AQA A Level Maths

Year 2

Series Editor
David Baker

Authors
Brian Jefferson, David Bowles, Eddie Mullan, Garry Wiseman, John Rayneau, Katie Wood, Mark Rowland, Mike Heylings, Paul Williams, Rob Wagner

Powered by MyMaths.co.uk

OXFORD UNIVERSITY PRESS

Great Clarendon Street, Oxford, OX2 6DP, United Kingdom

Oxford University Press is a department of the University of Oxford.

It furthers the University's objective of excellence in research, scholarship, and education by publishing worldwide. Oxford is a registered trade mark of Oxford University Press in the UK and in certain other countries.

First published in 2017

British Library Cataloguing in Publication Data
Data available

978 0 19 841296 0

10 9 8 7 6 5 4 3 2

Paper used in the production of this book is a natural, recyclable product made from wood grown in sustainable forests.
The manufacturing process conforms to the environmental regulations of the country of origin.

Printed and bound by CPI Group (UK) Ltd, Croydon, CR0 4YY

Acknowledgements

Series Editor: David Baker

Authors
Brian Jefferson, David Bowles, Eddie Mullan, Garry Wiseman, John Rayneau, Katie Wood, Mark Rowland, Mike Heylings, Paul Williams, Rob Wagner

Editorial team
Dom Holdsworth, Ian Knowles, Matteo Orsini Jones, Felicity Ounsted, Anna Clarke, Anna Gupta, Rosie Day

With thanks also to Anna Cox, Karuna Boyum, Katherine Bird, Keith Gallick, Linnet Bruce and Pete Sherran for their contribution.

Index compiled by Marian Preston, Preston Indexing

Although we have made every effort to trace and contact all copyright holders before publication, this has not been possible in all cases. If notified, the publisher will rectify any errors or omissions at the earliest opportunity.

p3, **p33**, **p63**, **p91**, **p128**, **p133**, **p163**, **p186**, **p223**, **p248**, **p259**, **p284**, **p294** Shutterstock; **p197** Vulkanov/iStockphoto; **p248** vkilikov/Shutterstock.com; **p280** Deborah Feingold/Getty Images; **p294** National Library Of Medicine/Science Photo Library; **p294** Nicku/Shutterstock.com

Message from AQA

This student book has been approved by AQA for use with our qualification. This means that we have checked that it broadly covers the specification and we are satisfied with the overall quality. We have not, however, reviewed the MyMaths and InvisiPen links, and have therefore not approved this content.

We approve books because we know how important it is for teachers and students to have the right resources to support their teaching and learning. However, the publisher is ultimately responsible for the editorial control and quality of this book.

Please note that mark allocations given in assessment questions are to be used as guidelines only: AQA have not reviewed or approved these marks. Please also note that when teaching the AQA A Level Maths course, you must refer to AQA's specification as your definitive source of information. While the book has been written to match the specification, it cannot provide complete coverage of every aspect of the course.

Full details of our approval process can be found on our website: www.aqa.org.uk

Contents

About this book 1

PURE

Chapter 12: Algebra 2
Introduction 3
12.1 Further mathematical proof 4
12.2 Functions 8
12.3 Parametric equations 16
12.4 Algebraic fractions 20
12.5 Partial fractions 24
Summary and review 28
Exploration 30
Assessment 31

Chapter 13: Sequences
Introduction 33
13.1 The binomial series 34
13.2 Introduction to sequences 40
13.3 Arithmetic sequences 44
13.4 Geometric sequences 50
Summary and review 56
Exploration 58
Assessment 59

Chapter 14: Trigonometric identities
Introduction 63
14.1 Radians 64
14.2 Reciprocal and inverse trigonometric functions 68
14.3 Compound angles 74
14.4 Equivalent forms for $a\cos\theta + b\sin\theta$ 80
Summary and review 84
Exploration 86
Assessment 87

Chapter 15: Differentiation 2
Introduction 91
15.1 The shapes of functions 92
15.2 Trigonometric functions 98
15.3 Exponential and logarithmic functions 102
15.4 The product and quotient rules 106
15.5 The chain rule 110
15.6 Inverse functions 114
15.7 Implicit differentiation 116
15.8 Parametric functions 120
Summary and review 124
Exploration 128
Assessment 129

Chapter 16: Integration and differential equations
Introduction 133
16.1 Standard integrals 134
16.2 Integration by substitution 140
16.3 Integration by parts 144
16.4 Integrating rational functions 148
16.5 Differential equations 152
Summary and review 158
Exploration 160
Assessment 161

Chapter 17: Numerical methods
Introduction 163
17.1 Simple root finding 164
17.2 Iterative root finding 168
17.3 Newton-Raphson root finding 174
17.4 Numerical integration 178
Summary and review 184
Exploration 186
Assessment 187

Assessment, chapters 12–17: Pure 190

MECHANICS

Chapter 18: Motion in two dimensions

Introduction 197
18.1 Two-dimensional motion with constant acceleration 198
18.2 Two-dimensional motion with variable acceleration 202
18.3 Motion under gravity 2 206
18.4 Motion under forces 212
Summary and review 218
Exploration 220
Assessment 221

Chapter 19: Forces 2

Introduction 223
19.1 Vectors in 3D 224
19.2 Statics 228
19.3 Dynamics 2 234
19.4 Moments 240
Summary and review 246
Exploration 248
Assessment 249

Assessment, chapters 18–19: Mechanics 253

STATISTICS

Chapter 20: Probability and continuous random variables

Introduction 259
20.1 Conditional probability 260
20.2 Modelling with probability 266
20.3 The Normal distribution 268
20.4 Using the Normal distribution as an approximation to the binomial 274
Summary and review 278
Exploration 280
Assessment 281

Chapter 21: Hypothesis testing 2

Introduction 283
21.1 Testing correlation 284
21.2 Testing a Normal distribution 288
Summary and review 292
Exploration 294
Assessment 295

Assessment, chapters 20–21: Statistics 297

Mathematical formulae 303

Statistical tables 305

Mathematical formulae – to learn 307

Mathematical notation 310

Answers 314

Index 383

About this book

This book has been specifically written for those studying the AQA 2017 Mathematics A Level. It's been written by a team of experienced authors and teachers, using a carefully selected range of features and exercises to build understanding and to help you get the most out of your course.

Every section starts by covering the basic **Fluency and skills** (A01), then builds on these techniques by looking at **Reasoning and problem-solving** (A02 and A03).

Strategy boxes help build problem-solving techniques.

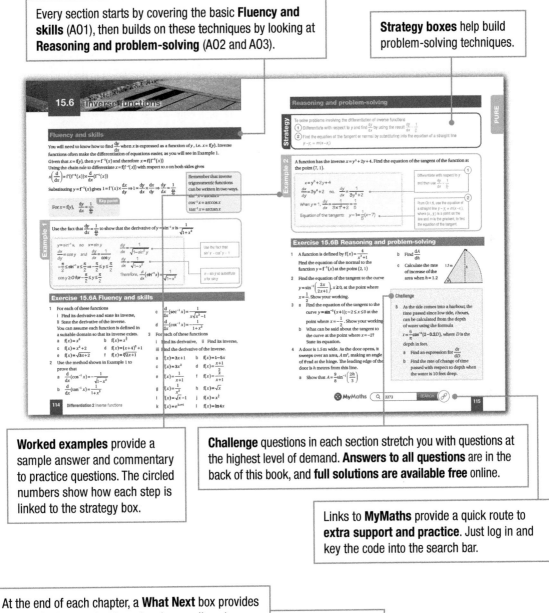

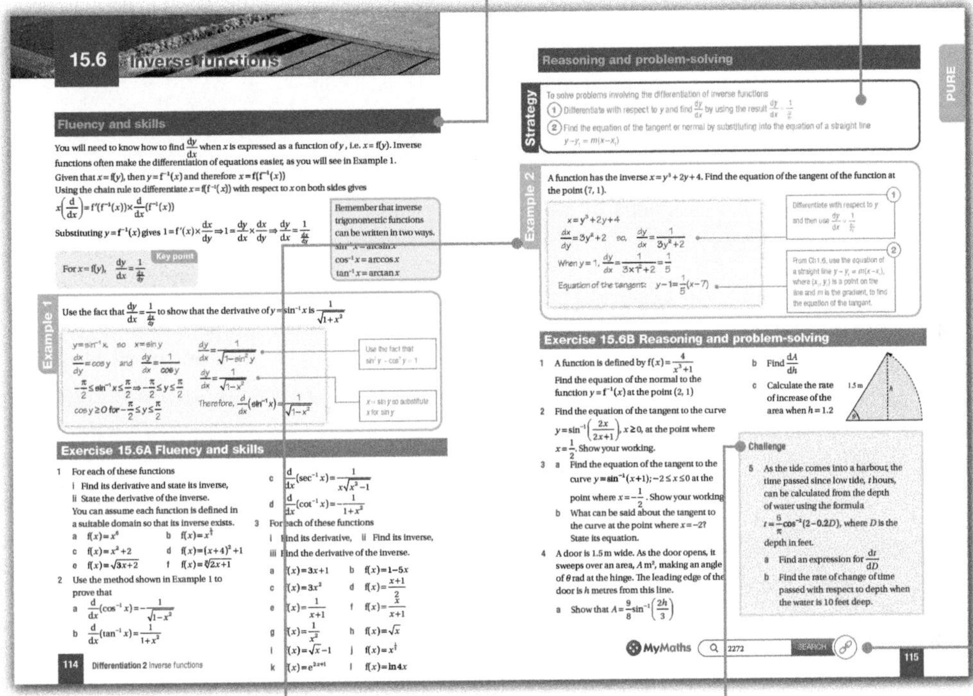

Worked examples provide a sample answer and commentary to practice questions. The circled numbers show how each step is linked to the strategy box.

Challenge questions in each section stretch you with questions at the highest level of demand. **Answers to all questions** are in the back of this book, and **full solutions are available free** online.

Links to **MyMaths** provide a quick route to **extra support and practice**. Just log in and key the code into the search bar.

At the end of each chapter, a **What Next** box provides links to further support based on how well you've understood the content.

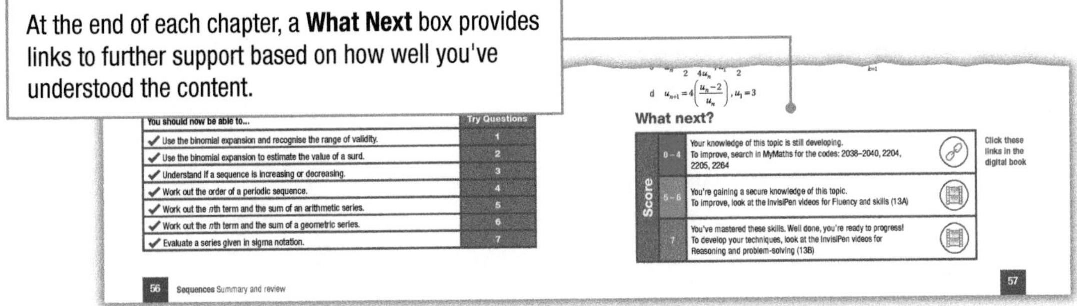

You should now be able to...	Try Questions
✔ Use the binomial expansion and recognise the range of validity.	1
✔ Use the binomial expansion to estimate the value of a surd.	2
✔ Understand if a sequence is increasing or decreasing.	3
✔ Work out the order of a periodic sequence.	4
✔ Work out the *n*th term and the sum of an arithmetic series.	5
✔ Work out the *n*th term and the sum of a geometric series.	6
✔ Evaluate a series given in sigma notation.	7

d $u_{n+1}=4\left(\frac{u_n-2}{u_n}\right)$, $u_1=3$

What next?

Score	
0–4	Your knowledge of this topic is still developing. To improve, search in MyMaths for the codes: 2038–2040, 2204, 2205, 2264
5–6	You're gaining a secure knowledge of this topic. To improve, look at the InvisiPen videos for Fluency and skills (13A)
7	You've mastered these skills. Well done, you're ready to progress! To develop your techniques, look at the InvisiPen videos for Reasoning and problem-solving (13B)

Click these links in the digital book

56 Sequences Summary and review

57

Links to **ICT resources** on Kerboodle show how technology can be used to help understand the maths involved.

ICT Resource online

To investigate reciprocal trigonometric functions, click this link in the digital book.

Support for when and how to use **calculators** is available throughout this book, with links to further demonstrations in the **digital book**. Unless otherwise stated, this book assumes that your calculator can do the required minimum according to specification guidelines. That is, it can perform an iterative function and it can compute summary statistics and access probabilities and values from standard statistical distributions.

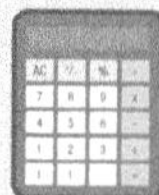

Try it on your calculator

You can use a calculator to evaluate the gradient of the tangent to a curve at a given point.

d/dx(5X² - 2X, 3)

28

Activity

Find out how to calculate the gradient of the tangent to the curve $y = 5x^2 - 2x$ where $x = 3$ on *your* calculator.

Assessment sections at the end of each chapter test everything covered within that chapter. Further **synoptic assessments** covering Pure, Mechanics and Statistics can be found at the end of chapters 17, 19 and 21

Dedicated questions throughout the statistics chapters will familiarise you with the **large data set.**

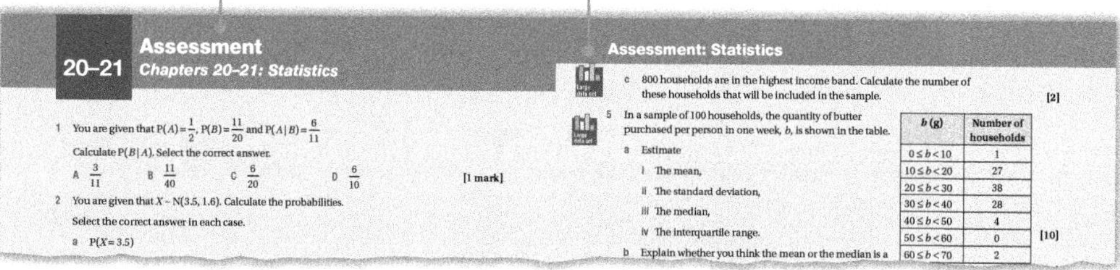

20–21 **Assessment**
Chapters 20–21: Statistics

1 You are given that $P(A)=\frac{1}{2}$, $P(B)=\frac{11}{20}$ and $P(A|B)=\frac{6}{11}$

Calculate $P(B|A)$. Select the correct answer.

A $\frac{3}{11}$ B $\frac{11}{40}$ C $\frac{6}{20}$ D $\frac{6}{10}$ [1 mark]

2 You are given that $X \sim N(3.5, 1.6)$. Calculate the probabilities.

Select the correct answer in each case.

a $P(X = 3.5)$

Assessment: Statistics

c 800 households are in the highest income band. Calculate the number of these households that will be included in the sample. [2]

5 In a sample of 100 households, the quantity of butter purchased per person in one week, b, is shown in the table.

a Estimate

i The mean,

ii The standard deviation,

iii The median,

iv The interquartile range. [10]

b Explain whether you think the mean or the median is a

b (g)	Number of households
$0 \le b < 10$	1
$10 \le b < 20$	27
$20 \le b < 30$	38
$30 \le b < 40$	28
$40 \le b < 50$	4
$50 \le b < 60$	0
$60 \le b < 70$	2

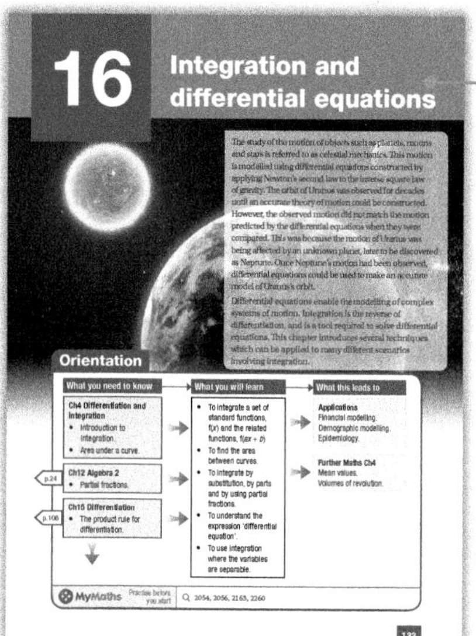

16 **Integration and differential equations**

The study of the motion of objects such as planets, moons and stars is referred to as celestial mechanics. This motion is modelled using differential equations constructed by applying Newton's second law to the inverse square law of gravity. The orbit of Uranus was observed for decades until an accurate theory of motion could be constructed. However, the observed motion did not match the motion predicted by the differential equations when they were compared. This was because the motion of Uranus was being affected by an unknown planet, later to be discovered as Neptune. Once Neptune's motion had been observed, differential equations could be used to make an accurate model of Uranus's orbit.

Differential equations enable the modelling of complex systems of motion. Integration is the reverse of differentiation, and is a tool required to solve differential equations. This chapter introduces several techniques which can be applied to many different scenarios involving integration.

Orientation

What you need to know	What you will learn	What this leads to
Ch4 Differentiation and integration • Introduction to integration. • Area under a curve. **Ch12 Algebra 2** (p.24) • Partial fractions. **Ch15 Differentiation** (p.106) • The product rule for differentiation.	• To integrate a set of standard functions, f(x) and the related functions, f(ax + b) • To find the area between curves. • To integrate by substitution, by parts and by using partial fractions. • To understand the expression 'differential equation'. • To use integration where the variables are separable.	**Applications** Financial modelling. Demographic modelling. Epidemiology. **Further Maths Ch4** Mean values. Volumes of revolution.

MyMaths Practise before you start Q 2054, 2056, 2163, 2260

133

On the chapter **Introduction page**, the **Orientation box** explains what you should already know, what you will learn in this chapter, and what this leads to.

At the end of every chapter, an **Exploration page** gives you an opportunity to explore the subject beyond the specification.

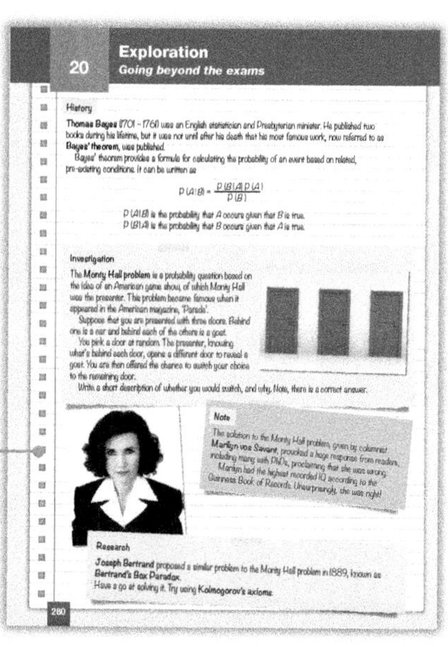

20 **Exploration**
Going beyond the exams

History

Thomas Bayes (1701 – 1761) was an English statistician and Presbyterian minister. He published two books during his lifetime, but it was not until after his death that his most famous work, now referred to as **Bayes' theorem**, was published.

Bayes' theorem provides a formula for calculating the probability of an event based on related, pre-existing conditions. It can be written as

$$P(A|B) = \frac{P(B|A)P(A)}{P(B)}$$

$P(A|B)$ is the probability that A occurs given that B is true.
$P(B|A)$ is the probability that B occurs given that A is true.

Investigation

The **Monty Hall problem** is a probability question based on the idea of an American game show, of which Monty Hall was the presenter. This problem became famous when it appeared in the American magazine, 'Parade'.

Suppose that you are presented with three doors. Behind one is a car and behind each of the others is a goat.

You pick a door at random. The presenter, knowing what's behind each door, opens a different door to reveal a goat. You are then offered the chance to switch your choice to the remaining door.

Write a short description of whether you would switch, and why. Note, there is a correct answer.

Note

The solution to the Monty Hall problem, given by columnist **Marilyn vos Savant**, provoked a huge response from readers, including many with PhDs, proclaiming that she was wrong.

Marilyn had the highest recorded IQ according to the Guinness Book of Records. Unsurprisingly, she was right!

Research

Joseph Bertrand proposed a similar problem to the Monty Hall problem in 1889, known as **Bertrand's Box Paradox**.

Have a go at solving it. Try using **Kolmogorov's axioms**.

280

12 Algebra 2

The secure transmission and storage of digital information requires strong forms of encryption. Public-key encryption is a vital part of online security today. It relies on trapdoor functions – these are functions which have no clear or easily-computable inverse. For example, calculating the prime factors of a number is notoriously difficult when the prime factors are very large, even for computers. The multiplication of two large prime numbers is therefore a trapdoor function: simple to perform, very hard to reverse. This forms the basis of RSA encryption, a widely used form of public-key encryption.

An understanding of functions and how to use them is vital in many areas of science, and they are a cornerstone of mathematics and, in particular, algebra. Functions are applicable to many topics, including parametric equations, partial fractions and other useful mathematical concepts covered in this chapter.

Orientation

What you need to know

Ch1 Algebra 1
- Direct proof, proof by exhaustion and counter examples.

Ch2 Polynomials and the binomial theorem
- Algebraic division and using the factor theorem.
- Transformations of graphs.

Ch3 Trigonometry
- Sine, cosine and tangent.

What you will learn

- To make logical deductions and prove statements directly.
- To understand and use functions, parametric equations and algebraic fractions.
- To decompose partial fractions.

What this leads to

Ch13 Sequences
Finding a binomial series by expanding an algebraic fraction.

Ch15 Differentiation
Differentiation of parametric functions.

Ch16 Integration and differential equations
Integration using partial fractions.

 MyMaths Practise before you start

2024, 2042, 2043, 2047, 2048, 2252, 2253, 2284

12.1 Further mathematical proof

Fluency and skills

See Ch1.1
For a reminder of proof.

In chapter 1 you learned how to prove a statement using **direct proof** and **proof by exhaustion** and also how to disprove a statement using a **counter example**.

You can also use **proof by contradiction** to show that, if some statement were true, a logical contradiction occurs, and so the statement must be false.

Example 1

Prove by contradiction that, for every real number $0° < x < 90°$, $\tan x - \sin x > 0$

Assume that $\tan x - \sin x \leq 0$ for some value of $0° < x < 90°$.

$\Rightarrow \frac{\sin x}{\cos x} - \sin x \leq 0$

$\Rightarrow \sin x\left(\frac{1}{\cos x} - 1\right) \leq 0$

$\Rightarrow$ Either $\sin x \leq 0$ or $\frac{1}{\cos x} \leq 1$ but not both.

$\sin x \leq 0 \Rightarrow 0° \not< x \not< 90°$ which contradicts the original statement.

$\frac{1}{\cos x} \leq 1 \Rightarrow 1 \leq \cos x \Rightarrow 0° \not< x \not< 90°$ which contradicts the original statement.

So, for every real number x between $0°$ and $90°$, $\tan x - \sin x > 0$

Write a statement that contradicts the given statement.

Substitute $\tan x = \frac{\sin x}{\cos x}$

Write the expression as a product.

Example 2

Prove that, if n^2 is even for $n \in \mathbb{Z}$, then n is even.

$\mathbb{Z}$ is the symbol for the set of integers.

Assume that there is an odd integer n such that n^2 is even.

Let $n = 2a + 1$ where a is an integer

$(2a + 1)^2 = 4a^2 + 4a + 1 = 2(2a^2 + 2a) + 1$

$2(2a^2 + 2a)$ is even.

So $2(2a^2 + 2a) + 1$ must be odd.

This means that n^2 must be odd which contradicts the original statement.

So, if n^2 is even, then n is even.

Write the opposite of the given statement.

You can write any odd number as $2a + 1$

Write the expression in the form $2k + 1$

This is a contradiction so the opposing statement is impossible.

Examples 3 and 4 show two important proofs that you need to know in your course of study.

An **irrational number** is one that you cannot express in the form $\frac{a}{b}$ where a and b are integers with no common factors.

PURE

Example 3

Prove by contradiction that $\sqrt{2}$ is irrational.

Assume that there exist integers a and b, with no common factors, such that $\sqrt{2}=\frac{a}{b}$

Write the opposing statement.

$\Rightarrow 2=\frac{a^2}{b^2}$ and so $2b^2=a^2$

Square both sides.

a^2 is a multiple of 2 so a must be even.

Use the result a^2 even $\Rightarrow$ a even.

Let $a=2m$ for an integer m, and so $a^2=4m^2$

So $2b^2=4m^2$, or $b^2=2m^2$

b^2 is a multiple of 2 so b must be even.

So $\sqrt{2}=\frac{\text{even}}{\text{even}}$

The numerator and denominator have a common factor of 2.

Both a and b have 2 as a factor. This contradicts the original statement that a and b have no common factors.

So the statement is false and $\sqrt{2}$ cannot be rational.

Example 4

Prove by contradiction that the number of prime numbers is infinite.

Assume that there are only n primes.

Write the opposing statement.

In this case the complete list of primes is

$2, 3, 5, 7, 11, 13, 17, 19, \ldots, p_n$

Let N be the product of all these primes + 1

So $N=(2\times3\times5\times7\times\ldots\times p_n)+1$

N is not divisible by 2 since it has a remainder of 1 when divided by 2

So 2 cannot be a factor of N

This also applies to all the other primes up to p_n

So none of the known primes is a factor of N and N must be a prime number.

Any number is either prime or divisible by a prime number.

This statement contradicts the original statement that the list of primes, $2, 3, 5, 7, 11, 13, 17, 19, \ldots, p_n$ was complete.

So the number of primes must be infinite.

Exercise 12.1A Fluency and skills

Use direct proof in questions 1 and 2

1 Prove that, for any integer n, $(5n+1)^4-(5n-1)^4$ is divisible by 40.

2 Show that for a 3-digit integer, if the sum of the digits is divisible by 9, then the number itself is divisible by 9

Use the method of exhaustion in questions 3 and 4

3 If p is an integer, prove that $p! \le 2^p$ for $0\le p\le 3$

4 Given that an integer can be even or odd, prove that the difference between the cubes of two consecutive integers is odd.

Use counter examples to disprove statements 5 to 7

5 If $a > b$ and $b > c$ then $ab > bc$

6 If p^2 is rational, then p is rational.

7 Cube numbers can end in any digit except 9

Use contradiction to prove statements 8 to 12

8 For any integer n, if n^2 is odd, then n is odd.

9 There are no values of a and b such that $a^2 - 4b = 2$. (Use the alternative statement $a^2 = 4b + 2$ and the fact that a^2 even $\Rightarrow a$ even.)

10 For every real number $0° < x < 90°$, $\sin x + \cos x \geq 1$. HINT: write down the opposing inequality and square both sides.

11 There are no integers m and n such that $\frac{m^2}{n^2} = 2$, where $\frac{m^2}{n^2}$ is a fully simplified fraction. You may assume that for an integer a, if a^2 is even then a is even.

12 Any integer greater than 1 has at least one prime factor.

Reasoning and problem-solving

Strategy

To prove a statement, P, using the method of contradiction

1. Assume P is not true.
2. Write the statement P′ which is the opposite of P.
3. Show that P′ leads to a contradiction.

Example 5

Natural numbers are integers greater than 0.

State why this proof by contradiction is flawed.

For any natural number n, the sum of all natural numbers less than n is not equal to n.

Opposing statement: For any natural number n, the sum of all smaller natural numbers is equal to n.

However, when $n = 5$, $1 + 2 + 3 + 4 = 10$ which is not equal to 5.

This is a contradiction.

So the assumption was false and the theorem must be true.

The contradiction of the universal statement "for all x, P(x) is true" is *not* the same as "for all x, P(x) is false".
The contradiction of the statement "for all x, P(x) is true" is the opposing statement "there exists an x such that P(x) is false".
So the statement "for any natural number n, the sum of all natural numbers less than n is not equal to n", has the opposing statement "there exists a natural number n such that the sum of all natural numbers smaller than n is equal to n".
In fact, for $n = 3$, $1 + 2 = 3$ so the statement can be *disproved* by counter example.

(2) Write the opposing statement.

Example 6

Saqib says that the curve $y = e^{2x}$ lies above the curve $y = e^x$ for all $x > 0$

Is he correct? Construct a proof to justify your answer.

Saqib is correct.

Assume the opposite: $y = e^{2x}$ lies below $y = e^x$ for some $x > 0$

$e^{2x} < e^x$ for some $x > 0$

$e^{2x} = (e^x)^2$

$e^x \times e^x < e^x \Rightarrow e^x < 1$

$\Rightarrow x < 0$ which contradicts the initial statement that $x > 0$

Therefore the curve $y = e^{2x}$ lies above the curve $y = e^x$ for all $x > 0$

Use your knowledge of the two graphs to choose a proof rather than a counterproof.

The statement uses 'for *all* $x > 0$' so the opposite statement uses 'for *some* $x > 0$'

Divide both sides of the inequality by e^x

Solve the inequality for x and state the contradiction.

State the conclusion clearly.

Exercise 12.1B Reasoning and problem-solving

1 Prove that, if m and n are both integers and mn is odd then m and n must both be odd.

2 Prove that, if $m^2 < 2m$ has any solutions, then $0 < m < 2$

3 Prove that if n is any odd integer, then $(-1)^n = -1$

4 Prove that the statement '$\sin(A - B) = \sin A - \sin B$, for all A and B' is false.

5 Prove that there is no integer k such that $(5m + 3)(5n + 3) = 5k$, where m and n are integers.

6 Emma says that the sum of the cubes of any two consecutive numbers always leaves a remainder of 1 when divided by 2. Is she correct? Construct a proof to support your answer.

7 Prove that, for $m < -\frac{1}{2}$, $1 - \frac{1}{m} < 3$

8 Prove that there is no smallest positive number.

9 A rule for multiplying by 11 in your head is shown.

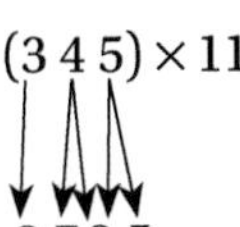

Working from the right:

1 Write down the 5

2 Add 5 + 4, add 4 + 3

3 Write down the 3

Prove algebraically that this rule works but state a 'catch'.

10 Prove that for an integer a, $(a^2 + 2) \div 4$ cannot be an integer.

11 Prove that there is no greatest odd integer.

12 Stephen is factorising integers. He thinks that, if b is a factor of a and c is a factor of b then c is a factor of a. Is he right? Use proof to justify your answer.

13 Prove that any positive rational number can be expressed as the product of two irrational numbers.

14 Prove that every cube number can be expressed in the form $9k$ or $9k \pm 1$, $k \in \mathbb{Z}$.

15 Prove that, for $a, b \in \mathbb{Z}$, there are no positive solutions to the equation $a^2 - b^2 = 1$

16 Prove that if a is an integer and is not a square number then $\sqrt{a}$ is irrational.

Challenge

17 State the error in this 'proof'.

Let $a = b$

So $a^2 = ab$

So $a^2 - b^2 = ab - b^2$

So $(a - b)(a + b) = b(a - b)$

So $\cancel{(a - b)}(a + b) = b\cancel{(a - b)}$

So $(a + b) = b$

So, since $a = b$, $2b = b$ and so $2 = 1$

MyMaths 🔍 2254 SEARCH

12.2 Functions

Fluency and skills

A **function**, f, is a mapping from a **domain**, set X, to a **range**, set Y, where each input value, $x \in X$, generates one and only one output value, $f(x) \in Y$. The range, Y, is the image of the domain, X, under the function, f.

A mapping from set X to set Y is a rule which associates each element of X with one element of Y.

Key point

For a function $f: X \rightarrow Y$, which maps elements of set X to elements of set Y

The set of all possible input values, X, is called the domain.

The set of all possible output values, Y, is called the range.

In a **one-to-one** function, each element in the range corresponds to exactly one element in the domain.

For example, $f(x) = 3x - 2$, $x \in \mathbb{R}$ is a one-to-one function.

$x = 3$ generates $f(x) = 7$

No other x value generates $f(x) = 7$

Its domain is the real numbers, $\mathbb{R}$. Its range is $\mathbb{R}$

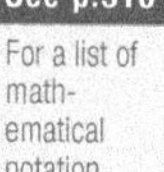

See p.310
For a list of mathematical notation.

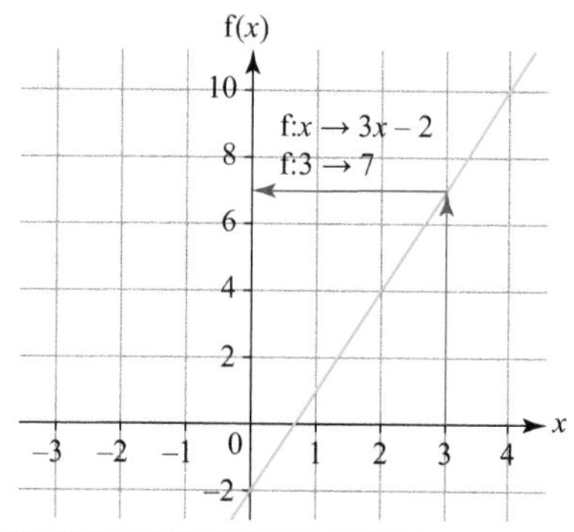

In a **many-to-one** function, an element in the range can be generated by more than one element in the domain.

For example, $f(x) = x^2$, $x \in \mathbb{R}$ is a many-to-one function.

Both $x = 3$ and $x = -3$ generate the value $f(x) = 9$

Its domain is $\{x: x \in \mathbb{R}\}$. Its range is $\{f(x): f(x) \in \mathbb{R}, f(x) \geq 0\}$

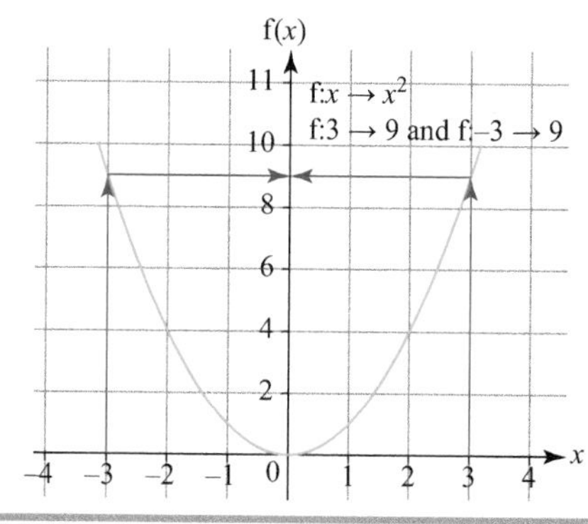

If any element in the domain generates more than one element in the range, then the relationship is not a function.

Example 1

For the function $f(x) = \frac{1}{\sqrt{x+6}}$

a State the maximum possible domain and the corresponding range. **b** Sketch its graph

a Maximum possible domain is $\{x: x \geq -6\}$, range is $\{f(x): f(x) > 0\}$

The denominator is strictly positive.

b

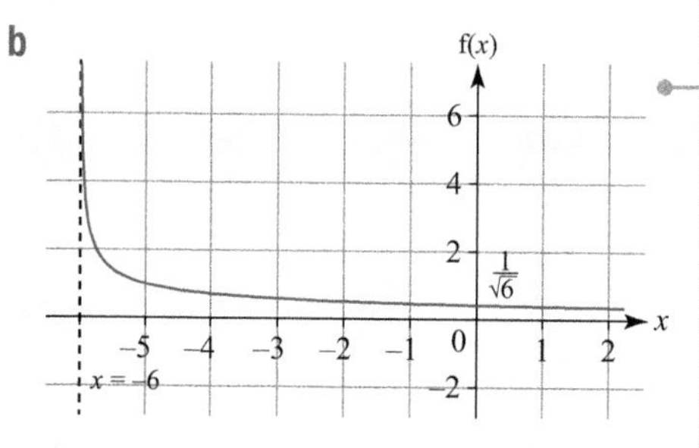

You can check your sketch using a graphical calculator.

A **composite function** is formed when you combine two or more functions. Given two functions f(x) and g(x):

- the composite function fg(x) means 'apply f to the results of g(x)'
- the composite function gf(x) means 'apply g to the results of f(x)'

For fg(x), the results, or range, of g(x) become the inputs, or members of the domain, of f. So the range of g(x) must be a subset of the domain of f.

The order is important. You always work out the 'inside' function first.

Composite functions are not commutative. fg(x) is not necessarily the same as gf(x).

Example 2

For the functions $f(x) = x^2, x \in \mathbb{R}$ and $g(x) = 2x + 1, x \in \mathbb{R}$

a write the composite functions fg(x), gf(x), ff(x) and gg(x)

b work out the values of fg(2), gf(2), ff(2) and gg(2)

a $fg(x) = (2x+1)^2$ $\quad gf(x) = 2x^2 + 1$

$ff(x) = (x^2)^2 = x^4$ $\quad gg(x) = (2(2x+1)+1) = 4x+3$

b $fg(2) = (2\times2+1)^2 = 25$ $\quad gf(2) = 2\times2^2+1 = 9$

$ff(2) = (2)^4 = 16$ $\quad gg(2) = 4(2)+3 = 11$

For fg(x), apply g first and then apply f to the result. For gf(x), apply f first and then apply g to the result.

For ff(x), apply f and then apply f to the result. For gg(x), apply g first and then apply g to the result.

You can check your answers by evaluating the composite functions at 2 on your calculator.

You can use composite functions in geometry to describe a transformation or a series of transformations.

See Ch2.4 For a reminder on transformations.

Example 3

a Describe the two transformations needed to transform the graph of $y = x^3$ into the graph of $y = 3 - x^3$

b The sequence of transformations applied to $y = x^3$ in part **a** can be described by the composite function $gf(x^3), x \in \mathbb{R}$. Write down

i $f(a)$ ii $g(a)$

c Sketch the curve of $fg(x^3)$ and describe the series of transformations needed to transform the graph of $y = x^3$ into the graph $y = fg(x^3)$

a

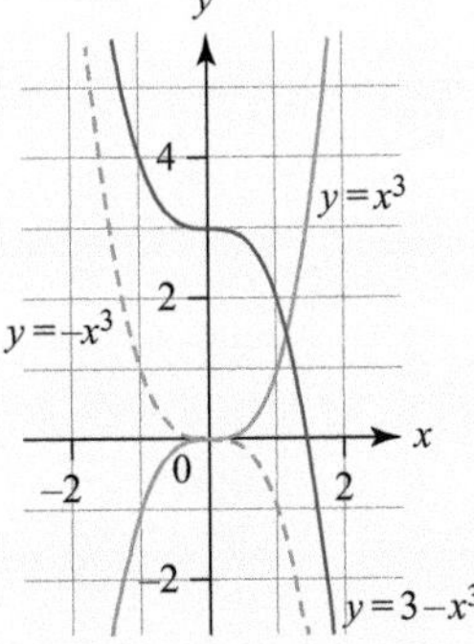

Reflection in the x-axis followed by translation by vector $\begin{pmatrix}0\\3\end{pmatrix}$

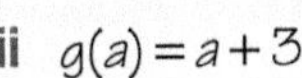

b i $f(a) = -a$

ii $g(a) = a + 3$

c

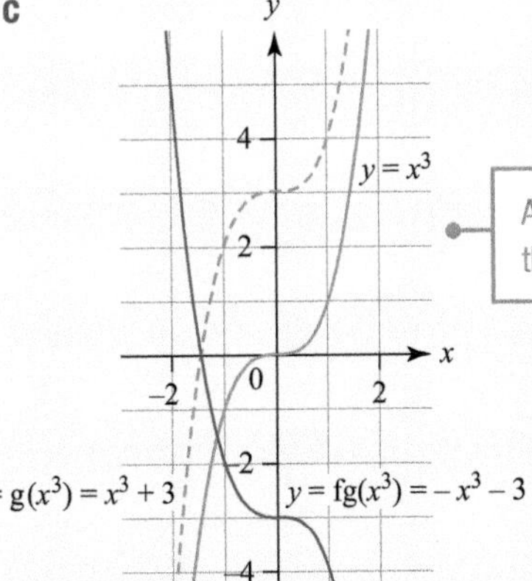

Translation by vector $\begin{pmatrix}0\\3\end{pmatrix}$ followed by reflection in the x-axis.

Apply g first, then apply f

The **inverse**, $f^{-1}(x)$, of a function, $f(x)$, reverses the effect of the original function.

You can find the inverse of a function by first rearranging the equation to make x the subject. You then replace the subject x with $f^{-1}(x)$ and replace $f(x)$ with x

f: 3 → 7
f⁻¹: 7 → 3

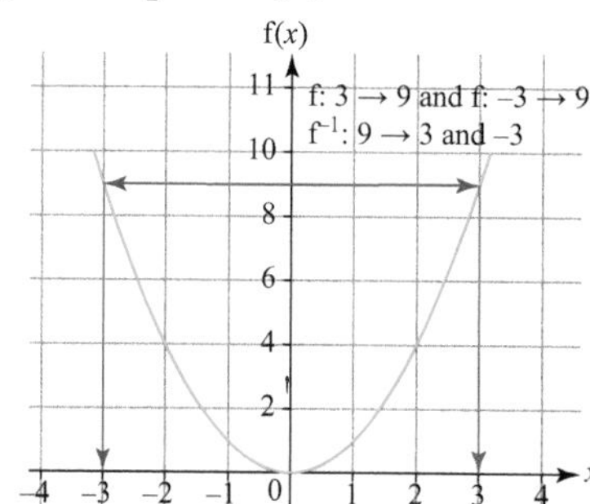

When inverted, many-to-one functions become one-to-many relationships.

$f(x) = 3x - 2, x \in \mathbb{R}$ maps elements from the domain $\mathbb{R}$ to the range $\mathbb{R}$

This can be rearranged to $x = \frac{f(x)+2}{3}$

$f^{-1}(x) = \frac{x+2}{3}, x \in \mathbb{R}$ is a one-to-one function which maps elements from the domain $\mathbb{R}$ to the range $\mathbb{R}$

$f(x) = x^2, x \in \mathbb{R}$ maps elements from the domain $\mathbb{R}$ to the range $f(x) \in \mathbb{R}, f(x) \geq 0$

This can be rearranged to $x = \pm\sqrt{f(x)}$

$f^{-1}(x) = \pm\sqrt{x}, x \in \mathbb{R}, x \geq 0$ is NOT a function because it maps some elements of the domain to more than one element in the range.

Key point

Only one-to-one functions have inverse functions.

Key point

If a function has an inverse, then
$ff^{-1}(x) = f^{-1}f(x) = x$
The domain of $f(x)$ is the range of $f^{-1}(x)$.

An inverse function, $f^{-1}(x)$, maps the elements in the range of $f(x)$ back onto the elements in the domain of $f(x)$.

Example 4

a Find the inverse of the function $f(x) = 8x^3 + 1, x \in \mathbb{R}$

b On one set of axes

i sketch the graph of $y = f(x)$ **ii** sketch the line $y = x$

iii reflect the graph of $y = f(x)$ in the line $y = x$

c Sketch the graph of $y = f^{-1}(x)$ on the same axes. What do you notice?

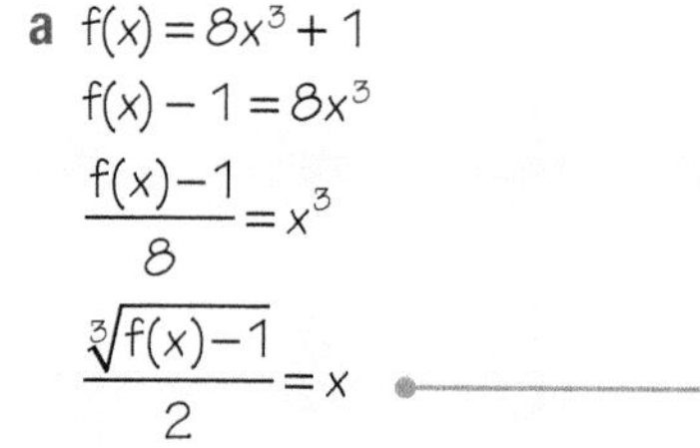

a $f(x) = 8x^3 + 1$

$f(x) - 1 = 8x^3$

$\frac{f(x)-1}{8} = x^3$

$\frac{\sqrt[3]{f(x)-1}}{2} = x$

Rearrange to make x the subject.

The inverse of $f(x) = 8x^3 + 1$ is

$f^{-1}(x) = \frac{1}{2}\sqrt[3]{x-1}$

To find the inverse function, swap $f(x)$ with x

b and **c**

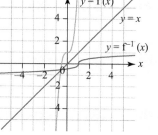

$y = f(x)$
$y = x$
$y = f^{-1}(x)$

The line $y = x$ is a mirror line of the function and its inverse.

Calculator

Try it on your calculator

You can sketch the inverse of a function on a graphics calculator.

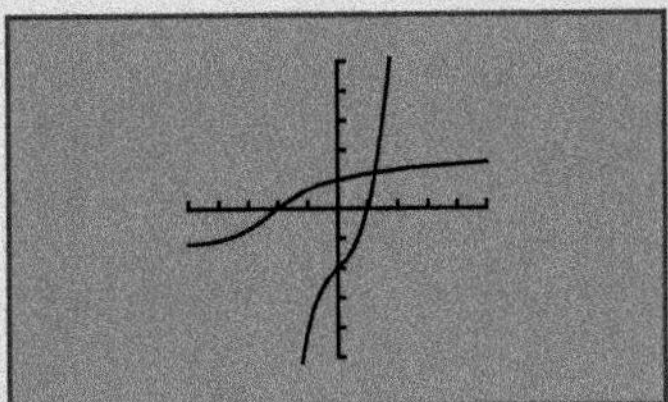

Activity

Find out how to sketch $f(x) = x^3 + x - 2$ and $f^{-1}(x)$ on *your* graphics calculator.

Key point

The graph of $f^{-1}(x)$ is a reflection of the graph of $f(x)$ in the straight line $y = x$

$|x|$ is called the **modulus** of x. It is also known as the **absolute value** of x.

The modulus of a real number is always positive. You can think of it as its distance from the origin.

For example, if $x = -3$, then $|x| = 3$

To sketch the graph of $y = |f(x)|$, you first sketch the graph of $y = f(x)$. You then take any part of the graph that lies below the x-axis and reflect it in the x-axis.

You can use your graphical calculator or computer software to sketch modulus functions.

Example 5

a On one set of axes **i** Sketch the graph of $y = 2x - 3$ **ii** Sketch the graph of $y = |2x - 3|$

b Sketch the graph of $y = 5 - |2x - 3|$

a

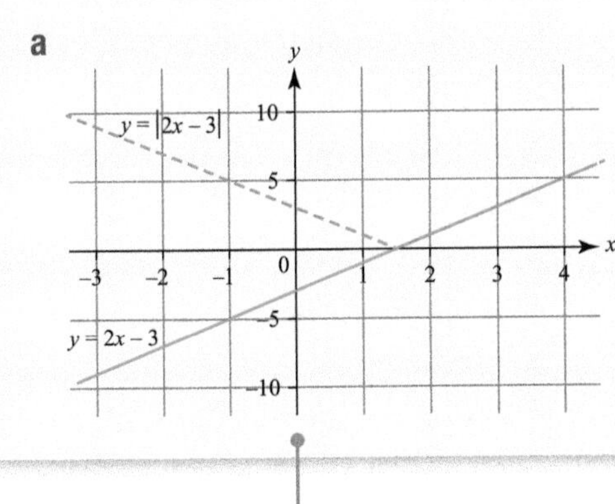

b

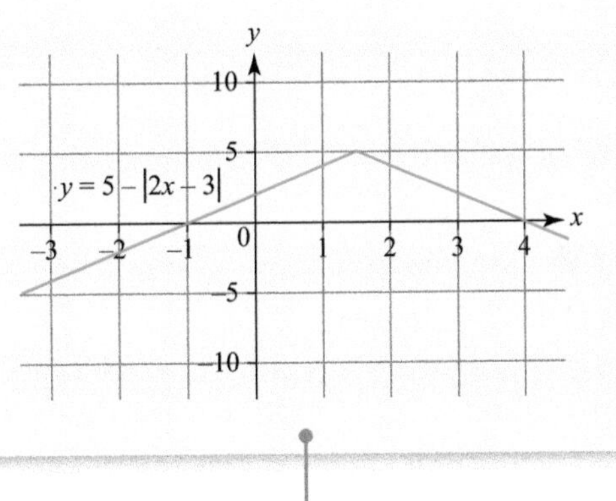

Only the absolute value of $2x - 3$ is taken. The negative portion of the graph is reflected in the line $y = 0$ and the minimum value of y is 0.

Reflect in the x-axis to get $-|2x - 3|$ then translate by $\begin{pmatrix} 0 \\ 5 \end{pmatrix}$ to get $5 - |2x - 3|$

Exercise 12.2A Fluency and skills

1 State the range of these functions.

a $f(x) = 5x - 2,\ x \in \mathbb{R}$

b $f(x) = 5x - 2,\ -6 < x < 6$

c $f(x) = (2x - 5)^2,\ x \in \mathbb{R}$

d $f(x) = (2x - 5)^2,\ 0 \le x \le 10$

2 State the maximum possible domain of these functions.

a $f(x) = 3x^3,\ f \in (-24, 81]$

b $f(x) = \ln x,\ f \in \mathbb{R}$

3 State the maximum possible domain and corresponding range of these functions.

a $f(x) = 4^x$ **b** $f(x) = \dfrac{1}{(x+3)^2}$

MyMaths 2049, 2135, 2138, 2139, 2142, 2261 SEARCH

4 State if each of these functions is one-to-one or many-to-one. Justify your answers.

a $f(x) = 2x^2, x \in \mathbb{R}$

b $f(x) = 3^{-x}, x \in \mathbb{R}$

c $f(x) = x^4, x \in \mathbb{R}$

d $f(x) = \sin^2 x, 90° \le x \le 270°$

e $f(x) = \frac{1}{x^2}, x \in \mathbb{R}, x \ne 0$

f $f(x) = -3x^3, x \in \mathbb{R}$

g $f(x) = \frac{1}{x-3}, x \in \mathbb{R}, x \ne 3$

h $f(x) = \cos x, 0° \le x \le 360°$

i $f(x) = \cos x, 0° \le x \le 180°$

j $f(x) = \cos 2x, 0° \le x \le 180°$

5 For each function

i Sketch the function

ii Find the point(s) of intersection between the function and $y = 3x, x \in \mathbb{R}$

a $f(x) = x^2, x \in [0,3)$

b $g(x) = x^2, x \in [0,5]$

c $h(x) = x^2, x \in (0,7]$

6 a Evaluate fg(1), gf(−2), $\text{ff}\left(\frac{-2}{3}\right)$ for these functions. Show your working.

i $f(x) = 3x + 7, x \in \mathbb{R}$

$g(x) = 3x^2 - 9, x \in \mathbb{R}$

ii $f(x) = \frac{1}{x}, x \in \mathbb{R}, x \ne 0$

$g(x) = \frac{-2}{x^2+1}, x \in \mathbb{R}$

b $f(x) = (x+1)^2, x \in \mathbb{R}$ and $g(x) = 1 - x$, $x \in \mathbb{R}$. Work out the values of fg(2) and gf(2), fg(−4) and gf(−4).

7 For each of these functions

a $f(x) = x^3$ b $f(x) = (x-4)^2 - 10$

i State the maximum possible domain and corresponding range.

ii Evaluate

f(0) f(−4) f(4)

8 Given that $|t| = 5$ work out all possible values of $|3t + 2|$.

9 In each case, the graph of $y = f(x)$ is a straight line passing through the points given. Sketch the graph.

i $y = f(x)$ ii $y = |f(x)|$

iii $|f(x-2)|$ iv $|f(x)| - 2$

a (−3, 0) and (0, 6)

b (−3, −2) and (2, −7)

10 Use the following functions to answer parts **i-iv**.

a $f(x) = x^2, 0 \le x < 8$

b $f(x) = (x-2)^3, x \in (2, 10)$

c $f(x) = 2^x, x < 0$

i State the range of $f(x)$.

ii Find the inverse function $f^{-1}(x)$, stating its domain.

iii Use a graphical calculator or a computer with appropriate software to sketch, on the same grid

1 $y = f(x)$

2 $y = f^{-1}(x)$

3 $y = x$

Check on your graph that the inverse is the reflection of the original function in the line $y = x$

iv State the range of $f^{-1}(x)$

11 a $f(x) = x^2, x \in \mathbb{R}$

i If $g(x) = 2x, x \in \mathbb{R}$ show that the composite function fg(x) is $4x^2, x \in \mathbb{R}$

ii Sketch the graph of $y = fg(x)$ and describe the transformation from $y = f(x)$ to $y = fg(x)$

b $f(x) = 4x^2, x \in \mathbb{R}$

i If $h(x) = x + 3, x \in \mathbb{R}$ show that the composite function fh(x) is $4x^2 + 24x + 36$

ii Sketch the graphs of f(x) and fh(x) and describe the transformation from f(x) to fh(x).

12 The curve $y = x^2, x \in \mathbb{R}$ is translated by $\begin{pmatrix} -3 \\ 0 \end{pmatrix}$ to create a new function, f(x). f(x) is then stretched parallel to the y-axis by scale factor 4 to create the composite function gf(x)

a Write an expression for f(x)
b Write an expression for g(x)
c Sketch the function gf(x)

13 a By writing each expression as a composition of simple functions, explain how you would transform the graph of $y = x^2$ into
i $x^2 - 6x + 13$ ii $4x^2 + 12x + 8$
b In the same way explain how you would transform the graph of $y = x^3$ into
i $(x+2)^3 - 7$ ii $(3x-5)^3 + 6$

14 Find an expression for f(x) when
a $fg(x) = e^{x^2}, x \in \mathbb{R}$ and $g(x) = x^2, x \in \mathbb{R}$
b $fg(x) = 3\log(x+1), x \in \mathbb{R}$ and $g(x) = x + 2, x \in \mathbb{R}$

15 For each pair of functions, sketch the graph of the composite function gf(x), stating its maximum possible domain.
a $f(x) = -x, x \in \mathbb{R}; g(x) = x^2, x \in \mathbb{R}$
b $f(x) = x^2, x \in \mathbb{R}; g(x) = x^{\frac{3}{2}}, x \in \mathbb{R}$

Reasoning and problem-solving

Strategy 1

To find the inverse of a function

1. Make sure the function is a one-to-one function.
2. Rearrange the function to make x the subject.
3. Interchange the variables: change x to $f^{-1}(x)$ and change f(x) to x
4. Check that the graphs of f(x) and $f^{-1}(x)$ are reflected in the line $y = x$

Example 6

Two functions exist only for $x \geq 0$

$f: x \rightarrow 7x - 2$ and $g: x \rightarrow ax^2 + b$ and $fg: x \rightarrow 28x^2 - 9$

$f: x \rightarrow 7x - 2$ is another way of writing $f(x) = 7x - 2$

a Explain why these functions are all one-to-one functions.

b Evaluate the values of a and b. c Work out the inverse of fg(x).

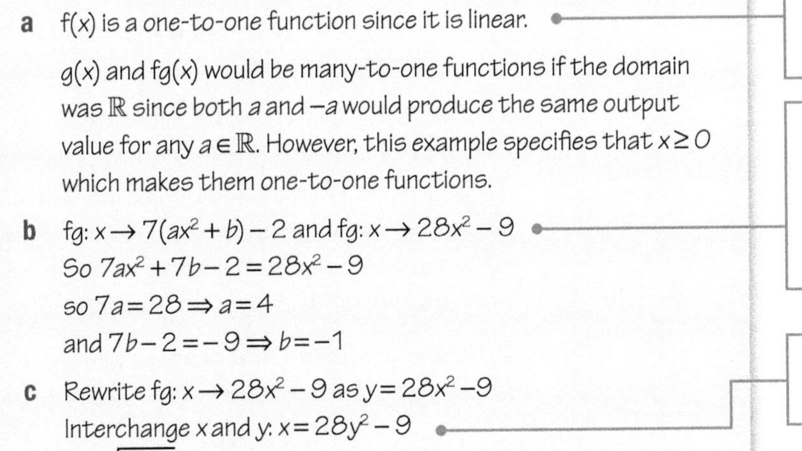

a f(x) is a one-to-one function since it is linear.

g(x) and fg(x) would be many-to-one functions if the domain was $\mathbb{R}$ since both a and $-a$ would produce the same output value for any $a \in \mathbb{R}$. However, this example specifies that $x \geq 0$ which makes them one-to-one functions.

b $fg: x \rightarrow 7(ax^2 + b) - 2$ and $fg: x \rightarrow 28x^2 - 9$
So $7ax^2 + 7b - 2 = 28x^2 - 9$
so $7a = 28 \Rightarrow a = 4$
and $7b - 2 = -9 \Rightarrow b = -1$

c Rewrite $fg: x \rightarrow 28x^2 - 9$ as $y = 28x^2 - 9$
Interchange x and y: $x = 28y^2 - 9$
$y = \sqrt{\frac{x+9}{28}}$
So $(fg)^{-1}: x \rightarrow \sqrt{\frac{x+9}{28}}$

1 All linear functions are one-to-one.

Combine f(x) and g(x) to give an expression for fg(x) and equate this to the expression given for fg(x).

2 Interchange the variables x and y.

3 Rearrange to make y the subject.

Strategy 2

To find the modulus, or absolute values, of a function for particular values of x

1. Substitute the values of x in the function $f(x)$
2. For any of the values of the function that are negative, replace the – sign with a + sign.
3. To sketch a modulus graph of a function reflect in the x-axis any parts of the graph which would be below the x-axis.

Example 7

Find, graphically, the set of values of x for which $|4x-3| < 2x$

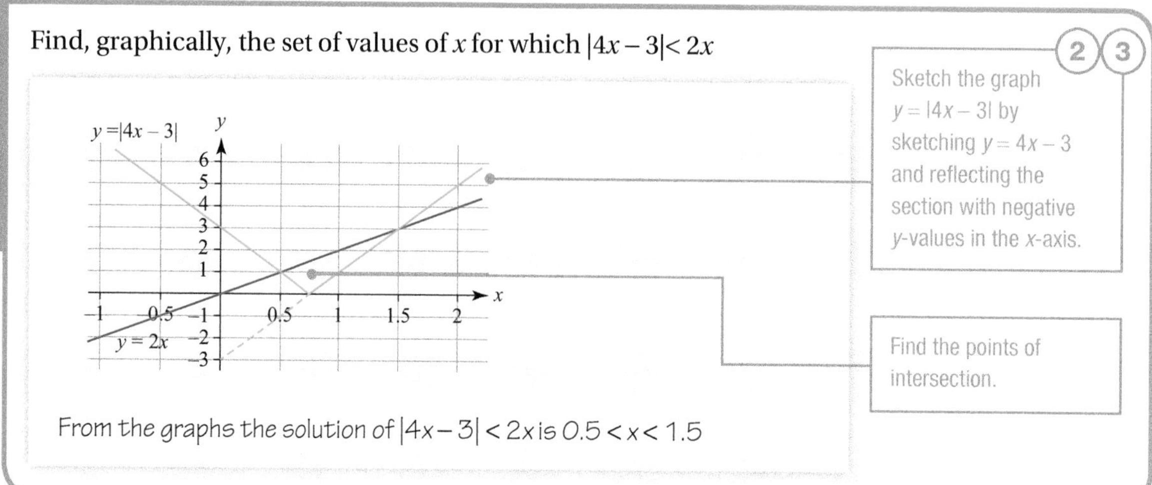

From the graphs the solution of $|4x-3| < 2x$ is $0.5 < x < 1.5$

Exercise 12.2B Reasoning and problem-solving

1 The graph of $y = f(x)$, $-1 < x < 5$ is shown.

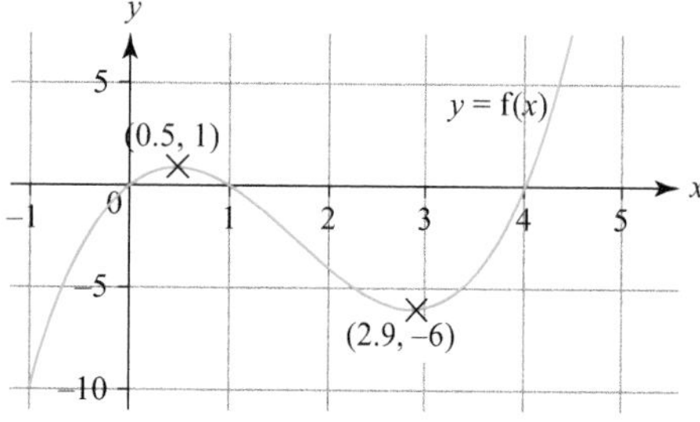

a Sketch the graph of $y = 3f(x) + 12$, $-1 < x < 5$ and indicate the coordinates of any turning points.

b The domain of $y = 3f(x) + 12$ is restricted to make the function invertible. State two different ways this could be done.

2 A company introduces a new product onto the market. Sales, S, (in thousands of pounds) initially increase steadily but then decrease, given by the function

$S = -2|w-15| + 30$ where w is the time (in weeks).

a Describe a sequence of transformations that maps $S = w$ onto $S = -2|w-15| + 30$

b Calculate the values of S for values of w every two weeks from 0 to 30

c Sketch the graph of the function.

d What was the maximum amount of sales in any one week?

3 The corners of a snooker table are (0, 0), (0, 6), (12, 0) and (12, 6). A player's cue ball is at (8, 4). He aims to bounce the ball off the bottom edge and send it into the pocket at (6, 6). Work out the equation he needs to pocket the ball and sketch the graph.

4 The shape of the roof of an art gallery is given by the equation

$$f(x) = \frac{-4}{3}|x-16.5| + 22$$

where $f(x)$ is the distance above ground level and x is horizontal distance in metres.

a Sketch the graph of this function.

b Write down the largest possible domain and corresponding range of this function.

c How high is the roof apex above its base?

d What is the length of the side of the square base on which the roof is mounted?

5 For each function, explain whether the inverse function exists and write an expression for the inverse if it exists.

a $f(x)=(x+2)^2$, $x \in \mathbb{R}$

b $g(x)=(x+2)^2$, $x \le 0$

c $h(x)=(x+2)^2$, $x \le -2$

6 a Work out the maximum possible domain and corresponding range of the function $\frac{1}{x-2}$

b Work out the inverse of the function $\frac{1}{x-2}$ and write down its domain and range.

c Compare your answers for the domain and range in parts **a** and **b**. What do you notice?

d Sketch a graph of the two functions on the same axes.

7 Use a graphical calculator or computer to check your answers in this question. Sketch the graphs of the functions in parts **a** to **d** on the same set of axes, for $-180° \le x \le 180°$

a $e(x)=\cos x$

b $f(x)=2\cos x$

c $g(x)=2\cos(x+90)°$

d $h(x)=2\cos(x+90)°-3$

e Describe the sequence of transformations that maps $e(x)$ to $f(x)$ to $g(x)$ to $h(x)$.

8 Describe a sequence of two transformations that maps the graph of $y=\sin x$ to $y=\sin(30-x)$

9 a Sketch the graph $y=\log_2 x$

b Transform the graph from part **a** by first translating it by $\begin{pmatrix}-2\\0\end{pmatrix}$ and then stretching it parallel to the y-axis by scale factor $\frac{1}{2}$

c Write the equation of the graph you drew in part **b**.

10 $f(x)=3x$, $x \in \mathbb{R}$ and $g(x)=x-4$, $x \in \mathbb{R}$

a i Describe a sequence of two transformations that maps $y=e^x$ to $y=fg(e^x)$

ii Write the expression $fg(e^x)$ in expanded form and sketch the graph $y=fg(e^x)$

b i Describe a sequence of two transformations that maps $y=e^x$ to $y=gf(e^x)$

ii Write the expression $gf(e^x)$ in expanded form and sketch the graph $y=gf(e^x)$

11 Calliste attempts to answer the following question. Explain and correct her mistakes.

$f(x)=4-x$, $x \in \mathbb{R}$. Sketch the graph of $y=|f(x)|$ and find the values of x for which $|f(x)|=\frac{1}{2}x$

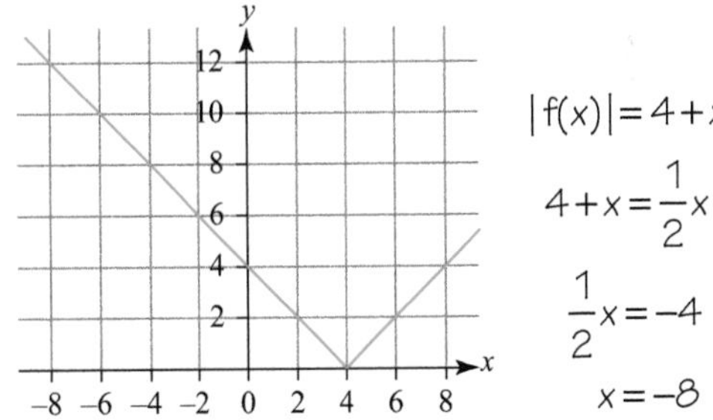

Challenge

12 $f(x)=x^3$, $x \in \mathbb{R}$ and $g(x)=\ln x$

a State the maximum domain and corresponding range for $g(x)$

b Saqib says that the composition $fg(x)$ is a function but the composition $gf(x)$ is not a function. Is he right? Explain your answer.

12.3 Parametric equations

Fluency and skills

A **Cartesian equation** can be written in the form $y = f(x)$.
A **parametric equation** expresses a relationship between two variables, x and y, in terms of a third variable, often named θ or t, which is known as the **parameter**.
x and y are given as functions of θ or t.

A Cartesian equation uses the variables x and y

Consider a circle with centre (3, 0) and radius 2. Any point, P, on the circle can be identified using the angle between the positive x-axis and the radius to the point P. The angle, θ, is the parameter in this case.

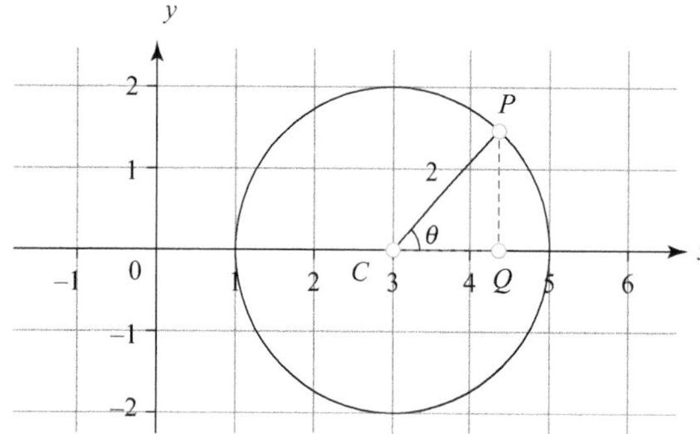

The point P can be expressed in terms of θ as $(3 + CQ, PQ)$

See Ch3.1
For a reminder of trigonometric functions.

$\cos\theta = \frac{CQ}{2} \Rightarrow CQ = 2\cos\theta$

$\sin\theta = \frac{PQ}{2} \Rightarrow PQ = 2\sin\theta$

Point P is $(3 + 2\cos\theta, 2\sin\theta)$

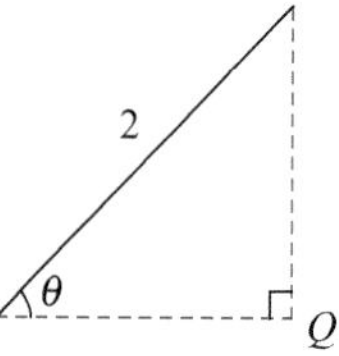

The parametric equations for the circle are

$x = 3 + 2\cos\theta$ and $y = 2\sin\theta$, for $0° \le \theta < 360°$

You can use the identity $\sin^2\theta + \cos^2\theta \equiv 1$ to write this as a Cartesian equation.

$$\begin{aligned}
x - 3 &= 2\cos\theta \\
(x-3)^2 &= 4\cos^2\theta \\
y^2 &= 4\sin^2\theta \\
(x-3)^2 + y^2 &= 4\cos^2\theta + 4\sin^2\theta \\
&= 4(\cos^2\theta + \sin^2\theta) \\
&= 4
\end{aligned}$$

The Cartesian equation of the circle is $(x-3)^2 + y^2 = 4$

Example 1

A curve is given by the parametric equations $x = t + 4$ and $y = t^2 - 10$

a Sketch the curve for $-4 \le t \le 4$

b Write the Cartesian equation for the curve.

a

t	−4	−3	−2	−1	0	1	2	3	4
x	0	1	2	3	4	5	6	7	8
y	6	−1	−6	−9	−10	−9	−6	−1	6

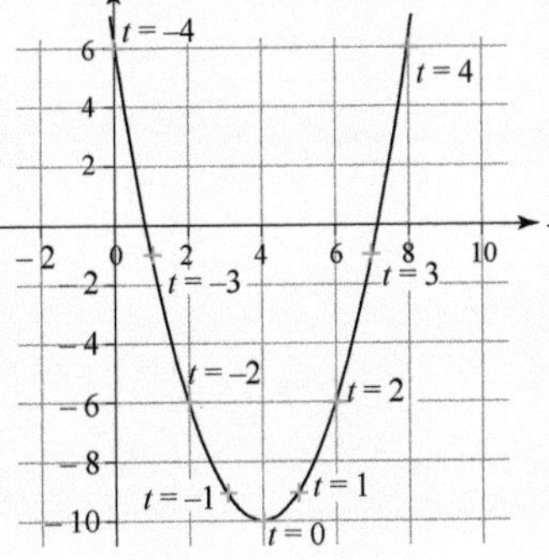

You can check your curve by drawing it on a graphics calculator.

b $x = t + 4$

$t = x - 4$

$y = (x - 4)^2 - 10$

Calculator

Try it on your calculator

You can sketch a curve given by parametric equations on a graphics calculator.

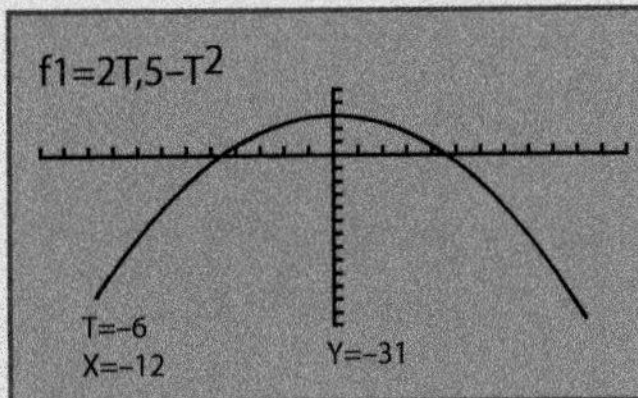

Activity

Find out how to sketch the curve given by $x = 2t$; $y = 5 - t^2$ on *your* graphics calculator.

Exercise 12.3A Fluency and skills

1 Work out the coordinates of the points on these parametric curves where $t = 5$, 2 and -3

a $x = t$; $y = \dfrac{4}{t}$ b $x = \dfrac{3}{t^2}$; $y = -2t$

c $x = \dfrac{1+t}{1-t}$; $y = \dfrac{2-t}{2+t}$ d $x = \dfrac{2t+5}{t+1}$; $y = \dfrac{t^3+4}{3}$

2 Work out the Cartesian equations given by these parametric equations.

a $x = t^2 + 2$; $y = 2t + 1$ b $x = 2t$; $y = -4t^3$

c $x = 3t$; $y = \dfrac{1}{2t}$ d $x = \dfrac{-2}{t}$; $y = 3t^4$

e $x = \dfrac{1+t}{1-t}$; $y = 2t$ f $x = 2\sin\theta$; $y = 2\cos\theta$

3 Use a table of values for t, x and y to sketch the graphs described by these parametric equations. Use a graphical calculator or computer software to check your sketches.

a $x = t$; $y = t^2$ b $x = t$; $y = 2t^3$

c $x = t^2 + 1$; $y = t + 2$ d $x = t^3$; $y = t^2$

e $x = 1 - t$; $y = \dfrac{1}{t}$ f $x = t$; $y = \dfrac{1}{t^2}$

4 A curve C is defined by parametric equations $x = \dfrac{4}{t-2}$, $y = \sqrt{t-2}$

a Theo says the largest possible domain is $t \ge 2$, Akeem says it is $t > 2$. Who is correct? Explain your answer.

b Write the Cartesian equation of the curve, stating the domain and range.

5 Show that any point on the parabola $y^2 = 4ax$ can be described by $(at^2, 2at)$ for a parameter t.

6 A curve C is defined by parametric equations

$x = e^{2t}$, $y = 5e^{-t}$ $t \in \mathbb{R}$

Write the Cartesian equation of the curve, stating the domain and range.

7 Show that the curve with parametric equations $x = 5\sin\theta$; $y = 5\cos\theta$ is a circle, centre O with radius 5.

8 Find the Cartesian equation of the curve given by the parametric equations

a $x = 7\cos\theta + 8$, $y = 7\sin\theta + 6$, $0° \le \theta < 360°$

b $x = 5\cos\theta + 3$, $y = 5\sin\theta - 1$, $0° \le \theta < 360°$

c $x = \frac{1}{2}\cos\theta - 4$, $y = \frac{1}{2}\sin\theta + 1$, $0° \le \theta < 360°$

d $x = \sqrt{2}\cos\theta - 4$, $y = \sqrt{2}\sin\theta - 3$, $0° \le \theta < 360°$

9 Work out the coordinates of the points on the curve $x = \frac{3+2t}{1-t}$, $y = \frac{2-t}{1+3t}$ where

a $t = -1$ b $t = 8$

10 By substituting $y = tx$, work out parametric equations for these curves.

a $y^3 = x^4$ b $y = x^2 - 3x$

c $y^2 = x^2 - 5x$ d $y^4 = x^3 + x^2$

e $y = x^3 + 3xy$

Reasoning and problem-solving

Strategy

To find a Cartesian equation from parametric equations

1. Make the parameter the subject of one of the equations.
2. Substitute the value for the parameter in the other equation and simplify.

Example 2

Point A moves across a coordinate grid in a straight line with speed $\begin{pmatrix} 6 \\ 8 \end{pmatrix}$ cm s^{-1}. Let t be the time in seconds. When $t = 0$, A is at (12, 0).

a Write down parametric equations in t for the position of A

b Find the Cartesian coordinates of the point where A crosses the line $y = x$

a $x = 12 + 6t$ $y = 8t$

Every second, the x-coordinate increases by 6 and the y-coordinate increases by 8.

b $12 + 6t = 8t \Rightarrow t = 6$ s, $(12 + 6 \times 6, 8 \times 6) = (8, 48)$

Exercise 12.3B Reasoning and problem-solving

1 A bullet is fired horizontally out to sea at 750 ms^{-1} from the top of a cliff 50 m high. Its position in relation to the foot of the cliff (in metres) is given by parametric equations $x = 750t$, $y = 50 - 5t^2$. Showing your working, work out

a When the bullet hits the sea

b How far from the base of the cliff it is at this time

c How far from the cliff the bullet is when it is 25 m above the sea.

2 A fairground roundabout has a radius of 10 m, with centre at the origin. A child gets on at the point (0, –10) and moves clockwise. Write parametric equations for the position of the child where the parameter is the angle $\theta°$ between the radius at any time and the negative direction of the y-axis. Give the coordinates of the child when θ is 90°, 135°, 180° and 270°.

3 An ellipse has the parametric equations $x = 5\cos t$, $y = \sin t$. Use a table of values to sketch this graph. Check your sketch using a graphical calculator or computer software.

4 A 'human cannonball' is fired from a cannon at a circus at a speed of $35\,\text{m s}^{-1}$ and an angle of 30° to the horizontal. His motion is described by $x = 35t\cos 30°$; $y = 35t\sin 30° - 5t^2$. He aims for a large safety platform with closest edge 90 m away horizontally and 7.5 m high.

Does he succeed?

5 A projectile passes through the air. Its passage can be modelled by the parametric equations

$y = -5t^2 + 20t + 105$, $x = 5t$, $t \geq 0$

where t is time (seconds), x is horizontal displacement (metres) and y is vertical displacement from the ground (metres). Show your working in parts **a** to **c**.

a What is the greatest height reached by the projectile?

b After how many seconds does the projectile hit the ground?

c How far does the projectile travel horizontally before hitting the ground?

6 Find the Cartesian equation of the locus given by $x = \sin\theta$, $y = 2\sin\theta\cos\theta$

7 A curve is given by parametric equations $x = \sin\theta$, $y = 3\cos\theta$, $0° < \theta \leq 360°$

The curve is first translated by the vector $\begin{pmatrix} 6 \\ 0 \end{pmatrix}$ and then stretched parallel to the y-axis by scale factor $\frac{1}{2}$

a Write parametric equations to describe the transformed curve.

b Write a Cartesian equation to describe the transformed curve.

8 Two curves are defined by parametric equations

Curve A: $x = 5 \times 3^{2t}$, $y = 3^{1-2t}$, $t \in \mathbb{R}$

Curve B: $x = \frac{3}{t}$, $y = 5t$, $t > 0$

a Show that curve A is identical to curve B.

b Write the Cartesian equation of the curve, stating its domain.

9 Find the point(s) of intersection of the curve given by $x = 5 - 2t$, $y = t^2 - 1$, $t \geq 0$ and the line given by $y = x - 2$, $x \in \mathbb{R}$. Show your working.

10 Find the points of intersection between the parabola $y = 2t$, $x = t^2$ and the circle $y^2 + x^2 - 3x - 90 = 0$. Show your working.

11 Work out the points of intersection of the curve $x = 3t^2 + 2$, $y = 4(t - 2)$ and the line $2x + 3y + 2 = 0$. Show your working.

12

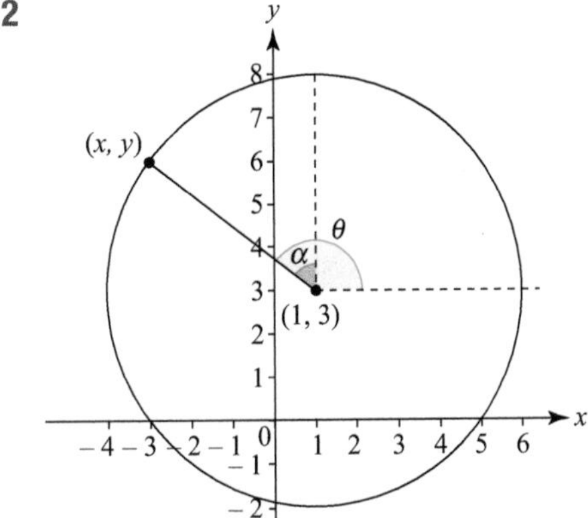

The diagram shows a circle with centre (1, 3) and radius 5. $0 < \theta \leq 360°$ is the angle measured anti-clockwise from the radius to (6, 3) and the radius to (x, y). $0 < \alpha \leq 360°$ is the angle measured anti-clockwise from the radius to (1, 8) and the radius to (x, y).

a Find parametric equations to represent the circle which use the parameter $0° < \theta \leq 360°$

b Find parametric equations to represent the circle which use the parameter $0° < \alpha \leq 360°$

Challenge

13 Show that the circle defined by $x = 3\cos\theta$, $y = 4 + 3\sin\theta$, $0° \leq \theta < 360°$ and the parabola defined by $x = -5t$, $y = 2t^2$, $t \in \mathbb{R}$ do not intersect.

12.4 Algebraic fractions

Fluency and skills

You can manipulate algebraic fractions using the same methods you use for arithmetical fractions.

Example 1

Simplify $\frac{x^2+x-6}{x^2-x-2} \times \frac{x^2-6x-7}{x^2+6x+9}$

$\frac{x^2+x-6}{x^2-x-2} \times \frac{x^2-6x-7}{x^2+6x+9}$

$= \frac{(x-2)(x+3)}{(x-2)(x+1)} \times \frac{(x-7)(x+1)}{(x+3)(x+3)}$ — Factorise the quadratic expressions first.

$= \frac{\cancel{(x-2)}\cancel{(x+3)}}{\cancel{(x-2)}\cancel{(x+1)}} \times \frac{(x-7)\cancel{(x+1)}}{\cancel{(x+3)}(x+3)}$ — Cancel common factors.

$= \frac{(x-7)}{(x+3)}$

See Ch2.3 For an introduction to algebraic division.

You can use algebraic long division for a divisor of the form $(ax+b)$ in the same way that you would for a divisor of the form $(x+b)$

Example 2

Calculate $(4x^3+3x-2)\div(2x-1)$

$$\begin{array}{r} 2x^2+x+2 \\ (2x-1)\overline{)4x^3+0x^2+3x-2} \\ \underline{4x^3-2x^2} \\ 2x^2+3x \\ \underline{2x^2-x} \\ 4x-2 \\ \underline{4x-2} \\ 0 \end{array}$$

$4x^3 \div 2x = 2x^2$

Use $0x^2$ to fill the place value for x^2

In Example 2 there is no remainder so $(2x-1)$ is a factor of $(4x^3+3x-2)$. You can find factors using a variant of the factor theorem.

In general, for a polynomial f(x) of degree $n \geq 1$ and any constants a and b

$f(x) \equiv (ax - b)g(x) + R$

where g(x) is a polynomial of order $n-1$ and R is a constant.

When $x = \frac{b}{a}$ this gives

$$f\left(\frac{b}{a}\right) \equiv \left(a\left(\frac{b}{a}\right) - b\right)g(x) + R$$

$$f\left(\frac{b}{a}\right) \equiv R$$

If $f\left(\frac{b}{a}\right) = 0$ there is no remainder when $f(x)$ is divided by $(ax - b)$ so $(ax - b)$ is a factor of $f(x)$

Key point

The factor theorem states that

$f\left(\frac{b}{a}\right) = 0$ if and only if $(ax - b)$ is a factor of $f(x)$

Example 3

Show that $(3x - 5)$ is a factor of $f(x) = (6x^3 - 7x^2 - 8x + 5)$

$(3x - 5)$ is a factor of $f(x)$ if $f\left(\frac{5}{3}\right) = 0$ — $3x - 5 = 0 \Rightarrow x = \frac{5}{3}$

$f\left(\frac{5}{3}\right) = 6\left(\frac{5}{3}\right)^3 - 7\left(\frac{5}{3}\right)^2 - 8\left(\frac{5}{3}\right) + 5 = 0$ — Substitute $x = \frac{5}{3}$

So $(3x - 5)$ is a factor of $f(x) = (6x^3 - 7x^2 - 8x + 5)$

Exercise 12.4A Fluency and skills

1 Simplify these fractions

a $\frac{x}{x-3} + \frac{2x}{x-5}$ b $\frac{2y}{y+8} - \frac{4y}{y-2}$

c $\frac{2x}{y-3} + \frac{3x}{2y-7}$ d $\frac{3z}{2z+9} - \frac{5z}{3z-4}$

2 a Simplify i $\frac{2x^2-5x+2}{x^2-4}$ ii $\frac{(x+2)^2}{2x^2+3x-2}$

b Use your answers from part **a** to write down the value of $\frac{2x^2-5x+2}{x^2-4} \times \frac{(x+2)^2}{2x^2-3x-2}$

3 Simplify these fractions

a $\frac{5z^2-10z}{8z+24} \times \frac{15z+45-10z}{12z^2-24z}$

b $\frac{3w+12}{w^2-7w} \div \frac{4w^2+16w}{w-7}$

c $\frac{2n^2-11n-6}{8n^2+22n+15} \div \frac{2n^2-11n-6}{12n^2-13n-35}$

d $\frac{4m^3-2m^2-12m}{9m^2+18m+5} \times \frac{3m^2-m-10}{2m^3-m^2-6m}$

4 Divide

a $8x^2 - 26x - 70$ by $(x - 5)$

b $3x^2 + 45x + 168$ by $(x + 8)$

5 Divide

a $x^3 + 6x^2 + 4x - 1$ by $(x + 1)$

b $2x^3 + x^2 - x - 63$ by $(2x - 3)$

c $3x^3 + 17x^2 - 11x - 33$ by $(3x + 2)$

d $6x^3 + x^2 - 19x - 12$ by $(3x + 1)$

e $x^3 + 27$ by $(x + 3)$

6 Work out the remainder when

a $2x^3 - 2x^2 + 7x + 14$ is divided by $(2x - 3)$

b $3x^4 - 13x^3 + 2x^2 - 5x - 10$ is divided by $(3x + 3)$

c $9x^3 + 5x^2 + 6x + 7$ is divided by $(3x - 1)$

d $4x^4 - x^3 - 12x^2 + 3x + 4$ is divided by $(2x + 11)$

7 Factorise $x^4 - 4x^2 + 3$

8 a Show that $(3x - 4)$ is a factor of $6x^3 - 5x^2 - 16x + 16$

b Show that $(2x + 1)$ is a factor of $2x^3 - x^2 + 7x + 4$

c Show that $(2x - 8)$ is a factor of $8x^3 - 36x^2 + 18x - 8$

9 a Use the factor theorem to show that $(3x+1)$ is a factor of $6x^3+5x^2+13x+4$

b Write $6x^3+5x^2+13x+4$ in the form $(3x+1)(Ax^2+Bx+C)$ where A, B and C are constants to be found.

10 The polynomial f(x) is given by $3x^3+10x^2-23x+10$

a Show that $(3x-2)$ is a factor of f(x)

b Factorise f(x) fully.

c Simplify the fraction $\frac{3x^3+10x^2-23x+10}{3x^2+4x-4}$

11 The expressions in parts **a** and **b** include a factor of the form $(x+a)$

a Factorise fully $6x^3+5x^2-12x+4$

b Factorise fully $4x^3-9x^2-x+6$

c Factorise fully $18x^4-27x^3-2x^2+10x$
This expression includes a factor of the form $(6x-b)$

Reasoning and problem-solving

Strategy

To solve a problem using the factor theorem

1. If the factor under consideration is of the form $(ax+b)$, solve the equation $ax+b=0$ for x
2. Substitute your solution into the expression f(x)
3. Interpret your solution. If the expression is equal to zero, $(ax+b)$ is a factor of f(x)

Example 4

Work out the values of p and q when $f(x)=6x^4+px^3-24x^2+qx+4$ is divisible by $(2x+1)$ and $(3x-1)$

$(2x+1)$ is a factor of f(x) $\Rightarrow f\left(-\frac{1}{2}\right)=0$ — ① ③ $2x+1=0 \Rightarrow x=-\frac{1}{2}$

$f\left(-\frac{1}{2}\right)=6\left(-\frac{1}{2}\right)^4+p\left(-\frac{1}{2}\right)^3-24\left(-\frac{1}{2}\right)^2+q\left(-\frac{1}{2}\right)+4$ — ② Substitute $x=-\frac{1}{2}$ into the expression

$=-\frac{p}{8}-\frac{q}{2}-\frac{13}{8}$

$=0$

$p+4q+13=0$ (1) — Simplify the equation by multiplying both sides by –8

$(3x-1)$ is a factor of f(x) $\Rightarrow f\left(\frac{1}{3}\right)=0$ — ① ③ $3x-1=0 \Rightarrow x=\frac{1}{3}$

$f\left(\frac{1}{3}\right)=6\left(\frac{1}{3}\right)^4+p\left(\frac{1}{3}\right)^3-24\left(\frac{1}{3}\right)^2+q\left(\frac{1}{3}\right)+4$ — ② Substitute $x=\frac{1}{3}$ into the expression

$=\frac{p}{27}+\frac{q}{3}+\frac{38}{27}$

$=0$

$p+9q+38=0$ (2) — Simplify the equation by multiplying both sides by 27

$9q-4q+38-13=0 \Rightarrow q=-5$ — Equation (2) – equation (1)

$p+4(-5)+13=0 \Rightarrow p=7$ — You can check your answer by solving these equations on your calculator.

Exercise 12.4B Reasoning and problem-solving

1 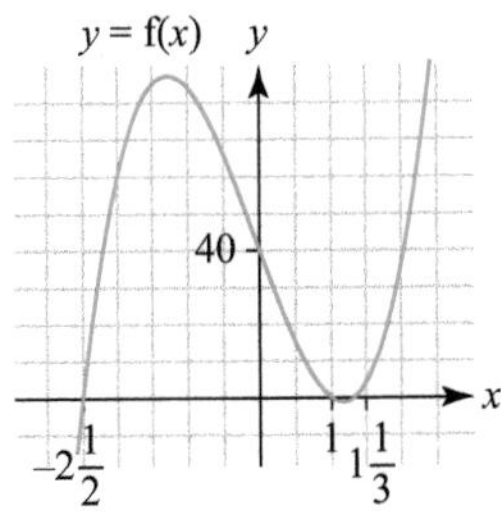

The graph shows the curve $y = f(x)$ where $f(x)$, $x \in \mathbb{R}$ is a cubic function. Write an expression for $f(x)$ in the form $Ax^3 + Bx^2 + Cx + D$

2 Work out the values of p and q when $12x^4 + 4x^3 + px^2 + qx + 8$ is divisible by $(3x - 2)$ and $(x + 1)$

3 Show there is only one vertical asymptote to the curve given by $y = \frac{x}{x-1} + \frac{1}{(x-1)(x-2)}$ and give its equation.

4 Solve for x, $\frac{1}{x} - \frac{1}{x-2} = \frac{1}{a} - \frac{1}{a-2}$

5 Prove that $(x^2 - 9)$ is a factor of the expression $x^4 + 6x^3 - 4x^2 - 54x - 45$

6 $x^4 - 13x^2 + 36$, $2x^3 + 3x^2 - 11x - 6$ and $3x^3 + x^2 - 20x + 12$ have a common quadratic factor. What is it?

7 Work out the HCF of $x^3 + 5x^2 + 2x - 8$ and $2x^3 + 3x^2 - 17x + 12$ in factorised form.

8 Show that $(x - 2)^2$ is a factor of $x^4 - 10x^3 + 37x^2 - 60x + 36$ and evaluate the other factors.

9 a Show that $(5x+2)$ is a factor of $30x^3 + 7x^2 - 12x - 4$

b Fully simplify the expression $\frac{30x^3 + 7x^2 - 12x - 4}{2x^2 - x} \times \frac{2x^2 - 9x + 4}{2x^2 - 7x - 4}$

c i State the maximum possible domain and corresponding range for the function $f(x) = \frac{30x^3 + 7x^2 - 12x - 4}{2x^2 - x} \times \frac{2x^2 - 9x + 4}{2x^2 - 7x - 4}$

ii Show that the gradient $f'(x)$ is positive at all points on the curve.

10 $(2x-5)$ is a factor of $f(x) = ax^4 + bx^2 - 75$

a Use the factor theorem to show that $(2x+5)$ is a factor of $f(x)$

b The third and final factor of $f(x)$ takes the form $(x^2 + c)$. Find the values of a, b and c

11 James attempts to complete the calculation $(4x^4 + 6x^3 + x - 8) \div (2x + 1)$. Explain and correct his mistakes.

$$
\begin{array}{r}
2x^3+2x^2-0.5 \\
2x+1 \overline{\big) \; 4x^4+6x^3+\;x-8} \\
-(4x^4+2x^3) \qquad\qquad \\
4x^3+\;x \qquad \\
-(4x^3+2x) \qquad \\
-x-8 \\
-(-x-0.5) \\
-7.5
\end{array}
$$

$(4x^4+6x^3+x-8) \div (2x+1) = 2x^3+2x^2-0.5-7.5$
$= 2x^3+2x^2-8$

12 For what values of b is $(x - b)$ a factor of $3x^3 + (b + 3)x^2 - (4b^2 + b - 7)x - 4$?

Challenge

13 a Prove that, if two polynomials $f(x)$ and $g(x)$ have a common linear factor $(x - a)$, then $(x - a)$ is a factor of the polynomial $f(x) - g(x)$

b Use the result from part **a** to prove that, if the equations $kx^3 + 3x^2 + x + 4 = 0$ and $kx^3 + 2x^2 + 9x - 8 = 0$ have a common linear factor, then $k = \frac{-9}{4}$ or $\frac{-59}{108}$

12.5 Partial fractions

Fluency and skills

The reverse process of adding algebraic fractions is splitting an algebraic fraction into its component parts. This technique is known as decomposing an algebraic fraction into **partial fractions**.

When adding fractions you find the lowest common multiple of the individual denominators. When decomposing a fraction you need to reverse this process, breaking the denominator into factors.

Example 1

Express $\frac{5x+2}{(x+4)(x-5)}$ in the form $\frac{A}{(x+4)}+\frac{B}{(x-5)}$ where A and B are integers.

Let $\frac{5x+2}{(x+4)(x-5)} \equiv \frac{A}{(x+4)}+\frac{B}{(x-5)}$

$(5x+2) \equiv \frac{A(x+4)(x-5)}{(x+4)}+\frac{B(x+4)(x-5)}{(x-5)}$

Multiply through by the denominator.

$(5x+2) \equiv A(x-5)+B(x+4)$

$5x+2 \equiv (A+B)x+(4B-5A)$

Look at both sides of the equation to compare the coefficient of the x term and the constant.

$A+B=5 \Rightarrow A=5-B$ (1)

$4B-5A=2$ (2)

$4B-5(5-B)=2 \Rightarrow B=3$

Solve the simultaneous equations by substituting the value of A from equation (1) into equation (2)

$A=5-3=2$

$\frac{(5x+2)}{(x+4)(x-5)} \equiv \frac{2}{(x+4)}+\frac{3}{(x-5)}$

If a denominator contains a squared factor, you need to consider the squared factor as a possible denominator, as well as all linear factors.

Example 2

a Find the sum of $\frac{3}{(x-5)}+\frac{7}{(x+4)}+\frac{10}{(x+4)^2}$

b Split $\frac{97x+1}{(x-5)(x+4)^2}$ into partial fractions.

a $\frac{3}{(x-5)}+\frac{7}{(x+4)}+\frac{10}{(x+4)^2}$

$\equiv \frac{3(x+4)^2+7(x-5)(x+4)+10(x-5)}{(x-5)(x+4)^2}$

$\equiv \frac{10x^2+27x-142}{(x-5)(x+4)^2}$

The distinct denominators $(x+4)$ and $(x+4)^2$ cannot be discerned from the final fraction.

(*Continued on the next page*)

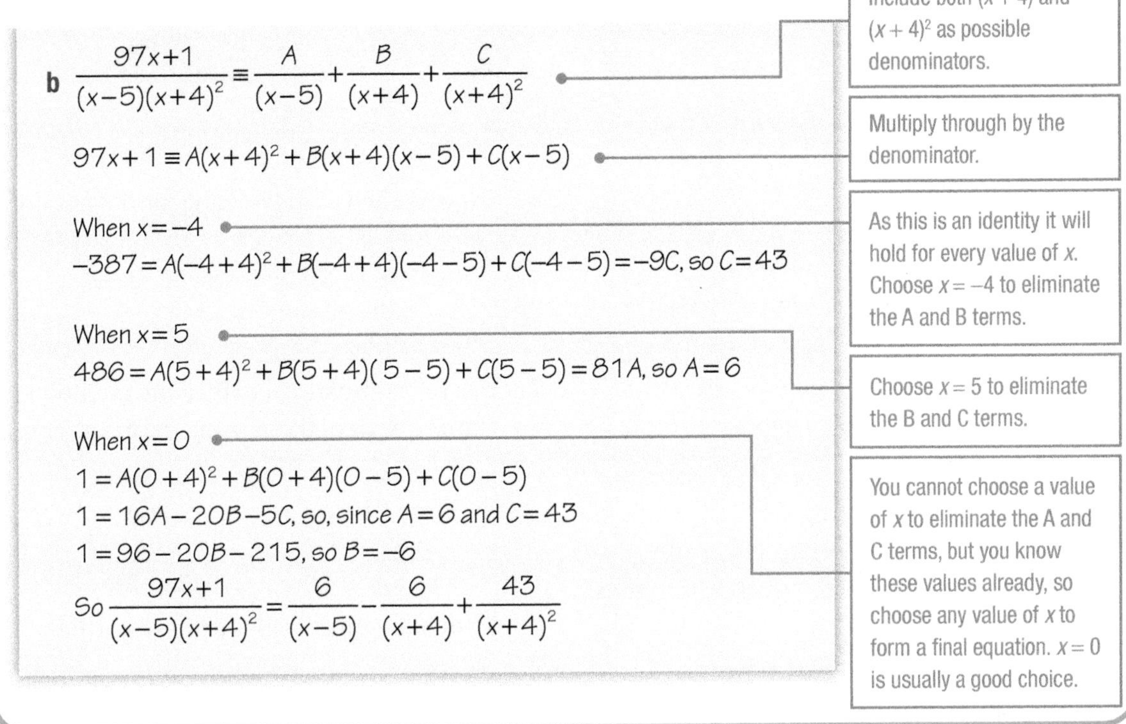

b $\frac{97x+1}{(x-5)(x+4)^2} \equiv \frac{A}{(x-5)} + \frac{B}{(x+4)} + \frac{C}{(x+4)^2}$

Include both $(x + 4)$ and $(x + 4)^2$ as possible denominators.

$97x+1 \equiv A(x+4)^2 + B(x+4)(x-5) + C(x-5)$

Multiply through by the denominator.

When $x = -4$

$-387 = A(-4+4)^2 + B(-4+4)(-4-5) + C(-4-5) = -9C$, so $C = 43$

As this is an identity it will hold for every value of x. Choose $x = -4$ to eliminate the A and B terms.

When $x = 5$

$486 = A(5+4)^2 + B(5+4)(5-5) + C(5-5) = 81A$, so $A = 6$

Choose $x = 5$ to eliminate the B and C terms.

When $x = 0$

$1 = A(0+4)^2 + B(0+4)(0-5) + C(0-5)$

$1 = 16A - 20B - 5C$, so, since $A = 6$ and $C = 43$

$1 = 96 - 20B - 215$, so $B = -6$

So $\frac{97x+1}{(x-5)(x+4)^2} = \frac{6}{(x-5)} - \frac{6}{(x+4)} + \frac{43}{(x+4)^2}$

You cannot choose a value of x to eliminate the A and C terms, but you know these values already, so choose any value of x to form a final equation. $x = 0$ is usually a good choice.

If the order of the numerator is equal to or greater than the order of the denominator, you divide the numerator by the denominator first and then express the remainder in partial fractions.

Example 3

Express $\frac{2x^2-3x-39}{(x+2)(x-3)}$ in the form $C + \frac{D}{(x+2)} + \frac{E}{(x-3)}$ where C, D and E are integers.

$\frac{2x^2-3x-39}{(x+2)(x-3)} \equiv C + \frac{D}{(x+2)} + \frac{E}{(x-3)}$

$2x^2-3x-39 \equiv C(x+2)(x-3) + D(x-3) + E(x+2)$

Multiply both sides by the denominator.

Let $x = -2$

$2(-2)^2 - 3(-2) - 39 \equiv C(0)(-5) + D(-5) + E(0)$

$-25 \equiv -5D$

$D = 5$

An identity is true for all values of x. Substitute a value which removes two unknowns.

Let $x = 3$

$2(3)^2 - 3(3) - 39 \equiv C(5)(0) + D(0) + E(5)$

$-30 \equiv 5E$

$E = -6$

Repeat the process to find E

Let $x = 0$

$2(0)^2 - 3(0) - 39 \equiv C(2)(-3) + 5(-3) - 6(2)$

$-39 \equiv -6C - 27$

$C = 2$

Use the values of D and E and any value of x to find the third unknown.

Hence $\frac{2x^2-3x-39}{(x+2)(x-3)} \equiv 2 + \frac{5}{(x+2)} - \frac{6}{(x-3)}$

MyMaths 2260 SEARCH

Exercise 12.5A Fluency and skills

1 Write $\frac{18x+26}{(3x+2)(x-4)}$ as the sum of two fractions with linear denominators.

2 Compare coefficients to work out the values of the constants in each identity.

a $12x-48 \equiv Ax(x-2)+B(x-3)(x+4)$

b $x^2-22x-35 \equiv C(x+3)(x-2) + D(x+1)(x-2) + E(x+1)(x+3)$

c $3x^3+11x^2-20 \equiv Fx(x+2)^2+Gx(x+2) - H(x+2)^2$

3 By substituting appropriate values of x, work out the values of the constants in these identities.

a $7x-15 \equiv A(x-1)+B(x-3)$

b $4x^2+24x+15 \equiv C(x+2)+D(x+2)^2 + E(x+1)$

c $24x-24 \equiv Fx(x+4)+G(x+4)(x-2) + Hx(x-2)$

4 Substitute $x=2$ to find the value of A in the identity $5x^2-5x-1 \equiv A(x^2-1)+(Bx+C)(x-2)$

Then substitute other values of x to find the values of B and C.

5 Express $\frac{1}{(1-x)^2}$ and $\frac{x}{(1-x)^2}$ in partial fractions. Compare your results.

6 Express each of these using partial fractions.

a $\frac{x-17}{(x+3)(x-2)}$ b $\frac{10x-2}{(x-1)(x+1)^2}$

c $\frac{450-20x}{(2x+5)(2x-5)^2}$

7 Use the comparing coefficients method to express each of these using partial fractions.

a $\frac{8}{x(x+4)}$ b $\frac{4}{(x+1)(x-3)}$

c $\frac{-36x+4}{(x+5)(x-7)^2}$ d $\frac{18x-12}{x(2x-3)(x+4)}$

8 By dividing first, evaluate these partial fractions.

a $\frac{3x^2-10x+11}{(x-1)(x-3)}$ b $\frac{5x^2+27x+26}{(x+1)(x+5)}$

9 Express using partial fractions

a $\frac{4x+2\sqrt{3}}{x^2-3}$ b $\frac{\sqrt{6}x+9\sqrt{5}}{6x^2-5}$

10 Express $\frac{x}{(x-a)(x-b)}$ in partial fractions.

11 Given that a, b, c, P and Q are constants, write expressions for P and Q in terms of a, b and c if $\frac{ax+b}{(x+c)^2}=\frac{P}{x+c}+\frac{Q}{(x+c)^2}$

Reasoning and problem-solving

Strategy

To find the partial fractions of an expression

1. Write a partial fraction for each linear or squared factor in the denominator.
2. Multiply through by the denominator and create an identity.
3. Either substitute suitable values or compare coefficients to find the constants and write the expression using partial fractions.

Example 4

a Express $\frac{1}{r(r+1)}$ in partial fractions.

b Deduce that $\frac{1}{1\times2}+\frac{1}{2\times3}+\frac{1}{3\times4}+\ldots+\frac{1}{n(n+1)}=\frac{n}{n(n+1)}$

(*Continued on the next page*)

a $\frac{1}{r(r+1)} \equiv \frac{A}{r} + \frac{B}{r+1}$

$1 \equiv A(r+1) + Br \Rightarrow 1 \equiv (A+B)r + A$

$A = 1$ and $A + B = 0 \Rightarrow B = -1 \Rightarrow \frac{1}{r(r+1)} \equiv \frac{1}{r} - \frac{1}{r+1}$

b $\frac{1}{1\times 2} + \frac{1}{2\times 3} + \frac{1}{3\times 4} + \ldots + \frac{1}{n(n+1)}$

$= \left(\frac{1}{1} - \frac{1}{2}\right) + \left(\frac{1}{2} - \frac{1}{3}\right) + \left(\frac{1}{3} - \frac{1}{4}\right) + \ldots + \left(\frac{1}{n-1} - \frac{1}{n}\right) + \left(\frac{1}{n} - \frac{1}{n+1}\right)$

$= \frac{1}{1} + \left(-\frac{1}{2} + \frac{1}{2}\right) + \left(-\frac{1}{3} + \frac{1}{3}\right) - \frac{1}{4} + \ldots + \frac{1}{n-1} + \left(-\frac{1}{n} + \frac{1}{n}\right) - \frac{1}{n+1}$

$= 1 - \frac{1}{n+1} = \frac{n+1}{n+1} - \frac{1}{n+1} = \frac{n}{n+1}$

(1) Write partial fractions with denominators r and $r + 1$

(2) Multiply through by the denominator and rearrange to compare coefficients.

Write each fraction as a pair of partial fractions. Include the first few terms and the last few terms to see the pattern.

Use the pattern to cancel zero terms.

Write 1 as $\frac{n+1}{n+1}$ and subtract to get one fraction.

Exercise 12.5B Reasoning and problem-solving

1 An athlete ran the first lap of a race in $\frac{3}{t-3}$ minutes and the second lap in $\frac{2}{t-7}$ minutes. Write a single fraction in t for her total time.

2 An arithmetic progression is of the form a, $a+d$, $a+2d$ etc. where a is the first term and d the common difference. The sixth term, $(a+5d)$, is $\frac{12x+4}{4x^2-1}$

Find d when $a = \frac{1}{2x+1}$

3 A lens has a focal length, $f = \frac{(p+2)(p-7)}{8p-11}$.

The object is at distance, $u = \frac{p+2}{\text{A}}$ and the image is at distance, $v = \frac{p-7}{\text{B}}$

Use the formula for a lens, $\frac{1}{u} + \frac{1}{v} = \frac{1}{f}$, to work out A and B.

4 Stephen says that, for real constants a and b,

$$\frac{a+b}{(x+c)(x+d)} \equiv \frac{a}{(x+c)} + \frac{b}{(x+d)} \text{ OR } \frac{b}{(x+c)} + \frac{a}{(x+d)}$$

a Show that this is not true for all real values of a, b, c and d.

b Find expressions for a and b for which the first identity holds.

5 Briana tries to decompose $\frac{14}{(x-5)(x+1)^2}$ into partial fractions. Explain and correct her mistakes.

$$\frac{14}{(x-5)(x+1)^2} \equiv \frac{A}{(x-5)} + \frac{B}{(x+1)^2}$$

$$14 \equiv A(x+1)^2 + B(x-5)$$

Let $x = -1 \Rightarrow 14 \equiv 0A - 6B$ so $B = -\frac{7}{3}$

Let $x = 5 \Rightarrow 14 \equiv 36A + 0B$ so $A = \frac{7}{18}$

$$\frac{14}{(x-5)(x+1)^2} \equiv \frac{7}{18(x-5)} - \frac{7}{3(x+1)^2}$$

Challenge

6 **a** Express $\frac{1}{(r+1)(r+2)}$ in partial fractions.

b Use your partial fractions to show that $\frac{1}{1(2)} + \frac{1}{2(3)} + \frac{1}{3(4)} + \frac{1}{4(5)} + \ldots\ldots + \frac{1}{(n+1)(n+2)} = 1 - \frac{1}{(n+2)}$

c Explain what happens to the sum if n approaches infinity.

12 Summary and review

Chapter summary

- You can use proof by contradiction to prove important results.
- A function is a relationship between two variables (usually x and y) where each input number (x) must generate only one output number (y). A function can be one-to-one or many-to-one.
- The sets of all possible values of x and y, are called the domain and the range respectively.
- A composite function is formed when two (or more) functions are combined.
- The inverse of a function is found by interchanging the variables x and y (or f(x)) and rearranging the equation to make y the subject. Only one-to-one functions have inverses.
- $|x|$ means the 'modulus of x' or the 'absolute value of x'.
- The modulus of a real number is always positive. You can think of it as its distance from the origin.
- A parameter is a single variable that can be used to express other variables.
- To obtain the Cartesian equation from the parametric equations you need to eliminate the parameter.
- Algebraic fractions use the same rules and techniques as arithmetical fractions.
- For a polynomial f(x), $f\left(\frac{b}{a}\right)=0$ if and only if $(ax-b)$ is a factor of f(x).
- The process of splitting a fraction into its component parts is called decomposing a fraction into partial fractions.
- When solving a partial fractions problem you can either compare coefficients or substitute appropriate values into the identity, or a mixture of the two.
- To split an expression in partial fractions, the degree of the numerator must be at least one lower than the degree of the denominator.

Check and review

You should now be able to...	Try Questions
✔ Make logical deductions and prove statements directly by exhaustion, by counter example and by contradiction.	1–3
✔ Understand and use functions.	4–9
✔ Understand and use parametric equations.	10, 11
✔ Understand and use algebraic fractions in all their forms.	12–15
✔ Decompose fractions into partial fractions.	16–17

1 Prove directly that, apart from 1! and 0!, all factorials are even.

2 Prove by exhaustion that, whichever way you factorise 385 you end up with the same factors.

3 Prove by contradiction that, if x is a rational number and y an irrational number, then $x-y$ is irrational.

4 $f(x)=3x+2,\ x\in\mathbb{R}$ and $g(x)=x^3,\ x\in\mathbb{R}$

Work out

a $f(4)$ b $g(-5)$ c $fg(2)$

5 a State a sequence of two transformations which map the graph of $y=x$ onto the graph of $y=5x-4$

b Sketch, on the same axes, the graphs of $y=5x-4$ and $y=|5x-4|$

c Given that $|x|=3$, work out all the possible values of $|5x-4|$. Use your sketch to confirm your solution.

6 Work out the inverse functions of

a $\frac{5}{x^2},\ x>1$ b $\frac{3x+7}{4},\ x\in\mathbb{R}$ c $\frac{x-2}{x-4},\ x\neq 4$

7 A function is defined by $f(x)=x^2-2,\ x\in\mathbb{R},\ x\geq 0$

a State the range of $f(x)$

b Write an expression for the inverse function $f^{-1}(x)$, stating its domain.

c Sketch the graphs of $f(x)$ and $f^{-1}(x)$ on the same set of axes.

d Find the value of x for which $f(x)=f^{-1}(x)$

8 Functions $f(x)$ and $g(x)$ are defined by

$f(x)=4-x,\ x\in\mathbb{R}$, and $g(x)=2x^2+5,\ x\in\mathbb{R}$

a State the range of $g(x)$

b Solve $gf(x)=7$

9 a Sketch the graph $y=2^x$

b The graph $y=2^x$ is transformed by first stretching it in the y-direction by a factor of 2 and then translating it by the vector $\begin{pmatrix}0\\-1\end{pmatrix}$

i Sketch the transformed curve.

ii Write the equation of the transformed curve.

10 Work out the Cartesian equation represented by $x=5t;\ y=\frac{-4}{t}$

11 Work out the points of intersection of the curve $x=2t+1,\ y=3-t^2$ and the straight line $x-y=6$. Show your working.

12 Divide $2x^4+x^3-5x^2-5x-3$ by $2x+3$

13 Simplify $\frac{n^2-n-6}{8n^2+4n+3}\div\frac{n^2-n-6}{n^2-4n-5}$

14 Fully factorise $4x^3-12x^2-x+3$

15 Show that $(3x-2)$ is a factor of $(3x^4-2x^3+12x^2-11x+2)$

16 Evaluate the partial fractions of $\frac{72}{(x+1)(x-5)^2}$

17 a Express each of these in partial fractions.

i $\frac{4x+1}{(x+1)(x-2)}$ ii $\frac{15-9x}{(x-1)(x-2)}$

b Hence solve the equation

$$\frac{4x+1}{(x+1)(x-2)}+\frac{15-9x}{(x-1)(x-2)}=1$$

What next?

Score		
0 – 8	Your knowledge of this topic is still developing. To improve, search in MyMaths for the codes: 2049, 2135, 2138, 2139, 2142, 2200, 2224, 2254, 2259–2262	
9 – 13	You're gaining a secure knowledge of this topic. To improve, look at the InvisiPen videos for Fluency and skills (12A)	
14 – 17	You've mastered these skills. Well done, you're ready to progress! To develop your techniques, look at the InvisiPen videos for Reasoning and problem-solving (12B)	

Click these links in the digital book

Exploration
Going beyond the exams

ICT

Some of the most interesting curves can be produced from **parametric equations** which use a combination of sines and cosines.

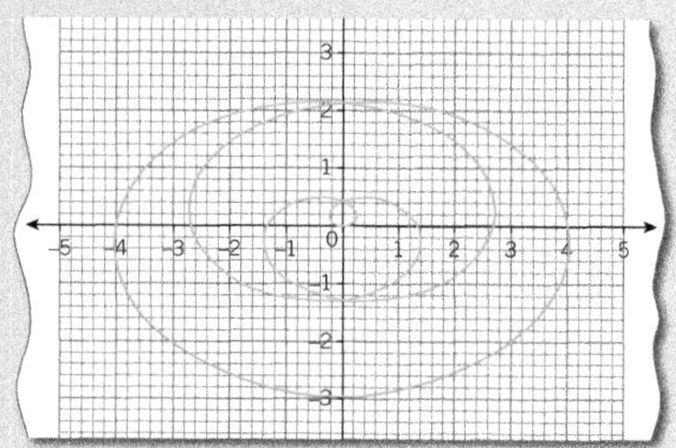

$x = t\cos(2.34t)$,
$y = t\sin(2.34t)$
$-4.7 \le t \le 4.7$

A small change to even just one of the functions can make a significant difference.

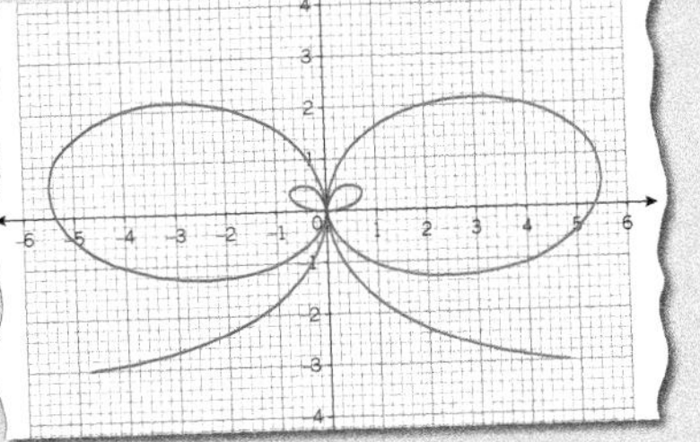

$x = t + t\cos(2.34t)$,
$y = t\sin(2.34t)$
$-4.7 \le t \le 4.7$

Other types of function can make different, even more interesting curves.

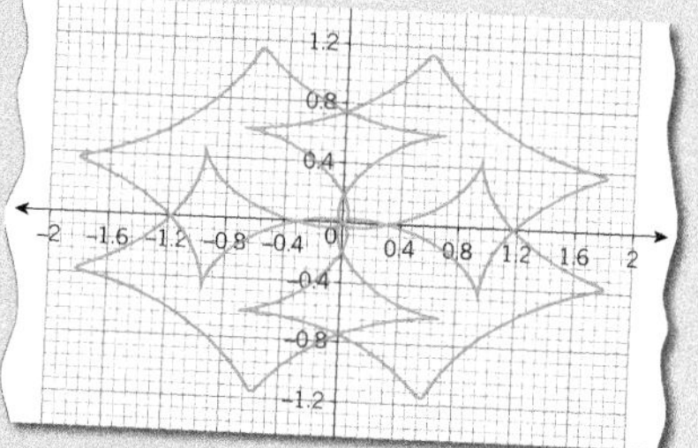

$x = \cos(2t) - \cos^3(10t)$
$y = \sin(2t) - \sin^3(10t)$
$-2 \le t \le 2$

Use a graph plotter or graphing software to plot different parametric curves. Try the equations above and vary the numbers – see what happens.

Applications

Lissajous curves (also known as **Bowditch curves**) represent complex harmonic motion. They take the form $x = A\sin(at + d)$, $y = B\sin(bt)$

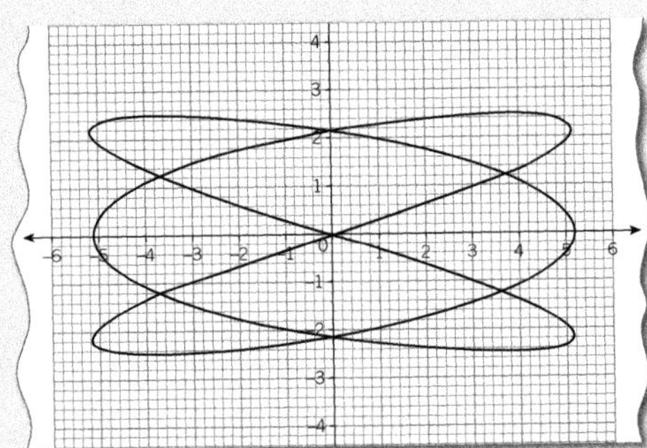

For example: $x = 6\sin(3t + \frac{\pi}{2})$
$y = 4\sin(2t)$
$-3.5 \le t \le 3.5$

In 2009 the European Space Agency launched the Herschel and Planck observatories. Their orbits were made to follow a Lissajous curve in order to save fuel.

Challenge

A ball is thrown at 20 m s^{-1} at an angle of 60° to the ground.
Write parametric equations to show the trajectory of the ball. Where does it land?

1 $f(x)=\frac{1}{x}, x\in\mathbb{R}$ and $g(x)=2x^2+5x, x\in\mathbb{R}$

Work out these values. Select the correct answer in each case.

a fg(3)

A $\frac{1}{33}$ B $\frac{17}{9}$ C $\frac{1}{51}$ D $15\frac{2}{9}$ **[1 mark]**

b gf(−1)

A −7 B $\frac{1}{3}$ C −3 D $-\frac{1}{3}$ **[1]**

2 a Work out the Cartesian equation represented by

$x=\frac{3}{t}$ and $y=2t^2-3$ $t\in\mathbb{R}, t\neq 0$

Select the correct answer.

A $y=6x^2-3$ B $x^2(y+3)=18$

C $x^2(y-3)=18$ D $y=18x^2-3$ **[1]**

b By substituting $y=tx$, work out parametric equations for $16y=5x^2$.

Select the correct answer.

A $x=\frac{4t}{\sqrt{5}}, y=\frac{4t^2}{\sqrt{5}}$ B $x=\frac{4t^2}{\sqrt{5}}, y=\frac{4t}{\sqrt{5}}$

C $x=\frac{16t}{5}, y=\frac{16t^2}{5}$ D $x=\frac{16t^2}{5}, y=\frac{16t}{5}$ **[1]**

3 Given that $\frac{3x^4+x^3-8x^2+3x+1}{x+2}\equiv Ax^3+Bx^2+Cx+D+\frac{E}{x+2}$, work out the values of the constants A, B, C, D and E **[6]**

4 Functions f(x) and g(x) are defined by

$$f(x)=\frac{x}{x-3}, x\in\mathbb{R}, x\neq 3, \text{ and } g(x)=\frac{5x-2}{x}, x\in\mathbb{R}, x\neq 0$$

a Work out an expression for the inverse function $f^{-1}(x)$ **[2]**

b Work out an expression for the composite function $gf(x)$ **[3]**

c Solve the equation $f^{-1}(x)=gf(x)$. Show your working. **[3]**

5 a Sketch the graph of $y=|2x-15|$ **[2]**

b Solve the equation $|2x-15|=3$ **[2]**

c Solve the inequality $|2x-15|\leq 3$ **[2]**

6 Determine which of the following functions are one-to-one, and which are many-to-one. Justify your answers.

A $y=3x+2, x\in\mathbb{R}$ B $y=x^2-5, x\in\mathbb{R}$

C $y=\frac{1}{x-3}, x\in\mathbb{R}, x\neq 3$ D $y=\sin x, x\in\mathbb{R}$ **[8]**

7 Use proof by contradiction to prove that, if n is an integer, and n^n is odd, then n is odd. **[5]**

8 The function $f(x)$ is defined by $f(x)=\frac{3x-7}{x^2-3x-4}-\frac{1}{x-4}$

a Show that $f(x)=\frac{2}{x+1}$ **[4]**

b What is the domain of $f(x)$? **[1]**

c Work out an expression for the inverse function $f^{-1}(x)$, stating its domain. **[2]**

d Solve the equation $f(x)=f^{-1}(x)$. Show your working. **[3]**

9 Decide which of the following statements are true and which are false. For those that are true, prove that they are true. For those that are false, give a counter example in each case.

A For $x\neq-1$, $\frac{4x}{(x+1)^2}\leq 1$

B $n!+1$ is prime for all positive integers, n

C The product of three consecutive odd integers is always a multiple of 15

D n^3-n is divisible by 6 for all positive integers, n **[13]**

10 Solve the equation $\frac{3}{x-2}-\frac{4}{x+1}=2$. Show your working. **[3]**

11 $f(x)=|3x|, x\in$, $g(x)=2x-1, x\in$

a Sketch the graph of $y=f(x)$ **[2]**

b Sketch the graph of $y=gf(x)$ **[2]**

c Describe the transformation from $f(x)$ to $gf(x)$ **[2]**

12 Given that $\frac{6x^4+5x^3-4x^2-3x+1}{2x+3}\equiv Ax^3+Bx^2+Cx+D+\frac{E}{2x+3}$, find the values of the constants A, B, C, D and E **[6]**

13 A curve, C, is given parametrically by $x=\sqrt{\sin t}$, $y=3\sin t\cos t$, $0°\leq t\leq 90°$

a Show that a Cartesian equation for C is $y=3x^2\sqrt{1-x^4}$ **[4]**

b Explain why there is no point on the curve for which $y=2$. **[5]**

14 Functions $f(x)$ and $g(x)$ are defined by

$$f(x)=e^{2x}, x\in\mathbb{R}, \quad \text{and} \quad g(x)=\ln(3x-2), x\in\mathbb{R}, x>\frac{2}{3}$$

a Write an expression for $fg(x)$ **[3]**

b Solve the equation $fg(x)=x^2$. Show your working. **[4]**

c Work out an expression for $f^{-1}(x)$ **[2]**

d Solve the equation $f(x)=5$. Show your working. **[2]**

15 a Work out the values of the constants A and B for which $\frac{3x-5}{(x-2)(x-1)}\equiv\frac{A}{x-2}+\frac{B}{x-1}$ **[4]**

b Hence show that $\frac{3x-5}{(x-2)(x-1)}$ is a decreasing function for $x>2$ **[2]**

16 a Prove that if a is an integer, and a^2 is a multiple of three, then a is also a multiple of three. **[4]**

b Use the method of proof by contradiction to prove that $\sqrt{3}$ is irrational. **[7]**

13 Sequences

The Fibonacci sequence (1 1 2 3 5 8 13...) is one of the most well known in mathematics, and it shows up time and time again in nature. The number of petals that a flower will have is frequently a Fibonacci number, and so is the number of spirals in pineapples, sunflower heads and pinecones. The ratio of any two adjacent terms tends towards the golden ratio as the values get larger. This is a ratio found in seashells, hurricanes, spiral galaxies, art and architecture.

Identifying and defining sequences is valuable for modelling situations which are expected to follow a pattern, rather than be random. They are applicable to financial analysis, research, computer science and engineering in determining the likelihood of different scenarios, and in analysing and predicting these scenarios. This chapter covers the groundwork for understanding different types of sequence and being able to describe them.

Orientation

What you need to know

Ch2 Polynomials and the binomial theorem

- The expansion of $(1 + x)^n$
- ${}^nC_r$ notation.

Ch12 Algebra 2

- p.20 Algebraic fractions.
- p.24 Partial fractions.

What you will learn

- To use the binomial expansion and recognise the range of validity.
- To use the binomial expansion to estimate the value of a surd.
- To understand if a sequence is increasing or decreasing.
- To find the order of a periodic sequence.
- To find the nth term and the sum of an arithmetic series and a geometric series.
- To evaluate a series given in sigma notation.

What this leads to

Ch17 Numerical methods
Iterative root finding.

Further Maths Ch2
Summing series and the method of differences.

Practise before you start

2041, 2200, 2260

13.1 The binomial series

Fluency and skills

See Ch2.2

For a reminder on the binomial theorem.

Key point

You know by the binomial theorem that the expansion of $(1+x)^n$, when n is a positive integer, is

$(1+x)^n = 1 + {}^nC_1x + {}^nC_2x^2 + {}^nC_3x^3 + ... + x^n$

You write $n \in \mathbb{Z}^+$ when n is a positive integer.

You can find the coefficients ${}^nC_1$, ${}^nC_2$, ${}^nC_3$ in this expansion by using Pascal's triangle or the $\boxed{nCr}$ button on your calculator.

Pascal's triangle

1
1 1
1 2 1
1 3 3 1

${}^3C_0 = \mathbf{1}$, ${}^3C_1 = \mathbf{3}$, ${}^3C_2 = \mathbf{3}$, ${}^3C_3 = \mathbf{1}$

For example, $(1+x)^3 = 1 + 3x + 3x^2 + x^3$

The sum on the right-hand side is a **series** of ascending powers of x. It is the result obtained when you expand $(1+x)(1+x)(1+x)$ and collect like terms.

When n is not a positive integer

- You cannot find $(1+x)^n$ by multiplying out $(1+x)$ n times,
- You cannot use the $\boxed{nCr}$ button to calculate coefficients. For example, when $n = -2$ ${}^{-2}C_1 = \text{Math ERROR}$

You write $n \notin \mathbb{Z}^+$ when n is *not* a positive integer.

Therefore, you must use a more general form of the binomial theorem.

Key point

When $n \notin \mathbb{Z}^+$, the binomial expansion of $(1+x)^n$ in ascending powers of x is

$$(1+x)^n = 1 + nx + \frac{n(n-1)}{2!}x^2 + \frac{n(n-1)(n-2)}{3!}x^3 + ...$$

The three dots mean the series has an **infinite** number of terms. You can also use this formula if n is a positive integer.

Remember that $3! = 3 \times 2 \times 1 = 6$

Example 1

Work out the binomial expansion of $(1+x)^{-2}$ up to and including the term in x^3

$$(1+x)^n = 1 + nx + \frac{n(n-1)}{2!}x^2 + \frac{n(n-1)(n-2)}{3!}x^3 + ...$$

$$n = -2,\ (1+x)^{-2} = 1 + (-2)x + \frac{(-2)(-3)}{2!}x^2 + \frac{(-2)(-3)(-4)}{3!}x^3 + ...$$

$$= 1 - 2x + \frac{6}{2!}x^2 - \frac{24}{3!}x^3 + ... = 1 - 2x + 3x^2 - 4x^3 + ...$$

Simplify coefficients.

You stop expanding at the x^3 term even though the series is infinite.

You can only use the binomial theorem in this way if the bracket is in the form $(1+x)^n$. If it is not, you need to re-write it in this form.

PURE

Example 2

Work out the binomial expansion of these expressions up to and including the term in x^2

a $(1+4x)^{\frac{1}{2}}$ **b** $(2+x)^{-1}$

a $(1+4x)^{\frac{1}{2}} = (1+y)^{\frac{1}{2}}$ where $y = 4x$

$= 1+\frac{1}{2}y+\frac{(\frac{1}{2})(-\frac{1}{2})}{2!}y^2+\ldots = 1+\frac{1}{2}(4x)+\frac{(\frac{1}{2})(-\frac{1}{2})}{2!}(4x)^2+\ldots$

$= 1+\frac{1}{2}\times 4x - \frac{1}{8}\times 16x^2+\ldots = 1+2x-2x^2+\ldots$

b $(2+x)^{-1} = \left[2\left(1+\frac{x}{2}\right)\right]^{-1} = 2^{-1}\left[1+\left(\frac{x}{2}\right)\right]^{-1} = 2^{-1}\left[1+y\right]^{-1}$ where $y = \frac{x}{2}$

$= \frac{1}{2}\left[1+(-1)y+\frac{(-1)(-2)}{2!}y^2+\ldots\right]$

$= \frac{1}{2}\left[1+(-1)\left(\frac{x}{2}\right)+\frac{(-1)(-2)}{2!}\left(\frac{x}{2}\right)^2+\ldots\right]$

$= \frac{1}{2}\left[1-\frac{x}{2}+\frac{x^2}{4}+\ldots\right] = \frac{1}{2}-\frac{1}{4}x+\frac{1}{8}x^2+\ldots$

Rewrite the expression in the form $(1+y)^n$. To do this, use $y = 4x$

$(4x)^2 = 16x^2$

Take out a factor of 2 so that the bracket starts with a 1

Keep the index -1 on the 2 when removed from the brackets.

The infinite sum $1+nx+\frac{n(n-1)}{2!}x^2+\frac{n(n-1)(n-2)}{3!}x^3+\ldots$ only makes sense if its terms get progressively smaller. To ensure this, you require $-1 < x < 1$

Key point

When $n \notin \mathbb{Z}^+$, the expansion $(1+x)^n = 1+nx+\frac{n(n-1)}{2!}x^2+\frac{n(n-1)(n-2)}{3!}x^3+\ldots$ is only **valid** if $-1 < x < 1$ (i.e. $|x| < 1$)

See Ch 12.2 For a reminder on domains and the modulus function.

This restriction is called the **range of validity** of the expansion. For values of x outside this range, the series has no value and the expansion is not valid.

Example 3

Evaluate the range of values of x for which the binomial expansion of $\frac{3}{1-2x}$ is valid.

$\frac{3}{1-2x} = 3(1+y)^{-1}$ where $y = -2x$

$= 3\left[1-y+y^2+\ldots\right]$

The expansion of $3(1+y)^{-1}$ is valid provided $-1 < y < 1$

In terms of x, the expansion of $\frac{3}{1-2x}$ is valid provided $-1 < -2x < 1$

$-1 < -2x < 1$ means $\frac{-1}{-2} > x > \frac{1}{-2}$

The range of validity is $-\frac{1}{2} < x < \frac{1}{2}$ or, equivalently, $|x| < \frac{1}{2}$

The multiplier 3 does not affect the range of validity.

Use $y = -2x$

You reverse an inequality sign when you divide through by a negative number.

See Ch 1.7 For a reminder on inequalities.

2204, 2205 SEARCH

Using similar methods to those used in Example 3, you can find the range of validity for the expansion of $(a+bx)^n$

$$(a+bx)^n = \left(a\left(1+\frac{b}{a}x\right)\right)^n$$

$$= (a(1+y))^n \qquad \text{where } y = \frac{b}{a}x$$

The expansion of $(a(1+y))^n$ is valid provided $-1 < y < 1$

So, in terms of x, the expansion of $(a+bx)^n$ is valid provided $-1 < \frac{b}{a}x < 1$

Key point

For $n \notin \mathbb{Z}^+$, the binomial expansion of $(a+bx)^n$ is only valid for $\left|\frac{b}{a}x\right| < 1$

Exercise 13.1A Fluency and skills

1 Work out the binomial expansions of these expressions up to and including the term in x^3

a $(1+x)^{-3}$ b $(1+x)^{\frac{1}{2}}$

c $(1+x)^{\frac{2}{3}}$ d $(1+4x)^{-1}$

e $(1-3x)^{-2}$ f $\left(1+\frac{1}{2}x\right)^{\frac{1}{3}}$

2 Work out the binomial expansions of these expressions up to and including the term in x^2

a $(2+x)^{-4}$ b $(3+2x)^{-1}$

c $(4-3x)^{\frac{3}{2}}$ d $\frac{3}{(3+x)^2}$

e $\sqrt{9+x}$ f $\frac{6}{3-4x}$

g $\frac{8}{\sqrt{4+9x}}$ h $\frac{27}{2(3-2x)^2}$

3 Work out the binomial expansions of these expressions, up to and including the term in x^2. Simplify coefficients in terms of the positive constant k

a $(1+kx)^{-3}$ b $\sqrt{(4+kx)}$

c $\left(\frac{k}{k+x}\right)^2$ d $\left(\sqrt{k}+x\right)^{-2}$

4 Evaluate the range of values of x for which the binomial expansion of these expressions is valid.

a $(1+4x)^{-3}$ b $(3+x)^{-2}$

c $(4-5x)^{\frac{3}{2}}$ d $\left(\frac{1}{2}+\frac{1}{3}x\right)^{\frac{1}{4}}$

5 Given that $(1+x)^{-2} = 1-2x+3x^2+\ldots$, work out the expansion of these expressions up to and including the term in x^2

a $x(1+x)^{-2}$

b $(2+x)(1+x)^{-2}$

c $(3-2x)(1+x)^{-2}$

d $(1-3x^2)(1+x)^{-2}$

6 By simplifying these expressions, or otherwise, work out their binomial expansions up to and including the term in x^2

a $\frac{1}{4+4x+x^2}$

b $\frac{4-6x}{4-9x^2}$

c $\frac{3}{\sqrt{1+4x+4x^2}}$

d $\frac{2}{\left(2-\sqrt{x}\right)} \times \frac{6}{\left(2+\sqrt{x}\right)}$

7 Given that the binomial expansion of $(a+4x)^{-3}$, where $a > 0$, is only valid for $-\frac{1}{2} < x < \frac{1}{2}$

a Calculate the value of a

b Hence evaluate the coefficient of x^3 in this expansion.

Reasoning and problem-solving

Strategy 1

To find an approximate value of an expression

1. Expand using the binomial expansion formula.
2. Equate coefficients if necessary.
3. Substitute the required value of x into the first few terms of the series.

Recall that, for $n \notin \mathbb{Z}^+$, $(1+x)^n = 1+nx+\frac{n(n-1)}{2!}x^2+\frac{n(n-1)(n-2)}{3!}x^3+...$

You can use the first few terms of this series to approximate $(1+x)^n$

Key point

If x is small enough that x^3 and higher powers can be ignored, then $(1+x)^n \approx 1+nx+\frac{n(n-1)}{2!}x^2$

$\approx$ is the approximation symbol.

For example, $(1+x)^{-2} = 1-2x+3x^2-4x^3+...$ so $(1+x)^{-2} \approx 1-2x+3x^2$ for suitably small values of x

If you substitute a low value of x, such as 0.1, then $1-2(0.1)+3(0.1)^2 = 0.83$

This is a reasonable approximation for 1.1^{-2} which is 0.8264...

Similarly, if you substitute $x = 0.01$, then $1-2(0.01)+3(0.01)^2 = 0.9803$, which is a very good approximation for 1.01^{-2} $(= 0.98029...)$

The greater number of terms you use, and the closer your value of x is to zero, the better the approximation. You can also use a binomial expansion to find estimates for surds.

Example 4

The first three terms in the binomial expansion of $\sqrt{1+px}$, where p is a negative constant, are $1-x-\frac{1}{2}x^2$

a Work out the value of p

b Use this series approximation and $x = 0.04$ to estimate $\sqrt{23}$

a $\sqrt{1+px} = (1+(px))^{\frac{1}{2}} = (1+y)^{\frac{1}{2}} = 1+\frac{1}{2}y+\frac{(\frac{1}{2})(-\frac{1}{2})}{2!}y^2+...$ — Rewrite the expression in the form $(1+y)^n$. To do this, use $y = px$

$= 1+\frac{1}{2}(px)+\frac{(\frac{1}{2})(-\frac{1}{2})}{2!}(px)^2+...$ — (1) Expand, leaving coefficients in terms of p

$= 1+\frac{1}{2}px-\frac{1}{8}p^2x^2+... \equiv 1-x-\frac{1}{2}x^2+...$

$\frac{1}{2}p = -1$ so $p = -2$ — (2) Equate coefficients of x

b $p = -2$ so $\sqrt{1-2x} \approx 1-x-\frac{1}{2}x^2$

When $x = 0.04$, $\sqrt{1-2(0.04)} = \sqrt{\frac{23}{25}}$, or $\frac{\sqrt{23}}{5}$ — (3) Substitute $x = 0.04$

and $1-(0.04)-\frac{1}{2}(0.04)^2 = 0.9592$ — Use the series approximation from part **a** as x is suitably small.

$\frac{\sqrt{23}}{5} \approx 0.9592$ so, $\sqrt{23} \approx 5 \times 0.9592 = 4.796$ — Check the result $\sqrt{23} = 4.7958...$

You can use partial fractions to find the expansion of an algebraic fraction.

Strategy 2

To find the expansion of an algebraic function involving fractions

1. Split the function into partial fractions.
2. Write each partial fraction in the form $(1+x)^n$
3. Work out the binomial expansion of each partial fraction.
4. Combine like terms.

Example 5

The curve C has equation $y = f(x)$, where $f(x) = \dfrac{3+5x}{(1+x)(1+2x)}$

a Work out the expansion of $f(x)$ up to and including the term in x^2

b State the range of values of x for which the full expansion of $f(x)$ is valid.

c Estimate the gradient of C at the point where $x = 0.05$

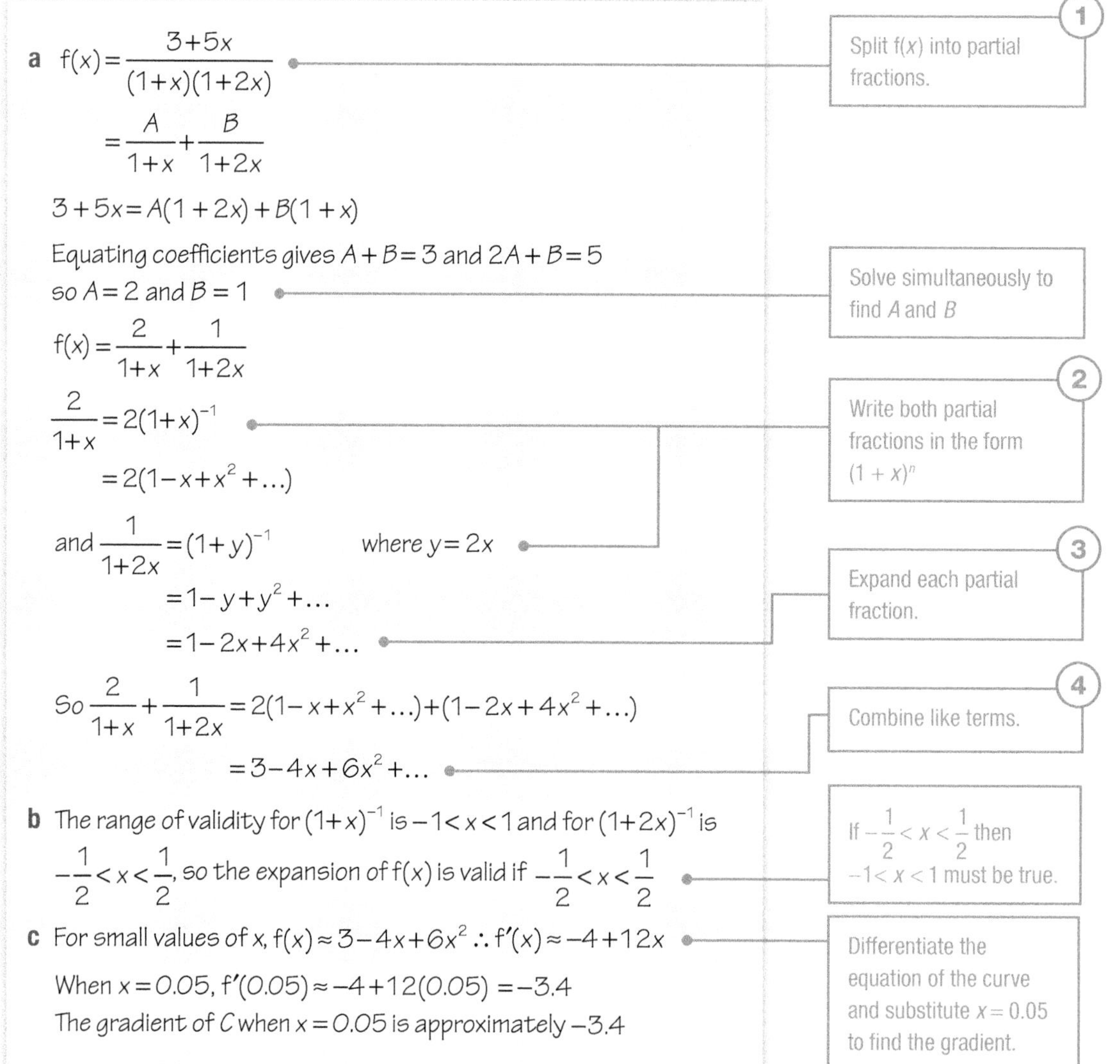

a $f(x) = \dfrac{3+5x}{(1+x)(1+2x)}$ — *Split f(x) into partial fractions.* (1)

$= \dfrac{A}{1+x} + \dfrac{B}{1+2x}$

$3+5x = A(1+2x) + B(1+x)$

Equating coefficients gives $A+B=3$ and $2A+B=5$

so $A=2$ and $B=1$ — *Solve simultaneously to find A and B*

$f(x) = \dfrac{2}{1+x} + \dfrac{1}{1+2x}$

$\dfrac{2}{1+x} = 2(1+x)^{-1}$ — *Write both partial fractions in the form $(1+x)^n$* (2)

$= 2(1-x+x^2+\ldots)$

and $\dfrac{1}{1+2x} = (1+y)^{-1}$ where $y = 2x$

$= 1-y+y^2+\ldots$

$= 1-2x+4x^2+\ldots$ — *Expand each partial fraction.* (3)

So $\dfrac{2}{1+x} + \dfrac{1}{1+2x} = 2(1-x+x^2+\ldots)+(1-2x+4x^2+\ldots)$

$= 3-4x+6x^2+\ldots$ — *Combine like terms.* (4)

b The range of validity for $(1+x)^{-1}$ is $-1<x<1$ and for $(1+2x)^{-1}$ is $-\frac{1}{2}<x<\frac{1}{2}$, so the expansion of $f(x)$ is valid if $-\frac{1}{2}<x<\frac{1}{2}$ — *If $-\frac{1}{2}<x<\frac{1}{2}$ then $-1<x<1$ must be true.*

c For small values of x, $f(x) \approx 3-4x+6x^2$ $\therefore f'(x) \approx -4+12x$ — *Differentiate the equation of the curve and substitute $x = 0.05$ to find the gradient.*

When $x = 0.05$, $f'(0.05) \approx -4+12(0.05) = -3.4$

The gradient of C when $x = 0.05$ is approximately -3.4

See Ch4.2 For a reminder on differentiation.

Use the quotient rule to work out the exact value of $f'(0.05)$ and compare your answer with this estimate. See Ch15.4 for a reminder on the quotient rule.

Exercise 13.1B Reasoning and problem-solving

1 a Work out the first three terms, in ascending powers of x, in the binomial expansion of $\sqrt{1+5x}$

b Use this series approximation and $x=0.05$ to show that $\sqrt{5}\approx\frac{143}{64}$

c Explain why substituting $x=0.8$ into the full expansion of $\sqrt{1+5x}$ does not give an answer equal to $\sqrt{5}$

2 a Work out the expansion of $(4+x)^{\frac{1}{2}}$, up to and including the term in x^2. State the range of values of x for which the full expansion of $(4+x)^{\frac{1}{2}}$ is valid.

b Use the series approximation found in part **a** and

i $x=0.25$ to show that $\sqrt{17}\approx\frac{2111}{512}$

ii $x=-0.25$ to calculate an estimate for $\sqrt{15}$

c Hence show that $\sqrt{17}-\sqrt{15}\approx\frac{1}{4}$

3 The first three terms in the binomial expansion of $(1+3x)^n$ are $1+kx-x^2$, where n and k are constants, $n>\frac{1}{2}$

a Work out the value of n and the value of k

b Use this series approximation and a suitable value of x to calculate an estimate for $\sqrt[3]{1.69}$

4 The first three terms in the binomial expansion of $(1+px)^n$ are $1+12x+24x^2$, where p and n are constants.

Using the series approximation and $x=0.01$, show that $\sqrt{3}\approx\frac{1403}{810}$

5 a Work out the expansion of $\frac{3+7x}{(1+x)(1+3x)}$ up to and including the term in x^3

b State the range of values of x for which this full expansion is valid.

6 The curve C has equation $y=\mathrm{f}(x)$, where $\mathrm{f}(x)=\frac{2-3x}{2-3x-2x^2}$

a Work out the expansion of $\mathrm{f}(x)$ up to and including the term in x^2

b Sketch a graph which approximates C for values of x close to zero.

7 $\mathrm{f}(x)=\frac{x+4}{(1+x)(2+x)^2}$

The curve C has equation $y=\mathrm{f}(x)$ and passes through point P with x-coordinate -0.1

a Calculate the y-coordinate of point P, giving your answer to 1 decimal place.

b Work out the first three terms in the expansion of $\mathrm{f}(x)$, giving your answer in the form $p+qx+rx^2$, stating the value of each constant p, q and r

c Use your answers to parts **a** and **b** to show that the tangent to this curve at point P can be approximated by the line with equation $y=0.98-2.2x$

Challenge

8 a By using $\cos^2 x+\sin^2 x\equiv 1$, show that $\frac{1}{\cos^2 x}=1+\sin^2 x+\sin^4 x+\sin^6 x+...$ and explain why this expansion is valid for $-90°<x<90°$

It is given that $\frac{1}{\sin^2 x}=1+\cos^2 x+\cos^4 x+\cos^6 x+...$ where $0°<x<90°$

b Read through this argument and explain where the error has occurred.

For $0°<x<90°$, $\frac{1}{\cos^2 x}\approx 1+\sin^2 x$

and $\frac{1}{\sin^2 x}\approx 1+\cos^2 x$

Adding these expressions

$\frac{1}{\cos^2 x}+\frac{1}{\sin^2 x}\approx(1+\cos^2 x)+(1+\sin^2 x)$

so $\frac{1}{\cos^2 x}+\frac{1}{\sin^2 x}\approx 3$

(as $\cos^2 x+\sin^2 x\equiv 1$)

But when $x=15°$, $\frac{1}{\cos^2 15°}+\frac{1}{\sin^2 15°}=16$ which is not close to 3

MyMaths 2204, 2205 SEARCH

13.2 Introduction to sequences

Fluency and skills

A **sequence** is a list of numbers, called **terms**. You can refer to each term by its position in the list. For example, for the sequence 3 , 6 , 9 , 12 , ..., 3 is the first term, 6 is the second term and so on.

You can use the notation u_n for the nth term of a sequence, where n is a positive integer. So for the sequence 3 , 6 , 9 , 12 , ... , you know that u_1 = 1st term = 3, u_2 = 2nd term = 6, and u_n = nth term.

Example 1

Work out the value of the first three terms of the sequence with nth term u_n where $u_{n+1} = 2u_n + 3$, $u_1 = 5$

$u_1 = 5$ — Write down the first term.

$n = 1$, $u_{1+1} = 2u_1 + 3$

$u_2 = 2(5) + 3 = 13$ — Apply the given formula to find the second term.

$n = 2$, $u_{2+1} = 2u_2 + 3$

$u_3 = 2(13) + 3 = 29$ — Use u_2 to find u_3

The first three terms are 5, 13 and 29

Example 1 uses a **recurrence relation** (or iterative formula), which can be simpler than a direct formula.

Calculator

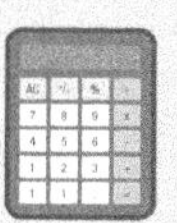

Try it on your calculator

You can calculate terms of a sequence defined by a recurrence relation on a calculator.

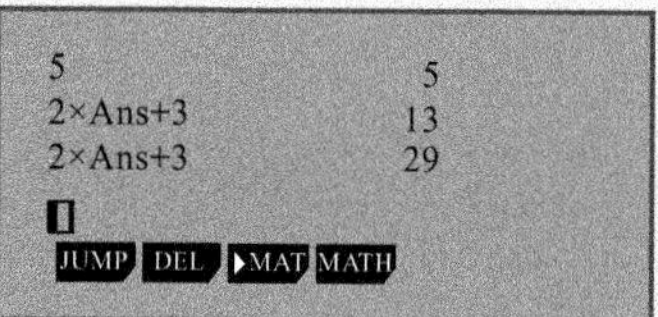

Activity

Find out how to calculate the first three terms of the sequence defined by $u_{n+1} = 2u_n + 3$, $u_1 = 5$ on *your* calculator.

Key point

A sequence with nth term u_n is

- **Increasing** if $u_{n+1} > u_n$ for all $n \in \mathbb{Z}^+$
- **Decreasing** if $u_{n+1} < u_n$ for all $n \in \mathbb{Z}^+$
- **Periodic** if the sequence consists of repeated terms.

Periodic sequences are neither increasing nor decreasing. For example, in the sequence 4, 3, 4, 3, 4, 3 ..., $u_{n+1} > u_n$ and $u_{n+1} < u_n$ are only sometimes true.

Example 2

Describe the list u_1, u_2, u_3, u_4 as either an increasing, decreasing or periodic sequence, where

a $u_n = n^2 + 2n + 3$ **b** $u_n = \cos(180n)°$ **c** $u_n = 0.5^n + 2$

a The terms 6, 11, 18, 27 form an increasing sequence. — e.g. $u_3 = (3)^2 + 2(3) + 3 = 18$

b The terms −1, 1, −1, 1 form a periodic sequence. — e.g. $u_2 = \cos(180 \times 2)° = 1$

c The terms 2.5, 2.25, 2.125, 2.0625 form a decreasing sequence. — e.g. $u_1 = 0.5^1 + 2 = 2.5$

Key point

A periodic sequence has **period** (or **order**) k, where k is the smallest number of terms that repeat as a block of numbers. For example, the periodic sequence 4, 1, 7, 4, 1, 7, 4, 1, 7, ... has period 3 since the first three terms 4, 1 and 7 repeat in blocks.

Key point

A sequence $u_1, u_2, u_3, \ldots$ **converges** to a number L if the terms get ever closer to L. L is called the **limit** of the sequence.

For example, if $u_n = \frac{n+1}{n}$, then $u_1 = 2$, $u_{10} = 1.1$, $u_{100} = 1.01$ and $u_{1000} = 1.001$

The terms of this sequence are getting ever closer to 1, which suggests that 1 is the limit of this sequence. You write $u_n \to 1$ as $n \to \infty$

The symbol $\to$ means 'tends to'. So $n \to \infty$ means 'n tends to infinity'.

You can also find the limit of a sequence by solving an equation.

Example 3

A sequence is defined by $u_{n+1} = 0.2u_n + 2$, $u_1 = 3$. The limit of u_n as $n \to \infty$ is L. Find the value of L

As $n \to \infty$, the terms u_n and u_{n+1} are approximately equal to L

Since $u_{n+1} = 0.2u_n + 2$, L must satisfy the equation $L = 0.2L + 2$ — Replace u_n and u_{n+1} with L in the equation $u_{n+1} = 0.2u_n + 2$

$0.8L = 2$ so $L = 2.5$ — Solve the equation $L = 0.2L + 2$

The limiting value of this sequence is 2.5

Use your calculator to verify that $u_n \to 2.5$ as $n \to \infty$

Exercise 13.2A Fluency and skills

In these questions, u_n represents the nth term of a sequence where $n \in \mathbb{Z}^+$

1 Work out the values of the first four terms of these sequences.

a $u_n = 5n+3$ b $u_n = n^2 - 3$

c $u_n = 2n^2(n-1)$ d $u_n = \frac{n}{n^2-6}$

2 Work out the values of u_2, u_3 and u_4 for these sequences.

a $u_{n+1} = 2u_n + 5$, $u_1 = 1$

b $u_{n+1} = u_n^2 + 2$, $u_1 = 1$

c $u_{n+1} = \frac{1}{u_n + 1}$, $u_1 = 3$

d $u_{n+1} = (u_n - 3)^2$, $u_1 = 5$

3 Describe the list u_1, u_2, u_3, u_4 as either an increasing sequence, a decreasing sequence or neither where

a $u_n = \frac{6}{n+1}$ b $u_n = n^2 + 4n - 3$

c $u_n = n^2 - 6n + 1$ d $u_n = \sqrt[n]{n}$

e $u_n = \log_{n+1}(n+2)$

f $u_n = \sin(30n)°\cos(30n)°$

4 Evaluate the order of these periodic sequences.

a $u_n = \tan(n \times 60°)$

b $u_{n+1} = 3 - \frac{9}{u_n}$, $u_1 = 1$

c $u_{n+1} = (-1)^n u_n$, $u_1 = 2$

d $u_{n+1} = \frac{2}{9}\left(3 - \frac{1}{u_n}\right)$, $u_1 = 3$

5 The nth term of a sequence is given by $u_n = 3n - 2$

a Calculate the value of n such that $u_n = 229$

b Work out the values of n for which $u_n > 1000$

6 The nth term of a sequence is given by $u_n = n^2 - 8n + 18$

a Calculate the value of n such that $u_n = 83$

b Work out the value of the smallest term in this sequence.

7 By solving an equation, find the limit L of these sequences as $n \to \infty$. Where appropriate, give answers in simplified surd form.

a $u_{n+1} = 0.4u_n + 3$ b $u_{n+1} = 0.6u_n - 3$

c $u_{n+1} = 3 - 0.25u_n$ d $u_{n+1} = \frac{3}{5}u_n - \frac{1}{2}$

e $u_{n+1} = 0.1(7u_n + 2)$ f $u_{n+1} = \frac{\sqrt{5}u_n + 4}{3}$

g $u_{n+1} = \left(\sqrt{3} - 2\right)u_n + 2\sqrt{3}$

h $u_{n+1} = \frac{1}{4}u_n^2 + 1$

Use your calculator or a spreadsheet with starting value $u_1 = 1$ to verify each answer.

Reasoning and problem-solving

Strategy

To solve a problem using a sequence

1. Define variables and translate words into mathematics.
2. Work out unknown values.
3. Give your answer in context, rounded appropriately.

You can describe, or **model**, real-life situations using sequences.

Example 4

The total number of people, p_n, who became infected with a virus n weeks after it started to spread can be modelled by the equation $p_{n+1} = 1.05p_n + c$, where c is a constant.

After week 1, 20 people were infected. During week 2, 41 people became infected.

Use this model to estimate the number of people who became infected during week 7

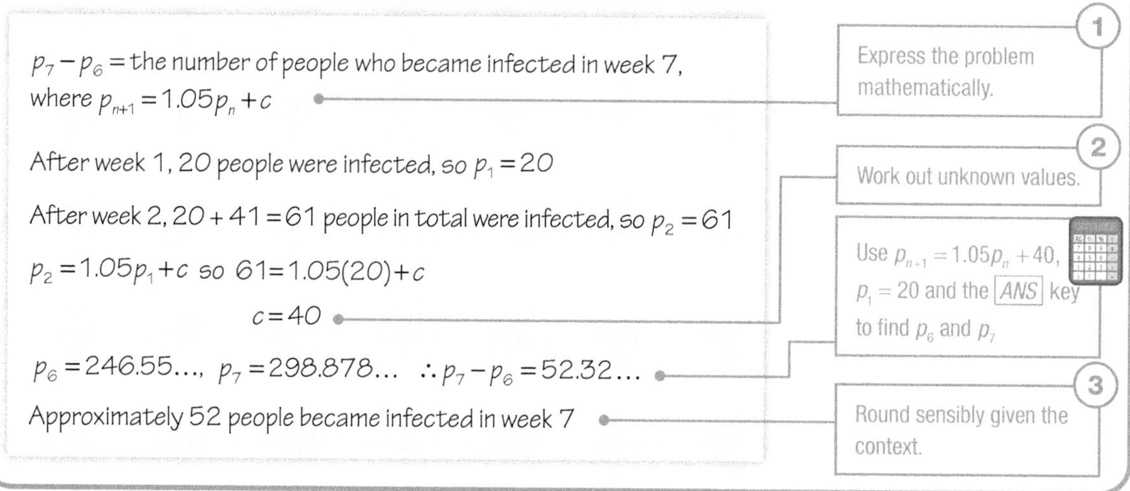

$p_7 - p_6 =$ the number of people who became infected in week 7, where $p_{n+1} = 1.05p_n + c$

After week 1, 20 people were infected, so $p_1 = 20$

After week 2, $20 + 41 = 61$ people in total were infected, so $p_2 = 61$

$p_2 = 1.05p_1 + c$ so $61 = 1.05(20) + c$

$c = 40$

$p_6 = 246.55...$, $p_7 = 298.878...$ $\therefore p_7 - p_6 = 52.32...$

Approximately 52 people became infected in week 7

1 Express the problem mathematically.
2 Work out unknown values.
Use $p_{n+1} = 1.05p_n + 40$, $p_1 = 20$ and the ANS key to find p_6 and p_7
3 Round sensibly given the context.

Exercise 13.2B Reasoning and problem-solving

In all questions, assume $n \in \mathbb{Z}^+$

1 A sequence is defined by the equation $u_{n+1} = 4u_n + k$, $u_1 = 1$, where k is a constant. Given that $u_3 = 31$

a Work out the value of u_4

b Evaluate the sum of the first four terms of this sequence.

2 A sequence is defined by the equation $u_{n+1} = au_n + 7$, $u_1 = -1$, where a is a constant. Given that $u_3 = 19$

a Show that $a^2 - 7a + 12 = 0$

b Work out the possible values of u_4

3 A sequence is defined by $u_{n+1} = au_n + b$ for constants a and b. The first three terms of this sequence are 15, 24 and 42, respectively.

a Calculate the values of a and b

Given that $u_{m+1} = 1.96u_m$, where $m \geq 2$

b Work out the value of u_{m+2}

4 A periodic sequence satisfies the equation $u_{n+1} = a - \frac{2a}{u_n}$, where $a > 0$ is a constant.
The sequence has period 3 and $u_3 = -2$

a Show that

i $u_2 = a - 1$ ii $a^2 - a - 2 = 0$

b Hence find the sum of the first 300 terms of this sequence.

5 Monthly observations at a wildlife sanctuary suggest that its population of otters is declining. The number of otters, p_n, seen during the $(n+1)$th observation was modelled by the equation $p_{n+1} = 0.75(p_n + 8)$. During the first observation 56 otters were seen.

a Show that there was a 12.5% decrease in the number of otters seen in the third observation compared to the second.

In one particular observation, at least 24 otters were observed.

b Show that the number of otters seen in the next observation was also at least 24

c Explain why, according to this model, the population of otters will never die out.

6 A code breaking competition consists of 10 rounds, each more difficult than the previous one. A round starts when the code is issued and contestants must break the code within two hours before being allowed to progress to the next round. It takes one of the contestants, Sam, m_n minutes to break the code in round n where $m_{n+1} = a(m_n - 1)$ and a is a positive constant. Sam takes 4 minutes to break the code in round 2 and 10 minutes to break the code in round 4

a Show that $3a^2 - a - 10 = 0$

b Work out the time, in minutes, that it takes Sam to break the code in round 1

c How many rounds of this competition does Sam successfully take part in?

7 In preparation to run a race, Paula undertakes weekly training sessions. In the nth session she runs e_n miles due East from her house, turns due South and runs s_n miles and then runs directly back to her house, so that the path she takes in each session is a right-angled triangle.

In the first session she runs 1.5 miles due East and 2 miles due South.

a Calculate the total distance she runs in session 1

For $n \geq 1$, it is given that $e_{n+1} = \frac{2}{3}e_n + 2$ and that $s_{n+1} = \frac{1}{2}s_n + k$ where k is a constant. In session 2, Paula runs 12 miles in total.

b Show that $\sqrt{10 + 2k + k^2} = 8 - k$ and hence evaluate the value of k

Paula correctly calculates that, to the nearest mile, the distance she runs in session 8 equals the length of the race.

c Calculate, to the nearest mile, the length of the race.

8 A sequence is defined by $u_{n+1} = pu_n + 6$, $u_1 = 5$, where p is a positive constant.

Given that $u_3 = 9.2$

a Show that $5p^2 + 6p - 3.2 = 0$

The limit of u_n as $n \to \infty$ is L

b Find the value of L

Challenge

9 A sequence is defined by $u_{n+1} = pu_n + q$, $u_1 = 12$, where p and q are constants. The second term of this sequence is 10 and the limit as $n \to \infty$ is 2

a Find the value of p and the value of q

b For these values of p and q, find the limit as $n \to \infty$ of the sequence defined by $v_{n+1} = qv_n + p$, $v_1 = 12$

13.3 Arithmetic sequences

Fluency and skills

Key point

A sequence is **arithmetic** if adding a constant to any term gives the next term.

For example, 4, 7, 10, 13, ... is an arithmetic sequence. You add 3 to each term to get the next term.

In this example, the constant 3 is called the **common difference** of the sequence. The difference $u_3 - u_2 = 10 - 7 = 3$

This means the difference $u_{n+1} - u_n = 3$ for all values of n

Key point

If an arithmetic sequence has first term a and common difference d, then

- For all n, $d = u_{n+1} - u_n$
- The terms are $a, a+d, a+2d, a+3d, \ldots$
- The nth term of the sequence is $u_n = a+(n-1)d$. For example the **4**th term is $a + (\mathbf{4} - 1)d$ or $a + 3d$
- If $d > 0$, the sequence is increasing and if $d < 0$, the sequence is decreasing.

Example 1

a Work out the nth term of the arithmetic sequence 3, 8, 13, 18, ...

b Hence evaluate the value of the 50th term of this sequence.

a First term a is 3. Hence, common difference d is $8-3=5$ — $d = u_2 - u_1$

n^{th} term is $u_n = a+(n-1)d = 3+(n-1)\times 5 = 5n-2$ — Put a and d into the nth term formula.

b When $n=50$, $u_{50} = 5(50)-2 = 248$ — Substitute $n = 50$

Key point

An **arithmetic series** is the sum of the terms of an arithmetic sequence.

For the arithmetic sequence $a, a+d, a+2d, a+3d, \ldots$

the corresponding arithmetic series is $a + (a+d) + (a+2d) + (a+3d) + \ldots$

An arithmetic sequence and its series share the same nth term and the same common difference.

You can use the notation S_n for the sum of the first n terms of an arithmetic series.

For example, for the arithmetic series $4+7+10+13+\ldots$

$S_2 = 4+7 = 11$ and $S_3 = 4+7+10 = 21$

You can work out a formula for the sum S_n. If you write out the series twice—once in order and once in reverse order—then a shortcut for finding the sum becomes clear.

$$S_n = u_1 + u_2 + u_3 + \dots + u_{n-1} + u_n$$

$$S_n = u_n + u_{n-1} + u_{n-2} + \dots + u_2 + u_1$$

$$\therefore S_n + S_n = (u_1 + u_n) + (u_2 + u_{n-1}) + (u_3 + u_{n-2}) + \dots + (u_{n-1} + u_2) + (u_n + u_1)$$

You can see that each of the bracketed terms in this sum are equivalent to $u_1 + u_n$. For example, $u_2 + u_{n-1} = (u_1 + d) + (u_n - d) = u_1 + u_n$ and $u_3 + u_{n-2} = (u_1 + 2d) + (u_n - 2d) = u_1 + u_n$

So $2S_n = \underbrace{(u_1 + u_n) + (u_1 + u_n) + (u_1 + u_n) + \dots + (u_1 + u_n) + (u_1 + u_n)}_{n \text{ brackets}}$

$= n(u_1 + u_n)$

Therefore, $S_n = \frac{1}{2}n(u_1 + u_n)$

Key point

If an arithmetic series has first term a and common difference d, then $S_n = \frac{1}{2}n(a + l)$, where $l = u_n$, the last term of the sum.

You can also write $S_n = \frac{1}{2}n[2a + (n-1)d]$, since $u_1 + u_n = a + (a + (n-1)d) = 2a + (n-1)d$. You can use this formula if you do not know the last term.

'Evaluate' means 'find the value of.'

Example 2

The arithmetic series $7 + 12 + 17 + \dots + 102$ has 20 terms.

a Evaluate this series. **b** Calculate the sum of the first 10 terms of this series.

a The sum has $n = 20$ terms, the first is $a = 7$ and the last is $l = 102$

$S_{20} = \frac{1}{2}(20)(7 + 102)$

$= 1090$

Use $S_n = \frac{1}{2}n(a + l)$

b The first term is $a = 7$ and the common difference, $d = 5$

$S_{10} = \frac{1}{2}(10)[2 \times 7 + (10 - 1) \times 5]$

$= 295$

Use $S_n = \frac{1}{2}n[2a + (n-1)d]$

You can use the symbol $\sum$ for a series. In the notation $\sum_{k=1}^{3} k^2$, the **index** k varies, taking in turn the values 1, 2, and 3

So $\sum_{k=1}^{3} k^2 = 1^2 + 2^2 + 3^2 = 14$

You read $\sum_{k=1}^{3} k^2$ as 'the sum of k^2 as the integer k varies from 1 to 3.'

Similarly, $\sum_{k=1}^{3} (4k - 3) = (4 \times 1 - 3) + (4 \times 2 - 3) + (4 \times 3 - 3) = 1 + 5 + 9 = 15$

Key point

For any function f, $\sum_{k=1}^{n} f(k) = f(1) + f(2) + \dots + f(n)$ In particular, $\sum_{k=1}^{n} 1 = n$

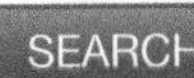

You may want to find the sum of a series where the value of the index k does not start at 1

e.g. $\sum_{k=3}^{6} k^2 = \underbrace{3^2+4^2+5^2+6^2}_{4\text{ terms}}$

In these instances, you can use the rule $\sum_{k=a}^{n} u_k = \sum_{k=1}^{n} u_k - \sum_{k=1}^{a-1} u_k$

Calculator

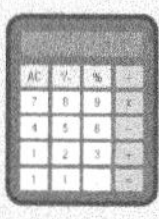

Try it on your calculator

You can calculate the sum of a series on a calculator.

$\sum_{x=2}^{7}(8x-3)$

198

Activity

Find out how to calculate $\sum_{x=2}^{7}(8x-3)$ on *your* calculator.

Example 3

Use the formula to evaluate the arithmetic series $\sum_{k=1}^{11}(6k+1)$

$\sum_{k=1}^{11}(6k+1) = (6\times1+1)+(6\times2+1)+(6\times3+1)+\ldots+(6\times11+1)$

$= \underbrace{7+13+19+\ldots+67}_{11\text{ terms}}$ — Write out the first few terms and the last term of the series.

There are 11 terms in this arithmetic series. The first is 7 and the last is 67

$S_{11} = \frac{1}{2}(11)(7+67) = 407$ — Use $S_n = \frac{1}{2}n(a+l)$

Exercise 13.3A Fluency and skills

In these questions u_n represents the nth term of a sequence where $n \geq 1$

1 Work out the values of the first four terms of these arithmetic sequences.

a $u_n = 3+2n$ b $u_n = 12-3n$

c $u_n = 7n-4$ d $u_n = \frac{3}{2}+\frac{5}{2}n$

2 Work out expressions for the nth terms of these arithmetic sequences, simplifying each answer as far as possible.

a $7, 11, 15, \ldots$ b $2, -1, -4, \ldots$

c $4, 5.5, 7, \ldots$ d $2, \frac{8}{3}, \frac{10}{3}, \ldots$

e $\frac{3}{5}, 1, \frac{7}{5}, \ldots$ f $\sqrt{2}, \sqrt{8}, \sqrt{18}, \ldots$

3 Work out the value of the 50th term in each of these sequences.

a $3, 6, 9, \ldots$ b $8, 5, 2, \ldots$

c $\frac{5}{4}, \frac{5}{2}, \frac{15}{4}, \ldots$ d $5.4, 3.2, 1.0, \ldots$

4 Evaluate these arithmetic series. The number of terms in each sum is shown in brackets.

a $60+57+\ldots+18$ (15 terms)

b $10+6+2+\ldots-34$ (12 terms)

c $2+2.5+3+\ldots$ (99 terms)

d $\sqrt{3}+\sqrt{27}+\ldots$ (10 terms)

5 Work out the number of terms in each of these arithmetic series.

a $4+\ldots+151$ when $S_n = 3875$

b $7+\ldots+52$ when $S_n = 295$

c $4+\ldots+44$ when $S_n = 2400$

d $\frac{1}{5}+\ldots+\frac{28}{5}$ when $S_n = 29$

6 Use the formula to evaluate these arithmetic series.

a $\sum_{k=1}^{20}(2k+3)$ b $\sum_{k=1}^{12}(45-3k)$

c $\sum_{k=3}^{9}(3k-1)$ d $\sum_{k=5}^{40}(1.2k+3)$

Reasoning and problem-solving

PURE

Strategy

To solve a problem using an arithmetic sequence or series

1. Define variables such as the first term a, the last term l and the common difference d using information given in the question.
2. Work out any unknown values.
3. Use appropriate formulae and give your answer in context, rounding appropriately.

Example 4

The third term of an arithmetic series is 15 and the seventh term is 31. Calculate the sum of the first 10 terms of this series.

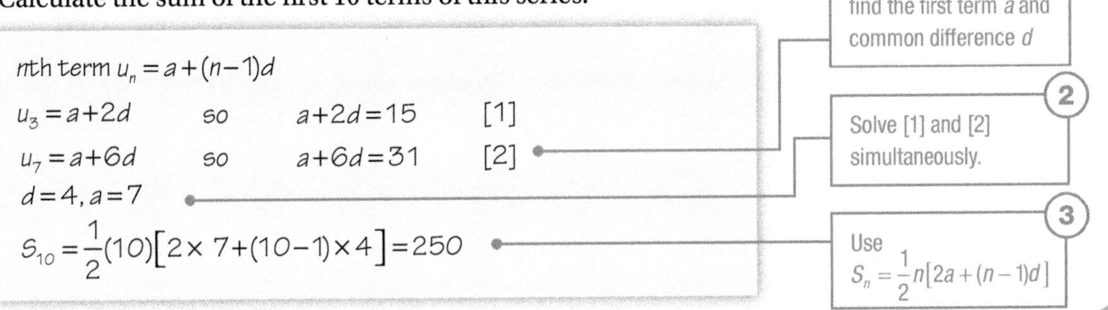

nth term $u_n = a + (n-1)d$

$u_3 = a + 2d$ so $a + 2d = 15$ [1]

$u_7 = a + 6d$ so $a + 6d = 31$ [2]

$d = 4, a = 7$

$S_{10} = \frac{1}{2}(10)[2 \times 7 + (10-1) \times 4] = 250$

1. Set up equations to find the first term a and common difference d
2. Solve [1] and [2] simultaneously.
3. Use $S_n = \frac{1}{2}n[2a + (n-1)d]$

You can create a mathematical model to approximate real-life situations involving an arithmetic sequence or series.

In this question you need to work with S_n, the **total** amount repaid after n working years.

Example 5

After graduating, Jane found a job and started to repay her student loan of £10 500

In her first working year, she repaid £550. In her second working year, she repaid £700.

The amount she repaid in each working year is modelled by an arithmetic sequence.

a Use this model to

  i Show that, by the end of her fourth working year, Jane owed £7400

  ii Work out the number of whole working years during which Jane owed more than £500

b Comment on the suitability of this model.

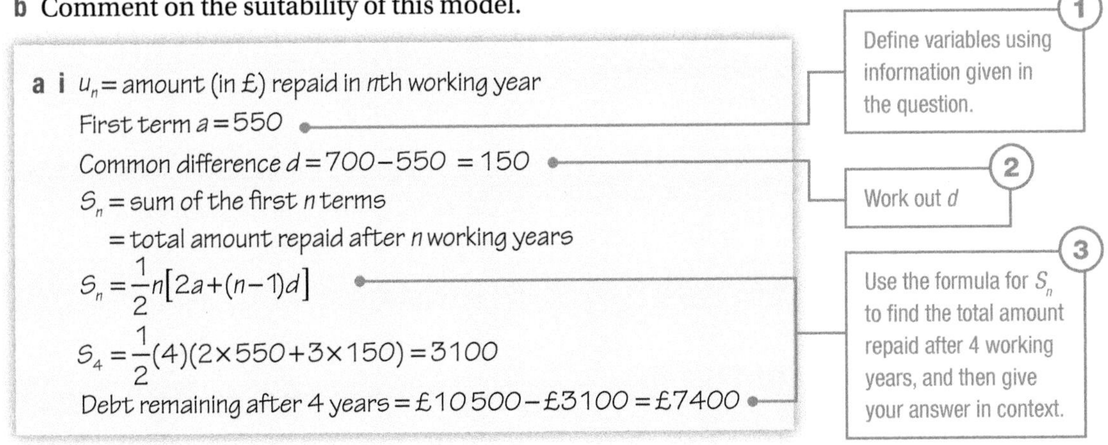

a i u_n = amount (in £) repaid in nth working year

First term $a = 550$

Common difference $d = 700 - 550 = 150$

S_n = sum of the first n terms

= total amount repaid after n working years

$S_n = \frac{1}{2}n[2a + (n-1)d]$

$S_4 = \frac{1}{2}(4)(2 \times 550 + 3 \times 150) = 3100$

Debt remaining after 4 years = £10 500 − £3100 = £7400

1. Define variables using information given in the question.
2. Work out d
3. Use the formula for S_n to find the total amount repaid after 4 working years, and then give your answer in context.

(Continued on the next page)

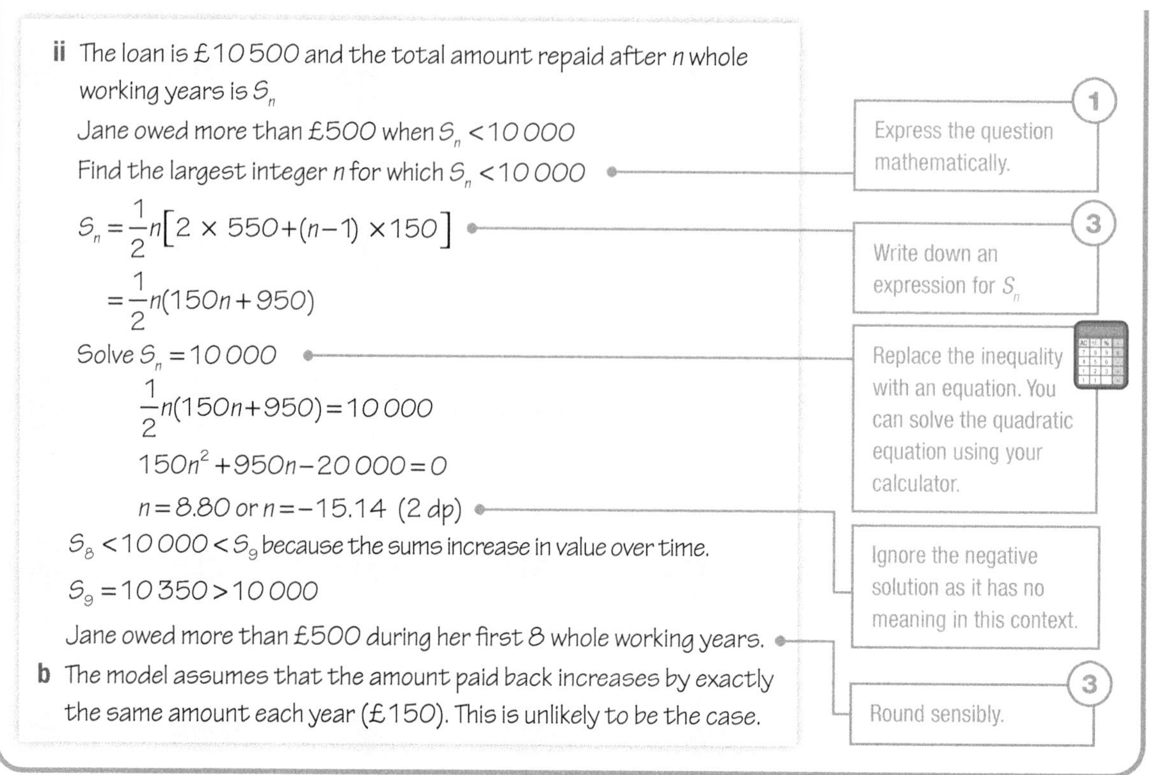

ii The loan is £10500 and the total amount repaid after n whole working years is S_n
Jane owed more than £500 when $S_n < 10000$
Find the largest integer n for which $S_n < 10000$
$S_n = \frac{1}{2}n[2 \times 550 + (n-1) \times 150]$
$= \frac{1}{2}n(150n + 950)$
Solve $S_n = 10000$
$\frac{1}{2}n(150n+950) = 10000$
$150n^2 + 950n - 20000 = 0$
$n = 8.80$ or $n = -15.14$ (2 dp)
$S_8 < 10000 < S_9$ because the sums increase in value over time.
$S_9 = 10350 > 10000$
Jane owed more than £500 during her first 8 whole working years.
b The model assumes that the amount paid back increases by exactly the same amount each year (£150). This is unlikely to be the case.
1
Express the question mathematically.
3
Write down an expression for S_n
Replace the inequality with an equation. You can solve the quadratic equation using your calculator.
Ignore the negative solution as it has no meaning in this context.
3
Round sensibly.

Exercise 13.3B Reasoning and problem-solving

1 The nth term of an arithmetic sequence is u_n, where $u_4 = 21$ and $u_7 = 36$

a Evaluate the first term a and the common difference d of this sequence.

b Calculate the value of N such that $u_N = 6 \times u_{10}$

2 The nth term of an arithmetic sequence is u_n, where $u_5 = 13$ and $u_{10} = 7u_2$

a Show that $u_{18} = 52$

b Work out the number of terms in this sequence which are factors of u_{18}

3 An arithmetic series has 20 terms. The first term is 4 and the sum of the first 10 terms of this series is 175

Calculate the sum of the last 10 terms of this series.

4 An arithmetic series has first term a and common difference d

The nth term is u_n and S_n is the sum of the first n terms of this series.

Given that $u_8 = 26$ and $S_5 = 205$

a Calculate the value of the smallest positive term of this series,

b Work out the greatest value of S_n

5 The first term of an arithmetic series is 5 and the common difference is 4

The nth term is u_n and S_n is the sum of the first n terms of this series.

a Show that $S_n = n(2n+3)$

b Work out the value of u_N, given that $S_N = 779$

6 An arithmetic series has 20 terms. The nth term is u_n and S_n is the sum of the first n terms of this series.

Given that $S_5 = 85$

a Work out the value of u_3

Given further that $S_{17} = 35 \times u_3$

b Evaluate the sum of all the terms of this series.

7 Quickline trains has invested in new timetabling software. The number of complaints received from their passengers each month in the following year was modelled using an arithmetic sequence. In the first month, 152 complaints were received. This decreased to 140 complaints in the second month.

a Use this model to show that a total of 1032 complaints were received during the year.

The train company claimed that there was roughly a 60% reduction in the number of complaints received between the first six months and the last six months of that year.

b Use this model to determine whether this claim is accurate.

c Comment on the suitability of this model in the following year.

8 The number of departures from a new airport each month was modelled using an arithmetic sequence. There were 450 departures in its first month of being operational. In the first three months of the airport being operational, there were 1470 departures.

a Use this model to work out the total number of departures in the first six months of the airport being operational.

If the number of departures in any complete month exceeded 1000 then the airport authority paid an environmental fine of £5000 for that month.

b Use this model to calculate how much the airport authority was fined in its first two years of being operational.

9 Jim has trained for a new job in sales. Each year he earns a basic salary of £24 000 plus 5% commission on the value of any sales made during that year. In his first working year, Jim earned £27 500. In his second working year, he made sales worth £80 000. The amount Jim earned each year was modelled using an arithmetic sequence.

a Use this model to work out

i The total amount Jim earned in his first five working years,

ii The total value of sales Jim made in his first five working years.

At the end of each working year, Jim paid 8% of the amount earned that year into a pension fund. The amount paid in each year formed an arithmetic sequence.

b Calculate the minimum number of whole years Jim will need to work for these pension contributions to have a total value of more than £30 000

c Give one criticism of using an arithmetic sequence to model the amount Jim earned each year.

Challenge

10 Without using a calculator, work out a relationship between the sums

$1+2+3+\ldots+n$ and $1^3+2^3+3^3+\ldots+n^3$

and hence find the value of the positive integer n for which

$1^3+2^3+3^3+\ldots+n^3=90\,000$

13.4 Geometric sequences

Fluency and skills

Key point

A sequence is **geometric** if multiplying any term by a fixed non-zero number gives the next term in the sequence.

For example, 3, 6, 12, 24, ... is a geometric sequence. You multiply each term by 2 to get the next term. In this example, the multiplier 2 is called the **common ratio** of the sequence. The ratio $\frac{u_3}{u_2}=\frac{12}{6}=2$

This means $\frac{u_{n+1}}{u_n}=2$ for all values of n

Key point

If a geometric sequence has first term a and common ratio r (both non-zero), then

- For all n, $r=\frac{u_{n+1}}{u_n}$
- The terms are $a, a\times r, a\times r^2, a\times r^3, \ldots$
- The nth term of the sequence is $u_n=a\times r^{n-1}$. For example, the **3**rd term is $a\times r^{3-1}$, or ar^2

Example 1

a Work out the nth term of the geometric sequence 2, 6, 18, 54, ...

b Calculate the value of the 10th term of this sequence.

a The first term $a=2$, the common ratio $r=\frac{u_2}{u_1}=\frac{6}{2}=3$

nth term is $u_n=a\times r^{n-1}=2\times 3^{n-1}$

b When $n=10$, $u_{10}=2\times 3^{10-1}=39\,366$

Key point

A **geometric series** is the sum of the terms of a geometric sequence.
For the geometric sequence $a, a\times r, a\times r^2, a\times r^3, \ldots$
the corresponding geometric series is
$a+(a\times r)+(a\times r^2)+(a\times r^3)+\ldots$
You say the series has nth term $a\times r^{n-1}$

A geometric sequence and its series share the same nth term and the same common ratio.

You can use the notation S_n for the sum of the first n terms of a geometric series.

For example, for the geometric series $3+6+12+24+\ldots$

$S_2=3+6=9$ and $S_3=3+6+12=21$

You can work out a formula for S_n

$$S_n = a + ar + ar^2 + \ldots + ar^{n-2} + ar^{n-1} \quad [1]$$

$$\therefore S_n \times r = (a + ar + ar^2 + \ldots + ar^{n-2} + ar^{n-1}) \times r$$

$$S_n \times r = ar + ar^2 + ar^3 + \ldots + ar^{n-1} + ar^n \quad [2]$$

[2]−[1]:

$$(S_n \times r) - S_n = (ar + ar^2 + ar^3 + \ldots + ar^{n-1} + ar^n) - (a + ar + ar^2 + \ldots + ar^{n-1})$$

$$\therefore S_n(r-1) = ar^n - a \qquad \text{so } S_n = \frac{a(r^n-1)}{r-1} \qquad \text{(provided } r \neq 1\text{)}$$

ICT Resource online

To investigate sequences, click this link in the digital book.

Key point

If a geometric series has first term a and common ratio r $(r \neq 1)$, then $S_n = \frac{a(r^n-1)}{r-1}$ or, equivalently, $S_n = \frac{a(1-r^n)}{1-r}$

When $r < 1$ use $S_n = \frac{a(1-r^n)}{1-r}$ to avoid unnecessary negative signs in your working.

Example 2

Calculate the sum of the first 12 terms of the geometric series 4+6+9+..., to 3 sf.

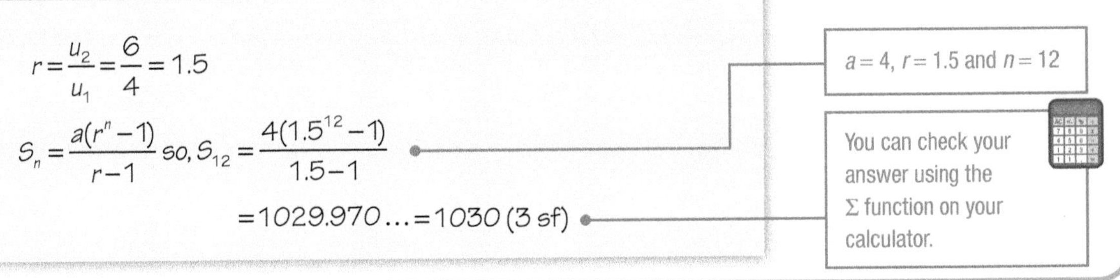

$$r = \frac{u_2}{u_1} = \frac{6}{4} = 1.5$$

$$S_n = \frac{a(r^n-1)}{r-1} \text{ so, } S_{12} = \frac{4(1.5^{12}-1)}{1.5-1}$$

$$= 1029.970\ldots = 1030 \text{ (3 sf)}$$

$a = 4$, $r = 1.5$ and $n = 12$

You can check your answer using the Σ function on your calculator.

You can evaluate a geometric series that is written using sigma notation.

Example 3

Use a formula to evaluate the geometric series $\sum_{k=1}^{8}(10 \times 0.2^k)$. Give your answer to 2 decimal places.

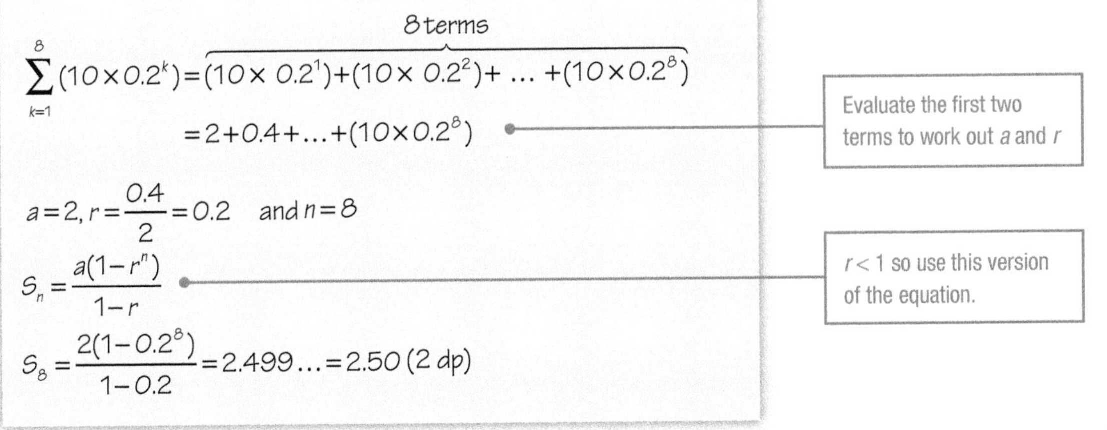

$$\sum_{k=1}^{8}(10 \times 0.2^k) = \overbrace{(10 \times 0.2^1) + (10 \times 0.2^2) + \ldots + (10 \times 0.2^8)}^{8 \text{ terms}}$$

$$= 2 + 0.4 + \ldots + (10 \times 0.2^8)$$

Evaluate the first two terms to work out a and r

$$a = 2,\ r = \frac{0.4}{2} = 0.2 \quad \text{and } n = 8$$

$$S_n = \frac{a(1-r^n)}{1-r}$$

$r < 1$ so use this version of the equation.

$$S_8 = \frac{2(1-0.2^8)}{1-0.2} = 2.499\ldots = 2.50 \text{ (2 dp)}$$

Exercise 13.4A Fluency and skills

1 Work out the values of the first four terms of the geometric sequences defined by

a $u_n = 4\times 3^{n-1}$ b $u_n = 3\times 4^{n-1}$

c $u_n = 8\times 0.5^n$ d $u_n = 6\times 3^{-n}$

e $u_n = 5\times 2^{-n-1}$ f $u_n = 2\times 0.5^{-(n-3)}$

2 Work out an expression for the nth term of these geometric sequences.

a $5, 10, 20, \ldots$ b $36, 24, 16, \ldots$

c $2, -6, 18, \ldots$ d $-8, 6, -4.5, \ldots$

e $7, -24.5, 85.75, \ldots$ f $1, \frac{1}{6}, \frac{1}{36}, \ldots$

3 Evaluate these geometric series. Give non-integer answers to 3 significant figures.

a $6+12+24+\ldots$ (10 terms)

b $64+96+144+\ldots$ (7 terms)

c $50+40+32+\ldots$ (15 terms)

d $4-6+9+\ldots$ (20 terms)

4 Use a formula to evaluate these geometric series.

a $\sum_{k=1}^{12}(2\times 3^k)$ b $\sum_{k=1}^{6}(192\times 0.5^k)$

c $\sum_{k=3}^{10}(5\times 2^k)$ d $\sum_{k=5}^{15}3\times(-2)^{k-1}$

Reasoning and problem-solving

Strategy

To solve a problem using a geometric sequence or series

1. Define variables such as the first term a, the common ratio r and the nth term u_n using information given in the question.
2. Work out any unknown values.
3. Use appropriate formulae and give your answer in context, rounding appropriately.

You can use simultaneous equations to solve a problem involving a geometric sequence or series.

Example 4

A geometric sequence has first term a, and common ratio r. The second term is 0.5 and the fifth term is 108. Work out the value of the ninth term of this sequence.

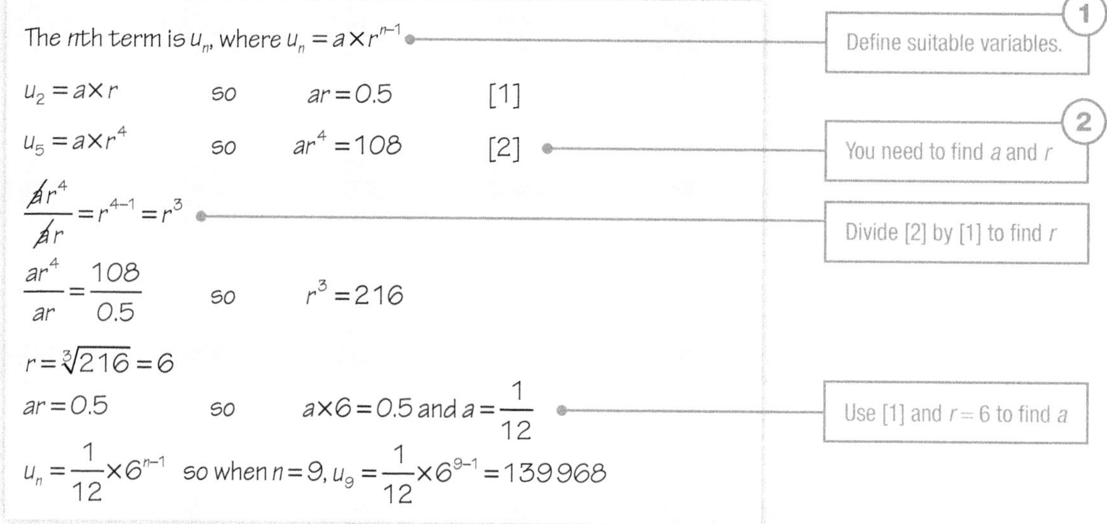

The nth term is u_n, where $u_n = a\times r^{n-1}$
Define suitable variables.
$u_2 = a\times r$ so $ar = 0.5$ [1]
$u_5 = a\times r^4$ so $ar^4 = 108$ [2]
You need to find a and r
$\frac{ar^4}{ar} = r^{4-1} = r^3$
Divide [2] by [1] to find r
$\frac{ar^4}{ar} = \frac{108}{0.5}$ so $r^3 = 216$
$r = \sqrt[3]{216} = 6$
$ar = 0.5$ so $a\times 6 = 0.5$ and $a = \frac{1}{12}$
Use [1] and r = 6 to find a
$u_n = \frac{1}{12}\times 6^{n-1}$ so when $n = 9$, $u_9 = \frac{1}{12}\times 6^{9-1} = 139968$

The geometric series $a + a\times r + a\times r^2 + a\times r^3 + ...$ has an **infinite** number of terms.

You know that the first n terms of this series has sum $S_n = \frac{a(1-r^n)}{1-r}$

If $-1 < r < 1$, then, as n increases, the term r^n in this formula approaches zero,

which means S_n approaches $\frac{a(1-0)}{1-r}$, or simply $\frac{a}{1-r}$

Key point

$\frac{a}{1-r}$ is the **sum to infinity** of the series. You can write S_∞ for this value.

If $-1 < r < 1$ then $S_\infty = \frac{a}{1-r}$

Example 5

a Evaluate the sum to infinity of the geometric series $48 + 12 + 3 + ...$

b Another geometric series has first term 36 and sum to infinity 20. Calculate the sum of the first six terms of this series. Give the answer to 1 decimal place.

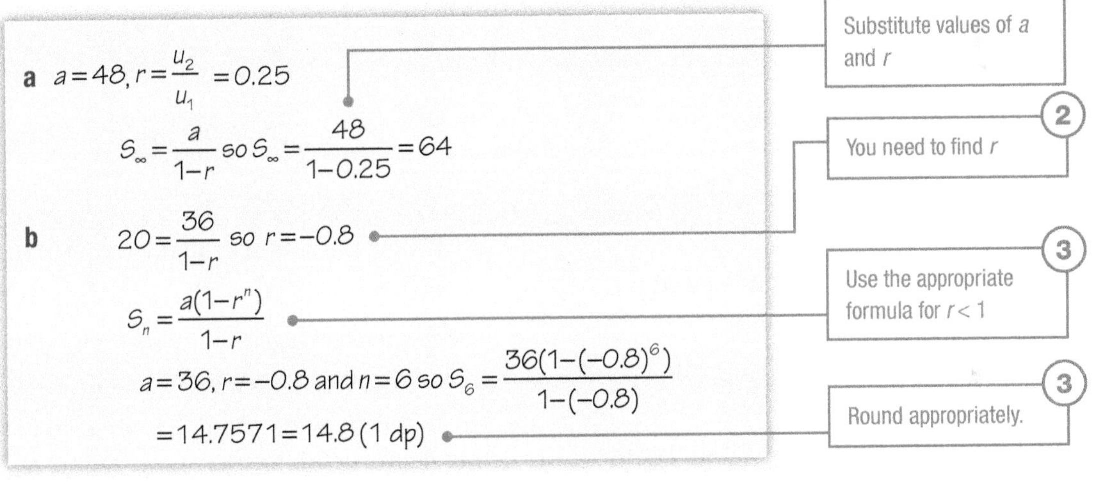

a $a = 48, r = \frac{u_2}{u_1} = 0.25$
$S_\infty = \frac{a}{1-r}$ so $S_\infty = \frac{48}{1-0.25} = 64$
Substitute values of a and r
2
b $20 = \frac{36}{1-r}$ so $r = -0.8$
You need to find r
$S_n = \frac{a(1-r^n)}{1-r}$
3
Use the appropriate formula for $r < 1$
$a = 36, r = -0.8$ and $n = 6$ so $S_6 = \frac{36(1-(-0.8)^6)}{1-(-0.8)}$
$= 14.7571 = 14.8$ (1 dp)
3
Round appropriately.

You can describe, or **model**, a real-life situation using a geometric sequence or series.

Example 6

At the start of year 1, £3000 was invested in a savings account.

The account paid 3% compound interest per year and no further deposits or withdrawals were allowed.

The balance of the account at the start of year n was modelled using a geometric sequence with common ratio r

a Work out the balance at the start of year 2 and state the value of r

b At the start of which year did the balance first exceed £3600?

(Continued on the next page)

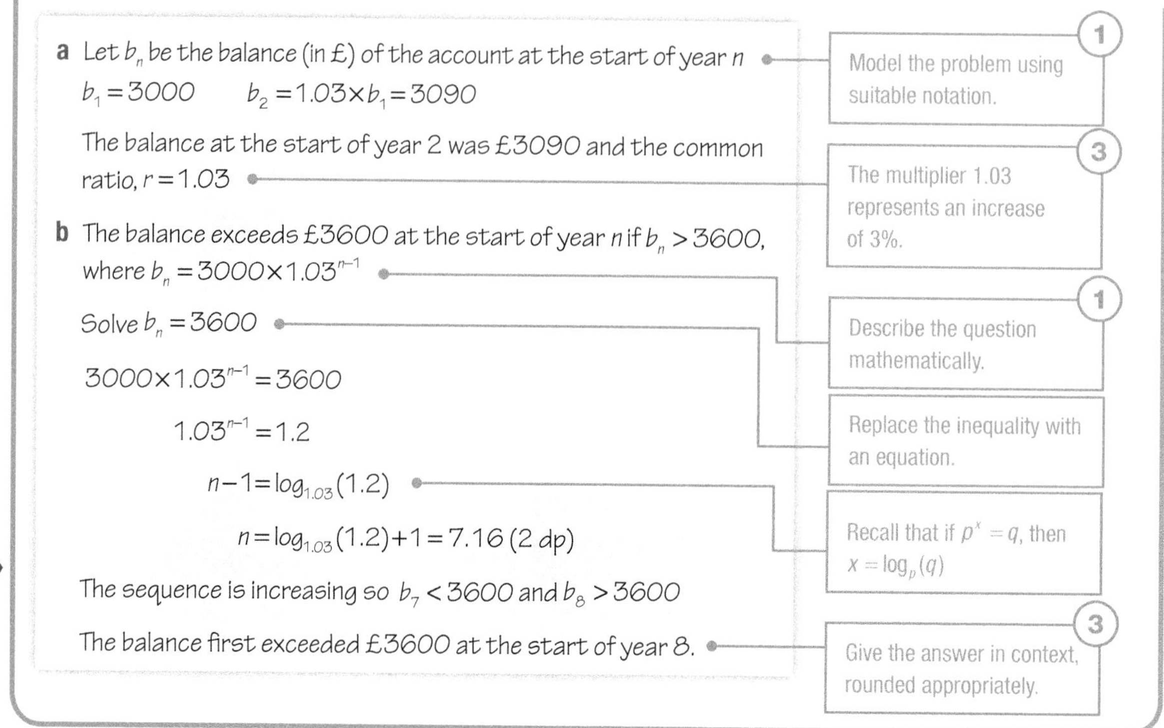

a Let b_n be the balance (in £) of the account at the start of year n

$b_1 = 3000 \qquad b_2 = 1.03 \times b_1 = 3090$

The balance at the start of year 2 was £3090 and the common ratio, $r = 1.03$

b The balance exceeds £3600 at the start of year n if $b_n > 3600$, where $b_n = 3000 \times 1.03^{n-1}$

Solve $b_n = 3600$

$3000 \times 1.03^{n-1} = 3600$

$1.03^{n-1} = 1.2$

$n - 1 = \log_{1.03}(1.2)$

$n = \log_{1.03}(1.2) + 1 = 7.16$ (2 dp)

The sequence is increasing so $b_7 < 3600$ and $b_8 > 3600$

The balance first exceeded £3600 at the start of year 8.

(1) Model the problem using suitable notation.

(3) The multiplier 1.03 represents an increase of 3%.

(1) Describe the question mathematically.

Replace the inequality with an equation.

Recall that if $p^x = q$, then $x = \log_p(q)$

(3) Give the answer in context, rounded appropriately.

See Ch5.2

For a reminder on logarithms and exponential functions.

Exercise 13.4B Reasoning and problem-solving

1 The nth term of a geometric sequence is u_n, where $u_3 = 2$ and $u_6 = 128$

a Work out the common ratio r and the first term a of this sequence.

b Calculate the value of the eighth term of this sequence.

c Express u_n in the form 2^{pn+q} stating the value of each constant p and q

2 The nth term of a geometric series is u_n, where $u_3 = 45$ and $u_5 = 405$. The series has common ratio r, where $r > 0$

a Calculate the sum of the first six terms of this series.

The sum of the first n terms of this series exceeds 1000 000

b Evaluate the least possible value of this sum.

3 A finite geometric series has first term 6 and common ratio r, where $r > 0$. The fifth term is 96 and the last term of this series has value 24 576

a Work out the value of r

b Evaluate the sum of all the terms of this series.

4 The nth term of a geometric series is u_n with first term a and common ratio 1.5

Given that $S_2 = 40$, where S_n is the sum of the first n terms of this series

a Calculate, to 3 significant figures, the value of S_n for which $9000 < S_n < 10000$

b Evaluate the sum to infinity of the geometric series $\frac{1}{u_1} + \frac{1}{u_2} + \frac{1}{u_3} + \ldots$

5 A geometric series has first term 50 and common ratio r, where $0 < r < 1$. The nth term is u_n and the sum of the first n terms of this series is S_n. Given that $S_3 = 98$

a Work out the value of S_∞

b i Evaluate the largest integer N for which $S_\infty - S_N > 0.5$

ii For this N evaluate $\sum_{k=N}^{\infty} u_k$, giving your answer to 3 decimal places.

6 At the start of year 1, £2500 is invested in a savings account. The account paid 2% compound interest per year and no further deposits or withdrawals were allowed. The balance (in £) of the account at the start of year n was modelled using a geometric sequence with common ratio r

a Write down the value of r

Use this model to calculate, to the nearest £

b The balance of the account at the start of year 4

c The amount of interest earned on the account in year 7

7 Toni bought a car for £24 000 which decreased in value by 25% each year after its purchase. The value of the car n years after its purchase was modelled by a geometric sequence with common ratio r

a Work out the value of r

b Use this model to show that three years after its purchase the value of the car was £10 125

Toni sold the car in the year its value fell below £2000

c Use this model to find the number of whole years Toni owned the car.

8 In his first year of driving, Tom drove 3125 miles. In his first two years of driving he drove 5625 miles. The distance (in miles) driven in Tom's nth year of driving was modelled using a geometric sequence.

a Use this model to

i Work out the distance Tom drove in his fourth year of driving,

ii Calculate the total distance Tom drove in his first six years of driving,

iii Show that the total distance Tom can drive in his lifetime is less than 15 625 miles.

b Comment on the suitability of this model in the long-term.

9 A rectangular piece of foil has length 50 cm, width 40 cm and is 0.4 mm thick. The foil was folded in half, and then in half again, and so on. The thickness (in mm) of the shape formed when the foil was folded n times was modelled using a geometric sequence.

a Calculate the thickness, in mm, of the shape when the foil is folded four times.

b Calculate the volume, in cm^3, of the shape when the foil is folded four times. You may assume after all folds the shape formed was a cuboid.

c Calculate the thickness of the shape if the foil were folded 40 times. Compare your answer with the distance from the Earth to the Moon (approximately 384 000 km).

10 At the end of the first month of opening, the electricity bill for a small business was £450. The amount due increased by 5% each month. The bills were analysed and it was correctly stated that, to the nearest £10, the average electricity bill per month for the first year was £600

Use a suitable model to determine which average (mean or median, or both) was used in this statement. Justify your answer.

Challenge

11 At 1 pm on a particular day, three people were told a rumour. At 2 pm that day, these three people had told three other people the rumour, who, at 3 pm that day, had each told three other people the rumour, and so on. Nobody was told the rumour more than once.

Given that the current estimate for the world's population is 7.4×10^9, how many hours will it take for the world's population to have heard this rumour?

13 Summary and review

Chapter summary

- When $n \notin \mathbb{Z}^+$, $1+nx+\frac{n(n-1)}{2!}x^2+\frac{n(n-1)(n-2)}{3!}x^3+\ldots$ is the binomial expansion of $(1+x)^n$
- The expansion of $(a+bx)^n$ when $n \notin \mathbb{Z}^+$ is only valid for $\left|\frac{b}{a}x\right|<1$
- A sequence with nth term u_n is increasing if $u_{n+1} > u_n$ for all $n \in \mathbb{Z}^+$, decreasing if $u_{n+1} < u_n$ for all $n \in \mathbb{Z}^+$, and periodic if the sequence consists of one block of terms repeated throughout.
- The period (or order) of a periodic sequence is the smallest number of terms that repeat as a block of numbers.
- If a sequence defined by $u_{n+1} = au_n + b$, where a and b are constants, has a limit L as $n \to \infty$ then L satisfies the equation $L = aL + b$. More generally, if the sequence defined by $u_{n+1} = \mathrm{f}(u_n)$ has limit L then $L = \mathrm{f}(L)$
- A sequence is arithmetic if for some constant d, $u_{n+1} - u_n = d$ for all $n \in \mathbb{Z}^+$
  d is the common difference of the sequence.
- $u_n = a+(n-1)d$ is the nth term of an arithmetic sequence with first term a
- A series is formed by adding the terms of a sequence together.
- $S_n = \frac{1}{2}n(a+l)$ is the sum of the first n terms of an arithmetic series, with first term a and last term l. This is also given by $S_n = \frac{1}{2}n[2a+(n-1)d]$
- Sigma notation can be used for series $\sum_{k=1}^{n}\mathrm{f}(k)=\mathrm{f}(1)+\mathrm{f}(2)+\ldots+\mathrm{f}(n)$
- A sequence is geometric if for some non-zero constant r, $\frac{u_{n+1}}{u_n} = r$ for all $n \in \mathbb{Z}^+$
  r is the common ratio of the sequence.
- $u_n = a \times r^{n-1}$ is the nth term of a geometric sequence with first term a
- $S_n = \frac{a(1-r^n)}{1-r}$, $r \neq 1$ is the sum of the first n terms of a geometric series.
- $S_\infty = \frac{a}{1-r}$, $-1 < r < 1$ is the sum to infinity of a geometric series.

Check and review

You should now be able to...	Try Questions
✔ Use the binomial expansion and recognise the range of validity.	1
✔ Use the binomial expansion to estimate the value of a surd.	2
✔ Understand if a sequence is increasing or decreasing.	3
✔ Work out the order of a periodic sequence.	4
✔ Work out the nth term and the sum of an arithmetic series.	5
✔ Work out the nth term and the sum of a geometric series.	6
✔ Evaluate a series given in sigma notation.	7

1 Work out the binomial expansion of these expressions up to and including the term in x^3. State the range of validity of each full expansion.

a $(1+x)^{\frac{4}{3}}$ b $\frac{1}{(1+2x)^3}$

c $(9-4x)^{\frac{3}{2}}$ d $\sqrt{\frac{64}{(4+x)^3}}$

2 Work out and use the first three terms in the binomial expansion of these expressions to estimate $\sqrt{3}$

a $(1+x)^{\frac{1}{2}}$ $x=0.08$

b $(4+3x)^{\frac{1}{2}}$ $x=-\frac{1}{3}$

3 Work out u_1, u_2, u_3 and u_4 for each of these sequences and describe as increasing, decreasing or neither.

a $u_{n+1}=3u_n-5,\ u_1=4$

b $u_{n+1}=\frac{1}{2}u_n^2-1,\ u_1=2$

c $u_{n+1}=u_n+\frac{1}{3},\ u_1=\frac{1}{2}$

d $u_{n+1}=5u_n-0.5,\ u_1=-1$

e $u_{n+1}=\frac{4}{u_n}-1,\ u_1=3$

f $u_{n+1}=\sqrt{2u_n+3},\ u_1=3$

4 Work out the order of these periodic sequences.

a $u_n=4(-1)^n+2(-1)^{n+1}$

b $u_n=2\sin(n\times 90)°+3\cos(n\times 90)°$

c $u_n=\frac{3}{2}-\frac{9}{4u_n},\ u_1=\frac{1}{2}$

d $u_{n+1}=4\left(\frac{u_n-2}{u_n}\right),\ u_1=3$

5 Work out the nth term and the sum of these arithmetic series.

a $7+13+19+...+73$ (12 terms)

b $15+11+7+...-81$ (25 terms)

c $9+14+19+...$ (30 terms)

d $\frac{1}{2}, 1, \frac{3}{2}, ...$ (50 terms)

e 18.5, 24, 29.5, ... (10 terms)

f $\frac{1}{2}+\frac{1}{3}+...$ (19 terms)

6 Work out the nth term and the sum of these geometric series. Give non-integer answers to 3 significant figures.

a $3+18+108+...$ (8 terms)

b $100+10+1+...$ (10 terms)

c $200-190+180.5+...$ (15 terms)

d $\log_2(3)+\log_2(9)+\log_2(81)+...$ (9 terms)

7 Use formulas to evaluate these series, each of which is either arithmetic or geometric. Give non-integer answers to 3 significant figures.

a $\sum_{k=1}^{30}(5k+3)$

b $\sum_{k=1}^{8}(25\times 1.8^k)$

c $\sum_{k=5}^{20}(40-3k)$

d $\sum_{k=1}^{\infty}(250\times 0.6^k)$

What next?

Score			
	0–4	Your knowledge of this topic is still developing. To improve, search in MyMaths for the codes: 2038–2040, 2204, 2205, 2264	
	5–6	You're gaining a secure knowledge of this topic. To improve, look at the InvisiPen videos for Fluency and skills (13A)	
	7	You've mastered these skills. Well done, you're ready to progress! To develop your techniques, look at the InvisiPen videos for Reasoning and problem-solving (13B)	

Click these links in the digital book

13 Exploration
Going beyond the exams

ICT

An **iterative** formula can be used to produce a sequence of numbers.

By using a spreadsheet, investigate the iterative formula $x_{n+1} = \frac{x_n}{a} + b$ for different values of a and b and for a range of starting values, x_0

In terms of a and b, what happens to these sequences as n increases?

B5 | fx =B4/$B$1+$B$2

	A	B	C	D	E
1	a =	2			
2	b =	2			
3					
4	x0	654			
5	x1	332			
6	x2	171			
7	x3	90.5			
8	x4	50.25			

Have a go

Using methods of calculus, **Isaac Newton** found an iterative formula for a sequence which will converge to the square root of any positive number, a

The iterative formula is given by
$$x_{n+1} = \frac{1}{2}\left(\frac{a}{x_n} + x_n\right)$$

Try using the formula to calculate $\sqrt{13}$

Note

A **convergent sequence** is a sequence for which the terms get closer and closer to a particular value.

For example, the sequence
$1, \frac{1}{2}, \frac{1}{3}, \frac{1}{4}, \frac{1}{5}, \ldots$ converges to 0

Information

Sequences can converge to their limits at different rates. Consider these sequences.

a $x_{n+1} = \frac{x_n}{10}$ this is equivalent to the sequence with nth term $\left(\frac{1}{10}\right)^n$

b $y_{n+1} = \frac{y_n}{10^{2n+1}}$ this is equivalent to the sequence with nth term $\left(\frac{1}{10}\right)^{n^2}$

Each of these sequences converges to 0, however sequence **b** has a higher rate of convergence. Despite the fact that there exists no value of n for which either sequence ever equals the limit, 0, you can think of sequence **b** as converging faster.

To get a sense of this, look at the size of the terms of each sequence for different values of n
For example, when $n = 3$
$x_3 = 1 \times 10^{-3} = 0.001$
$y_3 = 1 \times 10^{-9} = 0.000\,000\,001$

Even for small values of n, sequence **b** is already much closer to 0

Research

Find out what is meant by the terms **linear convergence, sublinear convergence** and **superlinear convergence.**

13 Assessment

1 Select the values of x for which the binomial expansion of $(5+3x)^{-2}$ is valid. **[1 mark]**

A $|x|<1$ B $|x|<3$ C $|x|<\frac{5}{3}$ D $|x|<\frac{3}{5}1$

2 Select the values of x for which the binomial expansion of $(x-7)^{-1}$ is valid. **[1]**

A $|x|<7$ B $|x|<-\frac{1}{7}$ C $|7x|<1$ D $|x|<1$

3 Find the value of $\sum_{k=1}^{n}(5k+3)$. Select the correct answer. **[1]**

A $5n+11$ B $5(n-1)$ C $8n+11$ D $\frac{1}{2}n(5n+11)$

4 Find the value of $\sum_{k=1}^{\infty}(4\times 0.2^k)$. Select the correct answer. **[1]**

A 1 B 4 C 5 D 20

5 A sequence is defined by the recurrence relation $u_{n+1}=1-\frac{1}{u_n}$, where $u_1=2$

a Write down the values of

i u_2 ii u_3 iii u_4 **[6]**

b Deduce the value of u_{50} **[2]**

6 Write the first four terms of each sequence, then describe the sequences as either increasing, decreasing or periodic.

a $u_n=2\cos(180n)°$ **[3]**

b $u_n=0.2^n+4$ **[3]**

c $u_n=n^2+4n-2$ **[3]**

7 Write down the first four terms in the binomial expansion, in ascending powers of x, of $(1-2x)^{-2}$, stating the values of x for which the expansion is valid. **[6]**

8 A car costs £30 000. Its value depreciates by 20% per annum. Work out

a Its value after 1 year, **[1]**

b Its value after 4 years, **[2]**

c The year in which it will be worth less than £5000 **[4]**

9 A sequence of terms is defined by the recurrence relation $u_{n+1}=4-ku_n$, where k is a constant.

Given that $u_1=3$

a Work out an expression in terms of k for u_2 **[2]**

b Work out an expression in terms of k for u_3 **[2]**

Given also that $u_1+u_2+u_3=9$

c Calculate the possible values of k **[4]**

10 The sum to infinity of a geometric series is 20. The first term is 4

a Calculate the common ratio of the series. **[3]**

b Evaluate the third term of the series. **[2]**

11 Adam plans to pay money into a savings scheme each year for 20 years. He will pay £800 in the first year, and every year he will increase the amount that he pays into the scheme by £100

a Show that he will pay £1000 into the scheme in year 3 **[1]**

b Calculate the total amount of money that he will pay into the scheme over the 20 years. **[2]**

c Over the same 20 years, Ben will also pay money into a savings scheme. He will pay £610 in the first year, and every year he will increase the amount that he pays into the scheme by £d. Given that Adam and Ben will pay in exactly the same total amounts over the 20 years, calculate the value of d **[3]**

12 When $(1+ax)^n$ is expanded the coefficients of x and x^2 are -4 and 20 respectively.

a Work out the value of a and the value of n **[8]**

b Evaluate the coefficient of x^3 **[2]**

13 The second term of a geometric series is 120 and the fifth term is 15. Work out

a The common ratio of the series, **[4]**

b The first term of the series, **[1]**

c The sum to infinity of the series. **[2]**

14 a Use a formula to evaluate $\sum_{r=1}^{40}(3r+1)$ **[3]**

b Calculate the value of n for which $\sum_{r=1}^{n}(3r+1)=9800$ **[4]**

15 a Write down the first three terms in the binomial expansion of $(1-2x)^{\frac{1}{2}}$, in ascending powers of x **[3]**

b Write down the first three terms in the binomial expansion of $(1+x)^{-\frac{1}{2}}$, in ascending powers of x **[3]**

c Use your answers to **a** and **b** to prove that $\sqrt{\frac{1-2x}{1+x}}=1-\frac{3}{2}x+\frac{3}{8}x^2+....$ **[4]**

16 The fourth term of an arithmetic series is 11 and the sum of the first three terms is -3

a Write down the first term of the series. **[4]**

b Work out the common difference of the series. **[1]**

c Given that the sum of the first n terms of the series is greater than 500, calculate the least possible value of n **[5]**

17 The first three terms of a geometric series are $(3p-1)$, $(p-3)$ and $(2p)$ respectively.

a Use algebra to work out the possible values of p **[5]**

b For the negative value of p, calculate the sum to infinity of the series. **[3]**

c For the positive value of p, evaluate the sum of the first 999 terms of the series. **[2]**

18 a Write down the first four terms in the binomial expansion $\sqrt{1-x}$, in ascending powers of x **[6]**

b By substituting $x=\frac{1}{4}$, work out a fraction that is an approximation to $\sqrt{3}$ **[4]**

19 A salesman sells vacuum cleaners for £120 each. In one week, he receives 2% commission on the first vacuum cleaner he sells, 4% commission on the second vacuum cleaner he sells, with commission increasing in steps of 2% , so that he receives commission of 30% on the sale of his fifteenth vacuum cleaner. Commission stays fixed at 30% for the sale of all vacuum cleaners, after the sale of his fifteenth vacuum cleaner in that week.

a Calculate how much commission he receives in a week for the sale of

i His first vacuum cleaner,

ii His fifth vacuum cleaner,

iii His twentieth vacuum cleaner. **[6]**

In one week he sells 40 vacuum cleaners.

b How much commission does he receive in total that week? **[5]**

20 The sum to infinity of a geometric series is 48, and the sum of the first two terms of the series is 45

The common ratio of the series is r

a Prove that r satisfies the equation $1-16r^2=0$ **[4]**

b Calculate the sum of the first four terms of the series. **[4]**

21 The training programme of a cyclist requires her to cycle 3 km on the first day of training.

Then, on each day that follows, she cycles 2 km more than she cycled on the day before.

a Calculate how far she cycles on the seventh day. **[1]**

b Calculate the total distance she has cycled by the end of the tenth day. **[2]**

c On which day of training will she cycle more than 100 km? **[3]**

d On which day of training will the total distance that she has cycled exceed 1000 km? **[5]**

22 a Write down the first three terms in the binomial expansion of $\sqrt{4-x}$, in ascending powers of x **[7]**

b Deduce an approximate value of $\sqrt{399}$, giving your answer to 3 decimal places. **[5]**

23 An investment scheme pays 3% compound interest per annum. The interest is paid annually.

A deposit of £1000 is invested in this scheme at the start of each year.

The initial investment of £1000 is made at the start of year 1

a Explain why the value of the investment at the start of year 2 is £2030 **[2]**

b Calculate the value of the investment at the start of year 3 **[2]**

c Work out the year in which the total value of the investment exceeds £50 000 **[4]**

24 The sum of the first two terms of an arithmetic series is 2. The sum of the first ten terms of the series is 330

a Work out the common difference of the series. **[5]**

b Write down the first term of the series. **[1]**

c Given that the sum of the first n terms of the series is equal to 1170, find the value of n **[4]**

25 Given that $f(x)=\frac{5x}{(2+x)(1-2x)}\equiv\frac{A}{2+x}+\frac{B}{1-2x}$

a Work out the values of the constants, A and B **[5]**

b Write down the series expansion of f(x), in ascending powers of x, up to and including the term in x^3 **[11]**

c State the values of x for which the expansion is valid. **[1]**

26 Given that $f(x)=\frac{13x-33}{(5-x)(1+3x)}$

a Work out the expansion of f(x) up to and including the term in x^3 **[14]**

b State the values of x for which the expansion is valid. **[1]**

27 When a ball is dropped from a height of h metres above a hard floor it rebounds to a height of $\frac{3}{4}h$

A ball is dropped from an initial height of 2 metres. Calculate

a The height to which the ball rises after the first bounce, **[2]**

b The total distance the ball has travelled when it hits the floor for the second time, **[2]**

c The total distance that the ball travels. **[3]**

28 Given that x, 15 and y are consecutive terms of an arithmetic series, and 1, x and y are consecutive terms of a geometric series, work out the possible values of x and y **[9]**

29 By solving an equation, find the limit of these sequences as $n\to\infty$. Where appropriate, give answers in simplified surd form.

a $u_{n+1}=0.2u_n+4$ **[2]**

b $u_{n+1}=9-0.2u_n$ **[2]**

c $u_{n+1}=\frac{1}{2}\left(\frac{1}{3}u_n-10\right)$ **[2]**

d $u_{n+1}=\left(\sqrt{2}-1\right)u_n+4$ **[2]**

e $u_{n+1}=\frac{1}{\sqrt{2}}u_n+\sqrt{2}$ **[2]**

f $u_{n+1}=0.5u_n^2+0.5$ **[2]**

14 Trigonometric identities

Tides are affected by the gravity of both the Moon and the Sun, and follow a predictable oscillatory pattern. Since tides are periodic, the sine and cosine functions can be used to model them, with components governed by the Sun and Moon. Predicting and modelling the tides has been essential to marine life for centuries, and now also plays an important role in calculating the effect and extent of climate change.

Trigonometric functions are suitable for modelling anything known to follow a periodic pattern, including tides, seasonal temperature fluctuations, hours of daylight, the motion of a Ferris wheel, the orbits of planets and satellites and music waves. The sine and cosine functions are very useful mathematical tools, and there are many trigonometric identities that aid in the manipulation and application of trigonometric expressions.

Orientation

What you need to know

Ch3 Trigonometry
- Sine, cosine and tangent.
- The sine and cosine rules.

Ch12 Algebra 2
- Transformations of functions.
- Inverse functions.

p.8

What you will learn
- To convert between degrees and radians and use radians in problems.
- To use reciprocal and inverse trigonometric functions.
- To use trigonometric formulae for compound angles, double angles and half angles.
- To find and use equivalent forms for $a\cos\theta + b\sin\theta$
- To simplify and solve equations using trigonometric formulae.

What this leads to

Ch15 Differentiation 2
Derivatives of trigonometric functions.

Ch18 Motion in two dimensions
Resolving forces.
Projectiles.

Practise before you start

2045–2048, 2138, 2142, 2284

14.1 Radians

Fluency and skills

As well as degrees, you can measure angles in **radians**. If you draw a circle with centre O and radius r, then 1 radian is the angle subtended at O by an arc that is equal in length to the radius.

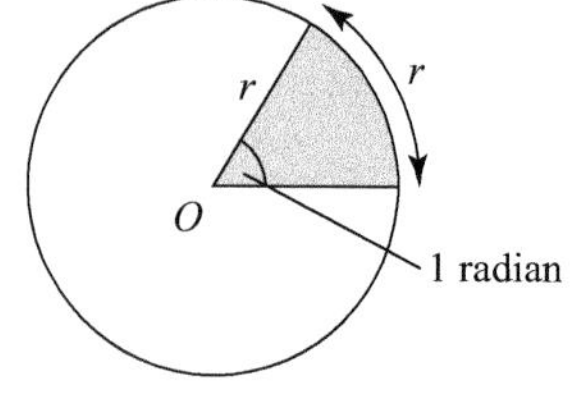

As the circumference is $2\pi \times r$, you can fit 2π of these arcs around the circumference. So, the total angle at O is 2π radians or 360°

Key point

2π radians = 360° π radians = 180°

1 radian $= \dfrac{180°}{\pi} \approx 57.3°$ 1 degree $= \dfrac{\pi}{180}$ radians

The symbol for radians is rad or, more rarely, ^{c} or ^{r}, as in 2^c and 2^r, where c means circular measure and rad is short for radians.

For sector POQ, the arc length, s, is a fraction of the circumference, and the sector area, A, is a fraction of the circle's area.

Radians are often used instead of degrees because they greatly simplify a number of circle formulae and the resulting calculations.

Key point

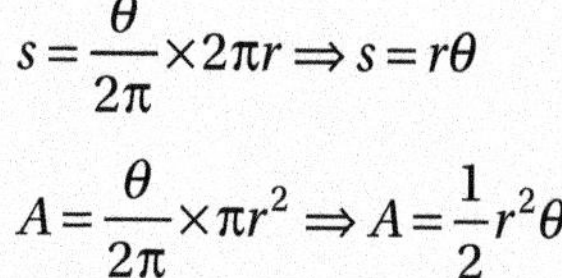

$$s = \frac{\theta}{2\pi} \times 2\pi r \Rightarrow s = r\theta$$

$$A = \frac{\theta}{2\pi} \times \pi r^2 \Rightarrow A = \frac{1}{2}r^2\theta$$

where θ is in radians.

Compare the formulae where θ is in radians with those where θ is in degrees: $s = \dfrac{\theta}{360°} \times 2\pi r$ and $A = \dfrac{\theta}{360°} \times \pi r^2$

Example 1

a Convert an angle of 30° into radians, writing it in terms of π

b Calculate the exact values of arc length PQ and sector area POQ

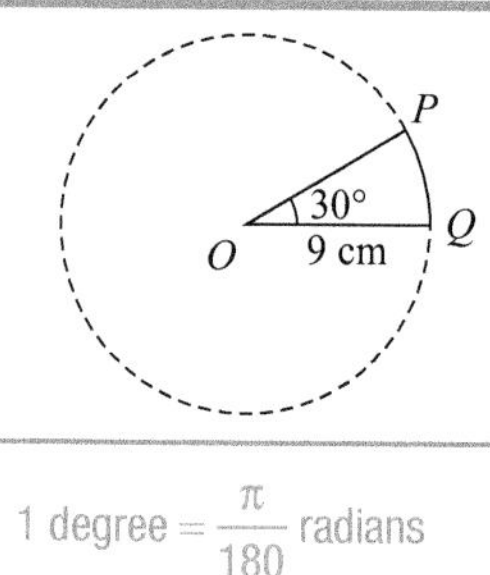

a $30° = 30 \times \dfrac{\pi}{180} = \dfrac{\pi}{6}$ radians

1 degree $= \dfrac{\pi}{180}$ radians

b Arc length $= r\theta = 9 \times \dfrac{\pi}{6} = \dfrac{3\pi}{2}$ cm

Sector area $= \dfrac{1}{2}r^2\theta = \dfrac{1}{2} \times 9^2 \times \dfrac{\pi}{6} = \dfrac{27\pi}{4}$ cm^2

When x is in radians, the graph of $y = x$ almost coincides with the graphs of $y = \sin x$ and $y = \tan x$ for small values of x. Similarly, the graph of $y = 1 - \dfrac{1}{2}x^2$ gives a close approximation to $y = \cos x$ for small x

Compare the graphs by plotting them on your graphics calculator, and check the results.

Key point

For small values of x in radians,

$\sin x \approx x$ $\tan x \approx x$ $\cos x \approx 1 - \dfrac{1}{2}x^2$

PURE

Example 2

Calculate the percentage error if 8° in radians is used as an approximation for sin 8°

$\sin 8° = 0.13917$ (to 5 dp)

$8° = \frac{8}{180} \times \pi = 0.13963$ radians (to 5 dp)

Percentage error $= \frac{\text{absolute error}}{\text{correct value}} \times 100\% = \frac{0.13963 - 0.13917}{0.13917} \times 100\%$

$= 0.3\%$ (to 1 sf)

These two special triangles can be used to find trigonometric ratios for angles in radians. It is useful to know these.

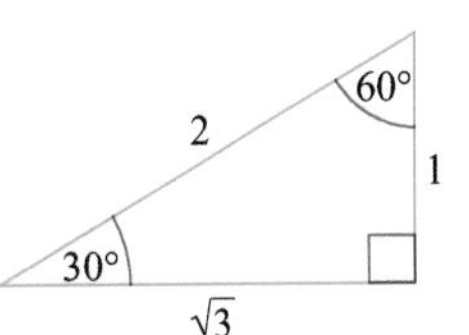

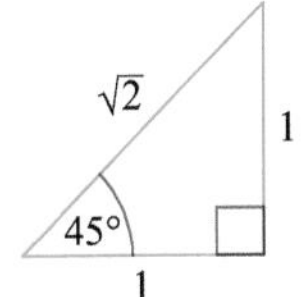

For example, $60° = \frac{1}{3}$ of 180°, which is $\frac{\pi}{3}$ radians

So, using the triangle, $\sin\frac{\pi}{3} = \sin 60° = \frac{\sqrt{3}}{2}$ and $\cos\left(\frac{\pi}{3}\right) = \cos 60° = \frac{1}{2}$

You can still use the sine rule, cosine rule and area of a triangle formula for triangles where angles are measured in radians.

You can change the set-up of your calculator from degrees to radians. Note that $\frac{\pi}{3} = 60$, so you use the special triangle here.

See Ch3.2
For a reminder on the sine and cosine rules.

Example 3

Calculate the length x and the area of the triangle, giving your answers as surds.

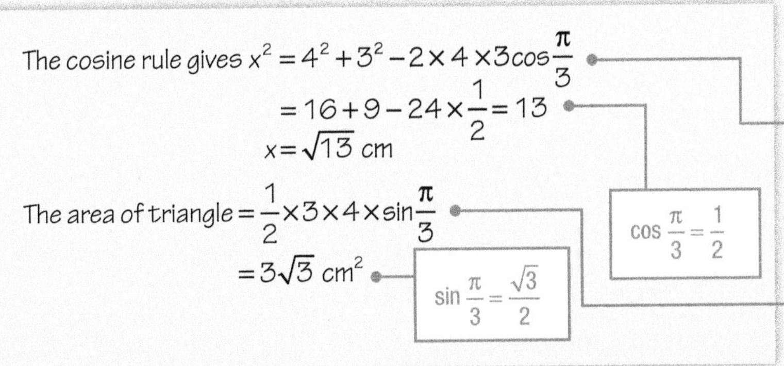

The cosine rule gives $x^2 = 4^2 + 3^2 - 2 \times 4 \times 3\cos\frac{\pi}{3}$

$= 16 + 9 - 24 \times \frac{1}{2} = 13$

$x = \sqrt{13}$ cm

The area of triangle $= \frac{1}{2} \times 3 \times 4 \times \sin\frac{\pi}{3}$

$= 3\sqrt{3}$ cm²

$\cos\frac{\pi}{3} = \frac{1}{2}$

$\sin\frac{\pi}{3} = \frac{\sqrt{3}}{2}$

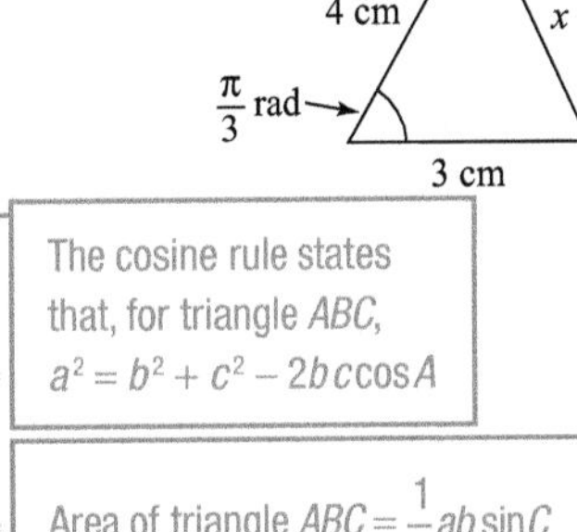

The cosine rule states that, for triangle ABC, $a^2 = b^2 + c^2 - 2bc\cos A$

Area of triangle $ABC = \frac{1}{2}ab\sin C$

Exercise 14.1A Fluency and skills

1 Convert these angles into degrees, correct to 1 dp.

a 2 rad b 3 rad c 0.5 rad

2 Convert these angles into radians. Give your answers in terms of π and also as decimals to 2 dp.

a 90° b 60° c 270°

3 Find the arc length and the sector area of a circle with radius 4 cm and an angle of 2 radians at the centre.

4 Copy and complete this table, giving radians in terms of π and exact values for the trigonometric ratios. Use the special triangles below Example 2 to do this.

θ	θ (radians)	$\sin\theta$	$\cos\theta$	$\tan\theta$
45°				
120°				
135°				
270°				
360°				

5 Calculate the unknown sides and angles in ΔDEF where $d = 72.4$ cm, $e = 43.2$ cm and angle $F = \frac{\pi}{4}$ rad

6 Giving your answers in terms of π, find the arc length and the sector area of a circle of radius 6 cm with an angle of $\frac{3}{4}\pi$ radians at the centre.

7 a Copy and complete this table, correct to 4 dp.

θ	θ (radians)	$\sin\theta$	$\tan\theta$
10°			
5°			
2°			

b At what value of $\theta°$ do $\sin\theta$, $\tan\theta$ and θ in radians have the same value correct to 4 dp?

8 If $\theta = 12°$, use the value of θ in radians as an approximation for $\tan\theta$ and calculate the percentage error in this approximation.

9 In this triangle, $\theta = 1°$ and $x = 2.5$ km

Calculate the height of the triangle y, in metres, using an approximation for $\tan\theta$

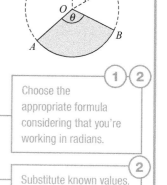

10 Find approximations for these expressions when θ and 2θ (in radians) are both small.

a $\frac{\sin 2\theta}{2\theta}$ b $\frac{1-\cos\theta}{2\theta^2}$ c $\frac{\theta\sin\theta}{1-\cos 2\theta}$

11 Solve these equations for $-\pi \le \theta \le \pi$. Show your working.

a $2\sin\theta - 1 = 0$ b $\sqrt{3} + 2\cos\theta = 0$

c $\sin\theta = \cos\theta$

Reasoning and problem-solving

Strategy

To solve problems that involve calculations with angles

1. Decide whether to work with angles in degrees or radians.
2. Choose appropriate formulae and, if necessary and if using radians, use values involving π or small-angle approximations.
3. Give your answer in the required form or, if you have a choice, choose the most appropriate form.

Example 4

A sector within a circle, centre O, has an arc of length 4π cm that makes an angle of $\frac{\pi}{3}$ radians at the centre. Another sector, AOB, has area 45π cm²

Calculate

a The radius, r, of the circle,

b The angle θ, in radians, in the sector AOB

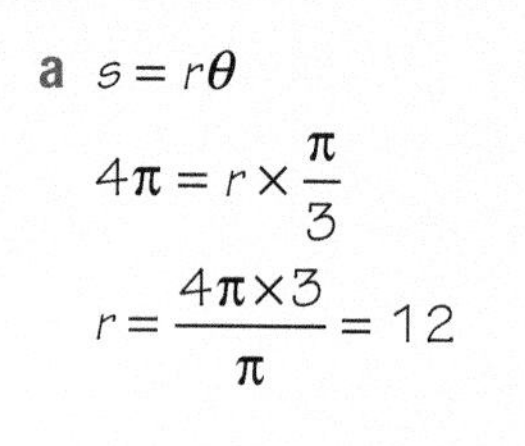
a $s = r\theta$

$4\pi = r \times \frac{\pi}{3}$

$r = \frac{4\pi \times 3}{\pi} = 12$

Radius $r = 12$ cm

b $A = \frac{1}{2}r^2\theta$

$45\pi = \frac{1}{2} \times 12^2 \times \theta$

$\theta = \frac{45\pi \times 2}{144} = \frac{5\pi}{8}$

Angle $\theta = \frac{5\pi}{8}$ radians

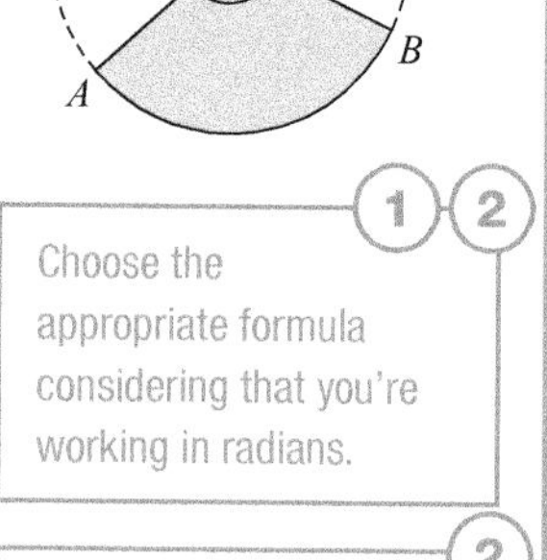

1 2 Choose the appropriate formula considering that you're working in radians.
2 Substitute known values.
3 Give your answer in radians.

Exercise 14.1B Reasoning and problem-solving

1 Through how many radians does the minute hand of a clock turn in 20 minutes?

2 Copy and complete the table showing values for the angle θ, arc length s, and area A, of sectors with radius r

r cm	θ radians	s cm	A cm^2
4		12	
5		2	
4			10
5			30

3 An aeroplane climbs at an angle of 5° for the first 2 km of its flight. Using an approximation for the angle, how high is it (in metres) after flying 2 km?

4 A hill is known to be 160 m high. The angle of elevation from a point x km away is 0.3°. Use an approximation to calculate the value of x

5 A comet is 10 million kilometres away from Earth, and its sides subtend an angle of 0.04° when viewed from Earth. Estimate the diameter of the comet, assuming it is spherical.

6 A triangle has sides of length 5 cm, 7 cm and 12 cm. Calculate the size of the largest angle in the triangle. Give your answer in radians.

7 A wheel of radius 2 m rolls 0.5 m along the ground. How many radians has it turned through?

8

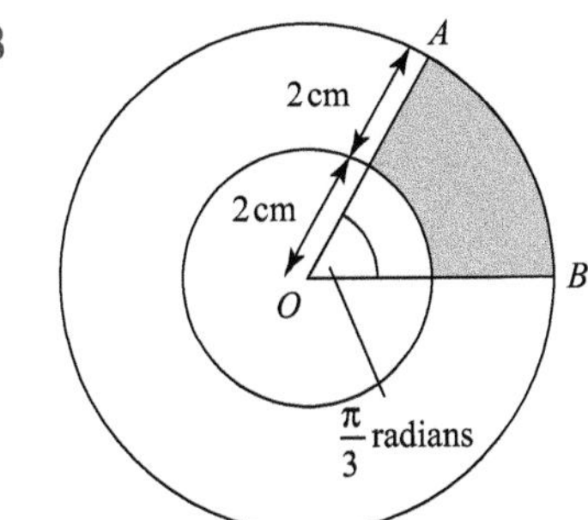

Find the perimeter and area of the shaded region in terms of π when $\angle AOB = \frac{\pi}{3}$.

9 Solve these equations for $-\frac{\pi}{2} \le \theta \le \frac{\pi}{2}$. Show your working.

a $2\sin 2\theta = \sqrt{3}$

b $\tan\left(\theta + \frac{\pi}{6}\right) = 1$

c $\sqrt{2}\sin 2\theta = 2\cos\theta$

d $2\tan\theta - 1 = \frac{1}{\tan\theta}$

10 A chord AB subtends an angle of 0.6 radians at the centre O of a circle of radius 8 cm. Calculate the area of the segment cut off by AB

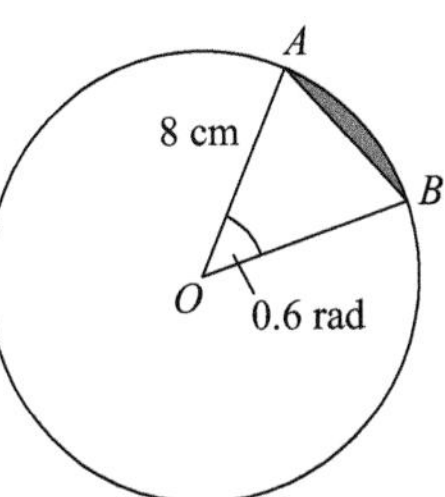

Challenge

11 Two equal circles of radius 7 cm have their centres 10 cm apart. Calculate the area of the overlap of the two circles.

12

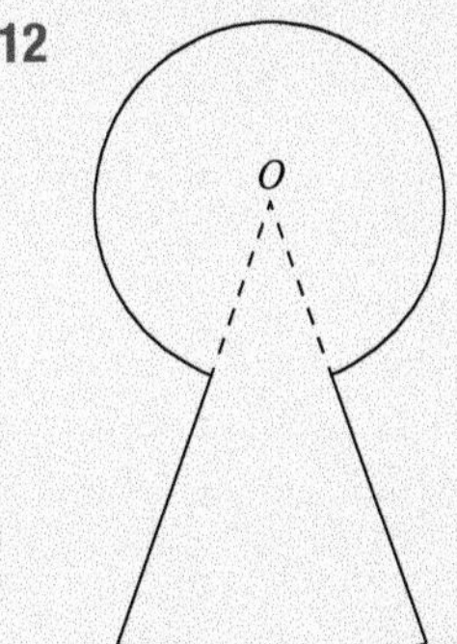

A keyhole is made from two sectors of two circles with a common centre O and radii 1 cm and 3 cm. If angle $AOB = \frac{\pi}{6}$ radians, calculate

a The perimeter of the keyhole,

b Its area,

giving answers in terms of π

13 Three equal circles of radius r are placed in the same plane so each one touches the other two. Find the area between them in terms of r

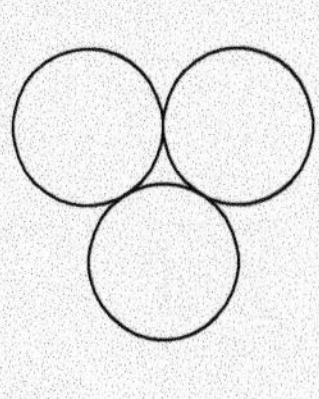

MyMaths 2050, 2266 SEARCH

14.2 Reciprocal and inverse trigonometric functions

Fluency and skills

Reciprocal functions

You know the **reciprocal** of x is $\frac{1}{x}$. The reciprocals of sine, cosine and tangent are cosecant (cosec), secant (sec) and cotangent (cot).

Be careful: cosec is related to sine (not cos) and sec is related to cosine.

Key point

$$\operatorname{cosec}\theta=\frac{1}{\sin\theta} \qquad \sec\theta=\frac{1}{\cos\theta} \qquad \cot\theta=\frac{1}{\tan\theta}$$

Your calculator may have buttons for cosec, sec and cot, but if not then you can calculate their values using the equivalent sin, cos and tan functions.

For example, $\sec 40° = \frac{1}{\cos 40°} = 1.305$

Example 1

Using your knowledge of trigonometric identities, and showing your working, find the exact values of

a cosec 225° **b** $\cot\frac{7\pi}{6}$

a $\operatorname{cosec} 225° = \frac{1}{\sin 225°} = \frac{1}{-\sin 45°} = -\sqrt{2}$

225° is in the 3rd quadrant in the CAST diagram, so sin and cosec are negative.

b $\frac{7\pi}{6}$ radians $= \frac{7\times 180°}{6} = 210°$

$\cot\frac{7\pi}{6} = \cot 210° = \frac{1}{\tan 210°} = \frac{1}{\tan 30°} = \sqrt{3}$

210° is in the 3rd quadrant so tan and cot are positive.

See Ch3.1 For a reminder on the CAST diagram.

You can use the graphs of sin θ, cos θ and tan θ to sketch the graphs of their reciprocal functions.

Sine and cosec

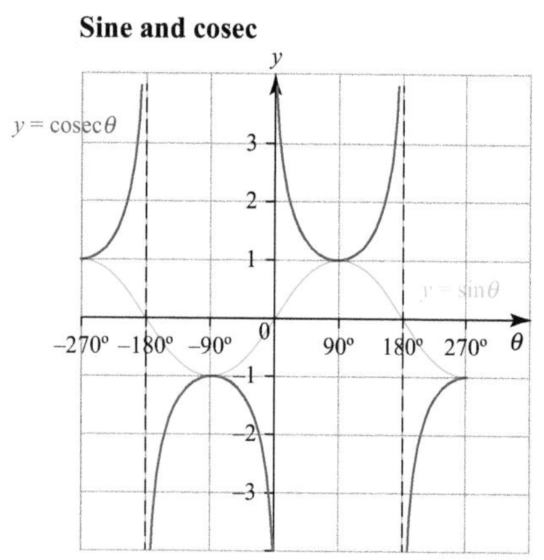

Cos and sec

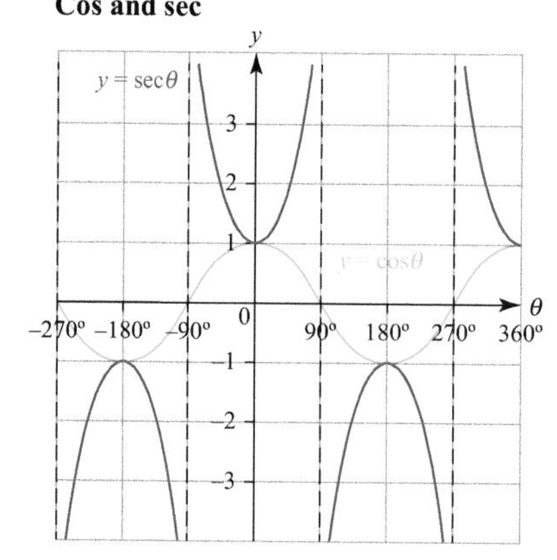

Tan and cot

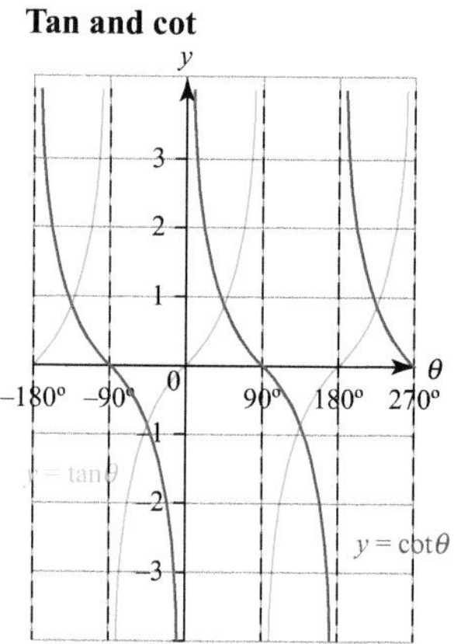

These graphs show the relationships between the original trigonometric functions and their reciprocals. They also show the **domains** and **ranges** of the reciprocal functions.

To investigate reciprocal trigonometric functions, click this link in the digital book.

PURE

	Domain	Range
$y = \operatorname{cosec}\theta$	$\theta \in \mathbb{R}, \theta \neq 0°, \pm180°, \pm360°, \ldots$	$y \in \mathbb{R}, y \geq 1, y \leq -1$
$y = \sec\theta$	$\theta \in \mathbb{R}, \theta \neq \pm90°, \pm270°, \ldots$	$y \in \mathbb{R}, y \geq 1, y \leq -1$
$y = \cot\theta$	$\theta \in \mathbb{R}, \theta \neq 0°, \pm180°, \pm360°, \ldots$	$y \in \mathbb{R}$

See Ch12.2
For a reminder on domains and ranges.

In Section 3.1 you derived $\sin^2\theta + \cos^2\theta \equiv 1$ by dividing Pythagoras' theorem by c^2

In a similar way, you can divide Pythagoras' theorem by a^2 or b^2 to derive results involving cosec, sec and cot.

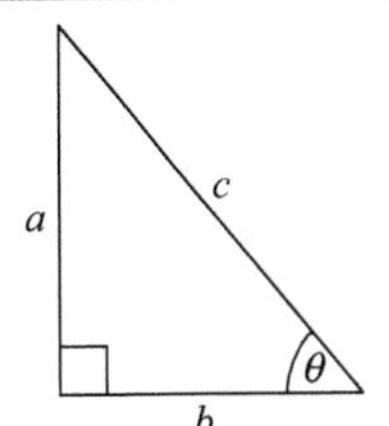

Key point

$\sec^2\theta \equiv 1 + \tan^2\theta$ and $\operatorname{cosec}^2\theta \equiv 1 + \cot^2\theta$

Example 2

Prove the identity $\frac{1}{\cos^2 x} - \frac{\sec^2 x}{\operatorname{cosec}^2 x} \equiv 1$

$$\frac{1}{\cos^2 x} - \frac{\sec^2 x}{\operatorname{cosec}^2 x} \equiv \sec^2 x - \frac{\left(\frac{1}{\cos^2 x}\right)}{\left(\frac{1}{\sin^2 x}\right)}$$

Use the definitions of sec x and cosec x to rearrange the expression and get rid of fractions.

$$\equiv \sec^2 x - \frac{\sin^2 x}{\cos^2 x} \equiv \sec^2 x - \frac{1}{\cos^2 x}\sin^2 x$$

$$\equiv \sec^2 x - \sec^2 x \sin^2 x$$

$$\equiv \sec^2 x(1 - \sin^2 x)$$

$$\equiv \sec^2 x \cos^2 x$$

Use $\sin^2 x + \cos^2 x \equiv 1$

$$\equiv \frac{1}{\cos^2 x} \times \cos^2 x \equiv 1$$

Use the definition of sec x

The identity is proved.

Inverse functions

The trigonometric functions have **inverse functions**. They can be written in two ways:

Key point

$\arcsin x$ or $\sin^{-1}x$ $\quad \arccos x$ or $\cos^{-1}x$ $\quad \arctan x$ or $\tan^{-1}x$

For example, $\sin\frac{\pi}{6} = 0.5$, so $\arcsin 0.5 = \frac{\pi}{6}$

Another way of thinking about arcsin 0.5 is 'the angle whose sine is 0.5'

Do not confuse the inverse $\sin^{-1}x$ with the reciprocal $(\sin x)^{-1}$

SEARCH

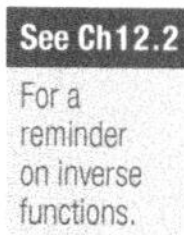
See Ch12.2 For a reminder on inverse functions.

The functions $\sin x$, $\cos x$ and $\tan x$ are many-to-one functions, so their inverses have many possible values (for example, $\arcsin 0.5 = 30°, 150°, 330°, \ldots$). However, inverse functions must have one-to-one mapping. This means that the domain of the original function must be restricted to ensure that it is one-to-one.

For the reflections below to work, axes must have the same scale and the angles must be in radians.

The graphs of the inverse functions are reflections of the original trigonometric graphs in the line $y = x$

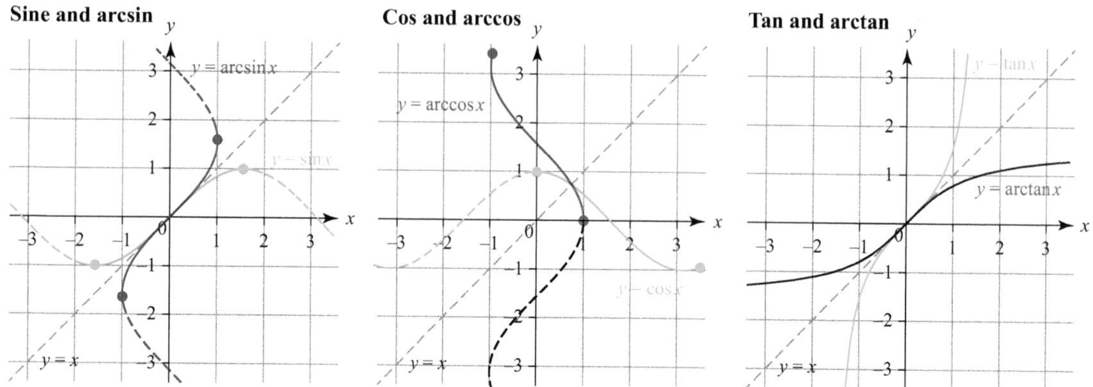

The graphs show the restricted domains and ranges as follows.

	$y = \sin x$	$y = \arcsin x$	$y = \cos x$	$y = \arccos x$	$y = \tan x$	$y = \arctan x$
Domain	$-\frac{\pi}{2} < y < \frac{\pi}{2}$	$-1 \le x \le 1$	$0 \le x \le \pi$	$-1 \le x \le 1$	$-\frac{\pi}{2} < y < \frac{\pi}{2}$	$x \in \mathbb{R}$
Range	$-1 \le y \le 1$	$-\frac{\pi}{2} \le y \le \frac{\pi}{2}$	$-1 \le y \le 1$	$0 \le y \le \pi$	$y \in \mathbb{R}$	$-\frac{\pi}{2} < y < \frac{\pi}{2}$

The **principal value** of $\arcsin x$, for a given value of x, is the unique value of $\arcsin x$ within the allowed range $-\frac{\pi}{2} \le y \le \frac{\pi}{2}$

Similarly, the principal values of $\arccos x$ are in the range $0 \le y \le \pi$ and the principal values of $\arctan x$ are in the range $-\frac{\pi}{2} \le y \le \frac{\pi}{2}$

Prove that $\arcsin\left(-\frac{1}{\sqrt{2}}\right) = -\frac{\pi}{4}$

Let $\arcsin\left(-\frac{1}{\sqrt{2}}\right) = \alpha$

You need to find the angle whose sine is $-\frac{1}{\sqrt{2}}$

So $\sin \alpha = -\frac{1}{\sqrt{2}}$

Angle α is a principle value of arcsin and so it lies between $-\frac{\pi}{2}$ and $\frac{\pi}{2}$ in the 1st or 4th quadrants.

See the graphs and table shown just before this example.

See Ch3.1 For a reminder on quadrants.

However, as $\sin \alpha$ is negative, α must lie in the 4th quadrant, not the 1st quadrant.

Using the special triangle, $\alpha = -\frac{\pi}{4}$ radians

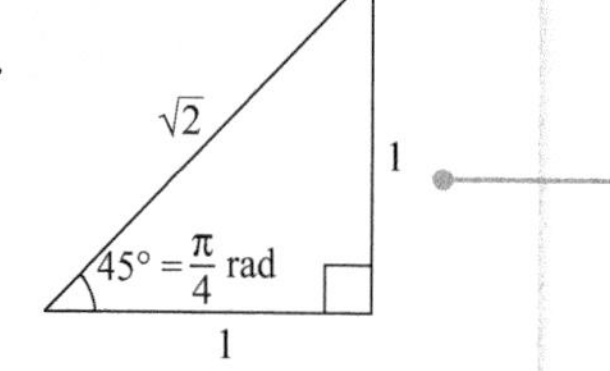

Recall this special triangle.

So, $\arcsin\left(-\frac{1}{\sqrt{2}}\right) = -\frac{\pi}{4}$

Exercise 14.2A Fluency and skills

1 Use the graphs following Example 1 to help you answer this question.
When the original trigonometric functions, $\sin\theta$, $\cos\theta$ and $\tan\theta$, have values of 0, +1 and −1, what can you say about the graphs of their reciprocal functions?

2 Use your calculator to find, to 3 sf, the values of

a $\sec 200°$ b $\cot 130°$ c $\text{cosec } 340°$
d $\sec\frac{3\pi}{5}$ e $\cot\frac{5\pi}{6}$ f $\text{cosec}\frac{2\pi}{9}$

3 Copy and complete this table, giving answers in exact form.

x, degrees	*x*, radians	cosec *x*	sec *x*	cot *x*
30°				
60°				
90°				

4 Without using a calculator, write down the exact values of

a $\cot 135°$ b $\sec 120°$ c $\text{cosec } 210°$
d $\cot\frac{4\pi}{3}$ e $\sec\frac{7\pi}{4}$ f $\text{cosec}\frac{3\pi}{2}$

5 Solve for $-180° \le x \le 180°$, showing your working.

a $\text{cosec } x = 2$ b $\sec x = 4$
c $\cot x = -1$ d $\sec(x - 10°) = 3$
e $\text{cosec }(x + 10°) = -4$ f $\cot(x + 30°) = -2$
g $\sec 3x = \sqrt{2}$ h $\text{cosec } 2x = 2$

6 Simplify these expressions.

a $\cot\theta\tan\theta$ b $\cot\theta\sin\theta\tan\theta$
c $\cot^2\theta\sec\theta\sin\theta$ d $\text{cosec }\theta\sec\theta\sin^2\theta$

7 Use your calculator to evaluate

a $\tan^{-1}3$ b $\cos^{-1}0.25$
c $\sin^{-1}\left(\frac{3}{4}\right)$ d $\cos^{-1}(-0.4)$
e $\tan^{-1}(-2)$ f $\sin^{-1}\left(-\frac{1}{4}\right)$

8 Giving your answers in terms of π, find the exact values of

a $\arcsin\frac{1}{\sqrt{2}}$ b $\arctan\sqrt{3}$
c $\arccos 1$ d $\arctan(-1)$
e $\arccos\frac{\sqrt{3}}{2}$ f $\arcsin 0$
g $\arccos\left(-\frac{1}{\sqrt{2}}\right)$ h $\arctan\left(-\frac{1}{\sqrt{3}}\right)$

9 Sketch the curves on two sets of axes and find algebraically the points of intersection of these pairs of curves for $0 < x < \pi$

a $y = 1 + 2\tan x$, $y = 1 + \sec x$
b $y = \sec x$, $y = 2 - \cos x$

Check your answers on a graphics calculator.

10 Solve these equations for $0 \le x \le 180°$, showing your wording.

a $\sin^{-1}\left(\frac{1}{2}\right) = x$
b $\sin^{-1}\left(\frac{\sqrt{3}}{2}\right) = x - 30°$
c $\cos^{-1}0.4 = x$
d $\cos^{-1}0.7 = x + 10°$
e $\tan^{-1}1 = x$
f $\tan^{-1}(-1) = x - 50°$
g $\sin^{-1}\left(\frac{1}{\sqrt{2}}\right) = x + 45°$
h $\cos^{-1}\left(\frac{1}{2}\right) = x - 45°$

11 Find the values, in degrees, of

a $\arccos\frac{\sqrt{3}}{2} + \arcsin\left(-\frac{1}{2}\right)$
b $\arctan 1 - \arctan(-1)$

12 Describe the transformations which map

a The graph of $y = \sec x$ onto
i $y = \sec(x - 45°)$ ii $y = 2\sec\left(\frac{1}{3}x\right)$

b The graph of $y = \text{cosec } x$ onto
i $y = 2\,\text{cosec}(x + 45°)$ ii $y = \text{cosec}\left(\frac{1}{2}x\right)$

c The graph of $y = \cot x$ onto
i $y = -\frac{1}{4}\cot x$ ii $y = 2 + \cot x$

13 Given that y is in radians, sketch the graphs of

a $y = \sin^{-1}\left(\frac{1}{2}x\right)$ b $y = \cos^{-1}(3x)$
c $y = 2\tan^{-1}\left(\frac{1}{3}x\right)$ d $y = 2 + \sin^{-1}(2x)$

In each case, write the exact coordinates of the end points of each graph.

Reasoning and problem-solving

Strategy

To solve problems involving reciprocal or inverse trigonometric functions

1. Look for opportunities to rearrange or simplify functions.
2. Use the appropriate definitions and trigonometric identities.
3. Find values using your calculator, graphs or special triangles.

Example 4

Solve the equation $2\tan^2\theta + 3\sec\theta = 0$ for $-\pi < \theta < \pi$. Show your working.

$2(\sec^2\theta - 1) + 3\sec\theta = 0$

$2\sec^2\theta + 3\sec\theta - 2 = 0$

$(2\sec\theta - 1)(\sec\theta + 2) = 0$

$\sec\theta = \frac{1}{2}$ or -2

$\cos\theta = 2$ or $-\frac{1}{2}$

$\theta = \pi \pm \frac{\pi}{3} = \frac{2\pi}{3}$ or $\frac{4\pi}{3}$

(1)(2) Substitute for $\tan^2\theta$ to give a quadratic equation.

Ignore $\cos\theta = 2$ as $\cos\theta \le 1$

(3) cos is positive in the 2nd and 3rd quadrants.

Example 5

Find the exact value of $\tan\left(\arcsin\frac{3}{4}\right)$

Find $\tan\theta$ given that $\sin\theta = \frac{3}{4}$

The principal value of $\arcsin\frac{3}{4}$ is positive, so θ is in the 1st quadrant.

$x^2 + 3^2 = 4^2$

$x = \sqrt{16-9} = \sqrt{7}$

$\tan\theta = \frac{3}{\sqrt{7}}$ and $\tan\left(\arcsin\frac{3}{4}\right) = \frac{3}{\sqrt{7}}$

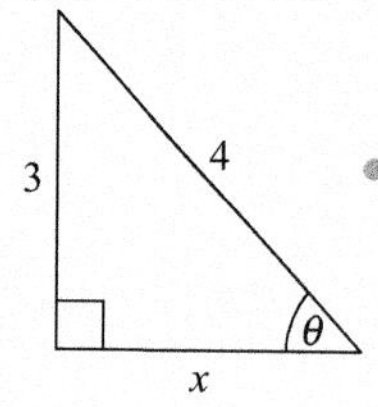

(Triangle with sides 3, 4, x and angle θ)

(2) Find the tangent of the angle whose sine is $\frac{3}{4}$

(3) Sketch a triangle with $\sin\theta = \frac{3}{4}$ and use Pythagoras' theorem.

Exercise 14.2B Reasoning and problem-solving

For all questions in this exercise you must show your working.

1 Solve these equations for $-\pi < \theta < \pi$

a $\sec^2\theta = 2$

b $\cot^2\theta = 3$

c $4\sin\theta = 3\operatorname{cosec}\theta$

d $\tan\theta = 4\sin\theta\cos\theta$

e $3\cos\theta = \sec\theta$

f $4\cot\theta = 3\tan\theta$

g $\cot 2\theta = \tan 2\theta$

h $\sec 3\theta - \operatorname{cosec} 3\theta = 0$

2 Solve these equations for $0 \le \theta \le 360°$

a $2 + \sec^2\theta = 4\tan\theta$

b $2\cot\theta = \tan\theta + 1$

c $3\sin\theta - 2\operatorname{cosec}\theta = 1$

d $2\sec\theta - 1 = \tan^2\theta$

3 Eliminate the trigonometric functions from these pairs of equations.

a $x = 4\sec\alpha$, $y = 2\tan\alpha$

b $x = 3\operatorname{cosec}\alpha$, $y = 2\cot\alpha$

c $x = 4\cos\alpha$, $y = 3\tan\alpha$

d $x = 1 - \cot\alpha$, $y = 1 + \operatorname{cosec}\alpha$

4 a Angle α is acute and $\sin\alpha = \frac{4}{5}$

Find the value of i $\operatorname{cosec}\alpha$ ii $\cot\alpha$

b Angle θ is obtuse and $\tan\theta = -\frac{8}{15}$

Find the value of i $\sec\theta$ ii $\cot\theta$

c Angle β is reflex and $\cos\beta = \frac{9}{41}$

Find the value of i $\operatorname{cosec}\beta$ ii $\cot\beta$

5 Write each of these expressions as a power of $\sec\alpha$, $\operatorname{cosec}\alpha$ or $\cot\alpha$

a $\dfrac{1}{\tan^2\alpha}$ b $\dfrac{\sec\alpha}{\cos^2\alpha}$ c $\dfrac{1-\cos^2\alpha}{\sin^4\alpha}$

6 Find the value of $\cot\theta$ when

a $2\sin\theta = 3\cos\theta$ b $4\tan\theta = 1$

7 Prove these identities.

a $\tan x\sin x + \cos x \equiv \sec x$

b $\dfrac{\cos x\tan x}{\operatorname{cosec}^2 x} \equiv \dfrac{\cos^3 x}{\cot^3 x}$

c $\tan\theta + \cot\theta \equiv \sec\theta\operatorname{cosec}\theta$

d $\sec\theta - \tan\theta \equiv \dfrac{1}{\sec\theta + \tan\theta}$

e $(1 + \sec\theta)(1 - \cos\theta) \equiv \tan\theta\sin\theta$

8 The force $F = 20\sin(3t - 1.5)$ N acts on an oscillating object where t is the time in seconds. Make t the subject of the formula and find its value when $F = 15$ N.

9 If $\theta = \arcsin x$, find $\tan\theta$ in terms of x

10 Given that $x = \tan\theta$, find $\operatorname{arccot} x$ in terms of θ

11 Show that, if $f(\theta) = \dfrac{1}{1+\sin\theta} + \dfrac{1}{1-\sin\theta} = 8$, then $\sec^2\theta = 4$

Hence, solve the equation $f\left(x - \dfrac{\pi}{6}\right) = 8$ where x is in radians and $0 < x < \pi$

12 a If $\arcsin x = \dfrac{\pi}{5}$, prove that $\arccos x = \dfrac{3\pi}{10}$

b If $2\arcsin x = \arccos x$, find the value of x

Challenge

13 Prove these identities.

a $\dfrac{\sin\theta}{1+\cos\theta} + \dfrac{1+\cos\theta}{\sin\theta} \equiv 2\operatorname{cosec}\theta$

b $\dfrac{1}{\cot\theta + \operatorname{cosec}\theta} \equiv \dfrac{1-\cos\theta}{\sin\theta}$

c $\dfrac{\tan^2\theta + \cos^2\theta}{\sin\theta + \sec\theta} \equiv \sec\theta - \sin\theta$

14 Given that $f(\theta) = 4\sec^2\theta - \tan^2\theta = k$, show that $\operatorname{cosec}^2\theta = \dfrac{k-1}{k-4}$

Hence, solve the equation $f\left(x + \dfrac{\pi}{4}\right) = 7$ where x is in radians and $0 < x < \pi$

14.3 Compound angles

Fluency and skills

A **compound angle** is the result of adding or subtracting two (or more) angles.

There are several identities that use compound angles.

You can prove that $\sin(A+B) \neq \sin A + \sin B$ by counter-example.

Substituting $A = B = 45°$ gives $\sin(45° + 45°) = \sin 90° = 1$ whereas $\sin 45° + \sin 45° = 2 \times \frac{1}{\sqrt{2}} = \sqrt{2}$ and $1 \neq \sqrt{2}$

To find the correct expansion of $\sin(A+B)$, you can use this diagram where $\angle ROP$ is the compound angle $A+B$

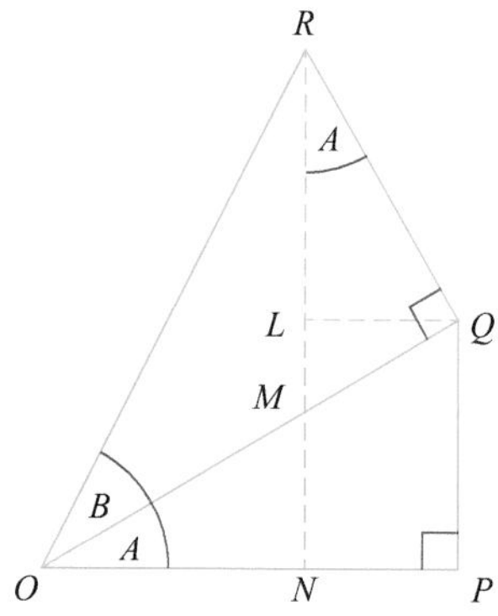

Triangles MON and MRQ are similar, so $\angle MRQ = \angle MON =$ angle A

In ΔORN, $\sin(A+B) = \frac{NR}{OR} = \frac{NL+LR}{OR}$

$$= \frac{PQ}{OR} + \frac{LR}{OR}$$

$$= \frac{PQ}{OQ} \times \frac{OQ}{OR} + \frac{LR}{RQ} \times \frac{RQ}{OR}$$

$$= \sin A \cos B + \cos A \sin B$$

This step introduces OQ and RQ to use in ΔOQP and ΔRQL

Although proved here for A and B as acute angles, this formula is true for all values of A and B

Replacing B by $-B$ gives $\sin(A-B) = \sin A \cos B - \cos A \sin B$

Angle $-B$ is in the 4th quadrant where sine is –ve and cosine is +ve, so $\sin(-B) = -\sin B$ and $\cos(-B) = \cos B$

There are also **compound angle formulae** for $\cos(A \pm B)$ and $\tan(A \pm B)$

Key point

$$\sin(A \pm B) \equiv \sin A \cos B \pm \cos A \sin B$$

$$\cos(A \pm B) \equiv \cos A \cos B \mp \sin A \sin B$$

$$\tan(A \pm B) \equiv \frac{\tan A \pm \tan B}{1 \mp \tan A \tan B}$$

Take care with the signs, especially the $\mp$ signs.

Example 1

Using special triangles and showing your working, find the value of cos 15° as a surd.

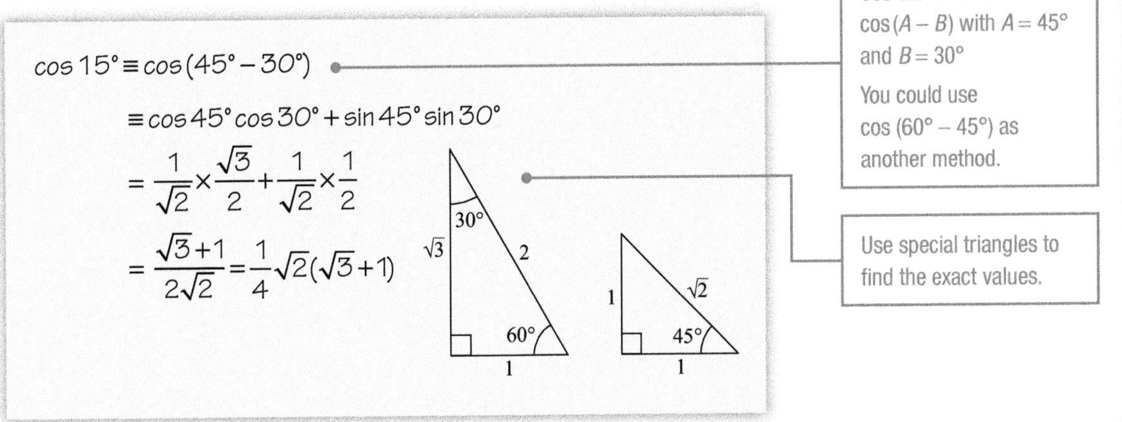

$\cos 15° \equiv \cos(45° - 30°)$
$\equiv \cos 45° \cos 30° + \sin 45° \sin 30°$
$= \frac{1}{\sqrt{2}} \times \frac{\sqrt{3}}{2} + \frac{1}{\sqrt{2}} \times \frac{1}{2}$
$= \frac{\sqrt{3}+1}{2\sqrt{2}} = \frac{1}{4}\sqrt{2}(\sqrt{3}+1)$
30°
√3
2
60°
1
1
√2
45°
1
Use the formula for cos (A – B) with A = 45° and B = 30°
You could use cos (60° – 45°) as another method.
Use special triangles to find the exact values.

Letting $A = B$ in the compound angle formulae creates the **double angle formulae**. $A + A = 2A$ is called a double angle.

Key point

$\sin 2A \equiv 2 \sin A \cos A \qquad \cos 2A \equiv \cos^2 A - \sin^2 A \qquad \tan 2A \equiv \frac{2\tan A}{1 - \tan^2 A}$

You can derive three versions of the formula for $\cos 2A$, by substituting the identity $\sin^2 A + \cos^2 A \equiv 1$

Key point

$\cos 2A =$ $\cos^2 A - \sin^2 A$ **or** $2\cos^2 A - 1$ **or** $1 - 2\sin^2 A$

You can rearrange the above identities for $\cos 2A$ to give expressions for $\cos^2 A$ and $\sin^2 A$

Key point

$\cos^2 A \equiv \frac{1}{2}(1 + \cos 2A) \qquad \sin^2 A \equiv \frac{1}{2}(1 - \cos 2A)$

Example 2

For the acute angle α, find the value of

a $\sin\alpha$ when $\cos 2\alpha = \frac{17}{25}$ **b** $\tan 2\alpha$ when $\sin\alpha = \frac{1}{3}$

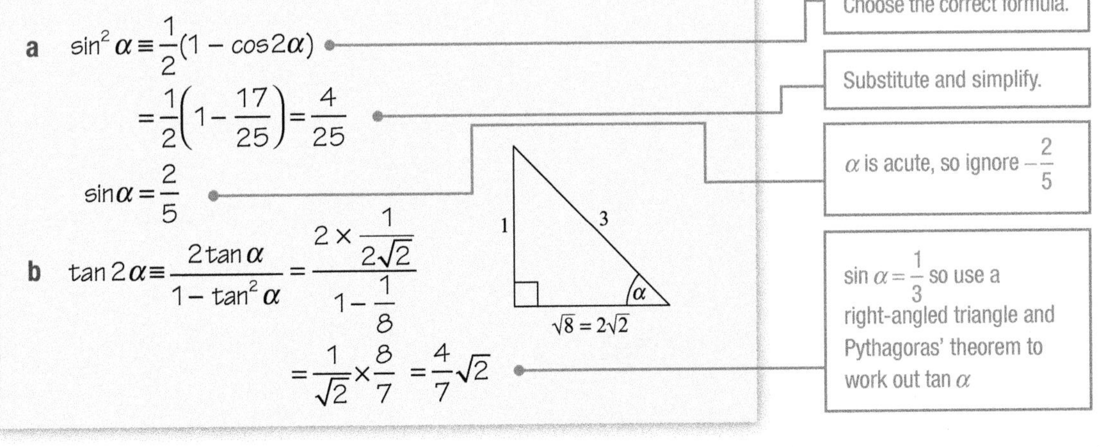

a $\sin^2\alpha \equiv \frac{1}{2}(1 - \cos 2\alpha)$
$= \frac{1}{2}\left(1 - \frac{17}{25}\right) = \frac{4}{25}$
$\sin\alpha = \frac{2}{5}$
b $\tan 2\alpha \equiv \frac{2\tan\alpha}{1 - \tan^2\alpha} = \frac{2 \times \frac{1}{2\sqrt{2}}}{1 - \frac{1}{8}}$
$= \frac{1}{\sqrt{2}} \times \frac{8}{7} = \frac{4}{7}\sqrt{2}$
1
3
α
√8 = 2√2
Choose the correct formula.
Substitute and simplify.
α is acute, so ignore $-\frac{2}{5}$
$\sin\alpha = \frac{1}{3}$ so use a right-angled triangle and Pythagoras' theorem to work out tan α

Exercise 14.3A Fluency and skills

For questions that ask for exact values, remember to show your working. You can check your final answer using a calculator.

1 a By writing $15° = 60° - 45°$, find the exact value of $\sin 15°$

b Find the exact values of

i $\cos 75°$ ii $\tan 75°$ iii $\tan 105°$

2 Write each expression as a single trigonometric ratio and find the exact value.

a $\sin 35° \cos 10° + \cos 35° \sin 10°$

b $\sin 70° \cos 10° - \cos 70° \sin 10°$

c $\cos 40° \cos 10° + \sin 40° \sin 10°$

d $\dfrac{\tan 75° - \tan 45°}{1 + \tan 75° \tan 45°}$

3 Simplify

a $\sin 2\theta \cos \theta + \cos 2\theta \sin \theta$

b $\cos 3\theta \cos \theta - \sin 3\theta \sin \theta$

c $\dfrac{\tan 2x + \tan x}{1 - \tan 2x \tan x}$ d $\dfrac{1 + \tan 3x \tan x}{\tan 3x - \tan x}$

4 Write each expression as a single trigonometric ratio.

a $\dfrac{\sqrt{3}}{2}\cos x + \dfrac{1}{2}\sin x$ b $\dfrac{\sqrt{3} + \tan x}{1 - \sqrt{3}\tan x}$

c $\dfrac{1}{\sqrt{2}}\cos x - \dfrac{1}{\sqrt{2}}\sin x$ d $\dfrac{1 + \tan x}{1 - \tan x}$

e $\dfrac{\cot x - \sqrt{3}}{1 + \sqrt{3}\cot x}$

5 Use compound angle formulae to show that

a $\sin(90° - A) = \cos A$

b $\sin(180° - A) = \sin A$

c $\cos(90° - A) = \sin A$

d $\cos(180° - A) = -\cos A$

6 Given acute angles α and β such that $\sin \alpha = \dfrac{12}{13}$ and $\tan \beta = \dfrac{3}{4}$, use trigonometric formulae to show that

a $\sin(\alpha + \beta) = \dfrac{63}{65}$ b $\tan(\alpha - \beta) = \dfrac{33}{56}$

7 Write each expression as a single trigonometric ratio.

a $2 \sin 23° \cos 23°$ b $\cos^2 42° - \sin^2 42°$

c $\dfrac{2\tan 70°}{1 - \tan^2 70°}$ d $2\cos^2 50° - 1$

e $2 \sin 3\theta \cos 3\theta$ f $\sin \theta \cos \theta$

g $\dfrac{1}{2}(1 + \cos 40°)$ h $1 - 2\sin^2 4\theta$

i $1 + \cos 2\theta$ j $\dfrac{1}{2}(1 - \cos 50°)$

k $\cos^2 \dfrac{\pi}{5} - \sin^2 \dfrac{\pi}{5}$ l $\sec \theta \operatorname{cosec} \theta$

m $\cot \theta - \tan \theta$ n $\dfrac{1 - \tan^2 4\theta}{2\tan 4\theta}$

8 Write each expression as a single trigonometric ratio and find the exact value.

a $2\sin\dfrac{\pi}{12}\cos\dfrac{\pi}{12}$ b $2\cos^2\dfrac{\pi}{8} - 1$

c $1 - 2\sin^2\dfrac{\pi}{4}$ d $\dfrac{1 - \tan^2 22\frac{1}{2}°}{2\tan 22\frac{1}{2}°}$

9 Use trigonometric formulae to find exact values of $\sin 2\theta$ and $\sin\dfrac{1}{2}\theta$ when θ is acute and

a $\cos \theta = \dfrac{3}{5}$ b $\tan \theta = \dfrac{5}{12}$

10 Find, in surd form, the values of $\sin\dfrac{x}{2}$, $\cos\dfrac{x}{2}$ and $\tan\dfrac{x}{2}$ when x is acute and

a $\cos x = \dfrac{1}{9}$ b $\sin x = \dfrac{3}{5}$

11 a Find the exact value of $\sin \theta$ when θ is acute and $\cos 2\theta = \dfrac{39}{49}$

b Find $\cos \theta$ when θ is acute and $\tan 2\theta = \dfrac{3}{4}$

12 a If $\tan(\alpha + \beta) = 4$ and $\tan \alpha = 3$, find $\tan \beta$

b If $\sin(\alpha + \beta) = \cos \beta$ and $\sin \alpha = \dfrac{3}{5}$, find $\tan \beta$

c If $\tan(\alpha - \beta) = 5$, find $\tan \alpha$ in terms of $\tan \beta$

Reasoning and problem-solving

Strategy

To solve problems involving compound trigonometric angles

1. Decide on the method you will use.
2. Use the appropriate angle formulae.
3. Simplify and manipulate the equation or expression.

Example 3

Solve the equation $\sin(\theta - 30°) = 3\cos\theta$ for $0 \le \theta \le 360°$. Show your working.

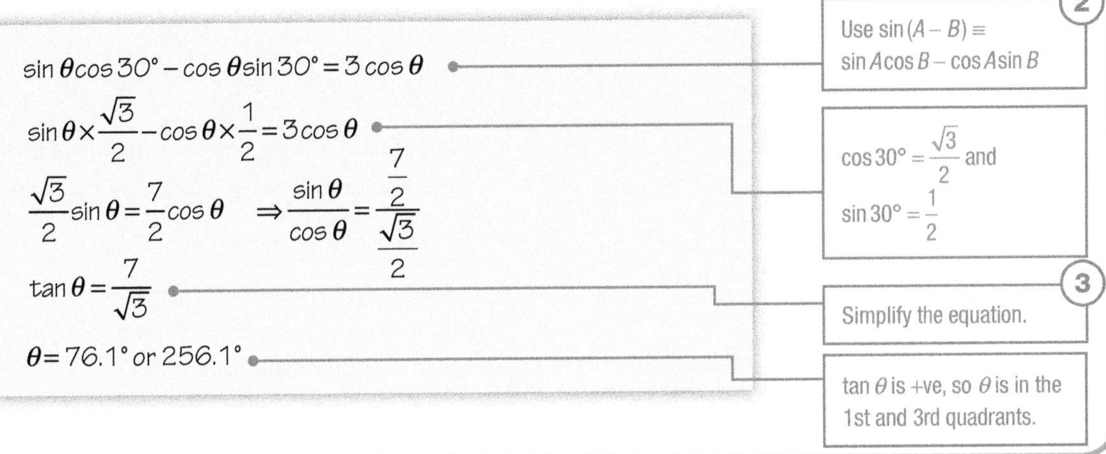

$\sin\theta\cos 30° - \cos\theta\sin 30° = 3\cos\theta$
$\sin\theta \times \frac{\sqrt{3}}{2} - \cos\theta \times \frac{1}{2} = 3\cos\theta$
$\frac{\sqrt{3}}{2}\sin\theta = \frac{7}{2}\cos\theta \Rightarrow \frac{\sin\theta}{\cos\theta} = \frac{\frac{7}{2}}{\frac{\sqrt{3}}{2}}$
$\tan\theta = \frac{7}{\sqrt{3}}$
$\theta = 76.1°$ or $256.1°$
2 Use $\sin(A - B) \equiv \sin A\cos B - \cos A\sin B$
$\cos 30° = \frac{\sqrt{3}}{2}$ and $\sin 30° = \frac{1}{2}$
3 Simplify the equation.
$\tan\theta$ is +ve, so θ is in the 1st and 3rd quadrants.

Example 4

Prove the identity $\cot\theta - \tan\theta \equiv 2\cot 2\theta$

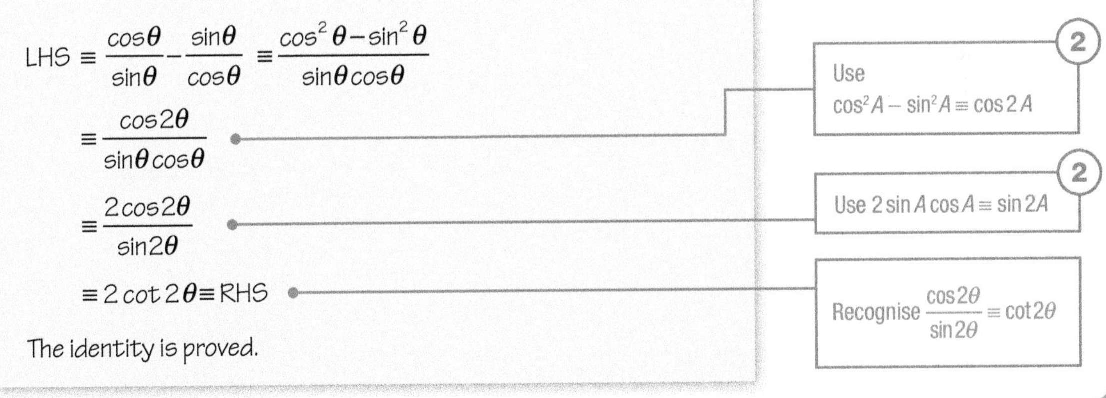

$\text{LHS} \equiv \frac{\cos\theta}{\sin\theta} - \frac{\sin\theta}{\cos\theta} \equiv \frac{\cos^2\theta - \sin^2\theta}{\sin\theta\cos\theta}$
$\equiv \frac{\cos 2\theta}{\sin\theta\cos\theta}$
$\equiv \frac{2\cos 2\theta}{\sin 2\theta}$
$\equiv 2\cot 2\theta \equiv \text{RHS}$
The identity is proved.
2 Use $\cos^2 A - \sin^2 A \equiv \cos 2A$
2 Use $2\sin A\cos A \equiv \sin 2A$
Recognise $\frac{\cos 2\theta}{\sin 2\theta} \equiv \cot 2\theta$

Example 5

Use the identity $\cos 2\theta \equiv \cos^2\theta - \sin^2\theta$ to find a Cartesian equation for the curve given by parametric equations $x - \sin\theta°$, $y = \cos 2\theta°$

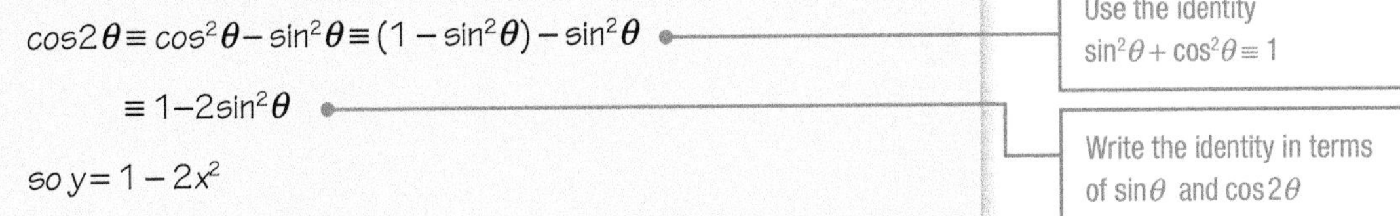

$\cos 2\theta \equiv \cos^2\theta - \sin^2\theta \equiv (1 - \sin^2\theta) - \sin^2\theta$
$\equiv 1 - 2\sin^2\theta$
so $y = 1 - 2x^2$
Use the identity $\sin^2\theta + \cos^2\theta \equiv 1$
Write the identity in terms of $\sin\theta$ and $\cos 2\theta$

MyMaths 2157, 2158, 2262 SEARCH

Exercise 14.3B Reasoning and problem-solving

1 Solve these equations for $0 \le \theta \le 180°$. Show your working.

a $3\sin\theta = \sin(\theta + 45°)$

b $2\cos\theta = \cos(\theta + 30°)$

c $2\sin\theta + \sin(\theta + 60°) = 0$

2

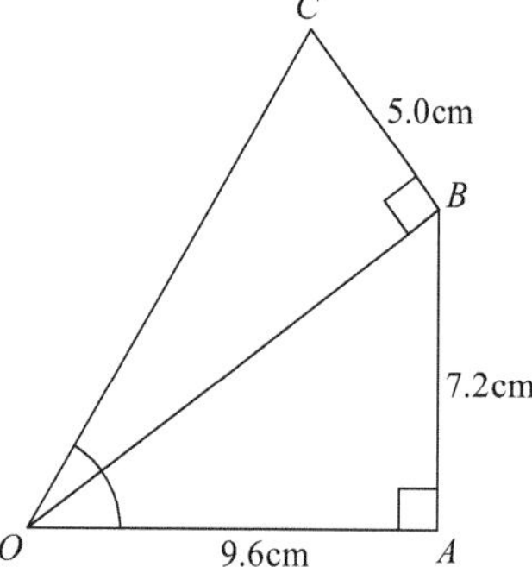

a Calculate lengths OB and OC in quadrilateral $OABC$

b Hence use compound angle formulae to calculate the sine, cosine and tangent of angle AOC

3 Prove these identities.

a $\tan A + \cot A \equiv 2\operatorname{cosec} 2A$

b $\operatorname{cosec} 2A + \cot 2A \equiv \cot A$

c $2\operatorname{cosec} 2A \equiv \sec A \operatorname{cosec} A$

d $\operatorname{cosec} A - \cot A \equiv \tan\frac{A}{2}$

4 If $\sin(\theta + \phi) = \cos\phi$, show that $\tan\theta + \tan\phi = \sec\theta$

5 Prove that

$$\arcsin\frac{1}{2} + \arcsin\frac{1}{3} = \arcsin\left(\frac{2\sqrt{2}+\sqrt{3}}{6}\right)$$

6 a Angle A is obtuse and angle B is acute such that $\tan A = -2$ and $\tan B = \sqrt{5}$. Use trigonometric formulae to find the values, in surd form, of

i $\cot(A + B)$ ii $\sin(A + B)$

b If $\tan\left(\theta + \frac{\pi}{3}\right) = \frac{1}{3}$, show that

$$\tan\theta = 2 - \frac{5}{3}\sqrt{3}$$

7 Solve these equations for $0 \le \theta \le 2\pi$. Show your working.

a $\tan 2\theta = 3\tan\theta$ b $\cos 2\theta + 3\sin\theta = -1$

c $\sin 2\theta - 1 = \cos 2\theta$ d $\sin 2\theta + \sin\theta = \tan\theta$

e $\cos 2\theta = \tan 2\theta$ f $2\sin 2\theta = \tan\theta$

8 Solve these equations for $0 \le \theta \le 360°$. Show your working.

a $\sin\theta = \sin\frac{\theta}{2}$ b $\tan\theta = 6\tan\frac{\theta}{2}$

c $3\cos\frac{\theta}{2} = 2 + \cos\theta$ d $\sin\theta = \cot\frac{\theta}{2}$

9 Find the Cartesian equation for the curve that has the following parametric equations.

$x = 4\cos\theta \quad y = 2\cos 2\theta$

10 A curve is defined by the parametric equations $x = 3\sec\theta$, $y = 6\tan\theta$. Find a Cartesian equation for the curve.

11 a Prove that $\tan(A + B) \equiv \frac{\tan A + \tan B}{1 - \tan A \tan B}$

b Find a similar formula for $\tan(A - B)$

12 Prove these identities.

a $\frac{\sin(\theta+\phi)}{\cos\theta\cos\phi} \equiv \tan\theta + \tan\phi$

b $\cot\theta \equiv \frac{\sin 2\theta + \cos\theta}{1 + \sin\theta - \cos 2\theta}$

c $\frac{\cos\theta + \sin\theta}{\cos\theta - \sin\theta} \equiv \sec 2\theta + \tan 2\theta$

d $\frac{1 + \sin 2\theta}{1 - \sin 2\theta} \equiv \tan^2\left(\theta + \frac{\pi}{4}\right)$

e $\frac{\cos^3\theta + \sin^3\theta}{\cos\theta + \sin\theta} \equiv 1 - \frac{1}{2}\sin 2\theta$

f $\frac{\sin 2\theta}{1 - \cos 2\theta} \equiv \cot\theta$

g $\sin 2\theta \equiv \frac{2\tan\theta}{1 + \tan^2\theta}$

h $\frac{\sin 2\theta + \sin\theta}{\cos 2\theta + \cos\theta + 1} \equiv \tan\theta$

13 Prove that $\cos^4\theta + 2\sin^2\theta - \sin^4\theta \equiv 1$

14 Prove that $\frac{1 - \tan 15°}{1 + \tan 15°} = \frac{1}{\sqrt{3}}$

15 Given that $\sin\theta\tan\theta = 2 - \cos\theta$, prove that $\cos\theta = 0.5$
Hence, solve the equation $\sin\phi\tan\phi = 2 - \cos\phi$ for $0 \le \phi < \pi$

16 Given that $3\cot\theta = 8\sec\theta$, prove that $3\sin^2\theta + 8\sin\theta - 3 = 0$

Hence solve the equation $3\cot\theta = 8\sec\theta$ for $0 < \theta < 180°$. Show your working.

17 **a** Prove that $(\cos x + \cos y)^2 + (\sin x + \sin y)^2 = 4\cos^2\left(\frac{x-y}{2}\right)$

b Use this result to prove that $\cos 15° = \frac{\sqrt{2}+\sqrt{6}}{4}$

18 **a** Use double angle formulae to prove that $\sin A + \sin B = 2\sin\left(\frac{A+B}{2}\right)\cos\left(\frac{A-B}{2}\right)$

b Solve the equation $\sin x + \sin y = 0$ for $0 \le x < \pi$ and $0 \le y < \pi$. Show your working.

19 Find all the solutions of the equation $\operatorname{cosec}^2\theta - \frac{1}{2}\cot\theta = 3$ for $0 \le \theta \le 2\pi$. Show your working.

20 **a** Show that the equation $\frac{1}{1-\sin x} + \frac{1}{1+\sin x} = 8$ can be rearranged as $\sec^2 x = 4$

b Use your result to solve the equation $\frac{1}{1-\sin(2\theta+10°)} + \frac{1}{1+\sin(2\theta+10°)} = 8$ for $0 \le \theta \le 180°$. Show your working.

21 The curve with the parametric equations $x = 4k\sin t$, $y = 3 + k\cos 2t$ intersects the y-axis at the point (0, 5). If k is a non-zero constant, find the value of k.

22 A curve has the Cartesian equation $y = \frac{4\sqrt{9-x^2}}{x}$. Two parametric equations also define this curve. If one of these equations is $x = 3\sin t$, find the other equation

23 **a** Given that $3\operatorname{cosec}^2 x - \cot^2 x = n$, use the identity $1 + \cot^2 x = \operatorname{cosec}^2 x$ to show that $\sec^2 x = \frac{n-1}{n-3}$, provided that $n \ne 3$

b Use your answer to **a** to solve the equation $3\operatorname{cosec}^2(\theta+30°) - \cot^2(\theta+30°) = 5$, giving your answers in degrees where $0 \le \theta \le 180°$. Show your working.

24 Two seas, with different tidal patterns, meet in a strait. The depth of each sea is given by $d_1 = 6 + 4\sin\left(t \times \frac{\pi}{6}\right)$ and $d_2 = 6 + 2\sin\left[(t-1) \times \frac{\pi}{6}\right]$, respectively, where t is the number of hours after midnight. The tides clash except when their depths are equal. At what times between 00:00 and 12:00 is there no clash? Show your working.

Challenge

25 **a** Prove these identities.

i $\sin 3A \equiv 3\sin A - 4\sin^3 A$

ii $\cos 3A \equiv 4\cos^3 A - 3\cos A$

b Express $\tan 3A$ in terms of $\tan A$

c Find identities for $\sin 4A$ and $\cos 4A$ in terms of $\sin A$ and $\cos A$

26 **a** Given that $2\arctan 3 = \operatorname{arccot} x$, substitute $\tan\alpha = 3$ and $\cot\beta = x$ to explain why $\tan 2\alpha = \tan\beta$
Hence use trigonometric formulae to show that $x = -\frac{4}{3}$

b Use similar substitutions to show the solution of $\arcsin x + \arccos\frac{12}{13} = \arcsin\frac{4}{5}$ is $x = \frac{33}{65}$

c Solve this equation for x
$x = \arcsin k + \arcsin\sqrt{1-k^2}$

14.4 Equivalent forms for $a\cos\theta + b\sin\theta$

Fluency and skills

Compare the graphs of $y = 3\cos\theta$, $y = 4\sin\theta$ and $y = 3\cos\theta + 4\sin\theta$

For each θ-value on the graph, you can add the two y-values on the blue and green curves to give the y-value on the black curve. Check this result by drawing these three graphs on a graphics calculator.

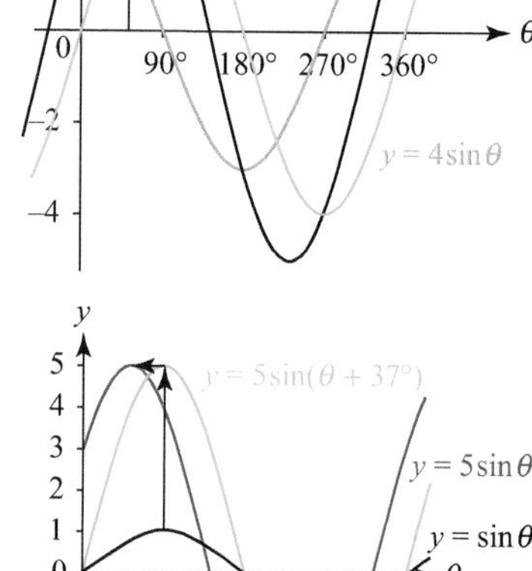

The graph of $y = 3\cos\theta + 4\sin\theta$ can also be created from the graph of $y = \sin\theta$ by

(a) Stretching, scale factor 5, parallel to the y-axis, then

(b) Translating by about 37° to the left. (You will work out the exact value of the translation in Example 1.)

So, $3\cos\theta + 4\sin\theta = 5\sin(\theta + 37°)$

You can check this result using a graphics calculator.

Similarly, you can use a graphics calculator to show that $3\cos\theta + 4\sin\theta = 5\cos(\theta - 53°)$

Key point

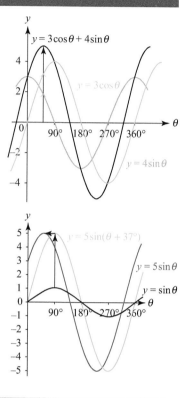

$a\cos\theta \pm b\sin\theta$ can be written as $r\sin(\theta \pm \alpha)$ or $r\cos(\theta \pm \alpha)$, where r is positive and angle α is acute.

Example 1

If $4\sin\theta + 3\cos\theta = r\sin(\theta + \alpha)$, find r and α such that $r > 0$ and α is acute.

$4\sin\theta + 3\cos\theta = r\sin\theta\cos\alpha + r\cos\theta\sin\alpha$ — Use a compound angle formula to expand the right-hand side.

$4 = r\cos\alpha$ [1]

$3 = r\sin\alpha$ [2] — Compare coefficients of $\sin\theta$ and $\cos\theta$

$\frac{3}{4} = \frac{r\sin\alpha}{r\cos\alpha} = \tan\alpha$ — Divide [2] by [1] and use $\tan\alpha \equiv \frac{\sin\alpha}{\cos\alpha}$

$\alpha = \arctan\left(\frac{3}{4}\right) = 36.9°$ — α is acute.

$3^2 + 4^2 = r^2(\sin^2\alpha + \cos^2\alpha)$ — Square [1] and [2] and add to find r

$= r^2$ — Use $\sin^2\alpha + \cos^2\alpha \equiv 1$

$r = \sqrt{25} = 5$ — Ignore $r = -5$ as r is +ve.

Hence, $4\sin\theta + 3\cos\theta = 5\sin(\theta + 36.9°)$

Exercise 14.4A Fluency and skills

1 In each case, find the values of r and α where $r > 0$ and α is acute.

Give r as a surd where appropriate and give α in degrees.

a $5\sin\theta + 12\cos\theta = r\sin(\theta + \alpha)$

b $4\cos\theta + 3\sin\theta = r\cos(\theta - \alpha)$

c $\cos\theta - 2\sin\theta = r\cos(\theta + \alpha)$

d $4\sin\theta - 2\cos\theta = r\sin(\theta - \alpha)$

e $3\sin\theta - 4\cos\theta = r\sin(\theta + \alpha)$

f $8\cos\theta + 15\sin\theta = r\cos(\theta + \alpha)$

g $2\cos 3\theta + 5\sin 3\theta = r\sin(3\theta + \alpha)$

2 Find the value of r so that

a $\cos\theta + \sin\theta = r\sin\left(\theta + \frac{\pi}{4}\right)$

b $\cos\theta - \sin\theta = r\cos\left(\theta + \frac{\pi}{4}\right)$

3 Find the value of α so that

a $\sqrt{3}\cos\theta + \sin\theta = 2\sin(\theta + \alpha)$

b $\cos\theta - \sqrt{3}\sin\theta = 2\cos(\theta + \alpha)$

4 a Use a graphics calculator or otherwise to draw the graph of $y = 12\sin\theta + 5\cos\theta$

If $y = r\sin(\theta + \alpha)$, write the values of r and α from your calculator display.

Check your answers algebraically.

b Repeat for $y = 4\sin\theta - 5\cos\theta$

5 The graph of $y = \sin\theta$ is stretched by a scale factor r parallel to the y-axis, and then translated by the vector $\alpha\mathbf{i} + 0\mathbf{j}$ to create the graph of $y = r\sin(\theta - \alpha)$, where α is in degrees. Describe the two translations which, when applied to the graph of $y = \sin\theta$, transform it into the graph of $y = 5\cos\theta + 12\sin\theta$

Reasoning and problem-solving

Strategy

To solve problems of the form $a\cos\theta + b\sin\theta = c$

1 Choose from $r\sin(\theta \pm \alpha)$ and $r\cos(\theta \pm \alpha)$ by matching the signs.

2 Expand using the compound angle formula.

3 Evaluate the values of r and α by matching coefficients.

3 Use $r\sin(\theta \pm \alpha)$ or $r\cos(\theta \pm \alpha)$ to solve equations or find stationary points.

You have four choices when converting $a\cos\theta + b\sin\theta$ to either $r\sin(\theta \pm \alpha)$ or $r\cos(\theta \pm \alpha)$

The algebra is easier if you match the signs. For example, since $2\sin\theta - 3\cos\theta$ has a negative sign, you should use $r\sin(\theta - \alpha)$ or $r\cos(\theta + \alpha)$ as they too have a negative sign when expanded. If instead you choose $r\sin(\theta + \alpha)$ or $r\cos(\theta - \alpha)$ for $2\sin\theta - 3\cos\theta$, you get a different sign when you expand and α will be negative.

Example 2

a Solve the equation $2\cos\theta + 3\sin\theta = 1$ for θ between 0 and 360°. Show your working.

b Find the maximum value of $2\cos\theta + 3\sin\theta$ and the smallest positive value of θ at which the maximum occurs.

a $2\cos\theta + 3\sin\theta = r\cos(\theta - \alpha)$ where α is acute

Choose $r\cos(\theta - \alpha)$ so that both sides have + signs. Alternatively, you could use $r\sin(\theta + \alpha)$ (1)

$= r\cos\theta\cos\alpha + r\sin\theta\sin\alpha$

Expand using the compound angle formula. (2)

$r\cos\alpha = 2$ [1]

$r\sin\alpha = 3$ [2]

Compare coefficients of $\sin\theta$ and $\cos\theta$ (3)

$r^2(\cos^2\alpha + \sin^2\alpha) = 2^2 + 3^2 = 13$

$r = \sqrt{13}$

Square and add [1] and [2] to find r

$\frac{r\sin\alpha}{r\cos\alpha} = \tan\alpha = \frac{3}{2}$

$\alpha = \arctan\frac{3}{2} = 56.3°$

Divide [2] by [1] and solve to find α

The equation is now $\sqrt{13}\cos(\theta - 56.3°) = 1$

$\cos(\theta - 56.3°) = \frac{1}{\sqrt{13}} = 0.2773...$

$\theta - 56.3° = 73.9°$ or $360° - 73.9° = 286.1°$

cos is +ve in 1st and 4th quadrants. (4)

$\theta = 73.9° + 56.3°$ or $286.1° + 56.3°$

$\theta = 130.2°$ or $342.4°$

b $2\cos\theta + 3\sin\theta = \sqrt{13}\cos(\theta - 56.3°)$

The maximum value of $2\cos\theta + 3\sin\theta$ is $\sqrt{13} \times 1 = \sqrt{13}$

The maximum value of $\cos(\theta - 56.3°)$ is 1 at 0° and 360° (4)

Maximum occurs when $\theta - 56.3° = 0°, 360°$

The smallest positive value of θ is 56.3°

Exercise 14.4B Reasoning and problem-solving

1 Solve these equations for $0 \le \theta \le 360°$. Show your working.

a $5\cos\theta + 12\sin\theta = 6$

b $2\cos\theta - 3\sin\theta = 1$

c $8\sin\theta + 15\cos\theta = 10$

d $3\sin\theta - 5\cos\theta = 4$

e $\sin\theta - \sqrt{3}\cos\theta = 0$

f $5\sin\theta + 8\cos\theta = 0$

2 Solve these equations for $-\pi \le \theta \le \pi$ Show your working.

a $\cos\theta + \sqrt{3}\sin\theta = 2$

b $\cos\theta + \sin\theta = \frac{1}{\sqrt{2}}$

c $\sin\theta + \sqrt{3}\cos\theta = 1$

d $\sqrt{3}\cos\theta - \sin\theta = \sqrt{2}$

3 Solve these equations for $0 \le \theta \le 180°$. Show your working.

a $2\sin 2\theta + \cos 2\theta = 1$

b $2\cos 3\theta - 6\sin 3\theta = 5$

c $6\cos\frac{\theta}{2}+8\sin\frac{\theta}{2}=9$

d $\sin\frac{\theta}{2}-4\cos\frac{\theta}{2}=1$

4 a Use a graphics calculator or otherwise to draw the graphs of $y=\sqrt{2}\sin(\theta+45°)$ and $y=\cos(\theta-30°)$ on the same axes. Find the points of intersection in the range $-180° \le \theta \le 180°$

b Use an algebraic method to calculate the same points of intersection.

5 Use a graphics calculator or otherwise to draw the graph of $y=3\sin x+4\cos x$

Find algebraically the values of x in the range $-90° \le x \le 90°$ for which $y \ge 3$. Show your working.

6 Use a graphics calculator or otherwise to draw the graph of $y=\sin 2x+2\cos 2x$. Showing your working, find algebraically the values of x in the range $0<x<180°$ for which

a $y>2$ b $y<-2$

7 An alternating electrical current i amps at a time t seconds (where $t \ge 0$) is given by $i=12\cos 3t-5\sin 3t$

a What is the initial value of the current?

b How many seconds does it take for the current to fall to 5 amps?

8 a Show that $\sqrt{2}\cos\theta+\sqrt{3}\sin\theta$ can be written in the form $r\cos(\theta-\alpha)$ where $r>0$ and α is acute. Hence, find the maximum value of $\sqrt{2}\cos\theta+\sqrt{3}\sin\theta$ and the values of θ between 0 and 360° at which the maximum value occurs.

b Find the minimum value of $\frac{1}{\sqrt{2}\cos\theta+\sqrt{3}\sin\theta}$ and the smallest value of θ at which it occurs. You can check your answer on a graphics calculator.

9 Use the compound angle formula to find the maximum and minimum values of each expression, giving your answers in surd form if necessary. In each case, state the smallest positive value of θ at which each maximum and minimum occurs.

a $\cos\theta-\sqrt{3}\sin\theta$ b $24\sin\theta-7\cos\theta$

c $3\sin\theta-2\cos\theta$ d $8\cos 2\theta-6\sin 2\theta$

You can check your answers using a graphics calculator.

10 a Express $7\cos\theta-24\sin\theta$ in the form $r\cos(\theta-\alpha)$, giving the values of r and θ

Show that $7\cos\theta-24\sin\theta+3 \le 28$ and find the minimum value of $7\cos\theta-24\sin\theta+3$

b Use a graphics calculator or otherwise to draw the graph of $f(\theta)=\frac{1}{7\cos\theta-24\sin\theta}$ for $0 \le \theta \le 2\pi$

Describe the transformations required to map the graph of $y=\sec\theta$ onto the graph of $y=f(\theta)$

Challenge

11 Use the compound angle formula to find the requested stationary values of these expressions and, in each case, state the smallest positive value of θ at which these values occur.

a The maximum value of

i $20+5\sin\theta-12\cos\theta$

ii $20-(5\sin\theta-12\cos\theta)$

b The minimum value of $\frac{65}{5\sin\theta-12\cos\theta}$

You can check your results on a graphics calculator.

12 Two alternating electrical currents are combined so that the resultant current I is given by $I=2\cos\omega t-4\sin\omega t$, where the constant $\omega=4$, and where $t \ge 0$ is the time in seconds.

a Use the compound angle formula to find the maximum value of I and the first time at which it occurs.

b For how many seconds in each cycle is the value of I more than half its maximum value?

14 Summary and review

Chapter summary

- Angles are measured in degrees or radians. $180° = \pi$ radians.
- Arc length $= r\theta$ and sector area $= \frac{1}{2}r^2\theta$, with θ in radians.
- For small θ in radians, $\sin\theta \approx \theta$, $\tan\theta \approx \theta$ and $\cos\theta \approx 1 - \frac{1}{2}\theta^2$
- The reciprocal trigonometric functions are $\operatorname{cosec}\theta = \frac{1}{\sin\theta}$, $\sec\theta = \frac{1}{\cos\theta}$ and $\cot\theta = \frac{1}{\tan\theta}$
- The inverse trigonometric functions are $\arcsin x$, $\arccos x$ and $\arctan x$, also known as $\sin^{-1} x$, $\cos^{-1} x$ and $\tan^{-1} x$
- The compound angle formulae give expansions for $\sin(A \pm B)$, $\cos(A \pm B)$ and $\tan(A \pm B)$
  - $\sin(A \pm B) \equiv \sin A\cos B \pm \cos A\sin B$
  - $\cos(A \pm B) \equiv \cos A\cos B \mp \sin A\sin B$
  - $\tan(A \pm B) \equiv \dfrac{\tan A \pm \tan B}{1 \mp \tan A\tan B}$
- The double angle formulae convert trigonometric functions for $2A$ into trigonometric functions for A
  - $\sin 2A \equiv 2\sin A\cos A$
  - $\cos 2A \equiv \cos^2 A - \sin^2 A$
  - $\tan 2A \equiv \dfrac{2\tan A}{1 - \tan^2 A}$
- The expression $a\cos\theta \pm b\sin\theta$ can be rewritten in the form $r\sin(\theta \pm \alpha)$ or $r\cos(\theta \pm \alpha)$ where $r > 0$ and angle α is acute.

Check and review

You should now be able to...	Try Questions
✔ Convert between degrees and radians and use radians in problems.	1, 2, 6, 14
✔ Use reciprocal and inverse trigonometric functions.	3, 7, 9, 14, 15
✔ Use trigonometric formulae for compound angles, double angles and half angles.	4, 8, 13
✔ Find and use equivalent forms for $a\cos\theta + b\sin\theta$	11, 12
✔ Solve equations using trigonometric formulae to simplify expressions.	5, 10, 15

1 a Convert these angles to degrees, giving your answers to 1 dp.

i 1 radian ii 2.5 radians

iii 3.5 radians

b Convert these angles to radians, giving your answers in terms of π

i 15° ii 72° iii 210°

2 Find approximations for these expressions when θ is a small angle.

a $\dfrac{\tan\theta}{\sin\theta}$ b $\dfrac{\sin 2\theta}{4\theta}$ c $\dfrac{1-\cos 2\theta}{\theta}$

3 Using special triangles, and showing any working, write the exact values of

a $\sec\frac{\pi}{3}$ b $\cot\frac{\pi}{6}$

c $\operatorname{cosec}\frac{3\pi}{4}$ d $\operatorname{cosec}\frac{3\pi}{2}$

e $\arccos\frac{1}{\sqrt{2}}$ f $\arctan\frac{1}{\sqrt{3}}$

g $\arccos 0$ h $\arcsin\left(-\frac{1}{\sqrt{2}}\right)$

4 Write each expression as a single trigonometric ratio.

a $\cos 40° \cos 10° - \sin 40° \sin 10°$

b $\frac{1-\tan 100° \tan 35°}{\tan 100° + \tan 35°}$

5 Solve for $0 \le x \le 180°$. Show your working.

a $\cot 2x = \frac{1}{2}$ b $\sec(2x + 40°) = 2$

c $3\cos x - \cot x = 0$ d $2\cot x = \operatorname{cosec} x$

e $4\cos x - \operatorname{cosec} x = 0$

f $\operatorname{cosec}^2 x = 4\cot x - 3$

g $\tan x + \cot x = 2$

6 A circle, of radius 4 cm, has a sector AOB subtending and angle of $\frac{\pi}{4}$ radians at the centre O

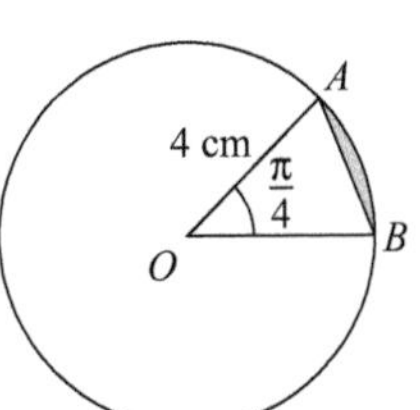

Calculate the exact area of

a The sector AOB

b The shaded segment.

7 Simplify

a $\sin x\left(\frac{\cot x}{\sec x} + \sin x\right)$ b $\frac{1}{\sin^2\theta} - \frac{\cot\theta}{\tan\theta}$

8 Find the values of tan 15° and sec 75°, writing your answers as surds.

9 Prove the identity $\arctan x \equiv \frac{\pi}{2} - \operatorname{arccot} x$

10 Solve these equations for $0 \le x \le 180°$. Show your working.

a $\sin(\theta - 60°) = 3\cos(\theta - 30°)$

b $4\sin 2\theta = \sin\theta$

c $\sin 2\theta + \sin\theta = \tan\theta$

d $\cos 2\theta = 3 - 5\sin\theta$

11 a Express $7\sin\theta - 24\cos\theta$ in the form $r\sin(\theta - \alpha)$, where $r > 0$ and α is acute.

b Hence, solve the equation $7\sin\theta - 24\cos\theta = 15$ for $0 \le \theta \le 360°$. Show your working.

12 a Write $7\cos\theta + 6\sin\theta$ in the form $r\sin(\theta + \alpha)$, where $r > 0$ and α is acute.

b Hence, find the maximum value of $7\cos\theta + 6\sin\theta$ and the values of θ at which it occurs in the interval $0 \le \theta \le 360°$

13 Prove these identities.

a $1 - \cos 2A \equiv \tan A \sin 2A$

b $\sin 2A(1 + \tan^2 A) \equiv 2\tan A$

c $\cot A - \tan A \equiv 2\cot 2A$

14 Sketch the graph of $y = \cos^{-1}\left(\frac{x}{2}\right)$ with y in radians. Write the exact coordinates of the end points of the graph.

15 If $f(x) = \frac{1-\sin x}{\cos x} + \frac{\cos x}{1-\sin x}$, show that $f(x) = 2\sec x$. Hence, solve the equation $f(x) = \tan^2 x - 2$ for $0 \le x < 360°$. Show your working.

What next?

Score		
0 – 7	Your knowledge of this topic is still developing. To improve, search in MyMaths for the codes: 2050, 2155–2160, 2262, 2266	
8 – 11	You're gaining a secure knowledge of this topic. To improve, look at the InvisiPen videos for Fluency and skills (14A)	
12 – 15	You've mastered these skills. Well done, you're ready to progress! To develop your techniques, look at the InvisiPen videos for Reasoning and Problem-solving (14B)	

Click these links in the digital book

Exploration

Going beyond the exams

History

The demands of navigation and astronomy in 16th century Europe required the solution of problems from a branch of geometry called **spherical trigonometry**, the study of sides and angles of **spherical polygons**.

The process demanded many calculations involving multiplication and division, and so a method was needed in order to make them easier.

Prior to the invention of **logarithms** in 1614, the most common method involved the use of trigonometric tables with an identity such as

$$\cos A \cos B \equiv \frac{1}{2}(\cos(A+B) + \cos(A-B))$$

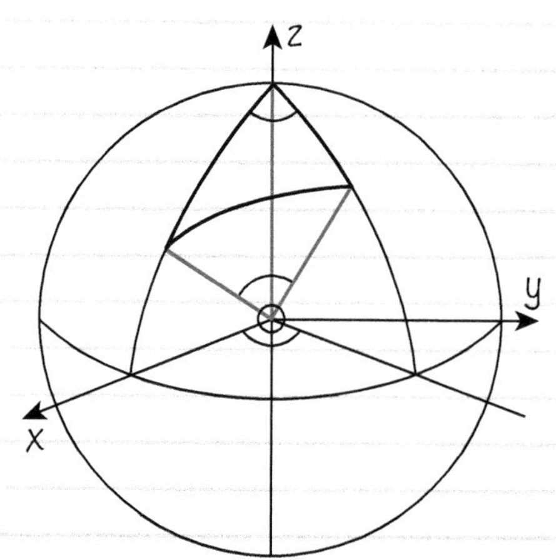

Tycho Brahe (1546–1601) made great use of this method over a twenty year period of observations and calculations, taking great pains to achieve accurate results. On his death, he left his entire work to **Johannes Kepler**, who used it to develop his theory of planetary motion.

Kepler's work influenced both **Galileo** and **Newton**.

Have a go

Suppose that you need to calculate 0.4379×0.9768

By considering the identity above, let $\cos A = 0.4379$ and $\cos B = 0.9768$

Use arccos (inverse cosine) to find A and B.
$A = 64.0300$ $B = 12.3659$

$A + B = 76.3959$ $\quad A - B = 51.6641$

$\cos(76.3959) = 0.2352$
$\cos(51.6641) = 0.6203$

$\frac{1}{2}(0.2352 + 0.6203) = ?$
(Work this out without a calculator.)

The exact answer to the original question is 0.42774072 – how does this compare?

Try calculating 87.14×3.519 using this method.

Note

The values of the six trigonometric ratios had, by this time, been worked out *by hand* to at least 10 decimal places.

"Geometry is the archetype of the beauty of the world."
- Johannes Kepler

Note

To work out 43.79×976.8, for example, the calculation is exactly the same, but the position of the decimal point is adjusted at the end.

14 Assessment

1 Given that $\arccos\left(x+\frac{1}{2}\right)=\frac{\pi}{4}$, find the exact value of x. Select the correct answer.

A 0.5 B $\frac{1}{\sqrt{2}}+\frac{1}{2}$ C 0.207... D $\frac{\sqrt{2}-1}{2}$ **[1 mark]**

2 A sector of a circle with radius 3 cm has an area of 5.4 cm^2. Calculate the angle θ, in radians, subtended at the centre of the circle. Select the correct answer.

A 3.6 B 1.2 C 0.3 D $\frac{\pi}{3}$ **[1]**

3 a Find the exact values of x for which $\tan x=\sqrt{3}$ in the interval $-180°\le x\le 180°$ **[3]**

b Sketch the graph of $y=\tan 2x$ for $-90°\le x\le 90°$. State the equations of the asymptotes. **[3]**

4 A sector of a circle of radius 5 cm has an angle of $\frac{2\pi}{3}$ at its centre. Calculate the exact value of

a The area of the sector, **[2]**

b The arc length of the sector. **[2]**

5 a State the exact value of i $\sec\left(\frac{\pi}{4}\right)$ ii $\cot\left(\frac{5\pi}{6}\right)$ **[4]**

b Sketch the graph of $y=\operatorname{cosec} x$ for the interval $-\pi\le x\le\pi$
Write the equations of the asymptotes. **[3]**

c State the range of $y=\operatorname{cosec} x$ **[1]**

6 Solve these equations for $0\le x\le 360°$. Show your working.

a $\operatorname{cosec} x=2$ **[3]** b $\sec x=-\sqrt{2}$ **[3]**

7 The velocity of a particle at time t seconds is given by $v=5\cos\left(\frac{t}{2}-\frac{\pi}{3}\right)\text{m s}^{-1}$

a State the maximum speed of the particle. **[1]**

b Give the times at which this maximum speed occurs for $0\le t<15$ **[3]**

8 a On the same set of axes, sketch the graphs of $y=\sin x$ for $-\frac{\pi}{2}\le x\le\frac{\pi}{2}$ and $y=\arcsin x$ for $-1\le x\le 1$ **[3]**

b Describe the geometric relationship between the two curves. **[1]**

c Work out the value of these expressions giving your answers as exact multiples of π

i $\arcsin\left(-\frac{\sqrt{3}}{2}\right)$ ii $\arcsin\left(\cos\frac{\pi}{3}\right)$ **[2]**

9 a Use the identity $\sin^2\theta+\cos^2\theta\equiv 1$ to show that $\sec^2\theta-\tan^2\theta\equiv 1$ **[3]**

b Given that $\tan\theta=\sqrt{5}$, find the exact value of

i $\sec\theta$ ii $\cos\theta$ **[5]**

10 By using the formula $\cos(A\pm B)\equiv\cos A\cos B\mp\sin A\sin B$, find the exact value of

a $\cos 75°$ **[3]** b $\cos 15°$ **[2]**

11 Given that $\sin(A+B)\equiv\sin A\cos B+\sin B\cos A$, show that $\sin 2x\equiv 2\sin x\cos x$ **[2]**

12 Write $6\sin\theta+8\cos\theta$ in the form $r\sin(\theta+\alpha)$, where $r>0$ and $0<\alpha<90°$ **[4]**

13 For $\theta=0.05$ radians, state the approximate value of

a $\sin\theta$ **[1]** b $\cos\theta$ **[1]** c $\tan\theta$ **[1]**

14 a Sketch the graphs of these equations for $0\le x\le 2\pi$. Label the x- and y-intercepts with their exact values.

i $y=\sin 3x$ ii $y=\cos\left(x+\frac{\pi}{3}\right)$ **[6]**

b Solve the equation $\sin 3x=\frac{\sqrt{2}}{2}$ for $0\le x\le\pi$. Show your working and give your answers in terms of π **[5]**

15 The area of an isosceles triangle is 100 cm^2. Calculate the perimeter of the triangle, given that one of the angles is $\frac{\pi}{6}$ rad. **[7]**

16 The population (in thousands) of a particular species of insect around a lake t weeks after a predator is released is modelled by

$P=6.5-4.1\sin\left(\frac{\pi t}{2.3}\right)$

a What was the initial population? **[2]**

b State the maximum possible population of the insect. **[1]**

c When does this maximum first occur? Give your answer to the nearest day. **[3]**

17 AB is the arc of a circle with radius 3.5 cm and centre C as shown.

a Calculate the

i Area of the sector, ii Perimeter of the sector. **[5]**

The segment S is bounded by the arc and the chord AB

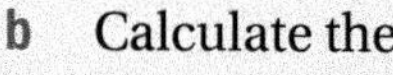

b Calculate the

i Area of S ii Perimeter of S **[6]**

18 a Sketch these graphs for $0\le x\le 2\pi$. Label the x- and y-intercepts and any asymptotes in terms of π

i $y=2\operatorname{cosec}x$ ii $y=\sec\frac{x}{2}$ **[6]**

b State the range of each of the graphs in part **a**. **[3]**

19 $f(x)=\cot(x-30°)$

a Sketch the graph of $y=f(x)$ for x in the interval $-180°\le x\le 180°$
Label the x- and y-intercepts and give the equations of any asymptotes. **[4]**

b Solve the equation $\cot(x-30°)=0.2$ for $-180°\le x\le 180°$, showing your working. **[3]**

20 Solve these equations for $0\le x\le 2\pi$. Show your working and give your solutions as exact multiples of π

i $\sec(x+\pi)=2$ ii $\operatorname{cosec}\left(x-\frac{\pi}{8}\right)=\sqrt{2}$ **[8]**

21 $f(x)=\arccos(x-1)$

a Sketch the graph of $y=f(x)$ for $0\le x\le 2$ **[3]**

b State the range of $f(x)$ **[2]**

c Work out the inverse of $f(x)$ and state its domain and range. **[4]**

22 Find the exact value of x for which

a $\arctan(2x-1)=\dfrac{\pi}{3}$ **[2]**   b $\operatorname{arccot}(x-5)=\dfrac{2\pi}{3}$ **[2]**

23 a Show that the equation $2\tan^2 x=\sec x-1$ can be written as $2\sec^2 x-\sec x-1=0$ **[3]**

b Hence solve the equation $2\tan^2 x=\sec x-1$ for x in the interval $0\le x\le 2\pi$, showing your working. **[4]**

24 a Given that $\sin^2\theta+\cos^2\theta\equiv 1$, show that $\operatorname{cosec}^2\theta-\cot^2\theta\equiv 1$ **[3]**

b Solve the equation $\operatorname{cosec}^2\theta-3\cot\theta-1=0$ for $0\le\theta<360°$, showing your working. **[6]**

25 a Show that $\sec^4 x-\tan^4 x\equiv\sec^2 x+\tan^2 x$ **[2]**

b Find the values of x in the range $-\pi\le x\le\pi$ that satisfy $\sec^4 x-\tan^4 x=5+\tan^2 x$, showing your working. **[4]**

26 a By writing $\cos 3x$ as $\cos(2x+x)$, show that $\cos 3x=4\cos^3 x-3\cos x$ **[5]**

b Hence solve the equation $8\cos^3 x-6\cos x=\sqrt{3}$ for x in the interval $0\le x\le 2\pi$

Show your working and give your answers as exact multiples of π **[5]**

27 a Use the identity $\cos(A+B)\equiv\cos A\cos B-\sin A\sin B$ to show that $\cos 2x\equiv 2\cos^2 x-1$ **[3]**

b Hence solve the equation $\cos 2x+3\cos x+2=0$ for $0\le\theta<360°$, showing your working. **[5]**

28 Prove by counter-example that $\cos(A+B)\ne\cos A+\cos B$ **[2]**

29 $f(x)=8\cos x+4\sin x$

a Write $f(x)$ in the form $r\cos(x-\alpha)$, where $r>0$ and $0<\alpha<\dfrac{\pi}{2}$ **[4]**

b Hence solve the equation $8\cos x+4\sin x=\sqrt{5}$ for $0<x\le 2\pi$, showing your working. **[4]**

30 Sketch these graphs for $0\le x\le 360°$

a $y=2\cos(x+60°)$ **[4]**   b $y=-\sin\left(\dfrac{x}{2}\right)$ **[2]**

31 $g(x)=\tan\left(\dfrac{\pi}{3}-x\right)$, for $0\le x\le 2\pi$

a Sketch the graph of $y=g(x)$, clearly labelling the x- and y-intercepts, and state the equations of the asymptotes. **[4]**

b Solve the equation $g(x)=\dfrac{\sqrt{3}}{3}$. Show your working and give your solutions in terms of π **[4]**

32 The area of sector ABC is 56.7 cm² and the length of the arc AB is 12.6 cm

Calculate the area of the shaded segment. **[8]**

33 The shape ABC is formed from a sector and a triangle as shown in the diagram.

a Find the length of AC **[2]**

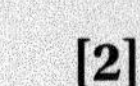

b Calculate

i The area of ABC ii The perimeter of ABC **[7]**

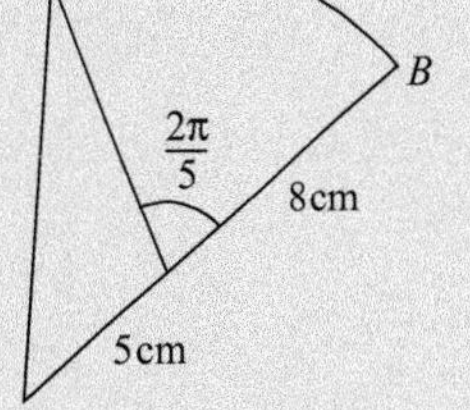

34 a Sketch the graphs of

i $y=\cot(2x-60)$ for $0 \le x \le 180°$

ii $y=1-\operatorname{cosec}\left(\frac{x}{2}\right)$ for $0 \le x \le 360°$

Give the equations of the asymptotes. **[7]**

b Solve the equation $\sec(3x+20)=\sqrt{2}$ for x in the interval $0 \le x \le 180°$. Show your working. **[5]**

35 a Solve the equation $\cot^2\theta=5$ for $0 \le \theta \le 2\pi$ **[4]**

b Find the exact solutions of $\sec^2\left(\theta+\frac{\pi}{6}\right)=2$ for $0 \le \theta \le 2\pi$. Show your working. **[5]**

36 Find the coordinates of the points of intersection of the curves $y=\operatorname{cosec} x$ and $y=\sin x$ in the interval $0 \le x \le 360°$ **[4]**

37 a Sketch the graph of $y=|\arcsin x|$ for $-1 \le x \le 1$ **[2]**

b Sketch the graph of $y=1+2\arccos x$ for $-1 \le x \le 1$ and label the y-intercept with its exact value. **[3]**

38 Show that the curve with Cartesian equation $\frac{x^2}{25}-\frac{y^2}{9}=1$ has parametric equations $x=5\sec\theta,\ y=3\tan\theta$ **[2]**

39 Solve the equation $3\operatorname{cosec}^2 x-\cot x=7$ for x in the interval $0 \le x \le 360°$. Show your working. **[8]**

40 Find all the solutions of $2\tan^2 2\theta+\sec 2\theta-4=0$ for $0 \le \theta \le \pi$. Show your working. **[9]**

41 Prove that $\cos\theta\cot\theta-\sin\theta \equiv \operatorname{cosec}\theta-2\sin\theta$ **[4]**

42 Solve the inequality $\cot^2 x>1+\operatorname{cosec} x$ for x in the interval $0 \le x \le \pi$. Show your working. **[6]**

43 a i Prove that $\frac{\cos x}{\sin x}-\frac{\sin x}{1-\cos x} \equiv -\operatorname{cosec} x$

ii For what values of x is this identity valid? **[5]**

b Solve the equation $\frac{\cos x}{\sin x}-\frac{\sin x}{1-\cos x}=3$ for $0 \le x \le 2\pi$. Show your working. **[4]**

44 a Prove that $\cos x \equiv 1-2\sin^2\left(\frac{x}{2}\right)$ **[3]**

b Find all the solutions of $\cos x+5\sin^2\left(\frac{x}{2}\right)=3$ for $0<x<360°$. Show your working. **[6]**

45 The height of a tide (in metres) in a harbour t hours after midnight is given approximately by the equation $h=2.8+\sqrt{3}\sin\left(\frac{t}{2}\right)-3\cos\left(\frac{t}{2}\right)$

A particular boat requires a depth of at least 3.5 m in order to safely leave or enter the harbour. The owners of the boat wish to depart the harbour in the afternoon.

Work out the earliest and latest times they can leave. **[8]**

46 Solve the equation $\sin 2x+\sin x=0$ for x in the interval $0<x<360°$. Show your working. **[7]**

47 Given that $\sin\theta+2\cos\theta$ can be written in the form $r\sin(\theta+\alpha)$, where $r>0$ and $0<\alpha<90°$,

a Find the value of r and the value of α **[4]**

b Calculate the minimum value of $\frac{1}{(\sin\theta+2\cos\theta)^2}$ and the smallest positive value of θ for which this minimum occurs, **[4]**

c Find the maximum value of $\frac{1}{3+\sin\theta+2\cos\theta}$ and express it in the form $a+b\sqrt{c}$. Find also the smallest positive value of θ for which this maximum occurs. **[5]**

15 Differentiation 2

Dido's problem refers to a conundrum that faced Queen Dido of Carthage in around 850 BC. According to legend, she was offered as much land as she could cover with an animal hide, and using intuition she cut it into thin strips and created a large circular area. She may not have used calculus at the time, but differentiation can be used to find solutions to problems like this: as it happens, the largest area that can be bound by a perimeter of given length is indeed a circle.

Differentiation is an incredibly powerful tool that crops up throughout mathematics, and has applications in subjects such as physics, engineering, biology and economics. It can be applied to simple examples, such as the one above, as well as more complex ones involving trigonometric functions, logarithms and so on. This chapter builds on what you studied in Chapter 4, and your method will depend on the expression you need to differentiate.

Orientation

What you need to know

Ch4 Differentiation and integration

- Introduction to differentiation.
- Turning points and the second derivative.

Ch5 Exponentials and logarithms

- The laws of logarithms.
- The natural logarithm.

Ch12 Algebra 2

- Parametric equations.

p.16

Ch14 Trigonometric identities

- Small angle approximations.
- Double angle formula.

p.64

p.74

What you will learn

- To find points of inflection and determine when a curve is convex or concave.
- To understand and use limits.
- To differentiate $\sin x$, $\cos x$, e^x, a^x, and $\ln x$
- To use the product, quotient and chain rules.
- To find the derivative of a function which is defined implicitly and of a function which is defined parametrically.
- To find the derivative of an inverse function.

What this leads to

Ch16 Integration and differential equations

Integration by substitution.
Integration by parts.

Further Maths Ch19

- Differentiate inverse trigonometric functions.
- Differentiate hyperbolic functions.

Practise before you start

2028, 2062, 2134, 2158, 2224, 2266, 2270

15.1 The shapes of functions

Fluency and skills

See Ch4.1 For an introduction to differentiation.

The first differential, or derivative, of a function gives you the gradient of that function, which tells you about the shape of its graph.

Key point

When $\frac{dy}{dx} > 0$, the function is increasing and the curve is **rising** in the positive x-direction.

When $\frac{dy}{dx} < 0$, the function is decreasing and the curve is **falling** in the positive x-direction.

When $\frac{dy}{dx} = 0$, the function is stationary (neither rising nor falling).

Remember that Leibniz notation for the first derivative is $\frac{dy}{dx}$, and $\frac{d^2y}{dx^2}$ for the second derivative. $\frac{dy}{dx}$ gives the rate of change of a function (i.e. the gradient), and $\frac{d^2y}{dx^2}$ gives the rate of change of the gradient.

See Ch4.5 For a reminder on turning points.

The graph shows the shape of a cubic function.

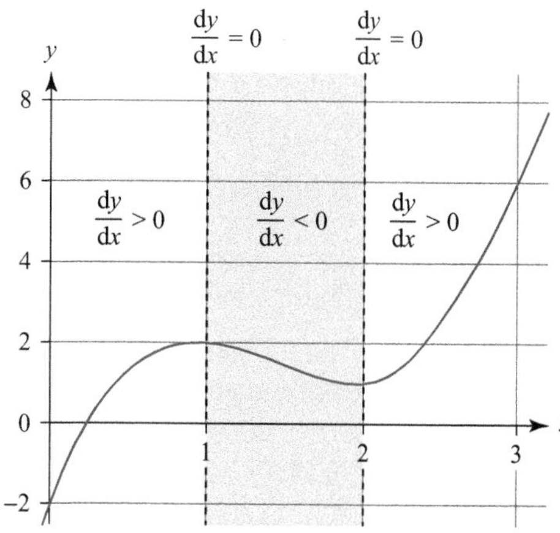

$\frac{dy}{dx} > 0$ in the regions where $x < 1$ or $x > 2$, so the curve is rising.

$\frac{dy}{dx} < 0$ in the regions where $1 < x < 2$, so the curve is falling.

$\frac{dy}{dx} = 0$ when $x = 1$ and $x = 2$, so the curve is neither rising nor falling.

The second derivative of a function, $\frac{d^2y}{dx^2}$, gives you further information about the shape of the curve. In the graph, at $x = 1$, there is a turning point where the gradient changes from positive to negative. So the second derivative is negative (it's decreasing). This point is a **maximum**, and the curve is described as **concave**. A curve is concave if any line segment joining two points on the curve stays below the curve.

Key point

If $\frac{dy}{dx} = 0$ and $\frac{d^2y}{dx^2} < 0$, the point is a maximum and the curve is concave.

At $x = 2$, there is a turning point where the gradient changes from negative to positive. So the second derivative is positive (it's increasing). This point is a **minimum**, and the curve is described as **convex**. A curve is convex if any line segment joining two points on the curve stays above the curve.

Key point

If $\frac{dy}{dx} = 0$ and $\frac{d^2y}{dx^2} > 0$, the point is a minimum and the curve is convex.

PURE

Between $x = 1$ and $x = 2$, the curve changes from concave to convex, that is, it changes **concavity**. The point at which a curve changes concavity is known as a **point of inflection**. At a point of inflection, the first derivative can take any value (the function does not need to be stationary), and the second derivative is *always* zero.

However, the reverse isn't true: a second derivative of zero does *not* necessarily show a point of inflection: you need to inspect the gradient at that point. If the gradient is non-zero, you have a point of inflection. If the gradient is zero, you must check that it has the same sign either side of the point for it to be a point of inflection. Alternatively, the second derivative must switch sign either side of the point to indicate a change in concavity.

Key point

At a point of inflection, it is always the case that $\frac{d^2y}{dx^2} = 0$

But if $\frac{d^2y}{dx^2} = 0$, further inspection is needed to determine the nature of the point.

The table shows what can be concluded from the value of the first and second derivatives.

Example of a curve	Lower x value	At turning point	Higher x value
Minimum	$\frac{dy}{dx} < 0, \frac{d^2y}{dx^2} > 0$ Convex	$\frac{dy}{dx} = 0, \frac{d^2y}{dx^2} > 0$ Convex	$\frac{dy}{dx} > 0, \frac{d^2y}{dx^2} > 0$ Convex
Maximum	$\frac{dy}{dx} > 0, \frac{d^2y}{dx^2} < 0$ Concave	$\frac{dy}{dx} = 0, \frac{d^2y}{dx^2} < 0$ Concave	$\frac{dy}{dx} < 0, \frac{d^2y}{dx^2} < 0$ Concave
	$\frac{dy}{dx} > 0, \frac{d^2y}{dx^2} < 0$ Concave	$\frac{dy}{dx} \geq 0, \frac{d^2y}{dx^2} = 0$ Point of inflection	$\frac{dy}{dx} > 0, \frac{d^2y}{dx^2} > 0$ Convex
	$\frac{dy}{dx} < 0, \frac{d^2y}{dx^2} > 0$ Convex	$\frac{dy}{dx} \leq 0, \frac{d^2y}{dx^2} = 0$ Point of inflection	$\frac{dy}{dx} < 0, \frac{d^2y}{dx^2} < 0$ Concave

Example 1

Use algebra to describe the shape of the curve $y = x^3 - 3x^2 + 2x$ at the following points.

a $(0, 0)$ **b** $(1, 0)$ **c** $(2, 0)$

$y = x^3 - 3x^2 + 2x \quad \frac{dy}{dx} = 3x^2 - 6x + 2 \quad \frac{d^2y}{dx^2} = 6x - 6$

Work out the first and second derivatives.

a At $x = 0$, $\frac{d^2y}{dx^2} = 0 - 6 = -6$

$\frac{d^2y}{dx^2} < 0$, so the curve is concave.

Evaluate the second derivative at that point.

b At $x = 1$, $\frac{d^2y}{dx^2} = 6 - 6 = 0$ and $\frac{dy}{dx} = 3 - 6 + 2 = -1$

So this is a point of inflection, on a decreasing section of the curve.

Given that $\frac{d^2y}{dx^2} = 0$, also evaluate the first derivative. As the first derivative is non-zero, you can conclude it is a point of inflection.

(Continued on the next page)

c At $x = 2$, $\frac{d^2y}{dx^2} = 12 - 6 = 6$

$\frac{d^2y}{dx^2} > 0$, so the curve is convex.

You can check your answers by drawing the curve on a graphics calculator.

Example 2

Use algebra to identify the points of inflection in each of these curves.

a $y = x^3$ b $y = x^3 - 3x$

a $y = x^3$ $\frac{dy}{dx} = 3x^2$ $\frac{d^2y}{dx^2} = 6x$

Work out the first and second derivatives.

When $\frac{d^2y}{dx^2} = 0$, $x = 0$. When $x = 0$, $\frac{dy}{dx} = 0$

Use the condition $\frac{d^2y}{dx^2} = 0$ to find the possible points of inflection. Also evaluate the first derivative. As the first derivative is zero, further inspection is required.

x	-0.1	0	0.1
$\frac{dy}{dx} = 3x^2$	+	0	+
tangent	/	—	/
$\frac{d^2y}{dx^2} = 6x$	−	0	+
convex or concave	concave	neither	convex

Use a table to inspect what is happening either side of the point on the curve where $x = 0$

So, the tangent is horizontal at $x = 0$ and the function is increasing either side of $x = 0$

Examine the first derivative.

The shape of the curve changes from concave to convex at $x = 0$

Examine the second derivative.

So, at $x = 0$, there is a horizontal point of inflection on the curve $y = x^3$

Draw a sketch of the curve at this point to check your conclusions.

b $y = x^3 - 3x$

$\frac{dy}{dx} = 3x^2 - 3$

$\frac{d^2y}{dx^2} = 6x$

Work out the first and second derivatives.

When $\frac{d^2y}{dx^2} = 0$, $x = 0$. When $x = 0$, $\frac{dy}{dx} = -3$

Given that $\frac{d^2y}{dx^2} = 0$, also evaluate the first derivative. As the first derivative is non-zero, you can conclude it is a point of inflection.

At $x = 0$, there is a point of inflection on a decreasing part of the curve $y = x^3 - 3x$. At the point of inflection, the gradient of the curve is -3

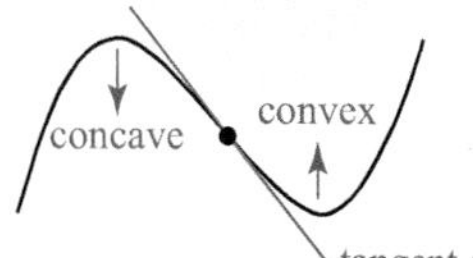

Use a graphics calculator to draw the curve and check your conclusions.

Exercise 15.1A Fluency and skills

1 For each of the following graphs, identify the regions in which the curve is concave and the regions in which the curve is convex.

a

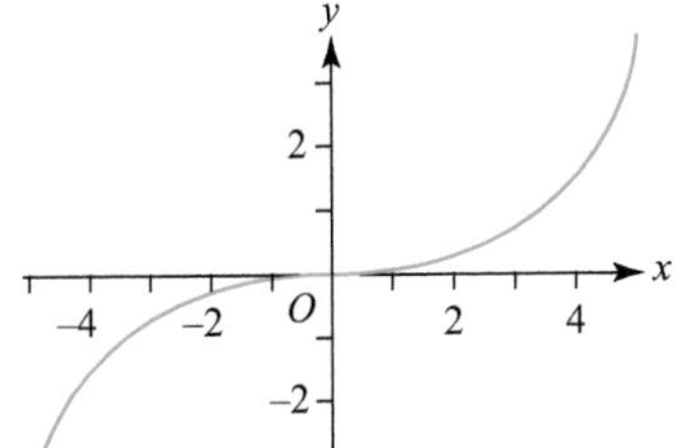

b

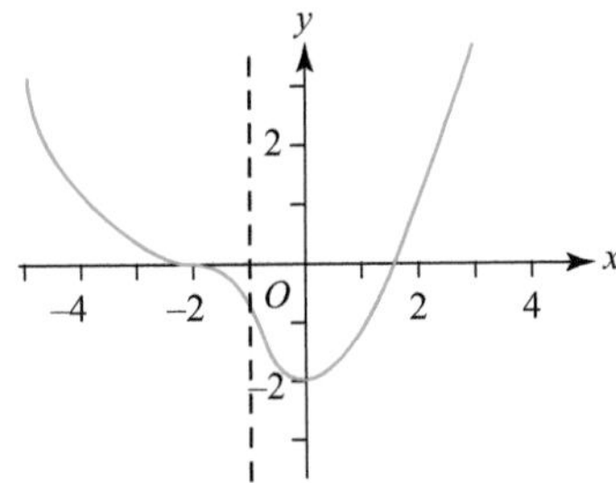

2 For each of the following graphs

a Identify the number of points of inflection,

b Describe the change of concavity at each point of inflection.

i

ii

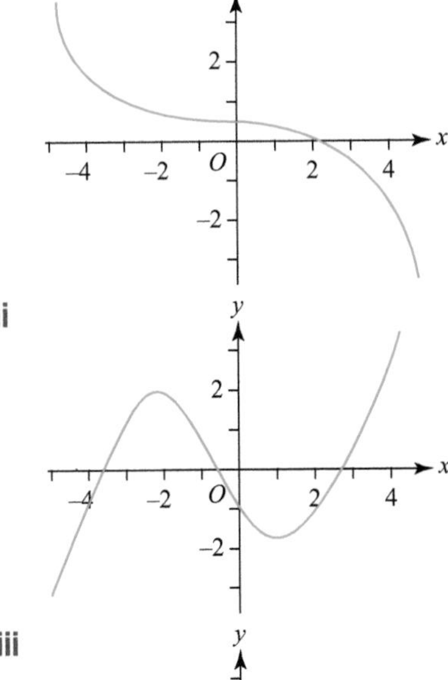

iii

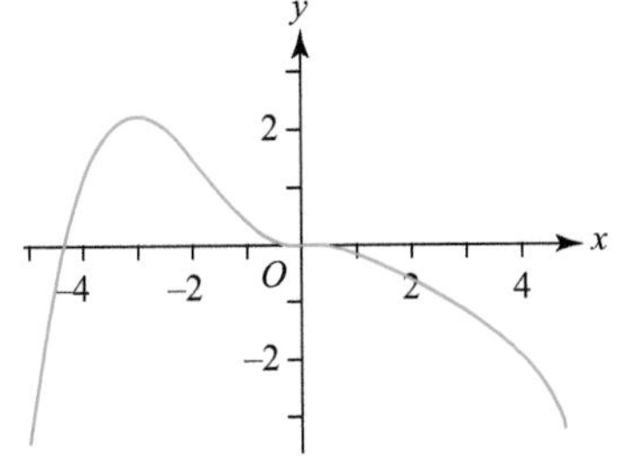

iv

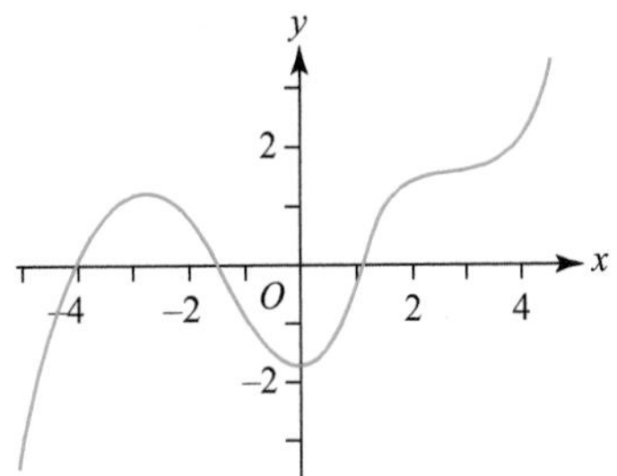

3 Use algebra to describe the shape of each curve at the given point. Show your working.

a $y=x^2+2x-1$ at $(1, 2)$

b $y=\frac{1}{x}$ at $(-1, -1)$

c $y=\frac{x+1}{x}$ at $(1, 2)$

d $y=1+x-x^2-x^3$ at $(1, 0)$

4 The curve $y=\frac{1}{3}x^3+2x^2+4x$ has a horizontal point of inflection.

a Determine the x-coordinate of the point.

b Show that the gradient of the curve at this point is zero.

c Determine whether the curve is concave or convex

i To the left of the point,

ii To the right of the point.

5 The curve $y=x^4-2x^3$ has two points of inflection. Identify them both and describe them.

6 Determine and describe the point of inflection on the curve $y=x^2+\frac{8}{x}$

7 Determine and describe the points of inflection on the curve $y=3x^5-10x^3$

8 a Use the second derivative to show that the curve $y=1+2x-x^2$ is always concave.

b Show that the curve $y=1-\sqrt{x}$, $x>0$ is always convex.

Reasoning and problem-solving

Strategy

To identify the key features of a curve

1. Work out both $\frac{dy}{dx}$ and $\frac{d^2y}{dx^2}$
2. Solve $\frac{d^2y}{dx^2}=0$ and, if necessary, $\frac{dy}{dx}=0$
3. Examine the possible values of x and interpret your findings.
4. If required, use the results to sketch the curve.

Example 3

$y=x^2(x+3)$

a Show that the curve has one point of inflection.

b Identify and classify any other turning points.

c Sketch the curve.

a $y=x^3+3x^2$ so $\frac{dy}{dx}=3x^2+6x=3x(x+2)$ and $\frac{d^2y}{dx^2}=6x+6=6(x+1)$

When $\frac{d^2y}{dx^2}=0,\ x=-1$

When $x=-1,\ \frac{dy}{dx}=(3\times(-1)^2)+(6\times-1)=-3$

So a point of inflection occurs when $x=-1$, on a decreasing part of the curve.

The curve is concave before $x=-1$ and convex after.

b When $\frac{dy}{dx}=0,\ x=0$ or $x=-2$

Therefore there are stationary points at (0, 0) and (−2, 4)

When $x=0,\ \frac{d^2y}{dx^2}=6$

so there is a minimum turning point at (0, 0)

When $x=-2,\ \frac{d^2y}{dx^2}=-6$

so there is a maximum turning point at (−2, 4)

c

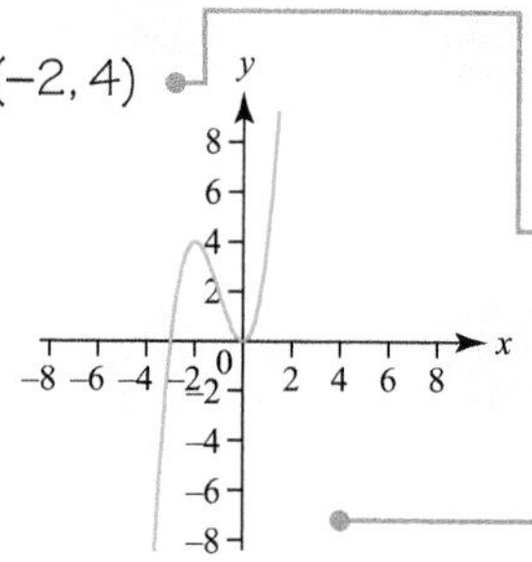

1 Work out the first and second derivatives.

2 Solve the second derivative equal to zero to find points of inflection.

3 Evaluate the gradient at this point. Check your answer by evaluating the gradient at $x=-1$ on your calculator.

2 Solve the first derivative equal to zero to find stationary points.

Substitute back into $y=x^2(x+3)$ to find the y-values.

4 Use the results to sketch the graph. You could check your sketch using a graphics calculator.

Example 4

Show that the graph of the function $y=x^4-4x^3+7x^2-12x-1$ is convex for all values of x

$y=x^4-4x^3+7x^2-12x-1$

So $\frac{dy}{dx}=4x^3-12x^2+14x-12$ and $\frac{d^2y}{dx^2}=12x^2-24x+14$

1 Work out the first and second derivatives.

(*Continued on the next page*)

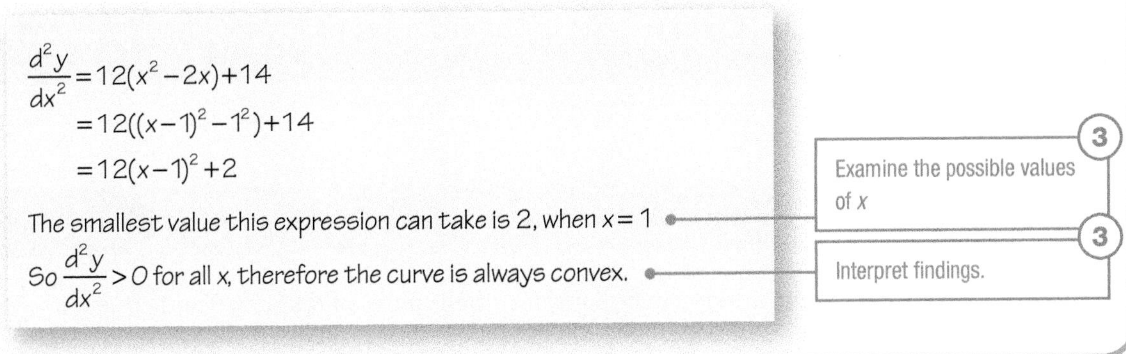

$$\frac{d^2y}{dx^2} = 12(x^2 - 2x) + 14$$
$$= 12((x-1)^2 - 1^2) + 14$$
$$= 12(x-1)^2 + 2$$

The smallest value this expression can take is 2, when $x = 1$ — **3** Examine the possible values of x

So $\frac{d^2y}{dx^2} > 0$ for all x, therefore the curve is always convex. — **3** Interpret findings.

Exercise 15.1B Reasoning and problem-solving

1 Sketch the following curves. Identify clearly any stationary points, points of inflection and intersections with the axes.

a $y = (x-4)(x-25)(x+20)$

b $y = x^4 - 24x^3 - 540$

c $y = x^3 - 9x^2 + 24x - 16$

2 Prove, using the second derivative, that the general quadratic $y = ax^2 + bx + c$, $a \neq 0$ is

a Always convex when $a > 0$

b Always concave when $a < 0$

3 a Show that the graph of the quartic function $y = x^4 + 8x^3 + 25x^2 - 5x + 10$ is convex for all values of x

b Show that the graph of the quartic function $y = 1 - 2x - 10x^2 + 4x^3 - x^4$ is concave for all values of x

c Show that the graph of the quartic function $y = 1 - 2x - 6x^2 + 4x^3 - x^4$ is never convex for any value of x

4 The general equation of the cubic function whose roots are a, b and c is $y = k(x-a)(x-b)(x-c)$, where k is a constant. Show that the point of inflection of the curve has an x-coordinate equal to the mean value of the roots.

5 The curve $y = 3x^5 - 5x^3$ has three stationary points. Identify the roots, the stationary points and the points of inflection, showing your working. Hence sketch the curve.

6 The curve $y = x^4 - 2x^3 - 12x^2$ has two points of inflection. Find the equation of the line that passes through both.

7 For what values of x is the graph of $y = x^3 - ax^2 - ax$ convex, where a is a constant? Give your answer in terms of a

Challenge

8 The general form of a cubic function is $f(x) = ax^3 + bx^2 + cx + d$ where a, b, c and d are constants and $a \neq 0$

a What conditions must be placed on the constants a, b and c so that the graph of $y = f(x)$ has

i No stationary points,

ii Exactly one stationary point,

iii Two distinct stationary points?

b In terms of a, b, c and d, for what values of x is the graph of $y = f(x)$

i Concave,

ii Convex,

iii At a point of inflection?

9 The general form of a quartic function is $f(x) = ax^4 + bx^3 + cx^2 + dx + e$ where a, b, c, d and e are constants and $a \neq 0$. What conditions must be placed on these constants so that there are exactly two changes of concavity on the curve $y = f(x)$?

15.2 Trigonometric functions

Fluency and skills

See Ch14.1 For more on small- angle estimations.

You may remember that, when working in radians, for small values of θ, $\sin\theta \approx \theta$ and $\cos\theta \approx 1 - \frac{1}{2}\theta^2$

These are known as small angle estimations. Hence, you can derive the following two results.

Key point

$$\lim_{\theta\to 0}\frac{\sin\theta}{\theta} = 1 \quad\text{and}\quad \lim_{\theta\to 0}\frac{1-\cos\theta}{\theta} = 0$$

See Ch14.3 To revise the double angle formula.

Using these results, and the compound angle formula for trigonometric functions, you can derive the first derivatives of $\sin x$ and $\cos x$ from first principles.

Proof

$\mathrm{f}(x) = \sin x$ so $\mathrm{f}'(x) = \lim_{h\to 0}\frac{\sin(x+h)-\sin x}{h}$

Applying the compound angle formula,

$$\mathrm{f}'(x) = \lim_{h\to 0}\frac{\sin x\cos h + \cos x\sin h - \sin x}{h} = \lim_{h\to 0}\frac{\sin x(\cos h - 1) + \cos x\sin h}{h}$$

$$= \lim_{h\to 0}\sin x\frac{(\cos h - 1)}{h} + \lim_{h\to 0}\cos x\frac{\sin h}{h} = \sin x\lim_{h\to 0}\frac{(\cos h-1)}{h} + \cos x\lim_{h\to 0}\frac{\sin h}{h}$$

Using the results derived from the small angle estimations, $\mathrm{f}'(x) = (\sin x\times 0) + (\cos x\times 1)$

Therefore, $\mathrm{f}'(x) = \cos x$

Key point

If $y = \sin x$, then $\frac{\mathrm{d}y}{\mathrm{d}x} = \cos x$ If $y = \cos x$, then $\frac{\mathrm{d}y}{\mathrm{d}x} = -\sin x$

You need to work in radians if you're differentiating trig functions.

Example 1

Determine the first and second derivatives of the function $y = 2x^3 + 3\sin x$

$y = 2x^3 + 3\sin x$ $\quad \frac{\mathrm{d}y}{\mathrm{d}x} = 6x^2 + 3\cos x$ $\quad \frac{\mathrm{d}^2y}{\mathrm{d}x^2} = 12x - 3\sin x$

Apply $\frac{\mathrm{d}}{\mathrm{d}x}\sin x = \cos x$

To find the second derivative, you must differentiate again, so apply $\frac{\mathrm{d}}{\mathrm{d}x}\cos x = -\sin x$

Example 2

Given that $\mathrm{f}(x) = \sin x + 2\cos x$, find the exact value of $\mathrm{f}'\left(\frac{\pi}{3}\right)$, showing your working.

$\mathrm{f}(x) = \sin x + 2\cos x$ $\quad \mathrm{f}'(x) = \cos x - 2\sin x$

$\mathrm{f}'\left(\frac{\pi}{3}\right) = \cos\frac{\pi}{3} - 2\sin\frac{\pi}{3} \quad = \frac{1}{2} - 2\times\frac{\sqrt{3}}{2} \quad = \frac{1}{2} - \sqrt{3}$

See Ch14.1 for a reminder on the special triangles which are useful when dealing with angles in radians.

Exercise 15.2A Fluency and skills

1 Work out the derivative of each of these functions.

a $3\sin x$

b $\frac{\cos x}{3}$

c $3-\cos x$

d $x+\sin x$

e $x^2+\cos x$

f $\sin x+\cos x$

g $3\sin x-4\cos x$

2 Find $\frac{dy}{dx}$ and $\frac{d^2y}{dx^2}$ for each of these functions.

a $y=2\sin x$

b $y=2x^2-\cos x$

c $y=2\sin x-3\cos x$

d $y=\sin x+\cos x+x^3$

e $y=\frac{\cos x-\sin x}{2}$

3 Show your working in parts **a**–**e**.

a Given that $f(x)=3\sin x$, calculate.

i $f'\left(\frac{\pi}{3}\right)$ ii $f''\left(\frac{\pi}{3}\right)$

b Given that $f(x)=2\cos x$, calculate

i $f'\left(\frac{\pi}{6}\right)$ ii $f''\left(\frac{\pi}{6}\right)$

c Given that $f(x)=\sin x+\cos x$, calculate

i $f'\left(\frac{\pi}{4}\right)$ ii $f''\left(\frac{\pi}{4}\right)$

d Given that $f(x)=1-\cos x$, calculate

i $f'\left(\frac{\pi}{2}\right)$ ii $f''\left(\frac{\pi}{2}\right)$

e Given that $f(x)=3\cos x-4\sin x$, calculate

i $f'\left(\frac{3\pi}{2}\right)$ ii $f''\left(\frac{3\pi}{2}\right)$

4 a By considering when $\frac{dy}{dx}=0$, find the turning points on the curve $y=1+\sin x$ in the interval $0\le x\le 2\pi$. Show your working.

b By considering when $\frac{d^2y}{dx^2}=0$, find the points of inflection on the curve $y=1-\cos x$ in the interval $0\le x\le 2\pi$

5 Find the gradient of the tangent to the curve $y=5\sin x-\sqrt{3}\cos x$ when $x=\frac{\pi}{3}$. Show your working.

6 Is the curve $y=x+\sin x$ convex or concave when $x=\frac{\pi}{3}$? Show your working.

7 Determine, from first principles, the derivative of

a $2\sin x$

b $-\cos x$

c $x-\sin x$

Reasoning and problem-solving

Strategy

To identify the features of a trigonometric function

1. Manipulate the function until it is the sum of terms of the form ax^n, $a\sin x$ or $a\cos x$, where a is a constant.
2. Find the first derivative, $\frac{dy}{dx}$, and/or the second derivative, $\frac{d^2y}{dx^2}$
3. Interpret their meaning in terms of the gradient or shape of the curve respectively.

MyMaths 2165 SEARCH

There is a change in concavity when the curve changes either from convex to concave or vice versa.

Example 3

a Show that the function $f(x) = x + \sin x$ is never decreasing.

b Show that its concavity changes at $x = 0, \pi, 2\pi, 3\pi, \ldots, n\pi$

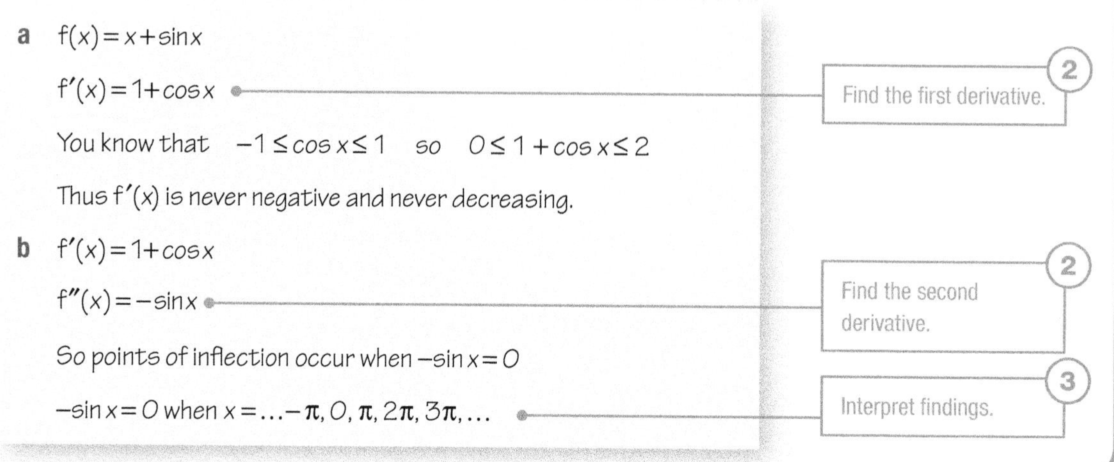

a $f(x) = x + \sin x$

$f'(x) = 1 + \cos x$

You know that $-1 \le \cos x \le 1$ so $0 \le 1 + \cos x \le 2$

Thus $f'(x)$ is never negative and never decreasing.

b $f'(x) = 1 + \cos x$

$f''(x) = -\sin x$

So points of inflection occur when $-\sin x = 0$

$-\sin x = 0$ when $x = \ldots -\pi, 0, \pi, 2\pi, 3\pi, \ldots$

Exercise 15.2B Reasoning and problem-solving

1 Find the first derivative of the following functions.

a $\dfrac{x\sin x - x}{x}$

b $\dfrac{\sin x}{\tan x}$

c $\dfrac{\sin x - \tan x}{\tan x}$

d $2\tan x\left(\cos x - \dfrac{\sin x}{2\tan x}\right)$

2 a Find the equation of the tangent to the curve $y = \sin x$ at the point where $x = \dfrac{\pi}{3}$ expressing your answer in the form $ax + by = c$ and leaving c as an exact value.

b Verify that the curve is concave at this point.

3 Differentiate the following functions.

a $\dfrac{x^2 - x\sin x}{2x}$

b $\dfrac{4 - x^3\cos x}{x^3}$

c $\cos x(\tan x + 1)$

d $\dfrac{\cos^2 x - \sin^2 x}{\cos x - \sin x}$

PURE

4 A curve has equation $y=\frac{x^2}{2\pi}-2\sin x$

a Showing your working, find its gradient when x is

i 0 ii $\frac{\pi}{2}$ iii π iv $\frac{3\pi}{2}$ v 2π

b Show that the function is always increasing when $x>2\pi$

c A related family of functions has equation $y=\frac{ax^2}{2\pi}-2\sin x$ where a is a constant.

For what values of a will the graph always be convex?

5 A Big Wheel is set up in a city as a tourist attraction. It takes just over 6 minutes (2π minutes) to make a complete turn. The vertical height of a passenger, y metres above the ground, is modelled by the function $y=21-20\cos t$, where t is the time in minutes since the passenger was at the bottom of the wheel.

a How high up is the passenger when $t=\frac{\pi}{3}$ minutes?

b What is the vertical speed of the passenger when $t=\frac{\pi}{6}$ minutes? Show your working.

c What is the vertical acceleration of the passenger when $t=\frac{\pi}{2}$ minutes?

6 It can be convenient to break the year into 2π units rather than 365 days.

The time of sunrise in Liverpool over a year can be modelled by $y=12-5\cos x$, where sunrise occurs at y hours after midnight, on the date given as x units.

a Find the rate at which y is changing when the date $x=0.5$ units.

b When is the rate of change of the time of sunrise equal to zero? Show your working.

c There are two points of inflection on the curve $y=12-5\cos x$ in a single year.

Find the values of x at which they occur.

7 The depth of water in a harbour basin over a day can be modelled by $y=2\sin\left(x+\frac{\pi}{3}\right)+5$, where y metres is the depth and x hours is the time since midnight.

a Find $\frac{dy}{dx}$, the rate of change of the depth with time.

b Find this rate at i 4 am ii 4 pm

c Is the water rising or falling at noon ($x=12$)?

Challenge

8 a Can you find the derivative of $\sin 2x$ from first principles? Hint: Remember that $\frac{1}{h}=\frac{2}{2h}$

b Can you find the derivative of $\sin ax$ from first principles, where a is a constant?

Hint: Remember that $\frac{1}{h}=\frac{a}{ah}$

15.3 Exponential and logarithmic functions

Fluency and skills

See Ch 5.1–5.3 For more about exponentials.

The mathematical constant e was first discovered by the mathematician Jacob Bernoulli whilst researching compound interest. Its value, to 6 significant figures, is 2.71828, although, like the value of π, it is irrational.

e is defined as $e = \lim_{n\to\infty}\left(1+\frac{1}{n}\right)^n$

Key point

It can be shown that if $y = e^x$ then $\frac{dy}{dx} = e^x$

And more generally that if $y = e^{ax}$ then $\frac{dy}{dx} = ae^{ax}$

Example 1

Find the gradient of the curve $y = x + 3e^{2x}$ where the curve crosses the y-axis. Show your working.

$y = x + 3e^{2x}$ $\quad \frac{dy}{dx} = 1 + 3\times 2e^{2x} = 1 + 6e^{2x}$ — Find the first derivative.

When $x = 0$, $\frac{dy}{dx} = 1 + 6e^0 = 7$ — The curve crosses the y-axis when $x = 0$

Check your answer by evaluating the derivative at $x = 0$ on your calculator.

See Ch5.2 For more about the natural log.

$\log_e x$ is known as the 'natural log' and is more commonly referred to as $\mathbf{\ln x}$

Key point

It can be shown that if

$y = \ln x$ OR $y = \ln ax$ then $\frac{dy}{dx} = \frac{1}{x}$

$\ln ax = \ln a + \ln x$. As $\ln a$ is a constant, it differentiates to zero and so the derivative of $\ln x$ is equal to the derivative of $\ln ax$

Example 2

Given that $y = \ln 2x^3$, find $\frac{dy}{dx}$

$y = \ln 2x^3 = \ln 2 + 3\ln x$ $\quad \frac{dy}{dx} = 3\times\frac{1}{x} = \frac{3}{x}$

Use the laws of logarithms to rearrange y including $\log_a(x^k) = k\log_a(x)$ and then differentiate.

You can differentiate the function $f(x) = a^x$ if you first express it in base e, rather than a

See Ch5.1 For a reminder on the laws of logarithms.

Let	$y = a^x$
Take natural logs on both sides	$\ln y = \ln a^x$
Use the log law $\log_a(x^k) = k\log_a x$	$\ln y = x\ln a$
Take exponentials on both sides	$y = e^{x\ln a}$ $\quad (a^x = e^{x\ln a})$
Differentiate	$\frac{dy}{dx} = \ln a \times e^{x\ln a}$
Since $e^{x\ln a} = a^x$, you have	$\frac{dy}{dx} = a^x \ln a$

Key point

If $y = a^x$, then $\frac{dy}{dx} = a^x \ln a$

PURE

Example 3

Differentiate 3^x

$y = 3^x$ $\quad \frac{dy}{dx} = 3^x \ln 3$ — y is in the form a^x where $a = 3$

Exercise 15.3A Fluency and skills

1 Differentiate each of these functions.

a $4e^x$ b $x - 2e^x$
c $5e^x - 3x^2$ d $x^{-1} + e^x$
e $\frac{1 + xe^x}{x}$ f $e^x - \frac{1}{x^2}$
g $x^3 + e^x - \cos x$ h $\frac{1}{3} - \frac{e^x}{4}$

2 Find $\frac{dy}{dx}$ and $\frac{d^2y}{dx^2}$ for each of these functions.

a e^{3x} b $2e^{-4x}$
c $e^x - e^{2x}$ d $6e^{0.5x}$
e $e^x + e^{-x}$ f $e^x - \frac{1}{e^x}$
g $e^{1.25x} + e^{0.5x}$ h $e^x(1 + e^{-3x})$

3 Find the first and second derivatives for each of these functions.

a $f(x) = 3\ln x$ b $f(x) = 1 - 2\ln x$
c $f(x) = \ln 2x$ d $f(x) = 3\ln x + \ln 3x$
e $f(x) = \ln x^5$ f $f(x) = \ln x^{-1}$
g $f(x) = \ln 3x^2$ h $f(x) = \ln \frac{3}{x}$
i $f(x) = \ln \sqrt{x}$

4 Find the derivative of each of these functions.

a 5^x b 2^x c 6^x
d $\frac{3^x}{5}$ e $8^x - 7^x$ f $5x - 5^x$
g 3×4^x

5 Showing your working, given that

a $f(x) = 3e^x$, find $f'(0)$
b $f(x) = x - 2e^x$, find $f'(1)$
c $f(x) = x^2 + 3e^{2x}$, find $f'(0.5)$
d $f(x) = \ln x$, find $f'(0.5)$
e $f(x) = \ln x^2$, find $f'(0.25)$
f $f(x) = \ln 4x$, find $f'(2)$

6 Find the gradient of each of these curves at the given point. Show your working.

a $y = 3e^x$ at $(0, 3)$ b $y = 5e^x$ at $(1, 5e)$
c $y = 4^x$ at $(0, 1)$ d $y = 5^x$ at $(1, 5)$
e $y = 2^x$ at $(-1, 0.5)$ f $y = x + 6^x$ at $(2, 38)$
g $y = 3\ln x$ at $(1, 0)$
h $y = x - 2\ln x$ at $(e, e - 2)$

Reasoning and problem-solving

Strategy

To solve problems involving the rate of change of exponential or logarithmic functions

1. Find the first and/or second derivative.
2. Use these to find the gradient of the curve or rate of change, as appropriate.
3. Interpret the solution within the context of the problem.

Example 4

A car initially has a value of £20 000

Its value after x years can be modelled by $y = 20000 \times e^{-0.357x}$ $(x \geq 0)$

Showing your working, find the annual rate of change of the car's value after

a 3 years, **b** 10 years.

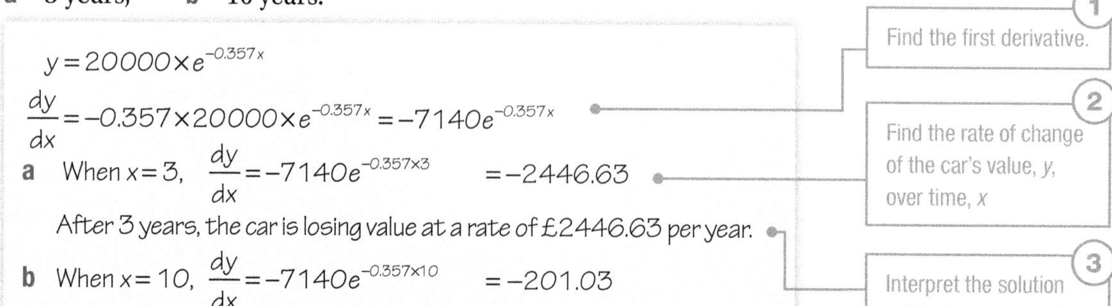

$y = 20000 \times e^{-0.357x}$

$\frac{dy}{dx} = -0.357 \times 20000 \times e^{-0.357x} = -7140e^{-0.357x}$

a When $x = 3$, $\frac{dy}{dx} = -7140e^{-0.357 \times 3} = -2446.63$

After 3 years, the car is losing value at a rate of £2446.63 per year.

b When $x = 10$, $\frac{dy}{dx} = -7140e^{-0.357 \times 10} = -201.03$

After 10 years, the car is losing value at a rate of £201.03 per year.

1
Find the first derivative.
2
Find the rate of change of the car's value, y, over time, x
3
Interpret the solution within the context of the problem.

Example 5

£100 is placed in an investment bond and its value increases.

The value of the bond is £x after y years. The time, y, and the value, x, are related by $y = 17.2\ln x - 79.2$

a After how many years would the value be £106?

b Find the rate of change of the value of the bond. Showing your working, calculate the time it takes to make £1 if there is

i £110 **ii** £200 in the bond.

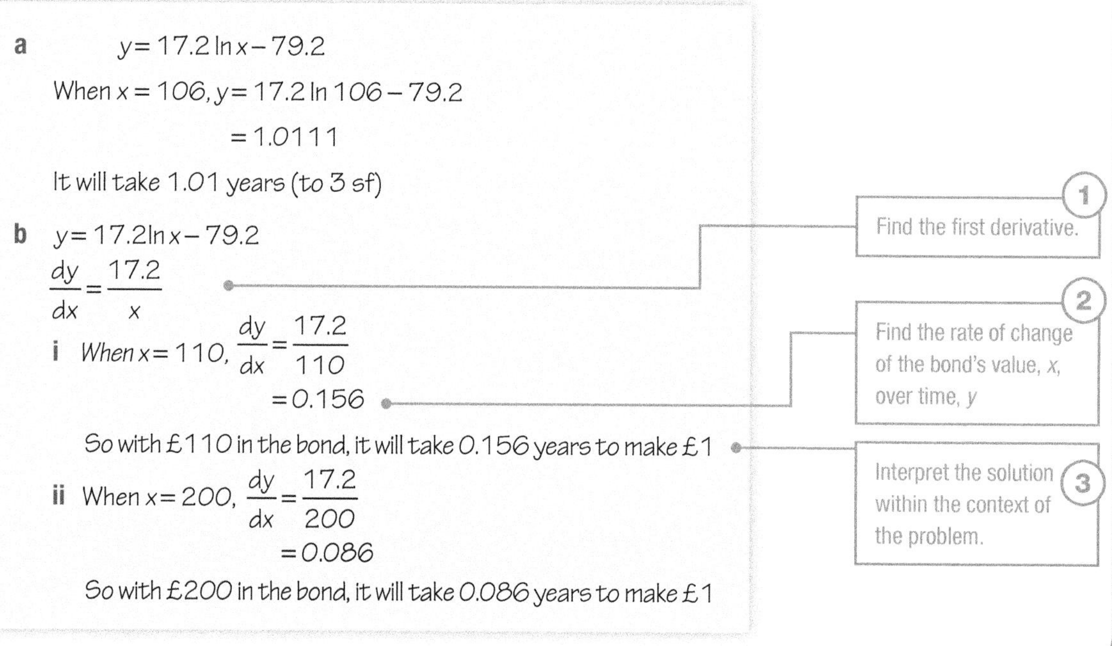

a $y = 17.2\ln x - 79.2$

When $x = 106$, $y = 17.2\ln 106 - 79.2$

$= 1.0111$

It will take 1.01 years (to 3 sf)

b $y = 17.2\ln x - 79.2$

$\frac{dy}{dx} = \frac{17.2}{x}$

i When $x = 110$, $\frac{dy}{dx} = \frac{17.2}{110}$

$= 0.156$

So with £110 in the bond, it will take 0.156 years to make £1

ii When $x = 200$, $\frac{dy}{dx} = \frac{17.2}{200}$

$= 0.086$

So with £200 in the bond, it will take 0.086 years to make £1

1
Find the first derivative.
2
Find the rate of change of the bond's value, x, over time, y
3
Interpret the solution within the context of the problem.

Exercise 15.3B Reasoning and problem-solving

1 In a managed moorland, the number of breeding pairs of pheasants is modelled by $P = 100e^{0.095t} - 50$, where P is the number of breeding pairs at the start of year t. At the beginning, $t = 0$

a How long will it take for the population to double?

b At what rate is the population changing at this time? Show your working.

c The population of pheasant pairs is plotted on a graph against time. Use the second derivative to prove that the curve is always convex.

2 A new dishwasher costs £170. Due to depreciation and other factors, its value drops. Its actual value, £y, in year t can be modelled by $y = 230e^{-0.134t} - 60$. Show your working in parts **a–d**.

a Verify that the dishwasher costs £170 at $t = 0$

b At what rate was the value changing when $t = 5$?

c In what year was the dishwasher dropping £12 a year in value?

d When the dishwasher reaches the end of its useful life, its value is zero.

i When does the model predict this will happen?

ii At what rate is it losing value at this point?

e The value of the dishwasher is plotted on a graph against time. Discuss the concavity of the curve.

3 A cup of tea is made from boiling water. Left alone, it will cool down according to the model $X = Ae^{-kt} + R$, where X°C is the temperature t minutes after the tea is made. R°C is the surrounding room temperature.

A and k are constants. Show your working in parts **a–d**.

a Calculate the values of the constants A and k given that, initially, the temperature of the tea is 100 °C and that, after 3 minutes, it had cooled to 80 °C. The room temperature was 18 °C.

b At what rate is the tea cooling

i The instant that it is poured,

ii After 10 minutes?

c After how many minutes will the rate of cooling drop below 1 °C per minute?

d At what rate is the rate of cooling changing when $t = 6$?

4 An altimeter works on the principle that altitude is a function of atmospheric pressure. When the altitude is a metres above sea level, the air pressure is P units, and $a = 54098 - 8155.5 \ln P$

a What should the altimeter read at the top of Snowdon where the pressure is 665.3 units?

b At what rate is the altitude changing as the pressure changes (in metres per unit) when the pressure is 700 units? Show your working.

c What is the rate of change of altitude with pressure, in metres per unit, when a walker is 1000 m up? Show your working.

5 A parasitic tick has infected a flock of sheep. The area is sprayed with an experimental insecticide. A control group is examined on day one and again on day seven after the spraying.

On day one, 250 ticks were discovered in the group. On day seven, there were 100 ticks found. The number of ticks, P, is related to the day, t, by a relationship of the form $P(t) = k \ln t + c$, where k and c are constants.

a Use this information to find the values of the constants.

b At what rate is the number of ticks changing on day four? Show your working.

c The flock will be deemed clear of ticks when P falls below one. Showing your working,

i Find the value of t when this occurs,

ii Work out the rate of change of the number of ticks at this time.

Challenge

6 When hatched ($t = 1$), an osprey chick weighs 80 g. It grows rapidly and, at 30 days, it is 1050 g, which is 75% of its adult weight. Over these 30 days, its mass w g can be modelled by $w = k \ln t + c$, where t is the time in days since hatching and k and c are constants.

a Find the values of k and c

b Showing your working, find the rate at which the chick gains mass on

i Day 7 ii Day 14.

c At what rate is it growing when it weighs 1000 g? Show your working.

d Show that the function $w(t)$, $1 \le t \le 30$, is an increasing function and that the rate of growth is slowing down over this interval.

15.4 The product and quotient rules

Fluency and skills

The **product rule** is used to differentiate two functions, u and v, that have been multiplied together.

Key point

In Leibniz notation, the product rule is written

$$\frac{d}{dx}(uv)=v\frac{du}{dx}+u\frac{dv}{dx}$$

The **quotient rule** is used to differentiate two functions, u and v, one of which has been divided by the other.

Key point

In Leibniz notation, the quotient rule is written

$$\frac{d}{dx}\left(\frac{u}{v}\right)=\frac{v\frac{du}{dx}-u\frac{dv}{dx}}{v^2}$$

A *quotient* is a quantity or function divided by another.

Example 1

Find $\frac{dy}{dx}$ when $y=x^4\ln x$

Let $u=x^4$ and $v=\ln x$ — Define u and v where y is in the form uv

So $\frac{du}{dx}=4x^3$ and $\frac{dv}{dx}=\frac{1}{x}$ — Differentiate u and v separately.

Therefore, $\frac{dy}{dx}=\ln x\times 4x^3+x^4\times\frac{1}{x}=4x^3\ln x+x^3$ — Substitute into the product rule.

Example 2

Find $\frac{dy}{dx}$ when $y=\tan x$

$y=\tan x=\frac{\sin x}{\cos x}$ — Write $\tan x$ as the quotient of $\sin x$ and $\cos x$

Let $u=\sin x$ and $v=\cos x$

So $\frac{du}{dx}=\cos x$ and $\frac{dv}{dx}=-\sin x$ — Define u and v and differentiate.

$\frac{dy}{dx}=\frac{\cos x\times\cos x-\sin x\times(-\sin x)}{\cos^2 x}$ — Substitute into the quotient rule.

$\frac{dy}{dx}=\frac{\cos^2 x+\sin^2 x}{\cos^2 x}=\frac{1}{\cos^2 x}=\sec^2 x$ — Simplify using the trigonometric identity $\sin^2 x+\cos^2 x=1$

Therefore, $\frac{d}{dx}(\tan x)=\sec^2 x$

See Ch3.1
For a reminder of the basic trigonometric identities.

Exercise 15.4A Fluency and skills

1 Differentiate each of these functions.

a $x\sin x$

b xe^x

c $e^x\cos x$

d $3x\ln x$

e $\sin x\ln x$

f $e^{3x}(x^2+1)$

g $\sqrt{x}\ln 3x$

h $\frac{1}{x}\cos x$

i $\left(x+\frac{1}{x}\right)\left(x-\frac{1}{x}\right)$

j $e^{3x}\ln x$

k $(2x^3+2x)(x-x^2)$

l $x^2\ln x^2$

m $(3-4x^2)\frac{1}{\sqrt{x}}$

n $\sqrt{x}\ln 3x$

o $(1+\sin x)(1-\sin x)$

p $e^{2x+1}(2x+1)$

2 a By writing $\sin^2 x=\sin x\sin x$, find the derivative of $\sin^2 x$

b Find the derivative of $\cos^2 x$

c Using the fact that $\sin 2x=2\sin x\cos x$, find the derivative of $\sin 2x$

d Using the fact that $\cos 2x=\cos^2 x-\sin^2 x$, find the derivative of $\cos 2x$

3 Show your working in parts **a–c**.

a $f(x)=x\sin x$

find i $f'\left(\frac{\pi}{2}\right)$ ii $f'\left(\frac{\pi}{6}\right)$

b $f(x)=x^3e^x$

find i $f'(0)$ ii $f'(1)$

c $f(x)=(x^2+2x+1)(1-3x-x^2)$

find i $f'(0)$ ii $f'(-1)$

4 Find the derivative of each of these functions.

a $\frac{x+1}{x-1}$

b $\frac{\sin x}{x}$

c $\frac{\sin x}{e^x}$

d $\frac{(x^2+1)}{x+1}$

e $\frac{\cos x}{\sin x}$

f $\frac{\ln x}{x}$

g $\frac{3}{\sin x}$

h $\frac{x}{\cos x}$

i $\frac{x^2}{1-\sqrt{x}}$

j $\frac{x^2-3x+1}{\sqrt{x}}$

k $\frac{e^{2x+3}}{x^3}$

l $\frac{1+\sin x}{\cos x}$

5 Show your working in parts **a–c**.

a Given that $f(x)=\frac{x}{\cos x}$

find i $f'(\pi)$ ii $f'(0)$

b Given that $f(x)=\frac{\ln x}{e^x}$

find i $f'(e)$ ii $f'(1)$

c Given that $f(x)=\frac{x^2+2x-1}{1-2x}$

find i $f'(0)$ ii $f'(-1)$ iii $f'(1)$

Reasoning and problem-solving

Strategy

To solve problems involving the differentiation of functions expressed as products or quotients

1. Separate the function into two appropriate parts, *u* and *v*, and differentiate these.
2. Apply the product rule or the quotient rule as appropriate.
3. Simplify the answer as much as possible.
4. Give the answer within the context of the problem.

MyMaths 2163, 2164 SEARCH

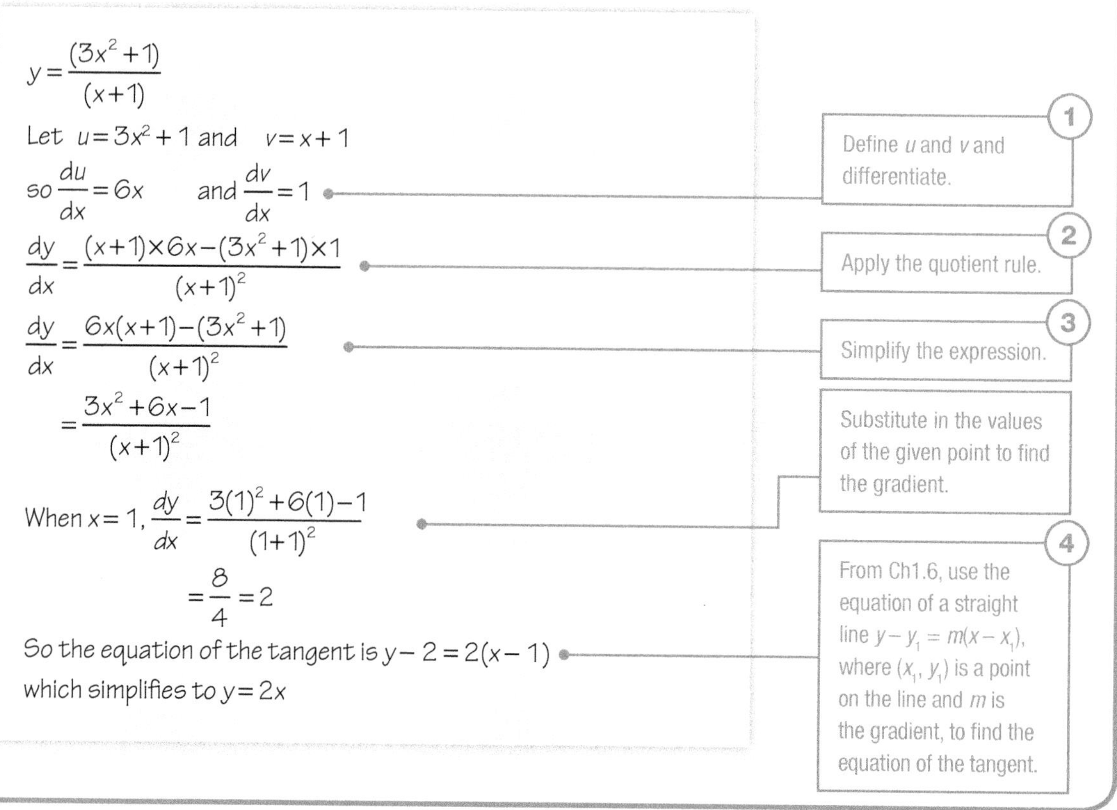

Example 3

Work out the equation of the tangent to $y=\dfrac{(3x^2+1)}{(x+1)}$ at the point (1, 2). Show all your working.

$y=\dfrac{(3x^2+1)}{(x+1)}$

Let $u=3x^2+1$ and $v=x+1$

so $\dfrac{du}{dx}=6x$ and $\dfrac{dv}{dx}=1$

$\dfrac{dy}{dx}=\dfrac{(x+1)\times 6x-(3x^2+1)\times 1}{(x+1)^2}$

$\dfrac{dy}{dx}=\dfrac{6x(x+1)-(3x^2+1)}{(x+1)^2}$

$=\dfrac{3x^2+6x-1}{(x+1)^2}$

When $x=1$, $\dfrac{dy}{dx}=\dfrac{3(1)^2+6(1)-1}{(1+1)^2}$

$=\dfrac{8}{4}=2$

So the equation of the tangent is $y-2=2(x-1)$ which simplifies to $y=2x$

1. Define u and v and differentiate.
2. Apply the quotient rule.
3. Simplify the expression.

Substitute in the values of the given point to find the gradient.

4. From Ch1.6, use the equation of a straight line $y-y_1=m(x-x_1)$, where (x_1, y_1) is a point on the line and m is the gradient, to find the equation of the tangent.

Example 4

A flu patient's temperature will quickly rise then slowly come back to normal over a week. The situation can be modelled by $X=\dfrac{12t}{e^t}$, $0\le t\le 7$ where X°C is the temperature above the patient's normal and t is the day since the flu began ($t=0$).

When will the fever peak? Show your working.

Let $u=12t$ and $v=e^t$

$\dfrac{du}{dt}=12$ and $\dfrac{dv}{dt}=e^t$

$\dfrac{dX}{dt}=\dfrac{(e^t\times 12)-(12t\times e^t)}{e^{2t}}=\dfrac{12(1-t)}{e^t}$

At a stationary point, $\dfrac{12(1-t)}{e^t}=0$, so $t=1$

When $t=1$, $X=\dfrac{12\times 1}{e^1}\approx 4.4$

$\dfrac{dX}{dt}=\dfrac{12(1-t)}{e^t}$

Let $u=12-12t$ and $v=e^t$

$\dfrac{du}{dt}=-12$ and $\dfrac{dv}{dt}=e^t$

1. Define u and v and differentiate.
2. Apply the quotient rule and simplify.

Identify when the first derivative equals zero to find any stationary points.

1. To determine the nature of this stationary point you need to inspect the sign of the second derivative, so you must use the quotient rule again. Separate the gradient function, $\dfrac{dx}{dt}$ into two appropriate parts.

(*Continued on the next page*)

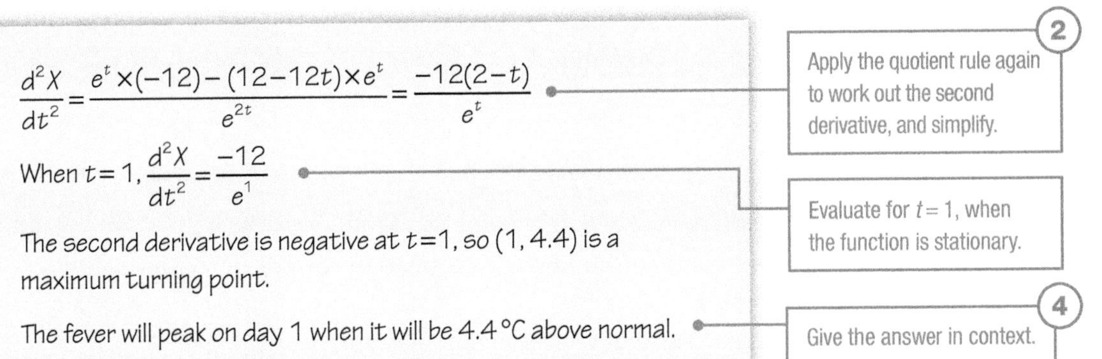
$\frac{d^2x}{dt^2} = \frac{e^t \times (-12) - (12 - 12t) \times e^t}{e^{2t}} = \frac{-12(2-t)}{e^t}$

Apply the quotient rule again to work out the second derivative, and simplify. 2

When $t = 1$, $\frac{d^2x}{dt^2} = \frac{-12}{e^1}$

Evaluate for $t = 1$, when the function is stationary.

The second derivative is negative at $t = 1$, so $(1, 4.4)$ is a maximum turning point.

The fever will peak on day 1 when it will be 4.4 °C above normal.

Give the answer in context. 4

Exercise 15.4B Reasoning and problem-solving

1 Work out the gradient of the tangent to the curve $y = \frac{2x+1}{x^2-1}$ at the point $(0, -2)$. Show your working.

2 Find the equation of the tangent to the curve $y = xe^x + 1$ at the point where it crosses the y-axis. Show your working.

3 Find the gradient of the tangent to the curve $y = x\cot x$ at the point where $x = \frac{\pi}{4}$ Show your working and leave your answer in terms of π

4 The sketch shows the function

$f(x) = \frac{\ln x}{e^x}$, $0 < x \le 8$

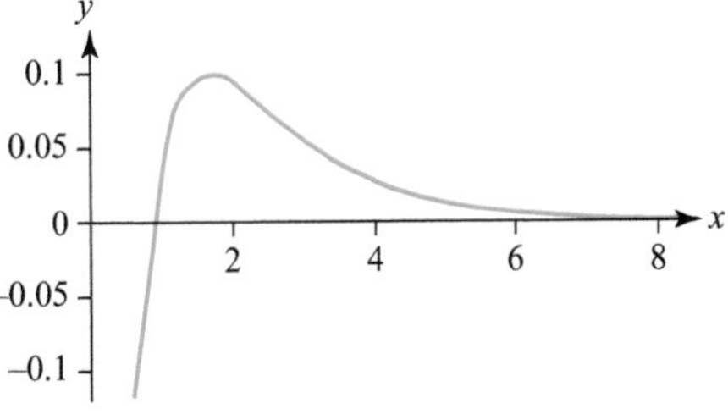

a Where does it cross the x-axis? Show your working.

b Show that there is a stationary point at $x^x = e$ and that this occurs when $x \approx 1.763$

c Show that a point of inflection occurs at $x = e^{\left(\frac{1+2x}{x^2}\right)}$ and that this occurs when $x \approx 2.55245$

5 Use the definitions

$\sec x = \frac{1}{\cos x}$; $\operatorname{cosec} x = \frac{1}{\sin x}$; $\cot x = \frac{1}{\tan x}$ to prove that

a $\frac{d}{dx}\sec x = \sec x \tan x$ b $\frac{d}{dx}\operatorname{cosec} x = -\operatorname{cosec} x \cot x$ c $\frac{d}{dx}\cot x = -\operatorname{cosec}^2 x$

Challenge

6 Differentiate a $\sin^3 x$ b $xe^x \sin x$

7 A river runs with a current of x miles per hour. A boat, which can reach 10 mph in still water, travels up-river for one mile, and then down-river for one mile, in T hours.

T is a function of x, the speed of the current, and can be expressed by the equation

$T(x) = \frac{20}{(10-x)(10+x)}$, $0 \le x < 10$

a Prove that, in the defined domain, T is an increasing function.

b What is the rate of change of T with respect to x when the current is 5 mph? Show your working.

c Showing your working, find the value of i $T(0)$ ii $T'(0)$

d In context, what happens as x approaches 10?

15.5 The chain rule

Fluency and skills

You will need to know how to differentiate a function of a function, known as a **composite function**. You can do this using the **chain rule**.

Take the function $y = \sin x^2$. You can say that y is a function of x, but, if you let $u = x^2$, then you can also say that y is a function of u, and u is a function of x

In other words, $y = f(u)$ and $u = g(x)$

Key point

The chain rule states that, for composite functions, $\frac{dy}{dx} = \frac{dy}{du} \times \frac{du}{dx}$

As you will see in Example 2, in certain contexts the chain rule is a significantly more efficient way of calculating a derivative.

Example 1

a Differentiate $y = \sin x^2$

b Differentiate $y = \sqrt{\cos x}$

a $y = \sin u \qquad u = x^2$ — Define y as a function of u, and u as a function of x

$\frac{dy}{du} = \cos u = \cos x^2 \qquad \frac{du}{dx} = 2x$ — Work out $\frac{dy}{du}$ and $\frac{du}{dx}$, and express both in terms of x

$\frac{dy}{dx} = \frac{dy}{du} \times \frac{du}{dx} = \cos x^2 \times 2x = 2x\cos x^2$ — Apply the chain rule and simplify.

b $y = \sqrt{u} = u^{\frac{1}{2}} \qquad u = \cos x$

$\frac{dy}{du} = \frac{1}{2}u^{-\frac{1}{2}} = \frac{1}{\sqrt{\cos x}} \qquad \frac{du}{dx} = -\sin x$ — Define each function and differentiate.

$\frac{dy}{dx} = \frac{dy}{du} \times \frac{du}{dx} = \frac{1}{\sqrt{\cos x}} \times -\sin x = -\frac{\sin x}{\sqrt{\cos x}}$ — Apply the chain rule and simplify.

You can also write the chain rule using a slightly different notation.

Key point

If you define a composite function as $y = f(g(x))$, then $\frac{dy}{dx} = f'(g(x)) \times g'(x)$

Example 2

Note that you could achieve the same result here by first expanding the brackets and differentiating each term.

Differentiate $(3x^2 + 1)^3$

$y = f(u)$ and $u = g(x)$ where $f(u) = u^3$ and $g(x) = 3x^2 + 1$ — This is a composite function, so define $g(x)$

$f'(u) = 3u^2$ and $g'(x) = 6x$

$\frac{dy}{dx} = f'(g(x)) \times g'(x)$

$= 3(3x^2 + 1)^2 \times 6x = 18x(3x^2 + 1)^2$ — Differentiate each function then multiply them together.

Exercise 15.5A Fluency and skills

1 For each of these functions, define y as a function of u and u as a function of x

a $y=\sin^2 x$ b $y=\tan 2x$

c $y=\sqrt{7x}$ d $y=\ln\cos x$

e $y=e^{7x}$ f $y=(2x+1)^4$

g $y=(3x-2)(3x-2)(3x-2)$

h $y=\frac{1}{\sqrt{\sin x}}$ i $y=\frac{2}{\sqrt{x^3}}$

j $y=\sqrt{x+\ln x}$ k $y=(\sin x+\ln x)^5$

l $y=\sqrt[3]{\cos x-\ln x}$

2 Differentiate each of these functions.

a $(3x+4)^5$ b $(2x-1)^7$

c $(x^2+1)^6$ d $(1-2x-3x^2)^3$

e $\sqrt{2x+1}$ f $\sqrt{3-5x}$

g $\sqrt{3x^2+4}$ h $\sqrt[3]{1-2x}$

i $\sqrt{x^2+3x+4}$ j $\frac{1}{(2x+3)^2}$

k $\frac{1}{\sqrt{1-3x}}$ l $\frac{3}{(x^2-2x+5)}$

m $\cos^2 x$ n $\sqrt{\cos x}$

o $\sin(3x+2)$ p $\tan(5x-1)$

q $\cos\sqrt{x+1}$ r $\sin(\cos x)$

s $e^{\sin x}$ t $e^{\sqrt{2x-1}}$

u $e^{(e^x)}$ v $\ln(\sin x)$

w $\ln\left(\sqrt{2x+3}\right)$ x $\ln(\ln x)$

y $\sin(\ln x)$ z $\frac{1}{\ln x}$

3 Find the gradient of each of these curves at the given point. Show your working.

a $y=(2x+1)^3$ at $x=0$

b $y=\sin^3 x$ at $x=\frac{\pi}{3}$

c $y=\sqrt{3x+1}$ at $x=5$

d $y=\ln(9-4x)$ at $x=2$

e $y=e^{\sqrt{10-3x}}$ at $x=2$

f $y=\ln(\cos x)$ at $x=\frac{\pi}{4}$

g $y=\frac{1}{(3x+1)^4}$ at $x=0$

h $y=\sin(x^2+x-2)$ at $x=1$

4 Find the derived function given that

a $f(x)=2^{\sin x}$

b $f(x)=\log_{10}(2x+1)$

Hint: $\log_a b=\frac{\ln b}{\ln a}$

c $f(x)=3^{x^2+2x-1}$

5 Find the equation of the tangent to each of these curves at $x=1$. Show your working.

a $y=(x^2-3x+1)^5$

b $y=e^{3x^2-2x}$

c $y=\sqrt{\sqrt{x}+3}$

d $y=\sin(x^2-1)$

e $y=\tan\left(2x^2-2x+\frac{\pi}{4}\right)$

f $y=e\ln(x^3-x+e)$

6 A semicircle with centre (4, 0) and radius 5 units has equation $y=\sqrt{9+8x-x^2}, -1\le x\le 9$

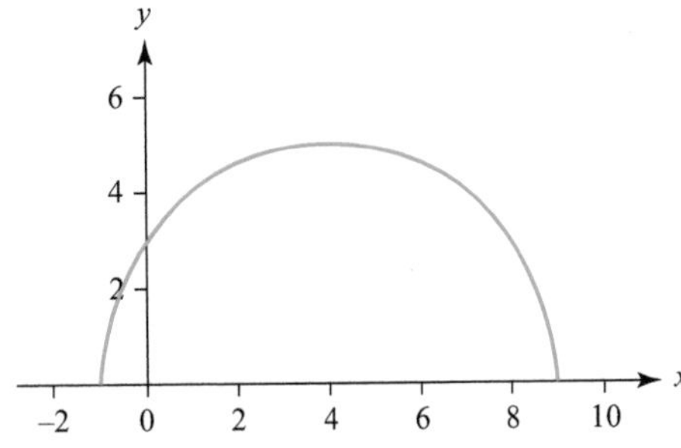

a Find the rate of change of y with respect to x when $x=7$. Show your working.

b Find the equation of the tangent to the circle when $x=1$ by using $\frac{dy}{dx}$. Show your working.

c What is the gradient of the curve as it crosses the y-axis? Show your working.

d What is the value of $\frac{d^2y}{dx^2}$ at this point?

Reasoning and problem-solving

Strategy

To solve problems involving differentiation of a composite function

1. Identify each function within the function and differentiate. Note that there may be more than one function within a function, and some of the functions may be products or quotients.
2. Apply the chain rule, and other necessary rules, to find the derivative of the composite function.
3. Simplify your answer and apply it to the problem.

Sometimes a composite function contains more than two distinct functions. You can still use the chain rule in these cases.

Key point

If $y = \mathrm{f}(u)$ where $u = \mathrm{g}(v)$ and $v = \mathrm{h}(x)$, then

$$\frac{dy}{dx} = \frac{dy}{du} \times \frac{du}{dv} \times \frac{dv}{dx}$$

You can use this principle for any number of functions within a composite function.

Example 3

Differentiate $\sin\sqrt{x^2+1}$

$y = \sin u,\ u = v^{\frac{1}{2}},\ v = x^2+1$

$\frac{dy}{du} = \cos u = \cos v^{\frac{1}{2}} = \cos\sqrt{x^2+1}$

$\frac{du}{dv} = \frac{1}{2\sqrt{v}} = \frac{1}{2\sqrt{x^2+1}}$ $\qquad \frac{dv}{dx} = 2x$

$\frac{dy}{dx} = \frac{dy}{du} \times \frac{du}{dv} \times \frac{dv}{dx} = \cos\sqrt{x^2+1} \times \frac{1}{2\sqrt{x^2+1}} \times 2x = \frac{x\cos\sqrt{x^2+1}}{\sqrt{x^2+1}}$

1 This is a function, within a function, within a function. First, define y, u and v

1 Calculate $\frac{dy}{du}$, $\frac{du}{dv}$ and $\frac{dv}{dx}$ and express in terms of x

2 3 Apply the chain rule and simplify your answer.

Example 4

A cylinder is placed in a hydraulic press and compressed, so that its height, h, decreases at a rate of $0.1\,\mathrm{m\,s^{-1}}$ and its radius squared, r^2, increases at a rate of $0.05\,\mathrm{m\,s^{-1}}$. At what rate is its volume changing when $h = 2\,\mathrm{m}$ and $r^2 = 0.25\,\mathrm{m^2}$?

Volume $V = \pi r^2 h = \pi u \qquad$ where $u = r^2 h$

$\frac{dV}{dt} = \frac{dV}{du} \times \frac{du}{dt}$

$\frac{dV}{du} = \pi \qquad \frac{du}{dt} = r^2\frac{dh}{dt} + h\frac{dr^2}{dt}$

$= (-0.1)r^2 + 0.05h$

$\frac{dV}{dt} = \frac{dV}{du} \times \frac{du}{dt} = \pi \times ((-0.1)r^2 + 0.05h) = 0.05\pi(h - 2r^2)$

$= 0.05\pi(2 - 2 \times 0.25) = 0.075\pi$

1 V is a function of r^2 and h, both of which are functions of t

2 You'll need the chain rule to calculate $\frac{dV}{dt}$

2 u is a product of two functions, r^2 and h, so use the product rule.

You're given the values of $\frac{dh}{dt}$ and $\frac{dr^2}{dt}$

3 Apply the chain rule and substitute in the values for h and r^2 given in the question.

Exercise 15.5B Reasoning and problem-solving

1 Find the derivative of

a $\sin(\cos 2x)$ b $e^{\sin 3x}$

c $(\sin x+\cos 2x)^5$ d $\sqrt{\sin(4x+1)}$

e $\sqrt{(x-1)(x+2)}$ f $\sin(xe^x)$

g $e^{x(3x+4)}$ h $\cos(x^2\sin x)$

i $\left(\frac{x+1}{x-1}\right)^{\frac{3}{2}}$ j $\sin\left(\frac{x}{x+1}\right)$

k $\ln\left(\frac{2x}{1-x}\right)$ l $\ln(xe^x)$

2 'Lighting up time' is defined as the time during which cars must use their headlights on public roads. Lighting up times can be calculated from the formula $T=2.5\cos\left(\frac{\pi D}{180}\right)+19$, where T hours is the lighting up time on day D after the longest day of the year.

a Find $\frac{dT}{dD}$, the rate at which lighting up time changes as the days pass.

b Use this to find the minimum value of T and the value of D when this occurs.

c Find $\frac{d^2T}{dD^2}$ and use it to verify that the answer to **b** is indeed the minimum.

3 A square, with side length x cm, is increasing in size such that its side length changes at a rate of 2 cm s^{-1}. At what rate is the area increasing when the side length is 10 cm?

4 Quadratic equations of the form $x^2+2bx+1=0$, where $b>1$, have two roots, one of which is $x=\sqrt{b^2-1}-b$

a Find $\frac{dx}{db}$

b Show that the graph of the function $x=\sqrt{b^2-1}-b$ is always increasing when $b>1$

c Find $\frac{d^2x}{db^2}$ and show that the curve of the function is always concave.

5 £1000 is deposited in a special account. The amount of money in the account, £P, can be calculated from the formula $P=1000e^{0.001y}$, where y is the number of years the money has been deposited.

a Find $\frac{dP}{dy}$ b At what rate is the money growing when $y=5$?

c Find $\frac{d^2P}{dy^2}$

6 The growth of a particular tree is modelled by $h=15\left(1-e^{-0.22t}\right)$, where h metres is the height of the tree after t years.

a Find $\frac{dh}{dt}$

b Show that $\frac{dh}{dt}=0.22(15-h)$

7 A boy is collecting stickers. There are 200 stickers to collect and he starts with 10

Each week he buys a new pack of stickers and discards duplicates. His number of stickers, N, at the end of week t is modelled by

$$N=\frac{200}{19e^{-0.7t}+1}$$

a Find $\frac{dN}{dt}$

b Express $19e^{-0.7t}$ in terms of N

c Show that $\frac{dN}{dt}=\frac{7}{2000}N(200-N)$

Challenge

8 Differentiate these functions.

a $\tan 3x=\sqrt{(b^2-1)}-b$ b $\sqrt{\sin x\cos x}$

c $\sqrt{\sin^2 x\cos 7x}$

9 A weather balloon can be modelled as a cube. As the balloon rises, the length of each side in metres is given by $x=2+1.5t$, where t is the time in minutes after the balloon is released. At what rate does the volume of the balloon increase when $x=8$ metres?

15.6 Inverse functions

Fluency and skills

You will need to know how to find $\frac{dy}{dx}$ when x is expressed as a function of y, i.e. $x = f(y)$. Inverse functions often make the differentiation of equations easier, as you will see in Example 1.

Given that $x = f(y)$, then $y = f^{-1}(x)$ and therefore $x = f(f^{-1}(x))$

Using the chain rule to differentiate $x = f(f^{-1}(x))$ with respect to x on both sides gives

$$x\left(\frac{d}{dx}\right) = f'(f^{-1}(x)) \times \frac{d}{dx}(f^{-1}(x))$$

Substituting $y = f^{-1}(x)$ gives $1 = f'(x) \times \frac{dx}{dy} \Rightarrow 1 = \frac{dy}{dx} \times \frac{dx}{dy} \Rightarrow \frac{dy}{dx} = \frac{1}{\frac{dx}{dy}}$

Remember that inverse trigonometric functions can be written in two ways.
$\sin^{-1} x = \arcsin x$
$\cos^{-1} x = \arccos x$
$\tan^{-1} x = \arctan x$

Key point

For $x = f(y)$, $\frac{dy}{dx} = \frac{1}{\frac{dx}{dy}}$

Example 1

Use the fact that $\frac{dy}{dx} = \frac{1}{\frac{dx}{dy}}$ to show that the derivative of $y = \sin^{-1} x$ is $\frac{1}{\sqrt{1+x^2}}$

$y = \sin^{-1} x$, so $x = \sin y$

$\frac{dx}{dy} = \cos y$ and $\frac{dy}{dx} = \frac{1}{\cos y}$

$-\frac{\pi}{2} \le \sin^{-1} x \le \frac{\pi}{2} \Rightarrow -\frac{\pi}{2} \le y \le \frac{\pi}{2}$

$\cos y \ge 0$ for $-\frac{\pi}{2} \le y \le \frac{\pi}{2}$

$\frac{dy}{dx} = \frac{1}{\sqrt{1-\sin^2 y}}$ — Use the fact that $\sin^2 y + \cos^2 y = 1$

$\frac{dy}{dx} = \frac{1}{\sqrt{1-x^2}}$ — $x = \sin y$ so substitute x for $\sin y$

Therefore, $\frac{d}{dx}(\sin^{-1} x) = \frac{1}{\sqrt{1-x^2}}$

Exercise 15.6A Fluency and skills

1 For each of these functions

i Find its derivative and state its inverse,

ii State the derivative of the inverse.

You can assume each function is defined in a suitable domain so that its inverse exists.

a $f(x) = x^6$ b $f(x) = x^{\frac{2}{3}}$

c $f(x) = x^2 + 2$ d $f(x) = (x+4)^2 + 1$

e $f(x) = \sqrt{3x+2}$ f $f(x) = \sqrt[3]{2x+1}$

2 Use the method shown in Example 1 to prove that

a $\frac{d}{dx}(\cos^{-1} x) = -\frac{1}{\sqrt{1-x^2}}$

b $\frac{d}{dx}(\tan^{-1} x) = \frac{1}{1+x^2}$

c $\frac{d}{dx}(\sec^{-1} x) = \frac{1}{x\sqrt{x^2-1}}$

d $\frac{d}{dx}(\cot^{-1} x) = -\frac{1}{1+x^2}$

3 For each of these functions

i Find its derivative, ii Find its inverse,

iii Find the derivative of the inverse.

a $f(x) = 3x + 1$ b $f(x) = 1 - 5x$

c $f(x) = 3x^2$ d $f(x) = \frac{x+1}{2}$

e $f(x) = \frac{1}{x+1}$ f $f(x) = \frac{x}{x+1}$

g $f(x) = \frac{1}{x^2}$ h $f(x) = \sqrt{x}$

i $f(x) = \sqrt{x} - 1$ j $f(x) = x^{\frac{1}{3}}$

k $f(x) = e^{2x+1}$ l $f(x) = \ln 4x$

Reasoning and problem-solving

Strategy

To solve problems involving the differentiation of inverse functions

1. Differentiate with respect to y and find $\frac{dy}{dx}$ by using the result $\frac{dy}{dx} = \frac{1}{\frac{dx}{dy}}$
2. Find the equation of the tangent or normal by substituting into the equation of a straight line $y - y_1 = m(x - x_1)$

Example 2

A function has the inverse $x = y^3 + 2y + 4$. Find the equation of the tangent of the function at the point (7, 1).

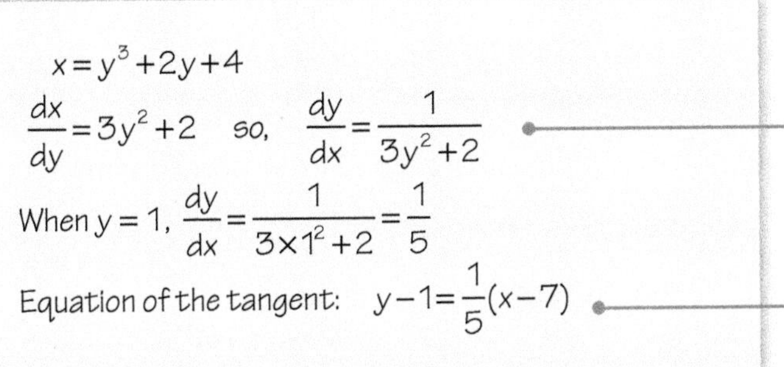

$x = y^3 + 2y + 4$
$\frac{dx}{dy} = 3y^2 + 2$ so, $\frac{dy}{dx} = \frac{1}{3y^2 + 2}$
When $y = 1$, $\frac{dy}{dx} = \frac{1}{3 \times 1^2 + 2} = \frac{1}{5}$
Equation of the tangent: $y - 1 = \frac{1}{5}(x - 7)$

(1) Differentiate with respect to y and then use $\frac{dy}{dx} = \frac{1}{\frac{dx}{dy}}$

(2) From Ch1.6, use the equation of a straight line $y - y_1 = m(x - x_1)$, where (x_1, y_1) is a point on the line and m is the gradient, to find the equation of the tangent.

Exercise 15.6B Reasoning and problem-solving

1 A function is defined by $f(x) = \frac{4}{x^3 + 1}$

Find the equation of the normal to the function $y = f^{-1}(x)$ at the point (2, 1)

2 Find the equation of the tangent to the curve $y = \sin^{-1}\left(\frac{2x}{2x+1}\right)$, $x \ge 0$, at the point where $x = \frac{1}{2}$. Show your working.

3 a Find the equation of the tangent to the curve $y = \sin^{-1}(x+1); -2 \le x \le 0$ at the point where $x = -\frac{1}{2}$. Show your working.

b What can be said about the tangent to the curve at the point where $x = -2$? State its equation.

4 A door is 1.5 m wide. As the door opens, it sweeps over an area, $A\,\text{m}^2$, making an angle of θ rad at the hinge. The leading edge of the door is h metres from this line.

a Show that $A = \frac{9}{8}\sin^{-1}\left(\frac{2h}{3}\right)$

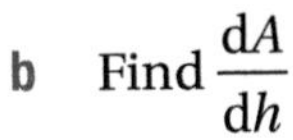

b Find $\frac{dA}{dh}$

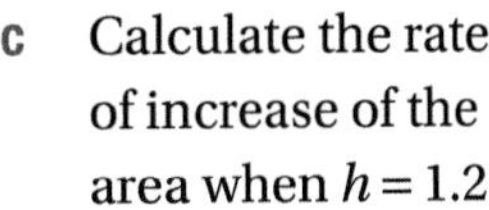

c Calculate the rate of increase of the area when $h = 1.2$

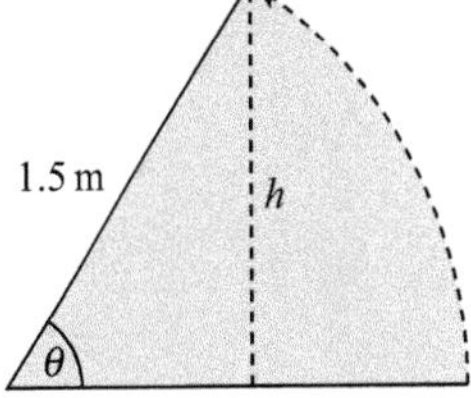

Challenge

5 As the tide comes into a harbour, the time passed since low tide, t hours, can be calculated from the depth of water using the formula $t = \frac{6}{\pi}\cos^{-1}(2 - 0.2D)$, where D is the depth in feet.

a Find an expression for $\frac{dt}{dD}$

b Find the rate of change of time passed with respect to depth when the water is 10 feet deep.

15.7 Implicit differentiation

Fluency and skills

In an equation of the form $y = f(x)$, y is called the dependent variable and x the independent variable. When the dependent variable is the subject of the equation, the function is said to be **explicit**.

However, equations are sometimes expressed in a form where the dependent variable is not explicitly given as the subject.

For example, the equation $2y + 3x = 5$ is the equation of a straight line and the fact that y is a function of x is only implied. This is an example of an **implicit function.**

The equation $(x-2)^2 + (y-1)^2 = 4$ is the equation of a circle and a second example of an implicit function.

Key point

When a function cannot be easily rearranged into the form $y = f(x)$, you can differentiate it implicitly.

If a term is expressed as a function of y, you must use the chain rule $\frac{d}{dx}f(y) = f'(y) \times \frac{dy}{dx}$

Example 1

A function is defined by $x^2 + 3xy + 4y^3 = 0$. Express $\frac{dy}{dx}$ in terms of x and y

Differentiating each term

Differentiating x^2 gives $2x$

Differentiating $3xy$ gives $\left(3x \times \frac{dy}{dx}\right) + (3y \times x^0) = 3x\frac{dy}{dx} + 3y$

Differentiating $4y^3$ gives $(4 \times 3y^2) \times \frac{dy}{dx} = 12y^2\frac{dy}{dx}$

Putting these terms together gives $2x + 3x\frac{dy}{dx} + 3y + 12y^2\frac{dy}{dx} = 0$

$\frac{dy}{dx}(3x + 12y^2) + (2x + 3y) = 0$

$\frac{dy}{dx} = -\frac{2x + 3y}{3x + 12y^2}$

This is an implicit function, so differentiate throughout with respect to x

The second term is a product of x and y, so you must use the product rule to differentiate. See Ch15.4 for a reminder.

The third term is a function of y, so you must use the chain rule to differentiate. See Ch15.5 for a reminder.

Collect terms multiplied by $\frac{dy}{dx}$

Exercise 15.7A Fluency and skills

1 For each expression, find $\frac{dy}{dx}$ in terms of x and y

a $x^2 = y^3$

b $(x+1)^2 + (y-3)^2 = 4$

c $2x^2 + y^2 = 4$

d $\frac{1}{x} + \frac{1}{y} = 1$

e $x^2 + 2xy + 3y = 0$

f $y + \frac{1}{y} = x^2$

g $x^2 + 2xy + y^2 = 6$

h $3e^x = \sqrt{y}$

i $\sin x = \cos 2y$

j $\ln(y+1) = x^2 + 1$

k $x\cos y = \tan x$

l $x^{\frac{3}{4}} + y^{\frac{3}{4}} = \pi$

m $x + \sin^{-1} x = y + \cos^{-1} y$

n $\ln(x+y) = x + xy$

o $e^{y+1} = x^2 + 2xy + 1$

p $\tan^{-1} y + xy = 0$

q $\frac{1}{x^2} + \frac{1}{y^2} = 144$

2 A hyperbola has equation $xy + 2x^2y^2 = 1$ Find the gradient of the curve at the point (1, 0.5). Show your working.

3 A circle has equation $x^2 + y^2 = 25$

a Find $\frac{dy}{dx}$ in terms of x and y

b Find the gradients of the tangents to the circle at the points where $x = 3$

c Find the gradient of the tangent to the circle at the point where $x = 0$ and $y = 5$

4 Find the stationary points on each of the following curves.

a $x^2 + 5y^2 = 20x$

b $x^2 + 2y^2 - 4x + 7y + 9 = 0$

c $x^3 + 3y^3 = 81$

5 A curve has equation $e^x + e^{2y} + 3x = 2 - 4y$

a Show that the point (0, 0) lies on the curve.

b Find $\frac{dy}{dx}$

c Find the gradient of the tangent at the point (0, 0)

6 A curve has equation $\ln(y-1) = x\ln x$

a Express $\frac{dy}{dx}$ in terms of x and y

b Show that the point (1, 2) lies on the curve.

c Find the gradient of the tangent to the curve at this point.

7 A circle has equation $x^2 + y^2 - 2x - 4y + 1 = 0$

a Find $\frac{dy}{dx}$ in terms of x and y

b The point (5, −1) lies on the circle. Determine the gradient of the tangent to the circle at this point.

8 St Valentine's Equation is $(x^2 + y^2 - 1)^3 = x^2y^3$

When the points which satisfy this equation are plotted, a heart is produced.

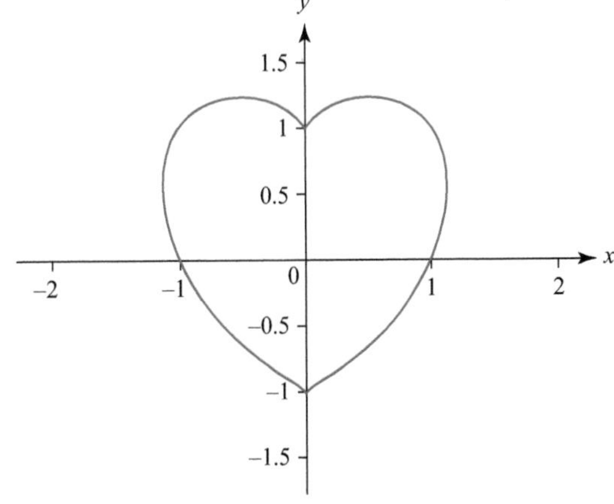

a Find an expression in x and y for $\frac{dy}{dx}$

b Find the gradient of the curve at the point (1, 1)

Reasoning and problem-solving

Strategy

To find the point where a tangent of given gradient touches the curve of an implicit function

1. Find the derivative of the implicit function.
2. Use the given conditions to express y in terms of x and obtain an equation for the tangent.
3. Substitute the expression into the implicit function and solve to find the x-values of the points of intersection.
4. Substitute the x-values into the equation of the tangent to find the corresponding y-values.

Example 2

The equation of a circle is $x^2+y^2+2x-7=0$

Find where the tangents to the circle with gradient 1 touch the circle.

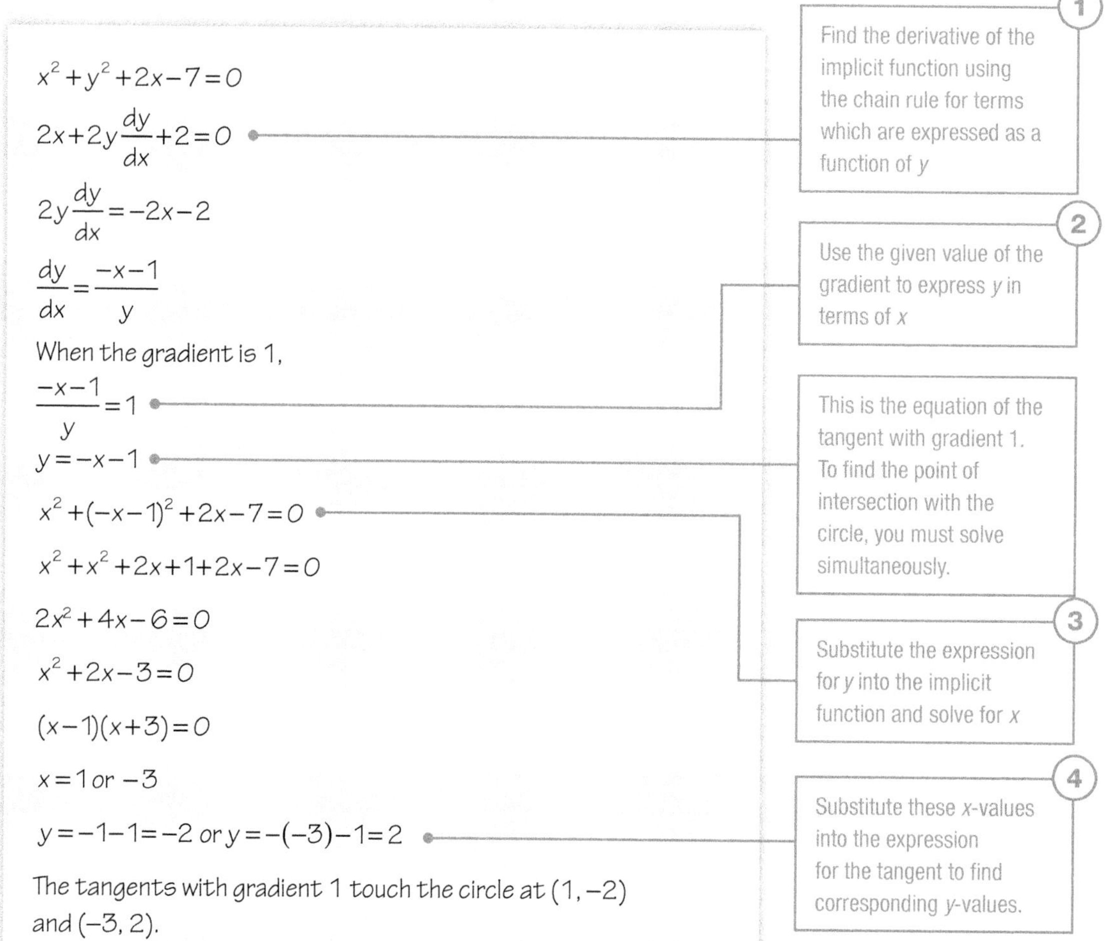

$x^2+y^2+2x-7=0$

$2x+2y\dfrac{dy}{dx}+2=0$ — ① Find the derivative of the implicit function using the chain rule for terms which are expressed as a function of y

$2y\dfrac{dy}{dx}=-2x-2$

$\dfrac{dy}{dx}=\dfrac{-x-1}{y}$

When the gradient is 1,

$\dfrac{-x-1}{y}=1$ — ② Use the given value of the gradient to express y in terms of x

$y=-x-1$ — This is the equation of the tangent with gradient 1. To find the point of intersection with the circle, you must solve simultaneously.

$x^2+(-x-1)^2+2x-7=0$ — ③ Substitute the expression for y into the implicit function and solve for x

$x^2+x^2+2x+1+2x-7=0$

$2x^2+4x-6=0$

$x^2+2x-3=0$

$(x-1)(x+3)=0$

$x=1$ or -3

$y=-1-1=-2$ or $y=-(-3)-1=2$ — ④ Substitute these x-values into the expression for the tangent to find corresponding y-values.

The tangents with gradient 1 touch the circle at $(1, -2)$ and $(-3, 2)$.

Exercise 15.7B Reasoning and problem-solving

1 An asteroid has the equation $x^{\frac{2}{3}}+y^{\frac{2}{3}}=5$

The asteroid passes through the point (1, 8). What is the equation of the tangent to the asteroid at this point?

2 A circle has the equation $x^2+y^2-8x-4y-5=0$

Find where the tangents to the circle with gradient $\frac{3}{4}$ touch the circle.

3 An ellipse has the equation $\frac{x^2}{a^2}+\frac{y^2}{b^2}=1$ where a and b are constants.

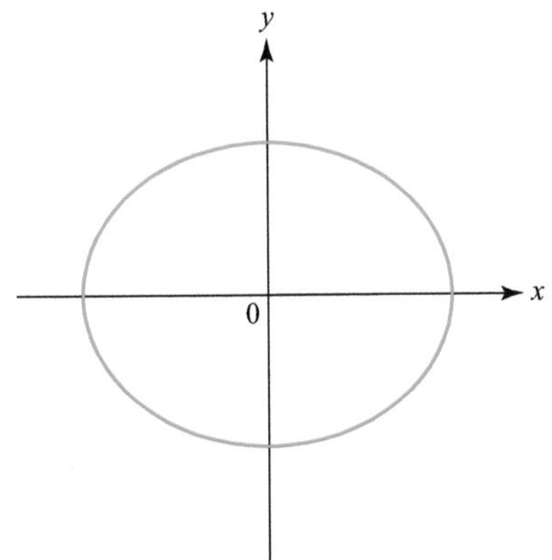

a Show that the horizontal width of the ellipse is $2a$

b Find the rate of change of y with respect to x

c Find the stationary points of the curve.

d What is the relationship between x and y when the gradient of the tangent to the ellipse is 1?

4 A hyperbola has the equation $(y+1)^2=(x+3)(2x+1)$

a Find $\frac{dy}{dx}$ in terms of x and y

b Find the equations of the tangents to the curve when $x=2$

c Find the point where the two tangents intersect. Show your working.

5 The ellipse shown has the equation $x^2+xy+y^2-9=0$

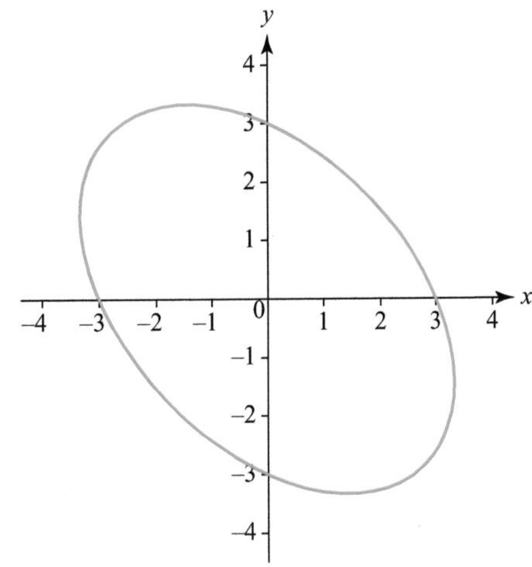

a Find the points at which the gradient of the ellipse is 1

b Find the maximum and minimum values of y, leaving your answer in surd form.

c Find the values of x where the tangents to the ellipse are parallel to the y-axis.

6 A circle has the equation $x^2+y^2-2x-10y+1=0$

a By differentiating, find the equation of the normal to the circle at

i (4, 9) ii (5, 8)

b Find where the two normals intersect, showing your working. What do you notice about this point of intersection?

Challenge

7 A curve has the equation $x^2-xy+2y^2=144$

a Find the equation of the tangent to the curve at the point where the curve intersects the line $y=3x$ in the first quadrant.

b Find the stationary points of the curve. Use graph plotting software to draw the curve and classify the stationary points.

15.8 Parametric functions

Fluency and skills

See Ch 12.3 For a reminder on parametric equations.

When two variables, x and y, are expressed as functions of a parameter, the derivative, $\frac{dy}{dx}$, can be found using the chain rule.

Key point

You can find $\frac{dy}{dx}$ from the parametric equations $x = f(t)$ and $y = g(t)$ by using the chain rule $\frac{dy}{dx} = \frac{dy}{dt} \times \frac{dt}{dx}$

Parametric equations are useful because they allow you to relate two variables with respect to a third variable. For example, in kinematics, you can represent the trajectory of an object using parametric equations which depend on time as the parameter.

Example 1

The parametric equations of a parabola are $x = 4t$; $y = 30t - 5t^2$

a Find $\frac{dy}{dx}$

b Find the turning point of the parabola. Show your working.

a $x = 4t$

$\frac{dx}{dt} = 4$

$y = 30t - 5t^2$

$\frac{dy}{dt} = 30 - 10t$ — Differentiate both parametric equations with respect to t

$\frac{dy}{dx} = (30 - 10t) \times \frac{1}{4}$ — Substitute into $\frac{dy}{dx} = \frac{dy}{dt} \times \frac{dt}{dx}$

$\frac{dy}{dx} = \frac{5}{2}(3 - t)$ — Simplify.

b At the turning point,

$\frac{dy}{dx} = 0$

$\frac{5}{2}(3 - t) = 0$ — Equate the first derivative to zero.

$\Rightarrow t = 3$ — Solve to find t

When $t = 3$,

$x = 4 \times 3 = 12$

$y = (30 \times 3) - (5 \times 3^2) = 45$

So the turning point is at $(12, 45)$ — Substitute $t = 3$ into the parametric equations to find the coordinates of the turning point.

Example 2

The curve shown has parametric equations $x = \sin 2\theta$; $y = 2\cos\theta$

Find an expression for its derivative in terms of the parameter θ

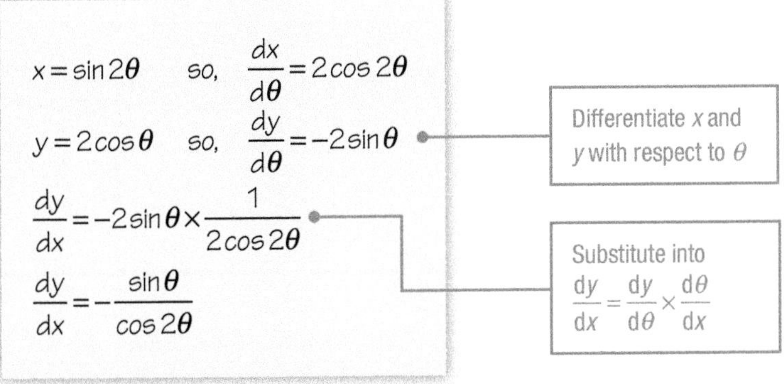

$x = \sin 2\theta$ so, $\frac{dx}{d\theta} = 2\cos 2\theta$
$y = 2\cos\theta$ so, $\frac{dy}{d\theta} = -2\sin\theta$
$\frac{dy}{dx} = -2\sin\theta \times \frac{1}{2\cos 2\theta}$
$\frac{dy}{dx} = -\frac{\sin\theta}{\cos 2\theta}$
Differentiate x and y with respect to θ
Substitute into $\frac{dy}{dx} = \frac{dy}{d\theta} \times \frac{d\theta}{dx}$

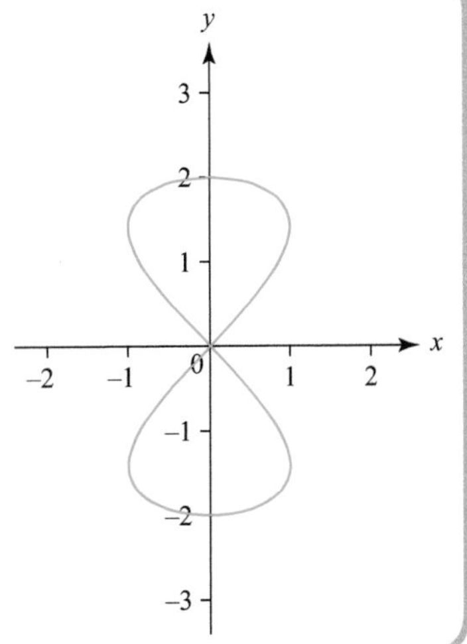

Exercise 15.8A Fluency and skills

1 Find $\frac{dy}{dx}$ for each pair of parametric equations.

a $x = 2t$; $y = t + 1$

b $x = t + 4$; $y = 2t - 1$

c $x = t^2$; $y = t$

d $x = \sin\theta$; $y = \cos\theta$

e $x = \sin(\theta + 3)$; $y = \cos(\theta + 2)$

f $x = 3 + \sin\theta$; $y = 1 + \cos\theta$

g $x = \sin(2\theta + 1)$; $y = \cos(2\theta + 1)$

h $x = e^t + e^{-t}$; $y = e^t - e^{-t}$

i $x = 10\ln t$; $y = 10e^{-t}$

j $x = 2t^3$; $y = 3t^2$

k $x = \sin t$; $y = t^2$

2 Find $\frac{dy}{dx}$ for each pair of parametric equations.

a $x = 9\cos\theta + 3\cos 3\theta$; $y = 9\sin\theta - 3\sin 3\theta$

b $x = 3\sin 2\theta$; $y = 4\cos 2\theta$

c $x = \sin^{-1}\theta$; $y = \cos^{-1}\theta$, $-1 \le \theta \le 1$

d $x = \sin^{-1}\theta$; $y = \tan^{-1}\theta$, $-1 \le \theta \le 1$

e $x = \sin^2\theta$; $y = \cos^2\theta$

f $x = \sqrt{3t}$; $y = \frac{1}{\sqrt{t}}$

g $x = \sqrt{\sin t}$; $y = \sqrt{\cos t}$

h $x = \cos 2\theta$; $y = 4\sin\theta$

i $x = e^{\sin t}$; $y = e^{\cos t}$

j $x = \frac{1}{\sin t}$; $y = \frac{\cos t}{\sin t}$

k $x = 1 - 5t$; $y = 1 + 20t - 5t^2$

3 A spiral has parametric equations $x = \theta\cos\theta$, $y = \theta\sin\theta$, $0 \le \theta \le 2\pi$

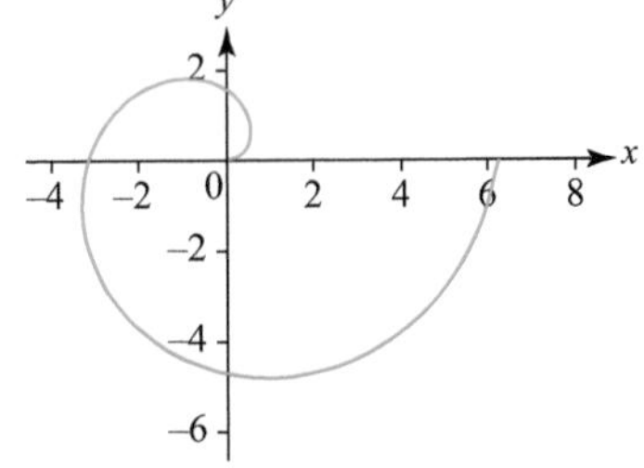

a Find the Cartesian coordinates of the point corresponding to a parameter value $\theta = \pi$

b Find the gradient of the tangent at this point. Show your working.

4 For each curve, find the coordinates of the point corresponding to the given parameter value. Find the gradient at that point, showing your working.

a $x = t^3$; $y = t$; when $t = 2$

b $x = \frac{t}{t+1}$; $y = \frac{t^2}{t+1}$, $t \ne -1$; when $t = 3$

c $x = \sqrt{2}\sin t$; $y = 2\sqrt{2}\cos t$; when $t = \frac{\pi}{4}$

d $x = \sqrt{2}\cos 3t$; $y = 3\sin 2t$; when $t = \frac{\pi}{12}$

e $x = 3\sec t$; $y = 4\tan t$; when $t = \frac{\pi}{3}$

f $x = 2t - 1$; $y = 4 + 3t$; when $t = 4$

Reasoning and problem-solving

Strategy

To solve problems involving differentiating parametric equations

1. Differentiate the equations with respect to the parameter.
2. Using the chain rule, substitute the results into $\frac{dy}{dx}=\frac{dy}{dt}\times\frac{dt}{dx}$
3. Find an expression relating only y and x
4. Use the given conditions to answer the question.

Example 3

An ellipse has parametric equations $x=2\sin\theta$; $y=3\cos\theta$

a Find the rate of change of y with respect to x when $x=1$. Show your working.

b State the relationship between the variables y and x alone.

a $x=2\sin\theta$ so, $\frac{dx}{d\theta}=2\cos\theta$

$y=3\cos\theta$ so, $\frac{dy}{d\theta}=-3\sin\theta$

$\frac{dy}{dx}=\frac{-3\sin\theta}{2\cos\theta}=-\frac{3}{2}\tan\theta$

When $x=1$, $\sin\theta=\frac{1}{2}$ so, $\theta=\frac{\pi}{6}$

$\frac{dy}{dx}=-\frac{3}{2}\tan\frac{\pi}{6}$

$=-\frac{3}{2}\times\frac{1}{\sqrt{3}}$

$=-\frac{\sqrt{3}}{2}$

So the rate of change is $-\frac{\sqrt{3}}{2}$ when $x=1$

b $x=2\sin\theta$ so, $\frac{x}{2}=\sin\theta$

$y=3\cos\theta$ so, $\frac{y}{3}=\cos\theta$

Squaring both equations: $\frac{x^2}{4}=\sin^2\theta$; $\frac{y^2}{9}=\cos^2\theta$

Adding both equations: $\frac{x^2}{4}+\frac{y^2}{9}=\sin^2\theta+\cos^2\theta$

The required relationship is: $\frac{x^2}{4}+\frac{y^2}{9}=1$

(1) Differentiate with respect to θ

(2) Substitute into $\frac{dy}{dx}=\frac{dy}{d\theta}\times\frac{d\theta}{dx}$

(4) Find the value of θ when $x=1$

(4) Substitute $\theta=\frac{\pi}{6}$ into the equation to find the rate of change.

(3) Manipulate the equations so that you can eliminate θ by using the identity $\sin^2\theta+\cos^2\theta=1$. This allows you to find an expression relating only y and x

Exercise 15.8B Reasoning and problem-solving

1 A function is defined by the parametric equations $x = \frac{t}{2}$; $y = 2t^3 - t^2 + 1$

a Find the rate of change of y with respect to x when $x = 4$. Show your working.

b Express y in terms of x alone.

2 An ellipse has parametric equations $x = \sin\theta$; $y = 5\cos\theta$

a Find an expression relating only the variables y and x

b Find the equations of the two tangents to the ellipse which are parallel to the x-axis.

3 The sketch shows the function with parametric equations $x = 2 - 4t^2$; $y = 3t^3 - t + 1$

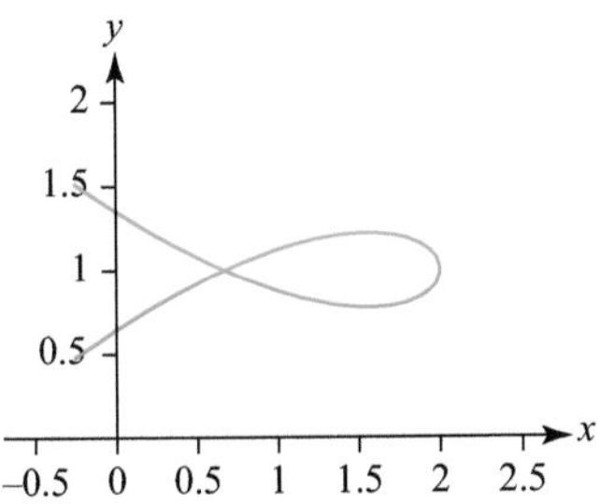

a Find $\frac{dy}{dx}$

b At what values of the parameter, t, will the curve be stationary?

c Find the stationary points and use the sketch to identify their nature.

d i Find the three values of the parameter, t, that correspond to $y = 1$

ii With reference to the sketch, comment on what's happening at these three points.

4 The parametric equations $x = 1 + 5\cos\theta$; $y = 2 + 5\sin\theta$ represent a circle.

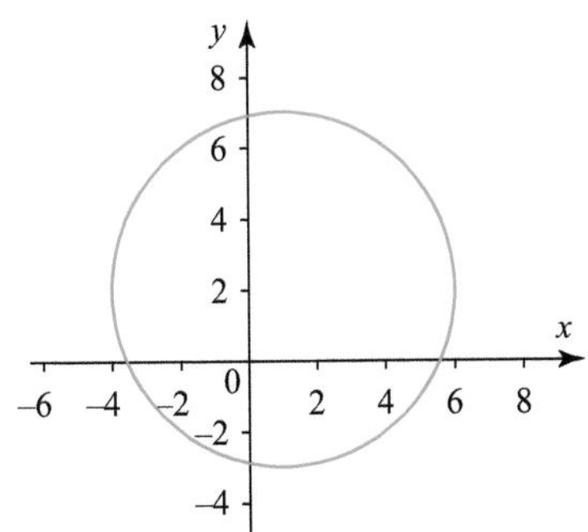

a Find $\frac{dy}{dx}$

b Find the gradient of the normal to the circle at the point that corresponds to a parameter value of $\theta = \frac{\pi}{3}$

c Using the parametric equations, show that the Cartesian equation of the circle is $(x-1)^2 + (y-2)^2 = 25$

5 A ball is thrown in the air. The flight-path follows a parabolic trajectory.

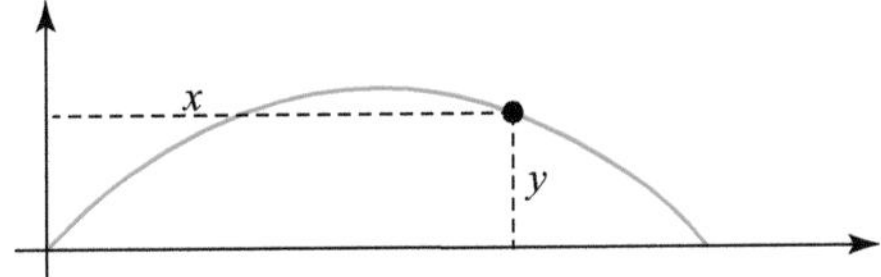

t seconds after release, the position of the ball (x, y) can be expressed by the parametric equations

$x = 10t$; $y = 10t - 5t^2$

a Find the rate of change of y with respect to x

b i What are the t-values that correspond to a y-value of zero?

ii What are the corresponding x-values?

c Using the parametric equations, show that the Cartesian equation of the flight-path is a quadratic of the form $y = ax^2 + bx + c$, where a, b and c are constants to be found.

Challenge

6 An ellipse has parametric equations $x = 2\cos\theta$; $y = 6\sin\theta$. Point A has coordinates $(\sqrt{2}, 3\sqrt{2})$ and lies on the ellipse. Find the point at which the normal to the ellipse at point A intersects the x-axis.

15 Summary and review

Chapter summary

- You can partially determine the shape of a curve by considering the derivative and second derivative.

  If $\frac{dy}{dx} > 0$, the function is increasing. If $\frac{dy}{dx} < 0$, the function is decreasing.

  If $\frac{d^2y}{dx^2} > 0$, the curve is convex. If $\frac{d^2y}{dx^2} < 0$, the curve is concave.
- A **point of inflection** is where the graph of the curve changes concavity, i.e. where the second derivative changes sign.
- At a point of inflection, $\frac{d^2y}{dx^2} = 0$, but $\frac{d^2y}{dx^2} = 0$ does not always imply a point of inflection.
- If $\frac{dy}{dx} = 0$ and there is a change in concavity, then you have a horizontal point of inflection.
- You can use small-angle approximations for $\sin\theta$ and $\cos\theta$ to derive the following results.

  $\lim_{\theta \to 0} \frac{\sin\theta}{\theta} = 1$ and $\lim_{\theta \to 0} \frac{1-\cos\theta}{\theta} = 0$

  Using these results, you can derive, from first principles, the derivatives of $\sin x$ and $\cos x$

  $\frac{d}{dx}(\sin x) = \cos x$; $\frac{d}{dx}(\cos x) = -\sin x$
- You need to know the following derivatives.
  - $\frac{d}{dx}(e^{kx}) = ke^{kx}$
  - $\frac{d}{dx}(\ln x) = \frac{1}{x}$
  - $\frac{d}{dx}(a^{kx}) = ka^{kx}\ln a$
- The product rule is used to differentiate two functions, u and v, that have been multiplied together.

  In Leibniz notation, it is written $\frac{d}{dx}(uv) = v\frac{du}{dx} + u\frac{dv}{dx}$
- The quotient rule is used to differentiate two functions, u and v, one of which has been divided by the other.

  In Leibniz notation, it is written $\frac{d}{dx}\left(\frac{u}{v}\right) = \frac{v\frac{du}{dx} - u\frac{dv}{dx}}{v^2}$
- If x is expressed as a function of y, then $\frac{dy}{dx} = \frac{1}{\frac{dx}{dy}}$

- The chain rule is used to differentiate a function of a function, known as a composite function. Consider the composite function $y = f(g(x))$. Let $g(x) = u$ so that $y = f(u)$.
Then $\frac{dy}{dx} = \frac{dy}{du} \times \frac{du}{dx}$
The chain rule allows you to solve problems involving connected rates of change.
- Equations presented in a form where the dependent variable is not explicitly given as the subject are known as implicit functions.
- To differentiate implicit functions, you must use the chain rule to differentiate terms expressed as a function of y and the product rule to differentiate terms which involve a product of x and y
- If x and y are defined parametrically, $x = f(t)$ and $y = g(t)$, you can use the chain rule to find $\frac{dy}{dx}$
$\frac{dy}{dx} = \frac{dy}{dt} \times \frac{dt}{dx}$

PURE

Check and review

You should now be able to...	Try Questions
✔ Determine when a curve is convex or concave.	1, 2
✔ Find points of inflection on a curve.	1, 2
✔ Use $\lim_{\theta \to 0} \frac{\sin\theta}{\theta} = 1$ and $\lim_{\theta \to 0} \frac{1-\cos\theta}{\theta} = 0$	3
✔ Differentiate $\sin x$ and $\cos x$	3, 4, 6, 7, 9, 10, 15
✔ Differentiate e^x, a^x, and $\ln x$	5–7, 9–11, 14
✔ Use the product and quotient rules for differentiation.	6, 7, 11
✔ Use the chain rule for differentiation.	4, 5, 8–11, 15
✔ Find the derivative of a function which is defined implicitly.	11–13
✔ Find the derivative of an inverse function.	14, 15
✔ Find the derivative of a function which is defined parametrically.	16–18

1 The curve $y = \frac{x^3}{3} + x^2 - 3x + 1$ has two turning points and a point of inflection.
Find each point and describe each fully. Show your working.

2 a Find the stationary points on the curve $y = x^4 + 8x^3 + 18x^2 + 1$. Show your working.

b Determine the nature of each of the points.

3 a Differentiate $y = \sin 2x$ from first principles.

b $f(x) = 2\sin x + 3\cos x$. Showing your working, calculate i $f'\left(\frac{\pi}{4}\right)$ ii $f''\left(\frac{\pi}{4}\right)$

4 A toy attached to a spring bobs up and down. Its position relative to its starting position is given by $y = 4\sin\left(3x + \frac{\pi}{4}\right)$ where the toy is y metres from its starting position after x seconds. When $y < 0$, the toy is below its starting position.

a Find the rate of change of the distance of the toy from its starting position at $x = 10$. Show your working.

b Is the toy rising or falling after 3 seconds? Show your working.

c When is *the rate of change of the rate of change* of y first equal to zero?

5 Find the derivative of each of the following.

a $4e^{2x+1}$ b $\ln(3x+1)$

c 2×3^x d 2^{-x}

6 Differentiate each of the following functions with respect to x

a $e^x \sin^2 x$ b $\frac{3\ln x}{2x}$

c $\sqrt{x}\tan x$ d $\frac{x^3 + 6x + 11}{\cos x}$

7 Use both the product rule and the quotient rule to differentiate these functions.

a $\frac{xe^x}{\ln x}$ b $\frac{\cos x}{x\sin x}$ c $\frac{x\cos x}{2x+1}$

8 $y = (2x^2 - 3)^5$

a Calculate $\frac{dy}{dx}$

b Hence, state the range of values of x for which y is increasing.

9 Find the gradient of the curve at the point given. Show your working.

a $y = (3x-1)^{\frac{1}{3}}$ at $x = 3$ b $y = \cos\left(2x^2 + \frac{3\pi}{2}\right)$ at $x = 0$

c $y = \ln(x^2 - 8)$ at $x = -3$ d $y = e^{\sqrt{2x+1}}$ at $x = 4$

10 Find the derived function when

a $f(x) = 3^{\cos x}$ b $f(x) = 4^{\ln x}$

11 Find $\frac{dy}{dx}$ when

a $\frac{3}{x^2} + \frac{1}{y^2} = 6$ b $\cos x \sin y = 5$

c $xe^y + ye^x = x$ d $e^y \ln x = x + 2y$

12 A curve has equation $3x + 2y^2 = 6y$

a Find an expression for $\frac{dy}{dx}$

b Find the gradient of the curve at the point $(-12, -3)$

13 A parabola has equation $x^2 + 6xy + 9y^2 + x + 3y + 1 = 0$. Find $\frac{dy}{dx}$

14 Given that $x = e^y + 2y$

a Find an expression for $\frac{dy}{dx}$ in terms of y

b Calculate the gradient at the point (1, 0)

15 Differentiate each of the following functions with respect to x

a $\sin^{-1}\left(\sqrt{1-x^4}\right)$ b $\cos^{-1}\left(1-e^{-\frac{x}{10}}\right)$ c $\tan^{-1}\sqrt{x^2-1}$

16 Determine $\frac{dy}{dx}$ for each pair of parametric equations.

a $x=3t-4;\ y=6t+1$ b $x=t^2+1;\ y=t-1$

c $x=3\sin\theta+1;\ y=4\cos\theta-5$ d $x=\ln t;\ y=t^2$

17 A curve has parametric equations

$x=2t, \quad y=\frac{3}{t^2}, \quad t\neq 0$

a Find an expression for $\frac{dy}{dx}$

b Calculate the gradient of the curve at the point where $t=4$

18 The sketch shows the function $x=\cos t-\sqrt{3}\sin t;\ y=2\sin t-\sqrt{3}\cos t$

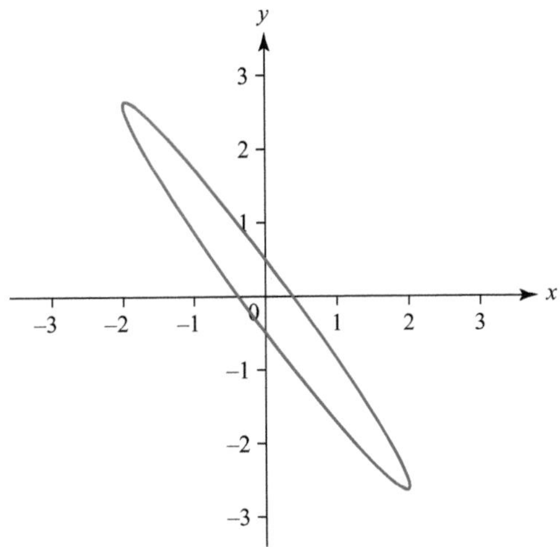

a Find where the curve cuts the x-axis. Show your working.

b Find the gradient of the ellipse at these points. Show your working.

What next?

Score			
	0–9	Your knowledge of this topic is still developing. To improve, search in MyMaths for the codes: 2161–2166, 2222, 2223, 2271, 2272	
	10–14	You're gaining a secure knowledge of this topic. To improve, look at the InvisiPen videos for Fluency and skills (15A)	
	15–18	You've mastered these skills. Well done, you're ready to progress! To develop your techniques, look at the InvisiPen videos for Reasoning and problem-solving (15B)	

Click these links in the digital book

Exploration

Going beyond the exams

History

Gottfried Leibniz (1640–1716) is generally credited with the product rule for differentiation. In **Leibniz notation**

$$\frac{d(uv)}{dx} = u\frac{dv}{dx} + v\frac{du}{dx}$$

Isaac Barrow (1630 – 1677) also contributed to the early development of calculus and may have independently discovered the product rule. **Isaac Newton**, who was once a student of Barrow's, then went on to develop the theory of calculus, causing a bitter dispute over who got there first between him and Leibniz.

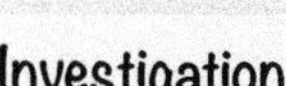

Investigation

Suppose that f and g are functions of x

Using $f(n)$ to represent the nth derivative of f with respect to x, we can write the rule for the derivative of a product as

$(fg)^{(1)} = fg^{(1)} + f^{(1)}g$

Differentiating a second time gives

$$(fg)^{(2)} = fg^{(2)} + f^{(1)}g^{(1)} + f^{(1)}g^{(1)} + f^{(2)}g$$
$$= fg^{(2)} + 2f^{(1)}g^{(1)} + f^{(2)}g$$

The process may be continued.

Differentiating a third time gives

$$(fg)^{(3)} = fg^{(3)} + f^{(1)}g^{(2)} + 2f^{(1)}g^{(2)} + 2f^{(2)}g^{(1)} + f^{(2)}g^{(1)} + gf^{(3)}$$
$$(fg)^{(3)} = fg^{(3)} + 3f^{(1)}g^{(2)} + 3f^{(2)}g^{(1)} + gf^{(3)}$$

Does this look familiar? Compare it to the binomial expansion formula

$$(a+b)^n = \sum_{r=0}^{n} {}^nC_r\, a^{n-r}b^r$$

Try to write a rule for the derivative of $(fg)^{(n)}$. This rule is known as **Leibniz's rule**.

Note

The brackets are used here to show that we are finding the nth derivative rather than the nth power.

This may be written as

$(f.g)^{(1)} = f.g^{(1)} + f^{(1)}g$

The dots indicate that we are taking the product of the functions rather than finding one as a function of the other.

"Nothing of worth or weight can be achieved with half a mind, with a faint heart, and with a lame endeavour."

\- Isaac Barrow

Challenge

Try using **Leibniz's rule** in reverse to integrate the expression

$$x^4e^x + 16x^3e^x + 72x^2e^x + 96xe^x + 24e^x$$

15 Assessment

1 Differentiate these expressions with respect to x

Select the correct answer in each case.

a $5\cos 2x$

A $-5\sin 4x$ B $-10\sin 2x$ C $5\cos 2$ D $10\sin 2x$ **[1 mark]**

b $x^3\ln x$

A $x^2(3\ln x+1)$ B $\frac{3x^2\ln x}{x}$ C $\frac{3x^2}{x}+\frac{\ln x}{x^3}$ D $3x^2\ln x+x^3$ **[1]**

c $\frac{e^x}{4x}$

A $\frac{4e^x(x-1)}{e^{2x}}$ B $\frac{e^x(1-x)}{4x^2}$ C $\frac{e^x(x+1)}{4x^2}$ D $\frac{e^x(x-1)}{4x^2}$ **[1]**

2 Calculate the exact gradient of the curve $y=\sin^2 x$ at the point $\frac{\pi}{3}$

Select the correct answer.

A 1.732 B $\frac{\sqrt{3}}{2}$ C 0.0365 D $\frac{-\sqrt{3}}{2}$ **[1]**

3 $y=3x^2-\frac{2}{\sqrt{x}}$

a Find $\frac{dy}{dx}$ **[2]**

b Calculate the rate of change of y with respect to x at the point where $x=4$ **[2]**

c Find the equation of the normal to the curve at the point where $x=1$ **[5]**

4 The volume (in m^3) of water in a tank at time t seconds is given by

$V=t-\frac{4}{t^2}$

At what time will the rate of change of the volume be $2\,m^3s^{-1}$? Show your working. **[4]**

5 Find the equation of the normal to $y=2\ln x$ at the point where $x=1$

Give your answer in the form $ax+by+c=0$ where a, b and c are integers. **[5]**

6 a Given that $y=xe^x$ find expressions for i $\frac{dy}{dx}$ ii $\frac{d^2y}{dx^2}$ **[4]**

b Write down an expression for the kth derivative of $y=xe^x$ **[1]**

7 $f(x)=3x\sin x$

a Calculate $f'(x)$ **[2]**

b Find an equation of the tangent to $y=f(x)$ at the point where $x=\pi$ **[3]**

8 a Differentiate these expressions with respect to x

i $\frac{x}{x+2}$ ii $\frac{3x^2}{\cos x}$ iii $(3x^3+5)e^x$ **[6]**

b Show that the derivative of $\frac{x^2+3x}{x-5}$ can be written as $\frac{ax^2+bx+c}{(x-5)^2}$, where a, b and c are constants to be found. **[3]**

9 Differentiate these expressions with respect to x

a $\cos 3x$ **[1]** b $2e^{3x}$ **[1]** c $\sin(2x-5)$ **[1]**

10 Given that $x = 2y^2 - 8\sqrt{y}$

a Find i $\frac{dx}{dy}$ ii $\frac{dy}{dx}$ in terms of y **[4]**

b Work out the equation of the normal at the point where $y = 4$ **[4]**

11 Find $\frac{dy}{dx}$ given that $5xy - y^3 = 7$ **[4]**

12 A curve is defined by the parametric equations $x = t^2, \quad y = 6t, \quad t > 0$

a Calculate the gradient of the curve when $y = 18$. Show your working. **[4]**

b Find the equation of the tangent at the point where $x = 4$ **[4]**

13 $y = 2x^2 e^x$

a Use calculus to find the stationary points of the curve. Show your working. **[6]**

b Classify each stationary point. **[4]**

14 A curve C has the equation $y = x^3 - 4x + 3$

a Identify and describe any points of inflection on the curve. **[4]**

b Sketch the curve C **[2]**

c For what values of x is the curve concave? **[1]**

15 A curve C has the equation $y = x^2e^x + e^x$

a Showing your working, find the stationary point on the curve and show that it is a point of inflection. **[5]**

b By considering $\frac{d^2y}{dx^2}$, show that the curve has another point of inflection. **[4]**

16 A cuboid has length twice its width as shown.

The volume of the cuboid is $192\,\text{cm}^3$

a Show that the surface area of the cuboid, S, is given by $S = 4x^2 + \frac{k}{x}$, where k is a constant to be found. **[5]**

b Find the minimum value of S, showing your working. **[5]**

c Use calculus to justify this is a minimum. **[3]**

17 Find and classify the stationary point on the curve $y = x^3 + 3x^2 + 3x - 5$. Show your working. **[5]**

18 Use differentiation from first principles to find the derivative of

a $\sin x$ **[4]** b $\cos 2x$ **[4]**

19 $f(x) = \tan x$

a Write down $f'(x)$ **[1]**

b Show that $f''(x) = 2\tan x \sec^2 x$ **[3]**

20 You are given that $y = 4^x$

a Write down $\frac{dy}{dx}$ **[1]**

b The tangent to the curve at the point $(-1, 0.25)$ cuts the x-axis at point A

Find the exact x-coordinate of A **[4]**

21 Given that $y = e^x \sin 5x$

a Calculate i $\frac{dy}{dx}$ ii $\frac{d^2y}{dx^2}$ **[6]**

b Find the smallest positive value of x to give a maximum value of y and prove it is a maximum. Show your working. **[5]**

22 a Differentiate the following with respect to t

i $\frac{e^{3t}}{t^2+1}$ ii $3t\ln t$ iii $e^{-t}\sin 4t$ **[8]**

b Find $g'(x)$, when $g(x)=2^x\tan x$ **[2]**

23 Given that $y=\sec x$

a Prove that $\frac{dy}{dx}=\sec x\tan x$ **[4]**

b Find $\frac{d^2y}{dx^2}$, giving your answer in terms of $\sec x$ only. **[3]**

24 The value of money in an account after t years is approximated by the formula

$V=ke^{0.03t}$, $k>0$ is a constant.

Given that £5000 is invested originally

a Work out the value in the account after 10 years, **[3]**

b Calculate the rate of increase in value when $t=5$. Show your working. **[3]**

25 A curve C has equation $x=y\ln y$

a Calculate $\frac{dy}{dx}$ at the point where $y=2$. Show your working. **[4]**

The normal to C at the point where $y=2$ intersects the y-axis at point A

b Find the exact y-coordinate of A. Show your working. **[5]**

26 Given that $x=\cos y$, show that $\frac{dy}{dx}=-\frac{1}{\sqrt{1-x^2}}$ **[4]**

27 Find an expression for the gradient in each case.

a $xe^y-\ln x=12$ **[4]** b $x\sin y-2x^2y=0$ **[4]**

28 A curve C has equation $\sin x+\frac{1}{2}\tan y=\sqrt{3}$, $0\le x, y\le\pi$

Find $\frac{dy}{dx}$ when $x=\frac{\pi}{3}$ **[7]**

29 A curve has parametric equations $x=3t+7$, $y=2+\frac{3}{t}$, $t\neq 0$

a Find the gradient of the curve at the point where $x=10$. Show your working. **[4]**

b Find a Cartesian equation of the curve in the form $y=f(x)$ where $f(x)$ is expressed as a single fraction. **[3]**

30 A curve is defined by the parametric equations $x=2\sin t$, $y=3\cos t$

a Find an expression for $\frac{dy}{dx}$ **[3]**

b Work out the equation of the tangent at the point when $t=\frac{2\pi}{3}$

Give your answer in the form $a\sqrt{3}x+by+c=0$, where a, b and c are integers. **[4]**

31 Find and classify the stationary point of the curve $y=x^4-3$. Show your working. **[5]**

32 Use calculus to determine if the following functions are convex or concave.

a $f(x)=3x^2-7x+8$ **[3]** b $g(x)=(2-x)^4$ **[4]**

33 Find the range of values of x for which the function $f(x)=\ln(x^2+1)$ is concave. Show your working. **[5]**

34 A sector of a circle has radius r and angle θ as shown. The arc length is 6 cm.

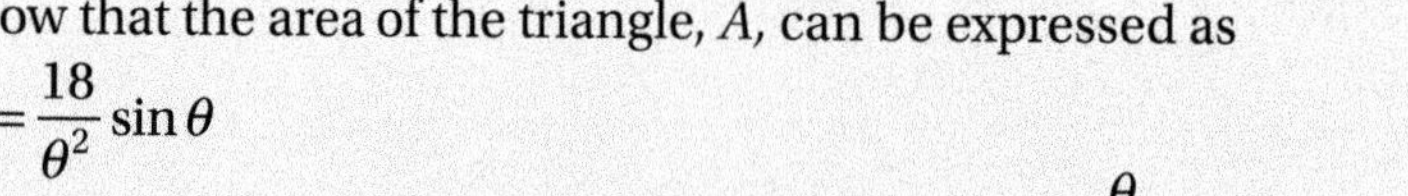

a Show that the area of the triangle, A, can be expressed as

$A=\dfrac{18}{\theta^2}\sin\theta$ **[5]**

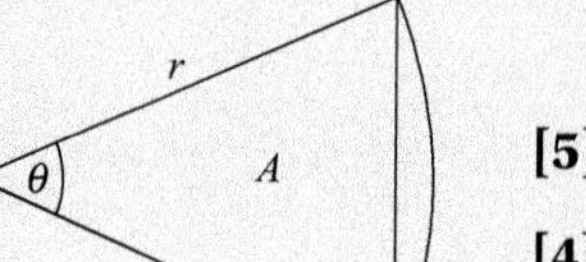

b Show that the stationary point occurs when $\tan\theta=\dfrac{\theta}{2}$ **[4]**

35 $f(x)=x-\ln(2x-3),\ x>\dfrac{3}{2}$

The tangent to the curve $y=f(x)$ at $x=3$ intersects the x-axis at point P

Find the x-coordinate of P, giving your answer in the form $a+\ln b$. Show your working. **[7]**

36 Given that $y=\operatorname{cosec}3x$

a Show that $\dfrac{dy}{dx}=A\cot3x\operatorname{cosec}3x$, where A is a constant to be found, **[4]**

b Find $\dfrac{d^2y}{dx^2}$ **[4]**

37 Given that $y=\arctan x$

a Find $\dfrac{dy}{dx}$ **[4]**

b Find an equation of the tangent to $y=\arctan x$ at the point where $x=1$ **[3]**

The tangent intersects the x-axis at point A and the y-axis at point B

c Show that the area of triangle OAB is $\dfrac{1}{16}(\pi^2-4\pi+4)$ **[5]**

38 A cube has side length x. The volume of the cube is increasing at a rate of 12 $cm^3 s^{-1}$

Find the rate at which x is increasing when the volume is 216 cm^3 **[6]**

39 The volume of a spherical balloon, V cm^3, is increasing at a constant rate of 6 $cm^3 s^{-1}$

Find the rate at which the radius of the sphere is increasing when the volume is 36π cm^3

Leave your answer in exact form. $\left[V=\dfrac{4}{3}\pi r^3\right]$ **[5]**

40 Prove that the derivative of $\arcsin2x$ is $\dfrac{2}{\sqrt{1-4x^2}}$ **[6]**

41 $xy^2+2y=3x^2$

a Find an expression in terms of x and y for $\dfrac{dy}{dx}$ **[4]**

b Calculate the possible rates of change of y with respect to x when $y=1$ **[5]**

42 Use implicit differentiation to prove that the derivative of a^x is $a^x\ln a$ **[4]**

43 Given that $x=\dfrac{2}{3-t},\quad y=\dfrac{t^2}{3-t},\quad t\neq3$

a Show that $\dfrac{dy}{dx}=\dfrac{6t-t^2}{2}$ **[5]**

b Find a Cartesian equation in the form $y=f(x)$. Simplify your answer. **[3]**

44 A curve C is defined by the parametric equations $x=\sec(\theta-4),\quad y=\tan(\theta-4)$

a Show that $\dfrac{dy}{dx}=\operatorname{cosec}(\theta-4)$ **[5]**

b Find a Cartesian equation of C **[2]**

c Show that the equation of the tangent to C at the point where $x=3$ and y is positive is given by $3x-2y\sqrt{2}=1$ **[6]**

16 Integration and differential equations

The study of the motion of objects such as planets, moons and stars is referred to as celestial mechanics. This motion is modelled using differential equations constructed by applying Newton's second law to the inverse square law of gravity. The orbit of Uranus was observed for decades until an accurate theory of motion could be constructed. However, the observed motion did not match the motion predicted by the differential equations when they were compared. This was because the motion of Uranus was being affected by an unknown planet, later to be discovered as Neptune. Once Neptune's motion had been observed, differential equations could be used to make an accurate model of Uranus's orbit.

Differential equations enable the modelling of complex systems of motion. Integration is the reverse of differentiation, and is a tool required to solve differential equations. This chapter introduces several techniques which can be applied to many different scenarios involving integration.

Orientation

What you need to know

Ch4 Differentiation and integration

- Introduction to integration.
- Area under a curve.

p.24

Ch12 Algebra 2

- Partial fractions.

p.106

Ch15 Differentiation

- The product rule for differentiation.

What you will learn

- To integrate a set of standard functions, $f(x)$ and the related functions, $f(ax + b)$
- To find the area between curves.
- To integrate by substitution, by parts and by using partial fractions.
- To understand the expression 'differential equation'.
- To use integration where the variables are separable.

What this leads to

Applications
Financial modelling.
Demographic modelling.
Epidemiology.

Further Maths Ch4
Mean values.
Volumes of revolution.

MyMaths Practise before you start 2054, 2056, 2163, 2260

16.1 Standard integrals

Fluency and skills

See Ch4.7 For a reminder on finding the area under a curve.

You can use definite integration to find the area under a curve.

Let A be the area between the curve $y = f(x)$, the x-axis, and the lines $x = a$ and $x = b$. The area of this region can be approximated by dividing it up into rectangles, finding the area of each rectangle, and adding the areas together.

See Ch4.2 For a reminder on Leibniz notation.

Suppose that the area is divided into n rectangles, each of width δx. The height of each rectangle is given by the value of y at the left-hand side of the rectangle.

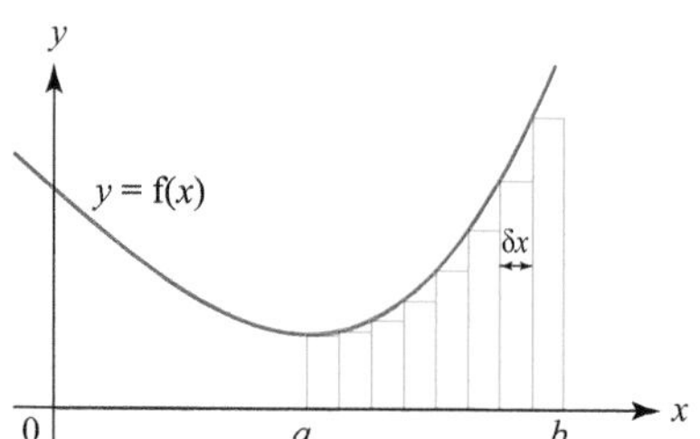

Summing these areas, you get $A \approx y_1\delta x + y_2\delta x + \ldots + y_n\delta x$

This can be written using sigma notation $A \approx \sum_{i=1}^{n} y_i \delta x$

To make this approximation more accurate, you must increase the number of rectangles between a and b. In other words, in order to find an accurate value of A, you let the number of rectangles, n, tend to infinity.

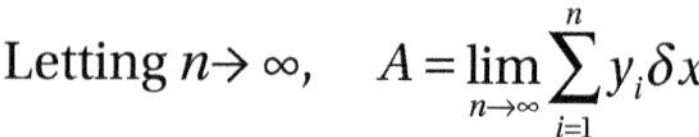

Letting $n \to \infty$, $A = \lim_{n\to\infty} \sum_{i=1}^{n} y_i \delta x$

As the number of rectangles is increased, the limit is equal to the definite integral.

Key point

$$\lim_{n\to\infty} \sum_{i=1}^{n} y_i \delta x = \int_a^b y \, dx$$

ICT Resource online

To investigate the integrals of trigonometric functions, click this link in the digital book.

Example 1

Consider the curve $y = 3x^2 + 2x + 1$. The area under the curve and between the lines $x = 1$ and $x = 3$ can be estimated by calculating and adding the area of 4 rectangles of equal width, δx

a What is the value of δx in this case?

b Use the total area of the four rectangles, $\sum_{i=1}^{4} y_i \delta x$, to find an estimate for this area.

c Calculate the exact area under the curve, showing your working.

a $\delta x = \dfrac{3-1}{4} = \dfrac{1}{2}$

To find δx, divide the total width of the interval by the number of rectangles.

b When $x = 1, y = 6$

When $x = 1.5, y = 10.75$

When $x = 2, y = 17$

When $x = 2.5, y = 24.75$

To find each rectangle height, calculate the y-value at the left-hand side of each rectangle.

$\sum_{i=1}^{4} y_i \delta x = \left(6 \times \frac{1}{2}\right) + \left(10.75 \times \frac{1}{2}\right) + \left(17 \times \frac{1}{2}\right) + \left(24.75 \times \frac{1}{2}\right)$

$= 29.25$

(*Continued on the next page*)

c $\int_1^3 (3x^2+2x+1)dx = [x^3+x^2+x]_1^3$

$= (3^3+3^2+3)-(1^3+1^2+1)$

$= 27+9+3-3$

$= 36$

Find the exact area using integration. Remember that you don't need to worry about the constant of integration when calculating a definite integral.

Substitute the limits and evaluate.

Integration is the reverse of differentiation. If the derivative of a function F(x) is f(x), then you can say that an indefinite integral of f(x) with respect to x is F(x)

Key point

$F'(x) = f(x) \Rightarrow \int f(x)dx = F(x) + c$

where c is the constant of integration.

An function to be integrated, i.e. $\int f(x)dx$, is sometimes called an integrand.

Using this definition, you can immediately work out a number of **standard integrals**.

Key point

1. $\frac{d}{dx}\left(\frac{1}{n+1}x^{n+1}\right) = x^n \Rightarrow \int x^n dx = \frac{1}{n+1}x^{n+1} + c,\ n \neq -1$
2. $\frac{d}{dx}(e^x) = e^x \Rightarrow \int e^x dx = e^x + c$
3. $\frac{d}{dx}(\ln x) = \frac{1}{x} \Rightarrow \int \frac{1}{x} dx = \ln|x| + c$
4. $\frac{d}{dx}(\sin x) = \cos x \Rightarrow \int \cos x dx = \sin x + c$
5. $\frac{d}{dx}(-\cos x) = \sin x \Rightarrow \int \sin x dx = -\cos x + c$
6. $\frac{d}{dx}(\tan x) = \sec^2 x \Rightarrow \int \sec^2 x dx = \tan x + c$

Example 2

Integrate $3\sin x + 2\cos x$ with respect to x

$\int 3\sin x + 2\cos x dx = 3\int \sin x dx + 2\int \cos x dx = -3\cos x + 2\sin x + c$

Example 3

Find **a** $\int \frac{1}{x} dx$ **b** $\int e^x + \sec^2 x dx$ **c** $\int \left(\cos x + \frac{3}{2x}\right) dx$

a $\int \frac{1}{x} dx = \ln|x| + c$

Use standard integral 3 from above. See Ch12.2 for a reminder on modulus notation.

b $\int e^x + \sec^2 x dx = \int e^x dx + \int \sec^2 x dx$

$= e^x + \tan x + c$

Use standard integrals 2 and 6 from above.

c $\int (\cos x + \frac{3}{2x}) dx = \int \cos x dx + \frac{3}{2}\int \frac{1}{x} dx$

$= \sin x + \frac{3}{2}\ln|x| + c$

Use standard integrals 3 and 4 from above.

Exercise 16.1A Fluency and skills

1 Find

a $\int 24x^5\,dx$ b $\int 7\sin x\,dx$

c $\int 4\cos x\,dx$ d $\int \frac{2}{3}\sec^2 x\,dx$

e $\int 5e^x\,dx$ f $\int 2x+\sqrt{x}\,dx$

g $\int \frac{3}{x^2}\,dx$ h $\int \frac{3}{x}\,dx$

i $\int \sin x+\sec^2 x\,dx$ j $\int 4\cos x-3\sin x\,dx$

k $\int 3\sec^2 x+2\cos x\,dx$ l $\int 5e^x\,dx$

2 Find

a $\int 3-4e^x\,dx$ b $\int \frac{1}{2x}\,dx$

c $\int \frac{1}{3x}+\frac{2}{5x}\,dx$ d $\int 1-3e^x\,dx$

e $\int 7-6\sin x\,dx$ f $\int 4x^3-\frac{5}{x}\,dx$

g $\int 1-2\cos x\,dx$ h $\int x-\frac{1}{x^3}\,dx$

i $\int \frac{1}{e}+\frac{e^x}{3}\,dx$ j $\int \frac{3}{4}+\frac{2}{x^3}-\frac{\sin x}{3}\,dx$

k $\int 4x^{-2}-5\sqrt{x}\,dx$ l $\int 1+x^{\frac{2}{3}}\,dx$

3 Evaluate the following integrals. Show your working.

a $\int_4^9 \frac{1}{\sqrt{x}}\,dx$ b $\int_{\frac{\pi}{2}}^{\pi} \sin x\,dx$

c $\int_{\frac{\pi}{6}}^{\frac{\pi}{3}} \cos x\,dx$ d $\int_{-\frac{\pi}{4}}^{\frac{\pi}{4}} \sec^2 x\,dx$

e $\int_0^1 e^x\,dx$ f $\int_e^{e^3} \frac{1}{x}\,dx$

g $\int_2^4 x+\frac{8}{x^2}\,dx$ h $\int_1^4 1-2x\,dx$

4 Evaluate the following integrals. Show your working.

a $\int_0^5 x^2+2x+1\,dx$ b $\int_{\frac{\pi}{2}}^{\pi} 1-\cos x\,dx$

c $\int_{\frac{\pi}{6}}^{\frac{\pi}{3}} 2\sin x+3\,dx$ d $\int_{-\frac{\pi}{4}}^{\frac{\pi}{4}} \sin x+\sec^2 x\,dx$

e $\int_0^1 x+e^x\,dx$ f $\int_{e^2}^{e^4} \frac{2}{3x}\,dx$

5 For each of the two following curves

i Split the area under the curve into four rectangles of equal width and find an approximation for each shaded area.

ii Write down a definite integral which, when evaluated, will give the exact shaded area.

iii Use integration to calculate the exact area. Show your working.

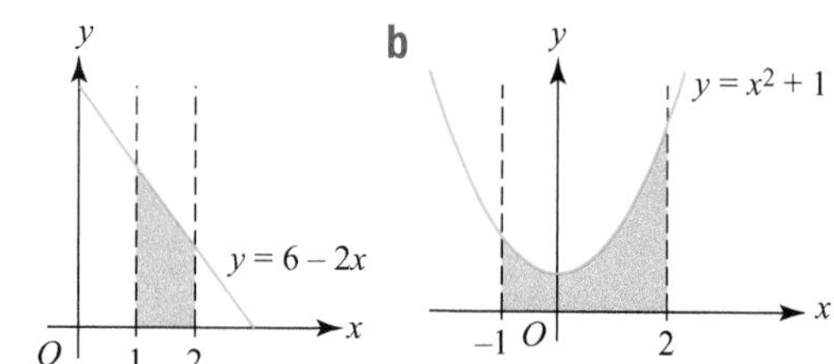

6 For each of the curves below

i Split the area under the curve into six rectangles of equal width and find an approximation for each shaded area.

ii Write down a definite integral which, when evaluated, will give the exact shaded area.

iii Use integration to calculate the exact area. Show your working.

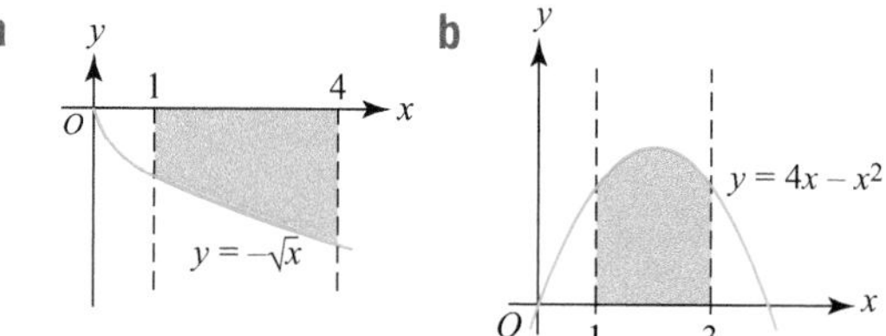

7 Find the shaded area between the curve $y=1+2\sin x$, the x-axis and the lines $x=\frac{\pi}{3}$ and $x=\pi$ as shaded in the diagram.

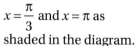

8 The graph shows the function $y=\frac{1}{2x}$ in the domain $1\le x\le 9$

Calculate the shaded area between the curve, the x-axis and the lines $x=3$ and $x=3e$. Show your working.

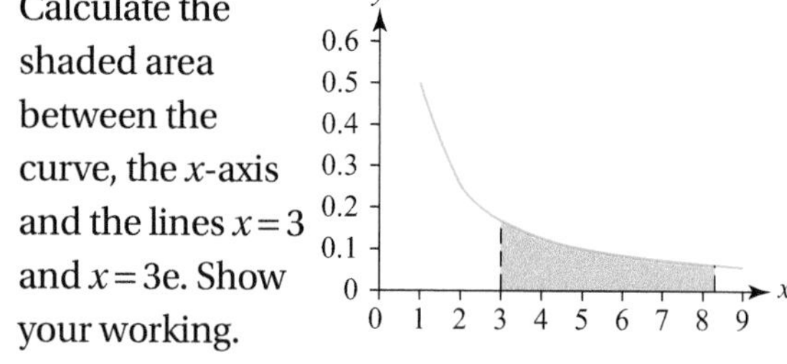

Reasoning and problem-solving

To find the area trapped between two curves, you subtract the area below the bottom curve from the area below the top curve.

If $f(x) \geq g(x)$ or $f(x) \leq g(x)$ for all x in the interval $a \leq x \leq b$, then the area between the two curves $y = f(x)$ and $y = g(x)$ and the lines $x = a$ and $x = b$ is given by

$$A = \left| \int_a^b f(x) - g(x)\,dx \right|$$

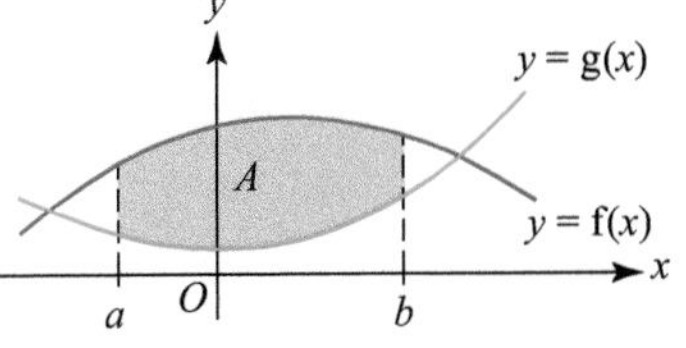

The formula for the area between two curves holds true even when the trapped area straddles the x-axis.

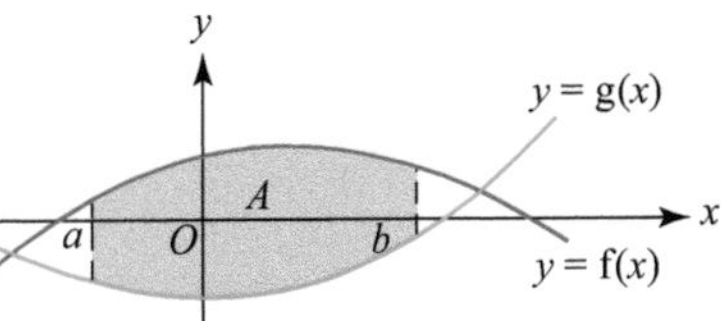

If the curves intersect in the interval, then each trapped area must be considered separately.

The total area is $A + B = \left| \int_a^c f(x) - g(x)\,dx \right| + \left| \int_c^b f(x) - g(x)\,dx \right|$

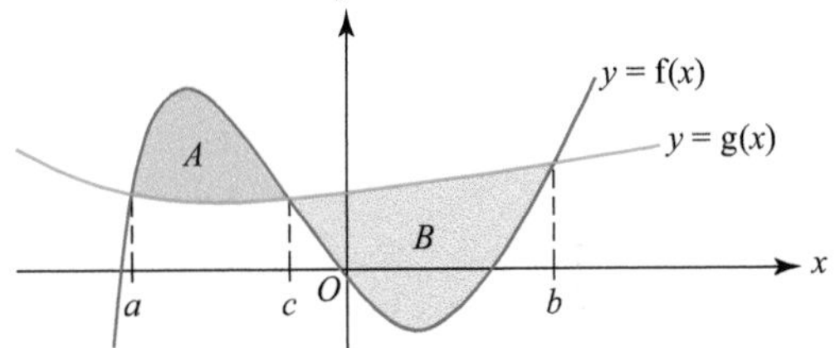

Strategy

To find the area trapped between two curves in an interval

1. Identify where the curves intersect. This will determine the limits of integration.
2. Draw a sketch to help you plan which integrals will need to be evaluated.
3. Integrate and substitute limits.
4. Where necessary, find the total area by addition.

Example 4

The curve $y = x^2 + x + 4$ intersects the curve $y = -x^2 + 2x + 5$ at the points $\left(-\frac{1}{2}, 3\frac{3}{4}\right)$ and $(1, 6)$

Find the exact area between the two curves. Show your working.

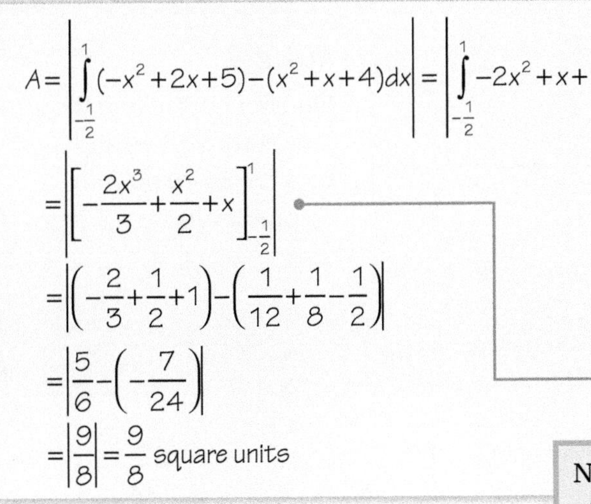

$$A = \left| \int_{-\frac{1}{2}}^{1} (-x^2 + 2x + 5) - (x^2 + x + 4)\,dx \right| = \left| \int_{-\frac{1}{2}}^{1} -2x^2 + x + 1\,dx \right|$$

$$= \left| \left[-\frac{2x^3}{3} + \frac{x^2}{2} + x \right]_{-\frac{1}{2}}^{1} \right|$$

$$= \left| \left(-\frac{2}{3} + \frac{1}{2} + 1\right) - \left(\frac{1}{12} + \frac{1}{8} - \frac{1}{2}\right) \right|$$

$$= \left| \frac{5}{6} - \left(-\frac{7}{24}\right) \right|$$

$$= \left| \frac{9}{8} \right| = \frac{9}{8} \text{ square units}$$

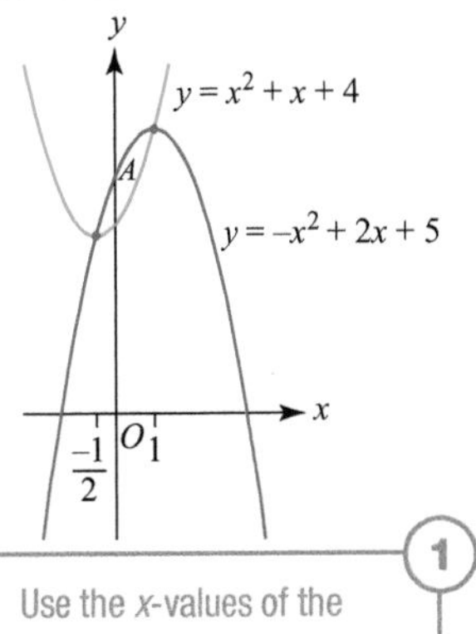

1 Use the x-values of the points of intersection as the limits of integration.

3 Integrate and substitute the limits.

Note that the same result is same regardless of which area is subtracted from which.

Example 5

Find the area trapped between the curves $y=x^3-2x^2+2x-2$ and $y=x^2-2$. Show each step of working.

$x^3-2x^2+2x-2=x^2-2$

$x^3-3x^2+2x=0$

$x(x-1)(x-2)=0$

$x=0$ or $x=1$ or $x=2$

$y=-2$ or $y=-1$ or $y=2$ respectively.

Equate the curves.

1 Identify where the curves intersect. These x-values will be used as the limits for integration.

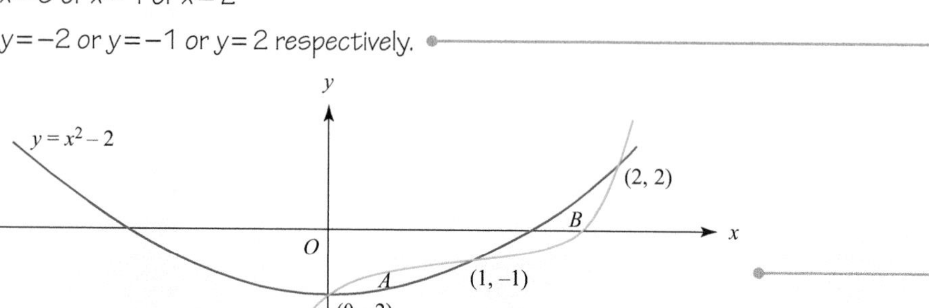

2 Draw a sketch. You could check your sketch using a graphics calculator.

$$A+B=\left|\int_0^1(x^3-2x^2+2x-2)-(x^2-2)dx\right|+\left|\int_1^2(x^3-2x^2+2x-2)-(x^2-2)dx\right|$$

$$=\left|\int_0^1 x^3-3x^2+2x\,dx\right|+\left|\int_1^2 x^3-3x^2+2x\,dx\right|$$

$$=\left|\left[\frac{1}{4}x^4-x^3+x^2\right]_0^1\right|+\left|\left[\frac{1}{4}x^4-x^3+x^2\right]_1^2\right|$$

$$=\left|\left(\frac{1}{4}-1+1\right)-0\right|+\left|(4-8+4)-\left(\frac{1}{4}-1+1\right)\right|$$

$$=\left|\frac{1}{4}\right|+\left|-\frac{1}{4}\right|=\frac{1}{2}$$

Trapped area = 0.5 square units

3 Integrate and substitute the limits.

4 Find the total area by addition.

Calculator

Try it on your calculator

You can find the area between two curves on a graphics calculator.

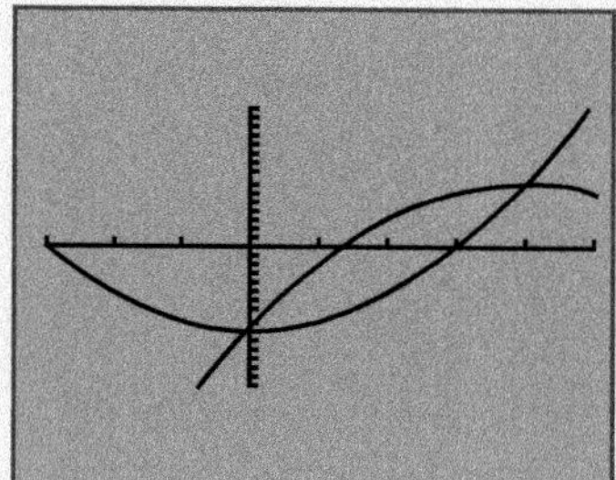

Activity

Find out how to find the area between $y=x^2-9$ and $y=-x^2+8x-9$ on *your* graphics calculator.

Exercise 16.1B Reasoning and problem-solving

1 Find

a $\int\frac{x^2}{\sqrt{x}}dx$ b $\int\frac{x+1}{3x}dx$

c $\int x^2(2x+3)dx$ d $\int\sqrt{x}\left(x+2\sqrt{x}\right)dx$

e $\int\frac{x^2+2x+1}{x}dx$ f $\int\frac{x+\sqrt{x}}{\sqrt{x}}dx$

g $\int\frac{1+\cos^3 x}{\cos^2 x}dx$ h $\int\frac{1+\sin x\cos^2 x}{\cos^2 x}dx$

i $\int e^x(e^x+1)dx$ j $\int (e^x+4)(e^x-1)dx$

k $\int (e^x+e^{-x})(e^x-e^{-x})dx$

l $\int \cos x \tan x\,dx$ m $\int \sin^2 x+\cos^2 x\,dx$

2 Each pair of curves traps a single region. Calculate the area of each region, showing your working.

a $y=x^2+2x+1$ and $y=x+3$

b $y=2x^2+12x-11$ and $y=x+10$

c $y=x^2+5x+6$ and $y=1-6x-x^2$

d $y=x^2$ and $y=\sqrt{x}$

3 a The curves $y=\sin x$ and $y=\cos x$ intersect at two points in the domain $0\le x\le 2\pi$. Find the two points, showing your working.

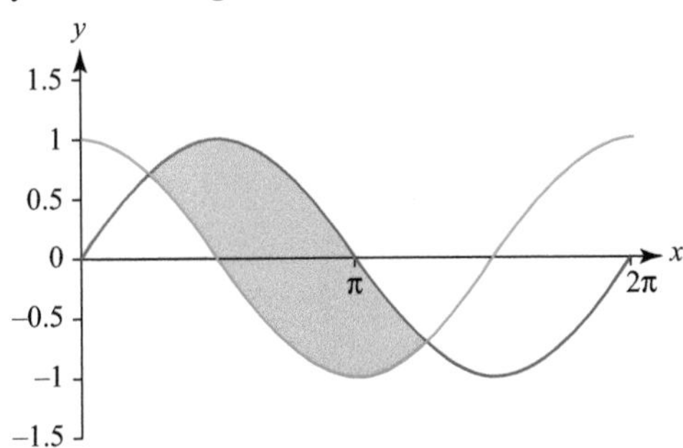

b Find the area trapped between the two curves between these two points, showing your working.

c What is the area of the region trapped between the curves and the y-axis?

4 Each pair of curves traps two distinct regions. Find the total area trapped in each case. Show your working.

a $y=x^3$ and $y=2x^2+x-2$

b $y=x^3+2x^2-3x$ and $y=x^2-x$

c $y=x^4+x^3-x^2$ and $y=2x^2-x^3$

d $y=x^2-6x+11$ and $y=\dfrac{6}{x}$

5 The cross-section of a riverbed can be represented by the equation $y=-2x(-x+4)$, for $0\le x\le 4$. The river flows at a rate of $0.4\ \text{m s}^{-1}$. Assuming the river water reaches the top of the bed, calculate the volume of water that flows past a given point in 1 minute. Show your working.

6 Two cars, A and B, enter a race to see which can travel the furthest in a straight line over 4 seconds. The speed, v, with respect to time, of each car during these 4 seconds is given by the equations

$v_A=4\sqrt{t}$

$v_B=t^2-\left(\dfrac{t}{2}\right)^3$

a Show that the velocity of each car is the same after 4 seconds.

b Show that car A wins the race by a distance of 8 metres. (Hint: the area under a velocity–time graph gives the distance travelled)

Challenge

7 a Expand

i $(x+1)^2$ ii $(x+1)^3$

iii $(x+1)^4$ iv $(x+1)^5$

b i Use the expansion of $(x+1)^2$ to find $\int (x+1)^2\,dx$

ii Rewrite c, the constant of integration, as $\dfrac{1}{3}+c_1$

iii Take out a common factor of $\dfrac{1}{3}$, collecting all terms except c_1

iv Simplify.

c i Use the expansion of $(x+1)^3$ to find $\int (x+1)^3\,dx$

ii Rewrite c, the constant of integration, as $\dfrac{1}{4}+c_1$

iii Take out a common factor of $\dfrac{1}{4}$, collecting all terms except c_1

iv Simplify.

d Suggest a general rule for $\int (x+1)^n\,dx$

e Investigate $\int (ax+1)^n\,dx$

16.2 Integration by substitution

Fluency and skills

Say you are presented with following integral.

$\int (2x+1)^5 \, dx$

Without expanding out 5 sets of brackets, the function would be difficult to integrate with the techniques you've learnt so far. But if you substitute the function inside brackets with a u, then integrating u^5 is actually fairly simple. This technique is known as **integration by substitution**.

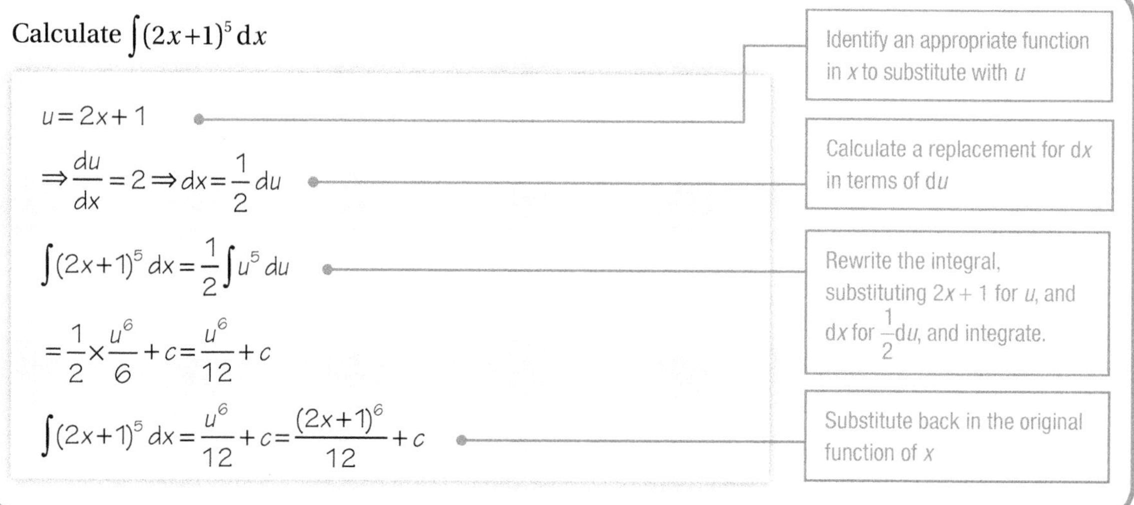

Example 1
Calculate $\int (2x+1)^5 \, dx$
$u = 2x+1$
$\Rightarrow \frac{du}{dx} = 2 \Rightarrow dx = \frac{1}{2} du$
$\int (2x+1)^5 \, dx = \frac{1}{2} \int u^5 \, du$
$= \frac{1}{2} \times \frac{u^6}{6} + c = \frac{u^6}{12} + c$
$\int (2x+1)^5 \, dx = \frac{u^6}{12} + c = \frac{(2x+1)^6}{12} + c$
Identify an appropriate function in x to substitute with u
Calculate a replacement for dx in terms of du
Rewrite the integral, substituting $2x+1$ for u, and dx for $\frac{1}{2} du$, and integrate.
Substitute back in the original function of x

See Ch15.5 For a reminder on the chain rule.

You may notice that this technique is essentially just the chain rule in reverse, so it's useful when you can spot a function and its derivative within the integral.

If the derivative of the function is, or contains, a constant, you can compensate by multiplying the integral by the reciprocal of that constant: in Example 1, the function is $2x+1$ and the derivative of that is 2, which is why you need to multiply the substituted integral by $\frac{1}{2}$

Try using the chain rule, where $u = 2x+1$, to differentiate $y = \frac{(2x+1)^6}{12}$:

$$\frac{dy}{dx} = \frac{dy}{du} \times \frac{du}{dx} = \frac{6u^5}{12} \times 2 = \frac{12}{12} u^5 = (2x+1)^5$$

This calculation shows Example 1 in reverse.

Some functions can be integrated quickly without the need for much working out. For example, the following result can be deduced using the integration by substitution technique.

Key point

In general, when $\int f(x) \, dx = F(x) + c$

$$\int f(ax+b) \, dx = \frac{1}{a} F(ax+b) + c$$

Example 2

Find **a** $\int \sin(3x+4)\,\mathrm{d}x$ **b** $\int \frac{1}{4x+3}\,\mathrm{d}x$ **c** $\int e^{1-5x}\,\mathrm{d}x$

a $\int \sin(3x+4)\,dx = -\frac{1}{3}\cos(3x+4)+c$

b $\int \frac{1}{4x+3}\,dx = \frac{1}{4}\ln|4x+3|+c$

c $\int e^{1-5x}\,dx = -\frac{1}{5}e^{1-5x}+c$

Use $\int f(ax+b)\,dx = \frac{1}{a}F(ax+b)+c$

When using substitution to find a **definite integral**, you substitute the corresponding values of u for the values of x in the limits. By doing this, you do not need to make the substitution back into the original variable after integration.

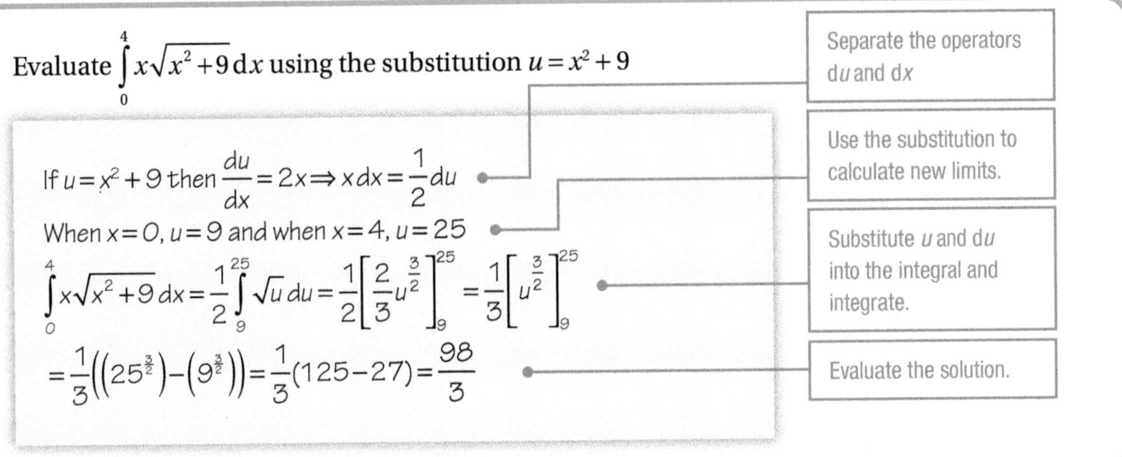

Example 3
Evaluate $\int_0^4 x\sqrt{x^2+9}\,\mathrm{d}x$ using the substitution $u=x^2+9$
If $u=x^2+9$ then $\frac{du}{dx}=2x \Rightarrow x\,dx=\frac{1}{2}du$
When $x=0$, $u=9$ and when $x=4$, $u=25$
$\int_0^4 x\sqrt{x^2+9}\,dx=\frac{1}{2}\int_9^{25}\sqrt{u}\,du=\frac{1}{2}\left[\frac{2}{3}u^{\frac{3}{2}}\right]_9^{25}=\frac{1}{3}\left[u^{\frac{3}{2}}\right]_9^{25}$
$=\frac{1}{3}\left(\left(25^{\frac{3}{2}}\right)-\left(9^{\frac{3}{2}}\right)\right)=\frac{1}{3}(125-27)=\frac{98}{3}$
Separate the operators du and dx
Use the substitution to calculate new limits.
Substitute u and du into the integral and integrate.
Evaluate the solution.

Exercise 16.2A Fluency and skills

1 Find

a i $\int \frac{1}{\sqrt{2x}}\,\mathrm{d}x$ ii $\int \frac{2}{\sqrt{4x-1}}\,\mathrm{d}x$

b i $\int \sin 4x\,\mathrm{d}x$ ii $\int \sin(5x-2)\,\mathrm{d}x$

c i $\int \cos 3x\,\mathrm{d}x$ ii $\int 3\cos(2-3x)\,\mathrm{d}x$

d i $\int \sec^2 5x\,\mathrm{d}x$ ii $\int \sec^2(4+3x)\,\mathrm{d}x$

e i $\int e^{7x}\,\mathrm{d}x$ ii $\int 4e^{-2x}\,\mathrm{d}x$

f i $\int \frac{1}{2x}\,\mathrm{d}x$ ii $\int \frac{2}{4x+5}\,\mathrm{d}x$

2 State the integral of

a $(3x+2)^{10}$ b $(5x-1)^8$

c $(7x-3)^{100}$ d $(3x-8)^{-8}$

e $(1-3x)^9$ f $(6-x)^7$

g $\frac{3}{(2x-1)^5}$ h $\frac{1}{(10-x)^5}$

i $\sin(3-5x)$ j $2\cos(4x-1)$

k $\cos(2x)$ l $\sec^2(4x+3)$

m $3\sec^2(2x+1)$ n e^{5x+2}

o $\frac{7}{3x+9}$ p $\frac{4}{8-x}$

3 Find each integral. A suitable substitution has been suggested.

a $\int 2x(x^2+2)^4\,\mathrm{d}x$; let $u=x^2+3$

b $\int (2x+1)(x^2+x-1)^3\,\mathrm{d}x$; let $u=x^2+x-1$

c $\int 6x^2\sqrt{2x^3-1}\,\mathrm{d}x$; let $u=2x^3-1$

d $\int x(2x^2-5)^3\,\mathrm{d}x$; let $u=2x^2-5$

e $\int \frac{x}{(x^2-7)^4}\,\mathrm{d}x$; let $u=x^2-7$

f $\int \frac{2x+3}{(x^2+3x-1)^2}\,\mathrm{d}x$; let $u=x^2+3x-1$

MyMaths 2167, 2169, 2216, 2219 SEARCH

g $\int \frac{3x}{\sqrt{1-x^2}}\,dx$; let $u = 1 - x^2$

h $\int \frac{1}{\sqrt[3]{x+4}}\,dx$; let $u = x + 4$

i $\int x\sin(3x^2+1)\,dx$; let $u = 3x^2 + 1$

j $\int \sec^2 x\tan^3 x\,dx$; let $u = \tan x$

k $\int \frac{\cos x}{\sin x - 1}\,dx$; let $u = \sin x$

l $\int \frac{\sin x}{3 - 2\cos x}\,dx$; let $u = 3 - 2\cos x$

m $\int x e^{2x^2+1}\,dx$; let $u = 2x^2 + 1$

n $\int \frac{e^{\sqrt{x}}}{\sqrt{x}}\,dx$; let $u = \sqrt{x}$

o $\int \sin x\, e^{\cos x}\,dx$; let $u = \cos x$

p $\int \frac{\ln x}{x}\,dx$; let $u = \ln x$

4 Find each integral using a suitable substitution.

a $\int x(2x^2-1)^3\,dx$

b $\int (x^2+x)\left(\frac{x^3}{3}+\frac{x^2}{2}\right)^{\frac{1}{2}}dx$

c $\int \frac{x^2-2x}{\sqrt{x^3-3x^2}}\,dx$

d $\int (x-1)\sin(x^2-2x+3)\,dx$

e $\int \frac{x^2+2x+3}{x^3+3x^2+9x+1}\,dx$

f $\int \frac{\sin\sqrt{x}}{\sqrt{x}}\,dx$ g $\int \frac{x+\cos\sqrt{x+1}}{\sqrt{x+1}}\,dx$

h $\int \frac{1}{\sqrt[3]{2x-1}}\,dx$ i $\int \sin x(1-\cos^3 x)\,dx$

j $\int \sec^2 x(1+\tan^2 x)\,dx$

k $\int \frac{\cos x}{\sqrt{1+\sin x}}\,dx$ l $\int \frac{e^x}{1+e^x}\,dx$

m $\int \sin 2x\, e^{\cos 2x}\,dx$ n $\int \frac{\sqrt{\ln(2x+3)}}{2x+3}\,dx$

5 Evaluate the following integrals. Show your working.

a $\int_1^7 2x\sqrt{x^2+15}\,dx$ b $\int_1^4 \frac{(\sqrt{x}-1)^2}{2\sqrt{x}}\,dx$

c $\int_0^1 (x-1)(x^2-2x-1)^3\,dx$

d $\int_1^2 \frac{2x+3}{x^2+3x}\,dx$ e $\int_5^{11} \frac{x}{\sqrt{x^2-21}}\,dx$

f $\int_1^2 \frac{x-2}{x^2-4x+1}\,dx$ g $\int_0^{\frac{\pi}{6}} \frac{\sin x}{\cos^2 x}\,dx$

h $\int_0^{\frac{\pi}{6}} \frac{\sin x}{\sqrt{\cos x}}\,dx$ i $\int_{\ln 3}^{\ln 8} e^x\sqrt{e^x+1}\,dx$

Reasoning and problem-solving

Strategy

To integrate functions with terms involving $\sin^2 x$ and $\cos^2 x$

1. Use the basic identities to express $\sin^2 x$ or $\cos^2 x$ in terms of $\cos 2x$
2. Integrate using the substitution method.

Example 4

Find $\int (1-3\sin^2 x)\,dx$

Recall that $\sin^2 x = \frac{1}{2}(1-\cos 2x)$

$$\int (1-3\sin^2 x)\,dx = \int\left(1-3\times\frac{1}{2}(1-\cos 2x)\right)dx$$

$$= \int\left(-\frac{1}{2}+\frac{3}{2}\cos 2x\right)dx$$

1 Express $\sin^2 x$ in terms of $\cos 2x$

$$= -\frac{1}{2}x+\frac{3}{2}\times\frac{1}{2}\sin 2x + c$$

2 Integrate using substitution. Remember that the integral of $f(ax+b)$ is $\frac{1}{a}F(ax+b)$, where in this case the function is cos, $a = 2$ and $b = 0$

$$= -\frac{1}{2}x+\frac{3}{4}\sin 2x + c$$

Example 5

Find $\int \cos^5 x\,dx$

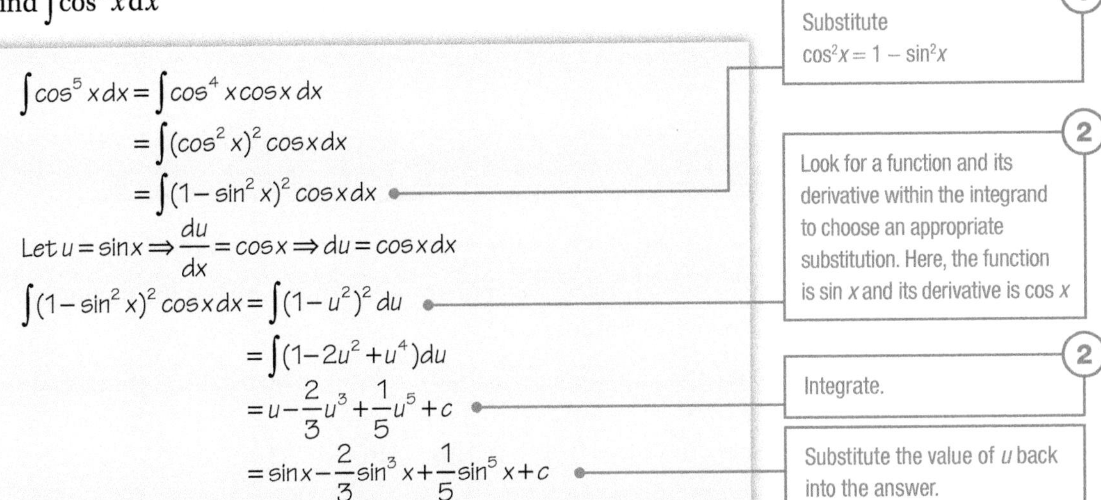

$\int \cos^5 x\,dx = \int \cos^4 x \cos x\,dx$
$= \int (\cos^2 x)^2 \cos x\,dx$
$= \int (1-\sin^2 x)^2 \cos x\,dx$
Let $u = \sin x \Rightarrow \frac{du}{dx} = \cos x \Rightarrow du = \cos x\,dx$
$\int (1-\sin^2 x)^2 \cos x\,dx = \int (1-u^2)^2\,du$
$= \int (1-2u^2+u^4)\,du$
$= u - \frac{2}{3}u^3 + \frac{1}{5}u^5 + c$
$= \sin x - \frac{2}{3}\sin^3 x + \frac{1}{5}\sin^5 x + c$
1
Substitute $\cos^2 x = 1 - \sin^2 x$
2
Look for a function and its derivative within the integrand to choose an appropriate substitution. Here, the function is $\sin x$ and its derivative is $\cos x$
2
Integrate.
Substitute the value of u back into the answer.

Exercise 16.2B Reasoning and problem-solving

1 a Find

i $\int \sin^2 x\,dx$ ii $\int \cos^2 x\,dx$

b Hence state

i $\int \sin^2 2x\,dx$ ii $\int \cos^2 2x\,dx$

iii $\int \sin^2(2x-1)\,dx$ iv $\int \cos^2(2x+3)\,dx$

v $\int \sin^2(1-x)\,dx$ vi $\int \cos^2(1-2x)\,dx$

2 Make use of trigonometric identities to find

a $\int 2\sin x\cos x\,dx$ b $\int 2\sin x(\cos x+1)\,dx$

c $\int (\cos x - \sin x)(\cos x - \sin x)\,dx$

3 a Transform the integral $\int \frac{1}{4+x^2}\,dx$ using the substitution $x = 2\tan u$

b Use the fact that $1+\tan^2 x = \sec^2 x$ to simplify the integral.

c Integrate with respect to u

d Substitute $u = \tan^{-1}\left(\frac{x}{2}\right)$ to complete the integration with respect to x

e Similarly, find

i $\int \frac{1}{9+x^2}\,dx$ ii $\int \frac{1}{1+4x^2}\,dx$

4 a Transform the integral $\int f(x)f'(x)\,dx$ using the substitution $u = f(x)$

b Perform the integration with respect to u

c Substitute $f(x) = u$ to the integral with respect to x

d Use your result to find the following integrals.

i $\int (2x+1)(x^2+x+5)\,dx$

ii $\int \sin x\cos x\,dx$

iii $\int \frac{\ln x}{x}\,dx$ iv $\int 2e^{2x}(e^{2x}+1)\,dx$

e Prove $\int \frac{f'(x)}{f(x)}\,dx = \ln|f(x)| + c$

f Find each of the following integrals.

i $\int \frac{\cos x}{\sin x}\,dx$ ii $\int \frac{\sec^2 x}{\tan x}\,dx$

iii $\int \frac{e^x}{e^x+3}\,dx$ iv $\int \frac{1}{x\ln x}\,dx$

Challenge

5 What is the integral of a^x?

a Use the identity that $a = e^{\ln a}$ and the rule that $\ln a^b = b\ln a$ to change the base of the exponential function from a to e

b Use the substitution $u = x\ln a$ to simplify the integral.

c Integrate and change the base back to a

d Use the technique to find

i $\int 4^x\,dx$ ii $\int 5^{2x}\,dx$ iii $\int 3^{x+1}\,dx$

16.3 Integration by parts

Fluency and skills

See Ch15.4 For a reminder of the product rule for differentiation.

Integration by parts is the technique used when you want to integrate the product of two functions.

You can derive a formula for integration by parts using the product rule, which is the technique used to *differentiate* the product of two functions.

Using the product rule for differentiation, $\frac{d}{dx}(uv)=v\frac{du}{dx}+u\frac{dv}{dx}$, where u and v are functions of x

Integrating throughout with respect to x gives $\int\frac{d}{dx}(uv)dx=\int v\frac{du}{dx}dx+\int u\frac{dv}{dx}dx$

Rearranging this gives $\int u\frac{dv}{dx}dx=uv-\int v\frac{du}{dx}dx$

Key point

To integrate by parts, consider the integral as being made up of two parts, u and $\frac{dv}{dx}$, and use

$$\int u\frac{dv}{dx}dx=uv-\int v\frac{du}{dx}dx$$

You usually select u as the function that becomes simpler when differentiated.

Example 1

Find **a** $\int 2x\cos x\,dx$ **b** $\int x^2\ln x\,dx$

a Let $u=2x$ Let $\frac{dv}{dx}=\cos x$

Here the two functions are $2x$ and $\cos x$. You know $2x$ becomes simpler when differentiated, so use this for u

$\Rightarrow\frac{du}{dx}=2$ $\Rightarrow v=\int\cos x\,dx=\sin x$

Differentiate to find $\frac{du}{dx}$ and integrate to find v

$\int u\frac{dv}{dx}dx=uv-\int v\frac{du}{dx}dx$

$\int 2x\cos x\,dx=2x\sin x-\int\sin x\times 2\,dx$

Substitute $u=2x$, $\frac{du}{dx}=2$, $v=\sin x$ and $\frac{dv}{dx}=\cos x$ into the equation and integrate.

$=2x\sin x-2\int\sin x\,dx$

$=2x\sin x+2\cos x+c$

b Let $u=\ln x$ Let $\frac{dv}{dx}=x^2$

Although x^2 becomes simpler on differentiating, you don't yet know how to integrate $\ln x$, so use $u=\ln x$

$\Rightarrow\frac{du}{dx}=\frac{1}{x}$ $\Rightarrow v=\int x^2\,dx=\frac{1}{3}x^3$

$\int u\frac{dv}{dx}dx=uv-\int v\frac{du}{dx}dx$

Differentiate to find $\frac{du}{dx}$ and integrate to find v

$\int\ln x\times x^2\,dx=\left(\ln x\times\frac{1}{3}x^3\right)-\int\left(\frac{1}{3}x^3\times\frac{1}{x}\right)dx$

$=\frac{1}{3}x^3\ln x-\frac{1}{3}\int x^2\,dx$

$=\frac{1}{3}x^3\ln x-\frac{1}{9}x^3+c$

Substitute $u=\ln x$, $\frac{du}{dx}=\frac{1}{x}$, $v=\frac{1}{3}x^3$ and $\frac{dv}{dx}=x^2$ into the equation and integrate.

Example 2

Evaluate the integral $\int_{-1}^{1}(x+1)e^x\,dx$. Show your working and give your answer in exact form.

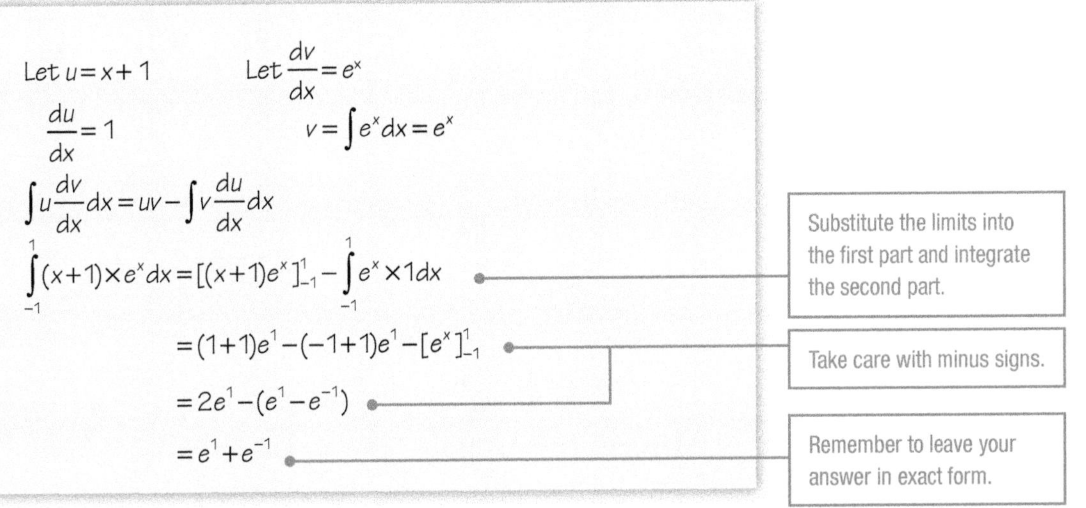

Let $u = x+1$ Let $\frac{dv}{dx} = e^x$

$\frac{du}{dx} = 1$ $v = \int e^x\,dx = e^x$

$\int u\frac{dv}{dx}dx = uv - \int v\frac{du}{dx}dx$

$\int_{-1}^{1}(x+1)\times e^x\,dx = [(x+1)e^x]_{-1}^{1} - \int_{-1}^{1} e^x \times 1\,dx$

$= (1+1)e^1 - (-1+1)e^1 - [e^x]_{-1}^{1}$

$= 2e^1 - (e^1 - e^{-1})$

$= e^1 + e^{-1}$

Substitute the limits into the first part and integrate the second part.

Take care with minus signs.

Remember to leave your answer in exact form.

Exercise 16.3A Fluency and skills

1 Use integration by parts to find

a $\int 3x\cos x\,dx$ b $\int 2x\sin x\,dx$

c $\int \frac{1}{2}x\cos x\,dx$ d $\int 3x\sin 2x\,dx$

e $\int x\cos(3x+1)dx$ f $\int \frac{x}{5}\cos(1-4x)dx$

g $\int (2x+3)\sin x\,dx$ h $\int (2x+1)\cos(x+1)dx$

i $\int (1-3x)\cos(4x)\,dx$

2 Use integration by parts to find

a $\int 3xe^x\,dx$ b $\int 2xe^{2x}\,dx$

c $\int \frac{1}{2}xe^{2x+1}\,dx$ d $\int (x+2)e^{2x}dx$

e $\int (3-5x)\,e^{1+2x}dx$ f $\int (3x-2)\,e^{1-3x}dx$

3 Use integration by parts to find

a $\int 2x\ln x\,dx$ b $\int 5x\ln x\,dx$

c $\int (x+1)\ln x\,dx$ d $\int (2x-1)\ln 2x\,dx$

e $\int \frac{1}{x^2}\ln(3x)dx$ f $\int (x+1)^2\ln x\,dx$

g $\int \frac{1}{\sqrt{x}}\ln(2x)dx$ h $\int x\sqrt{x}\,dx$

i $\int 2x\sqrt{x+1}\,dx$ j $\int x\sec^2 x\,dx$

4 Use integration by parts to evaluate the following integrals. Show your working.

a $\int_0^1 xe^x\,dx$ b $\int_1^3 (x^2-2x)\ln x\,dx$

c $\int_0^{\frac{\pi}{4}} (x+1)\cos x\,dx$ d $\int_1^5 \frac{\ln x}{x^2}dx$

e $\int_0^1 5xe^{-x}\,dx$ f $\int_{-\frac{\pi}{2}}^{\frac{\pi}{2}} x\cos x\,dx$

g $\int_3^8 x\sqrt{x+1}\,dx$ h $\int_0^{\frac{\pi}{2}} x\sin\left(x+\frac{\pi}{2}\right)dx$

5 The graph shows $y = x\sin x$ in the interval $0 \le x \le 2\pi$

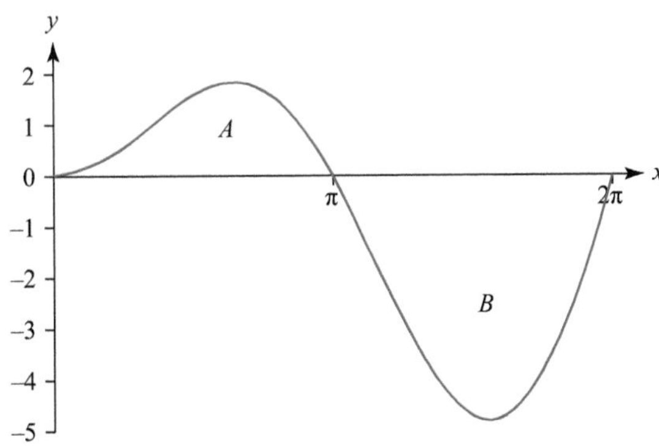

a Verify that it cuts the x-axis at 0, π and 2π

b Showing your working, calculate the area i A ii B

c Hence calculate the area trapped between the curve and the x-axis in the interval $0 \le x \le 2\pi$

Reasoning and problem-solving

Strategy

To integrate by parts when there appears to be only one function, f(x), and its integral is not known

1. Rewrite $\int f(x)\,dx$ as $\int f(x)\times 1\,dx$
2. Let $u = f(x)$ and let $\frac{dv}{dx} = 1$
3. Use integration by parts.

Example 3

Find $\int \ln x\,dx$

$\int \ln x\,dx = \int \ln x \times 1\,dx$ — (1) Rewrite the function in the expanded form.

Let $u = \ln x \Rightarrow \frac{du}{dx} = \frac{1}{x}$

Let $\frac{dv}{dx} = 1 \Rightarrow v = x$ — (2) Let $u = f(x)$ and let $\frac{dv}{dx} = 1$

$\int u\frac{dv}{dx}dx = uv - \int v\frac{du}{dx}dx$

$\int \ln x\,dx = (\ln x \times x) - \int\left(x \times \frac{1}{x}\right)dx$ — (3) Substitute $u = \ln x$, $\frac{du}{dx} = \frac{1}{x}$, $v = x$ and $\frac{dv}{dx} = 1dx$

$= x\ln x - \int 1\,dx$

$= x\ln x - x + c$

Strategy

To use integration by parts more than once

1. Use integration by parts to get $uv - \int v\frac{du}{dx}\,dx$
2. Use integration by parts on $\int v\frac{du}{dx}\,dx$

Example 4

Find $\int x^2 \sin x\,dx$

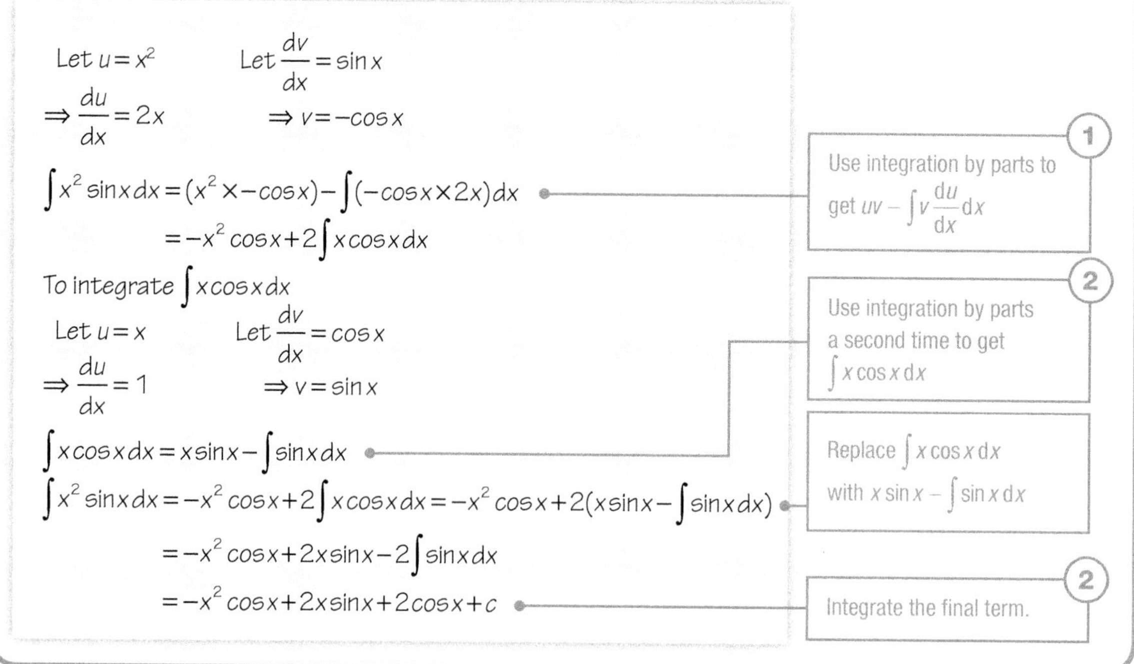

Let $u = x^2 \Rightarrow \frac{du}{dx} = 2x$

Let $\frac{dv}{dx} = \sin x \Rightarrow v = -\cos x$

$\int x^2\sin x\,dx = (x^2 \times -\cos x) - \int(-\cos x \times 2x)\,dx$ — (1) Use integration by parts to get $uv - \int v\frac{du}{dx}dx$

$= -x^2\cos x + 2\int x\cos x\,dx$

To integrate $\int x\cos x\,dx$

Let $u = x \Rightarrow \frac{du}{dx} = 1$

Let $\frac{dv}{dx} = \cos x \Rightarrow v = \sin x$

$\int x\cos x\,dx = x\sin x - \int \sin x\,dx$ — (2) Use integration by parts a second time to get $\int x\cos x\,dx$

$\int x^2\sin x\,dx = -x^2\cos x + 2\int x\cos x\,dx = -x^2\cos x + 2\left(x\sin x - \int \sin x\,dx\right)$ — Replace $\int x\cos x\,dx$ with $x\sin x - \int \sin x\,dx$

$= -x^2\cos x + 2x\sin x - 2\int \sin x\,dx$

$= -x^2\cos x + 2x\sin x + 2\cos x + c$ — (2) Integrate the final term.

Exercise 16.3B Reasoning and problem-solving

PURE

1 Use integration by parts to find each of the following.

a $\int \ln(4x)\,dx$ b $\int \ln(3x+1)\,dx$

c $\int x^7 \ln(x^3)\,dx$ d $\int \sin x \ln(\cos x)\,dx$

e $\int \ln(1-5x)\,dx$ f $\int \ln\left(\frac{1}{x}\right)dx$

g $\int (\ln x)^2\,dx$ h $\int (\ln x^2)\,dx$

i $\int x\sin^2 x\,dx$

2 Apply integration by parts to find

a $\int x^2 e^x\,dx$ b $\int x^2 \sin x\,dx$

c $\int x^2 \cos x\,dx$ d $\int (x+1)^2 \sin x\,dx$

e $\int (x^2+2x)\cos x\,dx$ f $\int (1-3x)^2 e^x\,dx$

g $\int (x^2+x+1)e^{-x}\,dx$ h $\int x^2(x+1)^7\,dx$

i $\int x^2(x+1)^{-2}\,dx$ j $\int (x+1)^2(x+3)^5\,dx$

k $\int x(x+1)\sin x\,dx$ l $\int x^2 \sin 2x\,dx$

m $\int x^2 \cos 3x\,dx$

3 Apply integration by parts twice to evaluate each of the following integrals. Show your working and give your answers in exact form.

a $\int_0^1 (x^2+5)e^{\frac{1}{3}x}\,dx$ b $\int_0^{\frac{\pi}{4}} x^2 \cos 2x\,dx$

c $\int_0^{\pi} (x-\pi)^2 \cos x\,dx$

4 A function is such that $\frac{dy}{dx} = x^2e^x$

a Express y in terms of x and c, the constant of integration.

b When $x=0$, $y=2$. Express y in terms of x alone.

c By considering the discriminant of a quadratic, prove that y is always positive.

5 a Find $\int x\ln x\,dx$

b Hence find $\int 4x(\ln x)^2\,dx$

6 The curve $y=(\pi-x)^2\sin x$, $0 \le x \le 2\pi$, exhibits half-turn symmetry about the point $(\pi, 0)$

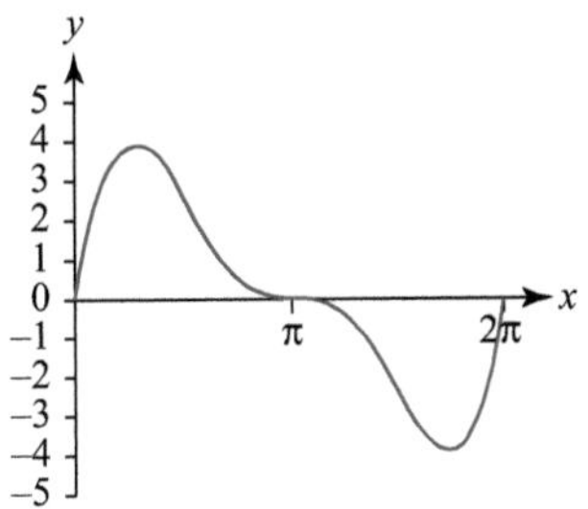

a Check that when $x = 0$, π, and 2π, the curve cuts the x-axis.

b Find the area trapped between the curve and the x-axis in the interval $0 \le x \le 2\pi$. Show your working.

7 The sketch shows the area below the curve $y=\frac{x^2}{\sqrt{2x-1}}$, bounded by $x=1$ and $x=2.5$

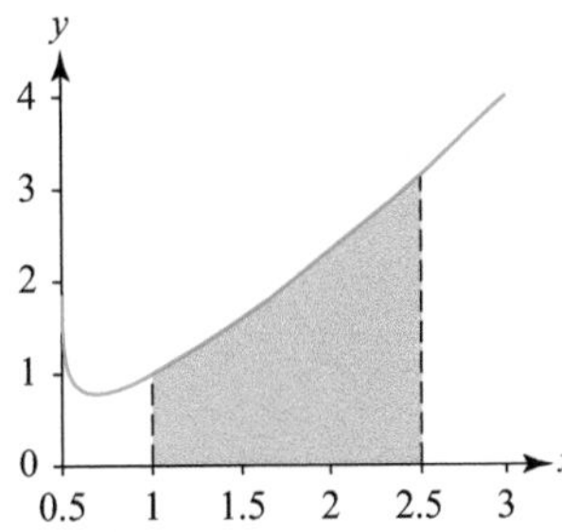

Use integration by parts twice to calculate the exact area. Show your working.

Challenge

8 Consider the integral $P=\int e^x \sin x\,dx$

a Let $u=e^x$ and $\frac{dv}{dx}=\sin x$ and perform integration by parts to get $P=$ (an expression which includes $\int e^x \cos x\,dx$)

b Let $u=e^x$ and $\frac{dv}{dx}=\cos x$ and perform integration by parts again. You should get $P=$ (an expression which includes $\int e^x \sin x\,dx$)

c Replacing this integral by P, you get $P=$ (an expression which includes P)

d Make P the subject of the equation to find $\int e^x \sin x\,dx$

16.4 Integrating rational functions

Fluency and skills

See Ch12.5 For a reminder on partial fractions.

You already know how to decompose a rational function of the form $\frac{px+q}{(ax+b)(cx+d)}$ into partial fractions of the form $\frac{A}{ax+b}+\frac{B}{cx+d}$ by a suitable choice of A and B. This technique often enables you to integrate rational functions. In the first form, no standard integral is evident. However, once resolved into partial fractions, each part is a standard integral.

The method can be extended to include $\frac{px^2+qx+r}{(ax+b)(cx+d)(ex+f)}$

Example 1

Find $\int \frac{x+1}{(x+2)(x+3)}\mathrm{d}x$

Let $\frac{x+1}{(x+2)(x+3)}=\frac{A}{x+2}+\frac{B}{x+3}$ — Express the integral using partial fractions and then multiply throughout by $(x+2)(x+3)$

$x+1=A(x+3)+B(x+2)$

Comparing coefficients for x: $Ax+Bx=x$ so $A+B=1$

Comparing coefficients for the constant: $3A+2B=1$

$A=-1$ and $B=2$ — Solve the simultaneous equations.

$\int\frac{x+1}{(x+2)(x+3)}dx=-\int\frac{1}{x+2}dx+\int\frac{2}{x+3}dx$ — Substitute your values for A and B to resolve into partial fractions and integrate.

$=-\ln|x+2|+2\ln|x+3|+c$

$=\ln k+2\ln|x+3|-\ln|x+2|$

$=\ln\left(k\frac{|x+3|^2}{|x+2|}\right)$ — Sometimes it can be convenient to express the constant of integration as a log.

Example 2

Find $\int \frac{2x+1}{x(x-1)(3x+1)}\mathrm{d}x$

$\frac{2x+1}{x(x-1)(3x+1)}=\frac{A}{x}+\frac{B}{x-1}+\frac{C}{3x+1}$

$2x+1=A(x-1)(3x+1)+Bx(3x+1)+Cx(x-1)$ — Express the integral using partial fractions and then multiply throughout by $x(x-1)(3x+1)$

Comparing coefficients for x^2: $3A+3B+C=0$

Comparing coefficients for x: $-2A+B-C=2$

Comparing coefficients for the constant: $A=-1$

$A=-1$ so $3B+C=3$ and $B=C$ — Substitute $A=-1$ into the simultaneous equations.

$A=-1$, $B=\frac{3}{4}$ and $C=\frac{3}{4}$ — Substitute $B=C$ into $3B+C=3$ to solve the simultaneous equations.

(*Continued on the next page*)

$$\int \frac{2x+1}{x(x-1)(3x+1)} = -\int \frac{1}{x}dx + \frac{3}{4}\int \frac{1}{x-1}dx + \frac{3}{4}\int \frac{1}{3x+1}dx$$

$$= -\ln|x| + \frac{3}{4}\ln|x-1| + \frac{3}{4}\times\frac{1}{3}\ln|3x+1| + \ln k$$

$$= \ln\left(k\frac{|x-1|^{\frac{3}{4}}|3x+1|^{\frac{1}{4}}}{|x|}\right)$$

Substitute your values for A, B and C to resolve into partial fractions and integrate.

Remember the constant of integration.

PURE

Exercise 16.4A Fluency and skills

1 Find

a $\int \frac{1}{x(x+1)}dx$ b $\int \frac{6}{x(x+3)}dx$

c $\int \frac{1}{x(3x+1)}dx$ d $\int \frac{x+1}{x(x+2)}dx$

e $\int \frac{8}{(x-2)(x+2)}dx$ f $\int \frac{5}{(x+1)(x-4)}dx$

g $\int \frac{2}{(x+2)(x+1)}dx$ h $\int \frac{21}{(x-4)(x+3)}dx$

2 Find

a $\int \frac{4}{(2x+1)(2x-3)}dx$ b $\int \frac{5}{(x+2)(3x+1)}dx$

c $\int \frac{7}{(x+5)(3x+1)}dx$ d $\int \frac{x+1}{(x-1)(x+3)}dx$

e $\int \frac{4x-2}{(x+4)(x-2)}dx$ f $\int \frac{5}{(2x+1)(3x-1)}dx$

g $\int \frac{6x}{(x-2)(x+1)}dx$ h $\int \frac{2x+1}{(2x+3)(x-1)}dx$

3 Find

a $\int \frac{4x-1}{(x-1)(2x+1)}dx$ b $\int \frac{3x}{(2x+1)(2x-1)}dx$

c $\int \frac{2-x}{(3x-1)(2x+1)}dx$ d $\int \frac{1}{x(1-x)}dx$

e $\int \frac{x+1}{x(4-x)}dx$ f $\int \frac{1}{2x(10-x)}dx$

4 Express each integrand as the sum of three rational functions, each of which has a linear denominator, and then integrate.

a $\int \frac{1}{x(x-1)(x+1)}dx$

b $\int \frac{10}{x(3x+2)(x-1)}dx$

c $\int \frac{x+3}{x(x-1)(4x-3)}dx$

d $\int \frac{6}{(x-1)(x-2)(x+1)}dx$

e $\int \frac{3x}{(2x-1)(x-2)(x-1)}dx$

f $\int \frac{x^2+2}{x(x-1)(x-2)}dx$

g $\int \frac{20x^2}{(x+3)(x+4)(x-1)}dx$

h $\int \frac{x^2+x-2}{(x+1)(x-3)(2x-1)}dx$

i $\int \frac{x^2+x+1}{x(x-1)(x+2)}dx$

j $\int \frac{x^2+x+1}{x(1-x)(x+2)}dx$

k $\int \frac{3-x^2}{x(1-2x)(x+3)}dx$

5 Evaluate each of the following integrals. Show your working and give your answers in exact form.

a $\int_1^3 \frac{5}{x(x-5)}dx$ b $\int_3^6 \frac{4}{x(x-2)}dx$

c $\int_2^4 \frac{2}{x(x+1)}dx$ d $\int_1^2 \frac{1}{x(4-x)}dx$

e $\int_4^5 \frac{2}{(x-3)(x-1)}dx$ f $\int_1^3 \frac{6}{(x-5)(x-2)}dx$

g $\int_0^{\frac{1}{2}} \frac{3}{(x-2)(x+1)}dx$

Strategy

To solve integration problems when dealing with improper rational functions

1. If necessary, expand the denominator.
2. Perform algebraic division to resolve the function into a polynomial and a proper rational function.
3. Resolve the rational function into partial fractions.
4. Integrate.
5. Simplify.

Example 3

This example uses the law of logarithms $k\log_a x = \log_a x_n$, which was covered in Section 5.1

Integrate $\dfrac{x^3+5x^2+1}{(x+1)(x+3)}$

$$f(x)=\frac{x^3+5x^2+1}{(x+1)(x+3)}=\frac{x^3+5x^2+1}{x^2+4x+3}$$

1 Expand the denominator.

$$\begin{array}{r} x+1 \\ x^2+4x+3\overline{\smash{)}x^3+5x^2+1} \\ \underline{x^3+4x^2+3x} \\ x^2-3x+1 \\ \underline{x^2+4x+3} \\ \underline{-7x-2} \end{array}$$

2 Perform algebraic division.

See Ch2.3

For a reminder on algebraic division.

$$f(x)=x+1+\left(\frac{-7x-2}{x^2+4x+3}\right)=x+1-\left(\frac{7x+2}{(x+1)(x+3)}\right)$$

2 Resolve the function into a polynomial and a proper rational function.

$$\frac{7x+2}{(x+1)(x+3)}=\frac{A}{x+1}+\frac{B}{x+3}$$

$$7x+2=A(x+3)+B(x+1)$$

Multiply through by $(x+1)(x+3)$

Comparing coefficents for x: $A+B=7$

Comparing coefficients for the constant: $3A+B=2$

$$A=-\frac{5}{2} \text{ and } B=\frac{19}{2}$$

Solve the simultaneous equations.

$$\frac{7x+2}{(x+1)(x+3)}=-\frac{5}{2(x+1)}+\frac{19}{2(x+3)}$$

3 Resolve into partial fractions.

$$\int\frac{x^3+5x^2+1}{(x+1)(x+3)}dx=\int(x+1)dx+\frac{5}{2}\int\frac{1}{x+1}dx-\frac{19}{2}\int\frac{1}{x+3}dx$$

$$=\frac{1}{2}x^2+x+\frac{5}{2}\ln|x+1|-\frac{19}{2}\ln|x+3|+c$$

4 Integrate.

$$=\frac{\ln|x+1|^{\frac{5}{2}}}{\ln|x+3|^{\frac{19}{2}}}+\frac{1}{2}x^2+x+c$$

5 Simplify.

Exercise 16.4B Reasoning and problem-solving

1 By first factorising the denominator, find

a $\int \frac{1}{x^2+2x}\,dx$ b $\int \frac{2}{x^2+5x}\,dx$

c $\int \frac{1}{2x^2+x}\,dx$ d $\int \frac{x+2}{x^2+x}\,dx$

e $\int \frac{1}{x^2-1}\,dx$ f $\int \frac{12}{x^2-9}\,dx$

2 By first factorising the denominator, find

a $\int \frac{2}{4x^2-1}\,dx$ b $\int \frac{3x}{9x^2-4}\,dx$

c $\int \frac{1}{x^2-4x+3}\,dx$ d $\int \frac{3}{x^2-5x+4}\,dx$

e $\int \frac{x}{2x^2+3x-2}\,dx$ f $\int \frac{2x}{x^2-7x+10}\,dx$

g $\int \frac{x+1}{x^2+3x-10}\,dx$ h $\int \frac{3x-1}{2x^2+7x-4}\,dx$

3 Find

a $\int \frac{x^2+x-1}{(x-1)(x+2)}\,dx$ b $\int \frac{x^2+3x+6}{(x+1)(x+2)}\,dx$

c $\int \frac{x^2+2x}{(x-1)(x+3)}\,dx$ d $\int \frac{2x^2-5}{(x-2)(x+2)}\,dx$

e $\int \frac{2x^2+8x+7}{(x+1)(x+3)}\,dx$ f $\int \frac{3x^2-8x+6}{(x-1)(x-2)}\,dx$

4 Evaluate the following integrals. Show your working and give your answers in exact form.

a $\int_0^2 \frac{4x-10}{(x+2)(x-4)}\,dx$ b $\int_4^5 \frac{2x-3}{(x-1)(x-2)}\,dx$

c $\int_3^4 \frac{1-2x}{(x-2)(x-5)}\,dx$ d $\int_0^2 \frac{-x-1}{(x-1)(x-3)}\,dx$

e $\int_1^3 \frac{x-8}{x(x-4)}\,dx$ f $\int_2^4 \frac{2x-14}{(x-1)(x-5)}\,dx$

5 This is a sketch of $y=\frac{2x-4}{(x-1)(x-3)}$

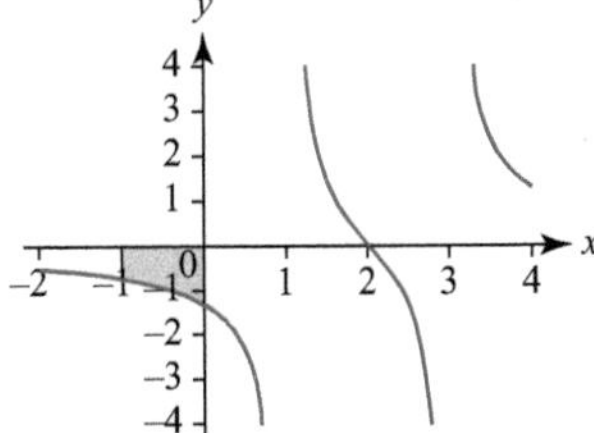

a Calculate the area bounded by the curve, the x-axis, $x=-1$ and $x=0$. Show your working.

b Calculate the area bounded by the curve, the x-axis, $x=4$ and $x=5$. Show your working.

c Why would it not be sensible to look for the area bounded by the curve, the x-axis, $x=0$ and $x=2$?

6 The curve shows sketches of $y=\frac{-8}{x(x-4)}$ and $y=\frac{-9}{x(x-3)}$

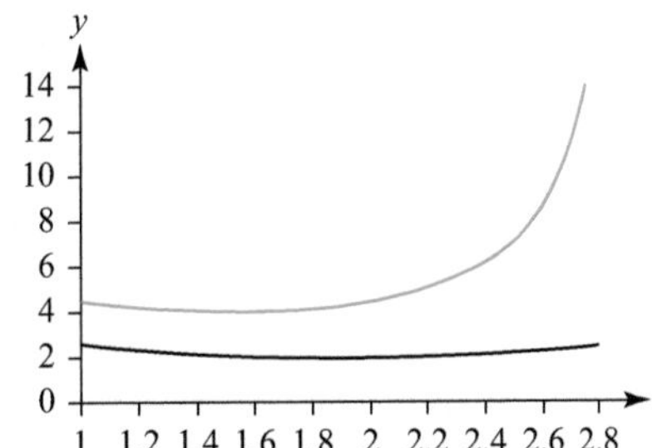

a Identify which curve goes with which equation.

b Showing your working, find the area bounded by

i $y=\frac{-8}{x(x-4)}$, the x-axis and $x=1$ and $x=2$

ii $y=\frac{-9}{x(x-3)}$, the x-axis and $x=1$ and $x=2$

c Hence find the area between the two curves in this interval, giving your answer in the form $a\ln 3+b\ln 2$

Challenge

7 The sketch shows the two curves $y_1=\frac{2x-3}{(x-1)(x-2)}$ and $y_2=\frac{1-2x}{(x-2)(x-5)}$

a Calculate the area between the curves in the interval $3\le x\le 4$ Show your working.

b Describe the intervals over which it would not be meaningful to consider the area between the curves.

16.5 Differential equations

Fluency and skills

A **differential equation** is an equation that expresses a relationship between functions and their derivatives. In order to find the **solution** of a differential equation you must **separate the variables**. That is, you must express the left-hand side purely in terms of y and the right-hand side purely in terms of x

Key point

Let y be a function of x, let f be a function of y, and let g be a function of x

If $f(y)\frac{dy}{dx} = g(x)$

Integrating both sides with respect to x gives $\int f(y)\frac{dy}{dx}dx = \int g(x)dx$

The chain rule then gives the result $\int f(y)dy = \int g(x)dx$

Integrating will result in a **general solution** that describes an infinite set of solutions. If you know at least one point (x, y) in the solution, then you can find the constant of integration and hence a **particular solution**.

The point (x, y) is usually referred to as **the initial condition.**

If a constant of proportion is also involved, then another point will be required to find its value.

Example 1

a Find the general solution of the equation $\frac{dy}{dx} = 2x(y+1);\ \ y > -1$

b Find the particular solution given that when $x = 3$, $y = 0$

a $\frac{1}{y+1}\frac{dy}{dx} = 2x$ — You can rewrite the equation in this form. This step is known as 'separating the variables'.

$\int \frac{1}{y+1}\frac{dy}{dx}dx = \int 2x dx$ — Integrate both sides with respect to x to find the general solution written as an implicit function.

$\Rightarrow \int \frac{1}{y+1}dy = \int 2x dx \Rightarrow \ln|y+1| = x^2 + c$

$\Rightarrow e^{\ln|y+1|} = e^{x^2+c} \Rightarrow y+1 = e^{x^2+c}$ — To get y in terms of x, take the exponential of each side of the equation.

$\Rightarrow \quad y = e^{x^2+c} - 1$ — This gives the general solution written as an explicit function.

b When $x = 3$ and $y = 0$, $\ln|0+1| = 3^2 + c$

$\Rightarrow c = -9$ — Substitute the initial conditions to find c

So the particular solution is given by $y = e^{x^2-9} - 1$

You also may be required to draw upon other integration techniques in order to solve a differential equation.

Example 2

Solve $\frac{dy}{dx} = \frac{2x}{y\sqrt{y^2-1}}$ given that when $x = 2$, $y = 1$

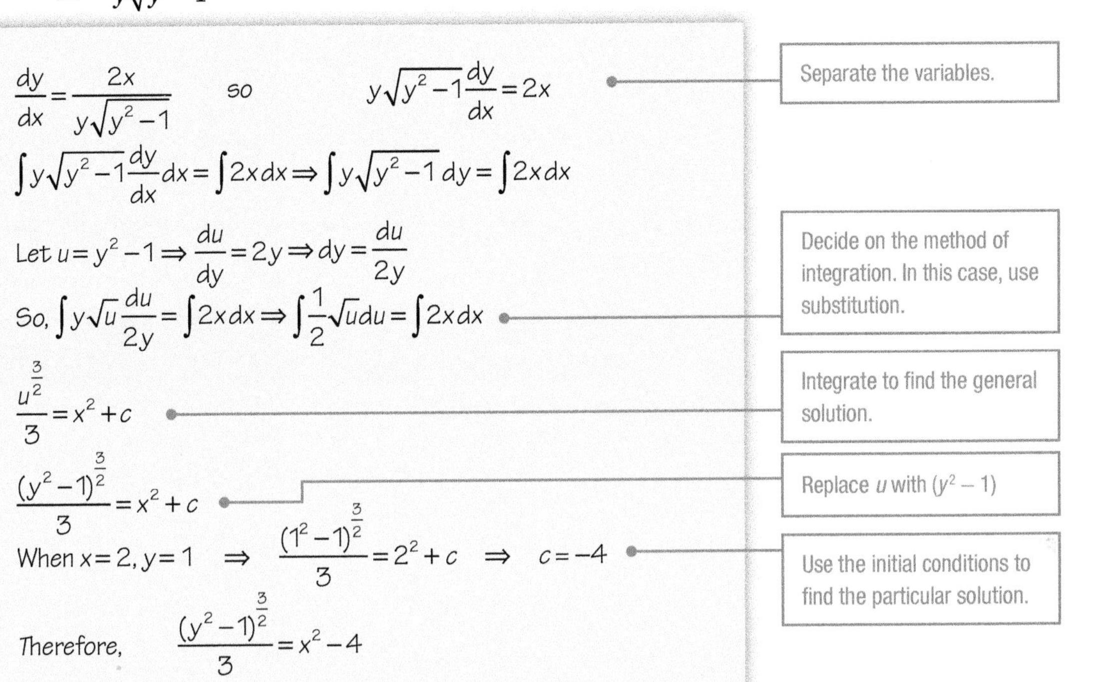

$\frac{dy}{dx} = \frac{2x}{y\sqrt{y^2-1}}$ so $y\sqrt{y^2-1}\frac{dy}{dx} = 2x$
$\int y\sqrt{y^2-1}\frac{dy}{dx}dx = \int 2x\,dx \Rightarrow \int y\sqrt{y^2-1}\,dy = \int 2x\,dx$
Let $u = y^2 - 1 \Rightarrow \frac{du}{dy} = 2y \Rightarrow dy = \frac{du}{2y}$
So, $\int y\sqrt{u}\frac{du}{2y} = \int 2x\,dx \Rightarrow \int \frac{1}{2}\sqrt{u}\,du = \int 2x\,dx$
$\frac{u^{\frac{3}{2}}}{3} = x^2 + c$
$\frac{(y^2-1)^{\frac{3}{2}}}{3} = x^2 + c$
When $x = 2, y = 1 \Rightarrow \frac{(1^2-1)^{\frac{3}{2}}}{3} = 2^2 + c \Rightarrow c = -4$
Therefore, $\frac{(y^2-1)^{\frac{3}{2}}}{3} = x^2 - 4$
Separate the variables.
Decide on the method of integration. In this case, use substitution.
Integrate to find the general solution.
Replace u with $(y^2 - 1)$
Use the initial conditions to find the particular solution.

Exercise 16.5A Fluency and skills

1 Find the general solution to each of the following differential equations.

a $\frac{dy}{dx} - 2x + 1 = 0$ b $\frac{dy}{dx} = 1 - x^2$ c $\frac{dy}{dx} = \sqrt{2x+1}$ d $\frac{dy}{dx} = \frac{3}{x^2}$

e $\frac{dy}{dx} = \frac{2}{x}$ f $\frac{dy}{dx} = 3y$ g $\frac{dy}{dx} = 5xy$ h $\frac{dy}{dx} = \frac{x}{y}$

i $\frac{dy}{dx} = \frac{x+1}{y-3}$ j $\frac{dy}{dx} = x(1+y)$ k $\frac{dy}{dx} + y = 5$ l $\frac{dy}{dx} = (x+2)(2y+1)$

m $2y\frac{dy}{dx} - 3x^2 = 0$ n $\frac{dy}{dx} = \frac{\sin x}{\cos y}$ o $\frac{dy}{dx} = e^{2x}\cos^2 y$ p $\frac{dy}{dx} = e^{x-y}$

q $\frac{dy}{dx} = e^{2x-3y+1}$

2 Find the particular solution to the differential equation that corresponds to the given initial conditions.

a $\frac{dy}{dx} - 4x + 1 = 0$; $(1, 4)$ b $\frac{dy}{dx} + 9x^2 - 2 = 0$; $(2, -10)$ c $\frac{dy}{dx} = \frac{1}{x}$; $(1, 3)$

d $\frac{dy}{dx} = 6y$; $\left(\frac{1}{3}, 1\right)$ e $\frac{dy}{dx} = x(y+3)$; $(2, -2)$ f $\frac{dy}{dx} = xy + 5y$; $(0, e)$

g $\frac{dy}{dx} = \frac{x+2}{2y-1}$; $(2, 1)$ h $\frac{dy}{dx} = \frac{\cos x}{\sin y}$; $\left(\frac{\pi}{5}, \frac{\pi}{3}\right)$ i $\frac{dy}{dx} = e^{2x-y}$; $\left(\frac{1}{2}, 0\right)$

j $2x\frac{dy}{dx} - y = 0$; $(1, 2)$

MyMaths 2226, 2227 SEARCH

3 Find the general solution, stated explicitly if possible.

a i $\dfrac{dy}{dx}=\dfrac{(y^2+1)^5}{y}$ ii $\dfrac{dy}{dx}=x\dfrac{\sin y}{\cos y}$ iii $e^{\sqrt{y}}\dfrac{dy}{dx}-2\sqrt{y}=0$

b i $\dfrac{dy}{dx}=\dfrac{\sin x}{y\sin y}$ ii $\dfrac{dy}{dx}=\dfrac{xe^x}{y}$ iii $\dfrac{dy}{dx}=\dfrac{9}{y^2\ln y}$

c i $\dfrac{dy}{dx}=\dfrac{6y}{x(2x-3)}$ ii $\dfrac{dy}{dx}=\dfrac{4y^2}{(x+1)(x-1)}$ iii $\dfrac{dy}{dx}=\dfrac{\sqrt{y}}{(x-2)(x-3)}$

4 The gradient of a curve at the point (x, y) is given by $\dfrac{\cos x}{e^y}$. The curve passes through the point $(\pi, 0)$. Find an expression for y in terms of x

5 The gradient of a curve at the point (x, y) is given by $\sin x\sqrt{y+5}$ and the curve passes through the point $\left(\dfrac{\pi}{2}, 2\right)$. Find an expression for y in terms of x

Reasoning and problem-solving

Strategy

To solve modelling problems involving differential equations

1. Form a differential equation using the information provided.
2. Separate the variables.
3. Integrate to find the general solution.
4. Use the initial conditions to find the constant and hence the particular solution.
5. Use the model to make predictions.
6. Consider any assumptions and any limitations of the model.

Example 3

In 1820, Thomas Malthus suggested that the rate of growth of a population, P, over time t, is proportional to the size of the population.

a Express this model using a differential equation.

b Find a general solution to your differential equation.

c Consider any limitations of the model.

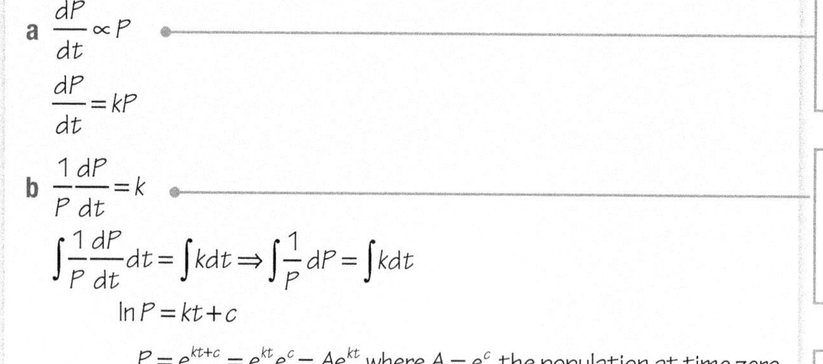

a $\dfrac{dP}{dt}\propto P$
$\dfrac{dP}{dt}=kP$
b $\dfrac{1}{P}\dfrac{dP}{dt}=k$
$\int\dfrac{1}{P}\dfrac{dP}{dt}dt=\int k\,dt \Rightarrow \int\dfrac{1}{P}dP=\int k\,dt$
$\ln P=kt+c$
$P=e^{kt+c}=e^{kt}e^{c}=Ae^{kt}$ where $A=e^{c}$, the population at time zero.
c Most natural populations do not grow exponentially fast since they are constrained by food supply, habitat or other external factors.

1 Form a differential equation. The rate of change of P is proportional to P. See Ch2.4 for a reminder on proportionality.

2 3 Separate the variables and integrate both sides with respect to t to find the general solution.

6 Consider any limitations of the model.

Example 4

A species of tree is expected to grow to 10 m in height.

The rate of growth in metres per year is directly proportional to the height it still has to grow to reach its potential. The rate of growth can be modelled by $\frac{dh}{dt} = k(10-h)$ where h is the height of the tree in metres, t is the time in years and k is a constant.

a Solve the differential equation to find the general solution.

b After 2 years, the tree is 1 m tall. Find the particular solution to the differential equation.

c Use the model to predict the age, to the nearest year, of a tree that is 6 m tall.

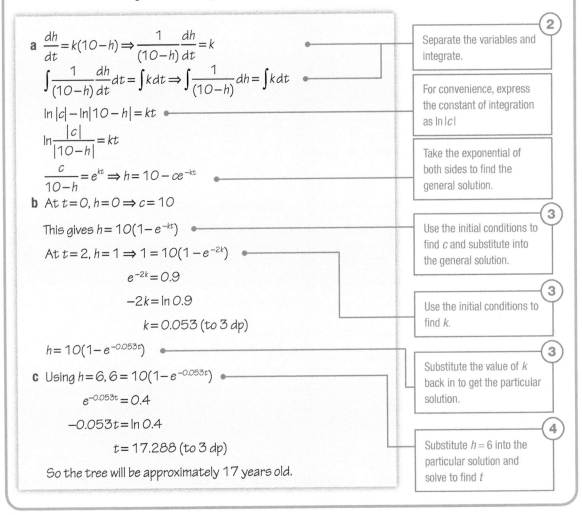

a $\frac{dh}{dt} = k(10-h) \Rightarrow \frac{1}{(10-h)}\frac{dh}{dt} = k$

$\int \frac{1}{(10-h)}\frac{dh}{dt}dt = \int k\,dt \Rightarrow \int \frac{1}{(10-h)}dh = \int k\,dt$

Separate the variables and integrate. (2)

$\ln|c| - \ln|10-h| = kt$

For convenience, express the constant of integration as $\ln|c|$

$\ln\frac{|c|}{|10-h|} = kt$

$\frac{c}{10-h} = e^{kt} \Rightarrow h = 10 - ce^{-kt}$

Take the exponential of both sides to find the general solution.

b At $t=0, h=0 \Rightarrow c=10$

This gives $h = 10(1-e^{-kt})$

Use the initial conditions to find c and substitute into the general solution. (3)

At $t=2, h=1 \Rightarrow 1 = 10(1-e^{-2k})$

$e^{-2k} = 0.9$

$-2k = \ln 0.9$

$k = 0.053$ (to 3 dp)

Use the initial conditions to find k. (3)

$h = 10(1-e^{-0.053t})$

Substitute the value of k back in to get the particular solution. (3)

c Using $h=6$, $6 = 10(1-e^{-0.053t})$

$e^{-0.053t} = 0.4$

$-0.053t = \ln 0.4$

$t = 17.288$ (to 3 dp)

So the tree will be approximately 17 years old.

Substitute $h=6$ into the particular solution and solve to find t (4)

Exercise 16.5B Reasoning and problem-solving

1 Express each sentence as a differential equation.

a When buying gold, the rate of change of the cost with respect to its weight is a constant. Let w grams cost £C

b A ball rolls down an inclined plane. The rate at which its distance from the top changes with respect to time is directly proportional to the time it has been rolling. Let S cm be the distance after t seconds.

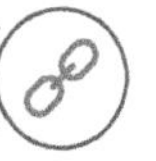

c Under compound interest conditions, the rate at which the amount in an account grows is directly proportional to the amount in the account. There are £A in the account after T years.

d The rate at which a cup of tea cools is directly proportional to the difference between the temperature of the tea and the temperature of its surroundings. After t minutes, with the room at a constant 20° C, the tea is at a temperature of T°C

e Earthquakes are measured using the Richter Scale. The rate of change of an earthquake's intensity, as its Richter scale number changes, is directly proportional to 10^R. R is the Richter scale number when the intensity is I units.

f A square metal plate expands when heated. The rate at which the length of the side increases with respect to the area is inversely proportional to the square root of the length of the side. A square of area x cm^2 has a side of y cm

g The growth of a particular plant is such that the rate at which its height is changing with respect to time is inversely proportional to $\sec^2 h$, where h cm is the height after t days.

2 Water is poured into a cistern which can hold 50 litres. The rate at which it fills can be modelled by $\frac{dV}{dt} = 2 + 0.6t$, where there are V litres in the cistern after t minutes.

a Solve the differential equation to express V in terms of t and a constant.

b Initially the tank is empty. Express V in terms of t

c The flow cuts off when the cistern is full. At what time will this occur?

3 Over a month, the rate at which the percentage of the moon which is visible changes with time can be modelled by $\frac{dP}{dt} = \frac{25\pi}{7}\cos\left(\frac{\pi t}{14}\right)$ where P is the percentage visible on day t of the month.

a Solve the differential equation to express P in terms of t and a constant.

b The month began with a half-moon. [Initially, $t = 0$, $P = 50$.] Express P in terms of t

c What percentage of the moon was visible on day 4?

d On which day was there a full moon?

4 A 10 metre ladder is held vertically against a wall. The foot of the ladder starts to slip away from the wall. The rate at which the height of the midpoint of the ladder changes with respect to its distance from the wall is modelled by $\frac{dy}{dx} = k\frac{x}{y}$ where the midpoint is y metres high when it is x metres from the wall and k is a constant.

a Solve the differential equation to express y in terms of x, k and c (the constant of integration).

b Initially the midpoint is 5 metres high and touching the wall. Find the value of c

When it is 4 metres high, it is 3 metres from the wall.

c Express the relationship between x and y as an implicit equation and describe the path traced out by the midpoint of the ladder as it falls.

5 Bacteria are growing in a petri dish. The rate at which they multiply is directly proportional to the number of bacteria. Initially there were 500 bacteria in the dish. On day t there are N bacteria.

a Form a differential equation.

b Find the general solution of the differential equation.

c Use the initial conditions to find the constant of integration.

d At the end of day 3, there were 4000 bacteria in the dish. Find constant of proportion and the particular solution to the equation, expressing N as an explicit function of t

e When will the number of bacteria exceed 50 000?

f Why might this model not be appropriate?

6 An osprey can be expected to reach an adult weight of 2000 g

On day zero, a chick will weigh 50 g on hatching. It fledges after 60 days when its weight is 1990 g

Its rate of growth is directly proportional to the difference between its weight and its expected adult weight. On day t, its weight is w grams.

a Form a differential equation to model the development of the osprey chick,

b Evaluate the constant of integration if at $t = 0$, $w = 50$,

c Find the constant of proportion,

d Express w explicitly as a function of t

e When is the chick's weight expected to exceed 1500 grams?

f Discuss any assumptions and any limitations of the model.

7 On a small island, it is estimated that there are enough resources to support a population of 500 breeding pairs of rabbits. Initially, 10 pairs are introduced on the island. By the end of year 1, there are 50 pairs. The rate of growth is found to be jointly proportional to both the population size and to the difference between the population size and the island's capacity for breeding pairs. Let P be the number of pairs on the island at the end of year t

a Form a differential equation to model the rabbit population in year t

b Express P explicitly in terms of t

Challenge

8 Use a spreadsheet to explore the models introduced in questions **3** and **4**

Draw graphs of each and see how they have been chosen to model each context.

How closely do they model the situation

a At the beginning,

b As time goes on?

Which one might best model the rate of uptake of a new idea in Britain (for example—the membership of a social media site)?

You'll need to get initial conditions from the web.

16 Summary and review

Chapter summary

- If the derivative of a function F(x) is f(x), then you can say that an indefinite integral of f(x) with respect to x is F(x). That is, $F'(x)=f(x) \Rightarrow \int f(x)\mathrm{d}x = F(x)+c$ where c is the constant of integration.
  This leads to the following set of standard integrals.
  1. $\int x^n\,\mathrm{d}x = \frac{1}{n+1}x^{n+1}+c;\ n\neq -1$
  2. $\int e^{kx}\,\mathrm{d}x = \frac{1}{k}e^{kx}+c$
  3. $\int \frac{1}{x}\mathrm{d}x = \ln|x|+c$
  4. $\int \cos kx\,\mathrm{d}x = \frac{1}{k}\sin kx + c$
  5. $\int \sin kx\,\mathrm{d}x = -\frac{1}{k}\cos kx + c$
  6. $\int \sec^2 kx\,\mathrm{d}x = \frac{1}{k}\tan kx + c$
- The area between the curves $y=$ f(x) and $y=$ g(x) is given by $A=\left|\int_a^b f(x)-g(x)\mathrm{d}x\right|$ if the curves do not intersect in the interval $a<x<b$. If the curves do intersect in the interval, then each trapped area must be considered separately.
- To integrate the product of a function and a derivate function, you can use integration by substitution.
  If $\int f(g(x))g'(x)\mathrm{d}x$, let $u=$ g(x) and d$u=$ g′(x)dx. This gives $\int f(u)\mathrm{d}u$
- To integrate a product of two functions, you can use integration by parts. To do this, consider the integral as being made up of two parts, u and dv, and use $\int u\frac{\mathrm{d}v}{\mathrm{d}x}\mathrm{d}x = uv - \int v\frac{\mathrm{d}u}{\mathrm{d}x}\mathrm{d}x$
- You can reduce a rational function of the form $\frac{px+q}{(ax+b)(cx+d)}$ into partial fractions of the form $\frac{A}{ax+b}+\frac{B}{cx+d}$ by a suitable choice of A and B. This technique often allows you to integrate rational functions more easily.
- An equation which relates functions to their derivatives is called a **differential equation**.
- The solution to a differential equation is the set of functions which satisfy the equation.
- If a differential equation can be expressed in the form $f(y)\frac{\mathrm{d}y}{\mathrm{d}x}=g(x)$ then you say the variables are separable.
- Integrating $\int f(y)\mathrm{d}y = \int g(x)\mathrm{d}x$ yields the *general solution* of the differential equation.
- If you know one point (x, y) in the solution, you can find a *particular solution*.

Check and review

You should now be able to...	Try Questions
✔ Integrate a set of standard functions, f(x) and the related functions, f($ax+b$)	1,2,3
✔ Find the area between two curves.	3
✔ Simplify an integral by changing the variable, referred to as *substitution*.	4,5
✔ Use integration by parts to integrate the product of two functions.	6,7,8
✔ Simplify an integral by decomposing a rational function into partial fractions.	9,10
✔ Understand the meaning of the expression 'differential equation'.	11,12
✔ Use integration where the variables are separable.	11,12

1 Find

a $\int 4x^6\,dx$ b $\int 2\sin(3x+1)\,dx$

c $\int \cos 2x\,dx$ d $\int x+\sec^2 x\,dx$

e $\int (e^x+e^{-x})\,dx$

2 Evaluate the following integrals. Show your working.

a $\int_1^4 \frac{2}{\sqrt{x}}\,dx$ b $\int_{\frac{\pi}{6}}^{\frac{\pi}{2}} \sin x\,dx$

c $\int_{-\frac{\pi}{4}}^{\frac{\pi}{4}} \cos 2x\,dx$ d $\int_0^{\frac{\pi}{4}} 1-\sec^2 x\,dx$

e $\int_{\ln 2}^{\ln 5} e^x\,dx$ f $\int_{e^2}^{e^4} \frac{3}{x}\,dx$

3 In each case the two functions trap a single region between them. Find the area of the region, showing your working.

a $y=x^2+x$ and $y=2x+2$

b $y=1+5x-x^2$ and $y=x^2-4x+5$

4 Find each integral. A suitable substitution has been suggested.

a $\int x\left(x^2+4\right)^5 dx$; let $u=x^2+4$

b $\int 8x^3(x^4-2)^3 dx$; let $u=x^4-2$

c $\int (x-2)(x^2-4x+1)\,dx$; let $u=x^2-4x+1$

d $\int \cos x\sqrt{\sin x}\,dx$; let $u=\sin x$

5 Evaluate the following integrals. Show your working.

a $\int_0^{\sqrt{3}} \frac{x}{\sqrt{x^2+1}}\,dx$ b $\int_0^{\frac{\pi}{3}} \frac{\sin x}{2\cos^2 x}\,dx$

6 Use integration by parts to find

a $\int 2x\sin x\,dx$ b $\int (3x+1)e^{2x}\,dx$

c $\int x^2\cos 3x\,dx$

7 Evaluate each of the following, showing your working.

a $\int_0^1 (x+3)e^{x+3}\,dx$ b $\int_1^e 9x^2\ln x\,dx$

8 Apply integration by parts twice to find

a $\int (x^2+2x-1)e^x dx$ b $\int 2x(\ln x)^2\,dx$

9 Find

a $\int \frac{8}{x(x-2)}\,dx$ b $\int \frac{2}{(x+2)(x+3)}\,dx$

10 Evaluate $\int_4^5 \frac{3x+1}{(x-1)(x-3)}\,dx$ expressing your answer in the form $a\ln b$ where a and b are integers. Show your working.

11 Find the particular solution to each differential equation.

a $\frac{dy}{dx}=y(x+1)$ given that when $x=1$, $y=1$

b $\frac{dy}{dx}=\frac{2x}{\cos y}$ given that when $x=1$, $y=\frac{\pi}{6}$

12 The rate at which a car loses value is directly proportional to the value of the car. The car is worth £V after T years.

Initially the car was worth £20 000
After 3 years it was worth £14 580

a Form a differential equation.

b Use the initial conditions to find the constants of proportion and integration.

c Express V as an explicit function of T

What next?

Score		
0–6	Your knowledge of this topic is still developing. To improve, search in MyMaths for the codes: 2057, 2167–2171, 2216–2221, 2226–2227, 2274	
7–9	You're gaining a secure knowledge of this topic. To improve, look at the InvisiPen videos for Fluency and skills (16A)	
10–12	You've mastered these skills. Well done, you're ready to progress! To develop your techniques, look at the InvisiPen videos for Reasoning and Problem-solving (16B)	

Click these links in the digital book

Exploration
Going beyond the exams

Have a go

Not all differential equations have **separable variables.**

Consider, for example

$$\frac{dy}{dx}+\frac{2y}{x}=\frac{1}{x^2}\cos x$$

Start by multiplying both sides of the equation by x^2

$$x^2\frac{dy}{dx}+2xy=\cos x$$

Recognise that the left-hand side of this equation is the derivative of x^2y with respect to x

$$\frac{d(x^2y)}{dx}=\cos x$$

Complete the solution to give y in terms of x

Note

No amount of rearranging this equation will leave all of the x terms on one side and the y on the other.

The trick here is to recognise that the left-hand side is the **derivative of a product.**

Information

The solution of the differential equation above relied on multiplying both sides by a factor that turned the left-hand side into the derivative of a product. Such a factor is called an **integrating factor.**

Use the chain rule to differentiate $e^{\int P dx}y$ with respect to x:

$$\frac{d(e^{\int P dx}y)}{dx}=e^{\int P dx}\frac{dy}{dx}+Pe^{\int P dx}y$$

$$=e^{\int P dx}(\frac{dy}{dx}dx+Py)$$

This result shows that a differential equation of the form $\frac{dy}{dx}+Py=Q$,

where P and Q are functions of x, has integrating factor $I=e^{\int P dx}$

Challenge

Use the integrating factor method to solve the differential equation

$$\frac{dy}{dx}+\frac{3}{x}y=\sqrt{x}$$

Given that $y=7$ when $x=1$, find the value of y when $x=4$

Challenge

Find the integrating factor for the equation

$$\frac{dy}{dx}+\frac{2y}{x}=\frac{1}{x^2}\cos x$$

Note

When working out $\int P dx$ it is not necessary to include a constant.

16 Assessment

1 Integrate $\frac{2}{3x+4}$. Select the correct answer.

A $6\ln(3x+4)+c$ B $\frac{2}{3(3x+4)^2}+c$ C $\frac{6}{(3x+4)^2}+c$ D $\frac{2}{3}\ln(3x+4)+c$ **[1 mark]**

2 $f'(x)=2+3\sin 6x$

Find the equation of the curve $y=f(x)$, which passes through the point (0, 1)

Select the correct answer.

A $y=2x+\frac{1}{2}\cos 6x+\frac{1}{2}$ B $y=2x+18\cos 6x+\frac{1}{2}$

C $y=2x-\frac{1}{2}\cos 6x+\frac{3}{2}$ D $y=2x-18\cos 6x+\frac{3}{2}$ **[1]**

3 Integrate each of these expressions.

a $\frac{1}{x}$ **[1]** b $\sin(x-3)$ **[2]** c e^{2x} **[2]** d $\cos 2x$ **[2]**

4 a Find these integrals.

i $\int\frac{1}{x+3}\,dx$ ii $\int\sin x\cos^2 x\,dx$ iii $\int 2x(x^2+4)^3\,dx$ **[5]**

b Find the exact value of $\int_0^1 \frac{x}{x^2+1}\,dx$. Show your working. **[4]**

5 a Use integration by parts to calculate each of these integrals.

i $\int xe^x\,dx$ ii $\int x\sin x\,dx$ **[5]**

b Show that the integral $\int_0^{\frac{\pi}{6}} x\sin 2x\,dx$ can be written in the form $\beta\sqrt{3}-\alpha\pi$

where α and β are constants to be found. **[4]**

6 a Find the general solution to the differential equation $\frac{dy}{dx}=\frac{y}{x+1}$

Give your answer in the form $y=f(x)$ **[4]**

b Find the particular solution for curve $y=f(x)$ which passes through the point (1, 8) **[2]**

7 a Express $\frac{10-13x}{(3+x)(1-2x)}$ in the form $\frac{A}{3+x}+\frac{B}{1-2x}$, where A and B are integers. **[3]**

b Hence find the area bounded by the curve $y=\frac{10-13x}{(3+x)(1-2x)}$, the x-axis and the lines $x=-1$ and $x=0$. Show your working. **[4]**

8 a Use the substitution $u=1+2x$ to find $\int\frac{2x}{1+2x}\,dx$. Give your answer in terms of x **[5]**

b Find the exact area of the region bounded by the curve $y=\frac{2x}{1+2x}$, the x-axis and the lines $x=0$ and $x=1$. Show your working. **[2]**

9 Find the value of $\int_1^2 \frac{4x+3}{2x^2+3x-2}\,dx$. Show your working. **[6]**

10 $f(x)=\frac{6x^3+14x^2+11x-1}{3x^2+7x+2}$

a Given that $f(x)$ can be expressed as $Ax+\frac{B}{x+2}+\frac{C}{3x+1}$, find the values of the constants A, B and C **[4]**

The gradient of a curve is given by $\frac{dy}{dx} = f(x)$

b Find the equation of the curve given that it passes through the point (0, 0) **[4]**

11 Find the general solution to the differential equation $\frac{dy}{dx} = \frac{5y}{2-3x-2x^2}$

Give your answer in the form $y = f(x)$ **[7]**

12 A radioactive material decays such that the rate of change of the number, N, of particles is proportional to the number of particles at time t days.

a Write down a differential equation in N and t **[1]**

Initially there are N_0 particles and it takes T days for N to halve.

b Solve your differential equation, giving the solution in terms of N_0, t and T, with N as the subject. **[6]**

13 a Sketch the graphs of $y = \sin 2x$ and $y = \cos x$ on the same axes for $0 \le x \le \pi$

Give the points of intersection with the coordinate axes. **[4]**

b Find the values of x in the interval $[0, \pi]$ of the points of intersection of the two curves. Show your working. **[4]**

c Calculate the total area enclosed between the two curves in the interval $[0, \pi]$. Show your working. **[4]**

14 The curve C has a tangent at the point P as shown.

The equation of C is $y = xe^{\frac{x}{3}}$ and P has x-coordinate 3

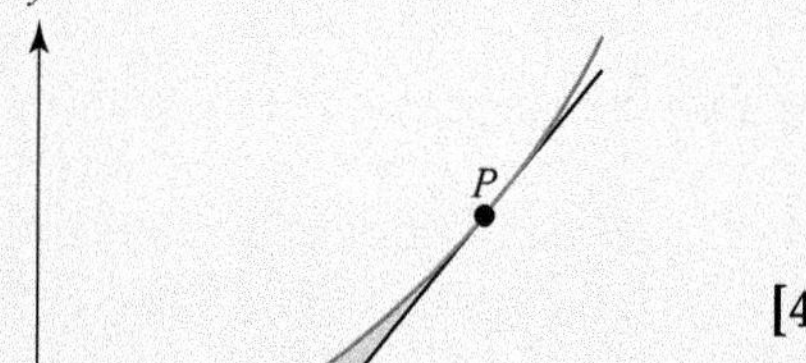

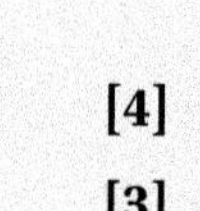

a Find the equation of the tangent to the curve at P. Show all your working. **[4]**

b Work out $\int xe^{\frac{x}{3}} dx$ **[3]**

The area bounded by the curve C, the tangent to the curve at P and the coordinate axes is shaded.

c Calculate the exact value of the shaded area. Show your working. **[5]**

15 $f(x) = \frac{4}{\sqrt{x}(x-4)}$

Use the substitution $u = \sqrt{x}$ to find $\int f(x)\,dx$

Give your answer as a single logarithm in terms of x **[8]**

16 Use an appropriate substitution to find $\int x(2x-5)^4\,dx$, give your answer in terms of x **[5]**

17 a Calculate $\int x^2 \sin x\,dx$ **[4]**

b The area R is bounded by the curve $y = x^2 \sin x$, the x-axis and the lines $x = 0$ and $x = 2\pi$

Calculate the area of R. Show your working and give your answer in terms of π **[4]**

18 At time t minutes, the volume of water in a cylindrical tank is $V\,m^3$. Water flows out of the tank at a rate proportional to the square root of V

a Show that the height of water in the tank satisfies the differential equation

$$\frac{dh}{dt} = -k\sqrt{h}$$ **[4]**

b Find the general solution of this differential equation.

Give your answer in the form $h = f(t)$ **[3]**

The tank is 2 m tall and is initially full. It then takes 2 minutes to fully empty.

c Show that the particular solution to the differential equation is $h = 2\left(1 - \frac{1}{2}t\right)^2$ **[4]**

17 Numerical methods

Atmospheric modelling is important for both understanding climate change and improving weather forecasting accuracy. Numerical analysis is an important tool for this kind of modelling. Partial differential equations, which express relationships between velocity, pressure and temperature, as well as laws for momentum, mass and energy, are needed. Each model will use a slightly different set of equations to place focus on different geographical locations and different atmospheric processes.

Equations like these can often be too complicated to solve algebraically, so mathematicians have developed a range of numerical techniques to find approximate solutions to problems. It is important to determine if these techniques are useful and if the solutions they provide are accurate enough for the purpose. As modern computing has advanced and computers have become more and more powerful, finding and evaluating these numerical methods has been made easier, and hence become a significant area of study.

Orientation

What you need to know

Ch1 Algebra 1

- Finding roots of an equation.

Ch4 Differentiation and integration

- Area under a curve.

Ch13 Sequences p.40

- Increasing, decreasing or periodic sequences.

Ch15 Differentiation 2 p.98

- Differentiation of trigonometric functions.

What you will learn

- To use the change of sign method to find and estimate the root(s) of an equation.
- To use an iterative formula to estimate the root of an equation.
- To recognise the conditions that cause an iterative sequence to converge.
- To use the Newton-Raphson method to estimate the root of an equation.
- To use the trapezium rule to find the area under a curve.

What this leads to

Applications

Weather forecasting.

Computer-aided engineering.

Software development.

MyMaths Practise before you start

2026, 2056, 2165, 2264

17.1 Simple root finding

Fluency and skills

See Ch1.4 For a reminder on finding the roots of an equation.

When you solve an equation you are finding its **roots**.

The roots of the equation $f(x) = 0$ are the x-coordinates of the points where the graph of $y = f(x)$ crosses the x-axis.

For example, the exact roots of the equation $x^2 - 2x - 2 = 0$ are $1 \pm \sqrt{3}$ and are shown on the diagram.

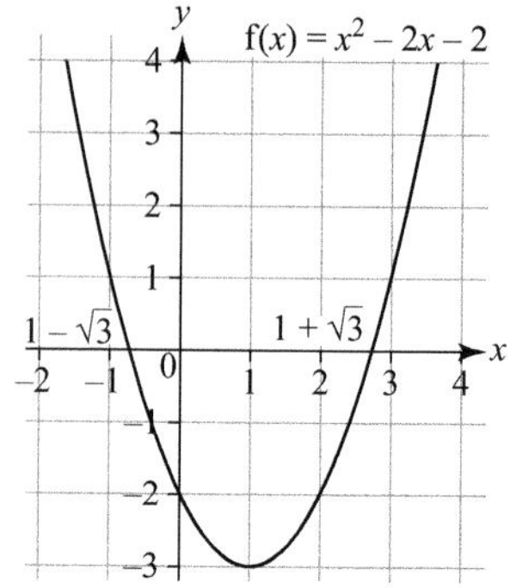

Often, it is difficult, or even impossible, to find the exact value of a root. When this happens you can use a **numerical method** to estimate its value. The simplest numerical method for detecting a root is the **change of sign** method.

Key point

If two real numbers c and d are such that $f(c)$ and $f(d)$ have opposite signs, you say $f(x)$ changes sign between $x = c$ and $x = d$

The diagram shows the graph of $y = f(x)$, which passes through points P and Q, with x-coordinates c and d, respectively.
$f(x)$ changes sign between $x = c$ and $x = d$

Since the curve joins P (which lies below the x-axis) to Q (which lies above it), the curve must cross the x-axis somewhere between $x = c$ and $x = d$. This is the point $R(\alpha, 0)$. Since $f(\alpha) = 0$, α is a root of the equation $f(x) = 0$

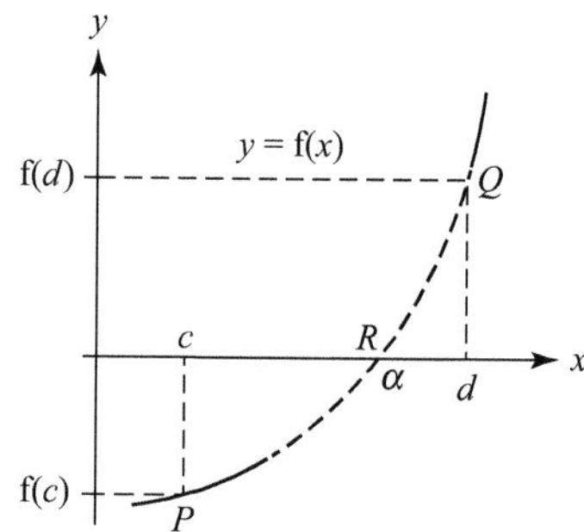

Key point

If $f(x)$ is **continuous** and changes sign between $x = c$ and $x = d$, then the equation $f(x) = 0$ has a root α, where $c < \alpha < d$

Example 1

$f(x) = x^3 - 4x - 4$ is a continuous function.

The equation $x^3 - 4x - 4 = 0$ has exactly one real root α

There is a cubic formula that gives exact roots but it can be difficult to use.

a Show that $2.35 < \alpha < 2.45$ **b** State the value of α, correct to 1 dp.

a $f(x) = x^3 - 4x - 4$

$f(2.35) = (2.35)^3 - 4(2.35) - 4$ and $f(2.45) = (2.45)^3 - 4(2.45) - 4$

$= -0.422125$ $= 0.906125$

Show that f(2.35) and f(2.45) have opposite signs.

$f(2.35) < 0$ and $f(2.45) > 0$ so $f(x)$ changes sign between $x = 2.35$ and $x = 2.45$

By the change of sign method, the equation $f(x) = 0$ has a root between 2.35 and 2.45

So $2.35 < \alpha < 2.45$ (as α is the only real root of this equation)

b Every number between 2.35 and 2.45 equals 2.4 when rounded to 1 dp, so $\alpha = 2.4$ to 1 dp.

2.35 **2.4** 2.45

Make a conclusion about the root.

Calculator

Try it on your calculator

You can use the change of sign method to find roots on a calculator.

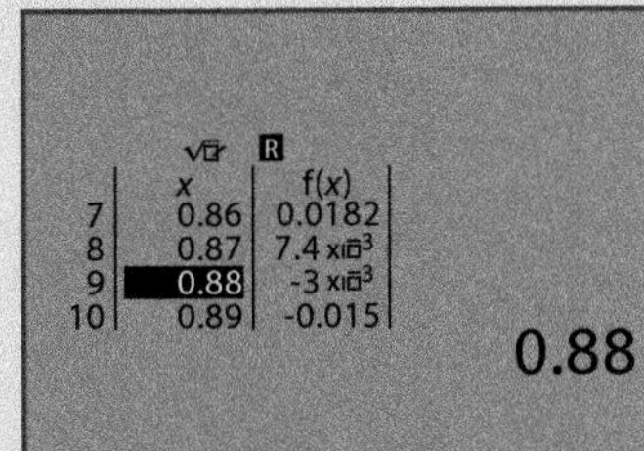

Activity

Find out how to estimate the root of $\sin x - x^2 = 0$ that lies between 0.8 and 0.9 on *your* calculator.

PURE

Exercise 17.1A Fluency and skills

For this exercise you can assume that all functions are continuous. Show your working in each question.

1 Use the change of sign method to show that these equations have a root between the given values c and d

a $7-3x-x^3=0, c=1, d=2$

b $x^2-\frac{1}{x}-4=0, c=2.1, d=2.2$

c $\sin(2x^c)-x^2+3=0, c=\frac{1}{2}\pi, d=\frac{2}{3}\pi$

d $e^x\ln x-x^2=0, c=1.69, d=1.71$

2 a Show that $e^x-x^3=0$ has a root

i α between 1.85 and 1.95

ii β between 4.535 and 4.545

b Hence write down the value of α correct to 1 dp and the value of β correct to 2 dp.

3 i Show that these equations have a root between the given values of c and d. Work in radians where appropriate.

ii Write down the value of this root to 1 dp.

a $\frac{2}{x^3}-\frac{1}{x}-2=0, c=0.85, d=0.95$

b $e^{-x}+2x-1=0, c=-1.35, d=-1.25$

c $x^2\sin x-0.5=0, c=3.05, d=\pi$

4 i Show that these equations have a root between the given values of c and d. Work in radians where appropriate.

ii Write down the value of this root to as many decimal places as can be justified.

a $x^4-3x^3+1=0, c=2.955, d=2.965$

b $e^{\frac{1}{x}}-x^2=0, c=1.414, d=1.424$

c $x^2-\sqrt{x}-2=0, c=1.8305, d=1.8315$

d $2\ln x-\sec x=0, c=\frac{8}{5}\pi, d=5.02725$

e $e^{\cos x}-\cos(e^x)=0, c=-\frac{3}{5}e, d=-\frac{1}{2}\pi$

5 Each of these equations has exactly one real root between 0 and 2

i For which equations does this root lie between 1.75 and 1.85?

a $x^2-\sin x^c-2=0$ b $x^3-\cos x^c-6=0$

c $\frac{1}{1+\tan x^c}-x+2=0$ d $2-x\operatorname{cosec} x=0$

ii For which of these equations is this root equal to 1.76 to 2 dp?

See Ch14.1 For a reminder on radian notation. A superscript c shows that x is in radians.

6 Each of the equations **i**, **ii** and **iii** can be paired, in some order, with exactly one of **A**, **B** and **C** to make a true statement.

a Find these pairings.

b Use these pairings to write down the value of each root to as many decimal places as can be justified.

i $x^2-\frac{1}{x}-2=0$

ii $e^{-x}-3\sin^2 x^c+1.8=0$

iii $x-\frac{1}{x^2}-2=0$

A Has a root between 2.205 and 2.215

B Has a root between 2.21 and 2.22

C Has a root between $\frac{1}{2}\pi$ and $\frac{3}{5}e$

7 Show that each of these equations has a root between 0 and the positive constant a

a $x^2+2x-2a=0$ b $ax^2+x-a^3=0$

c $\cos\left(\frac{\pi}{a}x\right)-\frac{a}{\pi}x=0$ d $x^3+(a+1)x^2-2a^3=0$

Reasoning and problem-solving

Strategy

To solve problems involving finding roots of equations

1. Sketch a graph to determine the number of real solutions in the interval.
2. Rearrange the equation into the form $f(x) = 0$
3. Use a suitable interval to test for any possible roots.
4. Test using the change of sign method and give any necessary conclusions.

Example 2

a Use a sketch to show that the equation $2^x = 4 - x$ has exactly one solution, α

b Show that $\alpha = 1.4$ to 1 dp.

The change of sign method works only for continuous functions – not when there is an asymptote.

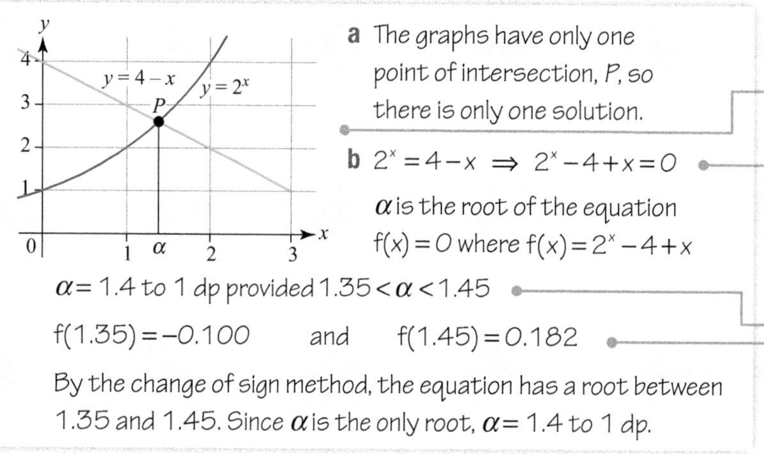

a The graphs have only one point of intersection, P, so there is only one solution.

(1) Sketch the graphs of $y = 2^x$ and $y = 4 - x$. You can check your sketch using a graphics calculator.

b $2^x = 4 - x \Rightarrow 2^x - 4 + x = 0$

α is the root of the equation $f(x) = 0$ where $f(x) = 2^x - 4 + x$

(2) Rearrange the equation into the form $f(x) = 0$

$\alpha = 1.4$ to 1 dp provided $1.35 < \alpha < 1.45$

(3) Use a suitable interval.

$f(1.35) = -0.100$ and $f(1.45) = 0.182$

(4) Test using the change of sign method.

By the change of sign method, the equation has a root between 1.35 and 1.45. Since α is the only root, $\alpha = 1.4$ to 1 dp.

You can describe a range of numbers using **interval** notation. For $c, d \in \mathbb{R}$, the **open interval** (c, d) is the set of real numbers x such that $c < x < d$. $f(x)$ is **continuous** on an interval (c, d) if you can draw the graph of $y = f(x)$ without taking your pen off the paper.

Example 3

If $f(x) = \dfrac{1}{x-2}$, $x \neq 2$, **a** Show that $f(x)$ changes sign across the interval $(1, 3)$

b Use a sketch to show that the equation $f(x) = 0$ has no real roots. Comment on this result.

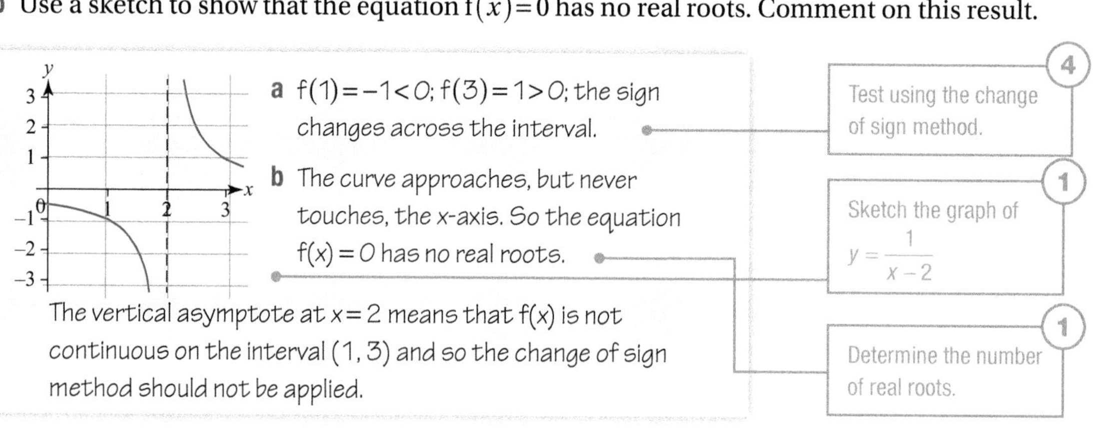

a $f(1) = -1 < 0$; $f(3) = 1 > 0$; the sign changes across the interval.

(4) Test using the change of sign method.

b The curve approaches, but never touches, the x-axis. So the equation $f(x) = 0$ has no real roots.

(1) Sketch the graph of $y = \dfrac{1}{x-2}$

(1) Determine the number of real roots.

The vertical asymptote at $x = 2$ means that $f(x)$ is not continuous on the interval $(1, 3)$ and so the change of sign method should not be applied.

Key point

A continuous function $f(x)$ does not change sign in an interval which contains an even number of roots (counting repetitions) of the equation $f(x) = 0$

Exercise 17.1B Reasoning and problem-solving

1 a Show that the equation $x^3 - x^2 = 1$ has a solution α in the interval $(1.4, 1.5)$

b By sketching on a single diagram the graph of $y = x^3 - x^2$ and the graph of $y = 1$, show that α is the only real solution of the equation $x^3 - x^2 = 1$

c Let β be the solution to the equation $8x^3 - 4x^2 = 1$. Use the result of part **a** to find the value of β to 1 dp.

2 a Find the coordinates of the stationary point on the curve $y = 2x - x^2$. Show your working.

b Show, by sketching a graph, that $2x - x^2 = 0.5^x$ has exactly two solutions.

c Given that these two solutions are α and β, where $\beta > 0.5$, show that

i α lies in the interval $(0.44, 0.48)$

ii β lies in the interval $(1.84, 1.88)$

d Hence find the value of $\beta - \alpha$ to 1 dp.

3 $f(x) = \tan(x-1) - 1$, where x is in radians.

a Solve $f(x) = 0$ for $0 \le x \le \pi$, to 1 dp. Show your working.

b Show that $f(x)$ changes sign across the interval $(2, 3)$

c Hence explain why $f(x)$ cannot be continuous in the interval $(2, 3)$

d Find, to 1 dp, the x-value at which $f(x)$ is not continuous for $2 \le x \le 3$

4 The diagram shows the curve of $y = \sqrt{x}$, $x \ge 0$, which is an increasing function.

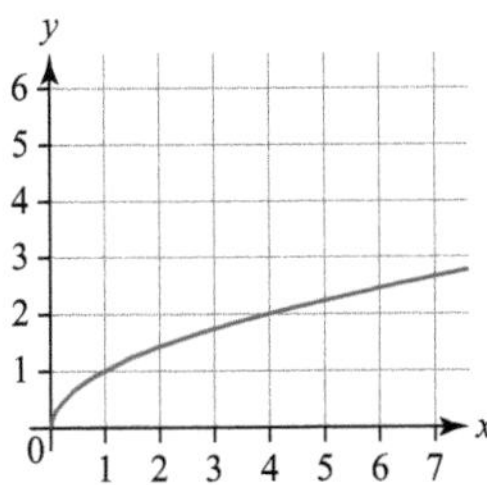

a On a copy of this diagram, draw a suitable straight line to show that the equation $2\sqrt{x} = 6 - x$ has exactly one real solution, α

b By applying the change of sign method to a suitable function and interval, show that $\alpha = 2.7$ correct to 1 dp.

c i Find the exact solution of the equation $2\sqrt{x} = 6 - x$, giving your answer in the form $a + b\sqrt{7}$. Show your working.

ii Hence show that $\sqrt{7} \approx \frac{53}{20}$

5 Given the continuous function $f(x) = 4\sin(\pi x) - 6x - 1$, where x is in radians,

a Show that $f(x)$ does not change sign across the interval $(0, 1)$,

b Evaluate $f\left(\frac{1}{6}\right)$ and hence explain why the equation $f(x) = 0$ must have at least two roots in the interval $(0, 1)$,

c Find all the roots of $4\sin(\pi x) - 6x - 1 = 0$. Justify that you have found all the roots by sketching two suitable graphs on a single diagram.

6 The continuous and differentiable function $f(x)$ changes sign across the intervals (a, b) and (b, c). Which of these statements are true? Support your answers with a reason.

a $f(x)$ does not change sign across (a, c)

b The equation $f(x) = 0$ has no roots in (a, c)

c The equation $f'(x) = 0$ has at least one root in the interval (a, c)

Challenge

7 Let $f(x) = x^3 - 2x - 1$

The equation $f(x) = 0$ has exactly one real root α in the interval $(1, 2)$ as shown in the diagram.

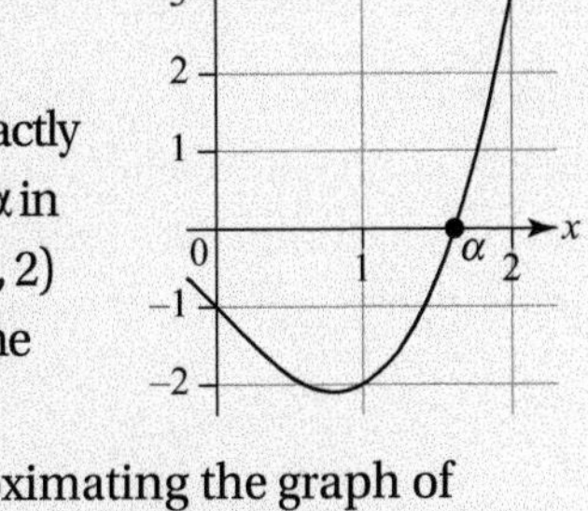

a By approximating the graph of $y = f(x)$ as a straight line across the interval $(1, 2)$, show that an approximate value of α is 1.4

b By repeating this method across $(1.4, 2)$, find another approximation for α and show it is accurate to 1 dp.

c Find the exact value of α

17.2 Iterative root finding

Fluency and skills

If an equation $x = g(x)$ has a solution α, then you can use an **iterative formula** written as $x_{n+1} = g(x_n)$ to solve the equation numerically. If the starting point x_1 is close to α, then the iterative formula produces a sequence $x_2, x_3, \ldots$ which can **converge** to α

See Ch 13.2
For a reminder on sequences.

Key point

A sequence $x_1, x_2, x_3\ldots$ converges to α if, as n increases, x_n gets ever closer to α

Example 1

A convergent sequence is defined by $x_{n+1} = 0.5x_n + 2$ with $x_1 = 3$

a Find the values of x_2, x_3 and x_4 **b** To what value does the sequence converge?

a $x_2 = 0.5x_1 + 2 = 0.5(3) + 2 = 3.5$

$x_3 = 0.5x_2 + 2 = 0.5(3.5) + 2 = 3.75$

$x_4 = 0.5x_3 + 2 = 0.5(3.75) + 2 = 3.875$

The terms $x_2, x_3\ldots$ are the iterates of the formula.

Substitute the values of x_1, x_2 and x_3

b Further calculations give $x_{10} = 3.998\ldots$ and $x_{20} = 3.999998\ldots$
The sequence appears to converge to 4. This can be confirmed by substituting 4 into the iterative formula and checking it is a solution.

As n increases, x_n gets ever closer to 4

Staircase or **cobweb** diagrams can be used to display iterates given by $x_{n+1} = g(x_n)$

To do this, draw the curve C with equation $y = g(x)$ and the straight line L with equation $y = x$
Label the point of intersection P and mark the x-coordinate of P as α. This is a solution to the equation $x = g(x)$

Go to curve C first – C comes before L in the alphabet.

Start at the point $(x_1, 0)$, given that x_1 is your initial guess for the root of the equation. Draw a vertical line until you reach the curve C, and then draw a horizontal line until you reach the line L. Repeat this process for the required number of iterations.

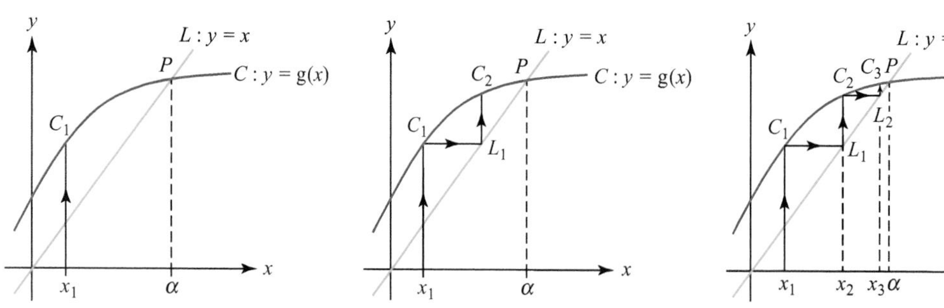

As the points $C_1, C_2, C_3, \ldots$ on the curve C converge to the point P, the sequence $x_1, x_2, x_3, \ldots$ converges to the solution α

This is an example of a **staircase** diagram.

Calculator

Try it on your calculator

You can draw a staircase diagram on a graphics calculator.

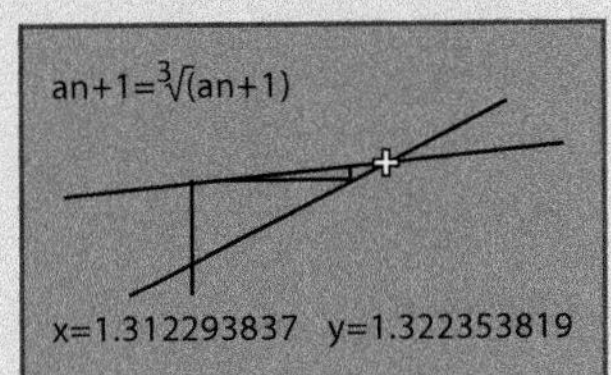

Activity

Find out how to draw a staircase diagram to display the iteration given by $x_{n+1} = \sqrt[3]{x_n + 1}$ with $x_0 = 1$ on *your* graphics calculator.

ICT Resource online

To investigate iterative root finding, click this link in the digital book.

PURE

Example 2

The diagram shows the curve with equation $y = g(x)$ and the line $y = x$. Also shown is a solution α to the equation $x = g(x)$ and an approximation to α, x_1

Display the iterates x_2, x_3 and x_4 found using the iterative formula $x_{n+1} = g(x_n)$

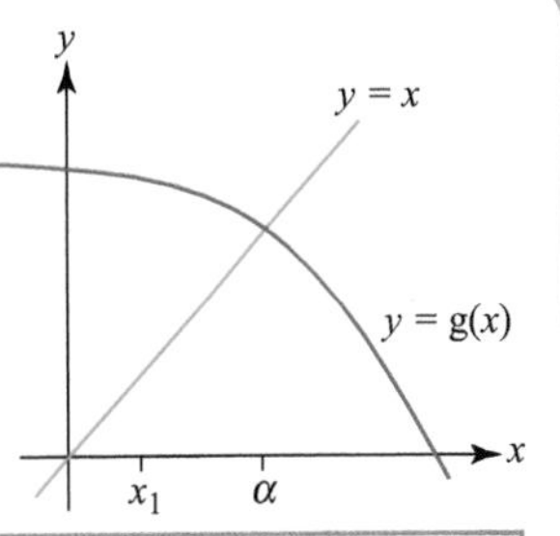

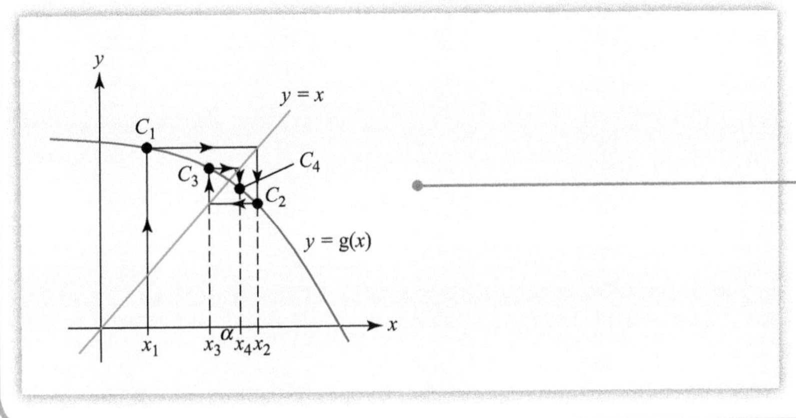

Start at the point $(x_1, 0)$. Draw a vertical line until you reach the curve $y = g(x)$, and then draw a horizontal line until you reach the line $y = x$. As you repeat this process, you obtain a more and more accurate approximation to the root.

This is a converging **cobweb** diagram.

Exercise 17.2A Fluency and skills

1 Each of these equations $x = g(x)$ has a solution α given to 3 dp.

i Use the iterative formula $x_{n+1} = g(x_n)$ and starting value $x_1 = 1$ to calculate x_2, x_3 and x_4, giving answers to 3 dp where appropriate.

ii Find the first iterate x_n such that $x_n = \alpha$ when both of these values are rounded to 3 dp.

a $x = 1 - 0.1x^3$, $\alpha = 0.922$

b $x = 2 + \sqrt[3]{x}$, $\alpha = 3.521$

c $x = \dfrac{2}{x^3 - 3}$, $\alpha = -0.618$

d $x = \cos(x - 1) + 0.6$, x in radians, $\alpha = 1.485$

2 For each equation $x = g(x)$, the sequence defined by the iterative formula $x_{n+1} = g(x_n)$ with the given starting value, converges to a solution α

i Use this iterative formula and starting value x_1 to find α correct to 2 dp, working in radians where appropriate.

2174

ii Write down an interval of width 0.01 with a mid-value of α

a $x=e^{-x}+2,\ x_1=1$ b $x=\dfrac{3x^2+2}{x^2-2},\ x_1=3$ c $x=\sqrt{2x+\ln x},\ x_1=2$

d $x=\left(4x^2-1\right)^{\frac{1}{3}},\ x_1=3$ e $x=2\sin x-\cos x,\ x_1=3$ f $x=\sqrt{x+e^{-x}},\ x_1=0$

3 The equation $x=\sin(\cos x)-\cos(\sin x)+1$, where x is in radians, has exactly one real root $\alpha=0.878$ to 3 dp. It is given that the sequence defined by the iterative formula $x_{n+1}=\sin(\cos x_n)-\cos(\sin x_n)+1$, with starting value $x_1=0$, converges to α

a Find the smallest number, N, of iterations required to produce an iterate which equals α when both numbers are rounded to 3 dp.

b Show that using this iterative formula with starting value $x_1=100$ requires *fewer* than N iterations to achieve this level of accuracy.

4 A student is asked to draw either a staircase or cobweb diagram on each of these sketches. Starting at x_1, the student draws three lines on each sketch, but in each case they make a mistake.

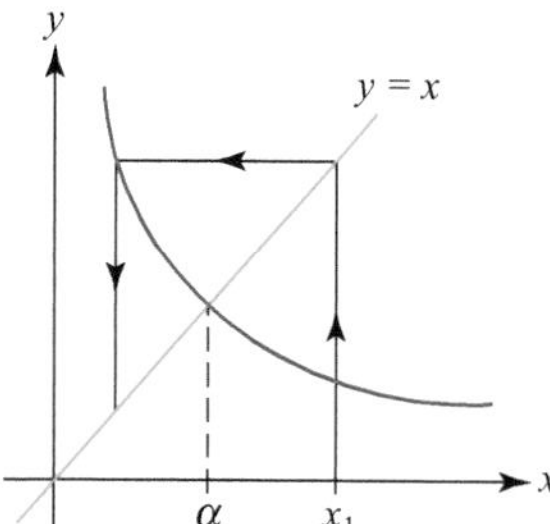

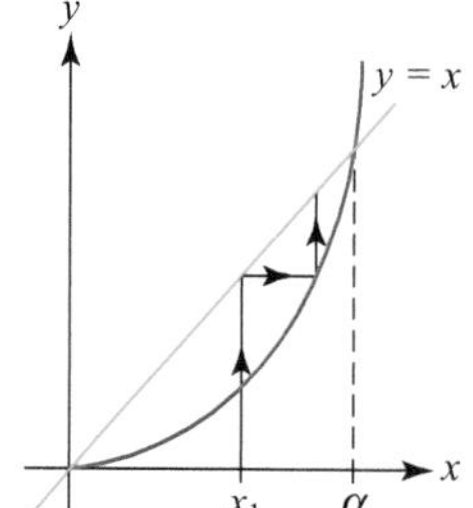

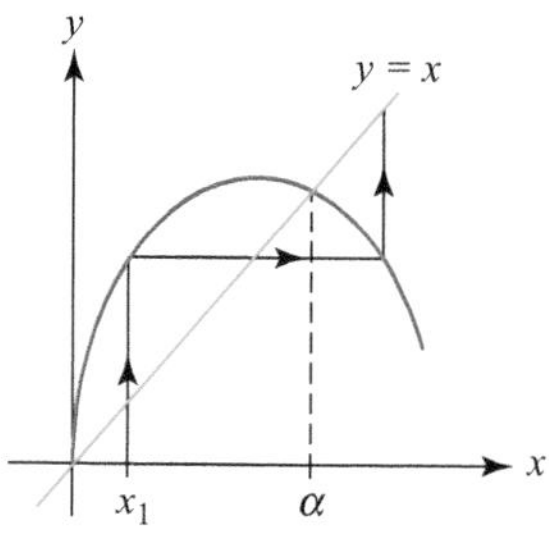

a On a copy of each sketch, draw the three arrowed lines that the student should have drawn.

It is given that, if continued, each diagram would show convergence.

b Which diagram will definitely not illustrate convergence to the solution α?

5 The diagram shows the curve with equation $y=g(x)$, the line $y=x$ and the points P and Q where the curve and line intersect. P is the stationary point on the curve $y=g(x)$ and the solutions of the equation $x=g(x)$ are α and β

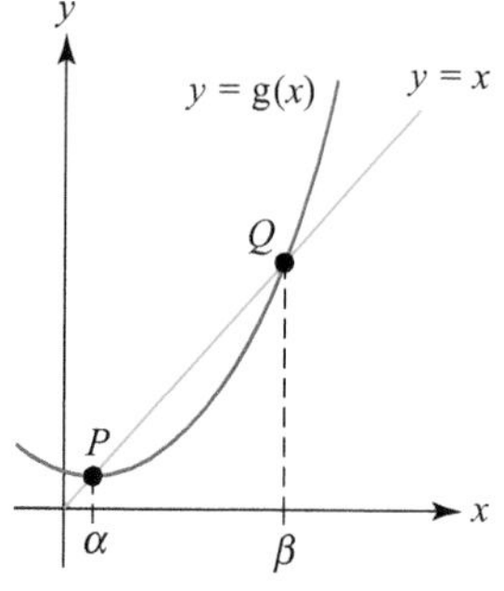

Using a copy of the diagram and the iterative formula $x_{n+1}=g(x_n)$

a Determine which of α or β these iterates converge to when x_1 is just less than β,

b Illustrate and describe the behaviour of these iterates when x_1 is just greater than β

Reasoning and problem-solving

Strategy

To solve problems that involve finding an approximation to the root of an equation

1. Select an appropriate way to rearrange the equation $f(x)=0$ into the required form $x=g(x)$
2. Apply the iterative formula $x_{n+1}=g(x_n)$ with a suitable starting value x_1
3. Use the change of sign method to prove accuracy levels have been achieved.
4. Examine the value of $g'(x)$ near a solution to determine if convergence to that root will occur.
5. Give your conclusion.

Example 3

The diagram shows the graph of $y = \text{f}(x)$, where $\text{f}(x) = x^3 - 6x + 2,\ x \geq 0$

The equation $\text{f}(x) = 0$ has exactly two positive roots α and β

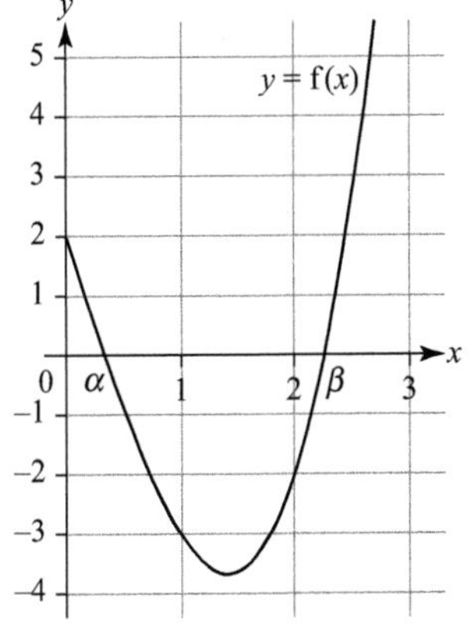

a **i** Show that the equation $\text{f}(x) = 0$ can be rearranged into the form $x = \dfrac{x^3 + 2}{6}$

ii Use the iterative formula $x_{n+1} = \dfrac{x_n^3 + 2}{6}$ with $x_1 = 0.5$ to find α to 2 dp, justifying your answer.

b **i** Show that the equation $\text{f}(x) = 0$ can be rearranged into the form $x = \sqrt{\dfrac{6x - 2}{x}}$

ii Comment on the suitability of the iterative formula $x_{n+1} = \sqrt{\dfrac{6x_n - 2}{x_n}}$ with $x_1 = 0.5$ for estimating α

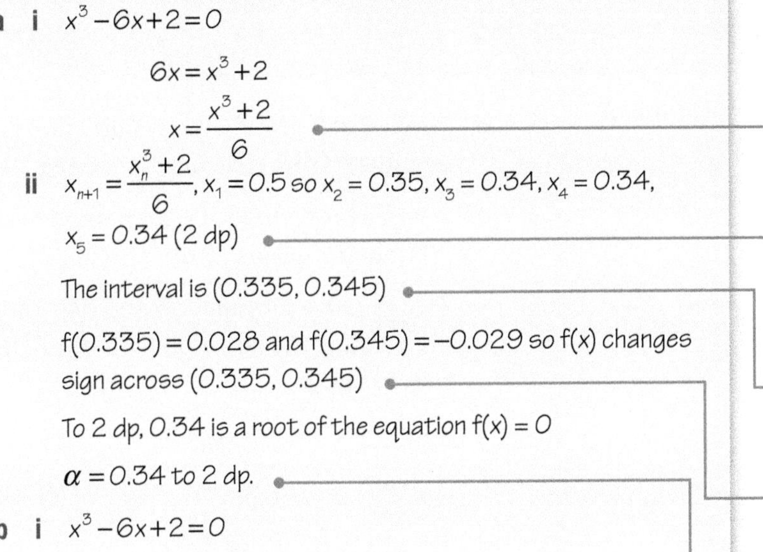

a **i** $x^3 - 6x + 2 = 0$

$$6x = x^3 + 2$$

$$x = \frac{x^3 + 2}{6}$$

Rearrange $x^3 - 6x + 2 = 0$ to make x the subject. (1)

ii $x_{n+1} = \dfrac{x_n^3 + 2}{6}$, $x_1 = 0.5$ so $x_2 = 0.35$, $x_3 = 0.34$, $x_4 = 0.34$, $x_5 = 0.34$ (2 dp)

Apply the iterative formula. Continue calculating iterates until their values agree to 2 dp. (2)

The interval is (0.335, 0.345)

All numbers in this interval round to 0.34 to 2 dp.

$\text{f}(0.335) = 0.028$ and $\text{f}(0.345) = -0.029$ so $\text{f}(x)$ changes sign across (0.335, 0.345)

Use the change of sign method to check accuracy. (3)

To 2 dp, 0.34 is a root of the equation $\text{f}(x) = 0$

$\alpha = 0.34$ to 2 dp.

Give your conclusion. (5)

b **i** $x^3 - 6x + 2 = 0$

$$x^3 = 6x - 2$$

$$\frac{x^3}{x} = \frac{6x - 2}{x}$$

$$x = \sqrt{\frac{6x - 2}{x}}$$

Rearrange $x^3 - 6x + 2 = 0$ to make x the subject. (1)

ii $x_{n+1} = \sqrt{\dfrac{6x_n - 2}{x_n}}$, $x_1 = 0.5$ so $x_2 = 1.41$, $x_3 = 2.14$, $x_4 = 2.25$, $x_5 = 2.26$, $x_6 = 2.26$ (2 dp).

Apply the iterative formula. (2)

This sequence converges to β so this formula is not suitable for estimating α

Give your conclusion. (5)

You can use the gradient of the curve to determine if $x_{n+1} = \text{g}(x_n)$ converges to the solution α

Key point

If $-1 < \text{g}'(x) < 1$ for all x in an interval which contains α and the starting value x_1, then $x_{n+1} = \text{g}(x_n)$ converges.

MyMaths 2174 SEARCH

Example 4

The equation $x = x^3 - 2$ has a solution α near 1.5

a Show that the sequence produced by $x_{n+1} = x_n^3 - 2$ with $x_1 = 1$ fails to converge to α

b Explain the cause of this failure.

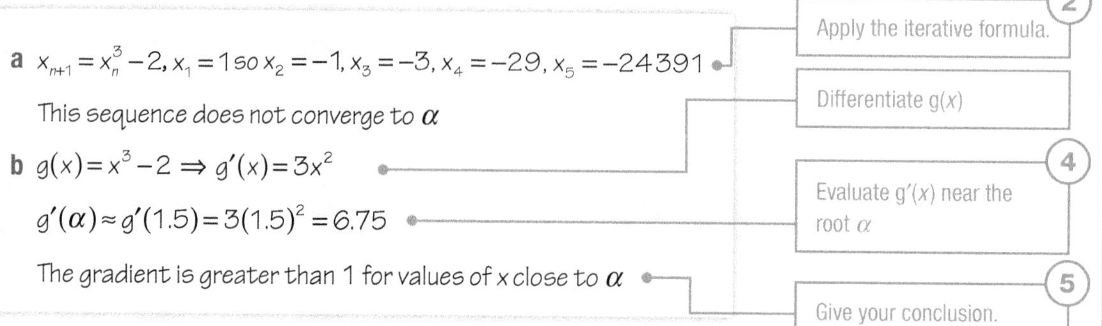

a $x_{n+1} = x_n^3 - 2$, $x_1 = 1$ so $x_2 = -1$, $x_3 = -3$, $x_4 = -29$, $x_5 = -24391$ — Apply the iterative formula. (2)

This sequence does not converge to α

b $g(x) = x^3 - 2 \Rightarrow g'(x) = 3x^2$ — Differentiate g(x)

$g'(\alpha) \approx g'(1.5) = 3(1.5)^2 = 6.75$ — Evaluate g′(x) near the root α (4)

The gradient is greater than 1 for values of x close to α — Give your conclusion. (5)

Exercise 17.2B Reasoning and problem-solving

1 The diagram shows part of the graph of $y = f(x)$ where $f(x) = x^4 - 2x - 1$

The equation $f(x) = 0$ has exactly two real roots α and β

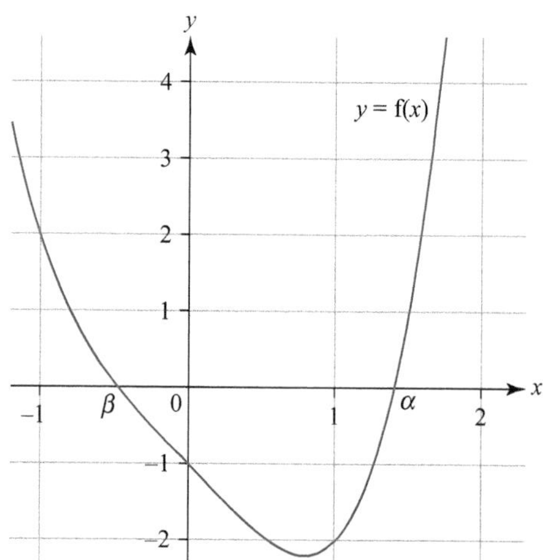

a Show that the equation $x^4 - 2x - 1 = 0$ can be rearranged into the form $x = \frac{\sqrt{2x+1}}{x}$

b Use the iterative formula $x_{n+1} = \frac{\sqrt{2x_n + 1}}{x_n}$ with $x_1 = 1$ to find α to 1 dp, justifying your answer.

c Show that $x_1 = -0.5$ is not an appropriate starting value when using the iterative formula $x_{n+1} = \frac{\sqrt{2x_n + 1}}{x_n}$ to estimate β

2 Given that $f(x) = x^3 - 3x^2 - 4$ and that the equation $f(x) = 0$ has exactly one real root, α

a Use the change of sign method to show that $\alpha = 3.355$ to 3 dp.

Two iterative formulae (I) and (II), found by rearranging the equation $f(x) = 0$, are used to estimate α

(I) $x_{n+1} = \sqrt[3]{3x_n^2 + 4}$, $x_1 = 3$

(II) $x_{n+1} = \sqrt{\frac{3x_n^2 + 4}{x_n}}$, $x_1 = 3$

b Compare how quickly each of these formulae produces an estimate for α which is correct to 3 dp.

c **i** Show that the equation $f(x) = 0$ can be rearranged into the form $x_{n+1} = g(x_n)$, where $g(x) = \frac{2}{\sqrt{x-3}}$

ii By finding $g'(x)$ determine whether or not this is a suitable rearrangement for estimating α. Support your answer by evaluating iterates.

3 At the start of an experiment substance A is being heated whilst substance B is cooling down. All temperatures are measured in °C. The equation $T_A = 10e^{0.1t}$ models the temperature T_A of substance A and the equation $T_B = 16e^{-0.2t} + 25$ models the

temperature T_B of substance B, t minutes from the start.

a Show that the time t at which the two substances have equal temperatures satisfies the equation $t = 10\ln(1.6e^{-0.2t} + 2.5)$

b Use the iterative formula $t_{n+1} = 10\ln(1.6e^{-0.2t_n} + 2.5)$ with $t_1 = 0$ to find this time, giving your answer to the nearest minute.

All the logarithm keys on Jane's calculator have stopped working.

c By letting $x = e^{0.1t}$ use a suitable iterative formula of the form $x_{n+1} = g(x_n)$ that will enable Jane to find the approximate time at which the two substances have equal temperatures.

4 **a** By sketching a pair of graphs on a single diagram, show that the equation $\theta = 2\sin\theta$, where θ is in radians, has exactly one solution between $\frac{1}{2}\pi$ and π

The diagram (not drawn to scale) shows a sector OAB of a circle, radius 4 cm. The angle subtended by the arc AB at the centre O is α radians.

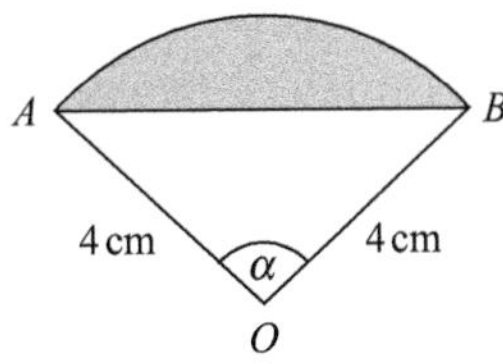

Given that the area of triangle OAB equals the area of the shaded segment

b Show that $\alpha = 2\sin\alpha$

c Use the iterative formula $\theta_{n+1} = 2\sin\theta_n$ and a suitable starting value to find angle α to 2 dp, justifying your answer.

5 Two points A and B on a circle, centre O, radius r cm are such that the length of arc AB is twice the length of the chord AB

a Show that $\alpha = \sqrt{8 - 8\cos\alpha}$, where α is the angle, in radians, subtended by this arc at the centre of this circle.

The equation $\theta = \sqrt{8 - 8\cos\theta}$ has exactly one positive solution.

b Use the iterative formula $\theta_{n+1} = \sqrt{8 - 8\cos\theta_n}$ with starting value $\theta_1 = 3$ radians to find angle α to 1 dp, justifying your answer.

c Given that $r = 2$ cm, calculate the area of triangle OAB to 1 decimal place.

Two students are investigating the convergent sequence defined by $\theta_{n+1} = \sqrt{2 - 2\cos\theta_n}$ with $\theta_1 = 1$ radian. One claims this sequence converges to a positive value, the other claims that it converges to zero. To determine who is correct, they are given this diagram which shows a circle with centre R and radius 1. The angle subtended by the arc PQ at the centre O is θ radians.

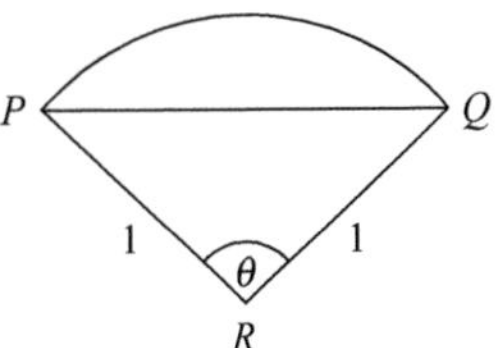

d **i** What does the equation $\theta = \sqrt{2 - 2\cos\theta}$ imply about the relationship between the arc length PQ and the length of the chord PQ?

ii Determine which student made the correct claim.

Challenge

6 Consider the equation $x^{\ln 2} - 2^{\ln x} = 0$, where $x > 0$

One possible iterative formula that comes from rearranging this equation is $x_{n+1} = \left(2^{\ln x_n}\right)^{\frac{1}{\ln 2}}$

a Investigate the behaviour of this sequence using different positive starting values.

b Comment on your observations and explain your findings.

17.3 Newton-Raphson root finding

Fluency and skills

See Ch13.2 For a reminder on sequences.

The **Newton-Raphson** method is another way to estimate a root of an equation.

See Ch15 For a reminder on differentiation.

Key point

If α is a root of the equation $f(x) = 0$, then the iterative sequence given by $x_{n+1} = x_n - \frac{f(x_n)}{f'(x_n)}$ converges to α, if it converges.

The starting value x_1 is a **first approximation** to α. In general, x_{n+1} is a better approximation to α than x_n

Example 1

The equation $x^3 - 2x - 3 = 0$ has exactly one real root α, where $1 < \alpha < 2$

a Taking $x_1 = 2$ as a first approximation to α, use the Newton-Raphson method to find the second and the third approximations to α. Give answers to 3 decimal places where appropriate.

b Use a change of sign to show that the iterates converge to 1.893 (3 dp).

a $f(x) = x^3 - 2x - 3$ — Define a suitable function f(x) which has a root α

$f'(x) = 3x^2 - 2$ — Differentiate.

$x_2 = x_1 - \frac{f(x_1)}{f'(x_1)} = 2 - \frac{f(2)}{f'(2)}$ — Substitute $x_1 = 2$

$f(2) = 2^3 - 2 \times 2 - 3 = 1$ and $f'(2) = 3 \times 2^2 - 2 = 10$ — Work out f(2) and f′(2)

$x_2 = 2 - \frac{1}{10} = 1.9$ — Substitute and evaluate x_2

A second approximation for α is $x_2 = 1.9$

$x_3 = x_2 - \frac{f(x_2)}{f'(x_2)}$ — Use x_2 to find x_3

$= 1.9 - \frac{f(1.9)}{f'(1.9)}$ — Substitute $x_2 = 1.9$

$f(1.9) = 0.059$ and $f'(1.9) = 8.83$

$x_3 = 1.8933\ldots$ — Substitute and evaluate x_3

A third approximation for α is $x_3 = 1.893$ (3 dp)

b $f(x) = x^3 - 2x - 3$

$f(1.8925) = -0.0069\ldots$ and $f(1.8935) = 0.0018\ldots$ — Apply the change of sign method across a suitable interval.

This change of sign shows that $\alpha = 1.893$ (3 dp)

Since $x_3 = 1.893$, the Newton-Raphson method has found α correct to 3 dp in just two iterations.

In practice, you can use the [Ans] key to calculate the iterates $x_2, x_3, x_4, \ldots$ To do this for the previous example you press [2] [=] to enter the starting value $x_1 = 2$, then key in the sequence

$$\boxed{\text{Ans}} - \left(\frac{\boxed{\text{Ans}}^3 - 2\boxed{\text{Ans}} - 3}{3\boxed{\text{Ans}}^2 - 2}\right)$$

to enter the iterative formula and then repeatedly press [=] to display each iterate.

Exercise 17.3A Fluency and skills

1 Use the Newton-Raphson method to find second and third approximations to a root α of the given equation, where x_1 is a first approximation to this root. Give answers to 2 dp where appropriate.

a $x^3-2x^2+x-3=0, 2<\alpha<3,\ x_1=2$

b $2x-x^3+5=0, 2<\alpha<3,\ x_1=2$

c $x^4-2x^3-1=0, -1<\alpha<0,\ x_1=-1$

d $x^2-2x^{\frac{1}{2}}-8=0, 3<\alpha<4,\ x_1=3$

e $x^3+\frac{4}{x}-6=0, 1<\alpha<2,\ x_1=2$

f $3\sqrt[3]{x}+\frac{1}{2x^2}-5=0, 4<\alpha<5,\ x_1=4$

2 Each of these equations has exactly one real root, α. Use the Newton-Raphson method with the given first approximation x_1 to find α to 3 dp. Justify that this level of accuracy has been achieved by using the change of sign method.

a $e^x+3x-4=0,\ x_1=1$

b $x^2-3e^{2x}=0,\ x_1=0$

c $x^2+3\ln x=0,\ x_1=1$

d $\sin x+x-3=0,\ x_1=0$ radians

e $x-\cos^2 x-3=0,\ x_1=0$ radians

f $x^2\ln x-2=0,\ x_1=0.7$

3 The equation $x^3-2x^2-7=0$ has exactly one real root α

a Show that two iterations of the Newton-Raphson method with first approximation $x_1=3$ are sufficient to produce an estimate for α which is accurate to 3 dp.

b Determine whether using the Newton-Raphson method with first approximation $x_1=1$ produces a reliable estimate for α

4 The equation $(x+\sin x)^2-1=0$ has exactly one positive root α

a Working in radians, show that two iterations of the Newton-Raphson method with first approximation $x_1=1$ produces an estimate for α which is

i Accurate to 2 dp,

ii Not accurate to 3 dp.

b Determine whether using the Newton-Raphson method with first approximation $x_1=2$ produces a reliable estimate for α

5 Given that $f(x)=\sin x+e^{-x}$, use the Newton-Raphson method (working in radians) with $x_1=2$ to find, correct to 3 dp, an approximate solution to the equation

a $f(x)=0$ b $f'(x)=0$

For each equation, justify that this level of accuracy has been achieved.

6 The Newton-Raphson method, with the given first approximation x_1, is used to solve these equations.

i For each equation, show that x_3 cannot be calculated, giving the reason.

ii Use the Newton-Raphson method to find all the solutions to each equation shown in the diagrams. Give your answers to 3 dp, justifying that this level of accuracy has been achieved.

a $2\sqrt{x}-\frac{1}{x}+1=0,\ x_1=1$

b $x^3-8\sqrt{x}+2=0,\ x_1=1$

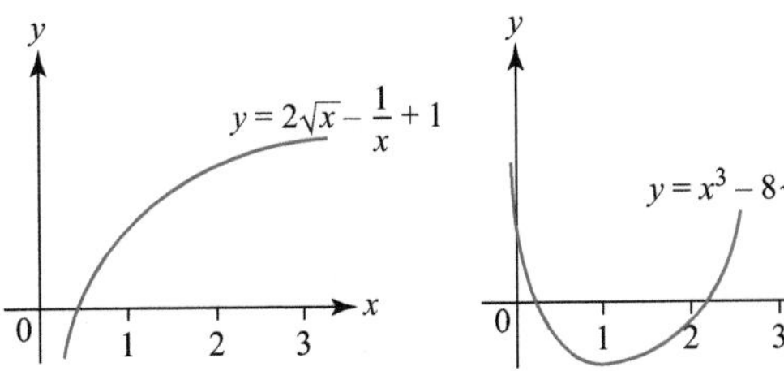

Reasoning and problem-solving

Strategy

To use the Newton-Raphson method to find solutions to practical problems

1. Rearrange the equation you are trying to solve into the form $f(x) = 0$
2. Use your calculator to efficiently calculate iterates.
3. Use the change of sign method to prove the accuracy of your estimate.

Example 2

The diagram shows the design for a sports field in the shape of a sector of a circle ABC, centre A, radius r. The acute angle $BAC = \alpha$ radians, and point D on AB is such that the line CD is perpendicular to AB

The length CB must be 30% longer than CD

Use the Newton-Raphson method to find α to 1 dp.

Arc length $CB = r\alpha$ — Find expressions for the lengths of CB and CD (1)

Using trigonometry, $\sin\alpha = \frac{CD}{AC}$ Length $AC = r$, so length, $CD = r\sin\alpha$

Length $CB = 1.3 \times$ length CD, so $r\alpha = 1.3r\sin\alpha \Rightarrow \alpha = 1.3\sin\alpha$

α is a solution to the equation $\theta = 1.3\sin\theta \Rightarrow \theta - 1.3\sin\theta = 0$ — Rearrange the equation.

$f(\theta) = \theta - 1.3\sin\theta$ so, $f'(\theta) = 1 - 1.3\cos\theta$ — Differentiate.

$\theta_{n+1} = \theta_n - \frac{f(\theta_n)}{f'(\theta_n)} = \theta_n - \frac{(\theta_n - 1.3\sin\theta_n)}{(1 - 1.3\cos\theta_n)}$, with $\theta_1 = 1$ radian — Find the iterative formula. 1 radian is a reasonable first approximation. (2)

$\theta_2 = 1.31...,\ \theta_3 = 1.22...,\ \theta_4 = 1.22...$ — The iterates converge to $\alpha = 1.2$ radians (to 1 dp).

$f(1.15) = -0.03659...$ and $f(1.25) = 0.01632...$

This change of sign proves that $\alpha = 1.2$ radians (to 1 dp) — Use the change of sign method across the interval (1.15,1.25) (3)

See Ch14.1 For a reminder on radians and the arc length formula.

See Ch15.2 For a reminder on differentiating trigonometric functions.

When using the Newton-Raphson method, you must choose a suitable value for the first approximation. In the example, $\theta_1 = 1$ radian was used as a first approximation to α

If, instead, $\theta_1 = 0.6$ radians, then $\theta_2 = -1.237...$, $\theta_3 = -1.221...$, $\theta_4 = -1.221...$

Clearly this sequence does not converge to the required root α. The reason for this failure is that the first approximation is close to the θ-coordinate of a stationary point on the curve $y = \theta - 1.3\sin\theta$

The diagram shows the point P on this curve where $\theta = 0.6$

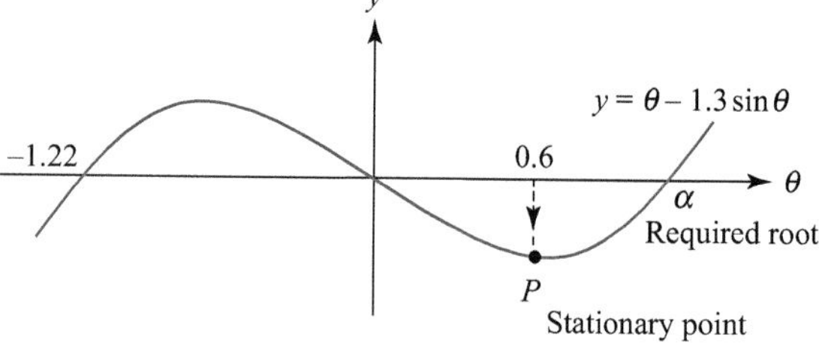

Key point

The Newton-Raphson method may not converge to the required root if the first approximation is close to the x-coordinate of a stationary point on the curve $y = f(x)$.
The Newton-Raphson method will also fail to converge if the derivative is zero at one of the values for x_n

Exercise 17.3B Reasoning and problem-solving

1 The equation $4x^3-12x^2+9x-1=0$ has exactly one real root α in the interval $(0,1)$

a Use the Newton-Raphson method with a first approximation $x_1=0$ to find α to 2 dp, justifying your answer.

b Explain why the Newton-Raphson method fails to find an estimate for α when the first approximation is

i $x_1=0.5$ ii $x_1=1$

c Using the factor theorem, find the exact value of α and hence find the number of iterations of the Newton-Raphson formula starting with $x_1=0$ that are required to find α correct to 8 dp.

2 The curve C has equation $y=\text{f}(x)$ where $\text{f}(x)=x^4+4x^2-6x$. C has exactly two real roots and exactly one stationary point.

a Write down one of the roots of C

b Use the Newton-Raphson method on a suitable function with starting value $x_1=1$ to find, to 1 dp,

i The other root of this curve,

ii The coordinates of the stationary point P on C

c Sketch the curve C

3 The curve C with equation $y=x\sin x+2\cos x$, where x is in radians, has exactly one stationary point P in the interval $(\pi, 2\pi)$. The x–coordinate of P is β

a Show that β is a root of the equation $\text{g}(x)=0$, where $\text{g}(x)=\tan x-x$

b Show that, when applied to $\text{g}(x)$, the Newton-Raphson formula can be written as $x_{n+1}=x_n-\left(\dfrac{\tan x_n-x_n}{\tan^2 x_n}\right)$

c Using this formula with first approximation $x_1=4.5$ radians, find the coordinates of point P to 2 dp.

d Use any appropriate technique to show that P is a minimum point.

4 a i Show with a sketch that $2e^x=6x+40$ has exactly one positive solution.

ii Apply the Newton-Raphson method to a suitable function to solve this equation to 1 dp, justifying your answer. Use $x_1=2$ as a first approximation.

For a particular country, the equation $S=2e^{0.5t}$ models the total S (£millions) paid in subscriptions for streaming music. The equation $D=3t+40$ models the total D (£millions) paid for music downloads t years after 1st January 2010.

b At the start of which year did the total amount paid in subscriptions for streaming first exceed the total amount paid for downloads? Show your working.

c Find, to the nearest million, the total amount paid in subscriptions and downloads from 1st January 2010 to the start of the year found in part **b**.

Challenge

5 For different values of $k>0$, investigate the behaviour of the sequence defined by $x_{n+1}=\dfrac{1}{2}\left(x_n+\dfrac{k}{x_n}\right)$, $x_1=1$

a How is the value to which this sequence converges related to the value of k?

b How can your findings be used to estimate the value of $\sqrt{54321}$ if the square-root button on your calculator is not working?

c Use the Newton-Raphson method to explain your findings.

17.4 Numerical integration

Fluency and skills

See Ch4.6

For a reminder on how to use integration to find the area under a curve.

The definite integral $\int_a^b f(x)\,dx$ can be used to calculate the area of the region **R** bounded by the curve $y=f(x)$ and the x-axis between $x=a$ and $x=b$

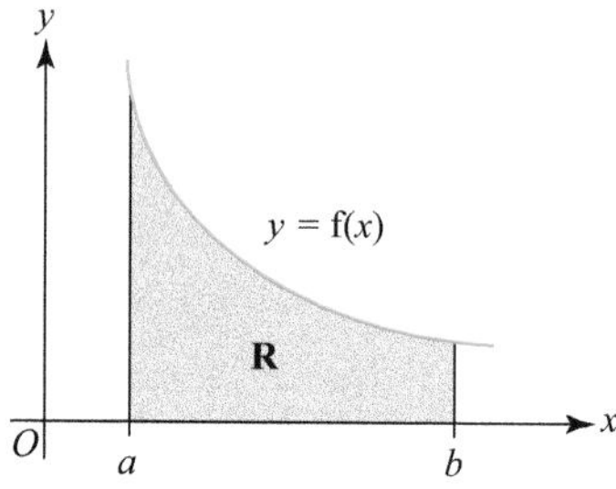

Often, $\int f(x)\,dx$ cannot be found, so you can't calculate the exact value of the definite integral. You can, however, use the trapezium rule to find an approximation to $\int_a^b f(x)\,dx$

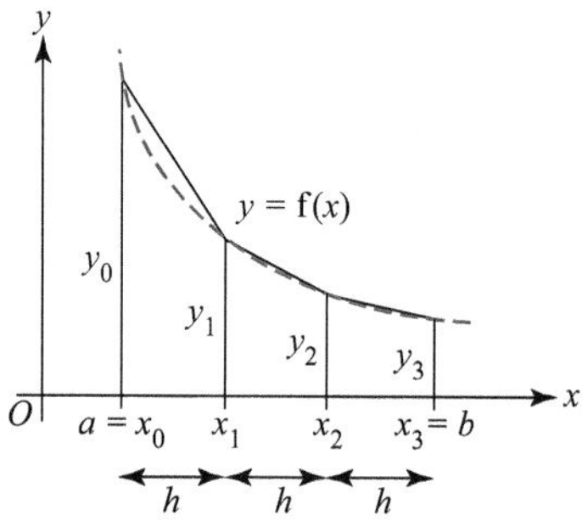

The diagram shows the region **R** split into three trapeziums of equal width. The x-coordinates between a and b are x_0, x_1, x_2 and x_3

Each trapezium has width h and their vertical sides are the y-values (or ordinates, where, $y_0=f(x_0)$, $y_1=f(x_1)$ and so on).

The area of the three trapeziums is

$$\frac{1}{2}h(y_0+y_1)+\frac{1}{2}h(y_1+y_2)+\frac{1}{2}h(y_2+y_3)$$

The area of $\mathbf{R} \approx \frac{1}{2}h\{(y_0+y_1)+(y_1+y_2)+(y_2+y_3)\}$

so $\int_a^b f(x) \approx \frac{1}{2}h\{(y_0+y_3)+2(y_1+y_2)\}$

More generally,

Recall that the area of a trapezium is found using $\frac{1}{2}h(a+b)$ where a and b are the lengths of the parallel sides and h is the distance between them.

Key point

The trapezium rule is $\int_a^b f(x) \approx \frac{1}{2}h\{(y_0+y_n)+2(y_1+y_2+...+y_{n-1})\}$

where the interval $a \le x \le b$ is divided into n intervals of equal width $h=\frac{b-a}{n}$ defined by the values $a=x_0, x_1, ..., x_{n-1}, x_n=b$

The ordinates are given by $y_0=f(x_0), y_1=f(x_1)$ etc.

Increasing the number of intervals improves the accuracy of the estimate for $\int_a^b f(x)$

See Ch4.6

For a reminder on how to use a calculator to evaluate a definite integral.

Remember that you can evaluate definite integrals using your calculator. You could use this to check answers found using the trapezium rule.

Example 1

The diagram shows the curve with equation $y=\sqrt{\ln x}$ for $x \geq 1$

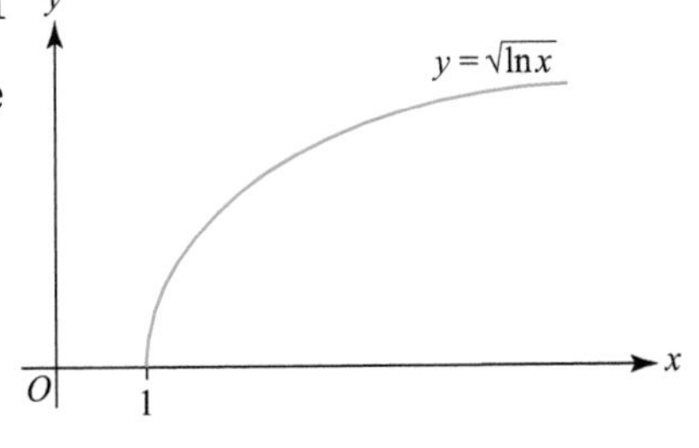

a Use the trapezium rule with four intervals to estimate the value of $\int_2^4 \sqrt{\ln x}\,dx$, giving the answer to 1 dp.

b Explain why the trapezium rule gives an underestimate for the value of $\int_2^4 \sqrt{\ln x}\,dx$

c Explain why increasing the number of intervals when using the trapezium rule gives a better estimate for this integral.

a The number of intervals $n=4$

The lower and upper limits of $\int_2^4 \sqrt{\ln x}\,dx$ are $a=2$ and $b=4$, respectively.

The width of each interval $h=\frac{b-a}{n}=\frac{4-2}{4}=0.5$

$x_0=2, x_1=2.5, x_2=3, x_3=3.5, x_4=4$

x_i	2	2.5	3	3.5	4
$y_i=\sqrt{\ln x_i}$	0.83	0.96	1.05	1.12	1.18

Draw a table to record ordinates to at least one more decimal place than that required in the final answer.

$$\int_2^4 \sqrt{\ln x}\,dx \approx \frac{1}{2}h[(y_0+y_4)+2(y_1+y_2+y_3)]$$

$$\approx \frac{1}{2}(0.5)[(0.83+1.18)+2(0.96+1.05+1.12)]$$

$$=2.0675$$

Substitute $h=0.5$ and the y-values into the formula.

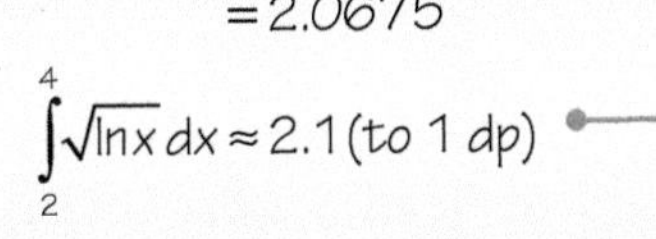

$$\int_2^4 \sqrt{\ln x}\,dx \approx 2.1 \text{ (to 1 dp)}$$

Give the answer produced by the trapezium rule to the required level of accuracy.

b The curve is concave so each trapezium lies entirely under the curve $y=\sqrt{\ln x}$ so the total area of the trapeziums is less than the exact area.

c Increasing the number of intervals decreases the width of each trapezium. The thinner each trapezium, the more accurately they approximate the curve.

Exercise 17.4A Fluency and skills

1 Use the trapezium rule with the stated number of intervals to find an estimate for these integrals. Give each estimate to two decimal places.

a $\int_1^5 e^{\sqrt{x}}\,dx$, 4 intervals

b $\int_1^3 8^{\frac{1}{x}}\,dx$, 4 intervals

c $\int_3^6 \ln\left(1+\sqrt{2x}\right)dx$, 5 intervals

d $\int_0^{\frac{\pi}{2}} \sin\left(x^2\right)dx$, 3 intervals

e $\int_{-1}^2 \frac{1}{x^3+5}\,dx$, 6 intervals

f $\int_1^2 x^x\,dx$, 5 intervals

MyMaths 2060 SEARCH

2 **i** Use the trapezium rule with five intervals to find an estimate for the area of the region **R** shown in each diagram. Work in radians where appropriate and give your answer to 3 dp.

ii Is your answer to part **i** an underestimate or an overestimate for the area of **R**? Why?

a

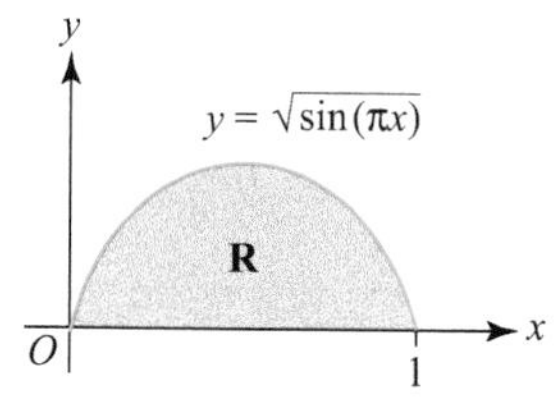

b

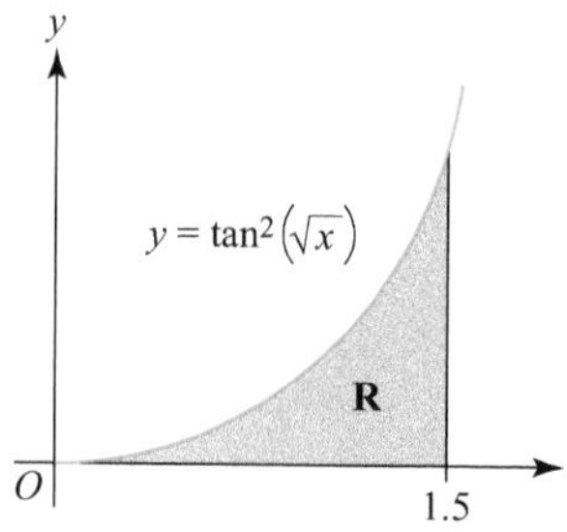

c

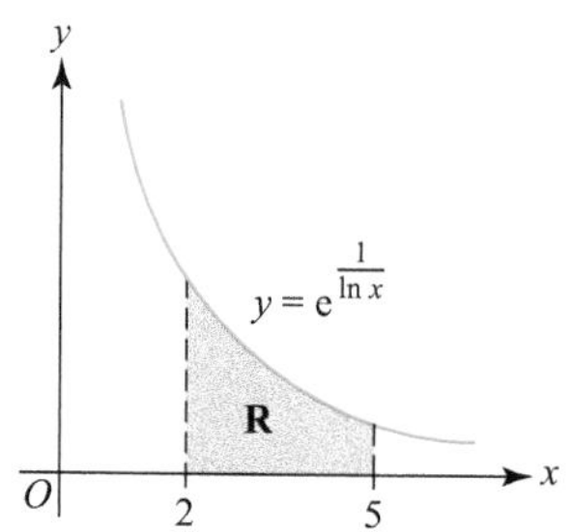

d

$y = \ln(10 - e^x)$

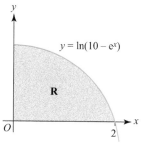

3 The table shows some values for x and y, where $y = 3^{x^2-1}$

The values of y have been rounded to 2 dp where appropriate.

x_i	1	1.1	1.2	1.3	1.4	1.5	1.6	1.7	1.8	1.9	2
y_i	1	1.26	1.62		2.87		5.55	7.98		17.59	27

a Copy and complete the table.

b Use the trapezium rule with all of the y-values in the completed table to find an estimate for $\int_1^2 3^{x^2-1}\,dx$. Give your answer to 1 dp.

4 **a** Use the trapezium rule with the stated number of intervals to find an estimate for these integrals. Give each estimate to 3 dp.

i $\int_1^4 \frac{e^x}{x} - 2\,dx$, 4 intervals **ii** $\int_1^4 6 - e^{0.1x^2}\,dx$, 5 intervals

The diagrams below show, in some order, part of the graphs with equations $y = \frac{e^x}{x} - 2$ and $y = 6 - e^{0.1x^2}$ for $1 \le x \le 4$

b Match each graph to its equation and hence state, with a reason, whether your answers to part **a** are underestimates or overestimates for each integral.

Graph 1

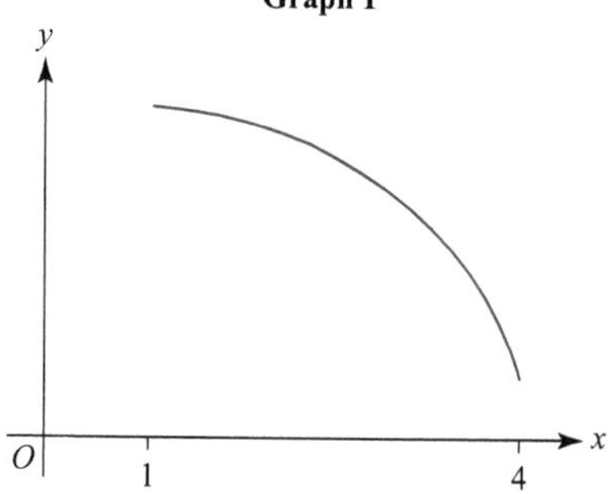

Graph 2

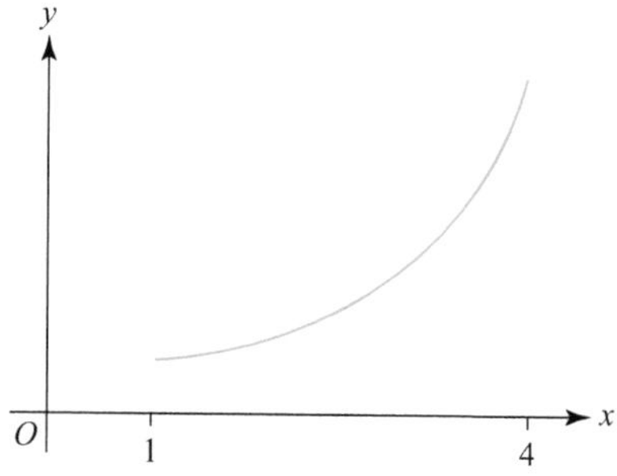

5 For each function f(x)

i Sketch the graph of $y = f(x)$ for $0 \le x \le 5$,

ii Use the trapezium rule with the stated number of intervals to estimate $R = \int_0^5 f(x)\,dx$, to 2 dp,

iii Is this answer an underestimate or an overestimate for the integral? Why?

iv Verify your answer to part **iii** by using integration to find the exact area of R. Show your working.

a $f(x) = \frac{1}{(x+1)} + 4$, 4 intervals

b $f(x) = e^{0.1x} + 2$, 5 intervals

Reasoning and problem-solving

Strategy

To use the trapezium rule to find numerical solutions to real-life problems

1. Calculate the width of each trapezium using the number of intervals.
2. Check that the number of ordinates is one more than the number of intervals used.
3. Substitute the values of h and the ordinates into the trapezium rule.
4. Use a sketch of a graph to determine if your answer is an underestimate or overestimate.
5. Interpret an estimate for the integral in context.

Example 2

The diagram shows the curve with the equation $v = 1.7e^{\sqrt{t}} - 0.5t^2$, where v models the speed (in ms^{-1}) of an athlete t seconds after he starts a warm-up sprint.

It takes the athlete six seconds to run the length of the track.

a Use the trapezium rule with five intervals to find an estimate for the value of $\int_0^6 1.7e^{\sqrt{t}} - 0.5t^2\,dt$, giving the answer to 1 dp

Is your answer an underestimate or overestimate? Why?

b State the minimum length of the track, to the nearest metre.

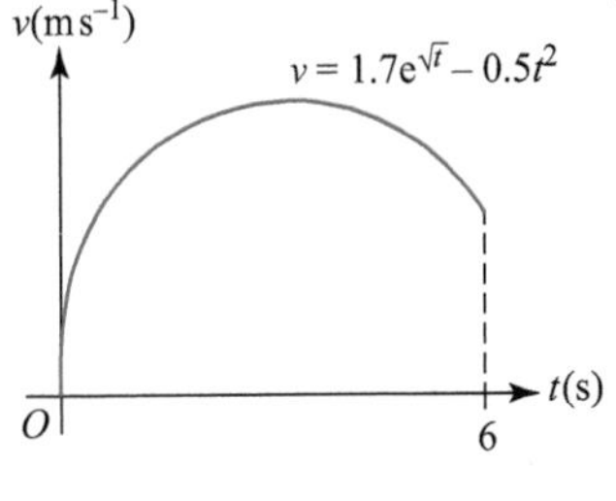

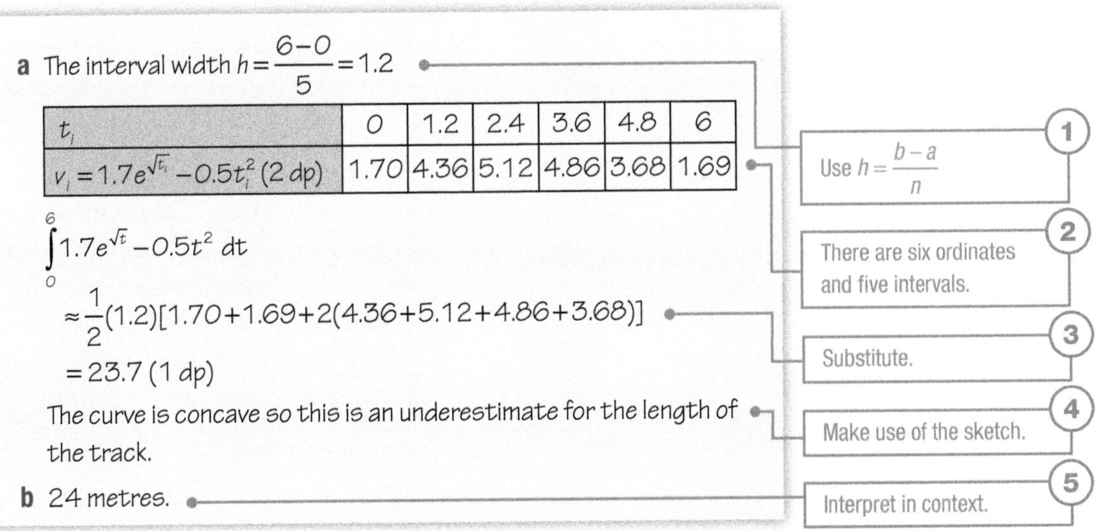

a The interval width $h = \frac{6-0}{5} = 1.2$

t_i	0	1.2	2.4	3.6	4.8	6
$v_i = 1.7e^{\sqrt{t_i}} - 0.5t_i^2$ (2 dp)	1.70	4.36	5.12	4.86	3.68	1.69

$$\int_0^6 1.7e^{\sqrt{t}} - 0.5t^2\,dt$$

$$\approx \frac{1}{2}(1.2)[1.70 + 1.69 + 2(4.36 + 5.12 + 4.86 + 3.68)]$$

$$= 23.7 \text{ (1 dp)}$$

The curve is concave so this is an underestimate for the length of the track.

b 24 metres.

1. Use $h = \frac{b-a}{n}$
2. There are six ordinates and five intervals.
3. Substitute.
4. Make use of the sketch.
5. Interpret in context.

Recall that the area under a speed–time graph equals the distance travelled, which in this case, is the length of the track.

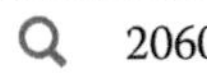

2060 SEARCH

Exercise 17.4B Reasoning and problem-solving

1 a Use the trapezium rule with five intervals to estimate the value of $\int_0^2 1-\sqrt{\cos(0.25\pi x)}\,dx$, where x is in radians. Give your answer to 2 sf.

The shaded region in the diagram shows the plan of a large field OAB where O is the origin. Point A has coordinates (0, 3) and point B has coordinates (2, 3). The curve OB has equation $y=3-3\sqrt{\cos(0.25\pi x)}$ and all distances are measured in kilometres.

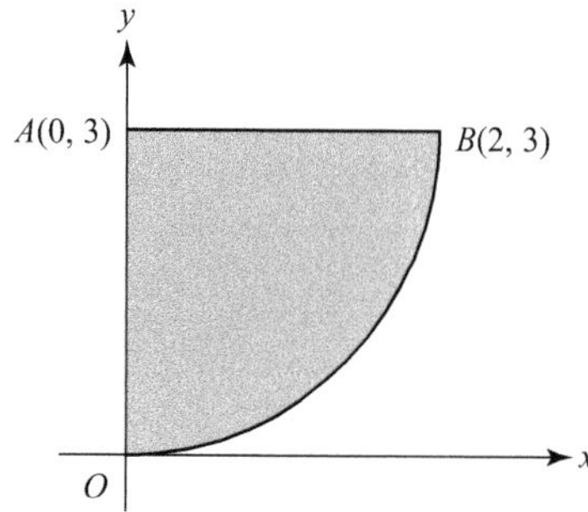

b i Use your answer to part **a** to find an estimate for the area of this field. Give your answer to 3 sf.

ii State, with a reason, whether this answer is an underestimate or an overestimate for the exact area of this field.

2 $I=\int_{4.5}^{6}\frac{x}{\sqrt{x^2-20}}\,dx$

a Use the trapezium rule with five intervals to estimate the value of I. Give your answer to 2 dp.

The trapezium rule with 20 intervals is used to estimate I. The sum of all the ordinates used is 52.725 (to 3 dp).

b Find this estimate for I, to 2 dp.

c Use integration to find I exactly. Show your working.

d Show that, as an estimate for I, your answer to part **b** is nine times more accurate than your answer to part **a**.

3 The diagram shows the cross section of a river from one side of its bank (point A) to the other (point B).

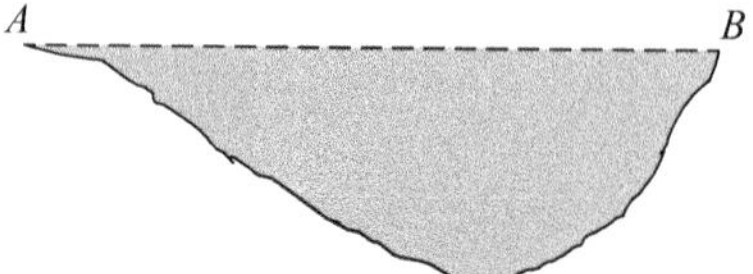

The depth h (in metres) of the river is modelled by the equation $h=4\sin\left(\frac{\pi}{36}x^2\right)$, where x is the distance (in metres) across the river from point A

a Find the width of the river from A to B

b Using the trapezium rule with six intervals, working in radians, find an estimate for the area of this cross section. Give your answer to 1 dp.

c A stick floating between A and B has speed 0.5 m s^{-1}. Given that 1 cubic metre = 1000 litres, find an estimate for the flow rate of the river. Give your answer in litres per second, to 1 sf.

4 a i Use the trapezium rule with five intervals to estimate the value of $\int_0^8 2^{0.25x}\,dx$ to 1 dp.

ii State, with a reason, whether your answer is an underestimate or an overestimate for this integral.

At the start of a race, runner B has a 10 metre head start on runner A. The equations $v_A=k+2^{0.25t}$ and $v_B=0.25t+1$ model the speeds (in m s^{-1}) of A and B, respectively, t seconds from the start, where k is a constant. After four seconds, A is running 1 m s^{-1} faster than B.

b Find the value of k

The diagram shows the speed–time graphs for runners A and B.

It takes B exactly eight seconds to complete the race.

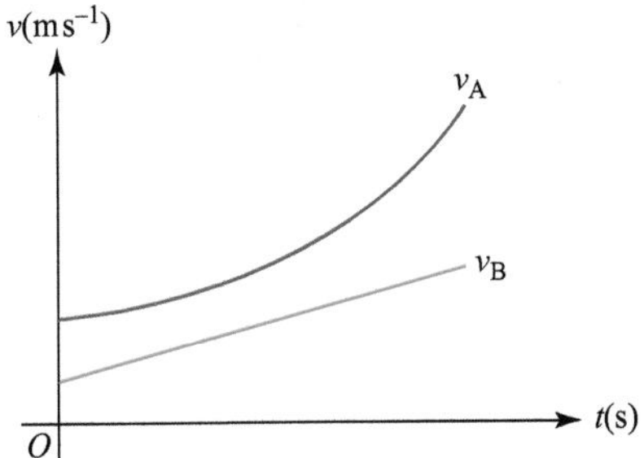

c Determine who wins the race, justifying your answer.

5 **a** Write down the formula for the sum of the first n terms of a geometric series with first term a and common ratio r

b Show that by using the trapezium rule with n intervals to estimate the value of $\int_0^1 2^{nx}\,\mathrm{d}x$ the answer given is $\frac{3}{2n}\left(2^n - 1\right)$

For a particular value of n, the estimate for $\int_0^1 2^{nx}\,\mathrm{d}x$ found using the trapezium rule with n intervals is 6144 to the nearest integer.

c **i** Use a suitable numerical method to find the value of n, justifying your answer.

ii For this n, find an estimate for the value of $\int_0^1 4^{\sqrt{nx}}\,\mathrm{d}x$, giving your answer to 1 dp.

6 $I = \int_2^3 \mathrm{e}^{-\frac{3}{x}}\,\mathrm{d}x$

a Use the trapezium rule with five intervals to estimate I to 3 dp.

b Using the substitution $u = -\frac{3}{x}$ show that

$$I = \int_{-1.5}^{-1} \frac{3\mathrm{e}^u}{u^2}\,\mathrm{d}u$$

c Use the trapezium rule with five intervals to estimate the value of $\int_{-1.5}^{-1} \frac{3\mathrm{e}^u}{u^2}\,\mathrm{d}u$ to 3 dp.

The diagram shows parts of the curves with equations $y = \mathrm{e}^{-\frac{3}{x}}$ and $y = \frac{3\mathrm{e}^x}{x^2}$

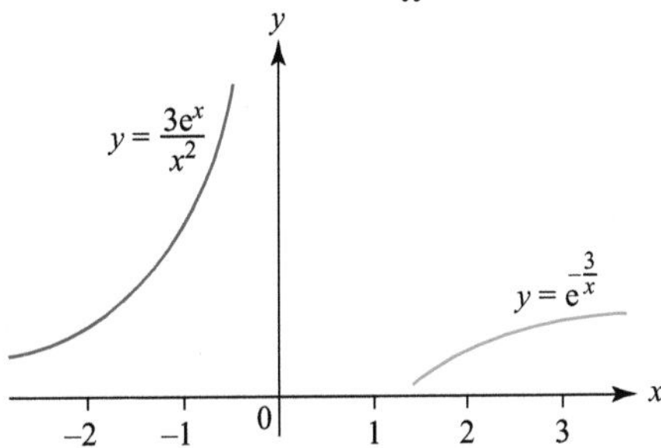

d Use your estimates and the diagram to find the value of I to as many decimal places as can be justified.

Challenge

7 **a** Use the trapezium rule to estimate $I = \int_0^1 \frac{4}{1+x^2}\,\mathrm{d}x$ using n intervals where

i $n = 4$

ii $n = 5$

iii $n = 8$

b Which famous value do these answers appear to approach as n increases?

c Use the substitution $x = \tan\theta$ to find the exact value of I

PURE

MyMaths 2060 SEARCH

17 Summary and review

Chapter summary

- An open interval (a,b) is the set of real numbers x such that $a < x < b$
- The solutions to the equation $f(x)=0$ are called the roots of the equation.
- $f(x)$ changes sign across (a,b) if $f(a)$ and $f(b)$ have opposite signs.
- $f(x)$ is continuous if you can draw the graph $y=f(x)$ without taking your pen off the paper.
- **The change of sign method:** If the continuous function $f(x)=0$ changes sign in the interval (c,d) then the equation $f(x)=0$ has a real root α, $c<\alpha<d$
- When the equation $f(x)=0$ is rearranged into the form $x=g(x)$, the sequence defined by the iterative formula $x_{n+1}=g(x_n)$ can converge to a root α of the equation $f(x)=0$
- If $|g'(x)|\geq 1$ for values of x close to a root α of the equation $f(x)=0$ then the sequence $x_1, x_2, x_3, \ldots$ will not converge to this root.
- Convergence and divergence can be displayed on a staircase or cobweb diagram.
- **The Newton-Raphson method:** If α is a root of the equation $f(x)=0$ then the iterative sequence given by $x_{n+1}=x_n-\frac{f(x_n)}{f'(x_n)}$ can converge to α
- If x_1 is close to an x-coordinate of a stationary point of the graph of $y=f(x)$ then the Newton-Raphson method may fail to provide a good estimate for α
- **The trapezium rule:** $\int_a^b f(x) \approx \frac{1}{2}h\{(y_0+y_n)+2(y_1+y_2+\ldots+y_{n-1})\}$ where the interval $a\leq x\leq b$ is divided into n intervals of equal width $h=\frac{b-a}{n}$ defined by the values $a=x_0, x_1, \ldots, x_{n-1}, x_n=b$

  The ordinates $y_0, y_1, \ldots, y_n$ are given by $y_i=f(x_i)$ for $i=1, 2, \ldots, n$
- Increasing the number of intervals (and therefore trapeziums used) improves the accuracy of the estimate for an integral.
- If a curve is convex, the trapezium rule overestimates an integral.
- If a curve is concave, the trapezium rule underestimates an integral.

Check and review

You should now be able to...	Try Questions
✔ Use the change of sign method to find and estimate the root(s) of an equation.	1, 2
✔ Use an iterative formula to estimate the root of an equation.	3
✔ Recognise the conditions that cause an iterative sequence to converge.	3, 4, 5
✔ Use the Newton-Raphson method to estimate the root of an equation.	4, 5
✔ Use the trapezium rule to find the area under a curve.	6

1 a Show that each equation has a solution in the given interval. Work in radians where appropriate.

i $x^3+3=5x^2$, $(0,1)$

ii $x^3+3=5x^2$, $(-0.8,-0.7)$

iii $2^x-e^{\sqrt{x}-1}=3$, $(2.15, 2.25)$

iv $x\sin x=\cos(\pi x)+1$, $(3, \pi)$

b Using these intervals, which of these solutions are known to 1 dp?

2 $f(x)=\dfrac{1}{x-2}-x$

a Use algebra to solve the equation $f(x)=0$

b Show that $f(x)$ changes sign between $x=1.5$ and $x=2.4$

c What information about the function $f(x)$ do the results of part **a** and part **b** give?

3 a Sketch graphs to show that the equation $2e^x=4-x^2$ has exactly one positive solution, α

b Show that the equation $2e^x=4-x^2$ can be rearranged into the form $x=\ln(2-0.5x^2)$

c Use the iterative formula $x_{n+1}=\ln(2-0.5x_n^2)$ with $x_1=1$ to find α to 3 decimal places, justifying your answer.

4 Use the Newton-Raphson method with first approximation $x_1=1$ to find a solution of these equations correct to 3 dp. Work in radians where appropriate.

a $x^3+6x=3$ b $x^2=2e^{\frac{1}{2}x}$ c $\ln(2x+1)=2x-1$ d $\sin(x^2+4)=e^{-x}$

5 $f(x)=x^2-4\sqrt{x}-1$

a Use the Newton-Raphson method with first approximation $x_1=2$ to find a root of the equation $f(x)=0$. Give your answer to 3 dp, justifying your answer.

b Explain why $x_1=1$ would not be an appropriate first approximation to use when applying the Newton-Raphson method to solve the equation $f(x)=0$

6 a Use the trapezium rule with five intervals to estimate the value of $\int_0^2 4^x\,dx$

Give your answer to 2 dp. State, with reason, whether this is an underestimate or overestimate.

b Use your answer to part **a** to find an estimate for the value of $\int_0^2 2^{2x+3}\,dx$

Give this estimate to 3 sf.

What next?

Score		
0 – 3	Your knowledge of this topic is still developing. To improve, search in MyMaths for the codes: 2060, 2173, 2174, 2176	
4 – 5	You're gaining a secure knowledge of this topic. To improve, look at the InvisiPen videos for Fluency and skills (17A)	
6	You've mastered these skills. Well done, you're ready to progress! To develop your techniques, look at the InvisiPen videos for Reasoning and problem-solving (17B)	

Click these links in the digital book

PURE

17 Exploration

Going beyond the exams

History

Leonhard Euler (1707–1783) was a Swiss mathematician, physicist, astronomer and engineer. He made important discoveries in many areas of mathematics and published 886 papers and books, many of which were in the last two decades of his life, when he was completely blind. One of the contributions that Euler made to calculus was a method for finding numerical solutions to differential equations.

Did you know?

Euler's identity, written as

$e^{i\pi} + 1 = 0$

is regarded by many as the most beautiful result in mathematics. In one equality, it links the five most fundamental mathematical constants.

Information

Given a differential equation such as $\frac{dy}{dx} + \frac{2y}{x} = \frac{4}{x}$ where $y = 4$ when $x = 2$, Euler's method allows you to find an approximate value for y, given a value for x, *without solving the equation.* This is particularly useful when there is no obvious algebraic method to solve the equation.

When $x = 3$, for example, you would proceed as follows. First substitute $x = 2$ and $y = 4$ into the equation to find that $\frac{dy}{dx} = -2$

Suppose that you now increase the value of x by some small amount, h, then the corresponding change in y is approximately $h\frac{dy}{dx} = -2h$. It follows that the point $(2 + h, 4 - 2h)$ will be close to the solution curve. For example, if you choose $h = 0.1$, then this gives the point (2.1, 3.8)

The whole process may now be repeated, this time starting from the new point (2.1, 3.8) and continuing in steps of 0.1 until the value of y is found when $x = 3$

Challenge

The numerical solution outlined above involves many calculations that take a while to complete manually. It is much easier to use a spreadsheet to carry out all the difficult calculations.

	A	B	C	D
1	xn	yn	y'	h
2	2	4	–2	0.1
3	=A2+$D$2	=B2+C2*$D$2	=4/A3–2/A3*B3	

Find y when $x = 3$ using $h = 0.1$

Experiment with smaller values of h and compare the results with the theoretical value of 2.89, correct to 2 dp.

17 Assessment

1 $f(x) = 3^x - 2x - 9$ is a continuous function.

The equation $3^x - 2x - 9 = 0$ has a real root α between 2 and 3

Use the change of sign method to calculate α to 1 dp.

Select the correct answer.

A 2.2 **B** 2.3 **C** 2.3... **D** 2.4 **[1 mark]**

2 $f(x) = x^3 - 7x + 4$ has root α between 0 and 1

Use the iterative formula $x_{n+1} = \dfrac{x_n^3 + 4}{7}$ with $x_1 = 1$ to find α to 2 dp.

Select the correct answer.

A 0.60 **B** 0.61 **C** 0.62 **D** 0.63 **[1]**

3 Show that the equation $x^3 + 2x^2 - 3x - 2 = 0$ has a root between $x = 1$ and $x = 2$ **[2]**

4 Given that $f(x) = x\sin x$, where x is in radians, show that $f(x) = 0$ has a root in the interval $3 < x < 3.5$ **[2]**

5 **a** Show that the equation $x^3 - 4x - 1 = 0$ has a root in the interval (2, 2.5) **[2]**

b Use the iterative formula $x_{n+1} = \sqrt{4 + \dfrac{1}{x_n}}$, starting with $x_1 = 2$ to find x_2 and x_3 to 2 dp. **[3]**

6 **a** Use the iterative formula $x_{n+1} = \ln(5 - x_n)$, starting with $x_1 = 1$ to find, to 2 decimal places, a root of the equation $e^x + x - 5 = 0$ **[4]**

b Prove that your solution is correct to 2 decimal places. **[3]**

7 **a** Show that the equation $x^3 - 3x + 1 = 0$ has a root between 1 and 2 **[2]**

b Taking 2 as a first approximation, use the Newton-Raphson process twice to find an approximation to the root of $x^3 - 3x + 1 = 0$, to 2 dp. **[4]**

8 You are given that a particle's motion is modelled by $f(x) = 2x^4 - 3x^3 + 4x$

a Use the Newton-Raphson process twice, taking $x = -1$ as the first approximation to find the negative root of the equation $f(x) = 0$ to 2 decimal places. **[4]**

b Prove that your solution is correct to 2 dp. **[2]**

9 Use the trapezium rule with four strips to estimate the integral $\int_1^3 \sin^3 x\,dx$, to 3 sf. **[4]**

10 Use the trapezium rule with four ordinates to estimate the integral $\int_0^3 \tan^3 x\,dx$, to 3 sf. **[4]**

11 An object's temperature is modelled by $f(x) = 5x - e^x$

a Prove that there is a root of $f(x) = 0$ between $x = 0$ and $x = 0.5$ **[2]**

The graph of $y = f(x)$ is shown.

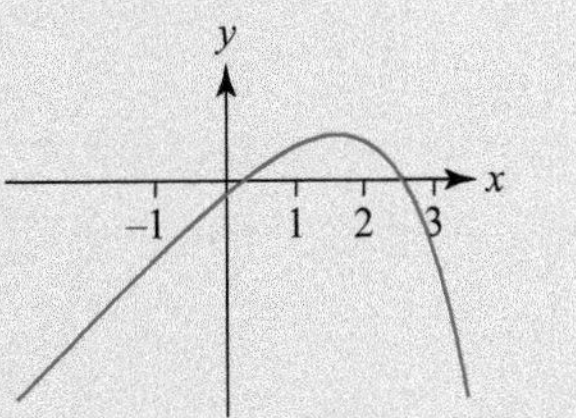

b Explain how you know from the graph that there is a root of $f(x) = 0$ between 2 and 3 **[1]**

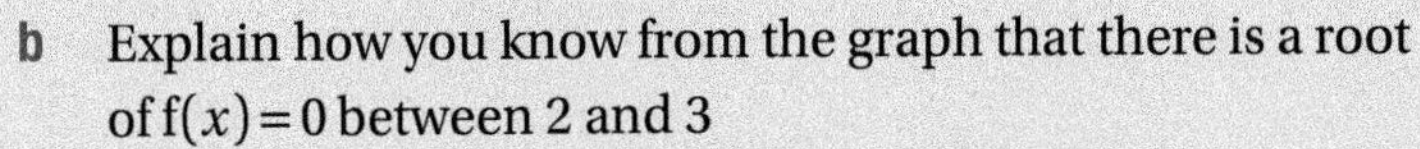

c Show that $x = 2.5$ is the root, correct to 1 dp. **[3]**

12 a Sketch, on the same axes, the graphs of $y=x+1$ and $y=\frac{4}{x}$ **[2]**

b Use your graphs to explain how many roots there are to the equation $x+1=\frac{4}{x}$ **[1]**

c Show that the equation $x+1=\frac{4}{x}$ has a root in the interval (1.5, 1.6) **[3]**

d Find the solutions to the equation $x+1=\frac{4}{x}$, give your answers to 3 significant figures. **[2]**

13 The graphs of $y=e^x\sin x$ and $y=x+2$ are shown, where x is in radians.

a Explain how many solutions there are to the equation $e^x\sin x=x+2$ **[1]**

b Show that one of the roots, α, is such that $1.2<\alpha<1.3$ **[3]**

c Find an interval of size 0.1 that the other positive root lies within. **[3]**

y
x
−π

14 a Show that the equation $x^3-3x^2-5=0$ can be written $x=\sqrt{\frac{5}{x}+3x}$ **[2]**

b Use the iteration formula $x_{n+1}=\sqrt{\frac{5}{x_n}+3x_n}$, starting with $x_1=3$ to find x_5

Give your answer to 2 decimal places. **[3]**

15 The graph of $y=2e^x+3x-7$ is shown.

a Use the iteration formula $x_{n+1}=\frac{7-2e^{x_n}}{3}$ with $x_1=0.8$ to find x_2, x_3, x_4, x_5 to 2 decimal places. **[3]**

b Explain what is happening in this case. **[1]**

c Now derive a different iteration formula and, again using $x_1=0.8$, calculate x_2, x_3, x_4, x_5 to 2 dp. **[4]**

y
x
−1 0 1 2

16 a Sketch the graphs of $y=x-3$ and $y=\sqrt{x}$ on the same axes. **[2]**

b Use an appropriate iteration formula with $x_1=2$ to find a root of $\sqrt{x}=x-3$ to 2 dp. **[4]**

c i Draw a suitable diagram to illustrate the results of the first two iterations. **[3]**

ii Write down the name of this diagram. **[1]**

17 a Sketch the graphs of $y=\ln x$ and $y=e^x-5$ on the same axes. **[3]**

b Explain how many roots the equation $\ln x=e^x-5$ has. **[1]**

c Show that one of the roots occurs between $x=1.6$ and $x=1.8$ **[3]**

d Use the Newton-Raphson process, to find this root correct to 2 decimal places. **[4]**

18 a Show that the equation $x^3+4x-3=0$ can be written $x=a(b-x^3)$, where a and b are constants to be found. **[2]**

b Use the iteration formula $x_{n+1}=a(b-x_n^3)$ for the values of a and b found in part **a** with $x_1=0.1$ to find x_5 correct to 2 significant figures. **[3]**

c i Draw a suitable diagram to illustrate the results of the first three iterations. **[3]**

ii Write down the name of this diagram. **[1]**

19 Use the Newton-Raphson method to find, to 3 significant figures, the solution of the equation $x\sin x=2\ln x$, where x is in radians, which is near 2 **[5]**

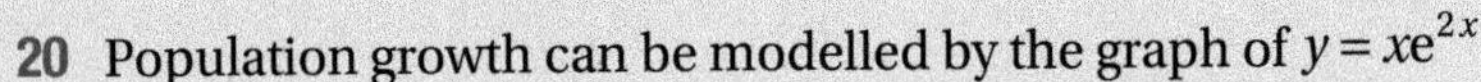

20 Population growth can be modelled by the graph of $y = xe^{2x}$

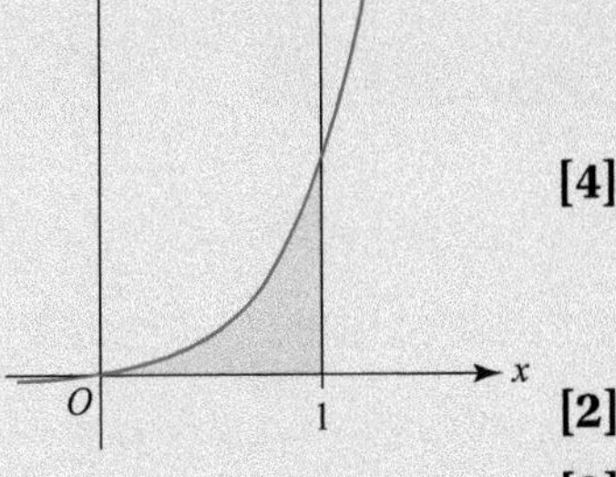

a Use the trapezium rule with five strips to estimate, to 4 significant figures, the area enclosed by the curve, the x-axis and the line $x = 1$ **[4]**

b State without further calculation whether this will be an overestimate or an underestimate of the actual area. Justify your answer. **[2]**

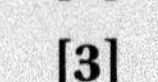

c **i** Use integration by parts to find the actual area. **[3]**

ii Calculate the percentage error in your approximation. **[2]**

21 The graph of $y = x^3 - 5x - 3$ is shown.

a How many solutions are there to the equation $x^3 - 5x - 3 = 0$? Justify your answer. **[1]**

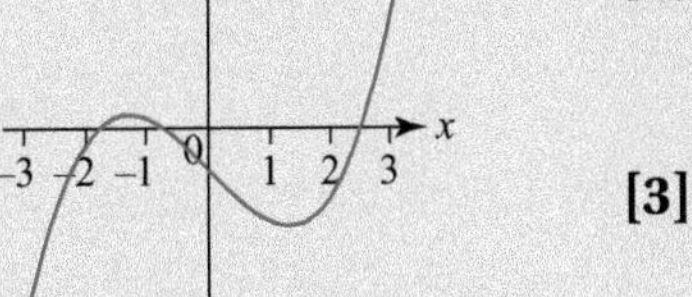

b Show that $x^3 - 5x - 3 = 0$ can be written in the form $x = \pm\sqrt{a + \frac{b}{x}}$, where a and b are constants to be found. **[3]**

c Use the iteration formula $x_{n+1} = \sqrt{a + \frac{b}{x_n}}$ with the values

you have found for a and b to calculate the positive root of the equation correct to 3 significant figures. **[3]**

d Use Newton-Raphson to find the largest negative root, correct to 2 significant figures. **[4]**

e Verify that the smallest negative root is -1.83 to 3 sf. **[2]**

22 Explain how the change of sign method will fail to find a root, α, to $f(x) = 0$ in these cases

a $f(x) = \frac{1}{x-3}$ for $2.5 < \alpha < 3.5$ **[2]**

b $f(x) = (3x-2)(2x-1)(x-4)$ for $0 < \alpha < 1$ **[2]**

23 Use the Newton-Raphson method to find a root, correct to 2 decimal places, to the equation $\sin^2 x = e^{-x}$, where x is in radians, using

a $x_1 = 1$ **[3]**

b $x_1 = 3$ **[3]**

24 **a** Use the trapezium rule with five ordinates to estimate, to 3 significant figures, the area enclosed by the curve with equation $y = \sqrt{\ln x}$, the x-axis and the line $x = 2$ **[4]**

b Comment on the suggestion that the actual area is close to 0.5 **[2]**

25. $f(x) = x\ln x - 1$, $x > 0$

a Find an interval of size 0.2 that contains the solution to $f(x) = 0$ **[3]**

b Use Newton-Raphson to approximate the root of the equation $f(x) = 0$

Ensure your answer is correct to 3 decimal places. **[8]**

12–17 Assessment

Chapters 12–17:

1 The area of a triangle is $\left(-1+3\sqrt{5}\right)$ cm². The height of the triangle is $\left(3+2\sqrt{5}\right)$ cm.

Calculate the length of the base of the triangle. Give your answer in the form $a+b\sqrt{5}$

Select the correct answer.

A $0+2\sqrt{5}$ B $0+1\sqrt{5}$ C $3-1\sqrt{5}$ D $6-2\sqrt{5}$ **[1 mark]**

2 Find the solutions to the equation $2^{2x+1}-7(2^x)+6=0$. Select the correct answer. **[1]**

A $x=2,\ 1.5$ B $x=1,\ 0.585$

C $x=0.693,\ 0.405$ D $x=-0.792,\ -0.158$

3 $f(x)=x^2+(k+1)x+2$

a Find the range of values of k for which the equation $f(x)=0$ has distinct real roots. **[4]**

b Find the solutions of the equation $x^4-3x^2+2=0$. Show your working. **[3]**

4 Calculate the points of intersection between a circle with radius 5 and centre (1, 2) and a line that passes through the points (1, 3) and (–2, 6). Show your working. **[8]**

5 Find the range of values of x that satisfy both $2-2x-3x^2 \geq 0$ and $4x+7>1$. Show your working. **[5]**

6 Solve the equation $5-\sin\theta-6\cos^2\theta=0$ for $0<\theta<360°$. Show your working. **[6]**

7 Solve the simultaneous equations $e^{x+y}=3$ $3x+2y=0$
Show your working and give each of your solutions as a single logarithm. **[5]**

8 $g(x)=6x^3-19x^2-12x+45$

a $y=g(x)$ and $y=0$ intersect at (3, 0) and at two other points. Calculate the remaining points of intersection, showing your working. **[5]**

b Sketch the curve $y=g(x)$, clearly labelling the points of intersection with the coordinate axes. **[3]**

c Calculate the area enclosed by the curve and the x-axis. Show your working. **[6]**

9 Find and classify all the stationary points of the curve with equation

$$y=\frac{1}{2}x^4-3x^3+2x^2+15x+1$$

Show your working. **[8]**

10 The number of cases of a viral infection in a school with 2000 students after t days is given by $N=Ae^{kt}$. There are initially two cases of the infection and this number doubles after three days.

a Calculate the exact values of A and k **[4]**

b According to this model, how many days until a quarter of the students have been infected? Show your working. **[3]**

The number of cases of a second type of viral infection after t days is given by $M=Br^t$. There are initially 10 cases of this infection and after five days there are 15 cases.

c After how many days will the number of cases of the first infection overtake the number of the second infection? Show your working. **[7]**

d Sketch on the same axes the graphs of N and M against t for $t>0$ **[4]**

e How realistic do you think these models are? Explain your answer. **[3]**

11 $f(x) = 6x^3 - 19x^2 - 51x - 20$

a Show that $2x+1$ is a factor of $f(x)$ **[3]**

b Find all the solutions to $f(x) = 0$, showing your working. **[3]**

12 Write each of these expressions in partial fractions.

a $\frac{3x+1}{(x+3)(2x+1)}$ **[4]** b $\frac{3x-5}{x^2-25}$ **[4]**

13 The function f is defined by $f : x \mapsto \frac{2x-14}{x^2-2x-3} + \frac{2}{x-3}, x > 3$

a Show that f(x) can be written as $\frac{k}{x+1}$, where k is an integer to be found. **[4]**

b Write down the

i Domain of f(x) ii Range of f(x) **[3]**

c Find the inverse function, $f^{-1}(x)$ and state its domain. **[4]**

14 Given that $g(x) = x^2 + 3$, $x \in \mathbb{R}$ and $h(x) = \frac{3}{x-2}, x \neq 2$

a Write down hg(−2), **[3]**

b Solve the equation $gh(x) = 12$. Show your working. **[4]**

c Is the range of gh(x) the same as the range of hg(x)? Explain how you know. **[3]**

15 Work out the first four terms of the binomial expansion of $(1+2x)^{\frac{1}{4}}, |x| < \frac{1}{2}$, in ascending powers of x **[4]**

16 A sequence is defined by $u_{n+1} = 2u_n + 1$, $u_1 = -2$. Showing your working, calculate

a u_2 **[2]** b $\sum_{1}^{5} u_r$ **[3]**

17 AB is the arc of a circle of radius 5 cm and centre C as shown.

The segment S is bounded by the arc and the line AB

a Calculate the area of S **[4]**

b Calculate the perimeter of S **[4]**

18 a Sketch each of these graphs on separate axes, for x in the range $0 \leq x \leq 2\pi$

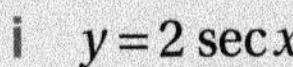

i $y = 2\sec x$ ii $y = \operatorname{cosec} 2x$ **[6]**

b Solve the equation $3\cot x + 2 = 4$ for x in the range $0 \leq x \leq 360°$. Show your working. **[4]**

19 Showing your working, find the exact solutions to each of these equations.

a $2\arcsin x = \frac{\pi}{2}$ **[3]** b $\arctan 4x = \frac{\pi}{3}$ **[2]**

20 a Differentiate with respect to x

i $3\sin x$ ii $x\ln x$ **[3]**

Given that $f(x) = (2x+1)\cos x$

b Find the exact gradient of the curve $y = f(x)$ when $x = \frac{\pi}{6}$. Show your working. **[4]**

21 a Find the coordinates of the minimum point of the curve $y = xe^{2x}$. Show your working. [5]

b Explain how you know this is a minimum point. [4]

22 a Work out each of these integrals

i $\int \sin x \, dx$ ii $\int \frac{3}{x} \, dx$ [3]

b Calculate the exact value of the integral $\int_2^6 \frac{2}{x-1} \, dx$. Show your working and give your answer in its simplest form. [4]

23 Calculate the area bounded by the x-axis and the curve $y = \cos x$ for $0 \le x \le \pi$. Show your working. [4]

24 $f(x) = x^3 - 6x - 12$

a Show that the equation $f(x) = 0$ has a root in the interval (3, 3.5). [2]

b Use the iterative formula $x_{n+1} = \sqrt{6 + \frac{12}{x}}$, starting with $x_1 = 3$ to find x_2 and x_3 to 2 decimal places. [2]

c Prove that your value of x_3 is a solution to $f(x) = 0$, correct to 2 decimal places. [3]

25 Use the trapezium rule with four strips to estimate the integral $\int_0^1 \cos^5 x \, dx$ to 3 significant figures. [5]

26 Prove by contradiction that if n^2 is odd then n is odd for all integers n [5]

27 Show that $\frac{2x^2 + 4x + 3}{2x^2 - x - 1}$ can be written $A + \frac{B}{x-1} + \frac{C}{2x+1}$ where A, B and C are integers to be found. [5]

28 The function f is given by $f : x \to |3 - 2x|$

a Sketch the graph of $y = f(x)$. [2]

b How many solutions will there be to the equation $|3 - 2x| = x$? Explain how you know. [2]

c Solve the inequality $|3 - 2x| \ge x$, showing your working. [4]

29 $f(x) = \ln(3x + 1)$, $x > -\frac{1}{3}$

a Find the inverse $f^{-1}(x)$. [3]

b Sketch $y = f(x)$ and $y = f^{-1}(x)$ on the same axes. [4]

c Write down the range and domain of $f^{-1}(x)$. [2]

30 a Express $\frac{6x+10}{(x-1)(x+3)^2}$ in partial fractions. [5]

b Integrate $\frac{6x+10}{(x-1)(x+3)^2}$ with respect to x [4]

31 A sequence is given by $x_{n+1} = (x_n)^2 - 2x_n$ where $x_1 = 1$

a Write down the value of x_2 and x_3 [3]

b Find an expression in terms of n for $\sum_1^n x_r$ [4]

32 The first term of a geometric series is 36 and the common ratio is $\frac{1}{3}$

a Find the difference between the second and third terms of the sequence. Show your working. [3]

b Calculate the difference between the sum to infinity and the sum of the first five terms of the series. Give your answer as a fraction. [5]

33 a Derive a formula for the sum of the first n terms of an arithmetic series with first term a and common difference d [4]

An arithmetic series has first term –3 and common difference 1.5

The sum of the first n terms of this an arithmetic series is 63

b Find the value of n [4]

34 a Sketch the graph in part **i** for $-1 \le x \le 1$ and the graph in part **ii** for $-2 \le x \le 0$

i $y = \arccos x$ ii $y = 2\arcsin(x+1)$ [6]

b State the range of each function in part **a**. [2]

c Write down the inverse of $f(x) = 2\arcsin(x+1)$, $-2 \le x \le 0$ and state its domain. [4]

35 a Sketch the graph of $y = 3\ln(x-1)$ and give the equation of any asymptotes. [3]

b Calculate the exact gradient of the curve at the point where $x = 3$. Show your working. [3]

36 a Using a small angle approximation, show that $\sec 2x \approx \frac{1}{(1-2x^2)}$ [4]

b Hence, find the first three terms of the binomial expansion for $\sec 2x$ in ascending powers of x [4]

c Use your expansion to find an approximate value for $\sec(0.2)$ [3]

37 Solve these equations for $0 \le x \le 2\pi$. Show your working and give your answers to 3 significant figures.

a $\sec 2x = 1.5$ [5] **b** $3\operatorname{cosec}\left(\frac{x+\pi}{2}\right) = 5$ [4]

38 a Prove that $3\sec^2\theta - 7\tan^2\theta \equiv 3 - 4\tan^2\theta$ [3]

b Hence solve the equation $1 + 14\tan^2\theta = 6\sec^2\theta$ for $0 \le \theta \le 2\pi$ [5]

39 a Use the formula for $\cos(A+B)$ to prove that $\cos 2x \equiv 1 - 2\sin^2 x$ [3]

b Hence solve the equation $\cos 2x + \sin x = 0$ for $-180° \le x \le 180°$ [5]

40 a Express $4\sin\theta + \cos\theta$ in the form $R\sin(\theta + \alpha)$, where $R > 0$ and $0 < \alpha < \pi$ [5]

b Write down the maximum value of $4\sin\theta + \cos\theta$ and the state the smallest positive value of θ at which this value occurs. [3]

41 Find the range of values of x for which the curve with equation $y = \frac{1}{12}x^4 + \frac{1}{3}x^3 - \frac{3}{2}x^2 + 5x$ is concave. [5]

42 A curve has equation $y = e^{2x}\cos 3x$

By first finding an expression for $\frac{dy}{dx}$, work out the equation of the tangent to the curve when $x = 0$ [7]

43 Given that $f(x) = \frac{\sin x}{x+2}$,

a Find $f'(x)$, [2]

b Prove that the x-coordinate of the minimum points of the curve $y = \mathrm{f}(x)$ satisfies the equation $x = \arctan(x+2)$ **[4]**

c Use the iterative formula $x_{n+1} = \arctan(x_n + 2)$ with $x_1 = 1.2$ to find the x-coordinate of a minimum point of the curve $y = \mathrm{f}(x)$ to 2 decimal places. **[3]**

44 A curve is defined by the parametric equations

$$x = t^2 - 3,\ y = 1 + 2t$$

a Find $\frac{\mathrm{d}y}{\mathrm{d}x}$ in terms of t **[3]**

b Find an equation of the normal to the curve at the point where $t = 2$ **[4]**

c Find a Cartesian equation of the curve. **[3]**

45 A curve is defined by the equation $x^2 + 2y^2 - 3xy = 1$

Find the gradient of the curve at each of the points where $x = 1$ **[7]**

46 Using small angle approximations, differentiate $\cos x$ from first principles. **[6]**

47 a Integrate these expressions.

i $x \sin x$ ii $\ln x$ **[6]**

b Use the substitution $u = x^2 + 1$ to find $\int \frac{x}{(x^2+1)^3}\,\mathrm{d}x$ **[4]**

48 Find the exact area enclosed by the curve $y = \frac{4x+5}{x^2+3x+2}$, the coordinate axes and the line $x = 1$. Show your working and give your answer as a single logarithm in its simplest form. **[8]**

49 Solve the differential equation $\frac{\mathrm{d}y}{\mathrm{d}x} = y\cos x$, given that $y = 1$ when $x = \frac{\pi}{6}$

Give your answer in the form $y = \mathrm{f}(x)$ **[5]**

50 a Sketch the graph of $y = 1 + \ln x$ and $y = x^2 - x$ on the same axes for $x > 0$ **[4]**

b Use an appropriate iteration formula to find, to 3 significant figures, a root of the equation $1 + \ln x = x^2 - x$. Show your working. **[5]**

c Verify that the solution is correct to 3 significant figures. **[2]**

51 This graph shows the curve $y = \mathrm{f}(x)$

where $\mathrm{f}(x) = x^3 - 2x^2 - 3x + 2$

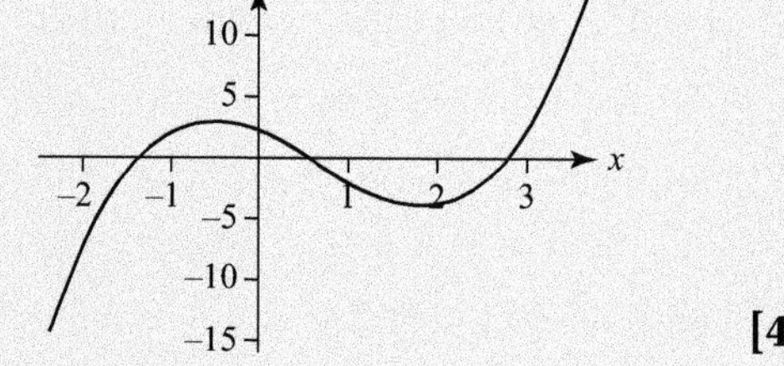

a Use the Newton-Raphson process twice, taking $x_1 = -1.3$ as the first approximation, to find a root of the equation $\mathrm{f}(x) = 0$

Give your answer correct to 3 decimal places. **[4]**

b Explain what happens if you use $x = 1.9$ as an initial approximation and illustrate on a copy of the graph. **[2]**

52 a Explain how many solutions there are to the equation $\sqrt{x} = \mathrm{e}^{-x}$ **[3]**

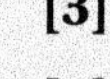

b Show that there is a solution between 0.4 and 0.5 **[2]**

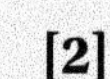

c Use the Newton-Raphson method to find this solution to 2 decimal places. Demonstrate how you know your answer is correct to this degree of accuracy. **[7]**

53 $f(x) = a - e^{2x}$ where $a > 0$ is a constant.

a Sketch the graph of $y = f(x)$. You should label the points where the curve crosses the coordinate axes and give the equation of any asymptotes. **[4]**

b Find the equation of $f^{-1}(x)$ and state its domain and range, in terms of a if appropriate. **[6]**

54 A woman plans to improve her fitness by running a miles in the first week, $a + d$ miles in the second week, and so on, with the number of miles forming an arithmetic sequence.

She runs 11 miles in the 5th week and a total of 208 miles after 13 weeks.

Calculate the values of a and d **[6]**

55 A geometric series has first term a and common ratio r. The 4th term of the series is $-\frac{3}{64}$ and the 7th term is $\frac{3}{4096}$

Find the sum to infinity of the series. **[7]**

56 a i Describe fully the series given by $u_n = 2\left(\frac{1}{3}\right)^r$

ii Calculate the exact value of $\sum_1^{\infty} 2\left(\frac{1}{3}\right)^r$ **[4]**

b State the type of series formed by $u_n = \ln 3^r$ and find an expression for $\sum_{r=1}^{n} \ln 3^r$ **[5]**

57 a Use the binomial expansion to expand $(8-5x)^{\frac{1}{3}}$, $|x| < \frac{8}{5}$, in ascending powers of x up to the term in x^2 **[4]**

b Use your expansion to find an approximation to $\sqrt[3]{7950}$ to 6 significant figures. **[5]**

58 Use compound angle formulae to prove that $\cos A - \cos B \equiv K \sin\left(\frac{A+B}{2}\right)\sin\left(\frac{A-B}{2}\right)$, state the value of K **[5]**

59 Solve the inequality $\operatorname{cosec}^2 x \le \cot x + 2$ for x in the interval $0 \le x \le \pi$

Show your working and give your limits to 3 significant figures. **[6]**

60 Solve the equation $2\cos 2x + 3\sin x + 1 = 0$ for $0 < x < 360°$

Show your working and give your answers to 1 decimal place. **[7]**

61 You are given that $f(x) = \left|\frac{1}{3\cos x + 4\sin x}\right|$, $0 < x < 180°$

Find the coordinates of a minimum point on the curve $y = f(x)$. Show your working. **[8]**

62 a Show that $\sin^4 x + \cos^4 x \equiv \frac{1}{4}(3 + \cos 4x)$ **[5]**

b Hence solve $3\sin^4 x + 3\cos^4 x = 2$ for $-\pi < x < \pi$. Show all your working and give your answers to 2 decimal places. **[6]**

63 The graph of $y = f(x)$ is shown.

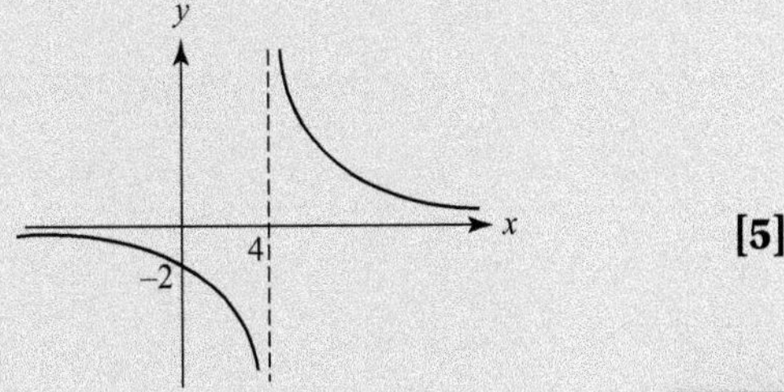

a Sketch the following graphs on separate axes, stating the equations of any asymptotes.

i $y = 2 + f(x-1)$ **ii** $y = f(3-x)$ **[5]**

Given that $f(x) = \frac{a}{x+b}$,

b Find the values of a and b **[3]**

c Write the equations of the graphs from part **a** with $f(x)$ as a single fraction in its simplest form. **[5]**

64 The curve, C is defined by the parametric equations

$$x = 4(t-5)^2,\ y = \frac{t^2}{t-5}$$

a Calculate the gradient of C at the point where $t = 2$. Show your working. **[6]**

b Find a Cartesian equation in the form $y = f(x)$, simplify your answer. **[4]**

65 You are given that $3x - 4x^2y + 3y^2 = 2$

Calculate the possible rates of change of y with respect to x when $y = 1$ **[9]**

66 A closed box of height h is a prism. The cross-section of the prism is a sector of a circle with radius r and angle $\frac{\pi}{6}$ as shown.

The volume of the box is 100 cm³

Find the minimum possible surface area of the box and the value of r that gives this surface area. **[11]**

67 Find the value of $\int_0^{21} \frac{x-5}{3x^2+10x+3}\,dx$. Show your working and give your answer in the form $k \ln 2$ **[9]**

68 Prove that the derivative of arccos 3 is $-\frac{3}{\sqrt{1-9x^2}}$ **[5]**

69 You are given that $f'(\theta) = \tan\theta$ and that $f\left(\frac{\pi}{3}\right) = \ln 6$

Prove that $f(\theta) = \ln|\sec\theta| + c$ where c is a constant to be found. **[7]**

70 a Find the general solution to the differential equation $\frac{dy}{dx} = \frac{e^{2x-y}}{y}$ **[4]**

Given that the curve passes through the point $(0, 1)$,

b Find the particular solution in the form $x = f(y)$. **[4]**

71 Use an appropriate substitution to find $\int x^2(3x-1)^5\,dx$, give your answer in terms of x **[6]**

72 Calculate the area bounded by the curve $y = x^2e^x$ and the line $y = 2xe^x$. Show your working. **[8]**

73 a Use the trapezium rule with 5 ordinates to estimate the area enclosed by the curve $y = \operatorname{cosec} x$, the x-axis and the ordinates $x = 1$ and $x = 3$ **[3]**

b Comment on the suggestion that the actual area is closer to 4 **[3]**

74 Calculate the area enclosed by the circle with equation $(x-5)^2 + (y-3)^2 = 16$ and the line with equation $2x - 5y = 3$. Show your working. **[10]**

75 Calculate the exact area bounded by the curve $y = \sin^2 x$, the x-axis and the lines $x = \frac{\pi}{3}$ and $x = \frac{5\pi}{6}$. Show your working. **[6]**

18 Motion in two dimensions

Many forces are involved in bungee jumping, the main ones being those due to gravity and the elasticity of the cord. Although a real-life jump takes place in three dimensions, it can be modelled in two dimensions by assuming there is no side-to-side motion. To calculate the motion of the bungee jumper in such a model, you would need to know the vertical and horizontal components of the fall and the force due to the elastic cord, as well as the length of the cord and its 'stretch' factor.

Scenarios which can be modelled in two dimensions include the motion of a snooker ball on a table, the movement of a boat on the ocean, the orbit of the Earth around the Sun and the motion of a thrown ball. The ability to understand and work in two dimensions is very useful for grasping the basic principles involved, and a stepping stone to working in three dimensions.

Orientation

What you need to know

Ch6 Vectors
- Resolving vectors into components.

Ch7 Units and kinematics
- Motion with variable acceleration.

Ch8 Forces and Newton's laws
- Motion under gravity.
- Resolving forces horizontally and vertically.

What you will learn

- To use the constant acceleration equation for motion in two dimensions.
- To use calculus to solve problems in two-dimensional motion with variable acceleration.
- To solve problems involving the motion of a projectile under gravity.
- To analyse the motion of an object in two dimensions under the action of a system of forces.

What this leads to

Ch19 Forces 2
Motion under a resultant force, including motion on a rough surface.
Finding forces and moments acting on a body.

Further Maths Ch8
Conservation of momentum.
Collisions.
Impulses.

Practise before you start — 2185, 2207, 2289

18.1 Two-dimensional motion with constant acceleration

Fluency and skills

See Ch6.2

For a reminder on components of a vector.

An object moving in two dimensions has position vector **r**, and its displacement, velocity and acceleration have components in the x- and y-directions. You use the following notation.

Key point

displacement $\mathbf{s} = x\mathbf{i} + y\mathbf{j}$ | acceleration $\mathbf{a} = a_x\mathbf{i} + a_y\mathbf{j}$

initial velocity $\mathbf{u} = u_x\mathbf{i} + u_y\mathbf{j}$ | final velocity $\mathbf{v} = v_x\mathbf{i} + v_y\mathbf{j}$

Provided a_x and a_y are constant, the motion in each direction will obey the equations for constant acceleration. So, for example,

$$v_x = u_x + a_x t \quad \text{and} \quad v_y = u_y + a_y t$$

You can combine these to give the vector equation

$$\mathbf{v} = \mathbf{u} + \mathbf{a}t$$

You can also write the other equations in vector form.

See Ch7.3

For a reminder on equations of motion for constant acceleration.

Key point

$$\mathbf{v} = \mathbf{u} + \mathbf{a}t \qquad \mathbf{s} = \frac{1}{2}(\mathbf{u} + \mathbf{v})t$$

$$\mathbf{s} = \mathbf{u}t + \frac{1}{2}\mathbf{a}t^2 \qquad \mathbf{s} = \mathbf{v}t - \frac{1}{2}\mathbf{a}t^2$$

Notice that the position vector **r** of a point is its displacement from the origin. You can sometimes write the relationships between displacement, velocity and acceleration with **r** in place of **s**

Expressing the equation, $v^2 = u^2 + 2as$, in vector form uses techniques beyond the scope of this book. You can still use it separately for components in the x- and y-directions.

Key point

$$v_x^{\,2} = u_x^{\,2} + 2a_x x \quad \text{and} \quad v_y^{\,2} = u_y^{\,2} + 2a_y y$$

Example 1

An object is initially at the point with position vector $\mathbf{r} = (6\mathbf{i} + 3\mathbf{j})\,\text{m s}^{-1}$ and travelling with velocity $(2\mathbf{i} + 15\mathbf{j})\,\text{m s}^{-1}$ when it undergoes an acceleration of $(\mathbf{i} - 2\mathbf{j})\,\text{m s}^{-2}$ for a period of four seconds.

Work out

a Its velocity at the end of the four seconds,

b The displacement it undergoes during the four seconds,

c Its position at the end of the four seconds,

d Its speed and direction of motion at the end of the four seconds.

(*Continued on the next page*)

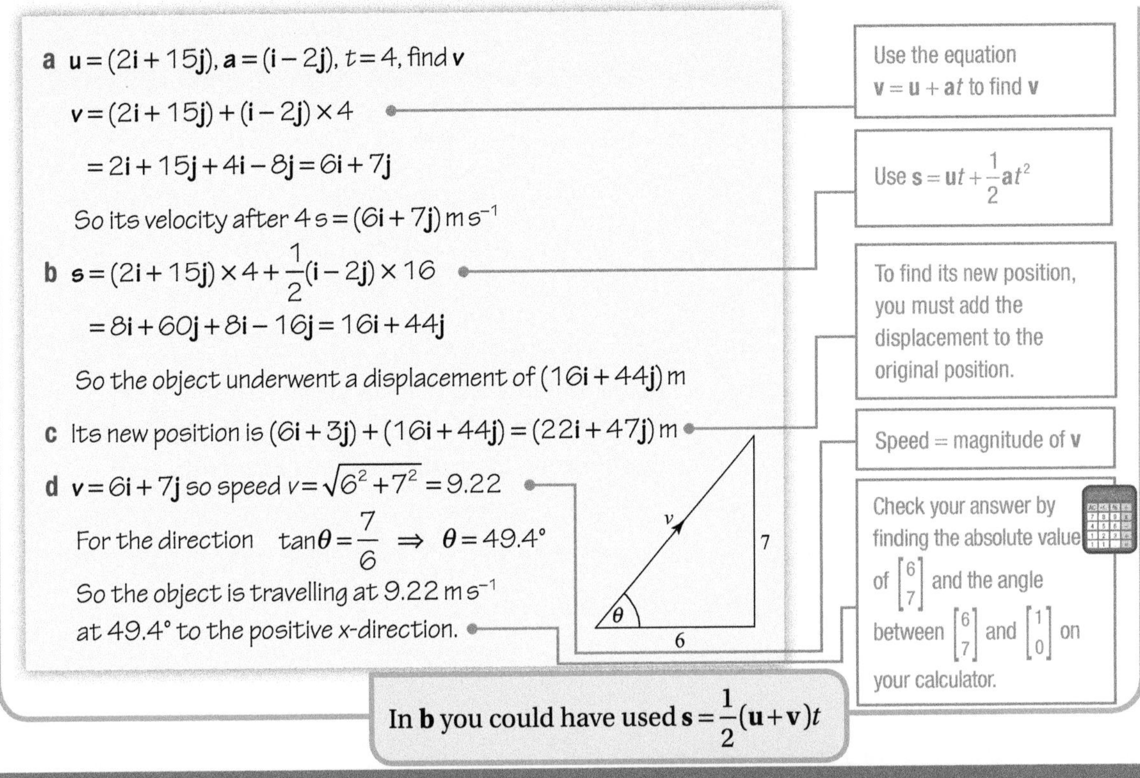

a $\mathbf{u} = (2\mathbf{i} + 15\mathbf{j})$, $\mathbf{a} = (\mathbf{i} - 2\mathbf{j})$, $t = 4$, find $\mathbf{v}$

$\mathbf{v} = (2\mathbf{i} + 15\mathbf{j}) + (\mathbf{i} - 2\mathbf{j}) \times 4$

$= 2\mathbf{i} + 15\mathbf{j} + 4\mathbf{i} - 8\mathbf{j} = 6\mathbf{i} + 7\mathbf{j}$

So its velocity after 4 s $= (6\mathbf{i} + 7\mathbf{j})\,\text{m s}^{-1}$

b $\mathbf{s} = (2\mathbf{i} + 15\mathbf{j}) \times 4 + \frac{1}{2}(\mathbf{i} - 2\mathbf{j}) \times 16$

$= 8\mathbf{i} + 60\mathbf{j} + 8\mathbf{i} - 16\mathbf{j} = 16\mathbf{i} + 44\mathbf{j}$

So the object underwent a displacement of $(16\mathbf{i} + 44\mathbf{j})$ m

c Its new position is $(6\mathbf{i} + 3\mathbf{j}) + (16\mathbf{i} + 44\mathbf{j}) = (22\mathbf{i} + 47\mathbf{j})$ m

d $\mathbf{v} = 6\mathbf{i} + 7\mathbf{j}$ so speed $v = \sqrt{6^2 + 7^2} = 9.22$

For the direction $\tan\theta = \frac{7}{6} \Rightarrow \theta = 49.4°$

So the object is travelling at $9.22\,\text{m s}^{-1}$ at 49.4° to the positive x-direction.

Use the equation $\mathbf{v} = \mathbf{u} + \mathbf{a}t$ to find $\mathbf{v}$

Use $\mathbf{s} = \mathbf{u}t + \frac{1}{2}\mathbf{a}t^2$

To find its new position, you must add the displacement to the original position.

Speed = magnitude of $\mathbf{v}$

Check your answer by finding the absolute value of $\begin{bmatrix}6\\7\end{bmatrix}$ and the angle between $\begin{bmatrix}6\\7\end{bmatrix}$ and $\begin{bmatrix}1\\0\end{bmatrix}$ on your calculator.

In **b** you could have used $\mathbf{s} = \frac{1}{2}(\mathbf{u} + \mathbf{v})t$

MECH

Exercise 18.1A Fluency and skills

1 A particle has initial velocity $(3\mathbf{i} + 2\mathbf{j})\,\text{m s}^{-1}$ and acceleration $(2\mathbf{i} - \mathbf{j})\,\text{m s}^{-2}$

  a Use $\mathbf{v} = \mathbf{u} + \mathbf{a}t$ to work out its velocity after 5 s

  b Use $\mathbf{s} = \mathbf{u}t + \frac{1}{2}\mathbf{a}t^2$ to work out its displacement after 3 s

2 A particle has initial velocity $(\mathbf{i} + 3\mathbf{j})\,\text{m s}^{-1}$ and travels for 3 s

  a Use $\mathbf{v} = \mathbf{u} + \mathbf{a}t$ to work out its final velocity if its acceleration is $(4\mathbf{i} + \mathbf{j})\,\text{m s}^{-2}$

  b Use $\mathbf{v} = \mathbf{u} + \mathbf{a}t$ to work out its acceleration if its final velocity is $(-2\mathbf{i} + 9\mathbf{j})\,\text{m s}^{-1}$

  c Use $\mathbf{s} = \mathbf{u}t + \frac{1}{2}\mathbf{a}t^2$ to work out its displacement if its acceleration is $(3\mathbf{i} - 2\mathbf{j})\,\text{m s}^{-2}$

3 A particle starts with velocity $(4\mathbf{i} - 3\mathbf{j})\,\text{m s}^{-1}$ and after 4 s its velocity is $(8\mathbf{i} + 5\mathbf{j})\,\text{m s}^{-1}$. Use $\mathbf{s} = \frac{1}{2}(\mathbf{u} + \mathbf{v})t$ to work out its displacement.

4 A particle accelerates at $(\mathbf{i} + 3\mathbf{j})\,\text{m s}^{-2}$ for 4 s and undergoes a displacement of $(16\mathbf{i} - 12\mathbf{j})$ m. Use $\mathbf{s} = \mathbf{v}t - \frac{1}{2}\mathbf{a}t^2$ to work out its final velocity.

5 For each of these situations, choose a suitable equation to work out the required value.

  a $\mathbf{u} = (2\mathbf{i} + 5\mathbf{j})$, $\mathbf{v} = (10\mathbf{i} - 7\mathbf{j})$, $t = 6$, work out $\mathbf{s}$

  b $\mathbf{u} = (-2\mathbf{i} + 3\mathbf{j})$, $\mathbf{a} = (3\mathbf{i} - 4\mathbf{j})$, $t = 8$, work out $\mathbf{s}$

  c $\mathbf{u} = (14\mathbf{i} + 6\mathbf{j})$, $\mathbf{v} = (4\mathbf{i} + 21\mathbf{j})$, $t = 5$, work out $\mathbf{a}$

  d $\mathbf{s} = (-4\mathbf{i} + 52\mathbf{j})$, $\mathbf{a} = (2\mathbf{i} - \mathbf{j})$, $t = 4$, work out $\mathbf{v}$

6 A particle is initially at the point with position vector $\mathbf{r} = (3\mathbf{i} - 2\mathbf{j})$ m and travelling with velocity $(5\mathbf{i} - \mathbf{j})\,\text{m s}^{-1}$ when it undergoes an acceleration of $(-\mathbf{i} + 2\mathbf{j})\,\text{m s}^{-2}$ for a period of 3 s. Work out its position at the end of this period.

7 A particle moving with velocity $(-\mathbf{i} + 2\mathbf{j})\,\text{m s}^{-1}$ undergoes a constant acceleration of $(2\mathbf{i} + \mathbf{j})\,\text{m s}^{-2}$ for 5 s. Work out its speed and direction at the end of this period. Show your working.

8 A particle undergoes a constant acceleration of $(2\mathbf{i} - 3\mathbf{j})\,\text{m s}^{-2}$. After 4 s it has velocity $(10\mathbf{i} - 4\mathbf{j})\,\text{m s}^{-1}$. If it was initially at the point with position vector $(\mathbf{i} + 5\mathbf{j})$ m, work out its position at the end of the period.

Reasoning and problem-solving

Strategy

To solve problems involving two-dimensional motion with constant acceleration

1. List the known values and the variable you need to find. This helps you decide which formula to use.
2. Be careful to distinguish between position, displacement and distance, and between velocity and speed.
3. Interpret your answers in the context of the question.

Example 2

Two particles, A and B, are moving in a plane. Initially A is at the point (0, 3) and B is at (2, 1). A has velocity $(2\mathbf{i}+\mathbf{j})\,\text{m s}^{-1}$ and acceleration $(\mathbf{i}-2\mathbf{j})\,\text{m s}^{-2}$, and B has velocity $(3\mathbf{i}-\mathbf{j})\,\text{m s}^{-1}$ and acceleration $2\mathbf{i}\,\text{m s}^{-2}$. Work out the distance between the particles after six seconds.

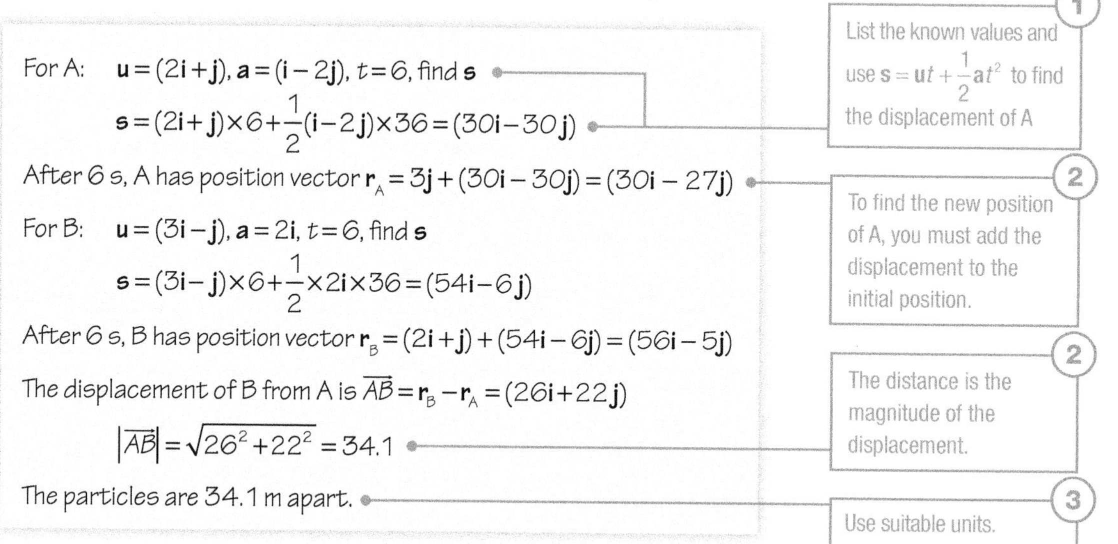

For A: $\mathbf{u}=(2\mathbf{i}+\mathbf{j}),\ \mathbf{a}=(\mathbf{i}-2\mathbf{j}),\ t=6$, find $\mathbf{s}$

$\mathbf{s}=(2\mathbf{i}+\mathbf{j})\times 6+\frac{1}{2}(\mathbf{i}-2\mathbf{j})\times 36=(30\mathbf{i}-30\mathbf{j})$

After 6 s, A has position vector $\mathbf{r}_A=3\mathbf{j}+(30\mathbf{i}-30\mathbf{j})=(30\mathbf{i}-27\mathbf{j})$

For B: $\mathbf{u}=(3\mathbf{i}-\mathbf{j}),\ \mathbf{a}=2\mathbf{i},\ t=6$, find $\mathbf{s}$

$\mathbf{s}=(3\mathbf{i}-\mathbf{j})\times 6+\frac{1}{2}\times 2\mathbf{i}\times 36=(54\mathbf{i}-6\mathbf{j})$

After 6 s, B has position vector $\mathbf{r}_B=(2\mathbf{i}+\mathbf{j})+(54\mathbf{i}-6\mathbf{j})=(56\mathbf{i}-5\mathbf{j})$

The displacement of B from A is $\overrightarrow{AB}=\mathbf{r}_B-\mathbf{r}_A=(26\mathbf{i}+22\mathbf{j})$

$|\overrightarrow{AB}|=\sqrt{26^2+22^2}=34.1$

The particles are 34.1 m apart.

1 List the known values and use $\mathbf{s}=\mathbf{u}t+\frac{1}{2}\mathbf{a}t^2$ to find the displacement of A

2 To find the new position of A, you must add the displacement to the initial position.

2 The distance is the magnitude of the displacement.

3 Use suitable units.

Example 3

A particle, P, is initially at the point with position vector $12\mathbf{j}$ m and moving with velocity $3\mathbf{i}\,\text{m s}^{-1}$. It undergoes an acceleration of $(-\mathbf{i}+2\mathbf{j})\,\text{m s}^{-2}$ for a period of 2 s. Show that it is then moving directly away from the origin O, and work out its speed.

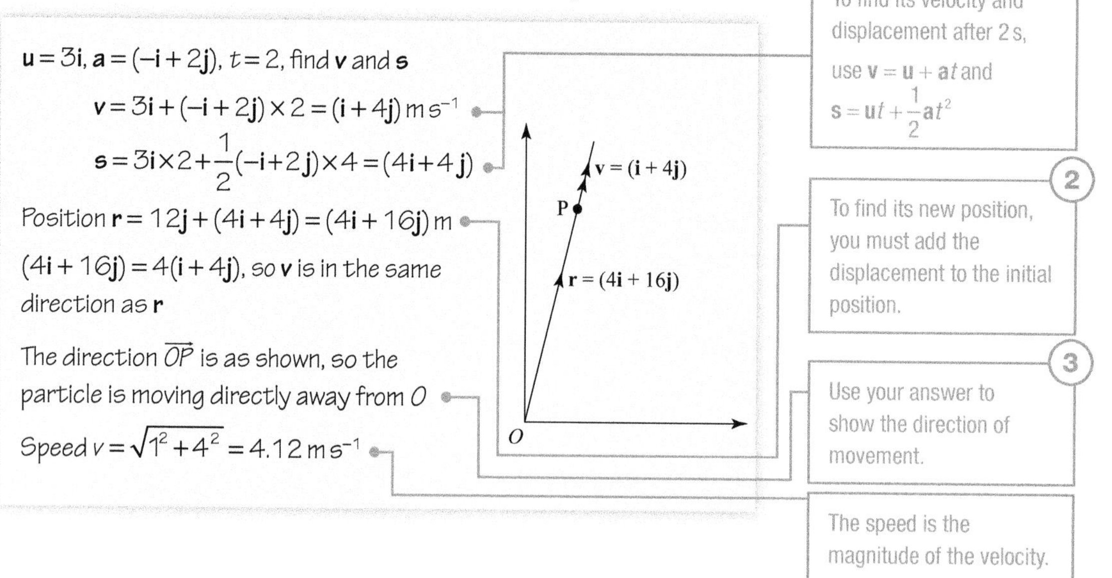

$\mathbf{u}=3\mathbf{i},\ \mathbf{a}=(-\mathbf{i}+2\mathbf{j}),\ t=2$, find $\mathbf{v}$ and $\mathbf{s}$

$\mathbf{v}=3\mathbf{i}+(-\mathbf{i}+2\mathbf{j})\times 2=(\mathbf{i}+4\mathbf{j})\,\text{m s}^{-1}$

$\mathbf{s}=3\mathbf{i}\times 2+\frac{1}{2}(-\mathbf{i}+2\mathbf{j})\times 4=(4\mathbf{i}+4\mathbf{j})$

Position $\mathbf{r}=12\mathbf{j}+(4\mathbf{i}+4\mathbf{j})=(4\mathbf{i}+16\mathbf{j})$ m

$(4\mathbf{i}+16\mathbf{j})=4(\mathbf{i}+4\mathbf{j})$, so $\mathbf{v}$ is in the same direction as $\mathbf{r}$

The direction $\overrightarrow{OP}$ is as shown, so the particle is moving directly away from O

Speed $v=\sqrt{1^2+4^2}=4.12\,\text{m s}^{-1}$

1 To find its velocity and displacement after 2 s, use $\mathbf{v}=\mathbf{u}+\mathbf{a}t$ and $\mathbf{s}=\mathbf{u}t+\frac{1}{2}\mathbf{a}t^2$

2 To find its new position, you must add the displacement to the initial position.

3 Use your answer to show the direction of movement.

The speed is the magnitude of the velocity.

See Ch6.1 For a reminder on collinear vectors.

Exercise 18.1B Reasoning and problem-solving

1 A particle is moving in a plane with acceleration $(4\mathbf{i}+3\mathbf{j})\,\text{m s}^{-2}$. Its initial velocity is $(2\mathbf{i}-6\mathbf{j})\,\text{m s}^{-1}$. After eight seconds it has velocity **v**. Work out

a Its speed and direction at this time, showing your working,

b The distance between the initial and final positions of the particle.

2 An object is moving in a plane with constant acceleration. Its initial velocity is $(3\mathbf{i}-2\mathbf{j})\,\text{m s}^{-1}$ and, six seconds later, its velocity is $(9\mathbf{i}+4\mathbf{j})\,\text{m s}^{-1}$. Its initial position is $(5\mathbf{i}+2\mathbf{j})\,\text{m}$. Work out its final position.

3 Two particles, A and B, are moving in a plane. Initially A is at the point (0, 5) and B is at (3, 7). A has velocity $(\mathbf{i}+3\mathbf{j})\,\text{m s}^{-1}$ and acceleration $(2\mathbf{i}+2\mathbf{j})\,\text{m s}^{-2}$, and B has velocity $2\mathbf{j}\,\text{m s}^{-1}$ and acceleration $(3\mathbf{i}+\mathbf{j})\,\text{m s}^{-2}$. Work out the distance between the particles after four seconds.

4 A boat moving with velocity $(6\mathbf{i}+2\mathbf{j})\,\text{m s}^{-1}$ undergoes an acceleration of $(-\mathbf{i}-7\mathbf{j})\,\text{m s}^{-2}$ for two seconds. Show that, at the end of that time, it is travelling in a direction perpendicular to its initial direction, and at twice the speed. Show your working.

5 A snooker ball is initially travelling with velocity $(4\mathbf{i}+5\mathbf{j})\,\text{m s}^{-1}$. It undergoes an acceleration of magnitude $2.5\,\text{m s}^{-2}$ in a direction given by the vector $(3\mathbf{i}-4\mathbf{j})$. Work out the velocity and displacement of the snooker ball from its initial position after four seconds.

6 A particle, P, is initially at the point (2, 6) in relation to an origin O. It is moving with velocity $(3\mathbf{i}+\mathbf{j})\,\text{m s}^{-1}$ and constant acceleration $(16\mathbf{i}+24\mathbf{j})\,\text{m s}^{-2}$. Show that, after two seconds, it is moving directly away from O. Work out its speed at that time. Show your working.

7 An object is moving in a plane. At time $t=0$ it is at the origin, O, and moving with velocity **u**. After two seconds, it is at A, where $OA=-2\mathbf{i}-4\mathbf{j}$. After a further three seconds, it is at B, where $OB=10\mathbf{i}-40\mathbf{j}$ Assuming the object has constant acceleration a, work out **a** and **u**

Challenge

8 A particle starts at the origin O, and then moves with constant velocity $(2\mathbf{i}+3\mathbf{j})\,\text{m s}^{-1}$ for four seconds. It then accelerates with constant acceleration $(\mathbf{i}-3\mathbf{j})\,\text{m s}^{-2}$ for six seconds, before slowing uniformly to rest in four seconds. Work out

a Its displacement during the first stage of the journey,

b Its velocity after the second stage of the journey,

c Its acceleration during the third stage of the journey,

d Its position at the end of the journey.

9 An aircraft is travelling horizontally in an easterly direction at a height of 400 m above level ground and at a speed of $120\,\text{m s}^{-1}$. When it is directly above an observer, O, it releases a package. In relation to the origin O, and to the x- and y- directions of east and up respectively, the package has acceleration $(-4\mathbf{i}-8\mathbf{j})\,\text{m s}^{-2}$. P is the position of the package at time t Work out

a An expression for the vector OP

b The time the package takes to reach the ground,

c How far from the observer the package lands.

MECH

2290

18.2 Two-dimensional motion with variable acceleration

Fluency and skills

See Ch7.4 For a reminder on motion with variable acceleration.

For motion in one dimension, the displacement s, velocity v and acceleration a are related by $v=\frac{ds}{dt}$ and $a=\frac{dv}{dt}$. You can extend these relationships to motion in which the displacement $\mathbf{s}$, velocity $\mathbf{v}$ and acceleration $\mathbf{a}$ are two-dimensional vectors.

Key point

$$\mathbf{v}=\frac{d\mathbf{s}}{dt} \quad \text{and} \quad \mathbf{a}=\frac{d\mathbf{v}}{dt}=\frac{d^2\mathbf{s}}{dt^2}$$

The differentiation of a vector involves parametric equations. The horizontal component, x, and the vertical component, y, of the displacement are expressed in terms of a third parameter. This third parameter is time, t

See Ch15.8 For a reminder on differentiating parametric equations.

To differentiate a vector, you differentiate its components, so

Key point

If $\mathbf{s}=x\mathbf{i}+y\mathbf{j}$, then $\mathbf{v}=\frac{dx}{dt}\mathbf{i}+\frac{dy}{dt}\mathbf{j}$ or $\mathbf{v}=\dot{x}\mathbf{i}+\dot{y}\mathbf{j}$

$$\mathbf{a}=\frac{d^2x}{dt^2}\mathbf{i}+\frac{d^2y}{dt^2}\mathbf{j} \quad \text{or} \quad \mathbf{a}=\ddot{x}\mathbf{i}+\ddot{y}\mathbf{j}$$

Example 1

The displacement of a particle at time t s is given by $\mathbf{s}=4t^2\mathbf{i}+(3t-5t^3)\mathbf{j}$ metres. Write an expression for its velocity at time t s

$\mathbf{v}=\frac{ds}{dt}=8t\mathbf{i}+(3-15t^2)\mathbf{j}\,\text{m s}^{-1}$

Differentiate displacement to find velocity. You need to differentiate each component with respect to t

Example 2

The velocity of a particle at time t s is given by $\mathbf{v}=\sin 2\pi t\mathbf{i}+t^2\mathbf{j}\,\text{m s}^{-1}$. Work out its acceleration when $t=1$

$\mathbf{a}=\frac{dv}{dt}=2\pi\cos 2\pi t\mathbf{i}+2t\mathbf{j}$

When $t=1$, $\mathbf{a}=(2\pi\mathbf{i}+2\mathbf{j})\,\text{m s}^{-2}$

Differentiate velocity to find acceleration. See Ch15.2 for a reminder on differentiating trigonometric functions.

You can also express the relationships between $\mathbf{s}$, $\mathbf{v}$ and $\mathbf{a}$ in terms of integration.

Key point

$$\mathbf{v}=\int\mathbf{a}\,dt \quad \text{and} \quad \mathbf{s}=\int\mathbf{v}\,dt$$

As with differentiation, to integrate a vector, you integrate each of its components.

Example 3

The acceleration of a particle at time t is given by $\mathbf{a} = 6t^2\mathbf{i} + (3t^2 - 8t)\mathbf{j}$. Write an expression for its velocity at time t, given that $\mathbf{v} = 5\mathbf{i} - 6\mathbf{j}$ when $t = 0$

$\mathbf{v} = \int 6t^2\mathbf{i} + (3t^2 - 8t)\mathbf{j}\, dt = 2t^3\mathbf{i} + (t^3 - 4t^2)\mathbf{j} + \mathbf{c}$ — Integrate acceleration to find velocity. The constant of integration is a vector.

When $t = 0$, $\mathbf{v} = 5\mathbf{i} - 6\mathbf{j}$, so $\mathbf{c} = 5\mathbf{i} - 6\mathbf{j}$ — Use the initial conditions to find **c**

$\mathbf{v} = 2t^3\mathbf{i} + (t^3 - 4t^2)\mathbf{j} + 5\mathbf{i} - 6\mathbf{j}$

So $\mathbf{v} = (2t^3 + 5)\mathbf{i} + (t^3 - 4t^2 - 6)\mathbf{j}$

Example 4

A particle is initially at the point with position vector $(2\mathbf{i} + \mathbf{j})$ m and has velocity $\mathbf{v} = (4\mathbf{i} + 6t\mathbf{j})\text{ m s}^{-1}$. Work out its position four seconds later.

$\mathbf{r} = \int 4\mathbf{i} + 6t\mathbf{j}\, dt = 4t\mathbf{i} + 3t^2\mathbf{j} + \mathbf{c}$ — Integrate velocity to find displacement, **r**, from the origin.

When $t = 0$, $\mathbf{r} = 2\mathbf{i} + \mathbf{j}$, so $\mathbf{c} = 2\mathbf{i} + \mathbf{j}$

$\mathbf{r} = 4t\mathbf{i} + 3t^2\mathbf{j} + 2\mathbf{i} + \mathbf{j} = (4t + 2)\mathbf{i} + (3t^2 + 1)\mathbf{j}$

When $t = 4$, $\mathbf{r} = 18\mathbf{i} + 49\mathbf{j}$, so the position after 4 s is $(18\mathbf{i} + 49\mathbf{j})$ m — Substitute $t = 4$ into your expression for the position vector.

Exercise 18.2A Fluency and skills

1 For each of these displacement or position vectors, express the velocity at time t

a $\mathbf{s} = 4t\mathbf{i} + (8 + 2t^2)\mathbf{j}$

b $\mathbf{r} = (t^2 - 4t)\mathbf{i} + (t^3 - 2t^2)\mathbf{j}$

c $\mathbf{s} = 2\cos t\,\mathbf{i} + 4\sin t\,\mathbf{j}$

d $\mathbf{r} = e^t\,\mathbf{i} + \ln(t+1)\,\mathbf{j}$

2 For each of these acceleration vectors, write an expression for the velocity vector consistent with the given initial condition.

a $\mathbf{a} = (3 - 2t)\mathbf{i} + (2t - 6t^3)\mathbf{j}$, $\mathbf{v} = 3\mathbf{i}$ when $t = 0$

b $\mathbf{a} = 4\cos 2t\,\mathbf{i} + 8\sin 2t\,\mathbf{j}$, $\mathbf{v} = 2\mathbf{i} + \mathbf{j}$ when $t = 0$

3 For each of these velocity vectors, write expressions for the acceleration and displacement or position vectors consistent with the given initial condition.

a $\mathbf{v} = 15\mathbf{i} + (20 - 10t)\mathbf{j}$, $\mathbf{s} = 4\mathbf{i} - 3\mathbf{j}$ when $t = 0$

b $\mathbf{v} = -8\sin t\,\mathbf{i} + 4e^{2t}\mathbf{j}$, $\mathbf{r} = \mathbf{i} + 3\mathbf{j}$ when $t = 0$

c $\mathbf{v} = 2t(1 - 3t)\mathbf{i} + t^2(3 - 4t)\mathbf{j}$, $\mathbf{r} = 2\mathbf{i}$ when $t = 0$

4 A particle has acceleration $\mathbf{a} = (4t\mathbf{i} + 3t^2\mathbf{j})\text{ m s}^{-2}$ and initial velocity $\mathbf{v} = (-3\mathbf{i} + \mathbf{j})\text{ m s}^{-1}$. Work out its velocity when $t = 2$ s

5 A particle has position vector $\mathbf{r} = (\sin \pi t\,\mathbf{i} + \cos \pi t\,\mathbf{j})$ m. Work out its velocity when $t = 3$ s. Show your working.

6 A particle starts from the point whose position vector is $(2\mathbf{i} + 3\mathbf{j})$ m. Its velocity at time t is given by $\mathbf{v} = ((4t - 2)\mathbf{i} + 3t^2\mathbf{j})\text{ m s}^{-1}$. Work out its position when $t = 2$ s

7 The acceleration of a particle at time t s is $\mathbf{a} = (6t\mathbf{i} + 2\mathbf{j})\text{ m s}^{-2}$. When $t = 0$, its velocity is $(3\mathbf{i} + 4\mathbf{j})\text{ m s}^{-1}$. Work out

a Its velocity at time t

b Its speed after 2 s

c Its direction of travel after 2 s, showing your working.

8 At time t the position of a particle is given by $\mathbf{r} = (t^3\,\mathbf{i} + 8\ln(1+t)\mathbf{j})$ m. What is its velocity and acceleration when $t = 2$? Show your working.

9 A particle has velocity at time t given by $\mathbf{v} = 6\cos 3t\,\mathbf{i} + 8\sin 2t\,\mathbf{j}$. It initially has position vector $\mathbf{r} = 2\mathbf{i} - 5\mathbf{j}$. Work out

a Its acceleration at time t

b Its position at time t

Reasoning and problem-solving

Strategy

To solve problems involving two-dimensional motion with variable acceleration

1. If the problem involves a variable force, use Newton's second law ($\mathbf{F} = m\mathbf{a}$) to find acceleration.
2. Identify whether you need to move from $\mathbf{s} \to \mathbf{v} \to \mathbf{a}$ (differentiation) or $\mathbf{a} \to \mathbf{v} \to \mathbf{s}$ (integration).
3. Be clear whether you are asked for displacement, position or distance, and velocity or speed.
4. Substitute any known values into your expressions to give the required solution.

Example 5

A particle of mass 4 kg is acted on by a force of $(8\mathbf{i} + 12t\mathbf{j})$ N. Initially, the particle has position vector $(3\mathbf{i} - \mathbf{j})$ m and velocity $(3\mathbf{i} - 5\mathbf{j})$ m s^{-1}. Work out

a Its speed after 4 s

b Its position at that time.

a $8\mathbf{i} + 12t\mathbf{j} = 4\mathbf{a}$ — (1) Use $\mathbf{F} = m\mathbf{a}$ to find the acceleration.

$\mathbf{a} = 2\mathbf{i} + 3t\mathbf{j}$ and $\mathbf{v} = \int 2\mathbf{i} + 3t\mathbf{j}\, dt = 2t\mathbf{i} + \frac{3}{2}t^2\mathbf{j} + \mathbf{c}_1$ — (2) Integrate acceleration to find velocity. Use the initial velocity to find $\mathbf{c}_1$

When $t = 0$, $\mathbf{v} = 3\mathbf{i} - 5\mathbf{j}$, so $\mathbf{c}_1 = 3\mathbf{i} - 5\mathbf{j}$

$\mathbf{v} = (2t + 3)\mathbf{i} + \left(\frac{3}{2}t^2 - 5\right)\mathbf{j}$

When $t = 4$, $\mathbf{v} = 11\mathbf{i} + 19\mathbf{j}$ so speed $= \sqrt{11^2 + 19^2} = 22.0$ m s^{-1} — (3) The speed is the magnitude of the velocity.

b $\mathbf{r} = \int (2t+3)\mathbf{i} + \left(\frac{3}{2}t^2 - 5\right)\mathbf{j}\, dt$ — (2) Integrate velocity to find displacement/position.

$= (t^2 + 3t)\mathbf{i} + \left(\frac{1}{2}t^3 - 5t\right)\mathbf{j} + \mathbf{c}_2$

When $t = 0$, $\mathbf{r} = 3\mathbf{i} - \mathbf{j}$, so $\mathbf{c}_2 = 3\mathbf{i} - \mathbf{j}$

$\mathbf{r} = (t^2 + 3t + 3)\mathbf{i} + \left(\frac{1}{2}t^3 - 5t - 1\right)\mathbf{j}$ — (4) Substitute $t = 4$ into your expression.

When $t = 4$, $\mathbf{r} = (31\mathbf{i} + 11\mathbf{j})$ m

Example 6

The displacement of a particle at time t s is given by $\mathbf{s} = 6t^2\mathbf{i} + (3t^2 - 12t)\mathbf{j}$ m. Work out the distance of the particle from its initial position at the instant it is travelling in the x-direction.

$\mathbf{s} = 6t^2\mathbf{i} + (3t^2 - 12t)\mathbf{j}$ and $\mathbf{v} = \frac{d\mathbf{s}}{dt} = 12t\mathbf{i} + (6t - 12)\mathbf{j}$ — (2) Differentiate to find velocity.

Travels in x-direction when $6t - 12 = 0 \Rightarrow t = 2$ — The y-component of the velocity must be zero.

$t = 2$ gives $\mathbf{s} = 24\mathbf{i} - 12\mathbf{j}$

The distance $= \sqrt{24^2 + (-12)^2} = 26.8$ m — (3) The distance is the magnitude of the displacement.

Exercise 18.2B Reasoning and problem-solving

1 A particle of mass 5 kg is acted on by a force $(10\mathbf{i} + 5t\mathbf{j})$ N. Initially the particle has position vector $(2\mathbf{i} + 4\mathbf{j})$ m. When $t = 2$, $\mathbf{v} = (5\mathbf{i} + 5\mathbf{j})\,\text{m s}^{-1}$. Work out

   a The initial speed of the particle,

   b The position of the particle at $t = 2$

2 The displacement of a boat at time t s is given by $\mathbf{s} = (5t^3 - 12t)\mathbf{i} + 10t\mathbf{j}$ m. Work out the distance of the boat from its initial position at the instant it is travelling in the y-direction. Leave your answer in surd form.

3 A particle moves on a plane such that its position at time t s is given by $\mathbf{r} = (3t - 2)\mathbf{i} + (4t - 2t^2)\mathbf{j}$ m

   a Write expressions for the velocity and acceleration of the particle at time t

   b Work out the initial speed of the particle.

   c At what time(s) is the particle moving parallel to the x-axis?

   d Is the particle ever stationary? Give a reason for your answer.

4 A particle moves with constant acceleration vector $\mathbf{a}$. Its initial velocity is $\mathbf{u}$ and, at time t, it has velocity $\mathbf{v}$ and displacement $\mathbf{s}$ from its starting position.

   a Show by integration that $\mathbf{v} = \mathbf{u} + \mathbf{a}t$

   b Show by integration that $\mathbf{s} = \mathbf{u}t + \frac{1}{2}\mathbf{a}t^2$

5 A ball of mass 5 kg is acted upon by a force of $(20t\mathbf{i} - 15\mathbf{j})$ N. Initially the ball is at the point with position vector $(2\mathbf{i} + 3\mathbf{j})$ m and is travelling with velocity $(-2\mathbf{i} + 12\mathbf{j})\,\text{m s}^{-1}$. Work out its velocity and position when $t = 6$

6 At time t a particle has position given by $\mathbf{r} = (2t - 1 + \cos t)\mathbf{i} + \sin 2t\mathbf{j}$. The particle starts at the origin.

   a Work out the value of t for which it next touches the x-axis.

   b For that value of t, work out its instantaneous velocity and acceleration. Show your working.

7 A force $\mathbf{F} = (2000t\mathbf{i} - 4000\mathbf{j})$ N acts on a particle of mass 500 kg at time t s. Initially the particle is at the origin and travelling with velocity $10\mathbf{i}\,\text{m s}^{-1}$. Work out

   a The speed of the particle when $t = 2$

   b The distance of the particle from the origin at that time.

8 An object moves on a plane so that its acceleration at time t s is given by $\mathbf{a} = (-4\cos 2t\mathbf{i} - 4\sin 2t\mathbf{j})\,\text{m s}^{-2}$. It is initially at the point $(1, 0)$ and travelling at $2\,\text{m s}^{-1}$ in the y-direction.

   a Show that the object moves at constant speed.

   b Work out the distance of the object from the origin at time t and hence describe the path of the object.

Challenge

9 Particles P and Q, each of mass 0.5 kg, move on a horizontal plane, with east and north as the **i** and **j** directions. Initially P has velocity $(2\mathbf{i} - 5\mathbf{j})\,\text{m s}^{-1}$ and Q is travelling north at $2\,\text{m s}^{-1}$. Each particle is acted on by a force of magnitude t N. The force on P acts towards the north-east, while that on Q acts towards the south-east. Work out the value of t for which

   a The two particles have the same speed,

   b The two particles are travelling in the same direction. Show your working.

10 The position of a particle at time t is given by $\mathbf{r} = (2 + 3t + 8t^2)\mathbf{i} + (6 + t + 12t^2)\mathbf{j}$. Work out

   a The times at which the particle is moving directly towards or directly away from the origin,

   b The position of the particle at the time(s) found in part **a**, and identify whether it is moving towards or away from the origin.

2291

18.3 Motion under gravity 2

Fluency and skills

See Ch8.3
For a reminder on motion under gravity.

When an object moves vertically under gravity you usually assume that it is a particle (it has no size and does not spin) and that there is no wind or air resistance. Its acceleration is $g\,\text{m s}^{-2}$ downwards.

However, an object thrown into the air is usually a projectile, with displacement and velocity in two dimensions. This means that to investigate the motion of a projectile, you will often need to use trigonometry to resolve velocity into its horizontal and vertical components.

See Ch6.2
For a reminder on components of a vector.

For example, consider a particle which is projected from a point O on a horizontal plane with a speed of $25\,\text{m s}^{-1}$ at an angle of 30° to the horizontal.

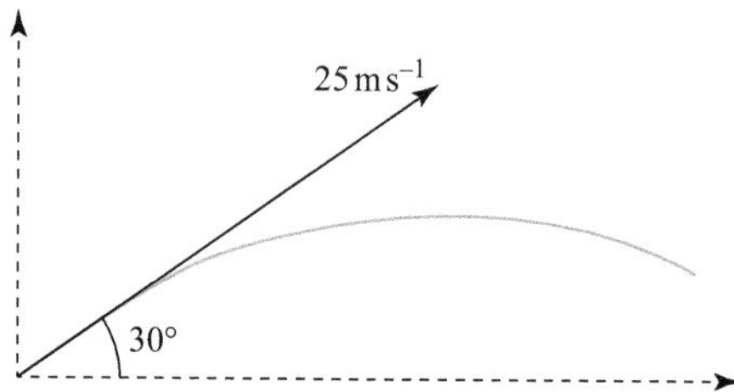

Resolving horizontally: $25\cos 30° \Rightarrow$ The initial horizontal component of velocity is $21.7\,\text{m s}^{-1}$

Resolving vertically: $25\sin 30° \Rightarrow$ The initial vertical component of velocity is $12.5\,\text{m s}^{-1}$

You will also need to use the equations of motion for constant acceleration. Remember that g will always act as the vertical component of acceleration in a downwards direction. Assuming there is no wind or air resistance, there will be no horizontal acceleration, so horizontal velocity is constant.

Example 1

A particle is projected from a point O on a horizontal plane with speed $40\,\text{m s}^{-1}$ at an angle of 30° to the horizontal. It lands on the plane at A. Taking $g = 9.81\,\text{m s}^{-2}$, work out

a How long it was in the air, **b** The distance OA **c** The greatest height it reached.

a The initial vertical component of velocity is $40\sin 30°$

The vertical acceleration is $-g$

At time t, the height $y = 40t\sin 30° - \frac{1}{2}gt^2$

> Use $s = ut + \frac{1}{2}at^2$ to find an expression for the height of the particle.

At A, $y = 0$, so $40t\sin 30° - \frac{1}{2}gt^2 = 0$

$\frac{1}{2}t(80\sin 30° - gt) = 0$

> When the particle lands, the vertical height will be zero.

So $y = 0$ when $t = 0$ (the start time) or

when $t = \frac{80\sin 30°}{g} = \frac{80\sin 30°}{9.81} = 4.08$ (to 2 dp)

The particle is in the air for 4.08 s

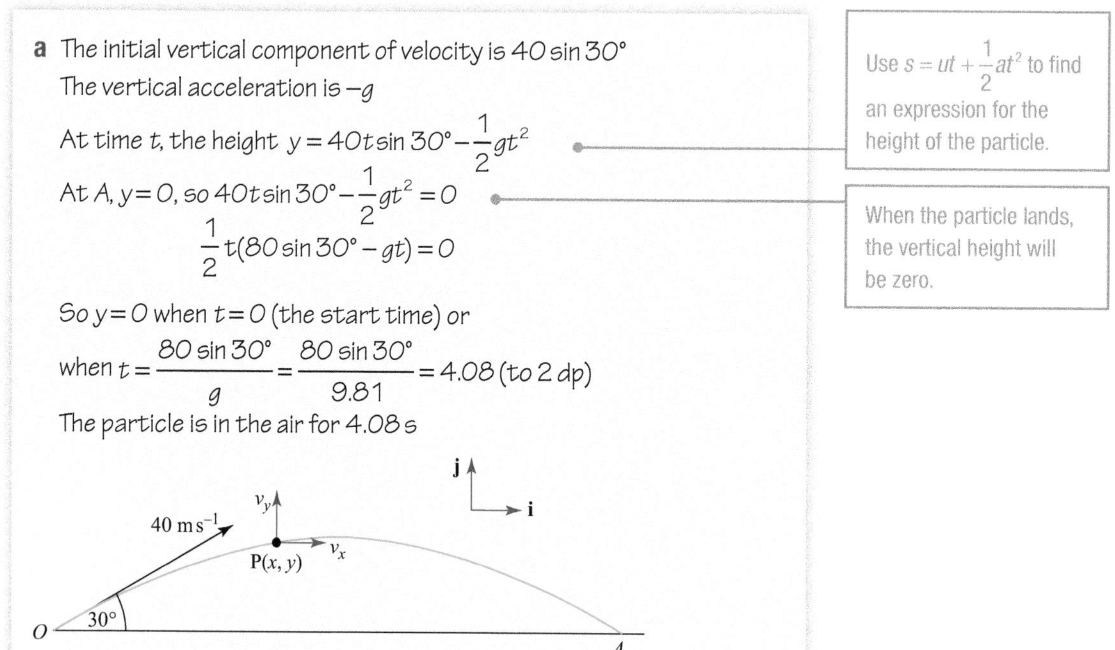

(Continued on the next page)

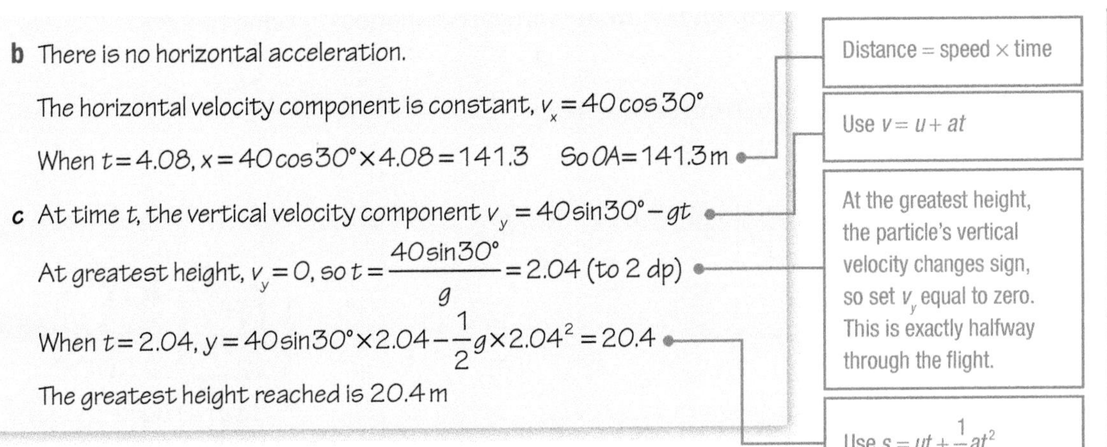

b There is no horizontal acceleration.

The horizontal velocity component is constant, $v_x = 40\cos 30°$

When $t = 4.08$, $x = 40\cos 30° \times 4.08 = 141.3$ So $OA = 141.3$ m

c At time t, the vertical velocity component $v_y = 40\sin 30° - gt$

At greatest height, $v_y = 0$, so $t = \dfrac{40\sin 30°}{g} = 2.04$ (to 2 dp)

When $t = 2.04$, $y = 40\sin 30° \times 2.04 - \dfrac{1}{2}g \times 2.04^2 = 20.4$

The greatest height reached is 20.4 m

Distance = speed × time

Use $v = u + at$

At the greatest height, the particle's vertical velocity changes sign, so set v_y equal to zero. This is exactly halfway through the flight.

Use $s = ut + \dfrac{1}{2}at^2$

You can also write the equations in vector form. You will need to know how to derive formulae for the time of flight, range and greatest height and the equation of the path of a projectile.

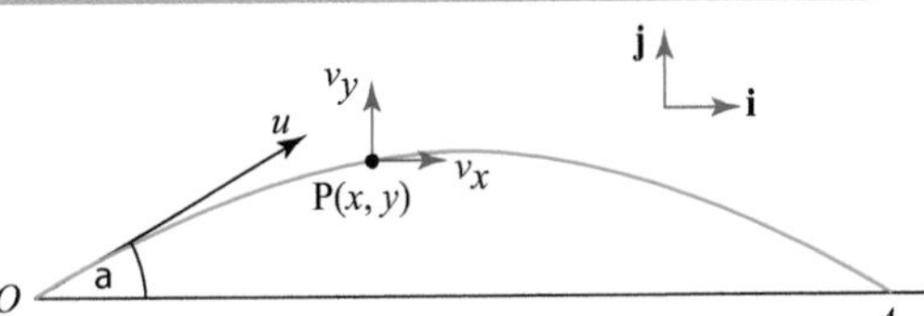

Particle, P, is projected from a point O on a horizontal plane with initial speed u and elevation α. Take **i** and **j** directions, as shown.

The horizontal and vertical components of initial velocity are $u_x = u\cos\alpha$ and $u_y = u\sin\alpha$, so the initial velocity vector is

$$\mathbf{u} = u\cos\alpha\mathbf{i} + u\sin\alpha\mathbf{j}$$

At time t, the particle is at the point with position vector $\mathbf{r} = x\mathbf{i} + y\mathbf{j}$ and has velocity $\mathbf{v} = v_x\mathbf{i} + v_y\mathbf{j}$. Its acceleration is $\mathbf{a} = -g\mathbf{j}$

From $\mathbf{v} = \mathbf{u} + \mathbf{a}t$: $\quad \mathbf{v} = u\cos\alpha\mathbf{i} + (u\sin\alpha - gt)\mathbf{j}$ ....[1]

From $\mathbf{r} = \mathbf{u}t + \frac{1}{2}\mathbf{a}t^2$: $\quad \mathbf{r} = ut\cos\alpha\mathbf{i} + \left(ut\sin\alpha - \frac{1}{2}gt^2\right)\mathbf{j}$ ....[2]

The horizontal distance the particle travels before landing (OA in the diagram above) is the **range**.

When the particle lands, $y = 0$ and so, from [2],

$$ut\sin\alpha - \frac{1}{2}gt^2 = 0$$

This gives $t = 0$ (the start time) and $t = \dfrac{2u\sin\alpha}{g}$, the **time of flight**. For this value of t, $x = \dfrac{2u^2\sin\alpha\cos\alpha}{g}$

You can simplify this using the formula $\sin 2\alpha = 2\sin\alpha\cos\alpha$

See Ch14.3 For a reminder on double angle formulae.

Key point

Range $x = \dfrac{u^2\sin 2\alpha}{g}$

As $\sin 2\alpha = \sin(180° - 2\alpha) = \sin 2(90° - \alpha)$, it follows that elevations of α and $(90° - \alpha)$ give the same range.

The range is greatest when $\sin 2\alpha = 1$. This gives $2\alpha = 90° \Rightarrow \alpha = 45°$

Key point

Elevations of α and $(90° - \alpha)$ give the same range.

The maximum range $= \dfrac{u^2}{g}$ when $\alpha = 45°$

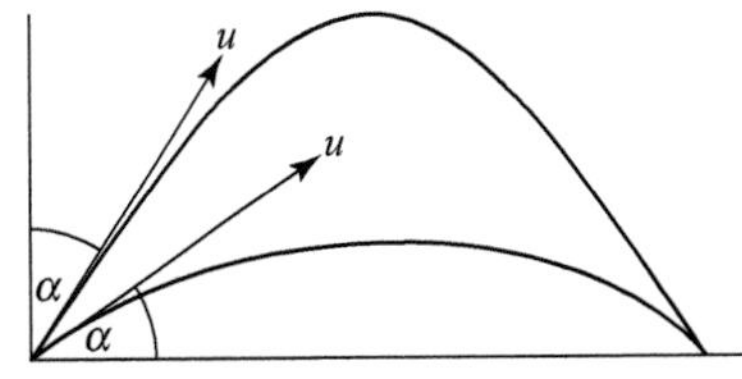

When the particle is at its **maximum height** the vertical component of velocity is zero, so, from [1],
$$u\sin\alpha - gt = 0$$

This gives $t = \frac{u\sin\alpha}{g}$, which is half the time of flight.

At this time, from [2],

$$y = u\left(\frac{u\sin\alpha}{g}\right)\sin\alpha - \frac{1}{2}g\left(\frac{u\sin\alpha}{g}\right)^2 = \frac{u^2\sin^2\alpha}{2g}$$

Key point

Maximum height $= \frac{u^2\sin^2\alpha}{2g}$

ICT Resource online

To investigate motion under gravity, click this link in the digital book.

You can find the **equation of the path** followed by the projectile.

At time t the particle is at the point (x, y), where, from [2],

$$x = ut\cos\alpha \quad .....[3]$$

$$y = ut\sin\alpha - \frac{1}{2}gt^2 \quad .....[4]$$

From [3], $t = \frac{x}{u\cos\alpha}$. Substituting this into [4] gives

$$y = u\left(\frac{x}{u\cos\alpha}\right)\sin\alpha - \frac{1}{2}g\left(\frac{x}{u\cos\alpha}\right)^2$$

This gives a quadratic function of x. The path is a parabola.

Key point

The equation of the path is $y = x\tan\alpha - \left(\frac{g\sec^2\alpha}{2u^2}\right)x^2$

While the formulae in these four Key point boxes are useful results, which you could be asked to derive, you ***should not*** quote or use them without justification.

Example 2

A particle is projected from a point O on horizontal ground with initial speed 12 m s^{-1} at an angle of 60° to the horizontal. Taking $g = 9.8$ m s^{-2}, work out

a Its height above the ground after travelling 7.2 m horizontally,

b Its speed and direction at that instant.

a Initial velocity is $\mathbf{u} = 12\cos60°\mathbf{i} + 12\sin60°\mathbf{j} = 6\mathbf{i} + 6\sqrt{3}\mathbf{j}$

Take **i** and **j** as the horizontal and vertical directions and resolve the velocity into these components.

Acceleration is $\mathbf{a} = -g\mathbf{j} = -9.8\mathbf{j}$

The position vector at time t is $\mathbf{r} = x\mathbf{i} + y\mathbf{j} = 6t\mathbf{i} + \left(6t\sqrt{3} - 4.9t^2\right)\mathbf{j}$

Use $\mathbf{s} = \mathbf{u}t + \frac{1}{2}\mathbf{a}t^2$ to find the time when the horizontal distance $x = 7.2$

When $x = 7.2$, $6t = 7.2$ and so $t = 1.2$

When $t = 1.2$, $y = 6\times1.2\sqrt{3} - 4.9\times1.2^2 = 5.41$ so its height above the ground is 5.41 m.

Use $s = ut + \frac{1}{2}at^2$ to find the vertical height when $t = 1.2$

b The velocity at time t is $\mathbf{v} = 6\mathbf{i} + (6\sqrt{3} - 9.8t)\mathbf{j}$

Use $\mathbf{v} = \mathbf{u} + \mathbf{a}t$

When $t = 1.2$, $\mathbf{v} = 6\mathbf{i} - 1.37\mathbf{j}$ and $v = |\mathbf{v}| = \sqrt{6^2 + (-1.37)^2} = 6.15$

$\tan\theta = \frac{1.37}{6} = 0.228$ and so $\theta = 12.8°$

6, θ, 1.37, v

The particle is travelling at 6.15 m s^{-1} at 12.8° below the horizontal.

Check your answer by evaluating the angle between **v** and $\begin{bmatrix}1\\0\end{bmatrix}$ and the absolute value of **v** on your calculator.

Exercise 18.3A Fluency and skills

1 A projectile is launched from a point O on horizontal ground with speed $15\,\text{m s}^{-1}$ at an angle of 35° to the horizontal. Take $g = 9.8\,\text{m s}^{-2}$

  a For how long is the projectile in the air?

  b What is the horizontal range of the projectile?

  c Work out the time taken for the projectile to reach its maximum height.

  d What is the greatest height reached by the projectile?

2 A projectile is launched from a point O on horizontal ground with speed $10\,\text{m s}^{-1}$ at an angle of 60° to the horizontal. Take $g = 9.81\,\text{m s}^{-2}$

  a For how long is the projectile in the air?

  b What is the horizontal range of the projectile?

  c Work out the time taken for the projectile to reach its maximum height.

  d What is the greatest height reached by the projectile?

3 A particle is projected from a point O on horizontal ground with speed $24\,\text{m s}^{-1}$ at an angle of 50° to the horizontal. Taking $g = 9.81\,\text{m s}^{-2}$, work out

  a The height of the particle after it has travelled 30 m horizontally,

  b The speed and direction of the particle at this instant. Show your working.

4 A ball is projected from a point O on horizontal ground. Its initial velocity is $(30\mathbf{i} + 40\mathbf{j})\,\text{m s}^{-1}$. Take $g = 9.8\,\text{m s}^{-2}$ and work out

  a Its position after

    i 1 s

    ii 2 s

  b The length of time it is in the air,

  c Its range,

  d The equation of its path.

5 A particle is projected at $196\,\text{m s}^{-1}$ at an angle α to the horizontal. It reaches a maximum height of 490 m. Work out the value of α. Take $g = 9.81\,\text{m s}^{-2}$

6 A particle is projected at 45° to the horizontal from a point O on horizontal ground. It strikes the ground again 100 m away. Take $g = 10\,\text{m s}^{-2}$ and work out the speed with which it was projected.

MECH

Reasoning and problem-solving

Strategy

To solve problems involving motion under gravity

1. Sketch a diagram.
2. Decide whether to work separately with the horizontal and vertical motion or in terms of vectors.
3. State what you know and choose the appropriate constant acceleration equation.
4. Interpret your solutions in the context of the question.

Example 3

A girl throws a stone horizontally from the top of a 40 m cliff at a speed which is just sufficient to reach the water's edge 50 m from the base of the cliff. Taking $g = 9.8\ \text{m s}^{-2}$, calculate

a The initial speed of the stone, **b** The speed and direction of the stone as it hits the water.

Take **i** and **j** as the horizontal and upward vertical directions, as shown.

U
j
i
40 m
50 m

(1) Sketch a diagram.

a For the vertical motion: $u = 0$, $a = -9.8$, $s = -40$, find t

(3) State what you know.

$-40 = 0 \times t - 4.9t^2 \Rightarrow t = \sqrt{\dfrac{40}{4.9}} = 2.86$

(3) Use $s = ut + \frac{1}{2}at^2$

The horizontal velocity is U and is constant.

$2.86U = 50 \Rightarrow U = 17.5$

Distance = speed × time

The initial speed of the stone is $17.5\ \text{m s}^{-1}$

b When the stone hits the water its horizontal velocity $v_x = 17.5$

Its vertical velocity is $v_y = -9.8 \times 2.86 = -28$

(3) Use $v = u + at$. Remember that $u = 0$ and $a = -9.8\ \text{m s}^{-2}$

Speed $v = \sqrt{17.5^2 + 28^2} = 33.0$

Direction is given by $\tan\theta = \dfrac{28}{17.5} \Rightarrow \theta = 58.0°$

17.5
θ
v
28

The stone hits the water at $33.0\ \text{m s}^{-1}$ in a direction 58.0° below the horizontal.

(4) Interpret your solutions in the context of the question.

Example 4

A projectile is fired at $50\ \text{m s}^{-1}$ from ground level and strikes a target 100 m away and 20 m above its point of projection. Work out the two possible angles of elevation at which it was fired.

Take $g = 9.81\ \text{m s}^{-2}$

Let the angle of elevation be θ

$\mathbf{u} = \begin{pmatrix} 50\cos\theta \\ 50\sin\theta \end{pmatrix}$, $\mathbf{a} = \begin{pmatrix} 0 \\ -9.81 \end{pmatrix}$ and, at time t, $\mathbf{r} = \begin{pmatrix} 100 \\ 20 \end{pmatrix}$

(2)(3) Use vectors and start by stating what you know.

$\begin{pmatrix} 100 \\ 20 \end{pmatrix} = \begin{pmatrix} 50\cos\theta \\ 50\sin\theta \end{pmatrix} t + \begin{pmatrix} 0 \\ -4.905 \end{pmatrix} t^2$

(3) Use $\mathbf{r} = \mathbf{u}t + \frac{1}{2}\mathbf{a}t^2$

$100 = 50t\cos\theta \Rightarrow t = \dfrac{2}{\cos\theta}$ [1]

Compare x components.

$20 = 50t\sin\theta - 4.905t^2$ [2]

Compare y components.

$20 = 100\tan\theta - 19.62\sec^2\theta$

Substitute [1] into [2]

$20 = 100\tan\theta - 19.62(1 + \tan^2\theta)$

$19.62\tan^2\theta - 100\tan\theta + 19.62 + 20 = 0$

$19.62\tan^2\theta - 100\tan\theta + 39.62 = 0$

So $\tan\theta = 4.66$ or 0.433

You could use your calculator to solve this quadratic equation.

The two angles of elevation are $\theta = 77.9°$ or $23.4°$

Exercise 18.3B Reasoning and problem-solving

1 A stone is thrown from the top of a 50 m high vertical cliff at 30 m s^{-1} at an angle of 20° below the horizontal.

Take $g = 10$ m s^{-2}

a Work out how far from the foot of the cliff the stone hits the sea.

b What assumptions have you made in your answer?

2 A ball is kicked from the floor along a corridor. The ball leaves the floor at 20 m s^{-1} at an angle of θ. The ceiling of the corridor is 3 m high.

Take $g = 9.81$ m s^{-2}

a Work out the maximum value of θ if the ball does not touch the ceiling.

b If you could not ignore the size of the ball, what effect would it have on your answer?

3 A ball is kicked from the floor of a sports hall at 40° to the horizontal and at a speed of V m s^{-1}. The ceiling of the hall is 10 m above the floor.

Take $g = 9.81$ m s^{-2}

a Work out in terms of V the maximum height to which the ball rises.

b Use your answer from part **a** to work out the greatest value of V for which the ball does not hit the ceiling.

c For this value of V work out the distance the ball travels before hitting the floor.

4 A man standing outside a football ground can see the ball if it rises above a height of 10 m. A ball is kicked from ground level at an angle of 50° to the horizontal and with speed 25 m s^{-1}. For how long is it visible?

Take $g = 9.8$ m s^{-2}

5 A tower of height h m stands on horizontal ground. An arrow is fired from the top of the tower with speed 40 m s^{-1} at an angle of 25° above the horizontal. It hits the ground 170 m from the base of the tower. Taking $g = 9.8$ m s^{-2}, calculate the value of h

6 A projectile has a range of 300 m, which takes it 8 s. Taking $g = 9.81$ m s^{-2}, calculate

a Its initial speed and direction,

b Its speed and direction after 2 s

c The length of time for which it was at a height greater than 40 m.

7 Show that a projectile fired with initial speed u at an angle of elevation α reaches a maximum height of $\frac{u^2 \sin^2 \alpha}{2g}$

Challenge

8 A tennis player serves the ball. At the moment that the racket strikes the ball, it is 11.6 m from the net and 3.2 m above the ground. The racket imparts a horizontal speed u m s^{-1} and a downward vertical speed v m s^{-1}. The ball hits the ground 6.4 m beyond the net after 0.3 s.

Take $g = 9.8$ m s^{-2}

a Calculate the values of u and v

b By how much does the ball clear the net, which is 0.91 m high?

9 A projectile is fired with initial speed u at an angle of elevation α

a Show that the equation of its path is $y = x\tan\alpha - \left(\frac{g\sec^2\alpha}{2u^2}\right)x^2$

b The projectile passes through the points (a, b) and (b, a). Show that its range is $\frac{a^2 + ab + b^2}{a + b}$

MECH

MyMaths 2198, 2199 SEARCH

18.4 Motion under forces

Fluency and skills

When motion is in two dimensions it is often necessary to resolve forces into components.

A 40 N force acts horizontally to the right and a 30 N force acts vertically upwards. Using Pythagoras' theorem and trigonometry, you can find a single force which is equivalent to these two, known as the **resultant**, R

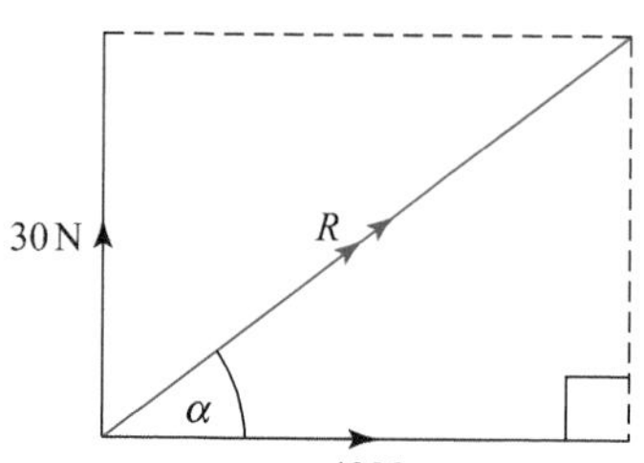

$R=\sqrt{30^2+40^2}=50\text{ N} \qquad \alpha=\tan^{-1}\left(\frac{30}{40}\right)=36.9°$ (to 1 dp)

So the resultant of the two forces is a single force of 50 N acting at an angle of 36.9° above the horizontal.

You will often need to reverse this process. You can use trigonometry to resolve the resultant 50 N force into its horizontal and vertical **components** X and Y

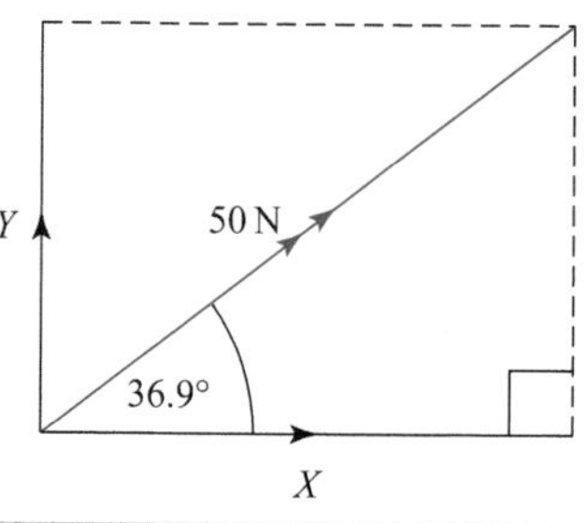

$X=50\cos 36.9°=40\text{ N}$ (to 2 sf)
$Y=50\sin 36.9°=30\text{ N}$ (to 2 sf)

Example 1

The diagram shows three forces acting on an object at a point O. The mass of the object is 4 kg

The effect of the forces accelerates the object from rest in the direction $\overrightarrow{OA}$. Work out

a The magnitude of the unknown force P

b The magnitude of the acceleration,

c How far the object travels in the first five seconds.

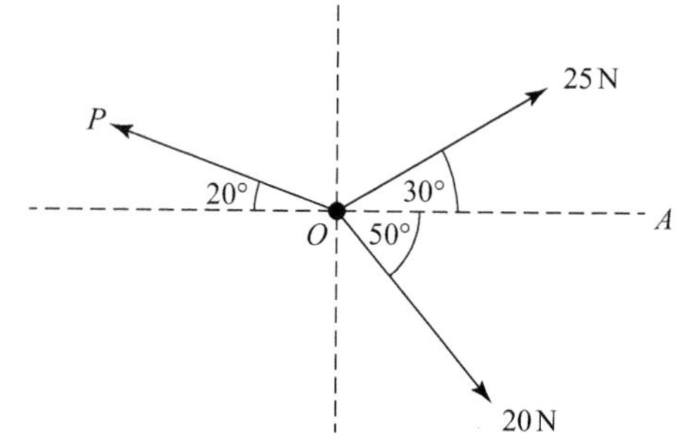

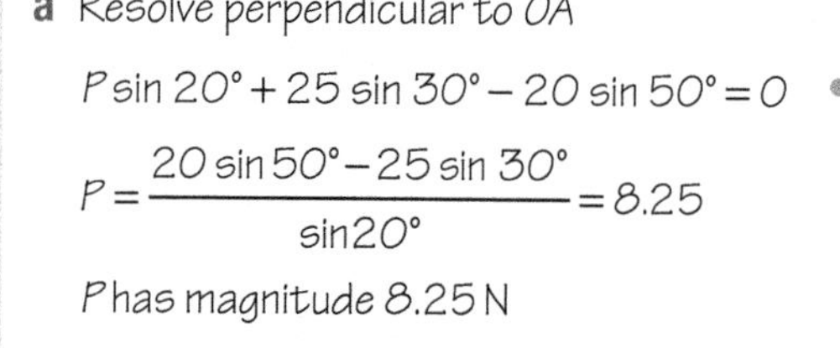

a Resolve perpendicular to OA

$P\sin 20°+25\sin 30°-20\sin 50°=0$ — There is no acceleration in the vertical direction so you can set this equation equal to zero..

$P=\dfrac{20\sin 50°-25\sin 30°}{\sin 20°}=8.25$

P has magnitude 8.25 N

b Let the magnitude of the acceleration be a — Resolve along OA

$25\cos 30°+20\cos 50°-8.25\cos 20°=4a$ — Use $F=ma$

$a=\dfrac{25\cos 30°+20\cos 50°-8.25\cos 20°}{4}=6.69$

The acceleration is 6.69 m s^{-2}

c $s=\frac{1}{2}\times 6.69\times 25=83.6$ The object travels 83.6 m — Use $s=ut+\frac{1}{2}at^2$. The object starts from rest so take $u=0$

You can also express your solution in terms of vectors.

Example 2

A block of mass 10 kg rests on a smooth horizontal plane. Forces of 50 N and 20 N, acting respectively at 20° and 40° to the horizontal, pull the block in opposite directions on the plane. Taking $g = 9.8\text{ m s}^{-2}$, work out

a The acceleration of the block,

b The normal reaction between the plane and the block.

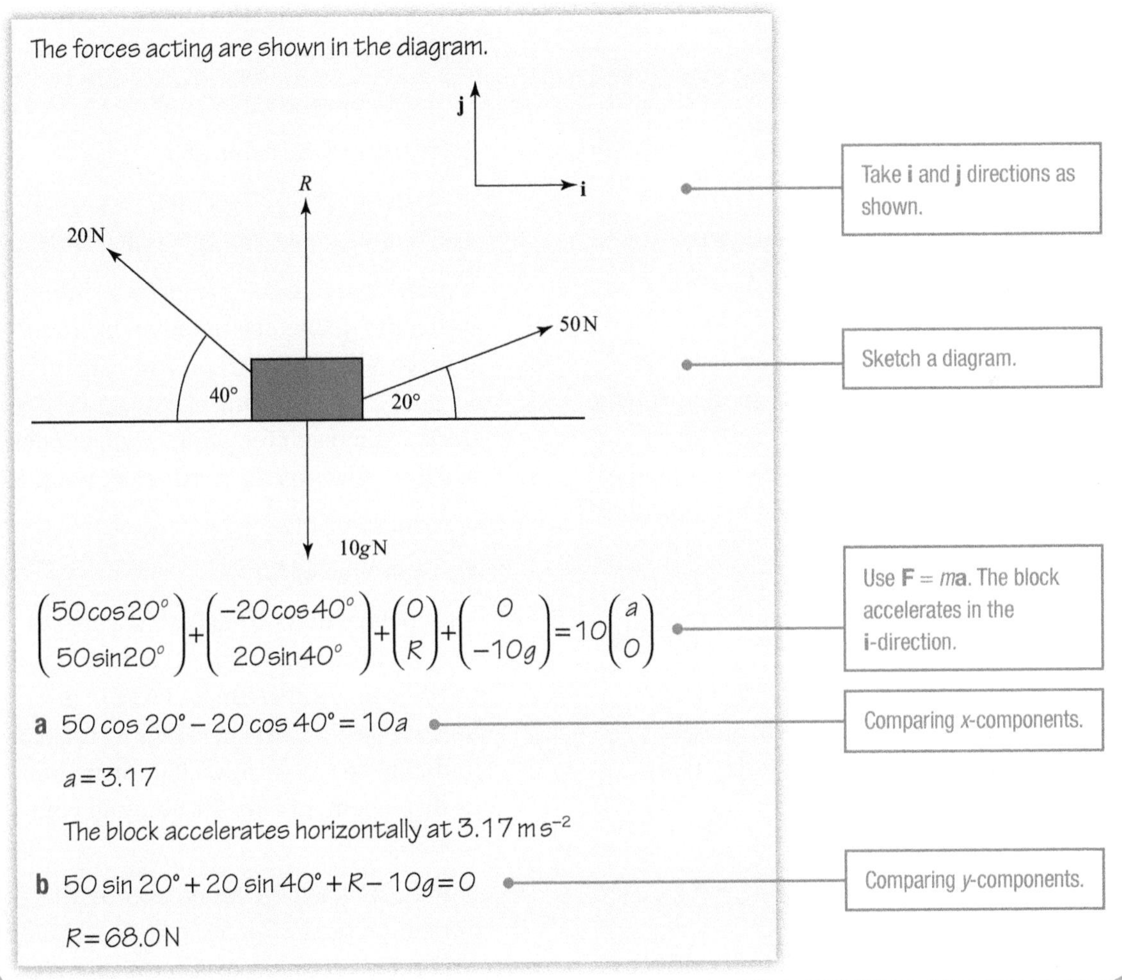

The forces acting are shown in the diagram.

Take **i** and **j** directions as shown.

Sketch a diagram.

$$\begin{pmatrix} 50\cos 20^\circ \\ 50\sin 20^\circ \end{pmatrix} + \begin{pmatrix} -20\cos 40^\circ \\ 20\sin 40^\circ \end{pmatrix} + \begin{pmatrix} 0 \\ R \end{pmatrix} + \begin{pmatrix} 0 \\ -10g \end{pmatrix} = 10\begin{pmatrix} a \\ 0 \end{pmatrix}$$

Use $\mathbf{F} = m\mathbf{a}$. The block accelerates in the **i**-direction.

a $50\cos 20^\circ - 20\cos 40^\circ = 10a$ — Comparing x-components.

$a = 3.17$

The block accelerates horizontally at 3.17 m s^{-2}

b $50\sin 20^\circ + 20\sin 40^\circ + R - 10g = 0$ — Comparing y-components.

$R = 68.0\text{ N}$

Exercise 18.4A Fluency and skills

1 A particle of mass 2 kg moves under the action of three forces: $(9\mathbf{i} + 4\mathbf{j})$ N, $(-2\mathbf{i} + 5\mathbf{j})$ N and $(5\mathbf{i} - \mathbf{j})$ N. Showing your working, work out

  a The magnitude of its acceleration,

  b The angle between its acceleration vector and the **i**-direction.

2 A particle of mass 1 kg moves under the action of three forces: $5\mathbf{i}$ N, $(2\mathbf{i} - 3\mathbf{j})$ N and $(-3\mathbf{i} + 5\mathbf{j})$ N. Showing your working, work out

  a The magnitude of its acceleration,

  b The angle between its acceleration vector and the **i**-direction.

3 A particle of mass 3 kg moves horizontally to the left under the action of its weight and two forces, P and Q, as shown in the diagram.

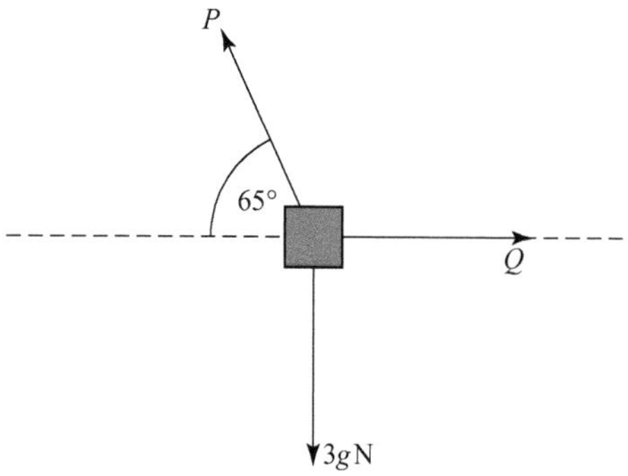

If the acceleration of the particle is $2\,\text{m s}^{-2}$, work out the values of P and Q. Take $g = 9.81\,\text{m s}^{-2}$

4 A particle of mass 5 kg rests on a smooth horizontal plane. It is acted upon by forces of 20 N, 12 N and P on bearings 035°, 160° and 295° respectively. As a result, the particle accelerates from rest in a northerly direction. Taking $g = 9.8\,\text{m s}^{-2}$, work out

a The value of P

b How far the particle moves in the first three seconds.

5 The diagram shows a particle of mass 8 kg which is accelerating horizontally at $a\,\text{m s}^{-2}$

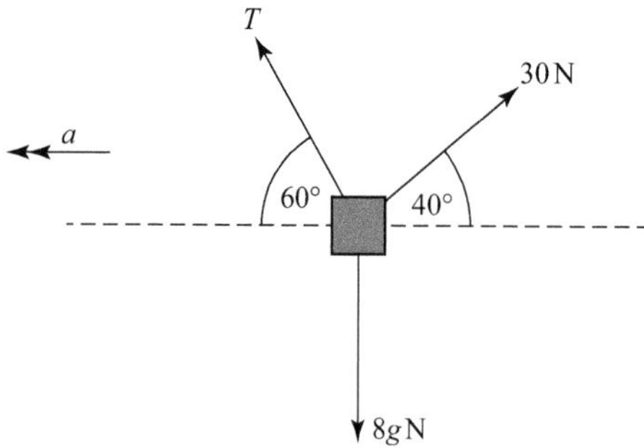

Taking $g = 9.81\,\text{m s}^{-2}$, work out

a The force T

b The value of a

6 The diagram shows a particle of mass 4 kg moving on a horizontal plane under the action of three forces. It has acceleration $a\,\text{m s}^{-2}$ in the direction shown.

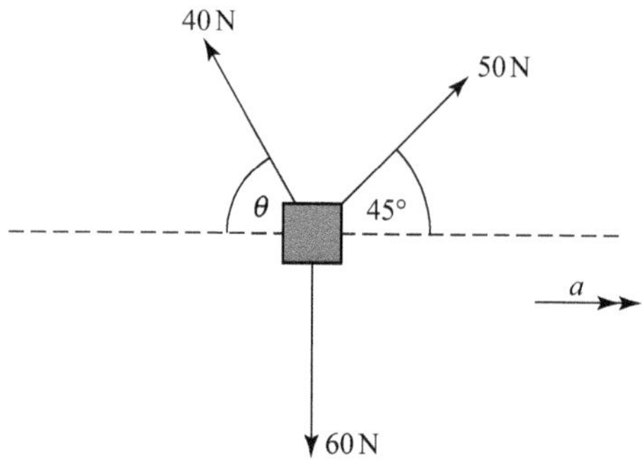

Taking $g = 9.8\,\text{m s}^{-2}$,

a Work out the value of θ

b Work out the value of a

c If the particle starts from rest, work out its speed after it has travelled 10 m

7 A particle of mass 4 kg rests on a smooth horizontal plane. It is acted on by three horizontal forces: 20 N on a bearing of 120°, 10 N on a bearing of 250° and P due north. As a result the particle accelerates at $a\,\text{m s}^{-2}$ towards the north-east. Work out

a The value of a

b The value of P

8 A particle of mass 10 kg is at rest on a smooth horizontal plane. It is acted upon by three horizontal forces: 60 N on a bearing of 065°, 45 N on a bearing of 320° and 22 N on a bearing of 200°. Work out its displacement after 6 s. Show all your working.

9 A body of mass 5 kg lies on a smooth horizontal plane. Forces of 60 N, 40 N and 50 N act in directions 030°, 120° and 200° respectively. Work out the magnitude and direction of the acceleration of the body. Show all your working.

10 A particle of mass 2 kg lies on a smooth horizontal plane. It is acted upon by a force of 20 N on a bearing of 040°, another force of 40 N on a bearing of 300° and a third force P on a bearing of 160°. As a result the particle accelerates in a westerly direction. Work out

a The magnitude of P

b The magnitude of its acceleration.

Reasoning and problem-solving

Strategy

To solve problems involving motion under forces

1. Draw a diagram showing all the forces in the problem.
2. Resolve each force into its two components and use $\mathbf{F} = m\mathbf{a}$
3. Interpret your solutions in the context of the question.

MECH

Example 3

A wedge is modelled as a prism of mass 10 kg, the cross-section of which is an isosceles right-angled triangle. It is placed with its largest face on a rough horizontal surface. A force of 60 N is applied to the wedge, perpendicular to one of its sloping faces. The normal reaction of the surface on the wedge is R and the motion is resisted by a frictional force of $0.2R$. Work out the acceleration of the wedge.

Take $g = 9.81\text{ m s}^{-2}$

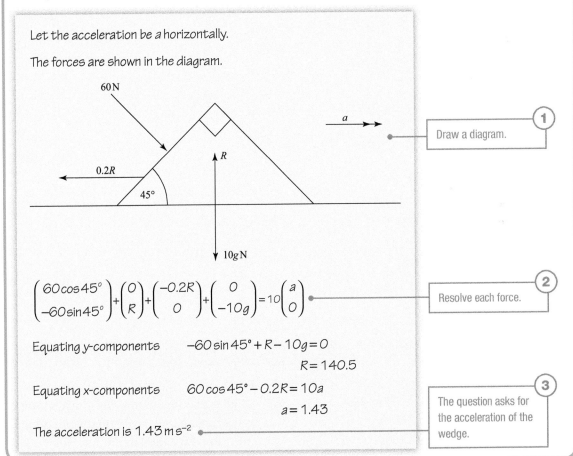

Let the acceleration be a horizontally.

The forces are shown in the diagram.

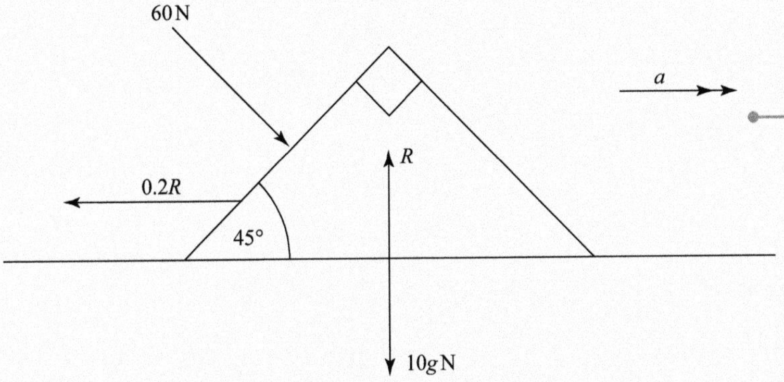

60 N

a

0.2R

R

45°

10g N

(1) Draw a diagram.

$$\begin{pmatrix} 60\cos 45° \\ -60\sin 45° \end{pmatrix} + \begin{pmatrix} 0 \\ R \end{pmatrix} + \begin{pmatrix} -0.2R \\ 0 \end{pmatrix} + \begin{pmatrix} 0 \\ -10g \end{pmatrix} = 10\begin{pmatrix} a \\ 0 \end{pmatrix}$$

(2) Resolve each force.

Equating y-components $\quad -60\sin 45° + R - 10g = 0$

$$R = 140.5$$

Equating x-components $\quad 60\cos 45° - 0.2R = 10a$

$$a = 1.43$$

The acceleration is 1.43 m s^{-2}

(3) The question asks for the acceleration of the wedge.

Example 4

A block of mass 5 kg rests on a rough horizontal plane. The block then accelerates uniformly from rest, pulled by a string inclined at 60° to the horizontal. The normal reaction of the plane on the block is R and the motion is resisted by a frictional force of $0.4R$. After 9 m, the block is travelling at $6\,\text{m s}^{-1}$. Taking $g = 9.8\,\text{m s}^{-2}$, work out the tension in the string.

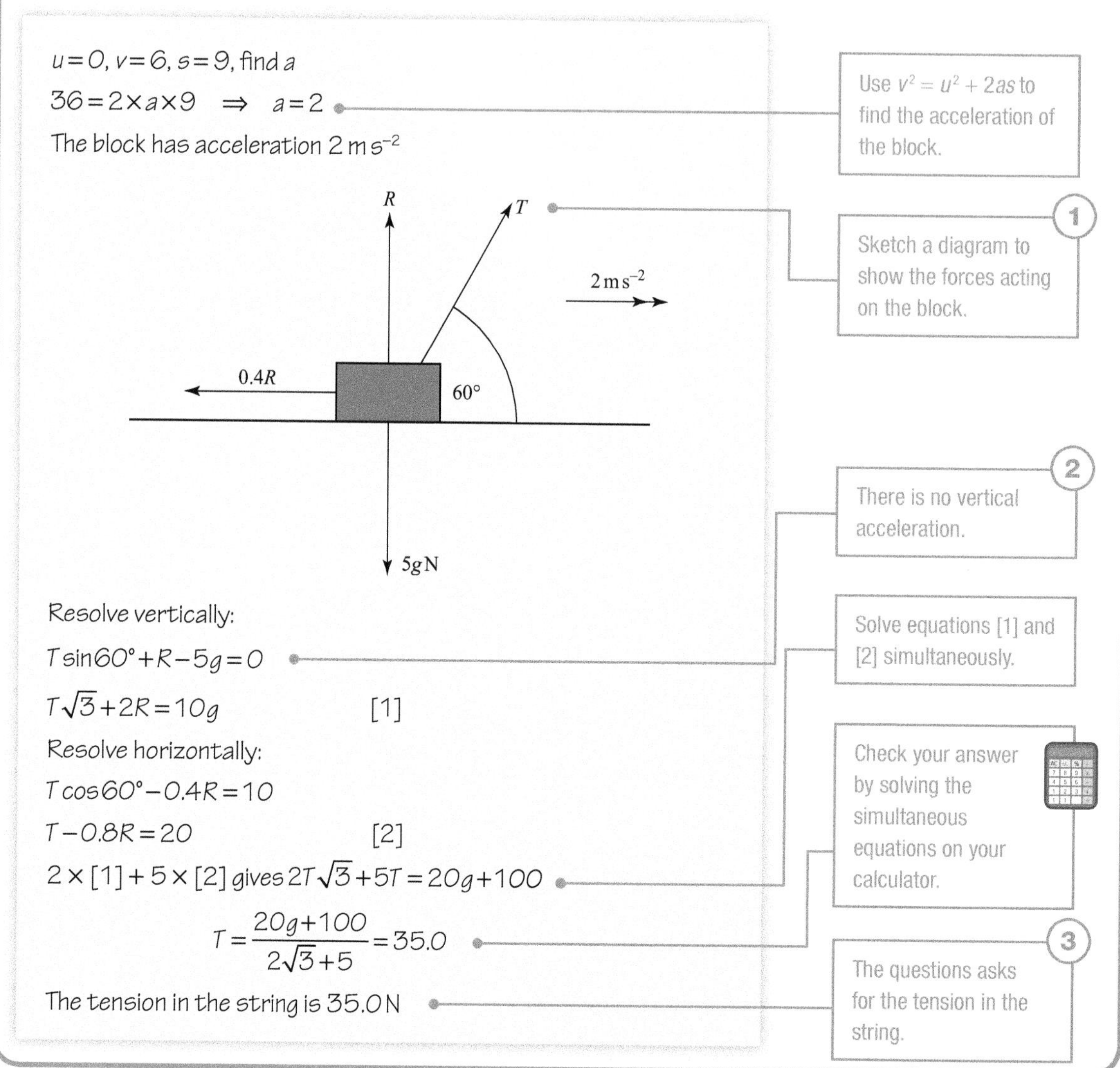

$u = 0, v = 6, s = 9$, find a

$36 = 2 \times a \times 9 \Rightarrow a = 2$

The block has acceleration $2\,\text{m s}^{-2}$

Resolve vertically:

$T\sin 60° + R - 5g = 0$

$T\sqrt{3} + 2R = 10g \qquad [1]$

Resolve horizontally:

$T\cos 60° - 0.4R = 10$

$T - 0.8R = 20 \qquad [2]$

$2 \times [1] + 5 \times [2]$ gives $2T\sqrt{3} + 5T = 20g + 100$

$$T = \frac{20g + 100}{2\sqrt{3} + 5} = 35.0$$

The tension in the string is 35.0 N

Use $v^2 = u^2 + 2as$ to find the acceleration of the block.

1 Sketch a diagram to show the forces acting on the block.

2 There is no vertical acceleration.

Solve equations [1] and [2] simultaneously.

Check your answer by solving the simultaneous equations on your calculator.

3 The questions asks for the tension in the string.

Exercise 18.4B Reasoning and problem-solving

1 A particle of mass 5 kg is pushed up a smooth slope inclined at 30° to the horizontal by a horizontal force of 40 N. Taking $g = 9.81\,\text{m s}^{-2}$, work out

a The reaction between the particle and the plane,

b The acceleration of the particle.

2 A rectangular block of mass 2 kg rests on a rough horizontal plane. The block is pulled by a string inclined at 30° to the horizontal. The normal reaction of the surface on the block is R and the motion is resisted by a frictional force of $0.2R$. If the tension in the string is 20 N, work out the acceleration of the block. Take $g = 9.8\,\text{m s}^{-2}$

3 The diagram shows a crate of mass 600 kg suspended from a helicopter by two cables, each inclined at 60° to the horizontal. The crate is being accelerated horizontally at $0.5\,\text{m s}^{-2}$ against air resistance of 400 N. Taking $g = 9.81\,\text{m s}^{-2}$, work out T_1 and T_2, the tensions in the two cables.

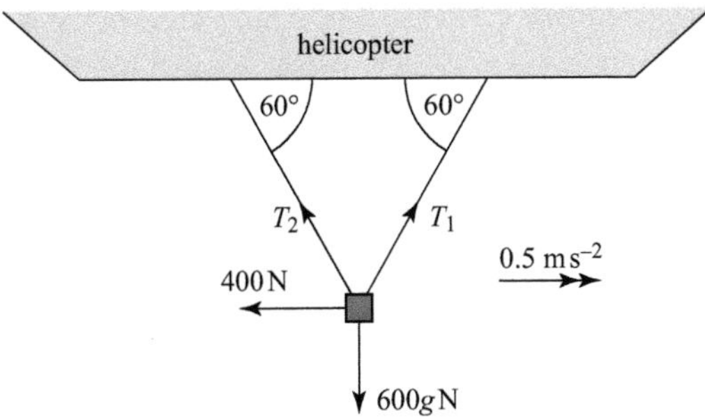

4 A block of mass m rests on a rough horizontal surface. The block is accelerated from rest, pulled by a string inclined at an angle θ to the horizontal. The normal reaction of the surface on the block is R and the motion is resisted by a frictional force of $0.75R$. The tension in the string equals the weight of the block.

a Show that the acceleration of the block is $a = \dfrac{g(3\sin\theta + 4\cos\theta - 3)}{4}$

b Hence show that for maximum acceleration $\theta = 36.9°$

5 A "kite buggy" is a wheeled cart pulled along by the force of the wind acting on a kite attached to it. The buggy has a mass of 100 kg (including the driver). It is travelling down a 10° slope towed by a kite whose tether makes an angle of 40° with the horizontal. The tension in the tether is 250 N and resistance forces total 50 N. Taking $g = 10\,\text{m s}^{-2}$, work out the acceleration of the buggy.

6 A ship of mass 800 tonnes is towed from rest by two tugs. The first exerts a force of 50 kN on a bearing of 020°, the second a force of 70 kN on a bearing of 100°. The resistance of the water is a force of magnitude 20 kN. Work out the distance travelled by the ship in the first 20 seconds.

Challenge

7 A rectangular block of mass 2 kg is at rest at a point A on a rough slope inclined at 30° to the horizontal. It is accelerated up the line of greatest slope, pulled by a string inclined at 20° to the slope. The normal reaction of the surface on the block is R and the motion is resisted by a frictional force of $0.6R$. The tension in the string is 30 N. After 2 s the string breaks and the block slows to rest at B. Taking $g = 9.8\,\text{m s}^{-2}$, work out the distance AB.

8 A cable car and its occupant have a mass of 100 kg. The cradle car is suspended from a trolley and travelling on a cable inclined at 10° to the horizontal. The cradle is supported by two ropes, AC and BC attached at A and B to the trolley as shown, where $AB = 4$ m. The ropes have lengths $AC = 4$ m and $BC = 5$ m. The car is accelerating up the cable at $0.2\,\text{m s}^{-2}$. Work out the tensions, T_1 and T_2, in the ropes. Take $g = 9.81\,\text{m s}^{-2}$

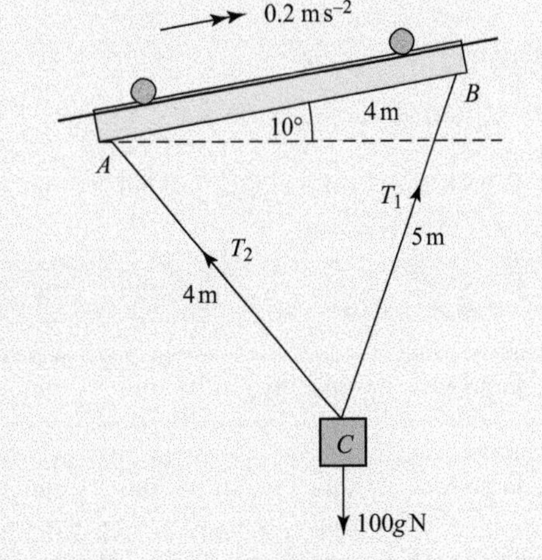

MECH

MyMaths 2192 SEARCH

18 Summary and review

Chapter summary

- For motion in two dimensions the position **r**, displacement **s**, initial and final velocities **u** and **v** and acceleration **a** are two-dimensional vectors.
- For constant acceleration: $\mathbf{v}=\mathbf{u}+\mathbf{a}t$ $\quad \mathbf{s}=\frac{1}{2}(\mathbf{u}+\mathbf{v})t$ $\quad \mathbf{s}=\mathbf{u}t+\frac{1}{2}\mathbf{a}t^2$ $\quad \mathbf{s}=\mathbf{v}t-\frac{1}{2}\mathbf{a}t^2$
- You can use the equation, $v^2=u^2+2as$, in component form to show the horizontal and vertical components of velocity, that is

  $v_x^{\,2}=u_x^{\,2}+2a_x x$ and $v_y^{\,2}=u_y^{\,2}+2a_y y$
- For variable acceleration $\mathbf{v}=\frac{\mathrm{d}\mathbf{s}}{\mathrm{d}t}$ and $\mathbf{a}=\frac{\mathrm{d}\mathbf{v}}{\mathrm{d}t}=\frac{\mathrm{d}^2\mathbf{s}}{\mathrm{d}t^2}$
- Differentiate a vector by differentiating components, so if $\mathbf{s}=x\mathbf{i}+y\mathbf{j}$ then $\mathbf{v}=\frac{\mathrm{d}x}{\mathrm{d}t}\mathbf{i}+\frac{\mathrm{d}y}{\mathrm{d}t}\mathbf{j}$, also written as $\mathbf{v}=\dot{x}\mathbf{i}+\dot{y}\mathbf{j}$, and $\mathbf{a}=\frac{\mathrm{d}^2x}{\mathrm{d}t^2}\mathbf{i}+\frac{\mathrm{d}^2y}{\mathrm{d}t^2}\mathbf{j}$, also written as $\mathbf{a}=\ddot{x}\mathbf{i}+\ddot{y}\mathbf{j}$

  You can express these relationships in terms of integration: $\mathbf{v}=\int\mathbf{a}\,\mathrm{d}t$ and $\mathbf{s}=\int\mathbf{v}\,\mathrm{d}t$
- For a projectile fired from a point on a horizontal plane with speed u and elevation α the initial velocity is $\mathbf{u}=u\cos\alpha\,\mathbf{i}+u\sin\alpha\,\mathbf{j}$ and acceleration is $\mathbf{a}=-g\mathbf{j}$. The velocity and position at time t are $\mathbf{v}=u\cos\alpha\,\mathbf{i}+(u\sin\alpha-gt)\mathbf{j}$ and $\mathbf{r}=ut\cos\alpha\,\mathbf{i}+\left(ut\sin\alpha-\frac{1}{2}gt^2\right)\mathbf{j}$
- You must be able to derive the following formulae:

  Time of flight $=\frac{2u\sin\alpha}{g}$ $\quad$ Range $=\frac{u^2\sin 2\alpha}{g}$ $\quad$ Maximum height $=\frac{u^2\sin^2\alpha}{2g}$

  Equation of path, $y=x\tan\alpha-\left(\frac{g\sec^2\alpha}{2u^2}\right)x^2$
- Elevations of α and $(90°-\alpha)$ give the same range.
- The maximum range $=\frac{u^2}{g}$ when $\alpha=45°$
- The equation of the path is $y=x\tan\alpha-\left(\frac{g\sec^2\alpha}{2u^2}\right)x^2$
- For motion under a system of forces, you resolve forces into components and apply Newton's second law, $\mathbf{F}=m\mathbf{a}$

Check and review

You should now be able to...	Try Questions
✔ Use the constant acceleration equations for motion in two dimensions.	1, 2
✔ Use calculus to solve problems in two-dimensional motion with variable acceleration.	3, 4, 5
✔ Solve problems involving the motion of a projectile under gravity.	6, 7
✔ Analyse the motion of an object in two dimensions under the action of a system of forces.	8

1 For each of these situations choose a suitable equation to work out the required value. Show all your working.

a $\mathbf{u} = (3\mathbf{i} - 2\mathbf{j})$, $\mathbf{v} = (7\mathbf{i} + 6\mathbf{j})$, $t = 5$, work out $\mathbf{s}$

b $\mathbf{u} = (2\mathbf{i} - 3\mathbf{j})$, $\mathbf{a} = (\mathbf{i} - 2\mathbf{j})$, $t = 6$, work out $\mathbf{s}$

c $\mathbf{u} = (12\mathbf{i} + 4\mathbf{j})$, $\mathbf{v} = (2\mathbf{i} + 19\mathbf{j})$, $t = 5$, work out $\mathbf{a}$

d $\mathbf{s} = (-8\mathbf{i} + 48\mathbf{j})$, $\mathbf{a} = (\mathbf{i} - 2\mathbf{j})$, $t = 10$, work out $\mathbf{v}$

2 A particle is initially at the point with position vector $\mathbf{r} = (5\mathbf{i} - \mathbf{j})$ m and travelling with velocity $(8\mathbf{i} - 3\mathbf{j})\,\text{m s}^{-1}$ when it undergoes an acceleration of $(-2\mathbf{i} + 7\mathbf{j})\,\text{m s}^{-2}$ for a period of 4 s. Work out its position at the end of this period.

3 A particle has position at time t given by $\mathbf{r} = (3t^2 - 2t)\mathbf{i} + (2t^3 - t^2)\mathbf{j}$. Work out its velocity and acceleration when $t = 2$. Show all your working.

4 A particle has velocity at time t given by $\mathbf{v} = t(4 - 9t)\mathbf{i} + t^2(6 - 2t)\mathbf{j}$. Its initial position is $\mathbf{r} = 3\mathbf{i} + 2\mathbf{j}$. Work out its position when $t = 2$

5 A particle has acceleration at time t s given by $\mathbf{a} = 4\sin 4t\mathbf{i} + 4\cos 2t\mathbf{j}$. At time $t = \frac{\pi}{4}$ s it has velocity $\mathbf{v} = \mathbf{i} + 2\mathbf{j}$ and position $\mathbf{r} = 3\mathbf{i} + \mathbf{j}$. Show that at time $t = \frac{\pi}{2}$ s its distance from the origin is $\sqrt{13}$

6 A projectile is launched from a point O on horizontal ground with speed $12\,\text{m s}^{-1}$ at an angle of 25° to the horizontal. Take $g = 9.8\,\text{m s}^{-2}$

a For how long is the projectile in the air?

b What is the horizontal range of the projectile?

c Work out the time taken for it to reach maximum height.

d What is the greatest height reached by the projectile?

7 A projectile is fired at $30\,\text{m s}^{-1}$ from a point on horizontal ground. It is directed so that its range is a maximum. Taking $g = 9.8\,\text{m s}^{-2}$, work out

a Its range,

b The equation of its path.

8 A particle of mass 4 kg rests on a smooth horizontal plane. It is acted upon by forces of 30 N, 18 N and P on bearings 030°, 150° and 305° respectively. As a result the particle accelerates in a northerly direction. Work out

a The value of P

b How far the particle moves in the first four seconds.

MECH

What next?

Score		
0 – 4	Your knowledge of this topic is still developing. To improve, search in MyMaths for the codes: 2192, 2198, 2199, 2290, 2291	
5 – 6	You're gaining a secure knowledge of this topic. To improve, look at the InvisiPen videos for Fluency and skills (18A)	
7 – 8	You've mastered these skills. Well done, you're ready to progress! To develop your techniques, look at the InvisiPen videos for Reasoning and problem-solving (18B)	

Click these links in the digital book

Investigation

The diagram shows a particle, P, moving with constant speed in a circular path of radius r and centre O

Does P have acceleration?

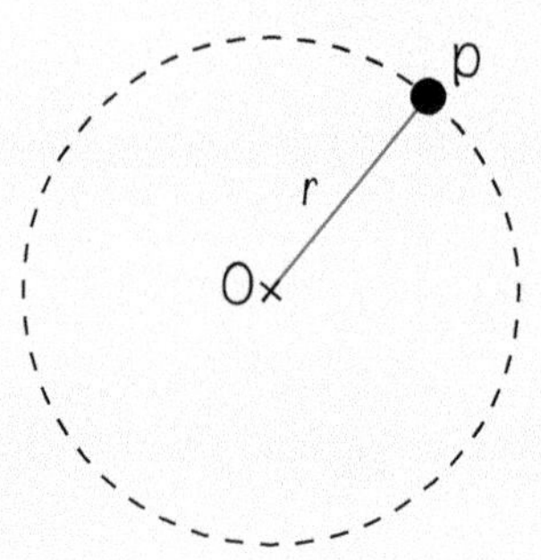

Note

Speed is a scalar quantity – it has magnitude but no direction.

Velocity is a vector quantity – it has both magnitude and direction.

Information

Vectors are a useful tool that can help to describe the motion of P

The first step is to find an expression for the position vector, **r**, of P, relative to O, in terms of the unit vectors **i** and **j**

$\mathbf{r} = r\cos\theta\,\mathbf{i} + r\sin\theta\,\mathbf{j}$

Differentiating with respect to time will give us the velocity vector, **v**, of P

$\mathbf{v} = -r\sin\theta\frac{d\theta}{dt}\mathbf{i} + r\cos\theta\frac{d\theta}{dt}\mathbf{j}$

$\frac{d\theta}{dt}$ is the angular velocity, which is constant.

Writing $\omega = \frac{d\theta}{dt}$ gives the velocity vector as

$\mathbf{v} = -\omega r\sin\theta\,\mathbf{i} + \omega r\cos\theta\,\mathbf{j}$

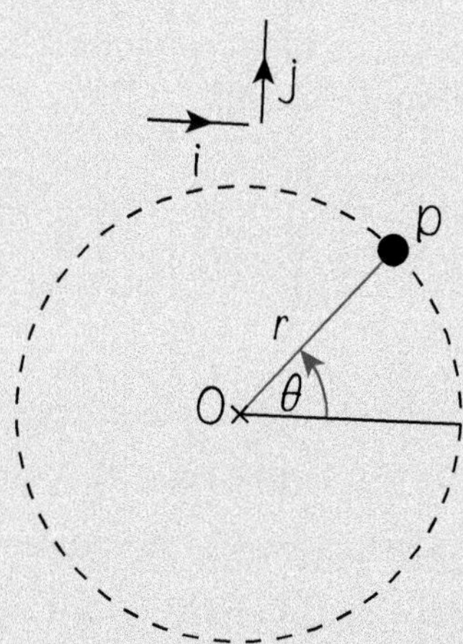

Note

The **chain rule** has been used to differentiate r as θ is also a function of t

Challenge

Differentiate the expression above found for **v** and write the acceleration, **a**, in terms of ω and r

Interpret the result to give the magnitude and direction of the acceleration of P.
What is the name given to this type of acceleration?

If P has mass m kg, what force is required to maintain the motion?

18 Assessment

1 A particle moves with an initial velocity of $(14\mathbf{i} - 4\mathbf{j})\,\text{m s}^{-1}$ and a constant acceleration of $(-\mathbf{i} + 7\mathbf{j})\,\text{m s}^{-2}$

a Work out its velocity at time $t = 4$. Select the correct answer. **[1 mark]**

A $(18\mathbf{i} + 24\mathbf{j})\,\text{m s}^{-1}$ B $(10\mathbf{i} - 32\mathbf{j})\,\text{m s}^{-1}$ C $6(3\mathbf{i} + 4\mathbf{j})\,\text{m s}^{-1}$ D $(10\mathbf{i} + 24\mathbf{j})\,\text{m s}^{-1}$

b Work out its speed at this time. Select the correct answer. **[1]**

A $26\,\text{m s}^{-1}$ B $33.5\,\text{m s}^{-1}$ C $30\,\text{m s}^{-1}$ D $12.2\,\text{m s}^{-1}$

2 A box of mass 6 kg moves up a smooth plane, at constant velocity, under action of a horizontal force of magnitude F Newtons. The plane is inclined at an angle of 30° to the horizontal. Work out the value of F **[5]**

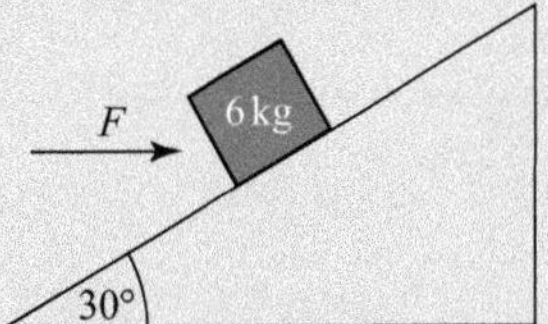

3 An object moving with constant acceleration has an initial velocity of $(-2\mathbf{i} + 5\mathbf{j})\,\text{m s}^{-1}$. Six seconds later it has a velocity of $(4\mathbf{i} - 7\mathbf{j})\,\text{m s}^{-1}$. Work out

a Its acceleration, **[3]**

b The distance of the object from its starting position after these six seconds. **[4]**

4 A particle is projected with a velocity of $58.8\,\text{m s}^{-1}$ at an angle of 30° to the horizontal from a point, X, on horizontal ground. Taking $g = 9.8\,\text{m s}^{-2}$, work out

a The highest point above the ground of the path of the particle, **[4]**

b The distance from X to the point where the particle lands, giving your answer correct to the nearest metre. **[5]**

5 A stone is thrown horizontally with a velocity of $20\,\text{m s}^{-1}$ from the top of a vertical cliff. Five seconds later the stone reaches the sea. Taking $g = 9.8\,\text{m s}^{-2}$, work out

a The height of the cliff, **[3]**

b The horizontal distance from the base of the cliff to the point where the stone lands, **[2]**

c The direction in which the stone is travelling at the instant before it hits the sea. **[4]**

6 Referred to a fixed origin, the position vector, $\mathbf{r}$ metres, of a particle, P, at a time t seconds is given by $\mathbf{r} = (\sin 2t)\mathbf{i} + (\cos 4t)\mathbf{j}$, where $t \geq 0$. At the instant when $t = \frac{\pi}{6}$, work out

a The velocity of P, showing your working, **[4]**

b The acceleration of P, showing your working. **[4]**

7 A particle moves through the origin with constant acceleration. It has an initial velocity of $(2\mathbf{i} - 3\mathbf{j})\,\text{m s}^{-1}$. After six seconds it is moving through the point with position vector $(-6\mathbf{i} + 18\mathbf{j})\,\text{m}$. Work out

a Its velocity at $t = 6$ **[3]**

b Its acceleration, **[2]**

c The time at which the particle is moving parallel to the x-axis. **[2]**

8 A parcel, P, of mass 5 kg, hangs from the horizontal roof of a lift, supported by two light inextensible strings, AP and BP. The lift is accelerating upwards at $\frac{1}{5}g\,\mathrm{m\,s^{-2}}$. Work out

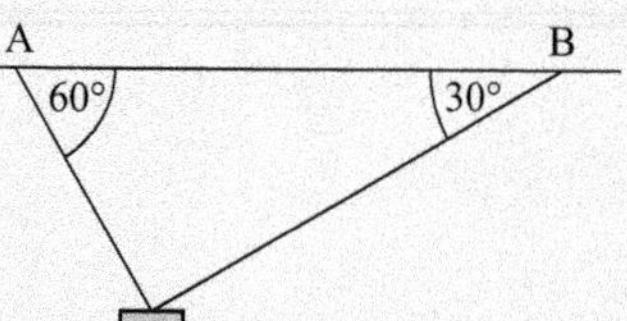

a The tension in BP, showing all your working, **[7]**

b The tension in AP **[1]**

9 A particle, P, of mass 0.4 kg moves under the action of a single force, **F** Newtons. At time t, the velocity, $\mathbf{v}\,\mathrm{m\,s^{-1}}$ of P is given by $\mathbf{v}=\begin{pmatrix}3t^2+5\\14t-2\end{pmatrix}\mathrm{m\,s^{-1}}$
At time $t=0$, P is at the origin. Work out

a **F** when $t=2$. Show all your working. **[5]**

b The distance of P from the origin when $t=2$ **[7]**

10 A ball is hit from a point that is one metre above horizontal ground, with a velocity of $20\,\mathrm{m\,s^{-1}}$ at an angle of elevation of α where $\tan\alpha=\frac{4}{3}$. The ball just clears a vertical wall, which is 12 metres horizontally from the point where the ball was hit.
Taking $g=9.8\,\mathrm{m\,s^{-2}}$, work out (to 3 sf)

a The height of the wall, **[7]**

b The speed of the ball at the instant when it passes over the wall, **[4]**

c The direction in which the ball is travelling at the instant when it passes over the wall. **[2]**

11 At time $t=0$, a particle, P, is at rest at the point (2, 0). At time t seconds, its acceleration, $\mathbf{a}\,\mathrm{m\,s^{-2}}$ is given by $\mathbf{a}=\begin{pmatrix}16\cos 4t\\\sin t-2\sin 2t\end{pmatrix}$. Work out

a The acceleration of P when $t=\frac{\pi}{2}$ **[2]**

b The velocity of P when $t=\frac{\pi}{4}$ **[7]**

c The position of P when $t=\pi$ **[7]**

12 Two boats, P and Q, are travelling with constant velocities $(3\mathbf{i}-8\mathbf{j})\,\mathrm{km\,h^{-1}}$ and $(-7\mathbf{i}+12\mathbf{j})\,\mathrm{km\,h^{-1}}$ respectively, relative to a fixed origin O. At noon, the position vectors of P and Q are $(4\mathbf{i}+11\mathbf{j})$ km and $(9\mathbf{i}+3.5\mathbf{j})$ km respectively. At time t hours after noon, the position vectors of P and Q, relative to O, are $\mathbf{S}_P$ and $\mathbf{S}_Q$. Write

a An expression in terms of t for $\mathbf{S}_P$ **[2]**

b An expression in terms of t for $\mathbf{S}_Q$ **[2]**

At a time, t hours after noon, the distance between the boats is given by d km

c Prove that $d^2=(-5+10t)^2+(7.5-20t)^2$ **[4]**

d Work out the time at which the boats are closest together. Show all your working. **[5]**

e Work out the distance between the boats at the time when they are closest together. **[2]**

13 A particle is projected from a point O, with an initial velocity of $u\,\mathrm{m\,s^{-1}}$, at an angle of α to the horizontal. In the vertical plane of projection, taking x and y as the horizontal and vertical axes respectively

a Show that $y=x\tan\alpha-\frac{gx^2}{2u^2}\sec^2\alpha$ **[5]**

Given that $u=42$ and that the particle passes through the point (60, 70)

b Find the two possible angles of projection. Take $g=9.8\,\mathrm{m\,s^{-2}}$ and show all your working. **[6]**

19 Forces 2

Tightening or loosening a bolt with a spanner is a simple action, but also involves the application of a particular type of force: the *turning force*, also called the *moment*. A turning force is the effect of a force around a fixed point (a pivot). Pushing a door open, removing a bottle cap with a bottle opener, steering a car, even bending your leg – these are all examples of turning forces.

This chapter introduces simple systems of forces: static equilibria, motion due to a force and turning forces. Objects are subject to forces all the time and an understanding of how these forces work and how they can be modelled is therefore vital in areas such as physics and engineering.

Orientation

What you need to know

Ch6 Vectors
- Resolving vectors into components.

Ch7 Units and kinematics
- The constant acceleration formulae.
- Equations of motion for variable acceleration.

Ch8 Forces and Newton's laws
- Equilibrium requires the resultant force to be zero.
- Newton's laws of motion.

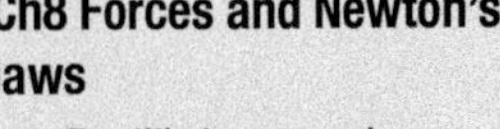

What you will learn

- To manipulate vectors in 3D and hence solve geometric problems.
- To understand that there is a maximum value that the frictional force can take when the object is moving or on the point of moving.
- To find unknown forces when a system is at rest or has constant acceleration.
- To use constant acceleration formulae.
- To solve differential equations arising from $F = ma$
- To take moments about points and resolve to find unknown forces.

What this leads to

Applications
Civil engineering.
Structural engineering.
Mechanical engineering.
Nuclear physics.
Aeronautical engineering.
Rocket science.

MyMaths Practise before you start 2184, 2186–2188, 2207, 2289

19.1 Vectors in 3D

Fluency and skills

To use vectors for three-dimensional problems, you use three perpendicular axes: the x-, y- and z-axes.

If you draw the x- and y-axes as usual, the convention is to draw the z-axis coming "out of" the page. So if the page is lying on your desk, the positive z-axis points towards the ceiling. Axes like these are a **right-hand set** because you can position your thumb, forefinger and second finger in the direction of the x-, y- and z-axis respectively.

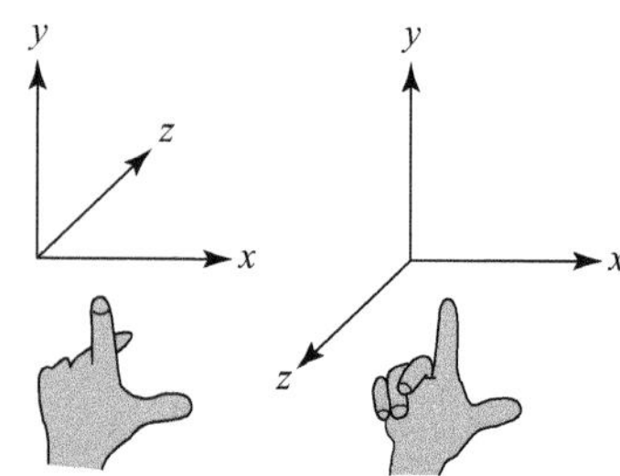

Left-hand set **Right-hand set**

See Ch6.2
For a reminder on vector notation.

A vector **r** in 3D has three components. You can write it as a column vector, or use **i**, **j** and **k**, which are the unit vectors in the x-, y- and z-directions. For example

$$\mathbf{r} = \begin{pmatrix} a \\ b \\ c \end{pmatrix} = a\mathbf{i} + b\mathbf{j} + c\mathbf{k}$$

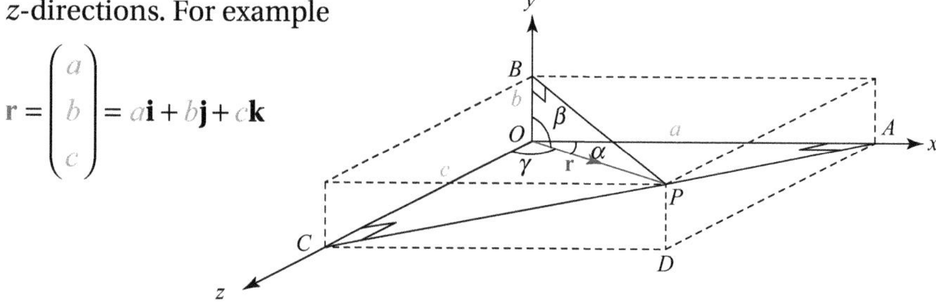

In this diagram a right-hand set of axes is used.

The **magnitude** (or **modulus**) of **r** is $|\mathbf{r}|$, the length of OP in the diagram.

From triangle PDC you have $PC^2 = a^2 + b^2$

From triangle OPC you have $OP^2 = PC^2 + c^2 = a^2 + b^2 + c^2$

Key point

The magnitude of vector $\mathbf{r} = a\mathbf{i} + b\mathbf{j} + c\mathbf{k}$ is

$$|\mathbf{r}| = \sqrt{a^2 + b^2 + c^2}$$

The direction of **r** makes angle α with the positive x-axis, β with the positive y-axis and γ with the positive z-axis, where

See Ch6.2
For a reminder on magnitude/direction form.

$$\cos\alpha = \frac{a}{|\mathbf{r}|},\ \cos\beta = \frac{b}{|\mathbf{r}|} \text{ and } \cos\gamma = \frac{c}{|\mathbf{r}|}$$

It follows that

$$\cos^2\alpha + \cos^2\beta + \cos^2\gamma = \frac{a^2 + b^2 + c^2}{|\mathbf{r}|^2} = 1$$

These cosine formulae aren't required in this course, but they're useful when working with vectors in 3D.

You combine or equate 3D vectors in component form by combining or equating their components.

Calculator

Try it on your calculator

You can find the angle between two vectors on a calculator.

Angle(VctA, VctB)
43.83017926

Activity

Find out how to calculate the angle between $\begin{pmatrix}6\\3\\-1\end{pmatrix}$ and $\begin{pmatrix}11\\-3\\1\end{pmatrix}$ on *your* calculator.

The position vector of a point A defines its position in relation to an origin O

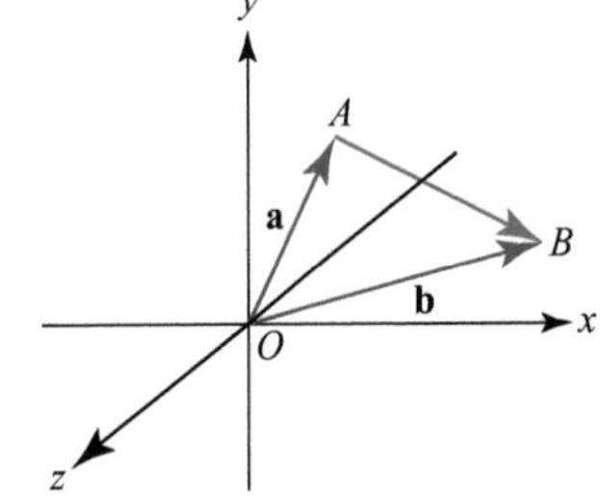

Key point

The position vector of A is $\overrightarrow{OA}$, you usually write this as **a**.

For points A and B, $\overrightarrow{AB}=\overrightarrow{AO}+\overrightarrow{OB}=\overrightarrow{OB}-\overrightarrow{OA}=\mathbf{b}-\mathbf{a}$

Example 1

Points A and B have position vectors $\mathbf{a}=3\mathbf{i}+2\mathbf{j}-5\mathbf{k}$ and $\mathbf{b}=4\mathbf{i}-7\mathbf{k}$ respectively.

Work out
a The length of AB
b The angle between $\overrightarrow{AB}$ and the y-direction, showing your working
c The unit vector in the direction of $\overrightarrow{AB}$

a $\overrightarrow{AB}=\mathbf{b}-\mathbf{a}=(4-3)\mathbf{i}-2\mathbf{j}+(-7+5)\mathbf{k}=\mathbf{i}-2\mathbf{j}-2\mathbf{k}$

$AB=|\overrightarrow{AB}|=\sqrt{1^2+(-2)^2+(-2)^2}=3$

b $\overrightarrow{AB}$ makes an angle of β with the positive y-direction

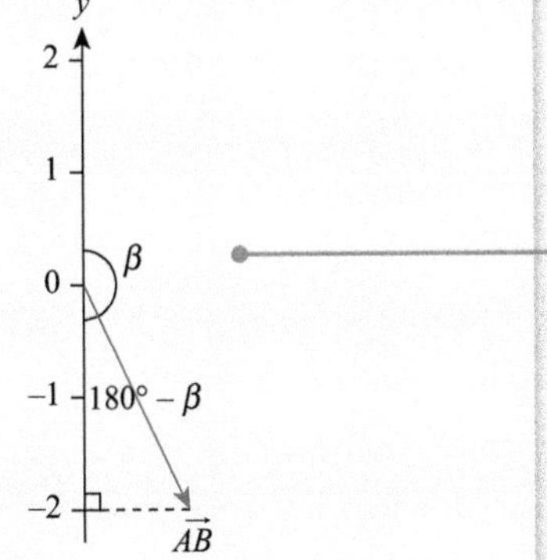

Simplify the question by moving $\overrightarrow{AB}$ to the origin without changing its direction.
Draw a simplified sketch.
The **j** component is −2

$\cos(180°-\beta)=\dfrac{2}{|\overrightarrow{AB}|}=\dfrac{2}{3}$

$\cos\beta=-\cos(180°-\beta)=\dfrac{-2}{3}$

This is equal to $\dfrac{\mathbf{j}\text{-component of }\overrightarrow{AB}}{|\overrightarrow{AB}|}$

$\beta=\cos^{-1}\left(-\dfrac{2}{3}\right)=131.8°$

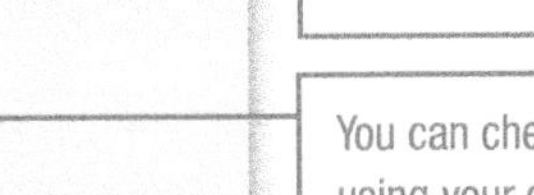

You can check this using your calculator.

c The unit vector in the direction of $\overrightarrow{AB}=\dfrac{\overrightarrow{AB}}{|\overrightarrow{AB}|}=\dfrac{1}{3}\overrightarrow{AB}=\dfrac{1}{3}\mathbf{i}-\dfrac{2}{3}\mathbf{j}-\dfrac{2}{3}\mathbf{k}$

You could state this problem in terms of coordinates $A(3, 2, -5)$ and $B(4, 0, -7)$

Example 2

Points P and Q have position vectors $\mathbf{p} = -3\mathbf{i} - 3\mathbf{j} + \mathbf{k}$ and $\mathbf{q} = 7\mathbf{i} + 4\mathbf{j} - 4\mathbf{k}$ respectively. The point R lies on PQ such that $PR:RQ = 2:3$. Work out the position vector, $\mathbf{r}$, of R

$\overrightarrow{PQ} = \mathbf{q} - \mathbf{p} = (7+3)\mathbf{i} + (4+3)\mathbf{j} + (-4-1)\mathbf{k} = 10\mathbf{i} + 7\mathbf{j} - 5\mathbf{k}$

$\overrightarrow{PR} = \frac{2}{5}\overrightarrow{PQ} = 4\mathbf{i} + 2.8\mathbf{j} - 2\mathbf{k}$

$\mathbf{r} = \mathbf{p} + \overrightarrow{PR} = (-3+4)\mathbf{i} + (-3+2.8)\mathbf{j} + (1-2)\mathbf{k} = \mathbf{i} - 0.2\mathbf{j} - \mathbf{k}$

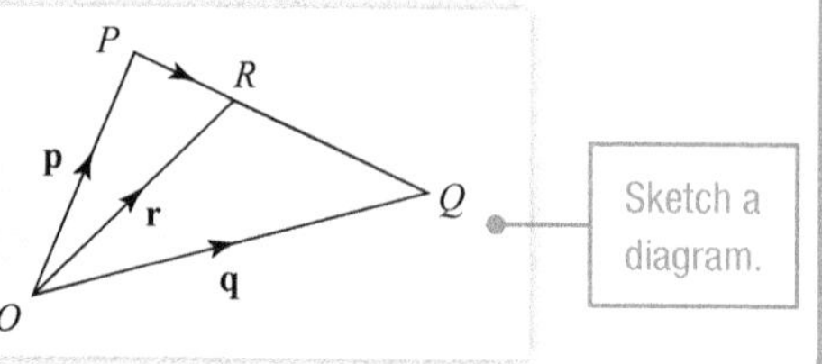

Sketch a diagram.

Exercise 19.1A Fluency and skills

1 Given vectors $\mathbf{a} = 3\mathbf{i} - \mathbf{j} + 2\mathbf{k}$, $\mathbf{b} = 6\mathbf{i} - 3\mathbf{j} - 2\mathbf{k}$ and $\mathbf{c} = \mathbf{i} + \mathbf{j} - 3\mathbf{k}$, work out

a i $\mathbf{a} - \mathbf{b}$ ii $2\mathbf{a} + 5\mathbf{c}$ iii $|\mathbf{b}|$

iv $|\mathbf{c} - \mathbf{a}|$ v Unit vector $\hat{\mathbf{b}}$

vi The angle between $\mathbf{b}$ and the positive x-, y- and z-directions. Show your working.

b A vector parallel to $\mathbf{b}$ with magnitude 28

c The values of p, q and r if $p\mathbf{a} + q\mathbf{c} = 3\mathbf{i} - 5\mathbf{j} + r\mathbf{k}$

2 The vector $\mathbf{r}$ has magnitude 8 and makes angles of 27°, 85° and 63.5° with the positive x-, y- and z-directions respectively. Express $\mathbf{r}$ in component form.

3 A vector $\mathbf{p}$ has magnitude 12 and makes angles of 68° and 75° with the positive y- and z-directions respectively.

a Work out the two possible angles between $\mathbf{p}$ and the positive x-direction.

b Use your answer to part **a** to express the two possible vectors $\mathbf{p}$ in component form.

4 Points A, B and C have position vectors $\mathbf{a} = \begin{pmatrix} 2 \\ 1 \\ 1 \end{pmatrix}$, $\mathbf{b} = \begin{pmatrix} 3 \\ 2 \\ 5 \end{pmatrix}$ and $\mathbf{c} = \begin{pmatrix} 6 \\ -1 \\ 5 \end{pmatrix}$ respectively.

a Work out the lengths of the sides of triangle ABC

b Deduce that the triangle is right-angled.

c State one other fact about the triangle.

5 Points A and B have position vectors $\mathbf{a} = 7\mathbf{i} + 2\mathbf{j} + 5\mathbf{k}$ and $\mathbf{b} = 5\mathbf{i} - 6\mathbf{j} + \mathbf{k}$ respectively. C is the midpoint of AB Work out

a The position vector, $\mathbf{c}$, of C

b The distance of C from the origin,

c The unit vector, $\hat{\mathbf{c}}$, in the direction of $\mathbf{c}$.

6 Points A and B have position vectors $\mathbf{a} = 9\mathbf{i} + 2\mathbf{j} - 4\mathbf{k}$ and $\mathbf{b} = 3\mathbf{i} + 5\mathbf{j} + \mathbf{k}$ respectively. The point C lies on AB, and $AC:CB = 7:3$. Work out the position vector of C

7 Given vectors $\mathbf{p} = \begin{pmatrix} 2 \\ 3 \\ -4 \end{pmatrix}$, $\mathbf{q} = \begin{pmatrix} 1 \\ 1 \\ -2 \end{pmatrix}$ and $\mathbf{r} = \begin{pmatrix} -2 \\ 2 \\ 1 \end{pmatrix}$, work out

a $\mathbf{p} + \mathbf{q} - 2\mathbf{r}$,

b $|\mathbf{p} - \mathbf{q}|$,

c A vector of magnitude 15 in the direction of $\mathbf{r}$,

d The angle between $\mathbf{r}$ and the positive x-direction, showing your working,

e The values of λ and μ if $\lambda\mathbf{p} + \mu\mathbf{q} = \begin{pmatrix} 1 \\ 3 \\ -2 \end{pmatrix}$

Reasoning and problem-solving

Strategy

To solve a problem using 3D vectors

1. Sketch a diagram. It does not need to be to scale, and it does not always need to be on 3D axes.
2. Express the required vectors in terms of the given ones.
3. Write a clear conclusion.

Example 3

A, B, C and D have position vectors $\mathbf{a}=\mathbf{i}+\mathbf{j}-2\mathbf{k}$, $\mathbf{b}=3\mathbf{i}+5\mathbf{j}+4\mathbf{k}$, $\mathbf{c}=-\mathbf{i}+7\mathbf{j}-2\mathbf{k}$ and $\mathbf{d}=7\mathbf{i}-3\mathbf{j}+6\mathbf{k}$ respectively.

P, Q, R and S are the midpoints of AB, BD, CD and AC respectively. Show that $PQRS$ is a parallelogram.

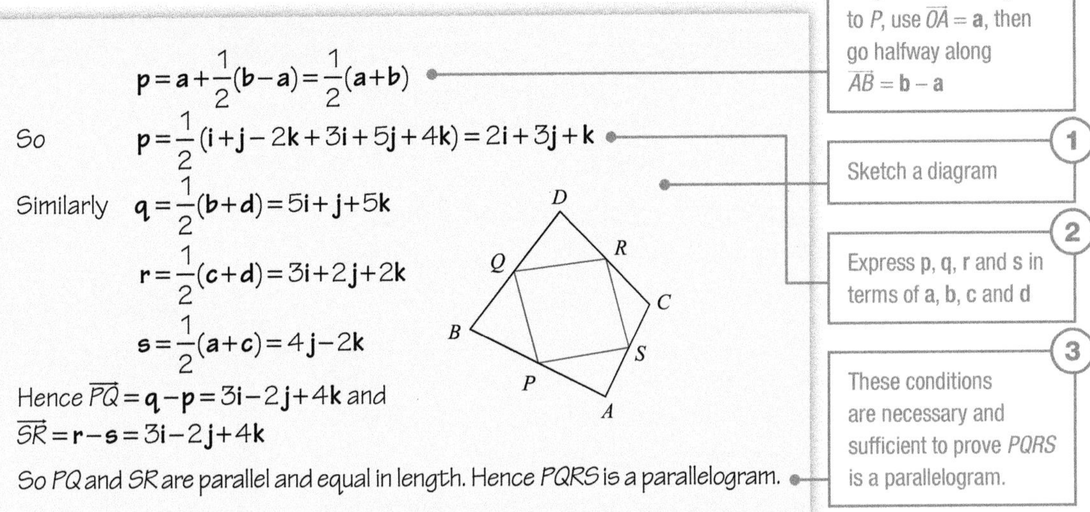

$$\mathbf{p}=\mathbf{a}+\frac{1}{2}(\mathbf{b}-\mathbf{a})=\frac{1}{2}(\mathbf{a}+\mathbf{b})$$

So $\mathbf{p}=\frac{1}{2}(\mathbf{i}+\mathbf{j}-2\mathbf{k}+3\mathbf{i}+5\mathbf{j}+4\mathbf{k})=2\mathbf{i}+3\mathbf{j}+\mathbf{k}$

Similarly $\mathbf{q}=\frac{1}{2}(\mathbf{b}+\mathbf{d})=5\mathbf{i}+\mathbf{j}+5\mathbf{k}$

$\mathbf{r}=\frac{1}{2}(\mathbf{c}+\mathbf{d})=3\mathbf{i}+2\mathbf{j}+2\mathbf{k}$

$\mathbf{s}=\frac{1}{2}(\mathbf{a}+\mathbf{c})=4\mathbf{j}-2\mathbf{k}$

Hence $\overrightarrow{PQ}=\mathbf{q}-\mathbf{p}=3\mathbf{i}-2\mathbf{j}+4\mathbf{k}$ and $\overrightarrow{SR}=\mathbf{r}-\mathbf{s}=3\mathbf{i}-2\mathbf{j}+4\mathbf{k}$

So PQ and SR are parallel and equal in length. Hence $PQRS$ is a parallelogram.

To get from the origin to P, use $\overrightarrow{OA}=\mathbf{a}$, then go halfway along $\overrightarrow{AB}=\mathbf{b}-\mathbf{a}$

1 Sketch a diagram

2 Express $\mathbf{p}$, $\mathbf{q}$, $\mathbf{r}$ and $\mathbf{s}$ in terms of $\mathbf{a}$, $\mathbf{b}$, $\mathbf{c}$ and $\mathbf{d}$

3 These conditions are necessary and sufficient to prove $PQRS$ is a parallelogram.

Exercise 19.1B Reasoning and problem-solving

1 Points A, B and C have positions vectors $2\mathbf{i}+3\mathbf{j}-3\mathbf{k}$, $\mathbf{i}-4\mathbf{j}+2\mathbf{k}$ and $3\mathbf{i}-5\mathbf{j}+\mathbf{k}$ respectively. Prove that ABC is a right-angled triangle.

2 The points A, B and C with position vectors $\mathbf{a}=\mathbf{i}+2\mathbf{j}$, $\mathbf{b}=2\mathbf{i}+\mathbf{j}-2\mathbf{k}$ and $\mathbf{c}=3\mathbf{i}-\mathbf{j}+\mathbf{k}$ are three vertices of a parallelogram. Work out all possible positions of the fourth vertex, D

3 The vector $\mathbf{V}$ has magnitude 6 and makes the same angle with each of the positive x-, y- and z-directions. Find the possible values of $\mathbf{V}$

4 In this question, east, north and upwards are the positive x-, y- and z-directions respectively. A child, standing at the origin O, flies a toy drone. She first sends it to A, 25 m north and at a height of 15 m, then for 35 m in the direction of the vector $6\mathbf{i}+3\mathbf{j}+2\mathbf{k}$ to B

a Work out the angle of elevation of B from O

b At B the drone's battery runs out and it falls to the ground. How far does she have to walk to retrieve it?

5 $ABCD$ is a tetrahedron. The position vectors of its vertices are $\mathbf{a}$, $\mathbf{b}$, $\mathbf{c}$ and $\mathbf{d}$ respectively. P, Q and R are the respective midpoints of AB, AD and BC. S divides PC in the ratio $1:2$. T is the midpoint of QR

a Show that D, T and S are collinear.

b Work out the ratio $DT:TS$

Challenge

6 The points A, B, C and D lie on a straight line, as shown.

$AC:CB=AD:BD=\lambda:\mu$, where $\lambda>\mu$

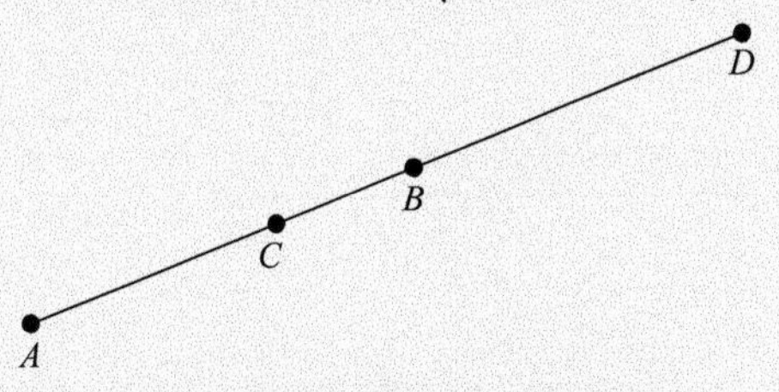

a If $\lambda=3$, $\mu=2$ and A and B have position vectors $2\mathbf{i}+\mathbf{j}+3\mathbf{k}$ and $7\mathbf{i}+3\mathbf{j}+9\mathbf{k}$ respectively, work out the length of CD

b Show that in general $CD:AB=2\lambda\mu:(\lambda^2-\mu^2)$

19.2 Statics

Fluency and skills

See Ch 8.1
For a reminder on equilibrium.

Statics is the study of forces in equilibrium. In the previous chapter, you learnt to resolve forces in motion. However, if an object is in equilibrium, then the resultant force acting on it is zero.

Imagine a book resting in equilibrium on a table. If you apply a horizontal force, P, to the book and it doesn't move, then friction F is acting to oppose P and $F = P$

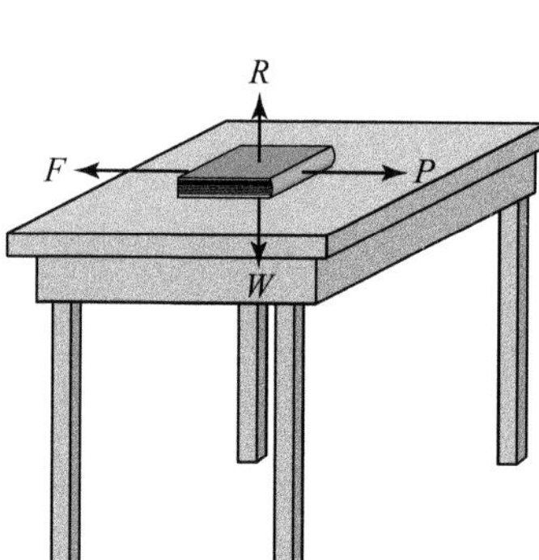

As you increase P and the book still does not move, F is increasing in line with P, but only up to the point when the book is on the point of moving. At this point, F has its maximum value which is called **limiting friction**. If P increases further, the book moves.

If you repeat this with different books, you'll find the limiting friction changes—it's proportional to the normal reaction, R. So, limiting friction $F = \mu \times R$, where μ is the **coefficient of friction**. The rougher the surface, the greater the value of μ. However, if the book is not on the point of moving, then $F < \mu R$

Key point

Friction acts to oppose any motion.
Friction, F, and normal reaction, R, are related by $F \le \mu R$, where μ is the coefficient of friction.
When the object is about to move, it is in **limiting equilibrium** and limiting friction $F = \mu R$

Example 1

Remember to convert the mass to a weight.

A crate of mass 50 kg lies on a rough horizontal surface. A string is attached to the crate and is pulled at 20° above the horizontal until the crate is just about to slip. At this point, the tension in the string is 200 N. Find the normal reaction, R, and the coefficient of friction, μ.
Use $g = 9.8\text{ m s}^{-2}$

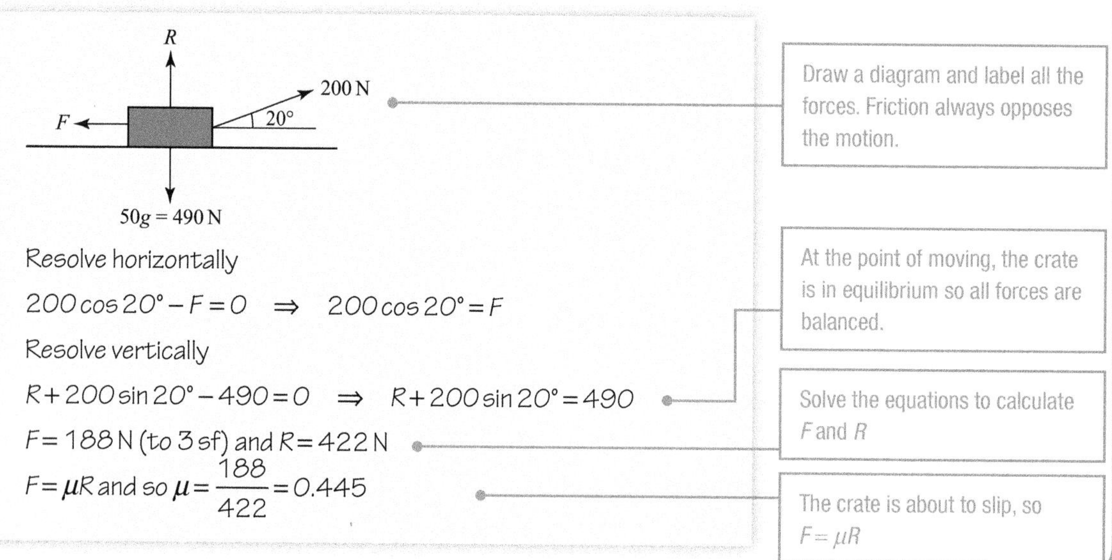

Draw a diagram and label all the forces. Friction always opposes the motion.

Resolve horizontally

$200\cos 20° - F = 0 \Rightarrow 200\cos 20° = F$

Resolve vertically

$R + 200\sin 20° - 490 = 0 \Rightarrow R + 200\sin 20° = 490$

At the point of moving, the crate is in equilibrium so all forces are balanced.

$F = 188$ N (to 3 sf) and $R = 422$ N

Solve the equations to calculate F and R

$F = \mu R$ and so $\mu = \dfrac{188}{422} = 0.445$

The crate is about to slip, so $F = \mu R$

Exercise 19.2A Fluency and skills

1 For each part **a** and **b**, calculate

i The horizontal component, X, and the vertical component, Y, (to 3 sf) of the resultant force,

ii The magnitude of the resultant force (to 2 sf) and the angle it makes with the upward vertical (to the nearest degree).

a

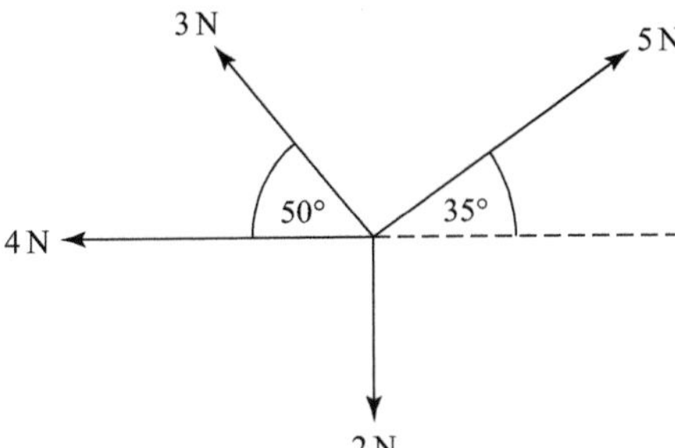

b

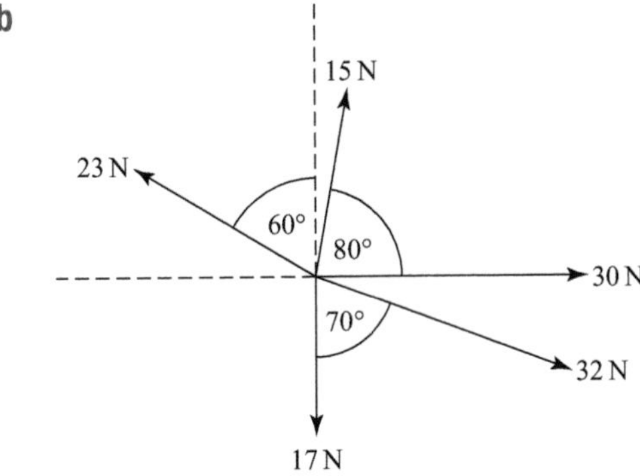

2 A crate of weight 800 N lies on a rough horizontal surface. A string is attached to the crate and is pulled at 30° above the horizontal until the crate is just about to slip. At this point the tension in the string is 350 N. Resolve to find the normal reaction of the surface on the crate, R, and the coefficient of friction, μ

3 A box of weight W lies on a rough horizontal surface. A string is attached to the box and pulled at an angle of 45° to the horizontal until the box is just about to slip. At this point the tension in the string is 600 N and the normal reaction of the surface on the box is R. The coefficient of friction between the box and the surface is $\frac{3}{4}$. Find W and R (to 2 sf).

4 A boy sits on a sledge on horizontal ground. The combined weight of the boy and the sledge is 750 N. A friend pulls the rope attached to the sledge so that it makes an angle of 10° to the horizontal.

When the sledge is just about to slip, the friend is pulling with a force T. The normal reaction of the ground on the block is R and the coefficient of friction between the sledge and the ground is $\frac{2}{3}$

Show that $T = 450$ N (to 2 sf) and find R (to 2 sf).

5 An object is free to move in a smooth horizontal plane. Four horizontal forces act on the object at the same time. One has magnitude 25 N and is directed due south, one has magnitude 60 N and is directed south-west, one has magnitude 40 N and is directed due west and the final force has magnitude 30 N and is directed north-west.

Find the magnitude of the resultant force and the bearing on which the object moves.

6 A box of weight 400 N lies on a rough horizontal surface. A string is attached to the crate and is pulled at 40° above the horizontal with a tension of magnitude T. The box is just about to slip. The normal reaction of the surface on the crate is 300 N and the coefficient of friction between the box and the surface is μ

Find μ and T

7 A box of weight W lies on a rough horizontal surface. A string is attached to the crate and is pulled at 20° above the horizontal with a tension of magnitude $\frac{1}{2}W$. The box is just about to slip. The normal reaction of the surface on the crate is 30 N and the coefficient of friction between the box and the surface is μ

Find μ and W

Reasoning and problem-solving

When an object is in equilibrium under several forces, you can consider three methods for calculating unknown forces and angles

1. Resolve in two perpendicular directions,
2. Draw a vector diagram and use trigonometry,
3. Use Lami's Theorem.

You have already used the first two of these methods in Chapter 18. You can use them in any problem. Lami's Theorem can only be used when you have just three forces and you know at least one of the angles between them. Although this is a specific case, it can be much faster than using vector diagrams or resolving in perpendicular directions.

Let three forces A, B and C, act on an object so that it is in equilibrium as shown.

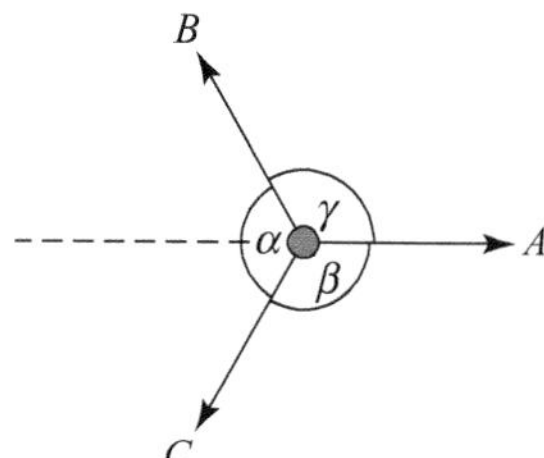

You can draw a vector diagram of the forces, nose-to-tail, to have zero resultant.

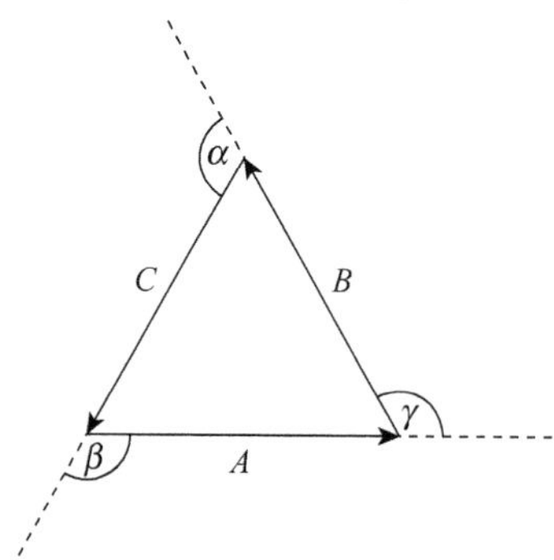

See Ch 3.2

For a reminder on the sine rule.

Applying the sine rule to the vector diagram gives

$$\frac{A}{\sin(180°-\alpha)}=\frac{B}{\sin(180°-\beta)}=\frac{C}{\sin(180°-\gamma)}$$

$\sin(180° - \alpha) = \sin\alpha$, and similarly for β and γ, so

$$\frac{A}{\sin\alpha}=\frac{B}{\sin\beta}=\frac{C}{\sin\gamma}$$

Lami's theorem isn't specifically required in this course, but it can be useful to know when solving equilibrium questions.

Key point

Lami's Theorem states that if there are three forces acting on an object and these forces are in equilibrium, each force is proportional to the sine of the angle between the other two forces.

You can use it when there are three forces acting on an object and you know at least one of the angles between them.

Strategy

To solve problems where objects are in equilibrium

1. Draw a diagram and mark all known and unknown forces and angles.
2. Decide which of these methods to use and form equations.
3. Solve the equations to find unknown forces and angles.

Example 2

A block of weight W is suspended by two ropes. One rope makes an angle of 30° with the vertical and has tension 60 N. The other rope makes an angle of 50° with the vertical and has a tension T

Find T and W (to 3 sf).

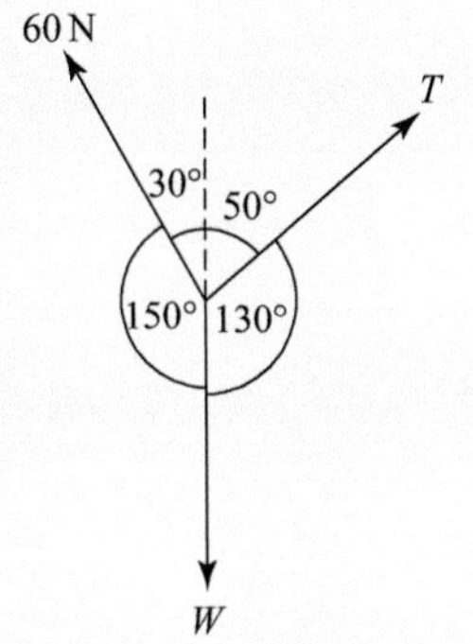

(1) Draw a diagram. Combine the 30° and 50° angles to give a single angle of 80° between the two ropes.

Using Lami's theorem

$$\frac{T}{\sin 150^\circ} = \frac{W}{\sin 80^\circ} = \frac{60}{\sin 130^\circ}$$

(2) There are three forces in equilibrium and the angles are known, so use Lami's theorem.

To find T, solve $\frac{T}{\sin 150^\circ} = \frac{60}{\sin 130^\circ}$

$$T = \frac{60 \sin 150^\circ}{\sin 130^\circ} = 39.2\text{ N (to 3 sf)}$$

To find W, solve $\frac{W}{\sin 80^\circ} = \frac{60}{\sin 130^\circ}$

$$W = \frac{60 \sin 80^\circ}{\sin 130^\circ} = 77.1\text{ N (to 3 sf)}$$

Calculate the missing forces T and W. Give your answers to the required degree of accuracy.

Example 3

A block of weight 60 N lies on a rough slope inclined at an angle 40° to the horizontal. A string is attached to the block and passes over a smooth pulley at the top of the slope. It is attached at its other end to a block of weight W which hangs vertically. The block on the slope is about to slip *down* the slope. The normal reaction of the slope on the block is R

The coefficient of friction between the block and the slope is $\frac{1}{2}$

Find W (to 3 sf).

(*Continued on the next page*)

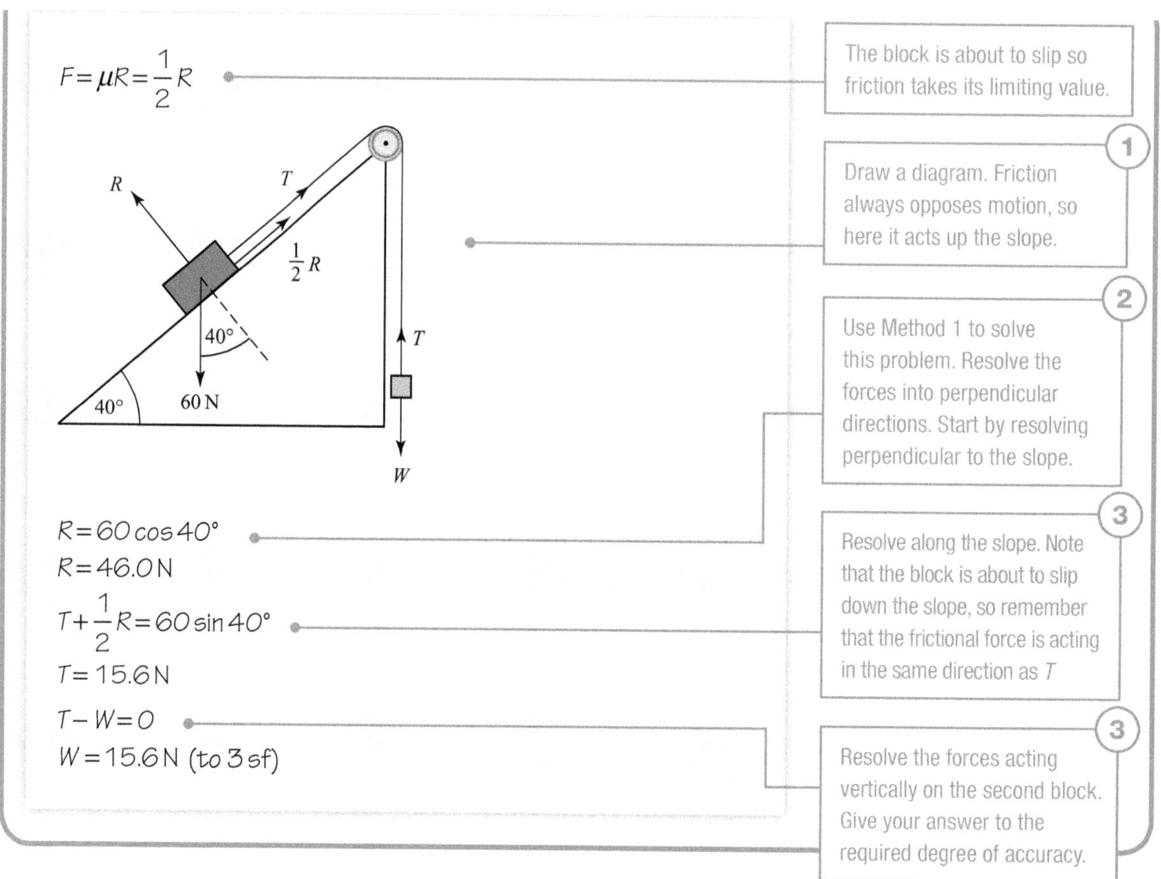

$F = \mu R = \frac{1}{2}R$
The block is about to slip so friction takes its limiting value.
1
Draw a diagram. Friction always opposes motion, so here it acts up the slope.
R
T
$\frac{1}{2}R$
40°
T
40°
60 N
W
2
Use Method 1 to solve this problem. Resolve the forces into perpendicular directions. Start by resolving perpendicular to the slope.
$R = 60\cos 40°$
$R = 46.0$ N
3
Resolve along the slope. Note that the block is about to slip down the slope, so remember that the frictional force is acting in the same direction as T
$T + \frac{1}{2}R = 60\sin 40°$
$T = 15.6$ N
$T - W = 0$
$W = 15.6$ N (to 3 sf)
3
Resolve the forces acting vertically on the second block. Give your answer to the required degree of accuracy.

Exercise 19.2B Reasoning and problem-solving

1 Two strings are attached to a box of weight 90 N so that the box is held in mid-air by the strings. Tension R in one string makes an angle of 30° with the upward vertical, whilst tension S in the the other string makes an angle of 20° with the upward vertical. Find R and S

2 A block of weight 120 N lies on a rough slope inclined at an angle of 25° to the horizontal. The coefficient of friction between the block and the slope is $\frac{3}{5}$
A string is attached to the block and is passed over a smooth pulley at the top of the slope. Its other end is attached to a block of weight W which hangs vertically. The block on the slope is about to slip *up* the slope. Find W (to 3 sf).

3 A rescue dinghy of weight 250 N is launched into water along a slope, which lies at an angle of 35° to the horizontal. The coefficient of friction between the two is 0.2. Before being launched, the dinghy is held by a light inextensible rope that is parallel to the slope, with tension T

a Find the smallest possible value of T that keeps the dinghy in equilibrium.

b Find the largest possible value of T that keeps the dinghy in equilibrium.

4 Two particles of weights 15 N and 10 N are connected by a light inextensible string which passes over a smooth pulley fixed at the end of a rough horizontal table. The 15 N weight rests on the table and the 10 N weight hangs freely below the pulley. Find the coefficient of friction between the 15 N weight and the table if the system is in limiting equilibrium.

5 A block of weight $6W$ lies on a rough slope inclined at an angle of 30° to the horizontal. The coefficient of friction between the block and the slope is $\frac{1}{3}$

A string is attached to the block and is passed over a smooth pulley at the top of the slope. Its other end is attached to a block of weight kW which hangs vertically. The block on the slope is about to slip *up*. Find k. Give your answer as an exact value using surds.

6 Three children kick a football at the same time. One kick is directed due north with magnitude 40 N. One kick is directed south-east with magnitude T and the third kick is directed due west with magnitude S. Find the magnitude of T and S if the magnitude of the resultant force is zero.

7 Two boxes of weights kW and W are connected by a light inextensible string which passes over a smooth pulley fixed at the end of a rough horizontal platform. The kW weight rests on the platform and the W weight hangs freely below the pulley. Find the coefficient of friction, in terms of k, between the kW weight and the table if the system is in limiting equilibrium.

8 A block of weight W lies on a rough slope inclined at an angle of 10° to the horizontal. The coefficient of friction between the block and the slope is μ. A string is attached to the block and is passed over a pulley at the top of the slope. It is attached at its other end to a block of weight $\frac{1}{2}W$ which hangs vertically. The block on the slope is about to slip *up* the slope.

Determine the value, of the coefficient of friction, μ

9 A block of mass 7 kg lies on a rough horizontal table. The coefficient of friction between the block and the table is $\frac{3}{7}$

Two light inextensible strings are attached to the block. One passes over a pulley at one end of the table and the other passes over a pulley at the other end of the table. The pulleys, the strings and the block are all in a straight line. Masses of m kg and 5 kg respectively are tied to the other ends of the string. These two masses are held with the strings taut.

a Calculate the value of m that will result in the 5 kg mass being on the point of moving downwards.

b Calculate the value of m that will result in the 5 kg mass being on the point of moving upwards.

Challenge

10 In a sports day event, a player wins a prize if they can move a box of weight 2000 N along a surface by pulling on a handle attached to it. The back of the box is also connected to two light inextensible strings that make an angle of 150° to the handle and 60° to each other, and pull on the box with a force of 300 N each.

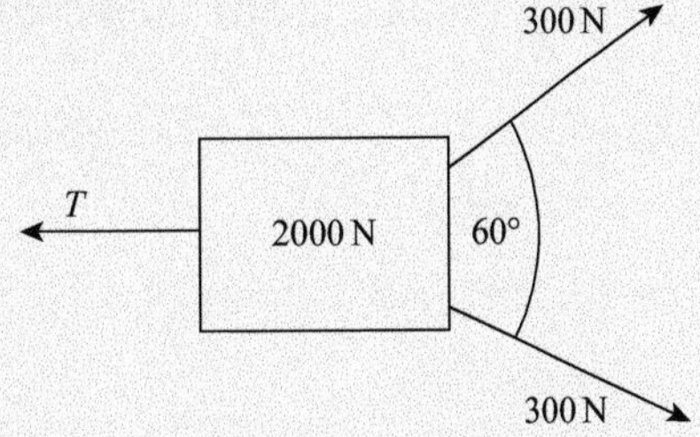

If the coefficient of friction between the box and the ground is 0.4, determine the minimum force with which a player would need to pull on the box for them to win the prize.

MECH

19.3 Dynamics

Fluency and skills

Dynamics is the study of forces in motion. From Newton's second law of motion, you know that, if a force F N acts on an object of mass m kg giving it an acceleration a m s^{-2}, then $F = ma$

See Ch8.2 For a reminder on Newton's second law of motion.

You used this equation in Chapter 18 to solve problems where the forces need to be resolved in perpendicular directions. In this chapter, you will recap resolving forces in two dimensions, and then learn to use Newton's second law to solve more complicated problems including the coefficient of friction, smooth pulleys and connected particles, and scenarios where acceleration is not constant.

ICT Resource online

To investigate resolving forces, click this link in the digital book.

Example 1

Three forces, of magnitudes 15 N, 20 N and 30 N, act on a particle of mass 4 kg as shown in the diagram.

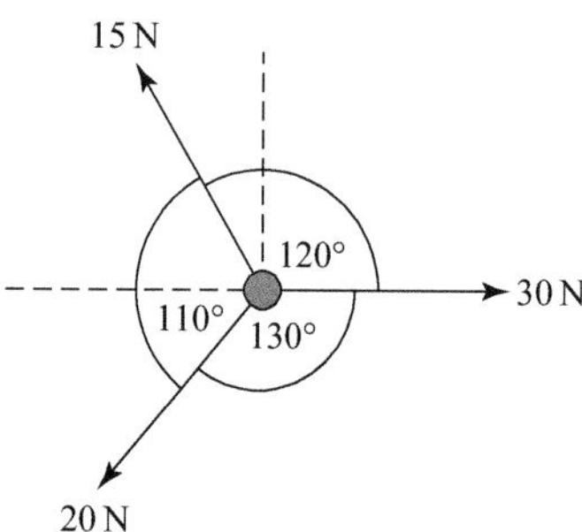

Find the magnitude and direction of the acceleration of the particle.

(→) $X = 30 - 15\cos 60° - 20\cos 50°$

$X = 9.64$ N

(↑) $Y = 15\sin 60° - 20\sin 50°$

$Y = -2.33$ N

Let X be the component of the resultant in the direction of the 30 N force and let Y be the component in the perpendicular direction.

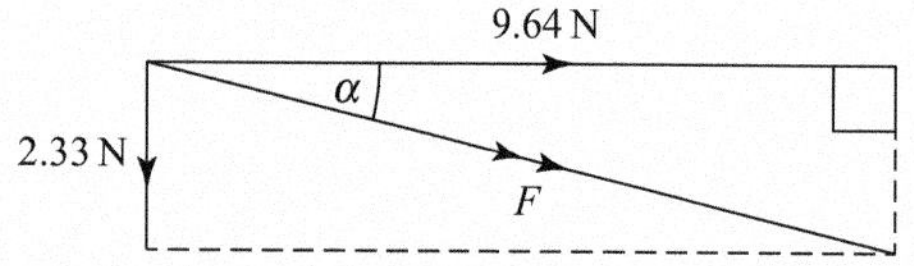

Draw a diagram. Y is in the downwards direction because it has a negative value.

$F = \sqrt{9.64^2 + 2.33^2} = 9.92$

$\alpha = \tan^{-1}\left(\frac{2.33}{9.64}\right) = 13.6°$

Calculate the magnitude and direction of the resultant force.

Using $F = ma$

$a = \frac{F}{m}$

$a = \frac{9.92}{4} = 2.48$ m s^{-2} at an angle of 13.6° below the 30 N force.

Example 2

A box of mass 60 kg is at rest on horizontal ground. A girl pulls a rope attached to the box so that it makes an angle of 35° to the horizontal. The coefficient of friction between the box and the floor is μ. When the girl is pulling with a force of 400 N, the box accelerates at 3 m s^{-2}. Find μ (to 2 sf). Use $g = 10\text{ m s}^{-2}$

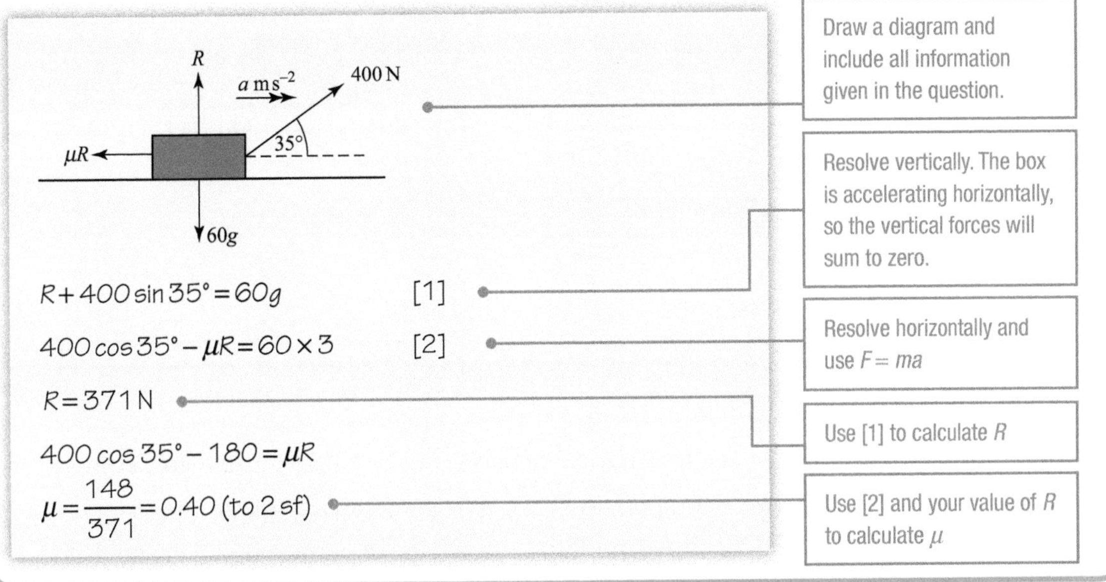

R
a ms⁻²
400 N
35°
μR
60g
Draw a diagram and include all information given in the question.
Resolve vertically. The box is accelerating horizontally, so the vertical forces will sum to zero.
Resolve horizontally and use F = ma
Use [1] to calculate R
Use [2] and your value of R to calculate μ

$R + 400\sin 35° = 60g$ [1]

$400\cos 35° - \mu R = 60 \times 3$ [2]

$R = 371\text{ N}$

$400\cos 35° - 180 = \mu R$

$\mu = \frac{148}{371} = 0.40$ (to 2 sf)

MECH

Exercise 19.3A Fluency and skills

1 Two perpendicular forces of magnitude X and 20 N are applied to an object of mass 2 kg. The mass ends up accelerating in a direction which makes an angle of 30° with the 20 N force. Find the value of X and the magnitude of the acceleration.

2 Three children kick a football of mass 0.45 kg. One kicks it with a magnitude of 30 N in the direction of due north, another kicks it with a magnitude of 40 N in the direction of due east and the third kicks it with a magnitude of 50 N in the direction of north-west. Find the magnitude and direction of the acceleration of the ball.

3 A box of mass 2 kg lies on a rough horizontal table. It is pulled by a tension of 10 N in a string acting at 25° to the horizontal. The coefficient of friction between the box and the table is 0.2. Find the acceleration of the box. Use $g = 10\text{ ms}^{-2}$

4 A book of mass 300 g lies on a rough horizontal desk. A boy pulls it along the table, applying a force of magnitude 2 N at an angle of 50° to the horizontal. The acceleration of the book is 0.8 m s^{-2}. Find the coefficient of friction between the book and the desk. Use $g = 9.8\text{ ms}^{-2}$

5 Three forces of magnitude 55 N, 40 N and 35 N act on a particle of mass 12 kg in the directions shown in the diagram. Find the magnitude and direction of the acceleration of the particle.

40 N, 55 N, 35 N, 100°, 145°, 115°

6 A boy sits on a sledge on rough horizontal ground. The combined mass is 80 kg. A friend pulls the rope attached to the sledge with a force of 500 N at an angle of 25° to the horizontal. The coefficient of friction between the sledge and the floor is $\frac{2}{3}$. Find the acceleration of the sledge to 2 sf. Use $g = 9.8\text{ m s}^{-2}$

7 A force of 100 N acting at an angle of 40° above the horizontal is applied to a crate of mass 20 kg lying on a rough horizontal surface.

The acceleration of the crate is $2\,\text{m s}^{-2}$. Find μ (to 2 sf). Use $g = 9.81\,\text{m s}^{-2}$.

Reasoning and problem-solving

Strategy

To solve questions involving acceleration

1. Draw a clear diagram, marking on the acceleration and all the forces which act on the object.
2. Resolve in suitable directions to form equations. You might need to use Newton's second law.
3. Solve these equations to find the unknown quantities.

Example 3

Particle A of mass 7 kg is connected to particle B of mass m kg by a light inextensible string.

Particle A rests on a rough slope (with $\mu = 0.4$) inclined at 30° to the horizontal. The string passes over a smooth pulley at the top of the slope and is fixed to particle B initially held at rest in mid-air.

Particle B is released and accelerates downwards at $2\,\text{m s}^{-2}$. Use $g = 9.81\,\text{m s}^{-2}$ to calculate m

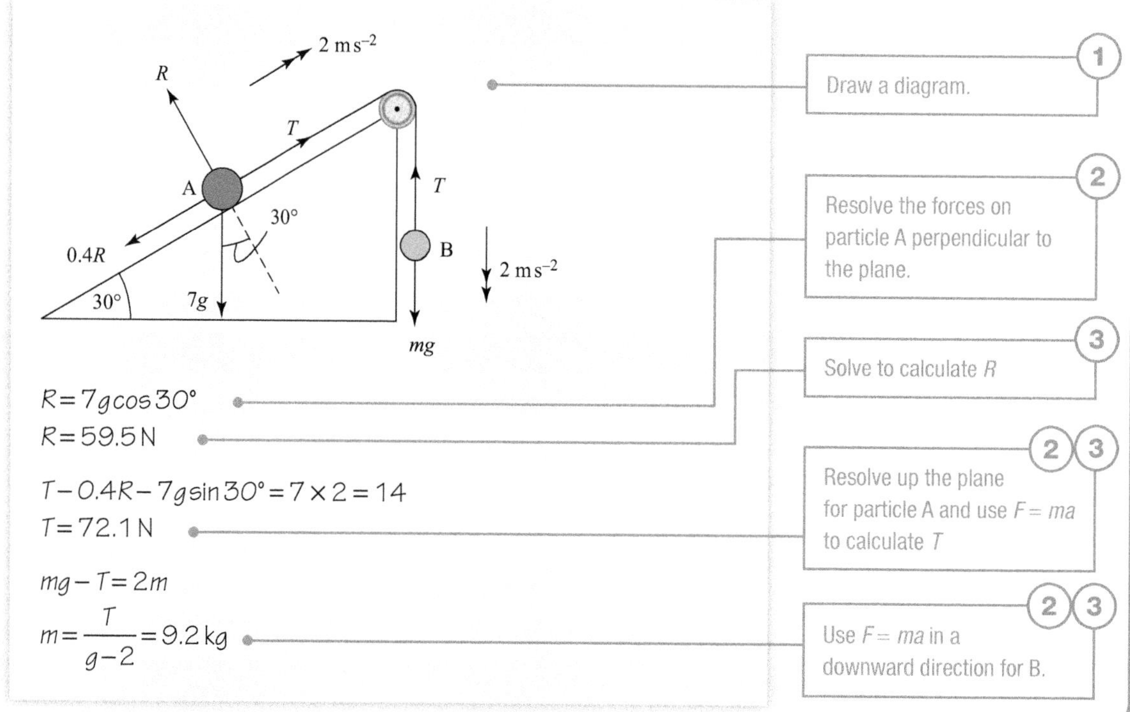

2 ms⁻²
R
T
A
30°
0.4R
30°
7g
T
B
2 ms⁻²
mg
1 Draw a diagram.
2 Resolve the forces on particle A perpendicular to the plane.
3 Solve to calculate R
2 3 Resolve up the plane for particle A and use F = ma to calculate T
2 3 Use F = ma in a downward direction for B.

$R = 7g\cos 30°$

$R = 59.5\,\text{N}$

$T - 0.4R - 7g\sin 30° = 7 \times 2 = 14$

$T = 72.1\,\text{N}$

$mg - T = 2m$

$m = \dfrac{T}{g-2} = 9.2\,\text{kg}$

In situations when the particles are moving with constant acceleration, you can use the constant acceleration formulae.

See Ch7.3
For a reminder on the constant acceleration formulae.

Example 4

Particle A, of mass 10 kg, sits on a rough table and is attached to one end of a light inextensible string. The string passes over a smooth pulley fixed on the edge of the table 3 m from A. The other end of the string is attached to particle B, of mass 5 kg, held in mid-air.

Mass B is released from rest and, after two seconds, particle A reaches the pulley.

Using $g = 9.8\,\text{m s}^{-2}$, show that

a The acceleration of the masses is $1.5\,\text{m s}^{-2}$

b The coefficient of friction is 0.27 (to 2 sf).

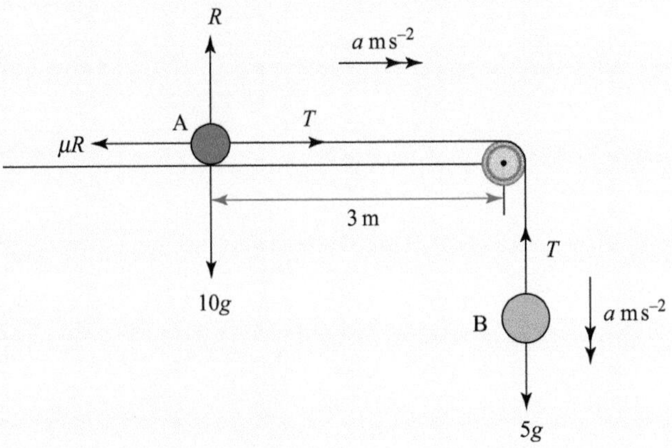

Draw a diagram. (1)

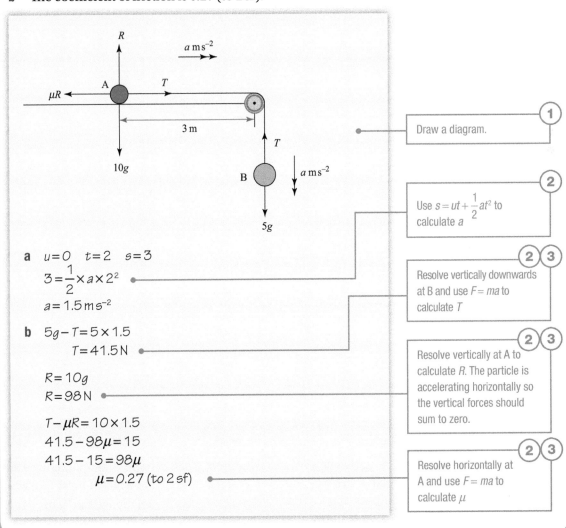

a $u = 0 \quad t = 2 \quad s = 3$

$3 = \frac{1}{2} \times a \times 2^2$

$a = 1.5\,\text{m s}^{-2}$

Use $s = ut + \frac{1}{2}at^2$ to calculate a (2)

b $5g - T = 5 \times 1.5$

$T = 41.5\,\text{N}$

Resolve vertically downwards at B and use $F = ma$ to calculate T (2)(3)

$R = 10g$

$R = 98\,\text{N}$

Resolve vertically at A to calculate R. The particle is accelerating horizontally so the vertical forces should sum to zero. (2)(3)

$T - \mu R = 10 \times 1.5$

$41.5 - 98\mu = 15$

$41.5 - 15 = 98\mu$

$\mu = 0.27$ (to 2 sf)

Resolve horizontally at A and use $F = ma$ to calculate μ (2)(3)

In situations where the acceleration is not constant, you can still use the equation $F = ma$ in the form $F = m\frac{dv}{dt}$

See Ch7.4
For a reminder on equations of motion for variable acceleration.

Example 5

Florence, mass 55 kg, parachutes from a helicopter. She falls vertically at 14.7 m s^{-1}. Once Florence opens her parachute, she experiences air resistance of magnitude $275v$ N, where $v\text{ m s}^{-1}$ is Florence's speed t seconds after she opens her parachute.

Using $g = 10\text{ m s}^{-2}$, show that $\frac{dv}{dt} = -5(v-2)$ and hence find v in terms of t

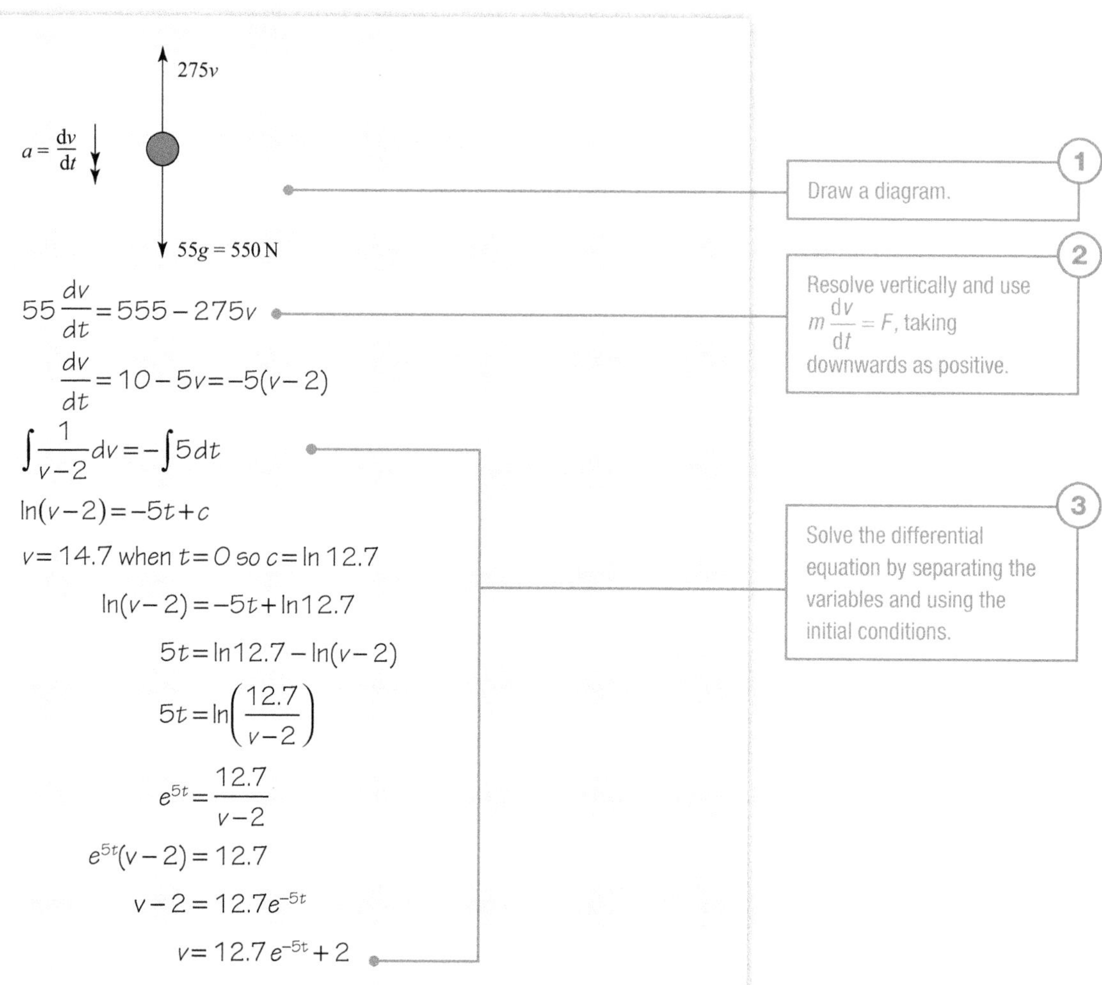

275v
a = dv/dt
55g = 550 N
55 dv/dt = 555 − 275v
dv/dt = 10 − 5v = −5(v − 2)
∫ 1/(v−2) dv = −∫ 5 dt
ln(v−2) = −5t + c
v = 14.7 when t = 0 so c = ln 12.7
ln(v − 2) = −5t + ln12.7
5t = ln12.7 − ln(v − 2)
5t = ln(12.7/(v−2))
e^5t = 12.7/(v−2)
e^5t(v − 2) = 12.7
v − 2 = 12.7e^−5t
v = 12.7e^−5t + 2
1
Draw a diagram.
2
Resolve vertically and use m dv/dt = F, taking downwards as positive.
3
Solve the differential equation by separating the variables and using the initial conditions.

See Ch16.5
For a reminder on solving differential equations.

Exercise 19.3B Reasoning and problem-solving

1 A block of mass 7 kg lies on a rough slope inclined at 35° to the horizontal. A string is attached to the block and is pulled with a force of 50 N up the slope. The coefficient of friction between the mass and the slope is 0.1. Find the acceleration of the block (to 2 sf). Use $g = 9.8\text{ ms}^{-2}$

2 A box of mass 5 kg lies on a rough slope inclined at 40° to the horizontal. A light inextensible string is attached to the box. The string passes over a smooth pulley fixed to the top of the slope. The other end of the string is attached to a box of mass 6 kg which hangs vertically, 1 m above the floor.

The 6 kg mass is released from rest and after two seconds it hits the floor. Find the coefficient of friction between the 5 kg mass and the slope. Use $g = 9.8\text{ ms}^{-2}$

3 A block of mass 75 kg lies on a rough slope inclined at 45° to the horizontal. The coefficient of friction between the block and the slope is 0.25. A cable is attached to the

block and is pulled with a force T parallel to the slope so that the block is accelerating up the slope at $4\,\text{m s}^{-2}$. Find T (to 2 sf).
Use $g = 9.8\,\text{m s}^{-2}$

4 Rachael, mass 50 kg, parachutes from a helicopter. She falls vertically at $12.8\,\text{m s}^{-1}$. Once Rachael opens her parachute, she experiences air resistance of magnitude $175v$ N, where $v\,\text{m s}^{-1}$ is her speed t seconds after she opens her parachute.

Using $g = 9.8\,\text{m s}^{-2}$, show that $\frac{dv}{dt} = -3.5(v - 2.8)$ and hence find v in terms of t

5 A block of mass 125 kg is at rest on a rough slope inclined at 40° to the horizontal. The coefficient of friction between the block and the slope is $\frac{1}{5}$. A cable is attached to the block and is pulled with a force of 1800 N so that the block is accelerating at $a\,\text{m s}^{-2}$ up the slope.

Show that $a = 6.6$ (to 2 sf).

Hence find how long it takes to travel 10 m (to 2 sf).

6 Three forces of magnitude 500 N, 600 N and X act on a particle of mass 80 kg as shown in the diagram. All forces act parallel to the ground. The particle accelerates in a direction 10° below the XN force. Find the value of X and the magnitude of the acceleration.

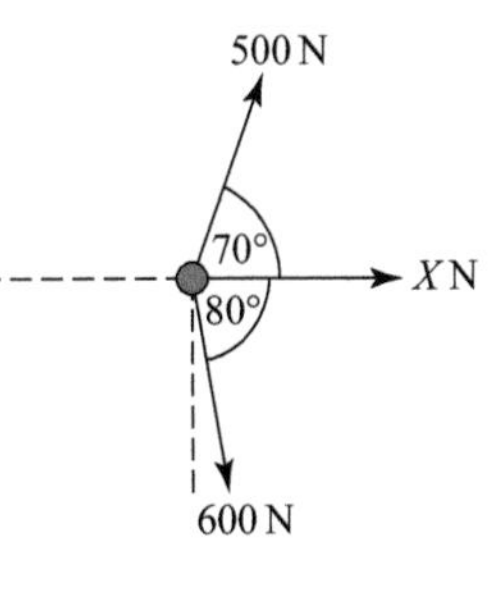

7 A boy kicks a block of mass 2 kg so that it slides up a rough plane inclined at 30° to the horizontal. The block has an initial speed of $3\,\text{m s}^{-1}$. The coefficient of friction between the block and the plane is $\frac{1}{2}$

a Show that, after it has been kicked, the deceleration of the block up the slope is $9.1\,\text{m s}^{-2}$ (to 2 sf).

b Find the distance travelled by the block before it comes to an instantaneous rest (to 2 sf).

c Find the acceleration of the block (to 2 sf) as it slides back down the slope.

d Show that the block will have a speed of $0.8\,\text{m s}^{-1}$ (to 1 sf) as it returns to the point where it was kicked.

8 A car of mass 800 kg drives along a horizontal road. The car's engine provides a driving force of 3600 N and, as the car accelerates from rest, it experiences a resistance force to its motion equal to $120v$ N, where v is its speed in m s^{-1}. Form an equation for the car's speed v in terms of t, and hence show that its maximum speed is $30\,\text{m s}^{-1}$ assuming the driving force does not change.

Challenge

9 Two particles, A and B, of mass $2m$ and $3m$ respectively, are attached to the ends of a light inextensible string. A is held at rest on a rough horizontal desk. The string passes over a small smooth pulley fixed on the edge of the desk. B hangs freely below the pulley.

The particles are released from rest with the string taut. They accelerate at $\frac{1}{2}g\,\text{m s}^{-2}$

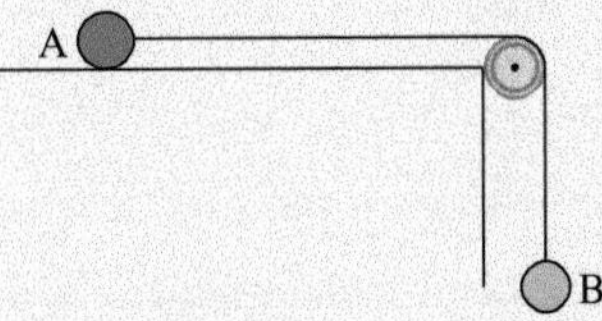

a Find the tension in the string in terms of m and g and show that $\mu = \frac{1}{4}$

When B has fallen h m it hits the ground and does not rebound.

In subsequent motion, particle A stops just before the pulley.

b How far, in terms of h, was A initially away from the pulley?

MECH

19.4 Moments

Fluency and skills

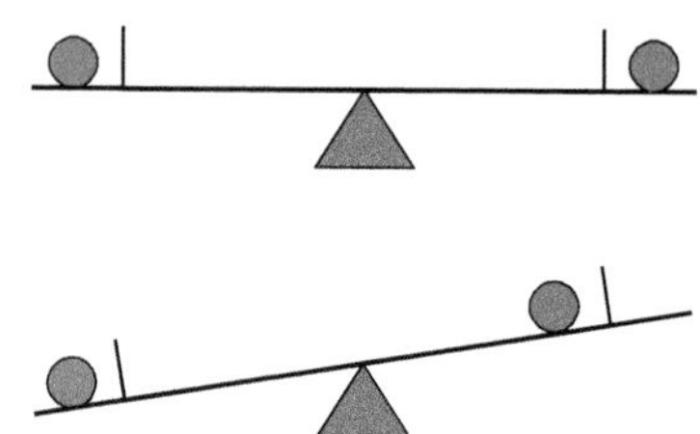

If two people with the same weight sit on either end of a see-saw then the see-saw will be balanced. If one person moves forwards, closer to the middle of the see-saw, then the see-saw will lift up at that end. This is because the turning effect of the person on the seat is greater than that of the person who is closer to the centre. These turning effects are called **moments**.

Key point

If a force F acts at a perpendicular distance d from a point P then the turning effect (or moment) of F about P is the product $F \times d$. Moments are measured in Newton-metres (Nm).

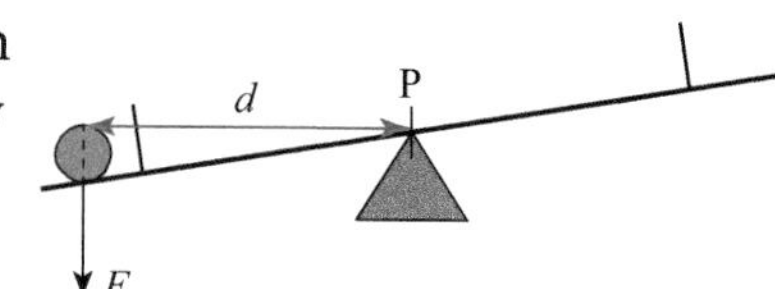

You draw the force F as a line and d is the perpendicular distance from this line to the point P. The diagram shows this for one person sitting on a see-saw. Note that d is not simply the length of see-saw between the pivot and the person.

The notation P $\curvearrowright$ is used when taking moments about P where the clockwise direction is chosen to be positive.

The notation P $\curvearrowleft$ is used when the anti-clockwise direction is chosen to be positive.

Example 1

For each set of forces, calculate the total clockwise moment about the point A.

a 3 m, A, 10 N

b 40 N, 30 N, 2 m, 5 m, A

c 30 N, 5 m, 50 N, 2 m, A

a A $\curvearrowright$ $10 \times 3 = 30$ Nm

Total clockwise moment is 30 Nm

b A $\curvearrowright$ $(40 \times 2) - (30 \times 5)$

$= -70$ Nm

c A $\curvearrowright$ $0 - (30 \times 5) - (50 \times 2)$

$= -250$ Nm

Choose the clockwise direction to be positive. Take moments using $F \times d$

There is only one moment and it acts clockwise.

Since the clockwise direction is taken as positive, any anti-clockwise moments are negative.

Both moments are anti-clockwise.

Example 2

The force $(2\mathbf{i} - 3\mathbf{j})$ N is applied at point A with position vector $(3\mathbf{i} + \mathbf{j})$ m. Calculate the magnitude of the moment of this force about the point B with position vector $(\mathbf{i} - \mathbf{j})$ m.

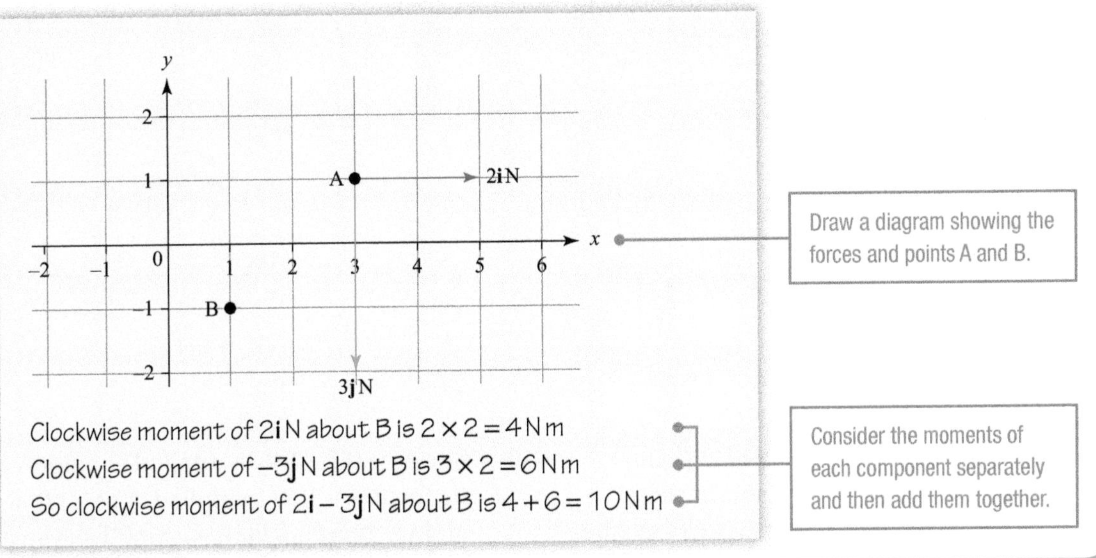

Consider the moments of each component separately and then add them together.

A beam, made from a material that is uniformly spread along its length, is called a **uniform beam.** You can assume that its weight acts at the midpoint of the beam. This point is called the **centre of mass.**

Key point

Any uniform shape (or lamina) that is symmetrical, such as a rectangle, has its centre of mass on every line of symmetry.

If an object is in equilibrium, then

- The resultant force in any direction is zero,
- The resultant moment of all forces about any point is zero, that is, the clockwise and anticlockwise moments balance.

Example 3

A uniform beam AB of weight 300 N has length 6 m. The beam rests on a support at A and is held horizontally in equilibrium by a vertical string at B. There is a vertical reaction R at A. Calculate the tension, T, in the string.

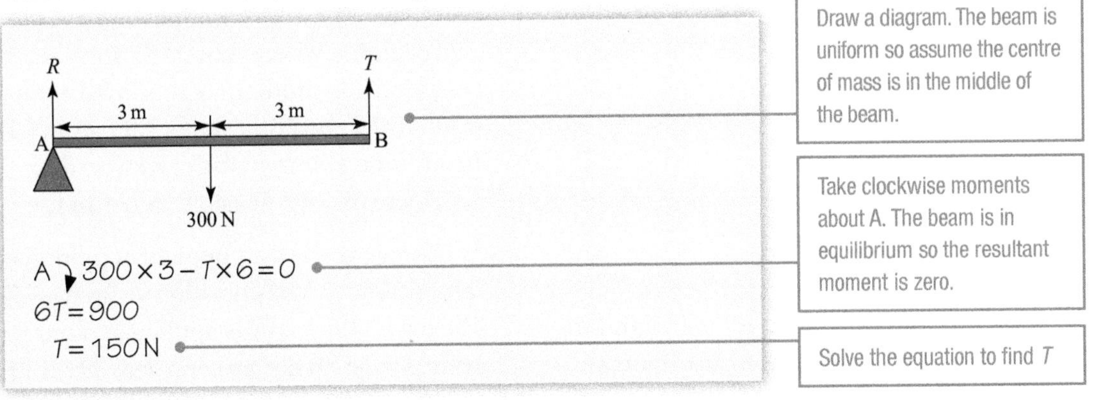

R
T
3 m
3 m
A
B
300 N
A ↻ 300 × 3 − T × 6 = 0
6T = 900
T = 150 N
Draw a diagram. The beam is uniform so assume the centre of mass is in the middle of the beam.
Take clockwise moments about A. The beam is in equilibrium so the resultant moment is zero.
Solve the equation to find T

By taking moments about A, the force R is not in the equation because it is at zero distance from A. This technique is important as it allows you to eliminate an unknown force from an equation.

  2197 SEARCH 

MECH

Exercise 19.4A Fluency and skills

1 Calculate the total clockwise moment of the forces about the point A.

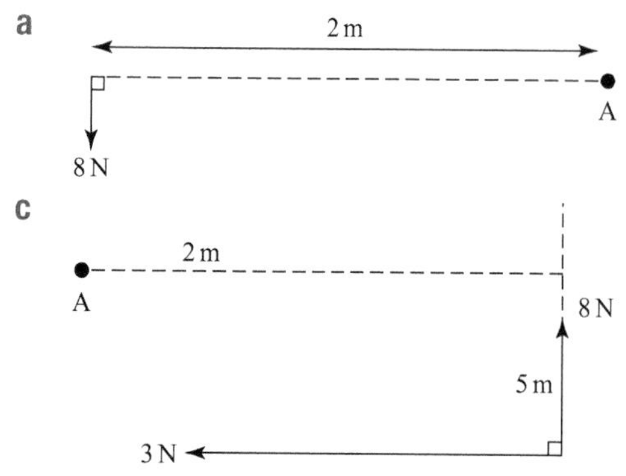

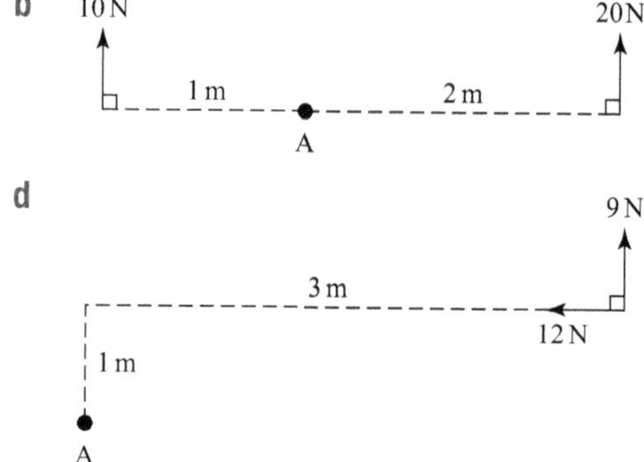

2 The force $(3\mathbf{i} - 4\mathbf{j})$ N is applied at point A of a lamina where A has position vector $(2\mathbf{i} - \mathbf{j})$ m relative to a fixed origin. Calculate the magnitude of the moment of this force about the point B with position vector $(\mathbf{i} + \mathbf{j})$ m

3 In each diagram, the circle acts as a pivot and the forces acting on the object are shown. Determine whether the object will turn, and if so, state the direction. Assume the objects have zero weight.

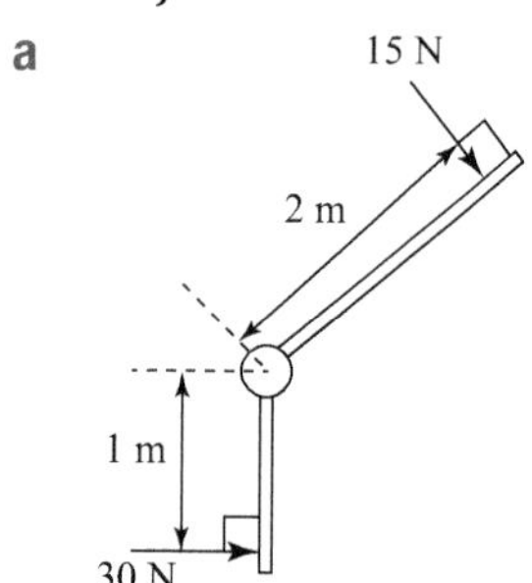

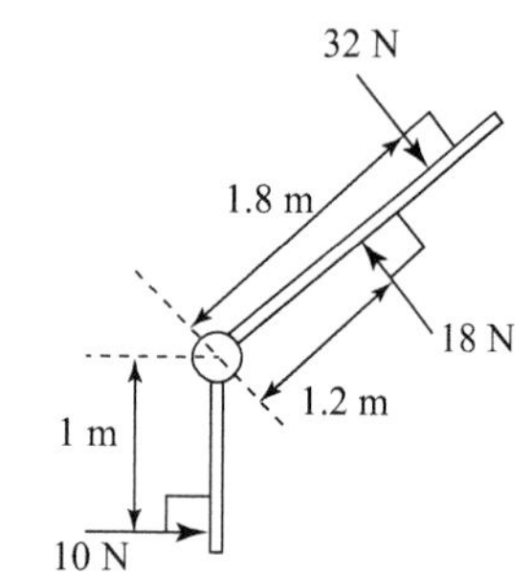

4 A horizontal uniform beam AB of length 4 m has weight 120 N. It rests in on a support at A and is held horizontally in equilibrium by a vertical string attached 3 m from A. There is a supporting force P acting at A and the tension in the string is T. Calculate T

5 The force $(5\mathbf{i} + 2\mathbf{j})$N is applied at point A and the force $(\mathbf{i} + 3\mathbf{j})$N is applied at point B of a lamina. A has position vector $(3\mathbf{i} - 2\mathbf{j})$m and B has position vector $(2\mathbf{i} + \mathbf{j})$m. Calculate the resultant moment of these forces about the point C with position vector $(\mathbf{i} - \mathbf{j})$m

6 A horizontal uniform beam with a length of 6 m has a weight of 100 N. The beam rests on a pivot at one end and is held up by a rope at the other. Three rocks, each with a weight of 200 N, are placed at 1.5 m intervals along the beam as shown.

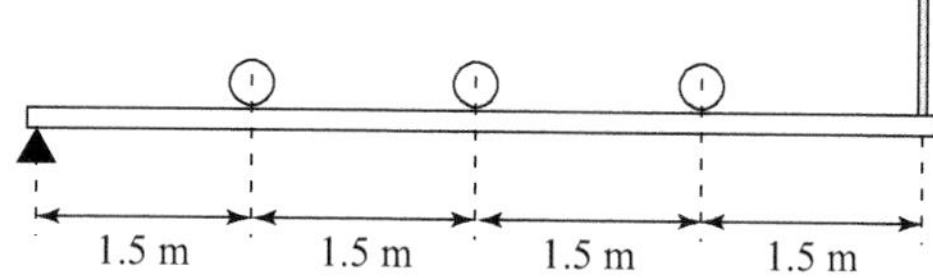

Calculate the tension in the rope.

7 The force $(p\mathbf{i} - 2p\mathbf{j})$N is applied at point A of a lamina where A has position vector $(\mathbf{i} + p\mathbf{j})$m. The moment of this force about the point B with position vector $(2\mathbf{i} - p\mathbf{j})$m has magnitude 12 N m in the clockwise direction. Calculate the two possible values of p

8 The force $(4\mathbf{i} - 2\mathbf{j})$N is applied at point A of a lamina where A has position vector $(\mathbf{i} + 3\mathbf{j})$m. The force $(10\mathbf{i} + x\mathbf{j})$N is applied at point B of a lamina where B has position vector $(-2\mathbf{i} + \mathbf{j})$m. The overall moment of these forces about the point C with position vector $(5\mathbf{i} - 2\mathbf{j})$m is zero. Calculate the value of x

Reasoning and problem-solving

Strategy

To solve questions involving moments

1. Draw a clear diagram, marking on all the forces that act on the object.
2. Take moments and/or resolve in suitable directions to create equations.
3. Solve the equations to find the unknown quantities.

MECH

Example 4

A uniform beam AB of weight 540 N has length 5 m. It rests in equilibrium on two supports C and D as shown in the diagram. Find the values of the vertical reactions P and Q acting at C and D respectively.

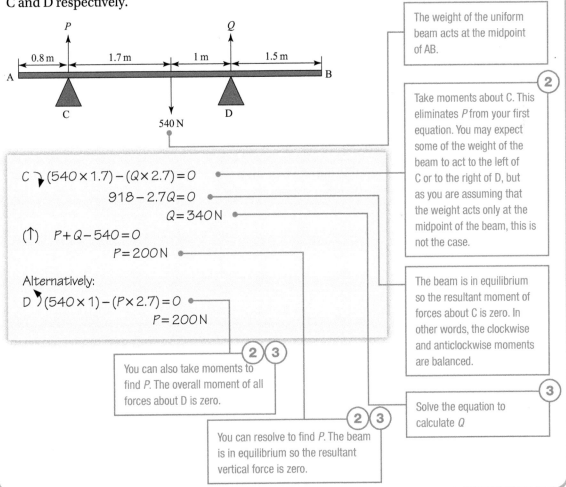

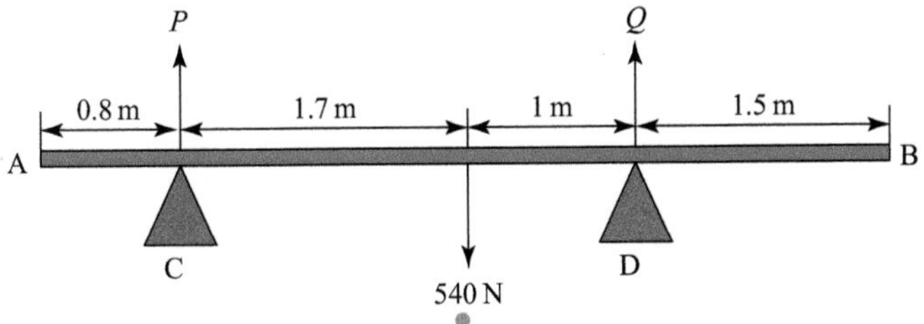

A beam AB with supports at C and D: AC = 0.8 m, C to midpoint = 1.7 m, midpoint to D = 1 m, DB = 1.5 m. Reaction P acts upward at C, reaction Q acts upward at D, and the weight 540 N acts downward at the midpoint.

The weight of the uniform beam acts at the midpoint of AB.

$C \curvearrowright (540 \times 1.7) - (Q \times 2.7) = 0$

$918 - 2.7Q = 0$

$Q = 340\text{ N}$

(2) Take moments about C. This eliminates P from your first equation. You may expect some of the weight of the beam to act to the left of C or to the right of D, but as you are assuming that the weight acts only at the midpoint of the beam, this is not the case.

The beam is in equilibrium so the resultant moment of forces about C is zero. In other words, the clockwise and anticlockwise moments are balanced.

(3) Solve the equation to calculate Q

$(\uparrow)\quad P + Q - 540 = 0$

$P = 200\text{ N}$

(2)(3) You can resolve to find P. The beam is in equilibrium so the resultant vertical force is zero.

Alternatively:

$D \curvearrowleft (540 \times 1) - (P \times 2.7) = 0$

$P = 200\text{ N}$

(2)(3) You can also take moments to find P. The overall moment of all forces about D is zero.

Example 5

A light, rigid rod, AB, rests in equilibrium on a support at C as shown in the diagram. A 20 N weight is attached to the rod at A and a 70 N weight is attached at D. A 50 N upward force is applied at B. Find the length y and the force P exerted by the support on the rod.

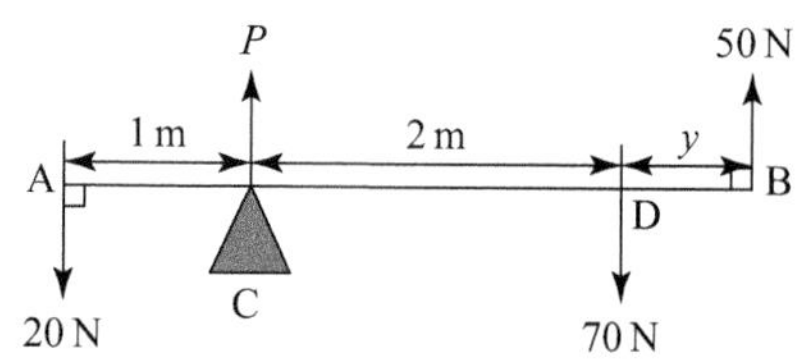

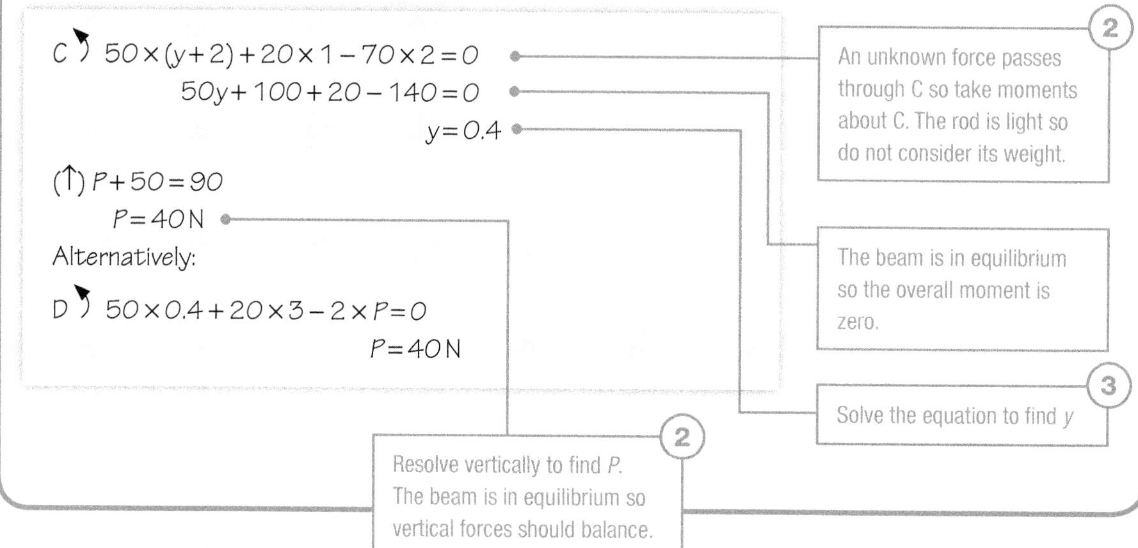

C ↻ $50 \times (y + 2) + 20 \times 1 - 70 \times 2 = 0$

$50y + 100 + 20 - 140 = 0$

$y = 0.4$

(↑) $P + 50 = 90$

$P = 40\text{ N}$

Alternatively:

D ↻ $50 \times 0.4 + 20 \times 3 - 2 \times P = 0$

$P = 40\text{ N}$

(2) An unknown force passes through C so take moments about C. The rod is light so do not consider its weight.

The beam is in equilibrium so the overall moment is zero.

(3) Solve the equation to find y

(2) Resolve vertically to find P. The beam is in equilibrium so vertical forces should balance.

If a horizontal beam is resting in equilibrium on two supports, and the beam is about to tilt about one of the supports, the reaction at the other support is zero.

Example 6

A uniform beam AB of length 6 m and weight 300 N rests in equilibrium on supports C and D where AC = 1 m and BD = 2 m

A 100 N weight is attached at A. Find the largest weight W which can be applied at B for the beam to remain in equilibrium.

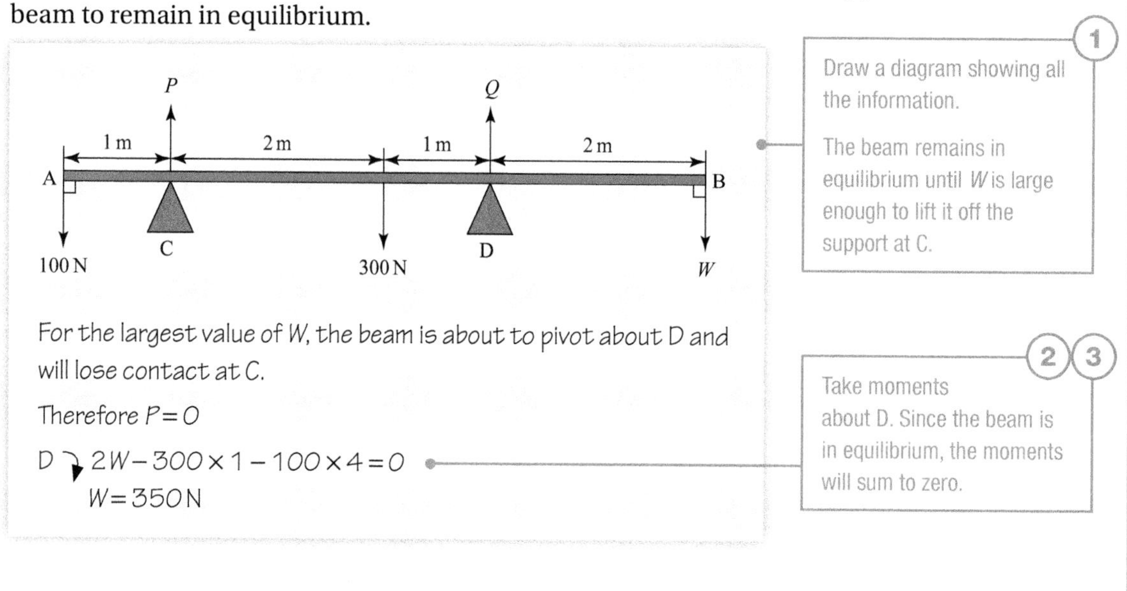

For the largest value of W, the beam is about to pivot about D and will lose contact at C.

Therefore $P = 0$

D ↻ $2W - 300 \times 1 - 100 \times 4 = 0$

$W = 350\text{ N}$

(1) Draw a diagram showing all the information.

The beam remains in equilibrium until W is large enough to lift it off the support at C.

(2)(3) Take moments about D. Since the beam is in equilibrium, the moments will sum to zero.

Exercise 19.4B Reasoning and problem-solving

1 A uniform rod of length 4 m weighs 50 N. 1 m of the rod lies on a horizontal table and the rest lies over the edge.

   a Where does the reaction act when the rod is about to tilt?

   b What downward force must be applied to the end of the rod to stop it from tilting?

   c What is the reaction of the table on the rod when it is about to tilt?

2 A screw-nut requires a turning force of 96 N m to be unscrewed. If a person can provide a force of 500 N, what is the minimum length of spanner they must use to turn the nut?

3 A rod, AB, of length 3 m and of weight 80 N, has its centre of mass 1 m away from A. It is hung from two vertical strings attached to A and to a point 0.5 m from B. Find the tensions in the two strings.

4 A uniform plank of wood, AB, has length 2 m and weight 30 N. It rests on a support which is 0.6 m away from A and a 50 N weight is placed on A. Calculate the weight that must be placed on the plank at B to balance it and the reaction at the support.

5 A uniform plank, AB, of mass 28 kg and length 9 m, lies on a horizontal roof in a direction at right angles to the edge of the roof. The end B projects 2 m over the edge. A man of mass 70 kg walks out along the plank.

   a Find how far along the plank he can walk without causing the plank to tip up.

   b Find the mass which must be placed on the end A so that the man can reach B without upsetting the plank.

6 A uniform beam AB of length 4 m and weight 200 N rests in equilibrium on supports C and D as shown in the diagram. A 500 N weight is attached at A and a weight W is attached at B.

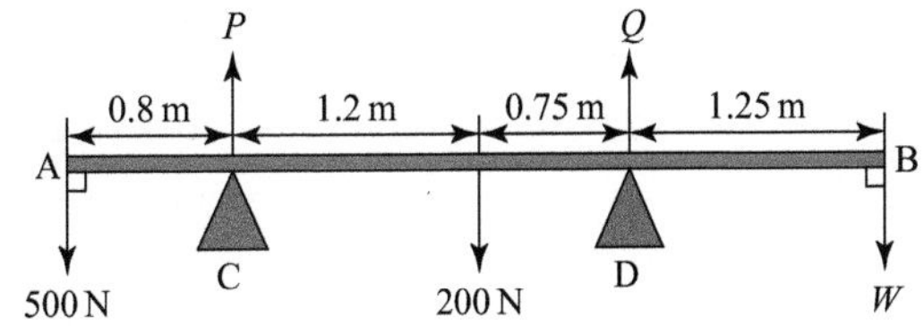

   a Find the largest possible value of W, for which the beam remains in equilibrium.

   b Find the smallest possible value of W for which the beam remains in equilibrium.

7 A uniform beam AB, of length 2.2 m and mass 150 kg, is supported in a horizontal position by two supports at C and D where AC = 0.75 m and AD = 1.5 m

   a Find, in terms of g, the force exerted by each support.

   b Find the mass of the heaviest man who can sit at

      i End A   ii End B

   of the beam without it tilting.

Challenge

8 A rectangular uniform drone, when viewed from above, is powered by four propellers with position vectors $(2\mathbf{i}+3\mathbf{j})$, $(4\mathbf{i}+2\mathbf{j})$, $(4\mathbf{i}+4\mathbf{j})$ and $(6\mathbf{i}+3\mathbf{j})$

   The first three propellers, in the order listed, are providing a driving force of $(-7\mathbf{i}+3\mathbf{j})$ N, $(3\mathbf{i}-5\mathbf{j})$ N and $(3\mathbf{i}+5\mathbf{j})$ N

   What force must the fourth propeller be providing if the drone is

   a Spinning but otherwise not moving,

   b Moving in the **j** direction but not spinning?

MECH

19 Summary and review

Chapter summary

- Axes in 3D form a right-hand set if conventional x- and y-axes are drawn on a horizontal surface and the positive z-axis points towards the ceiling.
- 3D vectors can be expressed in component form $\mathbf{r} = \begin{pmatrix} a \\ b \\ c \end{pmatrix} = a\mathbf{i} + b\mathbf{j} + c\mathbf{k}$
- The magnitude of a 3D vector $\mathbf{r}$ is $|\mathbf{r}| = \sqrt{a^2 + b^2 + c^2}$
- Forces can be resolved in any two perpendicular directions. If X and Y are the resolved parts or components of a force F, then F is the resultant of the forces X and Y
- If an object is in contact with a surface, then the frictional force F exerted by the surface on the object satisfies the inequality $F \le \mu R$, where μ is the coefficient of friction between the object and the surface, and R is the normal reaction of the surface on the object.
    - If the object is moving, or on the point of moving, then $F = \mu R$
- If an object in contact with a surface is about to move parallel to the surface, then it is said to be in limiting equilibrium and the friction is limiting friction.
- $F = ma$ can be used in any direction to find unknown forces acting on an object or the object's acceleration, where F is the component of the resultant force in that direction, and a is the acceleration in that direction.
- If a force F acts at a perpendicular distance d from a point P, then the turning effect or moment of F about P is the product $F \times d$. Moments are measured in Newton-metres (Nm).
- If a body is in equilibrium under a system of forces then
    - The vector sum of the forces is zero, i.e. the sum of the components of the forces in any given direction is zero,
    - The sum of the moments about any given point is zero, that is, the clockwise moments balance the anticlockwise moments.

Check and review

You should now be able to...	Try Questions
✔ Manipulate vectors in three dimensions, and solve geometrical problems in three dimensions by vector methods.	1
✔ Understand that there is a maximum value that the frictional force can take (μR) and that it takes this value when the object is moving or on the point of moving.	2, 5
✔ Resolve in suitable directions to find unknown forces when the system is at rest.	3
✔ Resolve in suitable directions to find unknown forces when the system has constant acceleration.	2, 5
✔ Use constant acceleration formulae for problems involving blocks on slopes or blocks connected by pulleys.	5
✔ Solve differential equations which arise from problems involving $F = ma$	4
✔ Take moments about suitable points and resolve in suitable directions to find unknown forces.	3

1 Points A and B have position vectors $\mathbf{a}=2\mathbf{i}+\mathbf{j}-3\mathbf{k}$ and $\mathbf{b}=5\mathbf{i}-2\mathbf{j}+3\mathbf{k}$. The point C, with position vector $\mathbf{c}$, lies between A and B

a Evaluate the vector $\overrightarrow{AB}$

b Work out the length of AB

c Work out the angle between $\overrightarrow{AB}$ and the positive z-direction, showing your working.

d Work out $\mathbf{c}$, given that $AC:CB=2:1$

2 A particle of weight $5g$ is pushed up a rough plane by a horizontal force of magnitude $3g$, giving it an acceleration of $\frac{g}{13}$ m s^{-2}. The plane is inclined to the horizontal at an angle α, where $\tan\alpha=\frac{5}{12}$

Find the coefficient of friction between the surface and the particle.

3 A uniform beam AB has mass 35 kg and length 6 m. The beam rests in equilibrium in a horizontal position on two smooth supports C and D where AC = 1 m and AD = 4.5 m

a Find, in terms of g, the magnitudes of the reactions on the beam at C and at D.

b Amanda wants to make the reactions at C and D equal and so she adds a mass of m kg at A. Find the value of m

4 A particle of mass 5 kg is travelling north along a horizontal surface with an initial velocity of 12.5 m s^{-1} when it is subject to a resistive force to the south of magnitude $12v$ N, where v m s^{-1} is the speed of the particle t seconds after the resistive force is applied. There is a constant force of 15 N acting on the particle in a northerly direction.

a Find the speed of the particle after two seconds.

b Find the minimum speed of the particle in the subsequent motion.

5 Two particles, A and B, of mass 5 kg and 2 kg respectively, are attached to the ends of a light inextensible string. A is held at rest on a rough horizontal table. The string passes over a small smooth pulley, P, fixed on the edge of the table which is $\frac{14}{15}$ m away from A. B hangs freely below the pulley. The particles are released from rest with the string taut. The coefficient of friction between A and the table is μ

a Find an inequality for μ if particle A moves towards the pulley.

b If $\mu=\frac{1}{3}$, calculate how long it takes for particle A to reach the pulley. Assume $g=9.8$ m s^{-2}

What next?

Score		
0 – 2	Your knowledge of this topic is still developing. To improve, search in MyMaths for the codes: 2190, 2191, 2193, 2194, 2197, 2208	
3 – 4	You're gaining a secure knowledge of this topic. To improve, look at the InvisiPen videos for Fluency and skills (19A)	
5	You've mastered these skills. Well done, you're ready to progress! To develop your techniques, look at the InvisiPen videos for Reasoning and problem-solving (19B)	

Click these links in the digital book

Exploration

Going beyond the exams

History

Albert Einstein (1879–1955) was born in Ulm, Germany.

His mass-energy equivalence, expressed as $E = mc^2$ has been called the most famous equation in the world. It was proposed as part of his **Special Theory of Relativity** in 1905.

In November 1915, Einstein published the equation of **General Relativity**, a theory of gravitation. The predictions of general relativity differ significantly from the classical mechanics of Newton, particularly in terms of time and space. Despite this, these predictions have been confirmed in all experiments and observations to date.

The one remaining prediction of the theory awaiting confirmation was the existence of ripples in the fabric of the Universe, so called **gravity waves**. It took 100 years, but on the 11th November 2016 it was confirmed that gravity waves had been detected for the first time.

Information

The differences in the predictions of Einstein's general relativity and Newton's classical mechanics only become apparent in extreme conditions. Newton's laws still remain the cornerstone of mechanics in everyday situations.

Newton considered that gravity was a force that would make a moving object deviate from a straight line.

Einstein's view was that gravity is a distortion of space-time.

"Imagination is more important than knowledge. Knowledge is limited. Imagination encircles the world."
-Albert Einstein

Applications

Global positioning systems work by using 24 satellites, each with a precise clock. A GPS receiver uses the time at which the signal from each satellite was received in order to determine position.

According to the theory of general relativity, gravity distorts space and time. This means that time on the satellites is moving slightly faster than time on Earth. Without taking into account this difference, a GPS tracker would become completely inaccurate within the course of about 2 minutes.

Research

In 1916, Albert Einstein proposed three tests of general relativity, subsequently called the **classical tests of general relativity**.

What were they? What was the outcome?

19 Assessment

1 Four forces, **P**, **Q**, **R** and **S**, act on an object. The object is in equilibrium.
P = 4**i** + 5**j**, **Q** = **i** – 8**j** and **R** = 3**i** – 12**j**

a Calculate **S**. Select the correct answer. **[1 mark]**

A –8**i** + 15**j** B 8**i** – 15**j** C –15**i** + 8**j** D 15**i** – 8**j**

b Calculate |**S**|. Select the correct answer. **[1]**

A $\sqrt{22}$ B 289 C 17 D $\sqrt{161}$

2 Points *P* and *Q* have position vectors **p** = 5**i** – 3**j** + **k** and **q** = –**i** + 7**j** + 5**k**.
The point *R* is the midpoint of the line *PQ*. Work out the position vector, **r**, of *R*.
Select the correct answer. **[1]**

A 7**i** – 8**j** + **k** B –**i** + 6.5**j** + 2.5**k** C –4**i** + 12**j** + 7**k** D 2**i** + 2**j** + 3**k**

3 A particle, P, of mass 20 kg, is attached to one end of a light inextensible string. The other end of the string is attached to a fixed point, *O*. A force, *F*, is applied to P, at right angles to the string. The system rests in equilibrium with the string at an angle of 30° to the vertical.

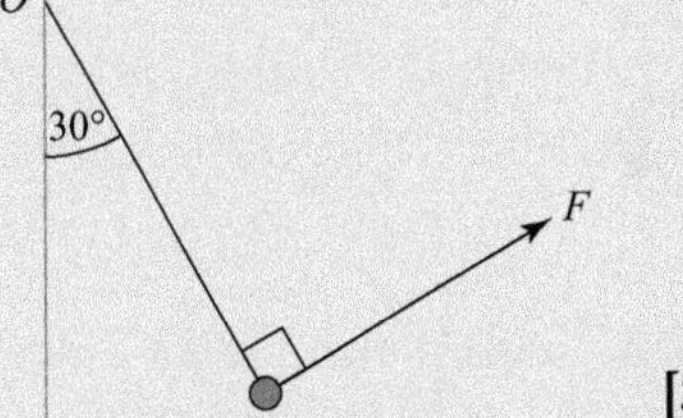

a Calculate the value of *F* in terms of *g* **[3]**

b Work out the tension in the string in terms of *g* **[3]**

4 A box of mass 20 kg is being pulled along a rough horizontal floor by a rope. The box is accelerating at 2 m s^{-2}. The tension in the rope is 148 N, and the rope makes an angle of 30° with the horizontal. The coefficient of friction between the box and the floor is μ. Using g = 9.8 m s^2 and modelling the rope as a light inextensible string,

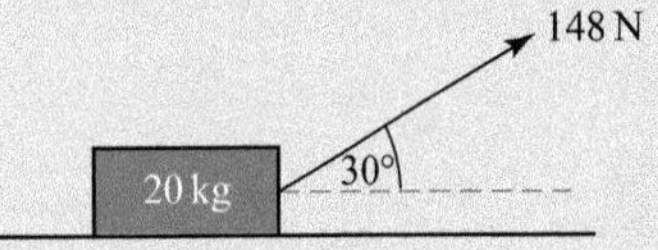

a Calculate the normal reaction between the box and the floor, **[3]**

b Calculate μ **[5]**

5 Two forces, $\mathbf{F}_1$ and $\mathbf{F}_2$, act on a body. $\mathbf{F}_1 = \begin{pmatrix} 5 \\ -6 \end{pmatrix}$ and $\boldsymbol{F}_2 = \begin{pmatrix} -1 \\ -2 \end{pmatrix}$. Find the magnitude and direction of their resultant. **[6]**

6 A uniform rod, AB, has weight 60 N and length 5 m. It rests in a horizontal position on two supports placed at P and Q, where AP = 1 m, and PQ = 2 m, as shown in the diagram. Calculate the magnitude of the force in the support

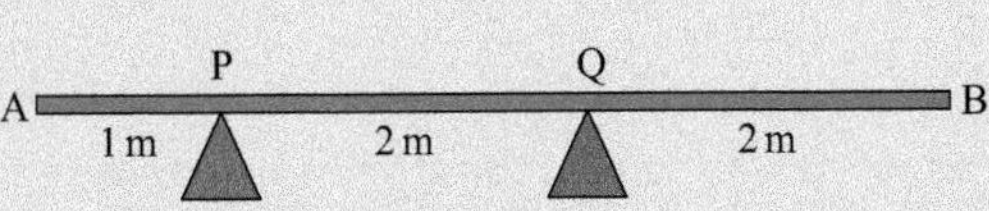

a At Q, **[3]**

b At P. **[2]**

7 A box of mass 5 kg rests in equilibrium on a rough plane. The plane is inclined at an angle of 30° to the horizontal. The coefficient of friction between the box and the plane is $\frac{\sqrt{3}}{5}$. The box is acted on by a force of P newtons, where P acts up the plane, along the line of greatest slope of the plane.

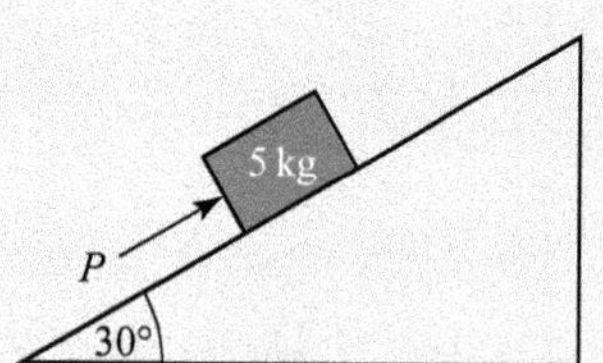

a Write an expression, in terms of g, for the normal reaction between the plane and the box. **[2]**

b Calculate, in terms of g, the range of possible values of P **[7]**

8 Two forces, **P** and **Q**, act on a particle. The force **P** is of magnitude 6 N and acts on a bearing of 035°. The force **Q** is of magnitude 8 N and acts on a bearing of 288°

a Draw a diagram and calculate the magnitude of the resultant of **P** and **Q** **[5]**

b Calculate the direction of the resultant of **P** and **Q** **[3]**

9 A parcel, P, of mass 5 kg, is attached to the ends of two light inextensible strings. The other ends of the strings are attached to two points, A and B, on a horizontal ceiling. The parcel hangs in equilibrium.

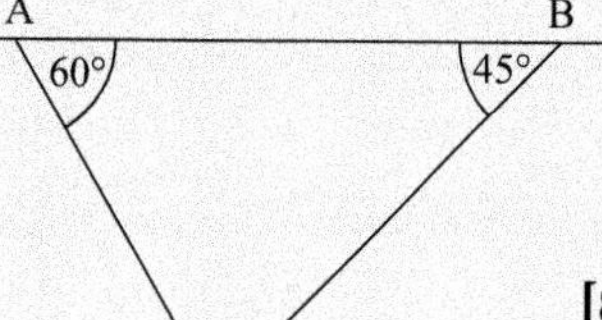

The string AP makes an angle of 60° with the ceiling, and the string BP makes an angle of 45° with the ceiling. Find, in exact form

a The tension in BP in terms of g **[8]**

b The tension in AP in terms of g **[1]**

10 A parcel of mass 8 kg slides down a ramp which is inclined at an angle of 40° to the horizontal. The parcel is modelled as a particle and the ramp as a rough plane. The coefficient of friction between the parcel and the ramp is 0.6. The ramp is 9 metres long and the parcel starts from rest at the top of the ramp. Using $g = 9.81 \text{ m s}^2$

a Calculate the acceleration of the parcel, **[7]**

b Calculate the speed of the parcel at the bottom of the ramp. **[2]**

11

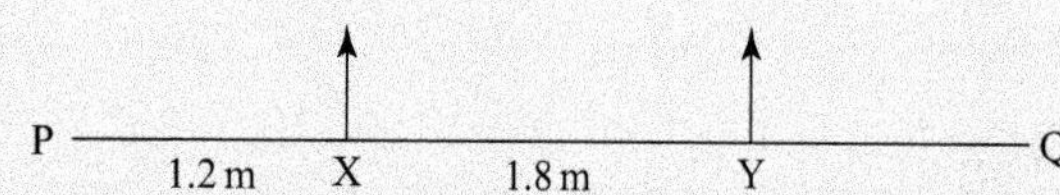

A uniform rod, PQ, has weight 200 N. It hangs in a horizontal position, supported by two light inextensible strings attached at X and Y, where PX = 1.2 m, and XY = 1.8 m as in the diagram. The tension in the string at Y is 120 N. Calculate

a The tension in the string at X, **[2]**

b The length of the rod. **[3]**

12 A particle, P, of mass 10 kg, is attached to one end of a light inextensible string. The other end of the string is attached to a fixed point, O. A horizontal force of 50 N is applied to P, so that the system rests in equilibrium with the string at an angle of $x°$ to the vertical. Using $g = 9.81 \text{ m s}^2$

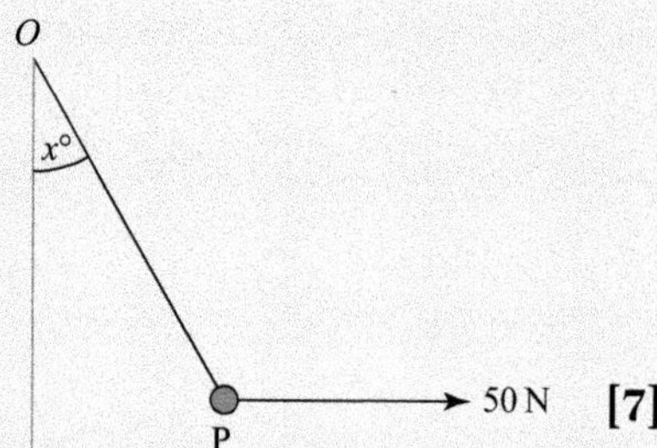

a Find the value of x **[7]**

b Find the tension in the string. **[2]**

13 Two forces $\mathbf{F}_1$ and $\mathbf{F}_2$ act on an object. $\mathbf{F}_1$ is of magnitude 5 N, and acts on a bearing of 041°. $\mathbf{F}_2$ acts on a bearing of 289°. Given that the resultant of $\mathbf{F}_1$ and $\mathbf{F}_2$ acts in a north-westerly direction, find the magnitude of $\mathbf{F}_2$ **[5]**

14 Two rings, A and B, each of mass 20 kg, are threaded on a rough horizontal wire. The coefficient of friction between each ring and the wire is μ. The rings are attached to the ends of a light inextensible string.

A smooth ring, C, of mass 30 kg, is threaded on the string, and hangs in equilibrium below the wire. The rings are in limiting equilibrium on the wire, on the point of slipping inwards. The angles between the strings and the wire are each $x°$, where $\tan x = \frac{3}{4}$

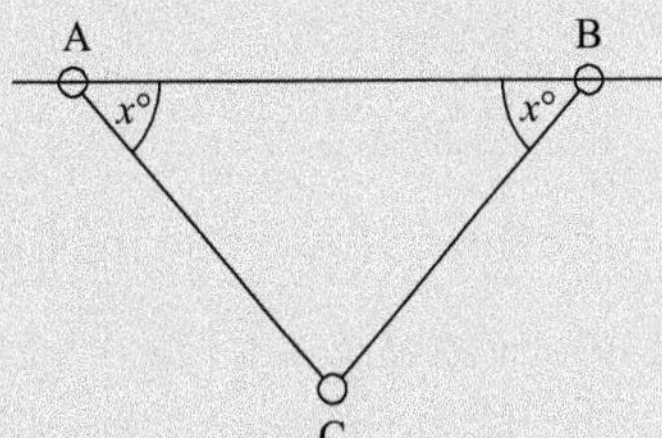

a Calculate the tension in the string in terms of g **[4]**

b Show that $\mu = \frac{4}{7}$ **[7]**

15 A box of mass 6 kg rests on a rough plane. The plane is inclined at an angle of 30° to the horizontal.

The coefficient of friction between the box and the plane is $\frac{\sqrt{3}}{2}$. A horizontal force of P newtons acts on the parcel, and it is in limiting equilibrium, on the point of moving up the plane. Calculate, in terms of g

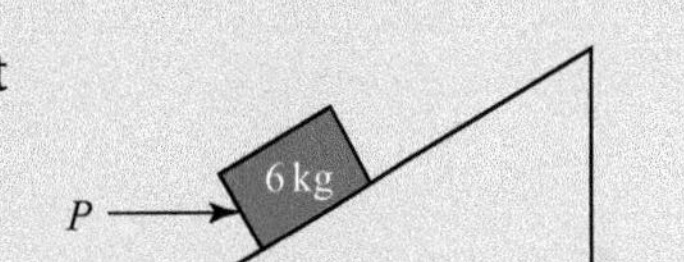

a The force P, **[5]**

b The normal reaction between the plane and the box, **[2]**

c The frictional force between the plane and the box. **[2]**

16 A particle is projected up the line of greatest slope of a rough plane with an initial velocity of 4 m s^{-1}. The plane makes an angle of 15° with the horizontal, and the coefficient of friction between the particle and the plane is $\frac{1}{10}$. Use $g = 9.81$ m s^2

a Calculate the acceleration of the particle. **[7]**

b Calculate the distance move by the particle before it comes instantaneously to rest. **[2]**

c Will the particle start to move back down the slope? Justify your answer. **[2]**

17

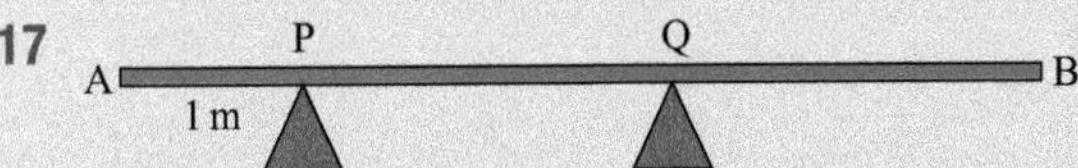

A uniform plank, AB, has mass 100 N and length 6 m. It rests in a horizontal position on two supports placed at P and Q, where AP = 1 m, as shown in the diagram. The reaction at Q is 20 N more than the reaction at P.

a Find the magnitude of the reaction at P. **[2]**

b Find the magnitude of the reaction at Q. **[1]**

c Find the distance AQ. **[3]**

18 The diagram shows three bodies, X, Y and Z, connected by two light inextensible strings, passing over smooth pulleys. Y lies on a rough horizontal table, and the coefficient of friction between Y and the table is $\frac{2}{5}$. X and Z hang freely. The system is released from rest. Calculate, in terms of g,

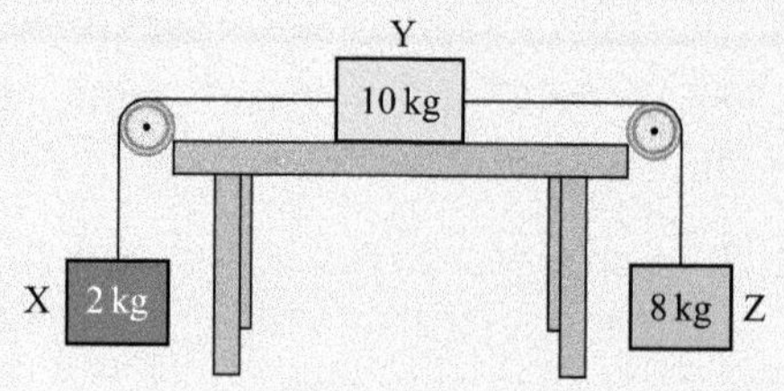

a The acceleration of Y, **[10]**

b The tension in the string joining X to Y, **[2]**

c The tension in the string joining Y to Z. **[2]**

19 A uniform plank, AB, has mass M kg and length 5 m. It rests in a horizontal position on two supports placed at P and Q, where AP = 1 m, and PQ = 3 m, as shown in the diagram. A mass of $\frac{M}{2}$ kg hangs from the rod at B.

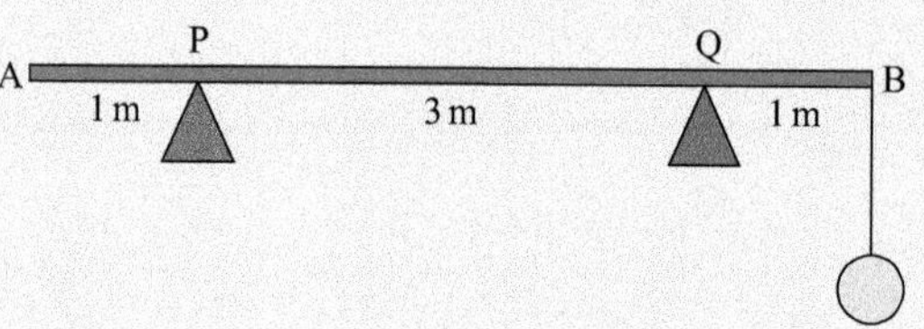

a Find expressions, in terms of M and g, for

i The magnitude of the force in the support at Q,

ii The magnitude of the force in the support at P. **[5]**

The mass at B is now replaced with a new mass of λM. The plank is on the point of tipping about Q.

b Find

i The value of λ

ii An expression, in terms of M and g, for the magnitude of the force in the support at Q. **[4]**

20 The diagram shows a particle, A, of mass 5 kg, on a rough plane, connected to a particle, B, of mass 10 kg, by a light inextensible string that passes over a smooth pulley. B hangs freely, and the string is parallel to the line of greatest slope of the plane. The coefficient of friction between A and the plane is $\frac{1}{4}$, and the plane is inclined at an angle of $x°$ to the horizontal, where $\sin x = \frac{3}{5}$

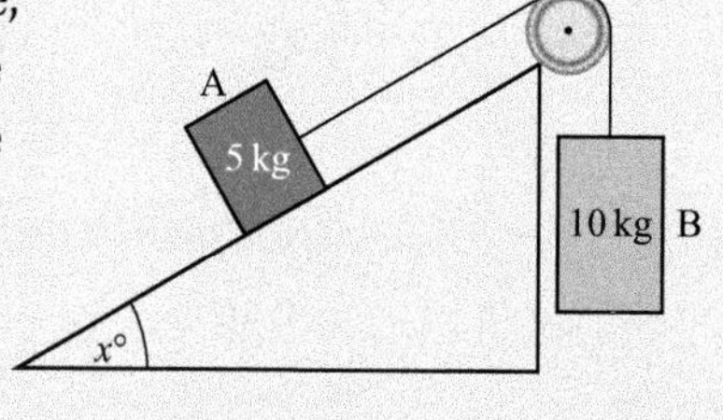

The particles are released from rest.

a Calculate, in terms of g, the acceleration of A. **[11]**

After B has dropped a distance of 1 metre, it hits the floor, and does not rebound.

b Calculate, in terms of g, the velocity with which B hits the floor. **[2]**

c Given that A does not hit the pulley, find the further distance travelled by A, until A is instantaneously at rest. **[4]**

Assessment 18–19

Chapters 18-19: Mechanics

1 A particle is held in equilibrium by a force of 3 N acting due east, a force of X acting on a bearing of 300° and a force of Y acting on a bearing of 210°

Calculate the values of X and Y.

Select the correct answer. [1]

A $X=\frac{3\sqrt{3}}{4}\text{N}, Y=\frac{3}{4}\text{N}$ B $X=\frac{3}{2}\text{N}, Y=\frac{3\sqrt{3}}{2}\text{N}$

C $X=\frac{3\sqrt{3}}{2}\text{N}, Y=\frac{3}{2}\text{N}$ D $X=\frac{3}{4}\text{N}, Y=\frac{3\sqrt{3}}{4}\text{N}$

2 A train of mass 50 tonnes is travelling at 35 m s^{-1} when the brakes are applied, causing a resultant braking force of 25 kN

Find the distance the train travels before coming to rest.

Select the correct answer. [1]

A 1225 m B 122.5 m C 35 m D 350 m

3 A car accelerates at 1.5 m s^{-2} from rest on a straight, horizontal road until it reaches a speed of 9 m s^{-2}. It then continues along the road at this constant speed for 3 s before the brakes are applied and the car comes to a stop 2 s later. After being stationary for 2 s, it then reverses for 2 s with a constant acceleration of 1 m s^{-2}. The car then stops instantly.

a Sketch a velocity-time graph. [4]

b Work out the displacement of the car from its starting point when it stops the second time. [3]

4 A ball is dropped from a window and hits the ground 1.5 s later.

a Calculate the height of the window in terms of g [2]

b Calculate the speed at which the ball hits the ground in terms of g [3]

At the same time as the ball is dropped, a second ball is thrown vertically upwards from the ground. The balls pass each other after one second.

c Calculate the possible speeds at which the second ball is thrown upwards in terms of g [4]

5 Two particles, A and B, of mass 30 g are 70 g respectively, are connected by a light inextensible string that passes over a smooth pulley. The particles are released from rest both 0.3 m from the pulley and 1 m from the ground. Take $g = 9.8$ m s^{-2}

a Calculate

i The tension in the string, ii The acceleration of the system. [5]

b Work out the height particle A travels to after particle B hits the ground. [4]

6 An object of mass 1.2 kg rests on a smooth horizontal table. It is attached to a second object of mass 0.8 kg that hangs over the edge of the table via a light, inextensible string that passes over a smooth pulley as shown in the diagram. The system is released from rest. Take $g = 9.81\ \text{m s}^{-2}$

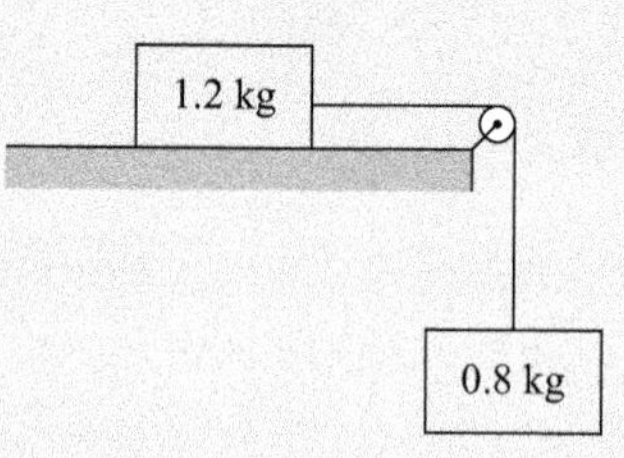

a Calculate

i The acceleration of the objects, ii The tension in the string. **[5]**

b Explain what is meant by each of these modelling assumptions,

i The string is light and inextensible, ii The pulley and table are smooth. **[3]**

7 The points A and B have position vectors $12\mathbf{i} + 7\mathbf{j} - 5\mathbf{k}$ and $3\mathbf{i} - 2\mathbf{j} - \mathbf{k}$ respectively.

Calculate the magnitude of the vector $\overrightarrow{AB}$ **[4]**

8 A particle passes through the origin at velocity $\begin{pmatrix} 2 \\ -4 \end{pmatrix}$ m s^{-1} then moves with constant acceleration for 5 seconds until it reaches a velocity of $\begin{pmatrix} -3 \\ 6 \end{pmatrix}$ m s^{-1}

a Find the acceleration over these 5 seconds. **[3]**

After 5 seconds, the particle is at point A

b Find $|\overrightarrow{OA}|$ **[4]**

9 A boat starts from rest then moves with acceleration $(2\mathbf{i} - 0.5\mathbf{j})$ m s^{-2}, where $\mathbf{i}$ and $\mathbf{j}$ are unit vectors directed due east and due north respectively.

a Find the speed of the boat 10 seconds after it starts moving. **[3]**

The boat stops accelerating after travelling 500 m.

b Calculate the time taken to travel to this point. **[4]**

10 A particle moves with velocity $v = t + \cos 2t$

a Calculate the acceleration of the particle when $t = \frac{\pi}{6}$ **[3]**

b Find an expression for the displacement of the particle from the origin at time t given that the particle is at the origin when $t = \pi$ **[4]**

11 The velocity of a particle is given by $\boldsymbol{v} = t^2\mathbf{i} + 2t\mathbf{j}$,

a Calculate the acceleration when $t = 3$ **[3]**

When $t = 0$, $\boldsymbol{r} = 3\mathbf{i} - 5\mathbf{j}$

b Find an expression for the displacement at time t **[4]**

12 A cyclist moves on a horizontal road. The position vector of the particle at t seconds is given by $\boldsymbol{r} = \begin{pmatrix} \frac{1}{3}t^3 - \frac{1}{2}t^2 - 3 \\ t^2 - \frac{1}{6}t^3 \end{pmatrix}$ m

a When $t = 3$, calculate

i The speed of the cyclist, ii The magnitude of the acceleration of the cyclist. **[7]**

The cyclist is due east of the origin at the points A and B

b Calculate the distance AB **[6]**

13 An object of mass 2 kg is acted on by a force of 12 N at 30° to the horizontal and a force of P at $\theta°$ to the horizontal as shown in the diagram. Take $g = 9.8\ \text{m s}^{-2}$

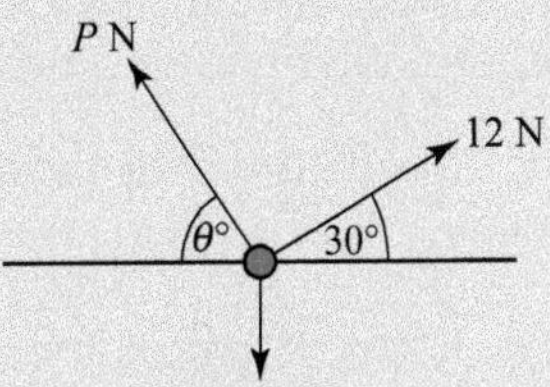

Find the values of P and θ when

a The system is in equilibrium, **[5]**

b The object is accelerating at $3\ \text{m s}^{-2}$ in the direction shown. 

[5]

14 A block of mass 7 kg rests on a rough horizontal plane, the coefficient of friction between the block and the plane is 0.7. A horizontal force P acts on the block as shown in the diagram. Take $g = 9.81\ \text{m s}^{-2}$

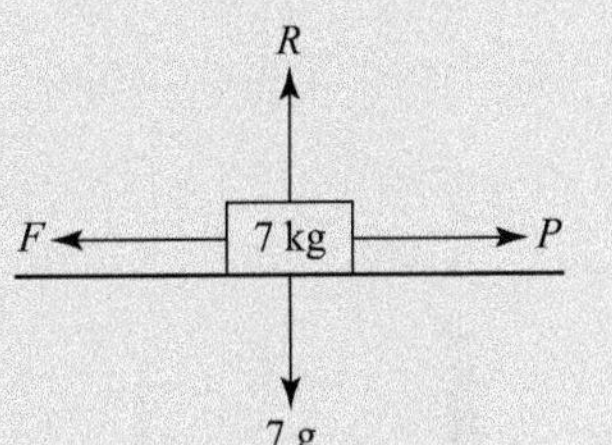

a Calculate the frictional force acting on the block when

i $P = 30$ N ii $P = 55$ N **[5]**

b Calculate the acceleration when $P = 60$ N **[2]**

15 A light beam, AB, of length 7 m is supported at points C and D which are positioned 1 m from the ends of the plank as shown. A mass of 3 kg is placed on the beam, 4 m from point C and the beam remains in equilibrium.

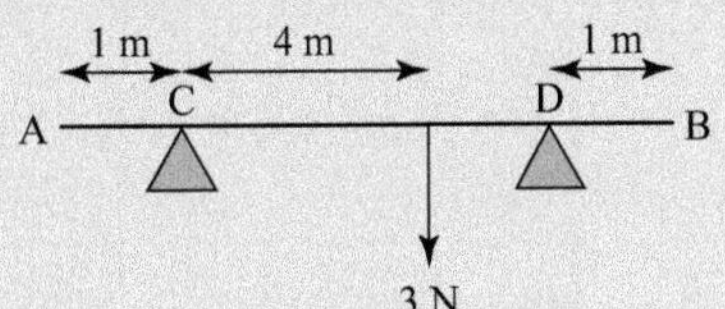

a Calculate the reaction force at points C and D. **[4]**

b Explain how you have used the assumption that the beam is light. **[1]**

16 A uniform rod, AB, of length 9 m and mass 30 kg is supported at each end. A mass of 12 kg is placed on the beam, 2 m from A. Take $g = 9.8\ \text{m s}^{-2}$

a Find the reaction at each of the supports. **[4]**

b Explain how you have used the assumption that the rod is uniform. **[1]**

17 You are given that $\overrightarrow{OA} = \begin{pmatrix} 2 \\ -1 \\ 5 \end{pmatrix}$ and $\overrightarrow{OB} = \begin{pmatrix} -3 \\ 2 \\ 0 \end{pmatrix}$

Calculate the magnitude of $\overrightarrow{OC}$ where $OACB$ is a rhombus. **[5]**

18 A jet ski is initially at rest at position vector $\boldsymbol{r} = \begin{pmatrix} 3 \\ -2 \end{pmatrix}$ m. It then starts moving with constant acceleration for four seconds until it reaches the point A where $\overrightarrow{OA} = \begin{pmatrix} -5 \\ 1 \end{pmatrix}$

a Calculate the acceleration over these four seconds. **[3]**

b Calculate the speed the jet ski is moving when it reaches point A **[4]**

c Work out the bearing the jet ski travels on. **[3]**

19 A particle of mass 2 kg has position vector at time t given by $\boldsymbol{r} = \begin{pmatrix} 1+\cos 2t \\ 3\sin 2t \end{pmatrix}$

a Calculate the speed of the particle when $t = \frac{5\pi}{6}$ **[5]**

b Calculate the force acting on the particle when $t = \frac{5\pi}{6}$ **[5]**

20 A ball is projected from point O with a velocity of 6 m s^{-1} at an angle of 60° to the horizontal. The ball lands at point A which is at the same horizontal level as point O

a Find the time taken for the ball to reach point A **[3]**

b Calculate the maximum height reached by the ball. **[3]**

c Calculate the distance OA **[3]**

d What assumptions have you made in modelling this situation? **[2]**

21 An object of mass 1.5 kg is at rest on a rough horizontal surface. The force $\boldsymbol{F} = \begin{pmatrix} 9 \\ 5 \end{pmatrix}$ N acts on the object causing it to accelerate at 2 m s^{-2} along the surface.

Calculate the coefficient of friction, μ, between the object and the surface. Take $g = 9.81\text{ m s}^{-2}$ **[5]**

22 A rough slope is inclined at 15° to the horizontal. A box of weight 60 N is on the slope and held in place by a rope which is parallel to the slope as shown.

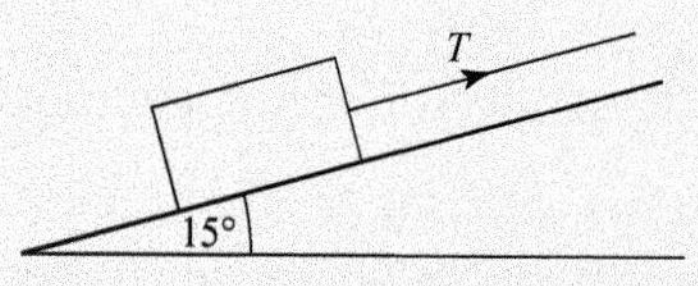

The coefficient of friction between the box and the slope is 0.13.

a Calculate the tension in the rope. **[5]**

The rope is detached and the box slides down the slope.

b Find the acceleration of the box. **[4]**

23 A uniform shelf of mass 16 kg and length 60 cm is supported by brackets at points A and B where B is at one end of the shelf and the distance between A and B is x cm. An object of mass 5 kg is placed on the other end of the shelf. The reaction force at A is double the reaction force at B. Take $g = 9.8\text{ m s}^{-2}$

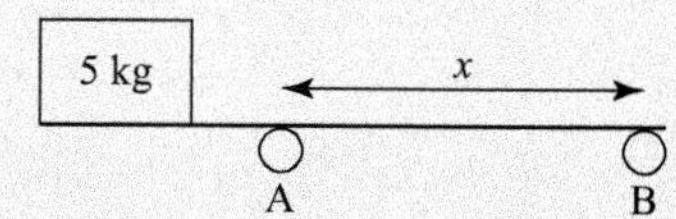

a By modelling the object as a particle, calculate

i The reaction forces at A and B, ii The length x **[7]**

The box is in fact a cube of side length 5 cm

b Without further calculation, explain the effect this will have on the value of x **[2]**

24 A particle of weight 7 N is hanging from two light, inextensible strings as shown in the diagram.

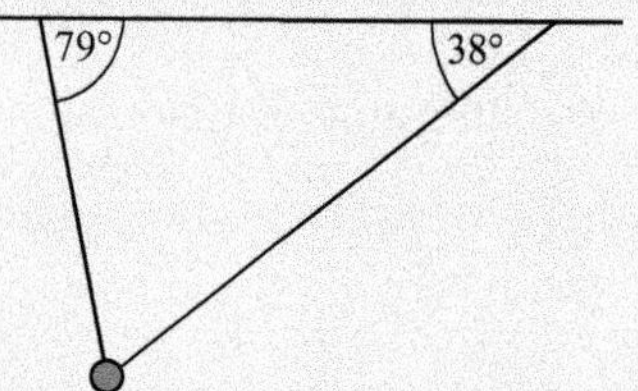

Find the tension in each string given that the system rests in equilibrium. **[5]**

25 A particle of mass 7 kg is initially at rest at the point with position vector $\begin{pmatrix}-1\\3\end{pmatrix}$ m.

It then moves with acceleration given by $\boldsymbol{a}=\begin{pmatrix}te^t\\1-2e^t\end{pmatrix}$ m s^{-2}

Calculate the distance of the particle from the origin after one second. **[10]**

26 A ball is hit at ground level on a horizontal surface and is caught 1.5 seconds later at a height of 2.1 m above the surface and a horizontal distance of 25 m from where it was projected.

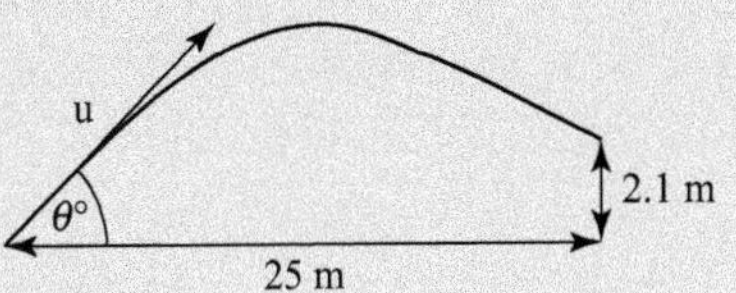

a Calculate

i The initial velocity, **ii** The angle of projection. **[6]**

b Calculate the speed and direction of the ball when it is caught. **[6]**

27 A car of mass 2500 kg is towing a caravan of mass 1000 kg up a hill which is inclined at 12° to the horizontal. The caravan is attached to the car using a light, inextensible tow bar.

The coefficient of friction between the vehicles and the road is 0.4. Initially the car and the caravan are at rest, and then they start to move with an acceleration of 4.5 m s^{-2}. Take $g = 9.8$ m s^{-2}

a Calculate

i The driving force, **ii** The tension in the tow bar. **[8]**

After four seconds, the driving force is removed.

b Calculate the length of time until the car and the caravan come to instantaneous rest. **[5]**

c Find the driving force required for the car and caravan to be on the point of moving up the hill. **[2]**

28 A particle is projected with a velocity of 21 m s^{-1} and lands 50 m away at the same height as it was projected from. Take $g = 9.8$ m s^{-2}

a Calculate

i The possible angles of projection,

ii The time taken for the ball to land in each case. **[10]**

Assuming the same initial velocity and the angles of projection calculated in part **a**,

b Explain how and why the horizontal displacement would change if the object being projected was a football. **[2]**

29 A uniform plank has mass 30 kg and length 3.6 m. It is held horizontal by two vertical strings attached at A and B which are 0.5 m and 2 m respectively from one end of the plank.

a Find the magnitudes of the tensions in the strings in terms of g **[5]**

When an object of weight of 40 N is attached x m from the end of the plank as shown, the plank is on the point of rotating about B.

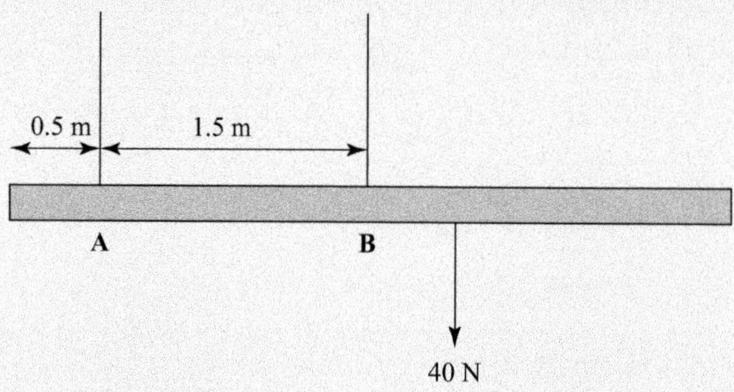

b Taking $g = 9.81\ \mathrm{m\,s^{-2}}$, calculate the length x **[5]**

30 A particle of mass 700 g is initially travelling at a constant speed of $5\ \mathrm{m\,s^{-1}}$ and passes through the origin when $t = 0$. It is then acted on by a horizontal force F, where $F = 3t + 6$, which causes it to accelerate in a straight line.

Calculate the distance of the particle from the origin when $t = 3$ s. **[8]**

31 A tennis player hits the ball directly towards the net at a height of 0.4 m above the ground and a perpendicular distance of 3.1 m from the net. The net is 1 m high and the velocity of the ball is $54\ \mathrm{km\,h^{-1}}$. The ball passes over the net. Take $g = 9.8\ \mathrm{m\,s^{-2}}$

a Calculate the range of possible angles of projection of the ball, giving your answer correct to the nearest tenth of a degree. **[8]**

The total length of the tennis court is 23.8 m and the net is exactly in the middle.

b Calculate the ranges of possible angles of projection of the ball given that is lands within the court on the other side of the net. Give your answer correct to the nearest degree. **[9]**

32 A particle of mass m kg moves in a straight line across a horizontal surface with speed $v\ \mathrm{m\,s^{-1}}$. As it moves it experiences a resultant resistive force of magnitude $\frac{1}{5}m\sqrt{v}$. When $t = 0$, the speed of the particle is $9\ \mathrm{m\,s^{-1}}$

Find the total distance travelled by the particle at its speed decreases from $9\ \mathrm{m\,s^{-1}}$ to zero. **[10]**

20 Probability and continuous random variables

Although light bulbs have an expected number of hours that they will last, they rarely last exactly the length of time printed on the box. Their lifespan varies depending on factors such as the environment, manufacture, extent of use and so on, and this can be modelled using a Normal distribution.

The Normal distribution is very useful for modelling outcomes which are influenced by a lot of small contributions, where individual outcomes are varied but the overall pattern is predictable. The distribution shows up in many places, from people's heights and IQs to measurement errors to exam marks. It is a useful tool for modelling the distributions of many random variables. It provides information on the probability of, say, a person being no taller than a certain height. Being able to model probability is very useful in many subject areas, including the social and natural sciences.

Orientation

What you need to know

Ch9 Collecting, representing and interpreting data

- Variance and standard deviation.

Ch10 Probability and discrete random variables

- Binomial distribution.

What you will learn

- To calculate conditional probabilities from data given in different forms.
- To apply binomial and Normal probability models in different circumstances.
- To use data to assess the validity of probability models.
- To solve problems involving both binomial and Normal distributions.

What this leads to

Ch21 Hypothesis testing 2
Testing a Normal distribution.

Applications
Weather forecasting.
Actuarial science.
Stochastic analysis.

Practise before you start

2111, 2281

20.1 Conditional probability

Fluency and skills

If a fair dice is thrown once, the probability of getting a six is $\frac{1}{6}$

See Ch10.1 For a reminder on the definition of **sample space**.

If you know the outcome of the dice roll is even, the probability it is a six changes to $\frac{1}{3}$ because there are now only 3 possible outcomes. **Conditional probability** is used when information about the outcome is known and it results in a smaller sample space.

Conditional probabilities can be calculated using a Venn diagram, a tree diagram, a two-way table, or formulae. You can also use **set notation** when calculating conditional probabilities.

Let $\varepsilon = \{1, 2, 3, ..., 10\}$, A = prime numbers, B = factors of 20

ε	The **universal set** consists of all elements under consideration	$\{1,2,3,4,5,6,7,8,9,10\}$
$A \cap B$	The **intersection** of A and B consists of elements common to *both* A and B	$\{2,5\}$
$A \cup B$	The **union** of A and B consists of elements that appear in *either* A or B or *both*	$\{1,2,3,4,5,7,10\}$
A'	The **complement** of A consists of elements that do *not* appear in A	$\{1,4,6,8,9,10\}$
$\subset$	A **subset** of ε is a set of elements which are all contained in ε	$A \subset \varepsilon$
$\varnothing$	The **empty set** contains no elements	$A' \cap A = \varnothing$

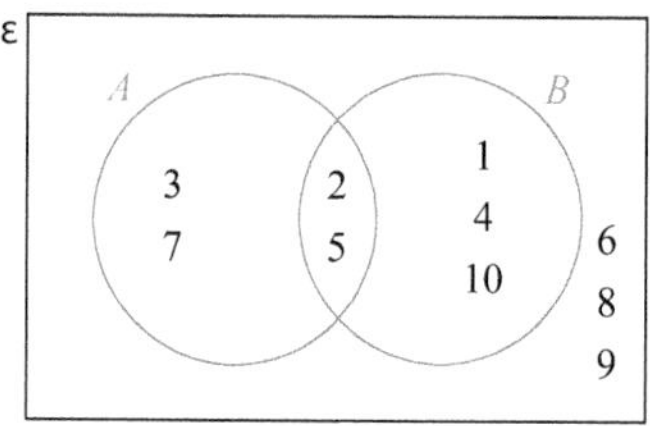

Key point

$P(A|B)$ represents the probability of A occurring *given that* B has occurred.

Example 1

25 consumers try three brands of mints, A, B and C, and say which of them, if any, they like. The Venn diagram shows the results.

Find the probability that a randomly chosen consumer

a Likes brand A

b Likes brand A and brand B

c Likes brand B given that they are known to like brand A

d Likes brand B or brand C

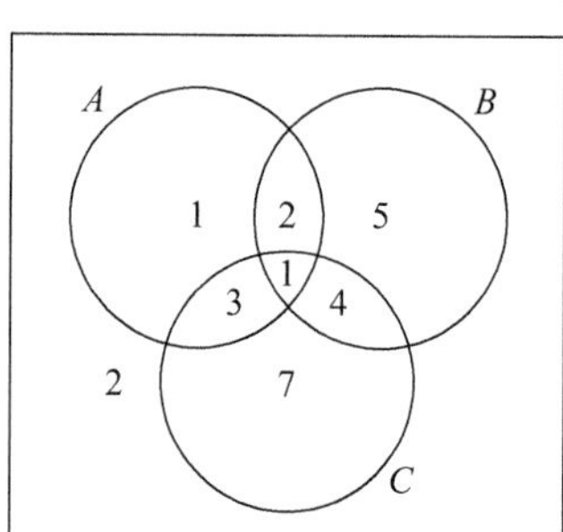

a $P(A) = \frac{n(A)}{N} = \frac{1+3+1+2}{25} = \frac{7}{25}$

b $P(A \cap B) = \frac{2+1}{25} = \frac{3}{25}$

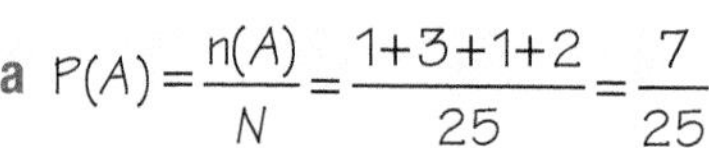

c $P(B \text{ given } A) = P(B|A) = \frac{n(A \cap B)}{n(A)} = \frac{2+1}{7} = \frac{3}{7}$

Out of the 7 people who like A, only 3 also like B

d $P(B \cup C) = \frac{2+5+1+4+3+7}{25} = \frac{22}{25}$

If it is 'given' that event A has occurred, the sample space is reduced to just those outcomes involving A

Key point

If A and B are two events associated with a random experiment,

$\mathrm{P}(B|A)=\dfrac{\mathrm{P}(A\cap B)}{\mathrm{P}(A)}$ and, rearranging, $\mathrm{P}(A\cap B)=\mathrm{P}(A)\times\mathrm{P}(B|A)$

This result can be used to solve tree diagram problems, including in cases where there is dependence between the stages.

In Example 1, $n(B\cup C)$ is the number of consumers who like at least one of B or C

$n(B\cup C)=2+5+1+4+3+7=22$

This is the same as $n(B)+n(C)-n(B\cap C)$. You subtract $n(B\cap C)$ to avoid double counting.

Key point

If A and B are two events associated with a random experiment, then

$\mathrm{P}(A\cup B)=\mathrm{P}(A)+\mathrm{P}(B)-\mathrm{P}(A\cap B)$

STATS

Example 2

Two art experts, S and T, view a painting. The probabilities that they correctly identify the artist are 0.8 and 0.65, respectively. One of the experts is chosen at random and asked to identify the artist.

a Calculate the probability that expert S is chosen and the artist is correctly identified.

b Calculate the probability that the artist is correctly identified.

a Define the events A: S chosen, B: T chosen and C: artist correctly identified.

$\mathrm{P}(A \text{ and } C)=\mathrm{P}(A\cap C)=\mathrm{P}(A)\times\mathrm{P}(C|A)=0.5\times0.8=0.4$ — 0.8 is the *conditional* probability of C given A

b $\mathrm{P}(C)=\mathrm{P}(A\cap C)$ or $\mathrm{P}(B\cap C)$ — Correct identification is by *either* S or T.

$=\mathrm{P}(A)\times\mathrm{P}(C|A)+\mathrm{P}(B)\times\mathrm{P}(C|B)$

$=0.5\times0.8+0.5\times0.65=0.725$

Example 3

In a certain population with equal numbers of males and females, 1 in every 100 males and 1 in every 10 000 females are colour blind. A person is chosen at random from the population. Draw a tree diagram to find the probability that the person is

a Male and colour blind, **b** Colour blind.

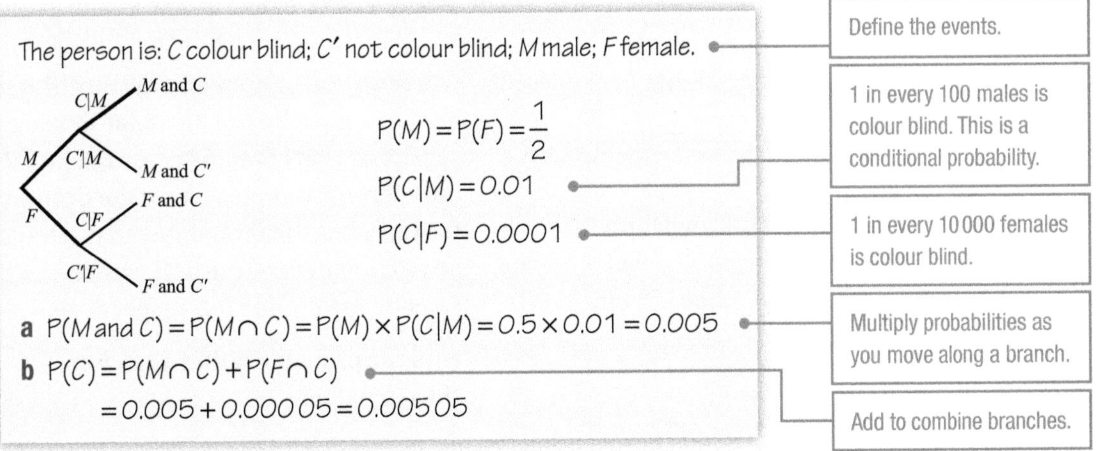

The person is: C colour blind; C′ not colour blind; M male; F female.
Define the events.
M
F
C|M
C′|M
C|F
C′|F
M and C
M and C′
F and C
F and C′
P(M) = P(F) = 1/2
P(C|M) = 0.01
P(C|F) = 0.0001
1 in every 100 males is colour blind. This is a conditional probability.
1 in every 10 000 females is colour blind.
a P(M and C) = P(M ∩ C) = P(M) × P(C|M) = 0.5 × 0.01 = 0.005
Multiply probabilities as you move along a branch.
b P(C) = P(M ∩ C) + P(F ∩ C)
= 0.005 + 0.000 05 = 0.005 05
Add to combine branches.

Conditional probability can be used to define independence between events.

2092, 2095 SEARCH

Two events, A and B, are independent if $P(A|B) = P(A)$ or if $P(A \cap B) = P(A) \times P(B)$

Key point

Exercise 20.1A Fluency and skills

1 In a survey, 32 people are asked if they like Indian, Bangladeshi or Thai foods. The Venn diagram shows the number of people who like each type of food.

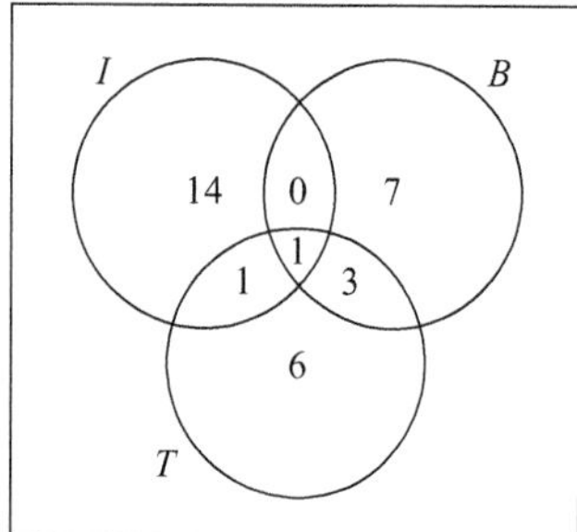

Find the probability that a randomly chosen customer

a Likes Indian food,

b Likes two types of food,

c Likes two types, given that they like Thai food.

2 In a sports club, 28 students choose from football, rugby and athletics. 7 choose only football, 6 choose only rugby and 7 choose only athletics. 6 choose football and rugby, and 2 choose football and athletics. No one chooses all three or rugby and athletics. A student is chosen at random. Use a Venn diagram to calculate the probability that the student chooses

a Rugby,

b Only football, only rugby or both,

c Football, given that they also choose athletics.

3 An ordinary six-sided dice is thrown once. A is the event 'the score is even' and B is the event 'the score is greater than 4'. Find

a $P(A \cap B)$ b $P(A|B)$ c $P(B|A)$

4 A box contains 8 red beads and 6 green beads. Two beads are chosen at random without being replaced and their colours are recorded. Draw a tree diagram to find the probability that the two chosen beads are different colours.

5 In a hospital survey, 40 patients were classified by diet (vegetarian or not) and blood pressure status (raised or not raised). The results are shown in the table.

Diet \ BP	R	R'	Total
V	3	5	8
V'	3	29	32
Total	6	34	40

Find the probability that a patient chosen at random

a Has raised blood pressure,

b Does not have raised blood pressure given that they are vegetarian.

6 If $P(A) = 0.5$, $P(B) = 0.7$ and $P(A \cap B) = 0.4$, find

a $P(A|B)$ b $P(A|B')$

c $P(A'|B)$ d $P(A'|B')$

7 A fair dodecahedral dice has sides numbered 1–12. Event A is rolling more than 9, B is rolling an even number and C is rolling a multiple of 3. Find

a $P(B)$ b $P(A \cap B)$

c $P(A|B)$ d $P((A \cap B)|C)$

8 In a drug trial, patients are given either the drug or a placebo (an identical-looking substitute that contains no drug). Equal numbers of patients receive the drug and the placebo. The probability that symptoms improve with the drug and with the placebo are 0.7 and 0.2, respectively.

Find the probability that a randomly chosen patient

a Is given the drug and improves,

b Improves.

Reasoning and problem-solving

Strategy

To solve a problem using conditional probability

1. Clearly define the events.
2. If it's helpful, display the data using a two-way table, Venn diagram or tree diagram.
3. Use formulae to find conditional probabilities.

Example 4

A bag contains the seeds of three different varieties of flower, R, S and T in proportions 1 : 3 : 6

The germination rates for these varieties are 74%, 45% and 60%, respectively.

One randomly-chosen seed is planted.

Given that it germinates, what is the probability that it was of variety S?

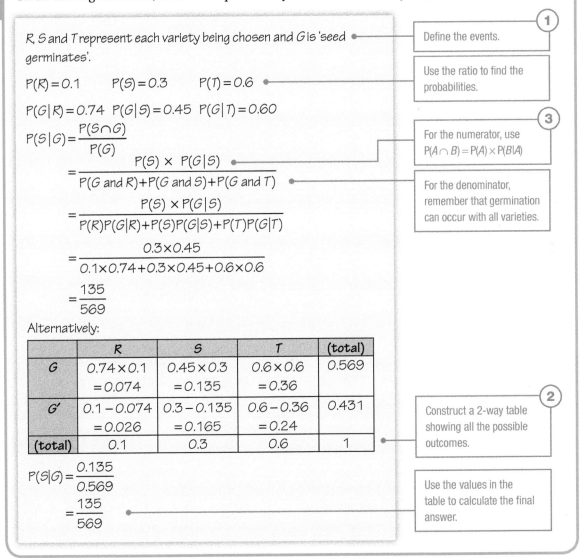

R, *S* and *T* represent each variety being chosen and *G* is 'seed germinates'. — Define the events. (1)

$P(R) = 0.1 \quad P(S) = 0.3 \quad P(T) = 0.6$ — Use the ratio to find the probabilities.

$P(G|R) = 0.74 \quad P(G|S) = 0.45 \quad P(G|T) = 0.60$

$$P(S|G) = \frac{P(S \cap G)}{P(G)}$$

$$= \frac{P(S) \times P(G|S)}{P(G \text{ and } R) + P(G \text{ and } S) + P(G \text{ and } T)}$$

(3) For the numerator, use $P(A \cap B) = P(A) \times P(B|A)$

For the denominator, remember that germination can occur with all varieties.

$$= \frac{P(S) \times P(G|S)}{P(R)P(G|R) + P(S)P(G|S) + P(T)P(G|T)}$$

$$= \frac{0.3 \times 0.45}{0.1 \times 0.74 + 0.3 \times 0.45 + 0.6 \times 0.6}$$

$$= \frac{135}{569}$$

Alternatively:

	R	*S*	*T*	(total)
G	$0.74 \times 0.1 = 0.074$	$0.45 \times 0.3 = 0.135$	$0.6 \times 0.6 = 0.36$	0.569
G'	$0.1 - 0.074 = 0.026$	$0.3 - 0.135 = 0.165$	$0.6 - 0.36 = 0.24$	0.431
(total)	0.1	0.3	0.6	1

(2) Construct a 2-way table showing all the possible outcomes.

$$P(S|G) = \frac{0.135}{0.569}$$

$$= \frac{135}{569}$$

Use the values in the table to calculate the final answer.

STATS

Example 5

S is a sample space for a random experiment where $X \cup Y = S$, $X \cap Y = \varnothing$ and $T \subset S$

$P(X) = P(Y)$, $P(T \mid Y) = \frac{1}{4}$ and $P(T \mid X) = \frac{2}{3}$

Find **a** $P(T)$ **b** $P(Y \mid T)$

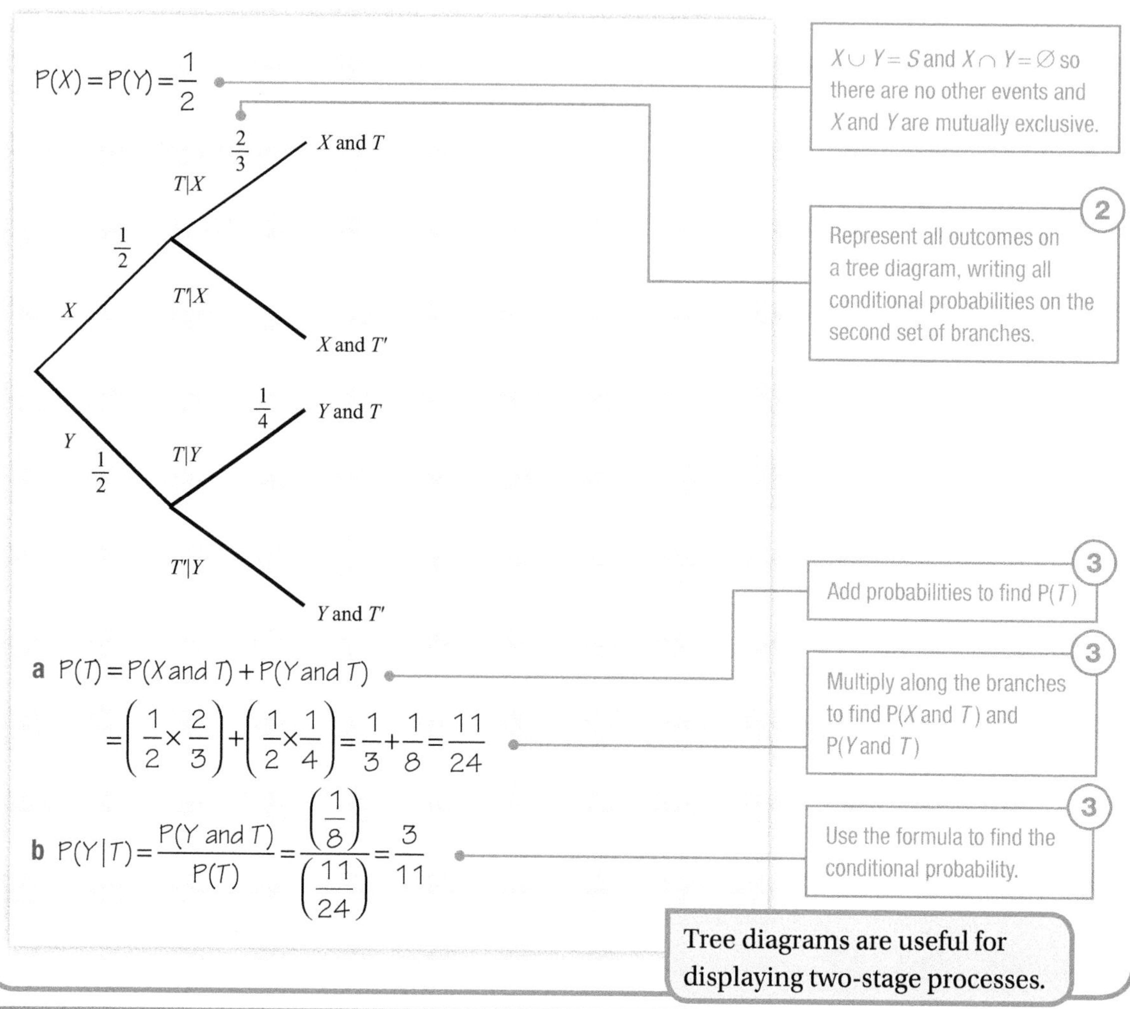

P(X) = P(Y) = 1/2
X and T
2/3
T|X
1/2
T'|X
X
X and T'
1/4
Y and T
Y
1/2
T|Y
T'|Y
Y and T'
X ∪ Y = S and X ∩ Y = ∅ so there are no other events and X and Y are mutually exclusive.
2
Represent all outcomes on a tree diagram, writing all conditional probabilities on the second set of branches.
3
Add probabilities to find P(T)
3
Multiply along the branches to find P(X and T) and P(Y and T)
3
Use the formula to find the conditional probability.

a $P(T) = P(X \text{ and } T) + P(Y \text{ and } T)$

$= \left(\frac{1}{2} \times \frac{2}{3}\right) + \left(\frac{1}{2} \times \frac{1}{4}\right) = \frac{1}{3} + \frac{1}{8} = \frac{11}{24}$

b $P(Y \mid T) = \frac{P(Y \text{ and } T)}{P(T)} = \frac{\left(\frac{1}{8}\right)}{\left(\frac{11}{24}\right)} = \frac{3}{11}$

Tree diagrams are useful for displaying two-stage processes.

Know your dataset

Large data set

In the LDS, purchased quantities of 'Yoghurt and fromage frais' will always be less than comparable purchased quantities of 'Other milk and cream' because the former is considered a *subset* of the latter. Click this link in the digital book for more information about the Large data set.

Exercise 20.1B Reasoning and problem-solving

1 The sample space for an experiment consists of two mutually exclusive and exhaustive events, P and Q. An event R is a subset of the same sample space. You are given that $P(P) = 0.2$, $P(R \mid P) = 0.4$ and $P(R \mid Q) = 0.7$

a Use a tree diagram to find $P(R)$

b Find $P(P \mid R)$

2 A test for a certain disease gives either a positive (disease present) or negative (no disease present) result. It correctly identifies 93% of cases where the disease is present. The proportion of cases where no disease is present but the test result is positive is 7%.

a Find the proportion of all cases where there is a positive result if 29% of the people being tested actually have the disease.

b Given that a person tests positive for the disease, what is the probability that she has the disease?

3 In a drug trial, patients are given either drug A or drug B. Equal numbers of patients are given each drug. The probability that symptoms improve is 0.71 with drug A and 0.43 with drug B

A patient is chosen at random. Find the probability that the patient

a Receives drug B and improves,

b Received drug B given she has improved.

Large data set

4 In 2012, the average purchased quantity of fully-skimmed milk in the East Midlands was 133 ml per person per week. Further investigation among residents of a town in the area showed that, among the 42% of people who purchased more than 133 ml of fully-skimmed milk, 55% considered themselves to be fit. The same statistic for those who purchased less than 133 ml was 38%. A resident of the town is chosen at random.

Find the probability that the person

a Purchased more than 133 ml of fully-skimmed milk and considered themselves fit

b Purchased more than 133 ml of fully-skimmed milk given that they considered themselves fit.

5 A large survey was carried out among people who identify as the main cook in their household. One quarter of the men questioned said that they purchased more olive oil than other vegetable or salad oil. Among women questioned, this increased to 31%. The ratio of women to men in the survey was 7 : 1

One person from the population is chosen at random. Find the probability that the person

a Is a woman who purchased more olive oil than other oils

b Purchased more olive oil than other oils

c Is a woman, given that she purchased more olive oil than other oils.

6 Three coins are used in a game. Two of them are fair and one has two heads. One coin is chosen at random and flipped.

a Find the probability that a heads is obtained.

b Given that a heads was obtained, find the probability that the coin with two heads was flipped.

7 A card is chosen at random from a set of twelve cards numbered 1–12

If the card shows a number less than 4, coin A, which is fair, is flipped.

If the card shows a number between 4 and 8 inclusive, coin B, for which the probability of a heads is $\frac{2}{3}$, is flipped.

If the number on the card is greater than 8, coin C, for which the probability of a heads is $\frac{1}{3}$, is flipped.

a Find the probability that the coin shows tails.

b If the coin shows tails, calculate the probability that coin B was flipped.

Challenge

8 A box contains 4 black and 8 red balls. A ball is drawn at random, its colour recorded, and the ball is placed back in the box along with *another* ball of the same colour.

a Find the probability that two black balls were drawn.

b If the second ball was black, find the probability that the first was black.

9 Suppose that 1 in every 100 males and 1 in every 10 000 females are colour blind. Assuming that there are equal numbers of males and females in the population, use conditional probability formulae to find the probability that a person chosen at random is female, given that they are not colour blind.

STATS

20.2 Modelling with probability

Fluency and skills

To model real-life situations mathematically, you often have to make simplifying assumptions. You can analyse and improve your model by

- comparing predicted results with actual data
- questioning any assumptions that have been made.

See Ch 10.2
For a reminder on the binomial distribution.

To test a binomial model you can use the mean and variance.

Key point

For $X \sim B(n, p)$, the mean (μ) and variance (σ^2) are given by $\mu = np$ and $\sigma^2 = np(1-p)$

Example 1

Jenny says she arrives at school before Claire about 90% of the time. At the end of each week she records how many days, X, out of five she arrives first, and says X follows a binomial distribution. The table shows her results for 40 weeks.

X	0	1	2	3	4	5
frequency	0	1	5	12	15	7

a State any assumptions Jenny has made and use the data to show that $X \sim B(5, 0.9)$ is not a good model for the data.

b Use the mean of the data to suggest a better value for p

a Assume that the probability remains constant every day, and the events are independent. *(Make simplifying assumptions.)*

	mean	variance
Using $X \sim B(n, p)$	$5 \times 0.9 = 4.5$	$5 \times 0.9 \times 0.1 = 0.45$
Using the data	3.55	0.9975

(Use your calculator to find the mean and variance.)

Values are not a close match, so the model is likely not a good one. *(Calculate and compare the mean and variance using the model and the actual data.)*

b A better value for p is $3.55 \div 5 = 0.71$ *(Use $\mu = np$)*

Exercise 20.2A Fluency and skills

1 Red and blue sweets are sold in randomly assorted packets of five. 40 packets are examined and the number of red sweets in each, X, is recorded.

X	0	1	2	3	4	5
frequency	0	3	10	12	13	2

a Calculate the mean and an unbiased estimate of the population variance.

b Bo suggests that X can be modelled by a binomial distribution B(5, 0.6). Use the sample to comment on the suitability of this model.

2 A dice is thrown 60 times. The number of scores greater than 2 is half the number of scores less than or equal to 2. Suggest a probability model for the number of outcomes greater than 2 when the dice is thrown 300 times, X. Justify your answer.

3 An experiment is modelled by a random variable X which is thought to have a binomial distribution, $n = 3$, $p = 0.2$. The experiment is performed 100 times.

X	0	1	2	3
frequency	49	40	9	2

Does the data support the suggested probability model?

Reasoning and problem-solving

Strategy

To solve a probability problem using modelling

1. Consider the assumptions that can be made in the context.
2. Where available, use real data to check the validity of the model.
3. Try to improve the model so it fits the data more closely.

Example 2

A town-centre bus service is scheduled to run six times an hour every weekday. The number of buses due between 7 pm and 8 pm which arrive on time, X, is modelled by $X \sim B(6, 0.72)$

a Give two reasons why a binomial model may not be suitable in this context.

b Explain why this model would not be appropriate for Y, the number of buses due between 4.30 pm and 5.30 pm which arrive on time. How will the model change?

a If one bus is late the increased passenger load may cause the following bus to be late, so events are not likely to be independent as assumed by a binomial model.

(1) Think about the conditions for a binomial model in context.

b Traffic is likely to be heavier during this time period because a high proportion of people travel home from work or school at this time. As a result, the chance of delays may be greater than at other times of day so the value of p is likely to be lower.

(1) Challenge the assumption that the risk of delays is constant throughout the day.

(3) Consider how the model could be improved.

Exercise 20.2B Reasoning and problem-solving

1 A coin is tossed 30 times, and yields X heads.

a Suggest a suitable probability distribution for X, explaining your reasoning.

b Give a reason why your model may not be suitable.

2 In a large 2001 survey of residents of a town in the north west of England, the total amount of mineral water purchased per person per week had $Q_1 = 155$ ml, $Q_2 = 170$ ml, $Q_3 = 187$ ml.

The survey was repeated in 2015 with a smaller sample of size 60.

a A 2015 respondent, A, is chosen at random. Using the 2001 data, find

i P(A buys less than 170 ml per week) **ii** P(A buys more than 155 ml per week)

Give one reason why this model might not be appropriate in 2015.

The results of the 2015 survey are given in the table.

a, amount (ml)	$a < 140$	$140 \le a < 150$	$150 \le a < 160$	$160 \le a < 170$	$170 \le a < 180$	$180 \le a < 190$	$190 \le a < 200$	$a \ge 200$	Total
frequency	4	6	5	9	11	8	12	5	60

b Were there significant changes in spending on mineral water between 2001 and 2015?

3 Sian plans to revise for 8 hours each day in the 3 weeks leading up to her exams. She reviews the first 15 pages of her notes in 90 minutes, so anticipates that the number of pages she will review each day, X, will have mean 80 and standard deviation 0.05

a Explain why Sian's model is based on unrealistic assumptions.

b Describe how more realistic assumptions would change Sian's model.

2113

20.3 The Normal distribution

Fluency and skills

When looking at the probability distribution of a **discrete random variable** (DRV), A, you assign a probability to each value that A could take.

A **continuous random variable** (CRV), X, could take any one of an infinite number of values on a given interval. Instead of assigning probabilities to individual values of X, you assign probabilities to *ranges* of values of X and the probability distribution is represented by a curve, or sequence of curves, called a **probability density function**, f(x).

$$P(a \leq X \leq b) = \int_a^b f(x)\,dx$$

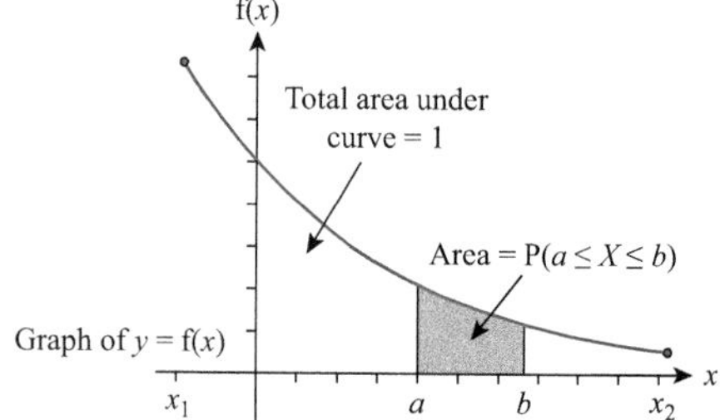

▲ Graph of $y = f(x)$ where X is a CRV taking values between x_1 and x_2 and f(x) is its probability density function

For a continuous distribution the probability of an individual value, a, is 0 because the area between a and a is 0. One result of this is that the signs < and ≤ become interchangeable.

Key point

If X is a continuous random variable

$P(a \leq X \leq b) = P(a < X \leq b) = P(a \leq X < b) = P(a < X < b)$

One of the most important and frequently used probability distributions is the **Normal distribution**. Continuous variables such as height, weight and error measurements are often Normally distributed.

ICT Resource online

To investigate the Normal distribution, click this link in the digital book.

The Normal **probability density function** has a bell-shaped curve. It is a **continuous function** and the area under the curve can be used to calculate probabilities. The total area under the curve equals 1

In a Normal distribution

- Mean ≈ median ≈ mode
- The distribution is symmetrical
- There are points of inflection one standard deviation (σ) from the mean
- ~68% of values lie within σ of the mean
- ~99.8% of values lie within 3σ of the mean

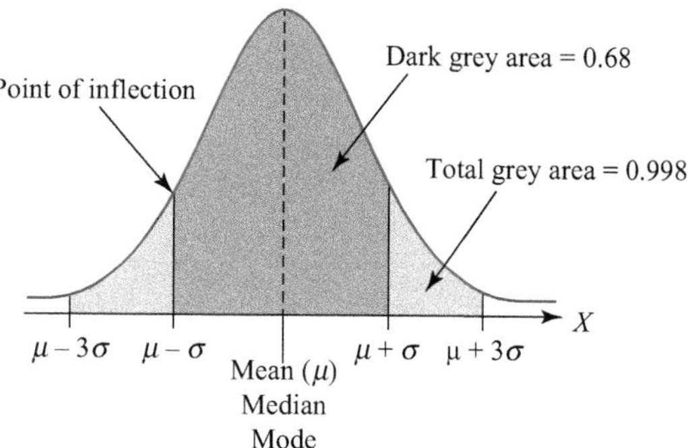

Each Normal distribution is distinguished by its mean, μ, and its variance, σ^2. If a variable X follows a Normal distribution with mean μ and variance σ^2 you write $X \sim N(\mu, \sigma^2)$

Calculator

Try it on your calculator

You can find probabilities from a Normal distribution on a calculator.

```
Normal    C.D
Data      :Variable
Lower     :4
Upper     :5.5
σ         :1.5
μ         :3
Save Res  :None
None LIST
```

```
Normal    C.D
P     =0.20470218
z:Low=0.66666666
z:Up  =1.66666667
```

Activity

Find out how to find $P(4 \le X \le 5.5)$ where $X \sim N(3, 1.5^2)$ on *your* calculator.

Example 1

All calculators work differently, make sure you know how to do this on your own calculator.

If $X \sim N(5, 3^2)$, find the probability that

a $P(2 < X < 11)$ **b** $P(X < 8)$ **c** $P(X > 8)$

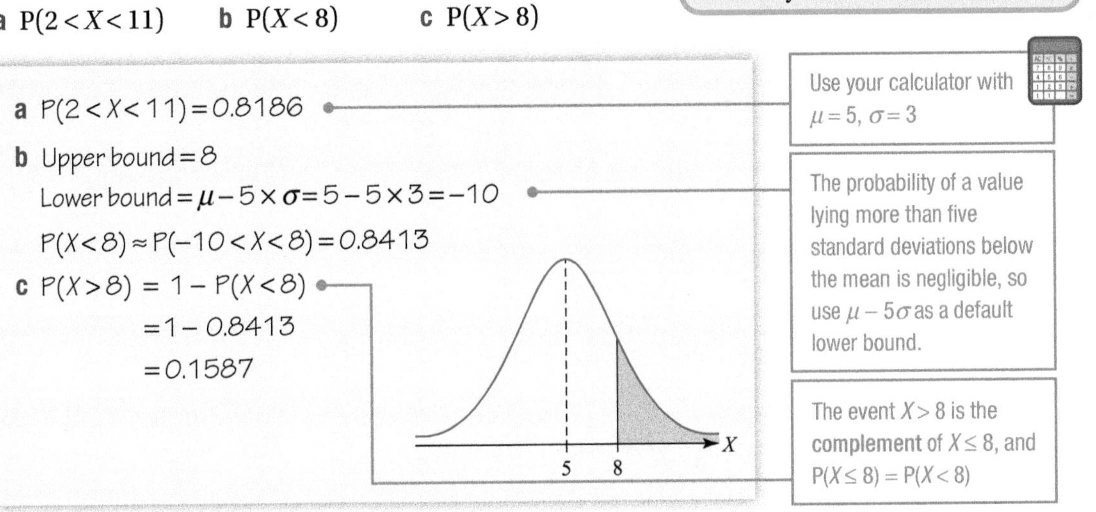

a P(2<X<11)=0.8186
b Upper bound=8
Lower bound=μ−5×σ=5−5×3=−10
P(X<8)≈P(−10<X<8)=0.8413
c P(X>8) = 1 − P(X<8)
=1− 0.8413
=0.1587
X
5
8
Use your calculator with μ = 5, σ = 3
The probability of a value lying more than five standard deviations below the mean is negligible, so use μ − 5σ as a default lower bound.
The event X > 8 is the complement of X ≤ 8, and P(X ≤ 8) = P(X < 8)

STATS

Since some calculators only give cumulative probabilities for $P(X < x)$ where $x \ge 0$, to calculate a probability you may first have to write it in this form. You can use the symmetry of the graph and the fact that the total area under the graph is 1 to do this.

Key point

$P(X > a) = 1 - P(X < a)$
$P(X < -a) = P(X > a) = 1 - P(X < a)$
$P(a < X < b) = P(X < b) - P(X < a)$

Some calculators allow you to skip this step and find $P(Z > z)$ or $P(Z < -z)$ directly.

To make calculations easier, you can transform any Normal distribution to the **standard Normal distribution**, which has a mean of 0 and a standard deviation of 1. This is a particularly useful instance of the Normal distribution and is usually given the symbol Z, written $Z \sim N(0, 1^2)$

Some calculators don't let you input the values of μ and σ directly. In this case, use the standard Normal distribution.

Key point

For any x-value in a Normal distribution $X \sim N(\mu, \sigma^2)$, the corresponding z-value in the distribution $Z \sim N(0, 1)$ is $z = \dfrac{x - \mu}{\sigma}$

Rather than finding probabilities corresponding to given values of a random variable, sometimes you need to find the value of a variable corresponding to a given probability. To do this you need to use inverse calculator functions.

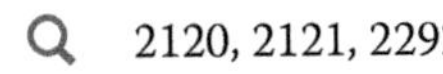

2120, 2121, 2292

Calculator

Try it on your calculator

You can solve problems using the inverse Normal distribution on a calculator.

Inverse Normal
Data : Variable
Tail : Left
Area : 0.69
σ : 4
μ : 11
Save Res : None
None LIST

Inverse Normal
xInv=12.9834014

Activity

Find out how to find a when $P(X \le a) = 0.69$ and $X \sim N(11, 4^2)$ on *your* calculator.

Example 2

The random variable, X, is randomly distributed with a mean of 20 and a variance of 8. Find the values of s and t given that

$P(X < s) = 0.7500$ and $P(t < X < s) = 0.5965$

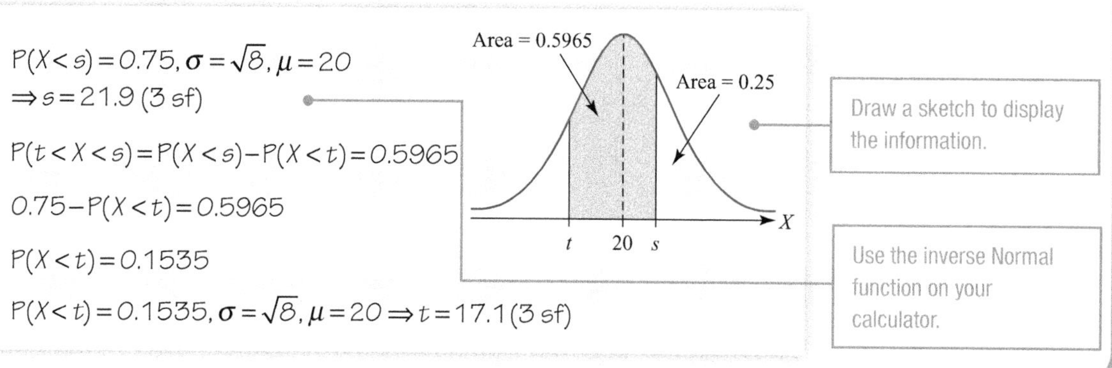

P(X<s)=0.75, σ=√8, μ=20
⇒s=21.9 (3 sf)
P(t<X<s)=P(X<s)−P(X<t)=0.5965
0.75−P(X<t)=0.5965
P(X<t)=0.1535
P(X<t)=0.1535, σ=√8, μ=20⇒t=17.1 (3 sf)
Area = 0.5965
Area = 0.25
X
t
20
s
Draw a sketch to display the information.
Use the inverse Normal function on your calculator.

Exercise 20.3A Fluency and skills

1 If $Z \sim N(0, 1)$, find

a $P(Z < 1)$

b $P(1 < Z < 1.5)$

c $P(-1 < Z < 2)$

2 If $X \sim N(6, 4^2)$, find

a $P(X < 9)$

b $P(5 < X < 8)$

c $P(2 < X < 14)$

3 X is a CRV with $X \sim N(3, \sigma^2)$ and $P(X > 8) = 0.35$

a Calculate $P(X < 8)$

$P(X < a) = 0.35$

b Find a

c Find $P(X > a)$

4 Find a in each of the following cases, where Z follows the standard Normal distribution $Z \sim N(0, 1)$

a $X \sim N(0, 16)$, $P(X \le 5) = P(Z \le a)$

b $X \sim N(4, 1)$, $P(X \le 6) = P(Z \le a)$

c $X \sim N(2, 9)$, $P(X \le 2) = P(Z \le a)$

d $X \sim N(0.5, 1.7^2)$, $P(X \le 3) = P(Z \le a)$

e $X \sim N(-1, 2.3^2)$, $P(X \le -4) = P(Z \le a)$

f $X \sim N(3, 0.25)$, $P(X \le -2) = P(Z \le a)$

5 The Normal random variable, X, has mean 12 and standard deviation 4. Find the value of d given that $P(X < d) = 0.35$

6 The Normal random variable, X, has mean 21 and variance 10. Find the value of c given that $P(c < X < 21) = 0.40$

7 The quantity of coffee dispensed from a drinks machine is Normally distributed with mean 350 ml and standard deviation 12 ml. Find the probability that a randomly chosen cup of coffee will have a volume between 340 ml and 370 ml.

Reasoning and problem-solving

Strategy

To solve a probability problem using a Normal distribution

1. Identify or find the mean and variance.
2. If necessary, calculate z-values and use the properties of the Normal distribution curve to write probabilities in the form $P(Z < z)$ where z is positive.
3. Use your calculator to find the probability or (using inverse functions) the value of the variable.
4. Compare the results of the model with the actual data.

Example 3

A machine produces metal rods whose lengths are intended to be 1 cm. The machine is set so that, on average, the rods are the correct length and the errors in the length have standard deviation 1 mm.

On inspection, it turns out that 12% of the rods are more than 2 mm from the correct length. Does the Normal distribution provide a good model to describe the length of the rods? Give reasons for your answer.

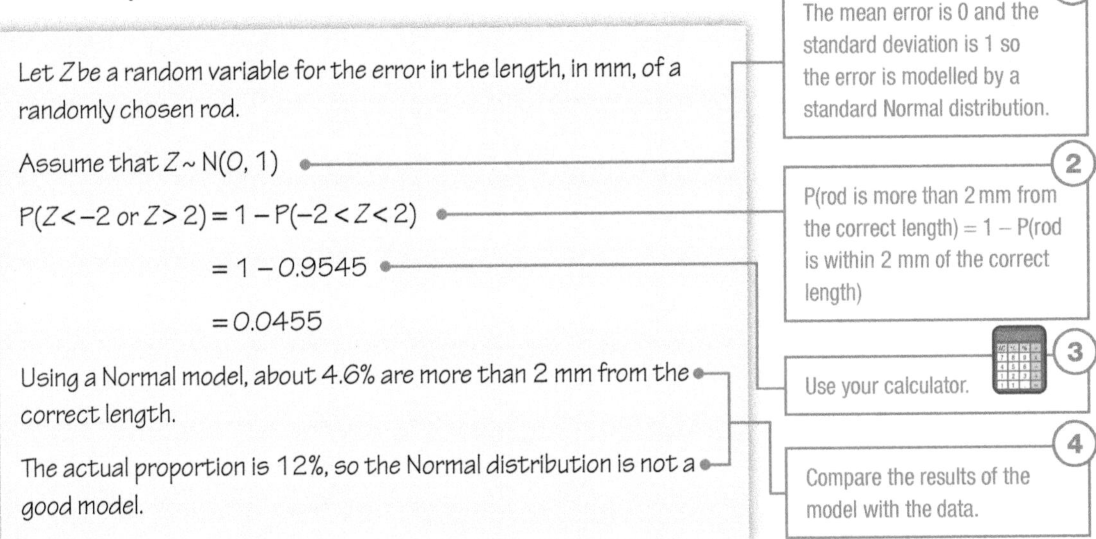

Let Z be a random variable for the error in the length, in mm, of a randomly chosen rod.

Assume that $Z \sim N(0, 1)$

$$P(Z < -2 \text{ or } Z > 2) = 1 - P(-2 < Z < 2)$$
$$= 1 - 0.9545$$
$$= 0.0455$$

Using a Normal model, about 4.6% are more than 2 mm from the correct length.

The actual proportion is 12%, so the Normal distribution is not a good model.

1. The mean error is 0 and the standard deviation is 1 so the error is modelled by a standard Normal distribution.
2. P(rod is more than 2 mm from the correct length) = 1 − P(rod is within 2 mm of the correct length)
3. Use your calculator.
4. Compare the results of the model with the data.

Sometimes you do not know the mean and the variance of a Normal distribution and you need to find them. You can work these out using known probabilities.

Example 4

A council records the waiting times at one of its centres over a one-week period. It finds that 8% of users wait less than 1 minute but 14% wait more than 5 minutes. Council guidelines state that no more than 2.5% of users should wait more than 6 minutes.

a Stating any necessary assumptions, show that this guideline is currently not being met.

b Assuming that the spread of waiting times remains unchanged, find the maximum mean waiting time that would meet the guidelines.

(*Continued on next page.*)

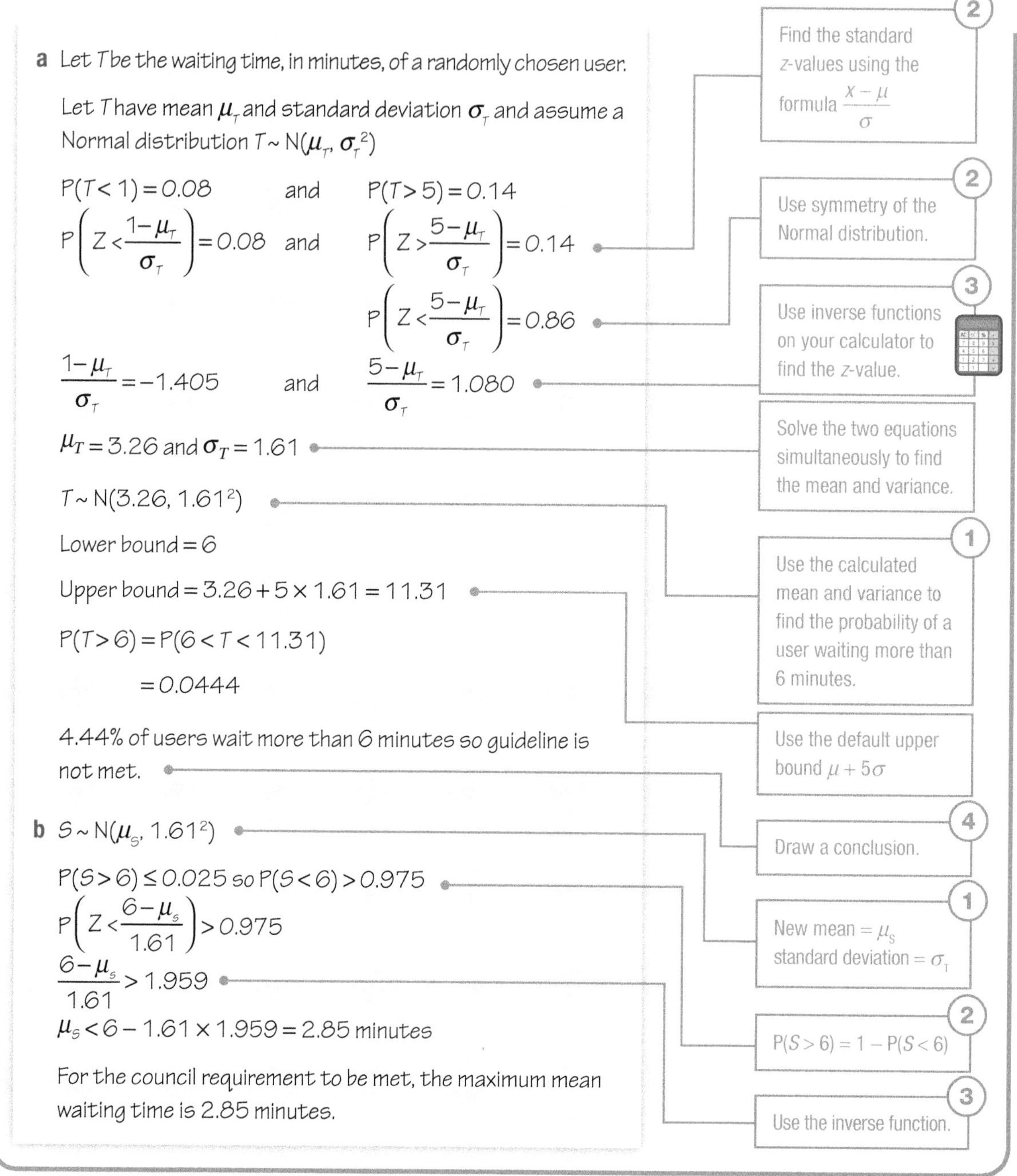

Find the standard z-values using the formula $\frac{x-\mu}{\sigma}$
Use symmetry of the Normal distribution.
Use inverse functions on your calculator to find the z-value.
Solve the two equations simultaneously to find the mean and variance.
Use the calculated mean and variance to find the probability of a user waiting more than 6 minutes.
Use the default upper bound $\mu + 5\sigma$
Draw a conclusion.
New mean $= \mu_S$ standard deviation $= \sigma_T$
$P(S > 6) = 1 - P(S < 6)$
Use the inverse function.

a Let T be the waiting time, in minutes, of a randomly chosen user.

Let T have mean μ_T and standard deviation σ_T and assume a Normal distribution $T \sim N(\mu_T, \sigma_T^2)$

$P(T < 1) = 0.08$ and $P(T > 5) = 0.14$

$P\left(Z < \frac{1-\mu_T}{\sigma_T}\right) = 0.08$ and $P\left(Z > \frac{5-\mu_T}{\sigma_T}\right) = 0.14$

$P\left(Z < \frac{5-\mu_T}{\sigma_T}\right) = 0.86$

$\frac{1-\mu_T}{\sigma_T} = -1.405$ and $\frac{5-\mu_T}{\sigma_T} = 1.080$

$\mu_T = 3.26$ and $\sigma_T = 1.61$

$T \sim N(3.26, 1.61^2)$

Lower bound $= 6$

Upper bound $= 3.26 + 5 \times 1.61 = 11.31$

$P(T > 6) = P(6 < T < 11.31)$

$= 0.0444$

4.44% of users wait more than 6 minutes so guideline is not met.

b $S \sim N(\mu_S, 1.61^2)$

$P(S > 6) \le 0.025$ so $P(S < 6) > 0.975$

$P\left(Z < \frac{6-\mu_S}{1.61}\right) > 0.975$

$\frac{6-\mu_S}{1.61} > 1.959$

$\mu_S < 6 - 1.61 \times 1.959 = 2.85$ minutes

For the council requirement to be met, the maximum mean waiting time is 2.85 minutes.

Exercise 20.3B Reasoning and problem-solving

1 Find the unknown parameters in each distribution.

a $R \sim N(\mu_R, 9)$ given $P(R < 15) = 0.7$

b $S \sim N(8, \sigma_S^2)$ given $P(S < 6) = 0.4$

c $T \sim N(\mu_T, \sigma_T^2)$ given $P(T < 15) = 0.7$ and $P(T < 18) = 0.9$

2 a Over several years, a school's cross-country running event was known to be completed in a mean time of 12 minutes 10 seconds with a standard deviation of 1 minute 20 seconds. One year 34 runners took part and a commendation was given to any runner who ran the course in less than 11 minutes. Estimate the number of runners receiving the commendation. State any distributional assumptions made.

b In the same event the following year, organisers wanted to specify a minimum time required to achieve a special commendation which would be awarded to the fastest 10% of runners. What should that time be? Give your answer to the nearest second.

3 a Show that, for a Normal population, approximately 95% of observations are within 2 standard deviations of the mean.

b In a reaction time test, 100 individuals were timed responding to a visual stimulus. The mean, median and modal times were 275 ms, 265 ms and 270 ms, and 93% of observations were within 2 standard deviations of the mean.

Is a Normal distribution a good probability model for these results? Give reasons for your answer.

4 The masses of 28 birds' eggs, in grams, are shown in the histogram.

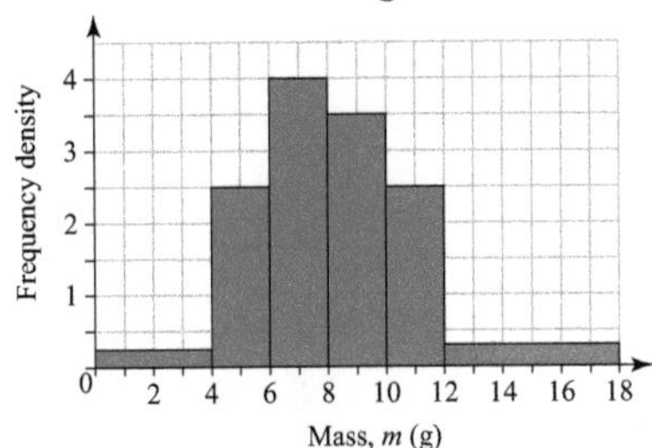

a Estimate the mean and standard deviation of this sample.

b Use linear interpolation to find the proportion of observations within 1 standard deviation of the mean.

c Comment on the distribution of egg weights in the light of your answers to parts **a** and **b**.

5 Following the installation of a new passenger scanner at an airport, the standard deviation of passenger waiting times remains at 3.4 minutes but the mean is increased from 8. If the probability of a passenger waiting less than 8 minutes is now 0.35, find the new mean waiting time.

6 Two mills produce bags of flour. Mill A produces bags with mass, X kg, $X \sim N(1.2, 0.05^2)$ Mill B produces bags with mass, Y kg, $Y \sim N(1.3, 0.1^2)$

a i Calculate the probability that a randomly chosen bag from Mill A has mass more than 1.25 kg

ii Calculate the probability that a randomly chosen bag from Mill B has mass more than 1.4 kg

iii Show that, for $W \sim N(\mu, \sigma^2)$, the probability of W taking a value more than one standard deviation above the mean is 0.1587

iv Show that, for $W \sim N(\mu, \sigma^2)$, the probability of W taking a value more than n standard deviations below the mean is $P(Z < -n)$

b The two mills are equally likely to produce a bag of mass less than a kg. Find a.

7 The average amount of skimmed milk purchased per person per week in Town A in 2012, X, follows the probability distribution $X \sim N(1279, 135^2)$ where values are in ml.

Find the probability that

a A randomly chosen person from this population bought less than 1 litre of skimmed milk in a given week.

b Three randomly chosen people all bought more than 1 litre of skimmed milk in a given week.

Challenge

8 A waste disposal company recorded waiting times over a one-week period at one of its centres. The results showed that 14% of users waited more than 5 minutes and that 8% of users waited less than 1 minute. The company requirement is that no more than 2.5% of users will wait more than 6 minutes.

a Making any necessary distributional assumptions, show that this requirement is currently not being met.

b Assuming that the spread of waiting times remains unchanged, find the maximum mean waiting time allowed to meet the requirement.

20.4 Using the Normal distribution as an approximation to the binomial

Fluency and skills

See Ch 10.2 For a reminder on the binomial distribution.

The binomial distribution models situations where a random variable takes only discrete values. The Normal distribution models continuous variables. If n is large enough (usually taken as $n > 30$ if p is roughly 0.5), you can use a Normal distribution to approximate a binomial distribution.

The random variables P, Q and R, with distributions B(10, 0.2), B(15, 0.2) and B(100, 0.2) respectively, are shown as probability histograms. As the number of trials, n, increases, the shape of the distribution becomes increasingly symmetric about its mean value and increasingly resembles a Normal distribution.

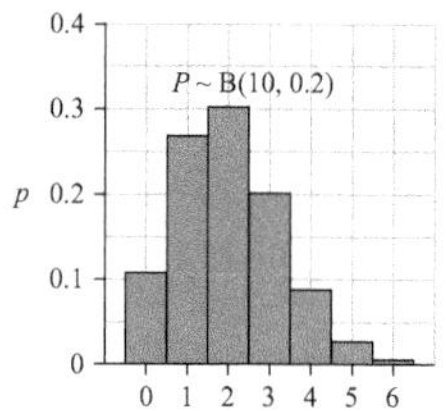

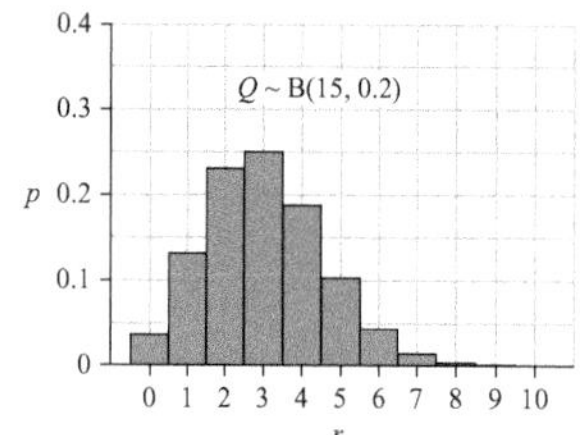

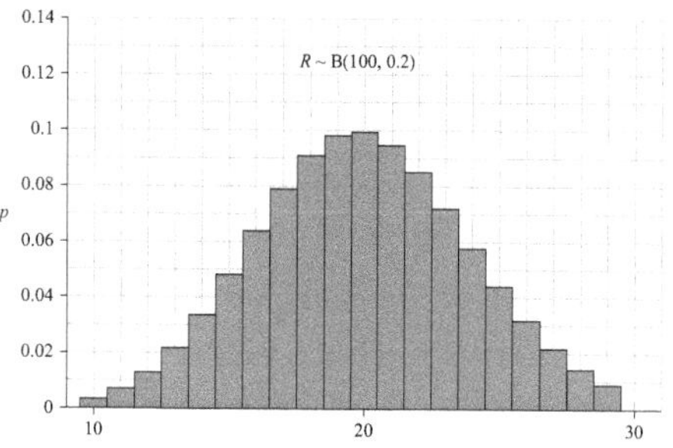

Key point

For $X \sim B(n, p)$, as n increases, the distribution of X tends to that of the random variable Y where $Y \sim N(np, np(1-p))$

The diagram shows the binomial distribution, $n = 30$, $p = 0.5$ in red. The blue curve shows the Normal distribution, which is used as an approximation. The area under the curve most closely approximating $P(X = 18)$ is between 17.5 and 18.5, and the area most closely approximating $P(X \le 18)$ is to the left of 18.5

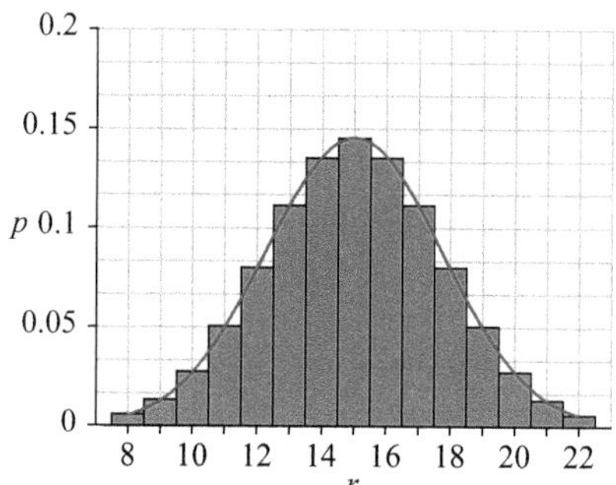

If n is large and p is close to 0.5 then $X \sim B(n, p)$ can be approximated by $Y \sim N(np, np(1-p))$

$$P(X = x) \approx P\left(x - \frac{1}{2} < Y < x + \frac{1}{2}\right)$$

Inclusion of the $\frac{1}{2}$ increases the accuracy of the approximation. It is known as a **continuity correction**. You should always use it when approximating a discrete distribution by a continuous distribution.

Continuity corrections are beyond the scope of this course, but are helpful in understanding how the approximation works.

Example 1

A data set is shown in the histogram

a Sketch a Normal distribution curve to fit the data.

b Identify the x-coordinates of the points of inflection on the curve.

c Use the symmetry of the distribution and your answer to part **b** to estimate the mean of the distribution.

d Use your answers to parts **b** and **c** to estimate the standard deviation of the distribution.

(Continued on next page.)

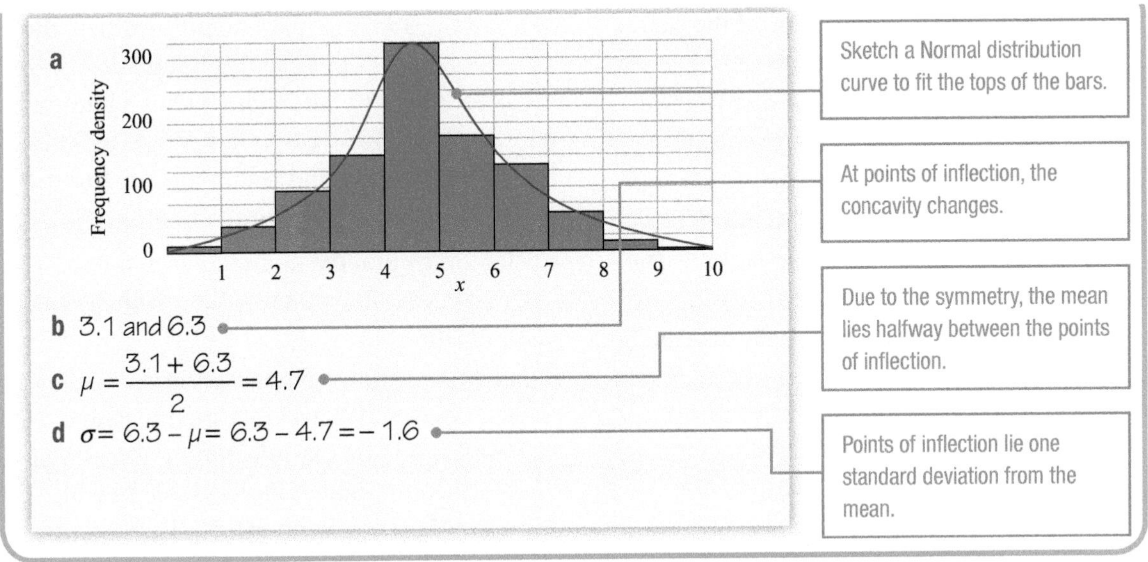

a
Frequency density
300
200
100
0
1 2 3 4 5 6 7 8 9 10
x
Sketch a Normal distribution curve to fit the tops of the bars.
At points of inflection, the concavity changes.
Due to the symmetry, the mean lies halfway between the points of inflection.
Points of inflection lie one standard deviation from the mean.
b 3.1 and 6.3
c $\mu = \frac{3.1 + 6.3}{2} = 4.7$
d $\sigma = 6.3 - \mu = 6.3 - 4.7 = -1.6$

The Normal distribution can be used to find p, the probability of success, in a binomial distribution.

Example 2

Waiting times at an airport security gate are Normally distributed with mean 10.2 minutes and standard deviation 3.4 minutes. Out of a randomly chosen 12 passengers, find the probability that more than five wait for less than 8 minutes.

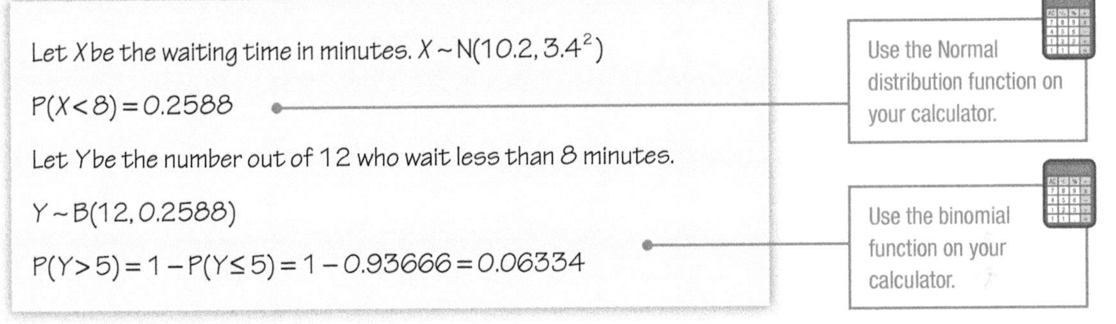

Let X be the waiting time in minutes. $X \sim N(10.2, 3.4^2)$
$P(X < 8) = 0.2588$
Let Y be the number out of 12 who wait less than 8 minutes.
$Y \sim B(12, 0.2588)$
$P(Y > 5) = 1 - P(Y \le 5) = 1 - 0.93666 = 0.06334$
Use the Normal distribution function on your calculator.
Use the binomial function on your calculator.

Exercise 20.4A Fluency and skills

1 The discrete random variable $X \sim B(60, 0.45)$ can be approximated by the continuous random variable $Y \sim N(27, 14.85)$

  i Apply a continuity correction to write down the equivalent probability statement for Y.

  ii Show that the two distributions yield roughly the same probability in each case.

  a $P(X = 23)$ b $P(X \le 50)$

  c $P(X \ge 12)$ d $P(32 \le X \le 51)$

  e $P(X > 17)$ f $P(X < 40)$

2 A student has used a Normal curve to model data given in a histogram.

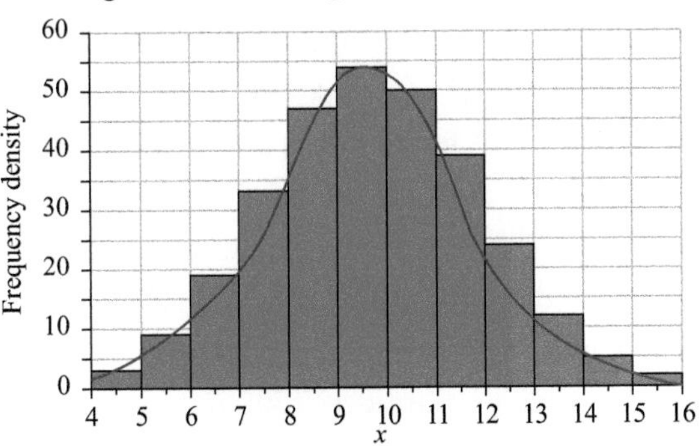

a Identify the x-coordinates of the points of inflection on the curve.

b Use the symmetry of the distribution and your answer to part **a** to estimate the mean of the distribution.

c Use your answers to parts **a** and **b** to estimate the standard deviation of the distribution.

3 A random variable X is Normally distributed with mean 26 and standard deviation 6. An independent random sample of size 6 is taken from the population. Find the probability that more than 3 of the observations are greater than 24

4 A plane carries 300 passengers. Usually, one third of passengers order a hot meal, and on a particular flight 115 meals are packed. Two flight attendants estimate the probability of not having enough meals to meet demand. Attendant A uses a binomial distribution and Attendant B uses a Normal approximation to this binomial distribution.

a Justify the use of both distributions.

b Show that the attendants get the same answer to 2 dp but different answers to 3 dp.

5 The mass shown on packets of red lentils is 1 kg. To satisfy weights and measures legislation, the manufacturer ensures that the mean weight of bags is 1.003 kg with a standard deviation of 0.004 kg. Find the probability that, out of 8 bags checked, less than a quarter of them are under 1 kg.

Reasoning and problem-solving

Strategy

To solve a probability problem by modelling with the binomial and Normal distributions

1. Use an appropriate approximating distribution if the necessary conditions are met.
2. For a binomial problem, the probability of success, p, may be found using a Normal distribution.
3. Use known data to check the validity of any model used.

If you are not told what distribution to use, you must make assumptions about what is appropriate and test against the given data.

Example 3

A machine pours mineral water into bottles. The bottles are labelled '500 ml' and the machine is set so that the mean volume of water in the bottles is 505 ml with a standard deviation of 7 ml.

a Assuming a Normal distribution for volumes, find the probability that a bottle chosen at random from the output will contain within 2 ml of 500 ml.

b Quality inspectors take samples of 5 consecutive bottles every 15 minutes for 10 hours and record how many bottles, Y, in each batch of 5 contain within 2 ml of 500 ml. The results are shown in the table.

Y	0	1	2	3	4	5
f	0	2	8	12	12	6

Find the mean and variance of Y from this data.

c Assuming a binomial model is appropriate for Y, does the result from part **b** suggest the Normal distribution $N(505, 7^2)$ is a suitable model for the distribution of volumes?

a Let X be the volume of a randomly chosen bottle. $X \sim N(505, 7^2)$

$P(503 < X < 507) = 0.2249$

1 Assume Normal distribution to find p

(Continued on next page.)

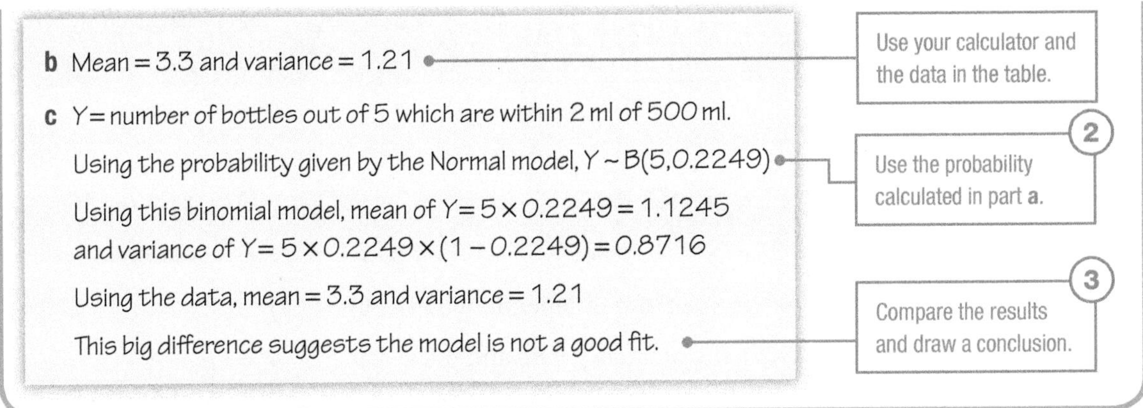
b Mean = 3.3 and variance = 1.21

Use your calculator and the data in the table.

c Y = number of bottles out of 5 which are within 2 ml of 500 ml.

Using the probability given by the Normal model, $Y \sim B(5, 0.2249)$

2 Use the probability calculated in part **a**.

Using this binomial model, mean of $Y = 5 \times 0.2249 = 1.1245$
and variance of $Y = 5 \times 0.2249 \times (1 - 0.2249) = 0.8716$

Using the data, mean = 3.3 and variance = 1.21

This big difference suggests the model is not a good fit.

3 Compare the results and draw a conclusion.

Exercise 20.4B Reasoning and problem-solving

1 A large number of batteries has a mean lifetime of 31 hours and a standard deviation of 21 hours. Half of all lifetimes are less than 20. Give two reasons why a Normal distribution would not provide a good probability model for this population.

2 $X \sim B(5, 0.05)$

Ava says X can be modelled by a continuous random variable $Y \sim N(0.25, 0.2375^2)$

Comment on the suitability of Ava's model. Consider the key features of the Normal distribution.

3 **a** A coin shows 7 heads in 10 tosses. Does this suggest the coin is biased?

b The same coin is tossed 10 000 times and a head appears 5450 times. By finding the probability of obtaining at least 5100 heads, decide whether this new evidence changes your view of the coin.

Large data set

4 **a** The Large data set gives no data for the quantities of school milk consumed and incomplete data for welfare milk consumed between 2001 and 2014. Explain why this is the case.

b The mean amount of milk (wholemilk plus skimmed milk) purchased per person per week in Yorkshire and the Humber in 2012 was 1625 ml.

i Explain why the sum of whole plus skimmed milk is more likely to be Normally distributed than either whole or skimmed milk separately.

ii The amount of milk bought per person per week in Town A in Yorkshire in 2012, X, follows the distribution $X \sim N(1625, 245^2)$. Find the probability that a randomly chosen person from this town bought more than 1800 ml milk in a randomly chosen week in 2012.

iii Find the probability that, out of 200 people in Town A, more than 50 purchased more than 1800 ml milk in a randomly chosen week in 2012.

5 The vegetable content of mushroom pies has $\mu = 98$ g, $\sigma = 10$ g. 10 pies are tested every hour for 24 hours and the number of pies with a vegetable content over 98 g, X, is noted. The mean and standard deviation of these X values are 3.1 and 2.1. Does the Normal distribution provide a good model for the vegetable content of the pies? Give reasons for your answer.

Challenge

6 Airlines sometimes sell more tickets than the number of seats available for the flight. For a particular airline, on average 6% of passengers fail to turn up for their flight. For a flight with 300 seats available, show that tickets sold should not exceed 309 if the probability that the flight cannot accommodate all passengers who turn up has to be less than 1%.

STATS

20 Summary and review

Chapter summary

- $P(A|B) = \dfrac{P(A \cap B)}{P(B)}$, and events A and B are independent if $P(A|B) = P(A)$
- For the distribution $X \sim B(n, p)$, the mean is $\mu = np$ and the variance is $\sigma^2 = np(1-p)$
- Normal probability density functions are symmetrical about a clearly defined central mean value so $P(X > \mu + a) = P(X < \mu - a)$ for any constant, a.
- A Normal distribution curve has points of inflection one standard deviation away from the mean.
- For a Normal distribution curve
    - Total area under the curve = 1
    - ~68% of values lie within σ of the mean
    - ~99.8% of values lie within 3σ of the mean
- If $X \sim N(\mu, \sigma^2)$, then the variable, Z, given by $Z = \dfrac{X-\mu}{\sigma}$, is also Normal with mean 0, variance 1
- $X \sim B(n, p)$ can be approximated by a Normal distribution $Y \sim N(np, np(1-p))$ when n is large.

Check and review

You should now be able to...	Try Questions
✔ Calculate conditional probabilities from data given in different forms.	1–3
✔ Apply binomial and Normal probability models in different circumstances.	4–6
✔ Use data to assess the validity of probability models.	4, 7
✔ Solve problems involving both binomial and Normal distributions.	8

1 You choose a card from a set of thirty cards numbered 1–30. If the card shows a multiple of 4 you flip a fair coin. If it is not a multiple of 4 you flip a coin for which the probability of a heads is $\frac{2}{3}$. Find the probability of obtaining

a A heads given the score was a multiple of 4

b A heads and not a multiple of 4

2 In a survey of travel arrangements to school, 40 students are questioned about their means of transport. Students who don't walk the whole distance either take the bus or cycle or do both. Of those who don't walk the whole distance, 10 take only the bus, 15 cycle at some point on their journey and 12 cycle the whole way.

a Show the results on a Venn diagram.

b Find the probability that a randomly chosen student

i Both cycles and takes the bus,

ii Takes the bus at some point given that they don't walk.

3 You have two fair dice. Dice 1 is 6-sided and numbered 1–6. Dice 2 is 10-sided and numbered 1–10. You choose a dice at random and roll it. What is the probability that dice 1 was chosen given that the score was 6?

4 A spinner has four sides coloured red, blue, green and yellow. You believe that the outcomes are equally likely. In an experiment, you spin it four times and record the number of times you get yellow.

a Copy and complete the following table with the theoretical probability of each outcome.

x	0	1	2	3	4
$P(X=x)$	0.3164	0.4219			

The experiment is repeated 80 times giving these results.

x	0	1	2	3	4
f	22	34	17	6	1

b Do these results support your belief that the spinner is fair? Give reasons for your answer.

5 a Given $X \sim N(12, 3)$, calculate

i $P(X<10)$ ii $P(9<X<14)$

b The random variable, X, has mean 19 and variance 8. Find the value of a given that $P(a<X<22)=0.65$

6 In a packing warehouse, peppers have weights that are Normally distributed with mean 150 g and standard deviation 12 g. Find the probability that a randomly chosen pepper will have a weight between 140 g and 170 g.

7 At a health spa, 75 people were timed to complete a fitness test. The mean and modal times were 281 s and 288 s respectively. Half of the observations were less than 279 s and 69% were within one standard deviation of the mean. Would a Normal distribution be a good probability model for this data? Give reasons for your answer.

8 In a call centre, waiting times are Normally distributed with mean 8.2 minutes and standard deviation 2.7 minutes. Out of a randomly chosen 18 callers, find the probability that fewer than 4 wait for more than 10 minutes.

STATS

What next?

Score		
0–4	Your knowledge of this topic is still developing. To improve, search in MyMaths for the codes: 2092, 2095, 2113, 2120–2121, 2286, 2292	
5–6	You're gaining a secure knowledge of this topic. To improve, look at the InvisiPen videos for Fluency and skills (20A)	
7–8	You've mastered these skills. Well done, you're ready to progress! To develop your techniques, look at the InvisiPen videos for Reasoning and Problem-solving (20B)	

Click these links in the digital book

Exploration

Going beyond the exams

History

Thomas Bayes (1701 – 1761) was an English statistician and Presbyterian minister. He published two books during his lifetime, but it was not until after his death that his most famous work, now referred to as **Bayes' theorem**, was published.

Bayes' theorem provides a formula for calculating the probability of an event based on related, pre-existing conditions. It can be written as

$$p(A|B) = \frac{p(B|A)\,p(A)}{p(B)}$$

$p(A|B)$ is the probability that A occurs given that B is true.
$p(B|A)$ is the probability that B occurs given that A is true.

Investigation

The **Monty Hall problem** is a probability question based on the idea of an American game show, of which Monty Hall was the presenter. This problem became famous when it appeared in the American magazine, 'Parade'.

Suppose that you are presented with three doors. Behind one is a car and behind each of the others is a goat.

You pick a door at random. The presenter, knowing what's behind each door, opens a different door to reveal a goat. You are then offered the chance to switch your choice to the remaining door.

Write a short description of whether you would switch, and why. Note, there is a correct answer.

Note

The solution to the Monty Hall problem, given by columnist **Marilyn vos Savant**, provoked a huge response from readers, including many with PhDs, proclaiming that she was wrong.

Marilyn had the highest recorded IQ according to the Guinness Book of Records. Unsurprisingly, she was right!

Research

Joseph Bertrand proposed a similar problem to the Monty Hall problem in 1889, known as **Bertrand's Box Paradox**.

Have a go at solving it. Try using **Kolmogorov's axioms**.

20 Assessment

1 For two events X and Y, it is know that $P(X) = 0.4$, $P(Y) = 0.5$ and $P(Y|X) = 0.3$

Calculate $P(X \cup Y)$. Select the correct answer. **[1 mark]**

A 0.7 B 0 C 0.2 D 0.78

2 For $X \sim N(12, 16)$ find $P(X > 15)$. Select the correct answer. **[1]**

A 0.7734 B 0.4256 C 0.2266 D 0.1797

3 For two events A and B, it is known that $P(A) = 0.3$, $P(B|A) = 0.4$ and $P(B' | A') = 0.35$

a Represent this information on a tree diagram showing all the individual probabilities. **[6]**

b Use your tree diagram to calculate $P(B)$ **[3]**

c Represent the information on a Venn diagram. **[6]**

4 Data are collected each year between 2001 and 2014 on how much cheese and how much sugar and preserves are purchased by people in England on average per person per week.

Large data set

	> 115 g Cheese	**≤ 115 g Cheese**	**Totals**
> 130 g Sugar & Preserves	2	3	5
≤ 130 g Sugar & Preserves	7	2	9
Totals	9	5	14

a Calculate the probability that, for a randomly-chosen year in the sample

i The average weekly purchase of cheese is over 115 g

ii The average amount of sugar and preserves purchased exceeds 130 g given that the average amount of cheese purchased exceeds 115 g

iii The average amount of sugar and preserves purchased exceeds 130 g

iv The average amount of cheese purchased exceeds 115 g given that the average amount of sugar and preserves purchased exceeds 130 g **[6]**

b If a new year is considered and the average amount of cheese purchased exceeds 115 g, would you assume that the average amount of sugar and preserves purchased exceeds 130 g? **[1]**

5 Between 05:30 and 22:30 inclusive, 170 number 1 buses arrive at a given bus stop.

a Calculate the average length of time between bus arrivals. **[3]**

b Assuming the buses arrive at regular intervals and never run late, what is the probability that a bus arrives between 08:30 and 08:40? **[2]**

c Discuss whether your answer to part **b** is reasonable for a real bus service, and state any assumptions that are likely to be wrong. **[1]**

6 For each of the following Normal distributions, calculate the probabilities to three significant figures.

a For $X \sim N(4,3)$ find $P(X < 2.3)$ **[2]**

b For $X \sim N(-4,21)$ find $P(X > -0.5)$ **[2]**

c For $X \sim N(17,4)$ find $P(X = 16)$ **[1]**

7 For $X \sim \mathrm{N}(4,4)$, $\mathrm{P}(X<2)=0.15866$ and $\mathrm{P}(X>7)=0.066807$ to five significant figures.

a Use this information to calculate $\mathrm{P}(2<X<7)$ to three significant figures. **[3]**

b Express the following probabilities in terms of X and use the information above to calculate their values to 3 significant figures.

i $\mathrm{P}(4<Y)$, where $Y \sim \mathrm{N}(8, 16)$

ii $\mathrm{P}(Z<-1)$, where $Z \sim \mathrm{N}(0, 1)$

iii $\mathrm{P}(W<-12 \text{ or } W>-2)$, where $W \sim \mathrm{N}(-8, 16)$ **[7]**

8 The distance an amateur archer lands an arrow from the centre of the target is modelled by a Normal distribution, with mean 0 cm and variance 100 cm^2

a Find the probability that the arrow lands within 3 cm of centre of the target. **[1]**

b The archer shoots ten arrows, one after the other. Assuming the arrows are shot independently, find the probability that at least three arrows land within 3 cm of the centre. **[3]**

c Is it reasonable to expect that the arrows have independent probabilities of landing within 3 cm of the centre? **[1]**

d In each round of a competition, the archers need to land three out of ten arrows within 3 cm of the centre to score a point for that round. The archer who scores the most points over five rounds wins. Assuming each round is independent of the others, find the expected number of points the archer in this question will score. **[2]**

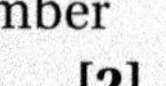

9 50 bags of flour are weighed and their masses are recorded in a histogram.

a Calculate estimates for the mean and variance of the data to 2 decimal places. **[4]**

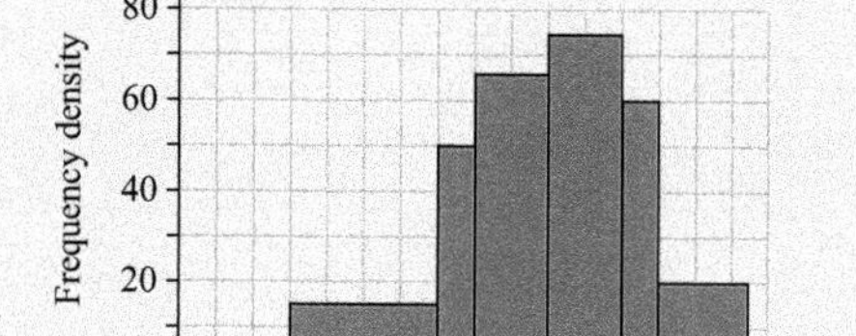

b Let X be the mass of a randomly chosen bag. Show that X can be modelled by a Normal distribution. **[5]**

c Use the Normal model from part **b** to calculate

i $\mathrm{P}(X<1)$ ii $\mathrm{P}(X\geq 0.7)$ iii $\mathrm{P}(X\leq 1.36)$ **[3]**

d Use the Normal model from part **b** to write the interval, centred at the mean, that 99.8% of the data lies in. **[1]**

10 A casino uses a dice testing machine to ensure a dice is rolling the right number of sixes.

a Assuming the dice is fair

i Calculate μ, the expected number of sixes in ten rolls,

ii Calculate the variance in the number of sixes over ten rolls,

iii Calculate $\mu+\frac{\sqrt{10}}{2}$ and $\mu-\frac{\sqrt{10}}{2}$

iv Find the probability that in ten rolls the number of sixes rolled is within $\frac{\sqrt{10}}{2}$ of the expected number. **[6]**

b If the dice is rolled n times, where n is a large number, state a suitable approximate distribution for the number of sixes rolled. **[2]**

c Calculate the probability that over n rolls, the number of sixes rolled is within $\frac{\sqrt{n}}{2}$ of the expected number, for

i $n=900$ ii $n=10000$ iii $n=1000000$ **[9]**

21 Hypothesis testing 2

Imagine a gymnast who wants to improve her success rate at achieving a score of over 12 for her beam routine. She spends three hours every day for a month practising this. Would you expect the number of points she scores for each performance to increase as the month progresses? This is a simple example of correlation: you would expect a clear link between the number of hours of practice and the number of points scored.

Correlation testing is an important part of statistics. While it does not prove causation, the ability to calculate the correlation between two variables helps us determine the likelihood of one having an influence on the other, particularly when the correlation is not as obvious as in the example above. It is a facet of hypothesis testing, along with techniques such as testing for a Normal distribution, and a vital part of statistics and research in many subject areas.

Orientation

What you need to know

Ch9 Collecting, representing and interpreting data

- Correlation and causation.

Ch11 Hypothesis testing 1

- Significance and p-value.

p.268

Ch20 Probability and continuous random variables

- Introduction to the Normal distribution.

What you will learn

- To state the null and alternative hypothesis when testing for correlation or when testing the mean of a Normal distribution.
- To compare a given PMCC to a critical value, or its p-value to a significance level, and use this to accept or reject the null hypothesis.
- To calculate the test statistic, compare to a critical value, or its p-value to a significance level, and use this to accept or reject the null hypothesis.
- To decide what the conclusion means in context about the correlation and about the mean of the distribution.

What this leads to

Applications
Scientific research.
Quality control.
Test engineering.

MyMaths Practise before you start 2115, 2120, 2283

21.1 Testing correlation

Fluency and skills

To find out how correlated two variables are, you need a way to test the strength of the **correlation**.

The population correlation coefficient takes values between −1 and 1. 0 indicates zero correlation, 1 and −1 indicate perfect correlation, so the points plotted on a scatter diagram lie on a straight line, with positive and negative gradients respectively.

See Ch9.4 For a reminder on correlation and causation.

Key point

The **population correlation coefficient**, ρ, describes how correlated two variables are.

Pearson's product moment correlation coefficient (PMCC), r, is a statistic that estimates ρ

The PMCC is a measure of correlation in a sample, and is used to estimate the population correlation coefficient. The estimate becomes better as the sample size increases, but it is likely to differ from the true value.

See Ch11.1 For a reminder on hypothesis testing.

In a hypothesis test, the **null hypothesis** states that the two variables have no correlation, i.e. H_0: $\rho = 0$. To start with, this is assumed to be true.

The **alternative hypothesis** is

- $H_1 : \rho \neq 0$ if you think there could be any kind of correlation,
- $H_1 : \rho > 0$ if you think there could be positive correlation,
- $H_1 : \rho < 0$ if you think there could be negative correlation.

$H_1 : \rho \neq 0$ is a two-tailed test.

$H_1 : \rho > 0$ or $H_1 : \rho < 0$ is a one-tailed test.

Key point

If the PMCC, r, is further from zero than the **critical value** then you have sufficient evidence to reject the null hypothesis and decide that $\rho \neq 0$.

Example 1

An art auction house measures the size and price of 10 portraits sold during one month, selected at random, to see if there is any correlation between the two. The null hypothesis is $H_0 : \rho = 0$ and the alternative hypothesis is $H_1 : \rho \neq 0$ with a 5% significance level.

a Explain why the alternative hypothesis takes this form.

The product moment correlation coefficient for the data is $r = 0.532$ and the critical value for this test is ± 0.6319

b State, with a reason, whether H_0 is accepted or rejected.

a The auction house is investigating to see if there is any correlation present. This is a two-tailed test.

'Any correlation' means there could be positive or negative correlation. A two-tailed test considers the possibility that the correlation coefficient is positive and the possibility that it is negative.

b Since the PMCC for the sample is closer to zero than the critical value is, there is insufficient evidence to reject H_0

The p-value is the probability of getting a result at least as extreme as the one obtained from the sample if the variables are not correlated.

Key point

Instead of using the critical value, you can compare the **p-value** for the PMCC to the significance level. You accept the null hypothesis if the p-value is greater than the significance level.

To investigate correlation using hypothesis tests, click this link in the digital book.

See Ch11.2
For a reminder on p-values.

Example 2

The number of hours spent training for a marathon and the number of hours taken to complete a marathon for a random sample of 30 marathon entrants are suspected to have a negative correlation. The hypotheses $H_0 : \rho = 0$ and $H_1 : \rho < 0$ are being considered at the 10% significance level. The PMCC for the sample is –0.3061, which has a p-value of 0.05 for a one-tailed test. State, with a reason, whether H_0 is accepted or rejected.

The p-value is 5%. Since 5% is less than 10%, the result is significant and so the null hypothesis is rejected.

If $\rho = 0$, the probability of this PMCC is less than the significance level.

STATS

Exercise 21.1A Fluency and skills

1 The average annual temperature and rainfall in 11 UK towns are measured to investigate whether there is any correlation between the two. The hypotheses $H_0 : \rho = 0$ and $H_1 : \rho \neq 0$ are being considered at the 5% significance level. The test statistic for the sample is 0.6137 which has a p-value of 0.0446 for a two-tailed test. State, with a reason, whether H_0 is accepted or rejected.

2 A jewellery auction house measures the masses and selling prices of a random sample of 10 pieces to see if there is any correlation between the two. The null hypothesis is $H_0 : \rho = 0$ and the alternative hypothesis is $H_1 : \rho \neq 0$ with a 5% significance level.

  a Explain why the alternative hypothesis takes this form.

  The product moment correlation coefficient for the data is $r = -0.148$. The critical value for this test is ± 0.6319

  b State, with a reason, whether H_0 is accepted or rejected.

3 The number of caffeinated drinks a person consumes during a day and the number of hours of sleep they get that night are suspected to have a negative correlation. A random sample of 30 people is surveyed to investigate whether any correlation is present. The hypotheses $H_0 : \rho = 0$ and $H_1 : \rho < 0$ are being considered at the 10% significance level. The critical value for the test is –0.2826 and the PMCC for the sample is $r = -0.837$. State, with a reason, whether H_0 is accepted or rejected.

4 The lengths and circumferences of 26 carrots are measured to see if there is any correlation between these properties. The hypotheses $H_0 : \rho = 0$ and $H_1 : \rho \neq 0$ are being considered at the 5% significance level. The PMCC of the sample is 0.189, which has a p-value of 0.355 for a two-tailed test. State, with a reason, whether H_0 is accepted or rejected.

5 A sample is taken across 40 towns to see if limiting alcohol sales after different times in the evening helps to reduce crime levels. The hypotheses $H_0 : \rho = 0$ and $H_1 : \rho < 0$ are tested at the 5% level where ρ measures the correlation between the number of hours before midnight that alcohol is limited and the number of crimes committed that night in the area. The sample is found to have a PMCC of −0.2935. Given that the critical value is −0.2605, state, with a reason, whether H_0 is accepted or rejected.

Reasoning and problem-solving

Strategy

To solve problems involving correlation

1. Conduct a test to decide if there is sufficient evidence to reject the null hypothesis.
2. Reject or accept the null hypothesis.
3. If the null hypothesis is rejected, then you conclude there is correlation and, if the null hypothesis is accepted, then you conclude there is no correlation.

Example 3

The average purchased quantities of fully skimmed and semi-skimmed milk in the North East are examined over time to see if there is any correlation. The hypotheses are being considered at the 5% significance level. 12 measurements are taken between 2003 and 2014 and the PMCC is found to be 0.6581, which has a p-value of 0.02 for a two-tailed test. State the hypotheses of the test and determine the conclusion in context.

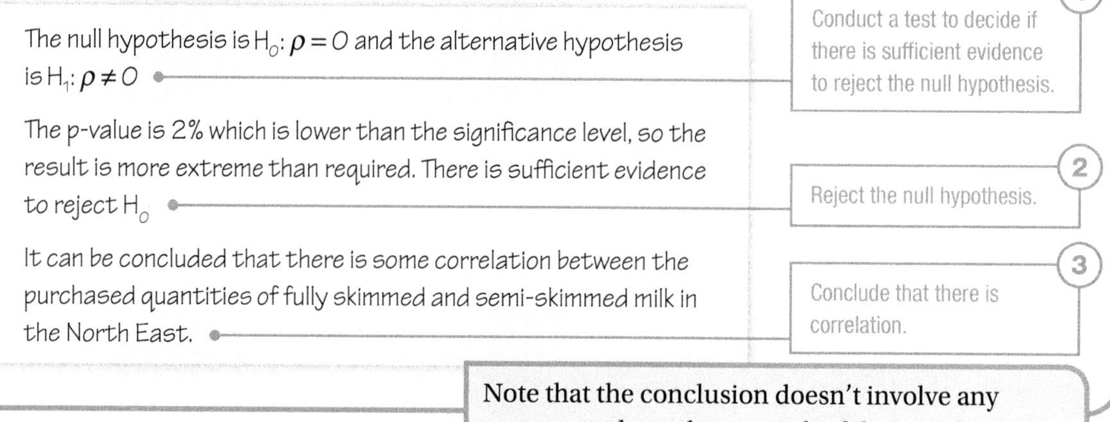

The null hypothesis is $H_0: \rho = 0$ and the alternative hypothesis is $H_1: \rho \neq 0$

(1) Conduct a test to decide if there is sufficient evidence to reject the null hypothesis.

The p-value is 2% which is lower than the significance level, so the result is more extreme than required. There is sufficient evidence to reject H_0

(2) Reject the null hypothesis.

It can be concluded that there is some correlation between the purchased quantities of fully skimmed and semi-skimmed milk in the North East.

(3) Conclude that there is correlation.

Note that the conclusion doesn't involve any statement about the strength of the correlation.

Example 4

The price of the set menu at a restaurant and the number of courses included are tested by a food blogger for positive correlation. The hypotheses $H_0 : \rho = 0$ and $H_1 : \rho > 0$ are being considered at the 5% significance level. A sample of 17 restaurants is taken and the PMCC is 0.295. The critical value is 0.412. State, with a reason, whether H_0 is accepted or rejected and determine the conclusion in context.

Since the PMCC is less than the critical value there is insufficient evidence to reject the null hypothesis.

(2) Accept the null hypothesis.

It can be concluded that there is no positive correlation between the price of a set menu and the number of courses included.

(3) Conclude that there is no positive correlation.

Exercise 21.1B Reasoning and problem-solving

1 Economists study the relationship between levels of taxation and the amount individuals give to charity to see if there is any correlation. The hypotheses $H_0 : \rho = 0$ and $H_1 : \rho \neq 0$ are considered at the 5% significance level. 24 countries are measured and the PMCC is −0.1601. The critical value for the test is ±0.4044. Explain whether H_0 is accepted or rejected and state your conclusion in context.

2 A teacher measures his students' test scores in Maths and English to see if there is any positive correlation between the two. The hypotheses $H_0 : \rho = 0$ and $H_1 : \rho > 0$ are considered at the 10% significance level. All 28 students' scores are recorded and the PMCC is found to be 0.2516, which has a p-value of 0.0983 for a one-tailed test. State, with a reason, whether H_0 is accepted or rejected and give your conclusion in context.

3 Food scientists investigate 20 different flavours of crisps to see if any show signs of being carcinogenic. For each flavour, the amount eaten is measured against a carcinogen index to check for positive correlation. Each flavour is tested at the 5% significance level.

a What does it mean for the test to have a 5% significance level?

b How many of the 20 flavours are expected to show positive correlation if in fact there is no correlation for any of the flavours?

It is found that exactly one of the 20 flavours has a PMCC which exceeds the critical value.

c What might the scientists conclude?

d Given your answer to part **b**, why would the scientists treat this conclusion suspiciously?

4 The average purchased quantities of infant milks and baby food per person per week in England from 2001–2014 are believed to be strongly positively correlated. The hypotheses H_0: $r = 0$ and H_1: $r > 0$ are considered at the 5% significance level, critical value 0.1216. Measurements are taken over five consecutive years to test this. The PMCC is 0.714. State, with a reason, whether H_0 is accepted or rejected and determine your conclusion in context.

Large data set

5 The amount of milk products (excluding cheese) and the amount of cheese consumed per person per week are investigated to check for correlation. The hypotheses H_0: $\rho = 0$ and H_1: $\rho \neq 0$ are being considered at the 5% significance level. Average amounts from 14 consecutive years in the South East are taken and the test statistic is −0.207. The critical value is ±0.264

a Explain whether H_0 is accepted or rejected.

b Give an example of two regions that suggest this relationship may not hold across the country.

STATS

6 Education specialists want to find out if there is any correlation between the geographical location of a school and the likelihood that a student attending the school will get at least one A at A-level. A random sample of 46 schools is taken to test the hypotheses $H_0 : \rho = 0$ and $H_1 : \rho \neq 0$ at the 5% level. The PMCC is 0.1319, which has a p-value of 0.382. State, with a reason, whether H_0 is accepted or rejected and determine your conclusion in context.

Challenge

7 A, B and C are properties which are tested for positive correlation at the 10% level. The critical value is 0.365

a The PMCC between A and B is 0.6 State, with a reason, whether there is reason to say that A and B are positively correlated.

b The PMCC between B and C is 0.6. State, with a reason, whether there is reason to say that B and C are positively correlated.

c The PMCC between C and A is 0.262. State, with a reason, whether you can say that C and A are positively correlated.

d Discuss how your answers to parts **a**, **b** and **c** relate to one another.

21.2 Testing a Normal distribution

Fluency and skills

See Ch20.3 For an introduction to the Normal distribution.

If a variable is Normally distributed, then you can perform a hypothesis test to identify whether the mean of the sample is sufficiently different to the hypothesised mean of the distribution.

$H_1 : \mu \neq \mu_0$ is a two-tailed test.

$H_1 : \mu < \mu_0$ or $H_1 : \mu > \mu_0$ is a one-tailed test.

The **null hypothesis** is the assertion that the population the sample is taken from has a particular mean. This is usually based on previous results.
You call this hypothesised mean μ_0, so $H_0 : \mu = \mu_0$

The **alternative hypothesis** is

- $H_1 : \mu \neq \mu_0$ if the mean of the sample is different to the hypothesised mean.
- $H_1 : \mu < \mu_0$ or $H_1 : \mu > \mu_0$ if the mean is lower or higher than the hypothesised mean.

If, $X \sim N(\mu, \sigma^2)$, then $\overline{X} \sim N\left(\mu, \frac{\sigma^2}{n}\right)$. To test these types of hypotheses you need a test statistic, z

Key point

$z = \frac{\overline{x} - \mu_0}{\frac{\sigma}{\sqrt{n}}}$, $\overline{x}$ is the mean of the sample, μ_0 is the hypothesised mean of the distribution, σ^2 is the variance of the distribution and n is the sample size.

You can perform a z-test using your calculator.

You can either compare this test statistic to the critical value, or you can find the p-value and compare that to the significance level. These calculations rely on the standardised Normal distribution $\Phi(z) = P(Z \leq z)$ where $Z \sim N(0,1)$

Key point

You accept the null hypothesis if the test statistic is smaller in size than the critical value, or if the p-value for the test statistic is greater than the significance level.

Example 1

A machine at a sausage factory is suspected of being faulty. The lengths of the sausages produced follow a Normal distribution. The variance is 16 cm^2, but the mean may have changed from the intended setting of 12 cm.

a State the null and alternative hypotheses.

A random sample of 25 sausages is taken and found to have a mean length of 13.2 cm. The test is carried out at the 10% significance level.

b Calculate the test statistic.

c Calculate the critical value from the significance level.

d State, with a reason, whether the null hypothesis is accepted or rejected.

The units of variance are cm^2. Standard deviation is often used in place of variance because it has the same units as the variable.

a $H_0 : \mu = 12$ and $H_1 : \mu \neq 12$

Even though the observed mean of the sample is higher than μ, a two-tailed test is used here since there is no reason to expect a shift in a particular direction.

(Continued on the next page)

b The test statistic $z=\frac{13.2-12}{\frac{4}{\sqrt{25}}}=1.5$

Substitute the values of $\bar{x}, \mu_0, \sigma$ and n into the formula for the test statistic.

c $\Phi^{-1}\left(1-\frac{10}{200}\right)=1.645$ so the critical value is ± 1.645

d Since the test statistic is less than the critical value, you accept the null hypothesis.

Compare the test statistic with the critical value.

Exercise 21.2A Fluency and skills

1 A Normal distribution is assumed to have variance 625 but its mean is unknown. A sample of size 49 is taken to investigate the claim that its mean could be 19. State the null and alternative hypotheses for this test.

2 A Normal distribution is known to have variance 196 but has an unknown mean. A sample of size 28 is collected to test whether or not the mean might be 43. State the null and alternative hypotheses for this test.

3 A machine at a spaghetti factory is suspected of being faulty. It is known that the length of each strand of spaghetti can be well-modelled by a Normal distribution with variance 2.25 cm^2, but it is thought that the mean may have changed from the intended setting of 32 cm.

a State the null and alternative hypotheses.

A random sample of 36 spaghetti strands is taken and is found to have a mean length of 31.5 cm. The test is done at the 10% significance level.

b Calculate the test statistic.

c Calculate the critical value at the given significance level.

d State, with a reason, whether the null hypothesis is accepted or rejected.

4 It is known that lengths of genetically-modified celery stalks follow a Normal distribution and the variance is 9 cm^2. It is believed that the mean may have changed from the original design of 25 cm.

a State the null and alternative hypotheses.

A random sample of 32 stalks is taken and is found to have a mean length of 25.9 cm. The test is done at the 10% significance level.

b Calculate the test statistic.

c Calculate the critical value from the significance level.

d State, with a reason, whether the null hypothesis is accepted or rejected.

5 A variable is Normally distributed with variance 196. The mean was known to be 29 at one point, but a sample of 36 is taken to see if the mean has increased. The mean of the sample is 33

a Calculate the test statistic.

b Calculate the p-value of the statistic.

c State, with a reason, whether the null hypothesis is accepted or rejected at the 10% level.

6 The heights of adult men in a large country are well-modelled by a Normal distribution with mean 177 cm and variance 529 cm^2. It is thought that men who live in a poor town may be shorter than those in the general population. The hypotheses $H_0: \mu=177$ and $H_1: \mu<177$ are tested at the 10% significance level with the assumption that the variance of heights is the same in the town as in the general population. A sample of 25 men is taken from the town and their heights are found to have a mean value of 171 cm.

a Calculate the test statistic.

b Calculate the p-value of the statistic.

c State, with a reason, whether the null hypothesis is accepted or rejected.

Reasoning and problem-solving

Strategy

To solve problems involving hypothesis testing

1. Define the null hypothesis and the alternative hypothesis.
2. Calculate the test statistic.
3. Calculate the critical value or the p-value.
4. Accept or reject the null hypothesis.
5. Give your conclusion.

Know your dataset

Large data set

The LDS shows lower purchased quantities of bread in London than in any other region of England. London is an outlier for several different measures in the LDS. Click this link in the digital book to access and explore the Large data set.

Example 2

A machine that makes copper rods is designed to produce rods with a mean length of 30 cm. An inspector is making sure that the mean is not less than this and assumes, from experience with similar machines, that the variance is 0.16 cm^2. The inspector measures the lengths of 32 random rods and finds that they have a mean length of 29.9 cm.

a State null and alternative hypotheses for this test.

b Calculate the test statistic and hence calculate the p-value.

c State, with a reason, whether the null hypothesis is accepted or rejected at the 10% significance level. Determine the conclusion of the hypothesis test.

a $H_0 : \mu = 30 \quad H_1 : \mu < 30$

(1) Since the inspector only wants to test if the mean is significantly lower than μ, you use a one-tailed test.

b $z = \dfrac{29.9 - 30}{\dfrac{\sqrt{0.16}}{\sqrt{32}}} = -1.414$ (4 sf)

The p-value is $1 - \Phi(1.414) = 7.87\%$

(2)(3) Calculate the test statistic and use this to calculate the p-value.

c Since the p-value is less than the significance level, the null hypothesis can be rejected. There is sufficient evidence to suggest that the machine might be producing rods which are too short.

(4)(5) Reject the null hypothesis and give your conclusion.

When testing a mean, you do not say with certainty that the hypothesised mean is incorrect or correct, simply that the evidence suggests the true mean is likely or unlikely to be different.

Example 3

Two scientists conduct an experiment. The results are modelled by a Normal distribution with variance 53.8, but the scientists think that the mean could be lower than the intended 167. They both separately take samples of size 41 to test the hypotheses $H_0 : \mu = 167$ and $H_1 : \mu < 167$

a One scientist tests at the 5% level and their sample has test statistic –2.53. The critical value is –1.64. State, with a reason, whether the null hypothesis is accepted or rejected.

b The other scientist tests at the 1% level and their sample has test statistic –2.27. The critical value is –2.33. State, with a reason, whether the null hypothesis is accepted or rejected.

c Determine the conclusion the two scientists should reach.

(Continued on the next page)

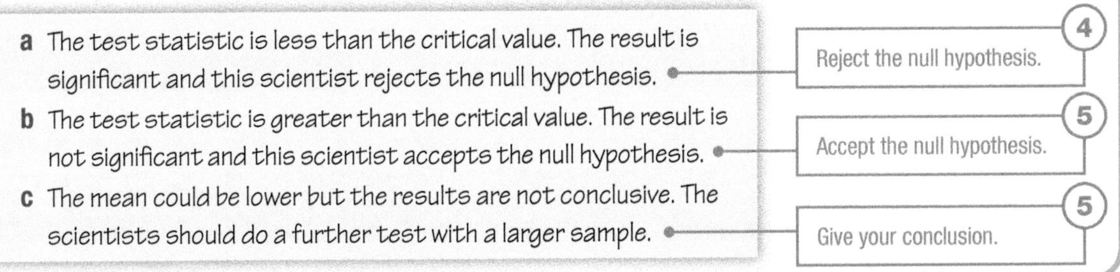

a The test statistic is less than the critical value. The result is significant and this scientist rejects the null hypothesis.
b The test statistic is greater than the critical value. The result is not significant and this scientist accepts the null hypothesis.
c The mean could be lower but the results are not conclusive. The scientists should do a further test with a larger sample.
4
Reject the null hypothesis.
5
Accept the null hypothesis.
5
Give your conclusion.

Exercise 21.2B Reasoning and problem-solving

1 Across England, people consume an average of 1845 ml of milk products (excluding cheese) per person per week. Assuming these are distributed Normally with variance 4288.3 ml^2, a survey is taken to see if the East consumes an average amount.

  a State null and alternative hypotheses.

  A sample of 48 people in the East are found to consume an average of 1824 ml.

  b Calculate the test statistic and the critical value if the test is at the 5% level.

  c State, with a reason, whether the null hypothesis is accepted or rejected. Determine the conclusion of the hypothesis test.

  d Using your understanding of the Large data set, comment on the reliability of this result.

2 A banana farm finds that the banana masses are well-modelled by a Normal distribution with mean 125 g and variance 150 g^2. During one summer, the bananas are thought to be larger than usual.

  a State the null and alternative hypotheses.

  A sample of 42 bananas is selected and their mean mass is 131 g.

  b Calculate the test statistic and hence calculate the p-value.

  c State, with a reason, whether the null hypothesis is accepted or rejected at the 1% significance level.

  d Determine the conclusion of the hypothesis test.

  e Discuss whether it's reasonable to say that the mean mass of that summer's bananas is 131 g.

Large data set

3 In England, the mean amount of margarine purchased per person per week in 2002 is well-modelled by a Normal distribution, with mean 13.6 grams and variance 71.9 grams^2. It is thought that people in London may be purchasing less margarine than the general population.

  a State the null and alternative hypotheses.

  A sample of 25 people is taken from London and their purchased quantities of margarine are found to have a mean value of 12.8 grams.

  b Calculate the test statistic and hence calculate the p-value.

  c Explain why a larger sample gives a better chance of the test showing a difference when the mean is lower.

  d State, with a reason, whether the null hypothesis is accepted or rejected at the 5% significance level. Write a conclusion for the test.

Challenge

4 A Normal distribution is known to have variance 9 but has an unknown mean. One statistician believes the mean is 16 but another believes that it is 17

  a State each person's null and alternative hypotheses.

  They take a sample of size 64 to decide at the 10% significance level who is correct. The observed mean of the sample is 16.6

  b Determine the conclusion of each statistician's hypothesis test and discuss the results.

21 Summary and review

Chapter summary

- The **population correlation coefficient**, ρ, describes how correlated two variables are.
- The product moment correlation coefficient (PMCC), r, is a statistic that estimates ρ. It is calculated from a sample and used to determine the presence of correlation.
- In a test to detect correlation, the null hypothesis is that there is none, i.e. $H_0 : \rho = 0$ and the alternative hypothesis can take any of the following forms.
  - There is some correlation, $H_1 : \rho \neq 0$, a 2-tailed test.
  - There is positive correlation, $H_1 : \rho > 0$, a 1-tailed test.
  - There is negative correlation, $H_1 : \rho < 0$, a 1-tailed test.

  The PMCC is the test statistic.
- A Normal distribution has mean μ and variance σ^2. Given a known or assumed variance you can use a sample to test if a value is a suitable estimate of the mean.
- The null hypothesis when testing your estimated mean for a Normal distribution is that the value you are testing is correct, i.e. $H_0 : \mu = \mu_0$ and the alternative hypothesis can take any of the following forms.
  - Your value is incorrect, $H_1 : \mu \neq \mu_0$, a 2-tailed test.
  - Your value is too small, $H_1 : \mu > \mu_0$, a 1-tailed test.
  - Your value is too large, $H_1 : \mu < \mu_0$, a 1-tailed test.

  For a sample of size n with mean value $\bar{x}$, the test statistic is $z = \dfrac{\bar{x} - \mu}{\frac{\sigma}{\sqrt{n}}}$
- When hypothesis testing:
  - If the test statistic is further from zero than the critical value, or
  - If the p-value associated with the PMCC is smaller than the significance level of the test, then you have sufficient evidence to reject the null hypothesis. Otherwise, you accept the null hypothesis. You should interpret this result in the context of the situation.

Check and review

You should now be able to...	Try Questions
✔ State null and alternative hypotheses when testing for correlation.	1, 2, 3
✔ Compare a given PMCC to a critical value or its p-value to the significance level, and use this comparison to decide whether to accept or reject the null hypothesis.	2, 3, 4, 5
✔ Decide what the conclusion means in context about the correlation.	3, 4, 5
✔ State null and alternative hypotheses when testing the mean of a Normal distribution.	6, 7
✔ Calculate the test statistic, compare it to a critical value or compare its p-value to the significance level, and use this comparison to decide whether to accept or reject the null hypothesis.	7
✔ Decide what the conclusion means in context about the mean of the distribution.	7

1 It's suspected that two variables are correlated but it is not known if the correlation would be positive or negative. State null and alternative hypotheses for a test to identify correlation.

2 The temperature in a seaside town is believed to be positively correlated with the number of ice cream cones sold by a van by the beach.

a State null and alternative hypotheses for a test with a 5% significance level.

In a sample of 30 days randomly selected over the course of the year, the data is found to have a PMCC of 0.6215. The critical value for this test is 0.3061

b State, with a reason, whether the null hypothesis is accepted or rejected.

3 A scientist believes there is negative correlation between the amount of a chemical in a petri dish and the number of bacteria present. To test this, she prepares 20 different petri dishes with varying amounts of the chemical in each. The null hypothesis is $H_0 : \rho = 0$ and the alternative hypothesis is $H_1 : \rho < 0$, which are tested at the 1% level. The test statistic is 0.4438, which has a p-value of 2.5%

a State, with a reason, whether the null hypothesis is accepted or rejected.

b Determine the scientist's conclusion.

4 Levels of annual household income and the amount spent per year on books are suspected to be positively correlated.

a State the null and alternative hypotheses.

40 households are used as a sample and the PMCC is found to be 0.198. The critical value at the 5% level is 0.264

b Determine the conclusion in context.

5 The heights and lengths heptathletes can jump in the high jump and long jump are tested for correlation. The hypotheses $H_0 : \rho = 0$ and $H_1 : \rho \neq 0$ are being considered at the 5% significance level. A sample of 29 competitors is taken and the PMCC is found to be 0.416, which has a p-value of 2.48% for a two-tailed test. State, with a reason, whether H_0 is accepted or rejected and determine the conclusion in context.

6 A Normal distribution is assumed to have variance 18 but has an unknown mean. It is believed that the mean could be –5, but could be larger. A sample is taken to test this. State the null and alternative hypotheses.

7 The lengths of some sticks of rock are well-modelled by a Normal distribution with variance 4 cm^2. Their mean length is unknown, but is supposed to be 30 cm. A sample of 28 sticks is taken to test their mean length at the 5% significance level.

a State the null and alternative hypotheses for this test.

The sample has a mean length of 30.45 cm and the test statistic gives a p-value of 0.2338

b State, with a reason, whether the null hypothesis is accepted or rejected.

c Determine the conclusion of this hypothesis test in context.

STATS

What next?

Score		
0 – 3	Your knowledge of this topic is still developing. To improve, search in MyMaths for the codes: 2287, 2288	
4 – 5	You're gaining a secure knowledge of this topic. To improve, look at the InvisiPen videos for Fluency and skills (21A).	
6 – 7	You've mastered these skills. Well done, you're ready to progress! To develop your techniques, look at the InvisiPen videos for Reasoning and problem-solving (21B).	

Click these links in the digital book

21 Exploration
Going beyond the exams

History

The **Pearson product-moment correlation coefficient** (PMCC) is named after **Karl Pearson** (1857–1936).

Pearson developed the mathematics in partnership with **Sir Francis Galton** (1822–1911), who originally developed the concepts of correlation and regression.

"All great scientists have, in a certain sense, been great artists; the man with no imagination may collect facts, but he cannot make great discoveries."

-Karl Pearson

ICT

B5 | *fx* =CORREL(A2:A8,B2:B8)

	A	B
1	X	Y
2	32	160
3	35	158
4	34	162
5	36	166
6	40	172
7	45	166
8	48	172
9		
10		PPMC = 0.762452

The PMCC can be calculated using spreadsheet software.

Collect data for two variables that you suspect may be connected, for example height and foot length.

Enter the data into two columns on a spreadsheet.

Use the *CORREL* function in Microsoft Excel on the two columns to find the PMCC.

Did you know?

Francis Galton was the half-cousin of **Charles Darwin**. The publication of Darwin's book, On the Origin of Species, in 1859, had a huge impact on Galton and it was Darwin who encouraged him to research inherited traits.

When studying human height, for example, Galton first thought that tall parents would tend to have children that were even taller. He found instead that the children tended to be slightly smaller.

Galton found similar results for other characteristics, which lead him to coin the term **'regression to the mean'**.

Note

Much of Galton's research pointed to the fact that a characteristic may not be the result of a single cause, but affected by multiple different causes, each with varying levels of influence.

This discovery led to the idea of **multiple regression**. Multiple regression is a way to model the data by considering various different factors and the amount of impact they have on the dependent variable.

21 Assessment

1 A machine producing metal rods is suspected of being faulty. The length of rods produced follow a Normal distribution and the variance is 25 cm but the mean may have increased from the intended setting of 40 cm

a State the null and alternative hypotheses. Select the correct answer.

A $H_0: \mu = 40$ $H_1: \mu \neq 40$ **B** $H_0: \mu = 40$ $H_1: \mu < 40$ **C** $H_0: \mu = 40$ $H_1: \mu > 40$ **D** $H_0: \mu = 40$ $H_1: \mu > 45$ **[1 mark]**

A random sample of 36 rods is taken and found to have a mean length of 48.3 cm.

b Calculate the test statistic. Select the correct answer.

A $z = 1.992$ **B** $z = 9.96$ **C** $z = 11.952$ **D** $z = 59.76$ **[1]**

2 A Normally distributed population has a variance of 8. A sample of size 25 was taken and had a mean of 38.1. Stating clearly your null and alternative hypotheses, test at the 3% significance level whether the mean is less than 40 **[4]**

3 A sample of size 10 is taken from a Normal population and gives the following values:

18.2, 19.6, 24.1, 19.3, 21.5, 22.6, 23.3, 20.9, 21.7, 20.3

Using a significance level of 3%, test whether the population mean is less than 22
Assume that the standard deviation of the population is 1.8 **[4]**

4 From extensive experience, a manufacturer knows that their halogen light bulbs have a mean lifetime of 1930 hours with a standard deviation of 245 hours. The firm introduces a new type of light bulb. The lifetimes of 20 of the new bulbs are given in the table.

Lifetime (l hours)	Frequency
$1400 \le l < 1800$	3
$1800 \le l < 2000$	5
$2000 \le l < 2100$	6
$2100 \le l < 2300$	4
$2300 \le l < 2700$	2

a Find the mean lifetime of these bulbs. **[1]**

b Assuming that the variation in the lifetimes of the bulbs remains unchanged, test at a 5% significance level whether the lifetimes have increased. **[3]**

5 The quantity of olive oil purchased per person per week, O, by a household is found to have variance 20 and a mean of 12. O is modelled by a Normal distribution. A sample of size 32 was taken and found to have a sample mean of 10.3. By finding the probability of the sample mean taking a value less than 10.3, test the hypothesis that the population mean is 12 against the alternative hypothesis that it is less than 12. You should use a significance level of 5%. **[4]**

Large data set

6 A survey measures the average purchased quantities of cheese and pickles per person per week in the North East to see if there is any correlation between the two.

a State null and alternative hypotheses for this test. **[1]**

The PMCC for this data is found to be 0.532. The critical value for this test is ±0.6319

b State, with a reason, whether H_0 is accepted or rejected and determine the conclusion in context. **[2]**

7 Two scientists have an experiment whose results are modelled by a Normal distribution with variance 17.6 but they think the mean could be different to the intended −13.6. They both

separately take samples of size 72 to test the hypotheses $H_0 : \mu = -13.6$ and $H_1 : \mu \neq -13.6$ at the 5% level. The critical value is ±1.96

a One scientist's sample has test statistic –2.04. State, with a reason, whether the null hypothesis is accepted or rejected. **[1]**

b The other scientist's sample has test statistic 2.13. State, with a reason, whether the null hypothesis is accepted or rejected. **[1]**

c Determine the conclusion the two scientists reach. **[2]**

8 According to EU legislation, one of the few products that is allowed to be labelled in Imperial measure is milk sold in returnable containers. 30 '1 pint' bottles filled at a dairy farm have the following volumes, given to the nearest ml.

569 567 568 570 569 568 563 568 572 571
568 573 570 572 569 571 568 574 567 569
570 572 564 566 568 572 576 566 571 570

It is required that the mean of the volume in each bottle must be greater than 1 pint (568 ml). Based on this sample, do you believe that this requirement is being met? You should assume that the sample standard deviation gives a good estimate of the population standard deviation and test at a significance level of 5%. **[5]**

9 The correlation coefficient for two variables, X and Y, is 0.31 based on a sample size of 19. Given that the critical value is ±0.468, test at the 5% significance level whether the population correlation coefficient is zero against the alternative hypothesis that it is not zero. **[2]**

10 The quantity of white bread and the quantity of brown and wholemeal bread is recorded for a sample of 30 households. A researcher looks to see if there is a positive correlation. The PMCC is found to be $r = 0.532$

a Write down the hypotheses for the test. **[1]**

b Given that the critical value is ±0.3061, determine the conclusion of the test. **[2]**

11 Given that the sample product moment correlation coefficient between variables X and Y is –0.17 based on a sample of size 46, and the critical value is –0.248, test the hypothesis that the population correlation coefficient, ρ, is less than zero at the 5% significance level. You should state your null and alternative hypotheses. **[3]**

12 A high-speed fabric weaving machine increases in temperature as it is operated. The number of flaws per square metre is measured at various temperatures and these variables are found to have a correlation coefficient of –0.42 based on a sample of size 30. The manufacturer claims that the number of flaws is independent of the temperature. Given that the critical value is ±0.367, test at a significance level of 5% the manufacturer's claim. **[3]**

13 a A Normal random variable X has an unknown mean μ and known standard deviation σ. A sample of size n is taken from the population and gives a sample mean of $\bar{x}$. A test at significance level 2% is to be carried out on whether the population mean has increased from a value μ_0. Find, in terms of μ_0, σ and n, the set of $\bar{x}$ values which would lead to the belief that the mean had increased. **[4]**

b Bars of steel of diameter 2 cm are known to have a mean breaking point of 80 kN with a standard deviation of 2.1 kN. An increase in the bars' diameter of 0.2 cm is thought to increase the mean breaking point. A sample of 40 bars with the greater diameter have a mean breaking point of 80.9 kN. Test at a significance level of 2% whether the bars with the greater diameter have a greater mean breaking point. State any assumptions used. **[3]**

Assessment

Chapters 20–21: Statistics

1 You are given that $P(A)=\frac{1}{2}$, $P(B)=\frac{11}{20}$ and $P(A|B)=\frac{6}{11}$

Calculate $P(B|A)$. Select the correct answer.

A $\frac{3}{11}$ B $\frac{11}{40}$ C $\frac{6}{20}$ D $\frac{6}{10}$ **[1 mark]**

2 You are given that $X \sim N(3.5, 1.6)$. Calculate the probabilities.

Select the correct answer in each case.

a $P(X=3.5)$

A 0 B 1 C 0.5 D 0.4 **[1]**

b $P(X>4.0)$

A 0.6227 B 0.3463 C 0.8944 D 0.1056 **[1]**

3 This two-way table shows the handedness of a group of men and women working in a large office. Not all the cells have been filled in.

Handedness	Men	Women	Total
Left	20		32
Right			
mixed/both	3	2	
Total	210		350

a Copy and complete the table. **[3]**

b A person is chosen at random. Find the probability they are

i Left-handed, ii A woman,

iii A right-handed man. **[3]**

c Are the following two events independent? Explain how you know.

"a randomly chosen person is mixed handed/ambidextrous"
"a randomly chosen person is a man" **[3]**

4 A town has 20 000 homes and a sample of 100 of these is to be surveyed to investigate the quantities of soft drinks purchased.

a Explain the difference between the parameter and the statistic in this case. **[2]**

A market researcher proposes to undertake the survey on a Monday morning and has chosen a road with over 100 houses. He plans to knock on every door until he has 100 responses.

b Explain any problems there might be with this proposal. **[2]**

A second researcher proposes taking a stratified random sample based on the income of the household.

c 800 households are in the highest income band. Calculate the number of these households that will be included in the sample. **[2]**

5 In a sample of 100 households, the quantity of butter purchased per person in one week, b, is shown in the table.

b (g)	Number of households
$0 \le b < 10$	1
$10 \le b < 20$	27
$20 \le b < 30$	38
$30 \le b < 40$	28
$40 \le b < 50$	4
$50 \le b < 60$	0
$60 \le b < 70$	2

a Estimate

i The mean,

ii The standard deviation,

iii The median,

iv The interquartile range. **[10]**

b Explain whether you think the mean or the median is a better measure of the average in this case. **[2]**

6 At a college, the probability a student studies Maths is 0.55, the probability they study Physics is 0.3, and the probability they study both is 0.25

a Calculate the probability that a randomly selected student does not study either Maths or Physics. **[3]**

b Draw a Venn diagram to illustrate this information. **[4]**

c Calculate the probability that a student studies Maths given that they study Physics. **[2]**

d Are the events “a student studies Maths” and “a student studies Physics” independent? Explain how you know. **[2]**

7 A sample of 20 households are questioned about their purchased quantities of cheese (C) and yoghurt (Y) during one week. The number that bought each item is given in the table.

	C	C'
Y	8	2
Y'	4	6

a Find each of these probabilities.

i $\mathrm{P}(C)$ ii $\mathrm{P}(Y|C)$ **[3]**

b Copy and complete the tree diagram with the probabilities on each branch. **[3]**

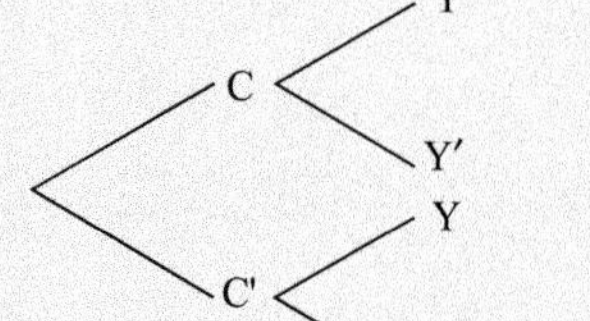

8 You are given that $\mathrm{P}(B) = \frac{1}{5}$, $\mathrm{P}(A|B) = \frac{1}{7}$ and $\mathrm{P}(A \cup B) = \frac{5}{7}$

a Calculate

i $\mathrm{P}(A \cap B)$ ii $\mathrm{P}(B \cap A')$

iii $\mathrm{P}(A|B')$ **[7]**

b The event C is independent to both A and B and $\mathrm{P}(C) = \frac{1}{3}$. Calculate

i $\mathrm{P}(A \cap C)$ ii $\mathrm{P}(C \cup B)$ **[5]**

9 The tree diagram shows the probabilities that a randomly chosen household purchases jam (J) in a particular week, depending on whether they also purchase bread (B).

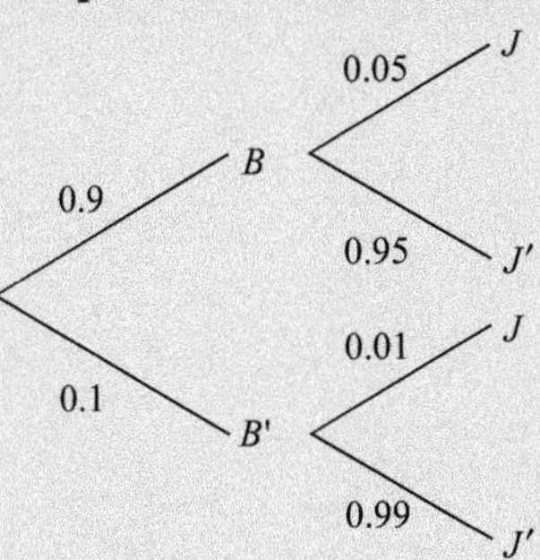

a Calculate

i $P(J \cap B)$ ii $P(J)$ **[5]**

b Draw a Venn diagram showing the probabilities. **[4]**

10 The continuous random variable X is modelled by a Normal distribution with mean 2.4 and standard deviation 0.3

a Calculate these probabilities.

i $P(X < 3)$ ii $P(X > 2.7)$

iii $P(1.8 < X < 2.2)$ **[6]**

b Find, to 4 significant figures, the value α such that $P(X < \alpha) = 0.95$ **[2]**

11 Two sets of times are recorded from 12 people taking part in an experiment.

The product moment correlation coefficient between the times is calculated to be $r = 0.782$. Investigate whether there is a positive correlation between the two times using a 1% significance level. State your hypotheses clearly. **[4]**

12 The quantity of processed cheese and the quantity of natural cheese purchased per week is recorded for a sample of 30 households, and the product moment correlation coefficient is found to be $r = -0.423$

a Investigate whether there is evidence of a negative correlation with a 5% significance level. State your hypotheses clearly. **[4]**

b Write down the critical region if you were to test at the 1% significance level. **[2]**

13 Find the range of test statistics that would lead to the rejection of the null hypothesis in a hypothesis test with significance level of 5% when there is a sample of size 20 and the hypotheses are defined as

$H_0: \rho = 0$ $H_1: \rho \neq 0$ **[4]**

14 The probability of a sports team winning a match in any weather is 0.36. If it is raining, the probability of them winning is 0.3. There is a 10% chance of it raining during the match.

a Calculate the probability of the team winning, given that it is not raining. **[4]**

b Calculate the probability that it was raining, given that the team won a match. **[4]**

15 The histogram shows the lengths (not including the tail), L, of a sample of fully-grown meerkats.

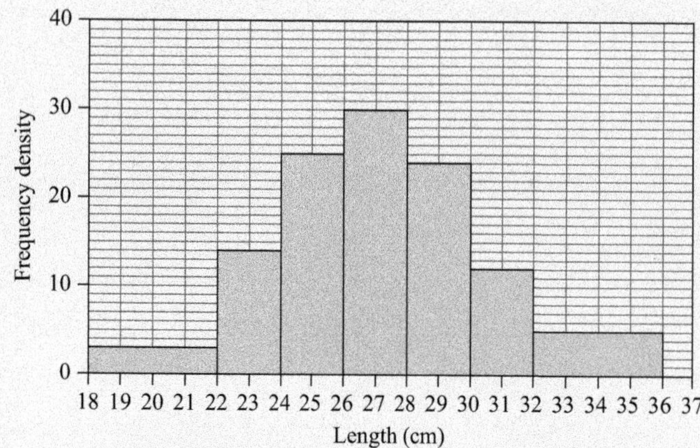

Given that there are 6 meerkats with lengths between 18 and 22 cm,

a Calculate the number of meerkats with a length between 22 and 24 cm, **[3]**

b Estimate

i The mean, ii The standard deviation. **[4]**

c Explain why the Normal distribution can be used to model the lengths, L, of the meerkats. **[2]**

d Using your answers to part **b** as approximations for μ and σ, use the Normal distribution to calculate each of these probabilities.

i $P(L < 31)$ ii $P(21.5 < L < 25.5)$ **[5]**

The product moment correlation coefficient between length and mass is calculated for n of these meerkats and found to be 0.6

A hypothesis test is carried out at the 1% level of significance with hypotheses $H_0: \rho = 0$ and $H_1: \rho > 0$

e What is the range of values of n that will lead to the rejection of the null hypothesis? **[2]**

16 You are given that $X \sim N(\mu, 1.5)$ and that $P(X > 10) = 0.9$

a Calculate the value of μ to 1 decimal place. **[5]**

b Find the x-coordinates of the points of inflection of this Normal distribution. **[3]**

17 The probability of a household purchasing more than 10 pints of milk in a week is found to be 0.35. Use the binomial distribution to model the number of households, H, purchasing more than 10 pints of milk in a week.

a Calculate the probability that, in a street with seven houses, none of them buy more than 10 pints of milk in a week. **[2]**

b Calculate the probability that, in a sample of ten households, more than four purchase more than 10 pints of milk in a week. **[3]**

c Use a Normal approximation to estimate the probability that, in a sample of 60 households, fewer than 15 purchase more than 10 pints of milk. **[4]**

18 The amount of bread, B g, purchased in a week by a sample of households is shown in the box and whisker diagram.

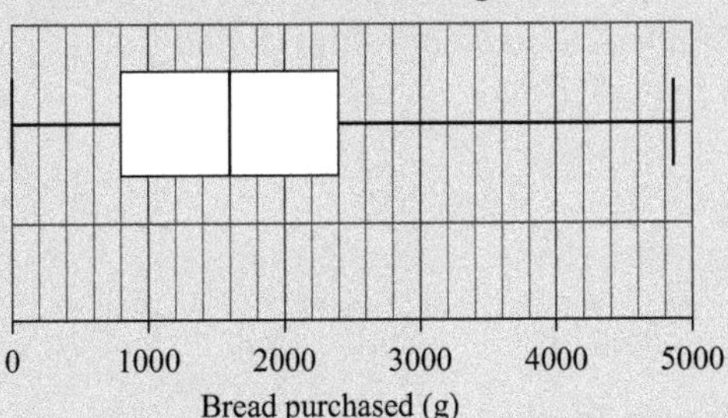

a Would a Normal distribution be suitable to model this data? Explain your answer. **[2]**

B is modelled by a Normal distribution with mean 1600 g and standard deviation 1100 g

b **i** Use the Normal distribution to estimate the probability that a household chosen at random purchases less than 2.4 kg of bread.

ii Compare this result to the data summarised in the box and whisker diagram. **[4]**

c Calculate these probabilities

i $P(B > 2000)$ **ii** $P(B > 1000)$

iii $P(1100 < B < 1850)$. **[6]**

19 For each of these random variables, decide if it could be modelled using the binomial distribution. If you think a binomial distribution is suitable, give the parameters. If not, explain why.

a A midwife delivers eight babies in a week, and X is the number that are girls. **[2]**

b A child rolls a dice repeatedly, and X is the number of throws until they obtain a six. **[2]**

c A bag contains 20 blue and 15 green counters, five counters are removed, and X is the number of blue counter removed. **[2]**

20 A group of 10 volunteers exercise strenuously and their maximum heart-rate is recorded along with their age. The product moment correlation coefficient between age and maximal heart-rate is found to be -0.42. Test whether there is a negative correlation between age and maximal heart-rate using a 10% significance level. State your hypotheses clearly. **[4]**

21 The amount of soup purchased per week, S, in the summer by a household is found to have a mean of 1.4 kg and a standard deviation of 0.6 kg. S is modelled by a Normal distribution.

Large data set

Rio believes that less soup is purchased in the summer so he records the amount of soup purchased over 4 weeks in the summer and calculates the mean to be 0.8 kg.

a Investigate Rio's claim using a 1% significance level. **[7]**

b Write down the acceptance region for the test statistic when using a 10% significance level. **[2]**

22 A continuous random variable is Normally distributed with mean μ and standard deviation σ

Using the facts that $P(X > 120) = 0.9868$ and $P(X > 140) = 0.0694$

a Calculate the values of μ and σ to 3 significant figures, **[9]**

b Find the value of α such that $P(\mu - \alpha < X < \mu + \alpha) = 0.4582$ **[4]**

23 In 2014, the average weekly quantity of soft drinks purchased per person S was 1517 ml. Assume that the quantity purchased can be modelled by a Normal distribution with mean 1517 ml and standard deviation σ. The probability of a household purchasing less than 1 litre of soft drink per person is estimated to be 0.4.

a Calculate the value of σ **[4]**

b Calculate the probability a household purchases more than 3 litres per person. **[2]**

c Calculate the probability that, in a sample of 5 households,

i None of them purchase more than 3 litres per person,

ii Fewer than 2 purchase more than 3 litres per person. **[5]**

24 $X \sim B(100, 0.2)$

a Estimate these probabilities.

i $P(X < 25)$ ii $P(17 \leq X \leq 21)$ **[8]**

b Explain why the Normal distribution is suitable to use as an approximation. **[2]**

Large data set

25 The weekly average quantity of soft natural cheese purchased per person is modelled by a Normal distribution with mean μ and standard deviation σ

The probability of a person purchasing on average less than 7g is 0.35 and the probability of purchasing on average more than 12g is 0.33

a Calculate the values of μ and σ. Give your answers to 3 significant figures. **[10]**

b Calculate the probability that a person purchases on average more than 15g **[2]**

c Estimate the probability that, in a sample of 40 people, there are at most 10 who purchase an average of more than 15g weekly. **[4]**

26 The lifetime of certain batteries is known to be Normally distributed with a mean of 30 hours of continuous use and a standard deviation of 5 hours. A customer purchases eight batteries and records their lifetimes, in hours, as shown.

26.6 25.7 30.5 27.3 20.1 29.5 28.2 25.3

The customer believes they have a faulty batch.

Test, at the 5% level, the customer's claim that the mean is less than 30 hours. **[7]**

Large data set

27 The average weekly purchase of salt, in grams, in a town in the UK is assumed to be Normally distributed with $S \sim N(13, 16)$.

Laura claims that she adds a lot more salt to her food than the average for her town.

She records the amount of salt she uses over n weeks and the mean is 15.2 g per week.

a State the hypotheses and the critical value for conducting a hypothesis test, with a 2.5% significance level, about the mean of the Normal distribution in this case. **[3]**

b Find the smallest value of n that will lead to the acceptance of Laura's claim. **[5]**

Mathematical formulae

For A Level Maths

The following mathematical formulae will be provided for you.

Pure Mathematics

Binomial series

$$(a+b)^n = a^n + \binom{n}{1}a^{n-1}b + \binom{n}{2}a^{n-2}b^2 + \ldots + \binom{n}{r}a^{n-r}b^r + \ldots + b^n \qquad (n \in \mathbb{N})$$

where $\binom{n}{r} = {}^nC_r = \dfrac{n!}{r!(n-r)!}$

$$(1+x)^n = 1 + nx + \frac{n(n-1)}{1.2}x^2 + \ldots + \frac{n(n-1)\ldots(n-r+1)}{1.2\ldots r}x^r + \ldots \qquad (|x|<1,\ n \in \mathbb{Q})$$

Arithmetic series

$$S_n = \frac{1}{2}n(a+l) = \frac{1}{2}n[2a+(n-1)d]$$

Geometric series

$$S_n = \frac{a(1-r^n)}{1-r} \qquad S_\infty = \frac{a}{1-r} \text{ for } |r|<1$$

Trigonometry: small angles

For small angle θ, measured in radians:

$$\sin\theta \approx \theta \qquad \cos\theta \approx 1 - \frac{\theta^2}{2} \qquad \tan\theta \approx \theta$$

Trigonometric identities

$$\sin(A \pm B) = \sin A\cos B \pm \cos A\sin B$$

$$\cos(A \pm B) = \cos A\cos B \mp \sin A\sin B$$

$$\tan(A \pm B) = \frac{\tan A \pm \tan B}{1 \mp \tan A\tan B} \qquad \left(A \pm B \neq \left(k+\frac{1}{2}\right)\pi\right)$$

Differentiation

$f(x)$	$f'(x)$
$\tan x$	$\sec^2 x$
$\operatorname{cosec} x$	$-\operatorname{cosec} x\cot x$
$\sec x$	$\sec x\tan x$
$\cot x$	$-\operatorname{cosec}^2 x$
$\dfrac{f(x)}{g(x)}$	$\dfrac{f'(x)g(x)-f(x)g'(x)}{(g(x))^2}$

Mathematical formulae for A Level Maths

Differentiation from first principles

$$f'(x) = \lim_{h \to 0} \frac{f(x+h) - f(x)}{h}$$

Integration

$$\int u \frac{dv}{dx} dx = uv - \int v \frac{du}{dx} dx$$

$$\int \frac{f'(x)}{f(x)} = \ln|f(x)| + c$$

$f(x)$	$\int f(x)dx$
$\tan x$	$\ln\lvert\sec x\rvert + c$
$\cot x$	$\ln\lvert\sin x\rvert + c$

Numerical solution of equations

The Newton-Raphson iteration for solving $f(x) = 0$: $x_{n+1} = x_n - \frac{f(x_n)}{f'(x_n)}$

Numerical integration

The trapezium rule: $\int_a^b y\,dx \approx \frac{1}{2} h\{(y_0 + y_n) + 2(y_1 + y_2 + \cdots + y_{n-1})\}$, where $h = \frac{b-a}{n}$

Mechanics

Constant acceleration

$s = ut + \frac{1}{2}at^2$	$\mathbf{s} = \mathbf{u}t + \frac{1}{2}\mathbf{a}t^2$
$s = vt - \frac{1}{2}at^2$	$\mathbf{s} = \mathbf{v}t - \frac{1}{2}\mathbf{a}t^2$
$v = u + at$	$\mathbf{v} = \mathbf{u} + \mathbf{a}t$
$s = \frac{1}{2}(u + v)t$	$\mathbf{s} = \frac{1}{2}(\mathbf{u} + \mathbf{v})t$
$v^2 = u^2 + 2as$	

Probability and Statistics

Probability

$P(A \cup B) = P(A) + P(B) - P(A \cap B)$

$P(A \cap B) = P(A) \times P(B|A)$

Discrete distributions

Distribution of X	$P(X = x)$	Mean	Variance
Binomial $B(n, p)$	$\binom{n}{x} p^x (1-p)^x$	np	$np(1 - p)$

Sampling distributions

For a random sample of n observations from $N(\mu, \sigma^2)$

$$\frac{\bar{X} - \mu}{\frac{\sigma}{\sqrt{n}}} \sim N(0, 1)$$

Statistical tables
For A Level Maths

The following statistical table will be provided for you.

Critical Values of the Product Moment Correlation Coefficient

The table gives the critical values, for different significance levels, of the product moment correlation coefficient, r, for varying sample sizes, n

One tail **Two tail**	**10%** **20%**	**5%** **10%**	**2.5%** **5%**	**1%** **2%**	**0.5%** **1%**	**One tail** **Two tail**
n						n
4	0.8000	0.9000	0.9500	0.9800	0.9900	**4**
5	0.6870	0.8054	0.8783	0.9343	0.9587	**5**
6	0.6084	0.7293	0.8114	0.8822	0.9172	**6**
7	0.5509	0.6694	0.7545	0.8329	0.8745	**7**
8	0.5067	0.6215	0.7067	0.7887	0.8343	**8**
9	0.4716	0.5822	0.6664	0.7498	0.7977	**9**
10	0.4428	0.5494	0.6319	0.7155	0.7646	**10**
11	0.4187	0.5214	0.6021	0.6851	0.7348	**11**
12	0.3981	0.4973	0.5760	0.6581	0.7079	**12**
13	0.3802	0.4762	0.5529	0.6339	0.6835	**13**
14	0.3646	0.4575	0.5324	0.6120	0.6614	**14**
15	0.3507	0.4409	0.5140	0.5923	0.6411	**15**
16	0.3383	0.4259	0.4973	0.5742	0.6226	**16**
17	0.3271	0.4124	0.4821	0.5577	0.6055	**17**
18	0.3170	0.4000	0.4683	0.5425	0.5897	**18**
19	0.3077	0.3887	0.4555	0.5285	0.5751	**19**
20	0.2992	0.3783	0.4438	0.5155	0.5614	**20**
21	0.2914	0.3687	0.4329	0.5034	0.5487	**21**
22	0.2841	0.3598	0.4227	0.4921	0.5368	**22**
23	0.2774	0.3515	0.4132	0.4815	0.5256	**23**
24	0.2711	0.3438	0.4044	0.4716	0.5151	**24**
25	0.2653	0.3365	0.3961	0.4622	0.5052	**25**
26	0.2598	0.3297	0.3882	0.4534	0.4958	**26**
27	0.2546	0.3233	0.3809	0.4451	0.4869	**27**
28	0.2497	0.3172	0.3739	0.4372	0.4785	**28**
29	0.2451	0.3115	0.3673	0.4297	0.4705	**29**
30	0.2407	0.3061	0.3610	0.4226	0.4629	**30**
31	0.2366	0.3009	0.3550	0.4158	0.4556	**31**
32	0.2327	0.2960	0.3494	0.4093	0.4487	**32**
33	0.2289	0.2913	0.3440	0.4032	0.4421	**33**

34	0.2254	0.2869	0.3388	0.3972	0.4357	**34**
35	0.2220	0.2826	0.3338	0.3916	0.4296	**35**
36	0.2187	0.2785	0.3291	0.3862	0.4238	**36**
37	0.2156	0.2746	0.3246	0.3810	0.4182	**37**
38	0.2126	0.2709	0.3202	0.3760	0.4128	**38**
39	0.2097	0.2673	0.3160	0.3712	0.4076	**39**
40	0.2070	0.2638	0.3120	0.3665	0.4026	**40**
41	0.2043	0.2605	0.3081	0.3621	0.3978	**41**
42	0.2018	0.2573	0.3044	0.3578	0.3932	**42**
43	0.1993	0.2542	0.3008	0.3536	0.3887	**43**
44	0.1970	0.2512	0.2973	0.3496	0.3843	**44**
45	0.1947	0.2483	0.2940	0.3457	0.3801	**45**
46	0.1925	0.2455	0.2907	0.3420	0.3761	**46**
47	0.1903	0.2429	0.2876	0.3384	0.3721	**47**
48	0.1883	0.2403	0.2845	0.3348	0.3683	**48**
49	0.1863	0.2377	0.2816	0.3314	0.3646	**49**
50	0.1843	0.2353	0.2787	0.3281	0.3610	**50**
60	0.1678	0.2144	0.2542	0.2997	0.3301	**60**
70	0.1550	0.1982	0.2352	0.2776	0.3060	**70**
80	0.1448	0.1852	0.2199	0.2597	0.2864	**80**
90	0.1364	0.1745	0.2072	0.2449	0.2702	**90**
100	0.1292	0.1654	0.1966	0.2324	0.2565	**100**

Mathematical formulae – to learn

For A Level Maths

You are expected to know the following formulae for A Level Mathematics.

Pure Mathematics

Quadratic equations

$ax^2 + bx + c = 0$ has roots $\dfrac{-b \pm \sqrt{b^2 - 4ac}}{2a}$

Laws of indices

$a^x a^y \equiv a^{x+y}$

$a^x \div a^y \equiv a^{x-y}$

$(a^x)^y = a^{xy}$

Laws of logarithms

$x = a^n \quad \Leftrightarrow \quad \log_a x$ for $a > 0$ and $x > 0$

$\log_a x + \log_a y \equiv \log_a xy$

$\log_a x - \log_a y \equiv \log_a \left(\dfrac{x}{y}\right)$

$k \log_a x \equiv \log_a (x)^k$

Coordinate geometry

A straight line graph, gradient m passing through (x_1, y_1) has equation $y - y_1 = m(x - x_1)$

Straight lines with gradients m_1 and m_2 are perpendicular when $m_1 m_2 = -1$

Sequences

General term of an arithmetic progression: $u_n = a + (n-1)d$

General term of a geometric progression: $u_n = ar^{n-1}$

Trigonometry

In the triangle ABC

Sine rule $\dfrac{a}{\sin A} = \dfrac{b}{\sin B} = \dfrac{c}{\sin C}$

Cosine rule $a^2 = b^2 + c^2 - 2bc \cos A$

Area $\dfrac{1}{2} ab \sin C$

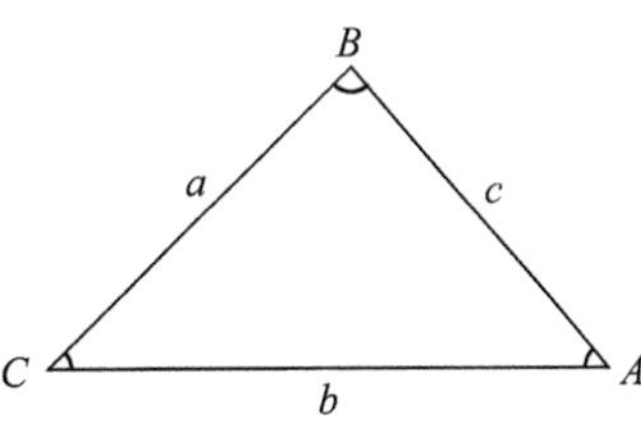

$\cos^2 A + \sin^2 A \equiv 1$

$\sec^2 A \equiv 1 + \tan^2 A$

$\text{cosec}^2 A \equiv 1 + \cot^2 A$

$\sin 2A \equiv 2 \sin A \cos A$

$\cos 2A \equiv \cos^2 A - \sin^2 A$

$\tan 2A \equiv \dfrac{2 \tan A}{1 - \tan^2 A}$

Mathematical formulae - to learn for A Level Maths

Mensuration

Circumference, C and area, A, of circle, with radius r and diameter d:

$C = 2\pi r = \pi d \qquad A = \pi r^2$

Pythagoras' Theorem:

In any right-angled triangle where a, b and c are the lengths of the sides and c is the hypotenuse, $c^2 = a^2 + b^2$

Area of a trapezium $= \frac{1}{2}(a+b)h$, where a and b are the lengths of the parallel sides and h is their perpendicular separation.

Volume of a prism = area of cross section × length

For a circle of radius r, where an angle at the centre of θ radians subtends an arc of length s and encloses an associated sector of area A:

$s = r\theta \qquad A = \frac{1}{2}r^2\theta$

Calculus and differential equations

Differentiation

Function	Derivative
x^n	nx^{n-1}
$\sin kx$	$k\cos kx$
$\cos kx$	$-k\sin x$
e^{kx}	ke^{kx}
$\ln x$	$\frac{1}{x}$
$f(x)+g(x)$	$f'(x)+g'(x)$
$f(x)g(x)$	$f'(x)g(x)+f(x)g'(x)$
$f(g(x))$	$f'(g(x))g'(x)$

Integration

Function	Integral	
x^n	$\frac{1}{n+1}x^{n+1}+c,$	$n \neq -1$
$\cos kx$	$\frac{1}{k}\sin kx + c$	
$\sin kx$	$-\frac{1}{k}\cos kx + c$	
e^{kx}	$\frac{1}{k}e^{kx} + c$	
$\frac{1}{x}$	$\ln\lvert x\rvert + c,$	$x \neq 0$
$f'(x)+g'(x)$	$f(x)+g(x)+c$	
$f'(g(x))g'(x)$	$f(g(x))+c$	

Area under a curve $= \int_a^b y\,dx \qquad (y \geq 0)$

Vectors

$\lvert x\mathbf{i} + y\mathbf{j} + z\mathbf{k}\rvert = \sqrt{x^2+y^2+z^2}$

Statistics

The mean of a set of data: $\bar{x} = \frac{\sum x}{n} = \frac{\sum fx}{\sum f}$

The standard Normal variable: $Z = \frac{X - \mu}{\sigma}$ where $X \sim N(\mu, \sigma^2)$

Mechanics

Forces and equilibrium

Weight = mass $\times g$

Friction: $F \le \mu R$

Newton's second law in the form: $F = ma$

Kinematics

For motion in a straight line with variable acceleration:

$$v = \frac{dr}{dt} \qquad a = \frac{dv}{dt} = \frac{d^2r}{dt^2}$$

$$r = \int v\,dt \qquad v = \int a\,dt$$

Mathematical notation

For AS and A Level Maths

You should understand the following notation without need for further explanation.
Anything highlighted is only used in the full A Level, and so will not be needed at AS Level.

Set Notation

$\in$	is an element of
$\notin$	is not an element of
$\subseteq$	is a subset of
$\subset$	is a proper subset of
$\{x_1, x_2, \ldots\}$	the set with elements $x_1, x_2, \ldots$
$\{x: \ldots\}$	the set of all x such that ...
$\mathrm{n}(A)$	the number of elements in set A
$\varnothing$	the empty set
ε	the universal set
A'	the complement of the set A
$\mathbb{N}$	the set of natural numbers, $\{1, 2, 3, \ldots\}$
$\mathbb{Z}$	the set of integers, $\{0, \pm1, \pm2, \pm3, \ldots\}$
$\mathbb{Z}^+$	the set of positive integers, $\{1, 2, 3, \ldots\}$
$\mathbb{Z}_0^+$	the set of non-negative integers, $\{0, 1, 2, 3, \ldots\}$
$\mathbb{R}$	the set of real numbers
$\mathbb{Q}$	the set of rational numbers, $\left\{\frac{p}{q} : p \in \mathbb{Z},\ q \in \mathbb{Z}^+\right\}$
$\cup$	union
$\cap$	intersection
(x, y)	the ordered pair x, y
$[a, b]$	the closed interval $\{x \in \mathbb{R} : a \le x \le b\}$
$[a, b)$	the interval $\{x \in \mathbb{R} : a \le x < b\}$
$(a, b]$	the interval $\{x \in \mathbb{R} : a < x \le b\}$
(a, b)	the open interval $\{x \in \mathbb{R} : a < x < b\}$

Miscellaneous Symbols

$=$	is equal to
$\neq$	is not equal to
$\equiv$	is identical to or is congruent to
$\approx$	is approximately equal to
∞	infinity
$\propto$	is proportional to
$<$	is less than
$\leqslant, \le$	is less than or equal to, is not greater than
$>$	is greater than
$\geqslant, \ge$	is greater than or equal to, is not less than
$\therefore$	therefore
$\because$	because
$\angle A$	angle A
$p \Rightarrow q$	p implies q (if p then q)
$p \Leftarrow q$	p is implied by q (if q then p)
$p \Leftrightarrow q$	p implies and is implied by q (p is equivalent to q)

a	first term for an arithmetic or geometric sequence
l	last term for an arithmetic sequence
d	common difference for an arithmetic sequence
r	common ratio for a geometric sequence
S_n	sum to n terms of a sequence
S_∞	sum to infinity of a sequence

Operations

$a+b$	a plus b
$a-b$	a minus b
$a\times b$, ab, $a\cdot b$	a multiplied by b
$a\div b$, $\frac{a}{b}$	a divided by b
$\sum_{i=1}^{n} a_i$	$a_1+a_2+\ldots+a_n$
$\prod_{i=1}^{n} a_i$	$a_1\times a_2\times\ldots\times a_n$
$\sqrt{a}$	the positive square root of a
$\lvert a\rvert$	the modulus of a
$n!$	n factorial: $n!=n\times(n-1)\times\ldots\times 2\times 1$, $n\in\mathbb{N}$; $0!=1$
$\binom{n}{r}$, ${}^nC_r$, ${}_nC_r$	the binomial coefficient $\frac{n!}{r!(n-r)!}$ for $n, r\in\mathbb{Z}_0^+$, $r\le n$ or $\frac{n(n-1)\ldots(n-r+1)}{r!}$ for $n\in\mathbb{Q}$, $r\in\mathbb{Z}_0^+$

Functions

$\mathrm{f}(x)$	the value of the function f at x
$\mathrm{f}: x\mapsto y$	the function f maps the element x to the element y
f^{-1}	the inverse function of the function f
gf	the composite function of f and g which is defined by $\mathrm{gf}(x)=\mathrm{g}(\mathrm{f}(x))$
$\lim_{x\to a}\mathrm{f}(x)$	the limit of $\mathrm{f}(x)$ as x tends to a
Δx, δx	an increment of x
$\frac{\mathrm{d}y}{\mathrm{d}x}$	the derivative of y with respect to x
$\frac{\mathrm{d}^n y}{\mathrm{d}x^n}$	the nth derivative of y with respect to x
$\mathrm{f}'(x)$ …, $\mathrm{f}^{(n)}(x)$	the first, …, nth derivatives of $\mathrm{f}(x)$ with respect to x
$\dot{x}$, $\ddot{x}$, …	the first, second, … derivatives of x with respect to t
$\int y\,\mathrm{d}x$	the indefinite integral of y with respect to x
$\int_a^b y\,\mathrm{d}x$	the definite integral of y with respect to x between the limits $x=a$ and $x=b$

Exponential and Logarithmic Functions

e	base of natural logarithms
e^x, $\exp x$	exponential function of x
$\log_a x$	logarithm to the base a of x
$\ln x$, $\log_e x$	natural logarithm of x

Trigonometric Functions

sin, cos, tan, cosec, sec, cot	the trigonometric functions
$\sin^{-1}$, $\cos^{-1}$, $\tan^{-1}$, arcsin, arccos, arctan	the inverse trigonometric functions
°	degrees
rad	radians

Vectors

$\mathbf{a}$, $\underline{a}$, $\underset{\sim}{a}$	the vector $\mathbf{a}$, $\underline{a}$, $\underset{\sim}{a}$
$\overrightarrow{AB}$	the vector represented in magnitude and direction by the directed line segment AB
$\hat{\mathbf{a}}$	a unit vector in the direction of $\mathbf{a}$
$\mathbf{i}, \mathbf{j}, \mathbf{k}$	unit vectors in the directions of the Cartesian coordinate axes
$\lvert\mathbf{a}\rvert$, a	the magnitude of $\mathbf{a}$
$\lvert\overrightarrow{AB}\rvert$, AB	the magnitude of $\overrightarrow{AB}$
$\begin{pmatrix} a \\ b \end{pmatrix}$, $a\mathbf{i} + b\mathbf{j}$	column vector and corresponding unit vector notation
$\mathbf{r}$	position vector
$\mathbf{s}$	displacement vector
$\mathbf{v}$	velocity vector
$\mathbf{a}$	acceleration vector

Probability and Statistics

A, B, C, etc.	events
$A \cup B$	union of the events A and B
$A \cap B$	intersection of the events A and B
$P(A)$	probability of the event A
A'	complement of the event A
$P(A \mid B)$	probability of the event A conditional on the event B
X, Y, R, etc.	random variables
x, y, r, etc.	values of the random variables X, Y, R etc.
$x_1, x_2, \ldots$	observations
$f_1, f_2, \ldots$	frequencies with which the observations $x_1, x_2, \ldots$ occur
$p(x)$, $P(X=x)$	probability function of the discrete random variable X
$p_1, p_2, \ldots$	probabilities of the values $x_1, x_2, \ldots$ of the discrete random variable X

E(X)	expectation of the random variable X
Var(X)	variance of the random variable X
~	has the distribution
B(n, p)	binomial distribution with parameters n and p, where n is the number of trials and p is the probability of success in a trial
q	$q = 1 - p$ for binomial distribution
N(μ, σ^2)	Normal distribution with mean μ and variance σ^2
$Z \sim \text{N}(0,1)$	standard Normal distribution
ϕ	probability density function of the standardised Normal variable with distribution N(0, 1)
Φ	corresponding cumulative distribution function
μ	population mean
σ^2	population variance
σ	population standard deviation
$\bar{x}$	sample mean
s^2	sample variance
s	sample standard deviation
H_0	null hypothesis
H_1	alternative hypothesis
r	product moment correlation coefficient for a sample
ρ	product moment correlation coefficient for a population

Mechanics

kg	kilograms
m	metres
km	kilometres
m/s, m s^{-1}	metres per second (velocity)
m/s^2, m s^{-2}	metres per second per second (acceleration)
F	Force or resultant force
N	Newton
N m	Newton metre (moment of a force)
t	time
s	displacement
u	initial velocity
v	velocity or final velocity
a	acceleration
g	acceleration due to gravity
μ	coefficient of friction

Answers

Full solutions to all of these questions can be found at the link in the page footer.

Chapter 12

Exercise 12.1A Fluency and skills

1 $(5n+1)^4-(5n-1)^4 \equiv [(5n+1)^2-(5n-1)^2][(5n+1)^2+(5n-1)^2]$
$\equiv [(25n^2+10n+1)-(25n^2-10n+1)]$
$[(25n^2+10n+1)+(25n^2-10n+1)]$
$\equiv [20n][50n^2+2]$
$\equiv [40n][25n^2+1]$
This has a factor of 40, so $(5n+1)^4-(5n-1)^4$ is divisible by 40

2 Let the three-digit number be $W=$ "xyz" with 100s digit x, 10s digit y, and units digit z
$W=100x+10y+z=99x+9y+(x+y+z)$
If the sum of the digits is divisible by 9, then $(x+y+z)=9P$ for some integer P
So $W=99x+9y+9P=9(11x+y+P)=9Q$ for some integer Q
Hence, W itself is divisible by 9

3 $p! \le 2^p$
$0!=0\le 1=2^0 \Rightarrow$ true for $p=0$
$1!=1\le 2=2^1 \Rightarrow$ true for $p=1$
$2!=2\le 4=2^2 \Rightarrow$ true for $p=2$
$3!=6\le 8=2^3 \Rightarrow$ true for $p=3$
So the statement is proved by exhaustion.

4 Let the numbers be n and $n+1$
$(n+1)^3-n^3\equiv n^3+3n^2+3n+1-n^3$
$\equiv 3n^2+3n+1$
Case 1: n is even.
Then n^2 is also even and hence $3n^2$ is even
$3n$ is also even
So $3n^2+3n+1$ would be even + even + odd, which is odd.
Case 2: n is odd.
Then n^2 is odd and hence $3n^2$ is odd
$3n$ is also odd
So $3n^2+3n+1$ would be odd + odd + odd, which is still odd.
So the difference between the cubes of two consecutive integers is odd.

5 If $a=5$, $b=-2$ and $c=-4$
Then $ab=5\times -2=-10$ and $bc=-2\times -4=8$
So, in this case $ab<bc$, which disproves the statement.

6 Suppose $p=\sqrt{5}$
$p^2=(\sqrt{5})^2=5$, which is rational
$p=\sqrt{5}$, which is irrational
This counterexample disproves the statement.

7 $9^3=729$, which disproves the statement.

8 Contradiction statement: There is an even integer, n such that n^2 is odd.
If n is even, it can be written as $n=2m$
Hence, $n^2=(2m)^2=4m^2$
But 4 times any integer is even and so the statement is contradicted.
So if n^2 is odd, then n is odd.

9 Contradiction statement: There are integers a and b for which $a^2-4b=2$
From this equation, $a^2=4b+2=2(2b+1)$
Any integer multiplied by 2 is even, so $2(2b+1)$ is even and so a^2 must be even.
Since a^2 is even, then a must also be even, so we can write $a=2m$ for some integer m
Substituting $a=2m$ back into the original equation $a^2-4b=2$ gives
$(2m)^2-4b=2$
$\Rightarrow 4m^2-4b=2$
$\Rightarrow 2m^2-2b=1$
$\Rightarrow 2(m^2-b)=1$
Since $2(m^2-b)$ is even, 1 must also be even
However, 1 is not even, and so the statement is contradicted.
So there are no integer values of a and b such that $a^2-4b=2$

10 For $0°<x<90°$, $\sin x>0$ and $\cos x>0$, so $\sin x+\cos x>0$
Alternative statement: Let us also assume that $\sin x+\cos x<1$ for $0°<x<90°$
$\Rightarrow (\sin x+\cos x)^2<1$
$\Rightarrow \sin^2 x+2\sin x\cos x+\cos^2 x<1$
But $\sin^2 x+\cos^2 x=1$, so $1+2\sin x\cos x<1$
However, for $0°<x<90°$, $\sin x>0$ and $\cos x>0$
$\Rightarrow 2\sin x\cos x>0$
Hence, $1+2\sin x\cos x>1$
This contradicts the statement.
Furthermore, if $x=0°$ or $90°$ then $\sin x+\cos x=1$
So for every real number x between $0°$ and $90°$,
$\sin x+\cos x\ge 1$

11 Alternative statement: m and n exist such that $\frac{m^2}{n^2}=2$, and $\frac{m^2}{n^2}$ is a fully simplified fraction.
If $\frac{m^2}{n^2}=2$, then $m^2=2n^2$
$2n^2$ must be even and so m^2 must also be even.
m^2 is even and so m must also be even.
so $m=2k$, $k\in\mathbb{Z}$
$\Rightarrow (2k)^2=2n^2$
$\Rightarrow 4k^2=2n^2$
So n^2 must be even and hence n must also be even.
Since both m and n are even, they both have a factor of 2
Hence both m^2 and n^2 have a factor of 4
This is a contradiction and disproves the original statement.

12 Alternative statement: There is at least one integer greater than 1 which has no prime factors.
Let the *smallest* such integer be n
Case 1: n is prime
Then n has a prime factor: itself, since $n=n\times 1$
This contradicts the assumption, so n is not prime.
Case 2: n is not prime
Then n has a factor, f, where $f\ne n$ and $f\ne 1$
Because f is a factor of n, $f<n$
Since n is the *least* integer with no prime factor, f *does* have a prime factor, p
Since p is a prime factor of f, and f is a factor of n, so p must also be a prime factor of n
This contradicts the assumption that n does not have a prime factor.
So every integer greater than 1 does have at least one prime factor.

Exercise 12.1B Reasoning and problem-solving

1 Alternative statement: Suppose that m and n are integers and mn is odd but m and n are not both odd, that is, at least one of them is even.
Without loss of generality, suppose that m is even.
m is even $\Rightarrow m = 2p$ for some integer p
Hence $mn = 2pn$
Now pn must be another integer, say q
Hence $mn = 2q$
But $2q$ must be an even number, and so mn is even
This contradicts the original statement.
So if m and n are both integers and mn is odd, then both m and n must be odd.

2 Alternative statement: Suppose that m exists such that $m^2 < 2m$, but m does not lie in the range $0 < m < 2$
If $m \geq 2$ and $m^2 < 2m$, then $m < 2$ (dividing by m), which is a contradiction.
So $m^2 < 2m$ cannot be true, for $m \geq 2$
If m is zero, then $m^2 < 2m$ is not true.
If m is negative, then m^2 is positive and $2m$ is negative.
So $m^2 < 2m$ cannot be true for $m \leq 0$
The alternative statement is therefore disproved.
Hence, if $m^2 < 2m$ has any solutions, then $0 < m < 2$

3 Let $n = 2m+1$ for integer m
Then $(-1)^n = (-1)^{2m+1}$
$= (-1)^{2m}(-1)^1$ [By laws of indices]
$= 1 \times (-1)^1$ [Since $(-1)^{2m} = 1$ as $2m$ is even]
$= -1$

4 Let $A = 60°$ and $B = 30°$
$\sin(A - B) = \sin(60 - 30) = \sin 30 = \frac{1}{2}$
$\sin A - \sin B = \sin 60 - \sin 30 = \frac{\sqrt{3}}{2} - \frac{1}{2} = \frac{\sqrt{3}-1}{2} \neq \frac{1}{2}$
Hence $\sin(A - B) \neq \sin A - \sin B$ for all A and B

5 Contradiction statement: Suppose that, for integers m and n, there exists an integer k such that $(5m + 3)(5n + 3) = 5k$
Then $5k = (5m + 3)(5n + 3)$
$\Rightarrow 5k = 25mn + 15m + 15n + 9$
$\Rightarrow 5k = 5(5mn + 3m + 3n + 1) + 4$
$\Rightarrow 5k = 5x + 4$ where $x = (5mn + 3m + 3n + 1)$, an integer
Hence, $4 = 5(k - x) = 5p$ for some integer p
5 is a factor of the RHS, but not a factor of the LHS, so this is a contradiction.
So the statement is disproved, and there is no integer k such that $(5m + 3)(5n + 3) = 5k$ for integers m and n

6 Let the numbers be $2m$ and $2m + 1$
$(2m)^3 + (2m + 1)^3 = 8m^3 + 8m^3 + 12m^2 + 6m + 1$
$= 16m^3 + 12m^2 + 6m + 1$
$= 2(8m^3 + 6m^2 + 3m) + 1$
2 is a factor of $2(8m^3 + 6m^2 + 3m)$
Hence, dividing $(2m)^3 + (2m + 1)^3$ by 2 leaves a remainder of 1

7 $m < -\frac{1}{2}$
$1 > -\frac{1}{2m}$ ($\div m$ on each side. m is negative, so reverses the inequality sign)
$2 > -\frac{1}{m}$ ($\times 2$ on each side)
$3 > 1 - \frac{1}{m}$ (+1 on each side)
$1 - \frac{1}{m} < 3$

8 Contradiction statement: There is a smallest positive number.
Let n be the smallest positive number.
If n is positive, then $\frac{n}{2}$ exists, and $0 < \frac{n}{2} < n$
So $\frac{n}{2}$ is a smaller positive number than n
This contradicts the statement that n is the smallest positive number.
So there is no smallest positive number.

9 Let the number, N, be 'abc'. Hence $N = 100a + 10b + c$
$11 \times (100a + 10b + c) = 1100a + 110b + 11c$
$= 1000a + 100a + 100b + 10b + 10c + c$
$= 1000a + (a + b)100 + (b + c)10 + c$
Written as a four-digit number, this is '$a\langle a+b\rangle\langle b+c\rangle c$', proving the rule.
The catch is when $a + b$ or $b + c$ comes to 10 or more, in which case a 'carry' has to be inserted.

10 Case 1: a is even $\Rightarrow a = 2m$ for some m
then $a^2 + 2 = (2m)^2 + 2$
$= 4m^2 + 2$
$= 2(2m^2 + 1)$
This has a factor of 2, but $2m^2$ is even and so $2m^2 + 1$ is odd
So $a^2 + 2$ cannot have the other required factor of 2
Hence, if a is even, it cannot be divided equally by 4 to give an integer.
Case 2: a is odd $\Rightarrow a = (2m + 1)$ for some m
then $a^2 + 2 = (2m + 1)^2 + 2$
$= 4m^2 + 4m + 3$
Both $4m^2$ and $4m$ are even and so $4m^2 + 4m + 3$ is odd
So $a^2 + 2$ cannot be divided equally by 4 to give an integer.

11 Alternative statement: Suppose there is a greatest odd integer, n
Then $n + 2 > n$ and $(n + 2)$ is also odd.
This is a contradiction, since n is the greatest odd integer.
Hence, there is no greatest odd integer.

12 If b is a factor of a then $a = rb$ for some integer, r
If c is a factor of b then $b = sc$ for some integer, s
So $a = rb = rsc = tc$ where t is an integer, equal to rs
Hence, c is a factor of a and Stephen is right.

13 Any positive rational number can be written as $\frac{a}{b}$ where a and b are either both positive or both negative
We can write $\frac{a}{b} = c\sqrt{2}$ for some positive number c
Since $\sqrt{2}$ is an irrational number, we must prove that c is also an irrational number.
Now $c = \frac{a}{b\sqrt{2}} = \frac{a\sqrt{2}}{2b}$ which is an irrational number.
Hence any positive rational number can be expressed as the product of two irrational numbers.

14 Every cube number is the cube of an integer.
Every integer, m, is either:
i a multiple of 3
ii or one less than a multiple of 3
iii or one more than a multiple of 3.
Hence these three cases are exhaustive. We will prove each in turn:
Case 1: Let m be a multiple of 3 $\Rightarrow m = 3p$ for some integer p
$m^3 = 27p^3 = 9(3p^3) = 9k$ for some integer k
Case 2: Let m be one more than a multiple of 3 $\Rightarrow m = 3p + 1$ for some integer p
$m^3 = (3p + 1)^3$
$= 27p^3 + 27p^2 + 9p + 1$
$= 9(3p^3 + 3p^2 + p) + 1$
$= 9k + 1$ for some integer k
Case 3: Let m be one less than a multiple of 3 $\Rightarrow m = 3p - 1$ for some integer p
$m^3 = (3p - 1)^3$
$= 27p^3 - 27p^2 + 9p - 1$
$= 9(3p^3 - 3p^2 + p) - 1$
$= 9k - 1$ for some integer k

So every cube number can be expressed in the form $9k$ or $9k$ 1 where k is an integer.

15 Alternative statement: There is a solution and $a^2 - b^2 = 1$ for some positive integers a and b
If this is true, then $a^2 - b^2 = 1$ so $(a - b)(a + b) = 1$
Now, since a and b are integers, then:
either $a - b = 1$ and $a + b = 1$
or $a - b = -1$ and $a + b = -1$
Solving $a - b = 1$ and $a + b = 1$ leads to the solution $a = 1$ and $b = 0$ which contradicts the original statement that a and b are both positive.
Solving $a - b = -1$ and $a + b = -1$ leads to the solution $a = -1$ and $b = 0$ which again contradicts the original statement that a and b are both positive.
So there are no positive integer solutions to the equation $a^2 - b^2 = 1$

16 Assume that a is a non-square integer and that $\sqrt{a} = \frac{m}{n}$, where $m, n \in \mathbb{Z}$ are co-prime, $n \neq 0$
Then $\left(\frac{m}{n}\right)^2 = a$
Thus $m^2 = an^2$ and a divides m^2
So a must also divide m
So $m = ak$ for some integer k
Hence, $a^2k^2 = an^2$
$\Rightarrow ak^2 = n$ and a divides n^2
So a must also divide n
So $n = aj$ for some integer j
Hence, $\frac{m}{n} = \frac{ak}{aj}$ which implies m and n share a common factor, a
This contradicts our original assumption that m and n share no common factors.
Hence $\sqrt{a}$ is irrational.

17 The error occurs in the 5th line: "So ~~$(a - b)$~~$(a + b) = b$~~$(a - b)$~~"
Dividing by $(a - b)$ is the same as dividing by 0, since $a = b$
Division by zero is undefined and so the argument becomes invalid.

Exercise 12.2A Fluency and skills

1 **a** $f(x) \in \mathbb{R}$
b $f(x) \in \mathbb{R} : -32 < y < 28$
c $f(x) \in \mathbb{R} : y > 0$
d $f(x) \in \mathbb{R} : 0 \leq y \leq 225$

2 **a** $x \in \mathbb{R} : -2 < x \leq 3$
b $x \in \mathbb{R} : x > 0$

3 **a** Domain $\{x \in \mathbb{R}\}$; range $\{f(x) \in \mathbb{R} : f(x) > 0\}$
b Domain $\{x \in \mathbb{R} : x \neq -3\}$; range $\{f(x) \in \mathbb{R} : f(x) > 0\}$

4 **a** $f(x) = 2x^2$ is many-to-one
E.g. $f(1) = 2$ and $f(-1) = 2$
b $f(x) = 3^{-x}$ is one-to-one
No two values of x give the same value for $f(x)$
c $f(x) = x^4$ is many-to-one
E.g. $f(1) = 1$ and $f(-1) = 1$
d $f(x) = \sin^2 x$ is many-to-one in this interval
E.g. $f(90) = 1$ and $f(270) = 1$
e $f(x) = \frac{1}{x^2}$ is many-to-one
E.g. $f(1) = 1$ and $f(-1) = 1$
f $f(x) = -3x^3$ is one-to-one
No two values of x give the same value for $f(x)$
g $f(x) = \frac{1}{x-3}$ is one-to-one
No two values of x give the same value for $f(x)$
h $f(x) = \cos x$, $0° \leq x \leq 360°$ is many-to-one
E.g. $f(0)$ and $f(360) = 1$
i $f(x) = \cos x$, $0° \leq x \leq 180°$ is one-to-one
No two values of x give the same value for $f(x)$
j $f(x) = \cos 2x$, $0° \leq x \leq 180°$ is many-to-one
E.g. $f(0)$ and $f(180) = 1$

5 **a** **i**

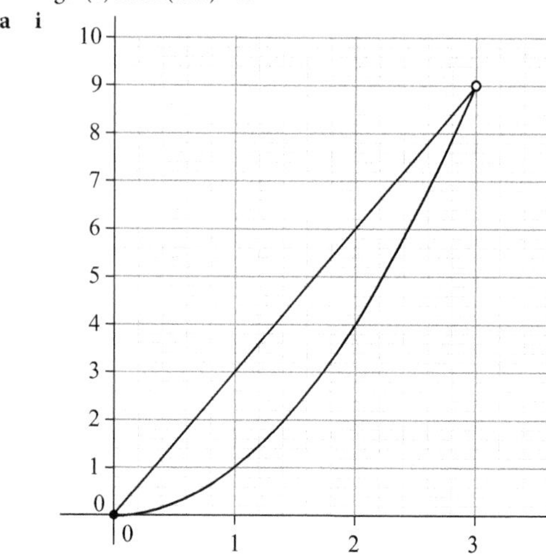

ii Point of intersection (0,0)

b **i**

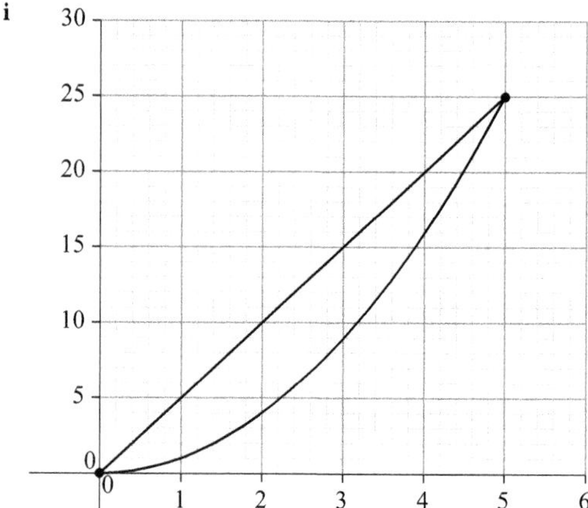

ii Points of intersection (0,0) and (3,9)

c **i**

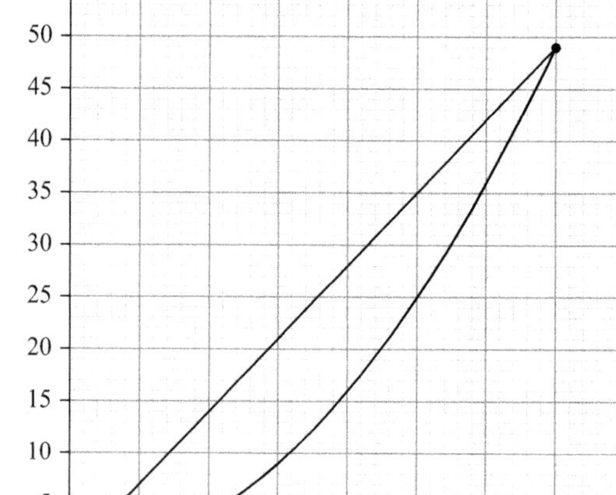

ii Point of intersection (3,9)

6 a i $fg(1) = -11$
$gf(-2) = -6$
$ff\left(\frac{-2}{3}\right) = 22$
ii $fg(1) = -1$
$gf(-2) = \frac{-8}{5}$
$ff\left(\frac{-2}{3}\right) = \frac{-2}{3}$
b $fg(2) = 0$
$gf(2) = -8$
$fg(-4) = 36$
$gf(-4) = -8$
7 a i The domain is $\{x \in \mathbb{R}\}$
The range is $\{f(x) \in \mathbb{R}\}$
ii $f(0) = 0$
$f(-4) = -64$
$f(4) = 64$
b i The domain is $\{x \in \mathbb{R}\}$
The range is $\{f(x) \in \mathbb{R}: y \geq -10\}$
ii $f(0) = 6$
$f(-4) = 54$
$f(4) = -10$
8 13 or 17
9 a $y = 2x + 6$

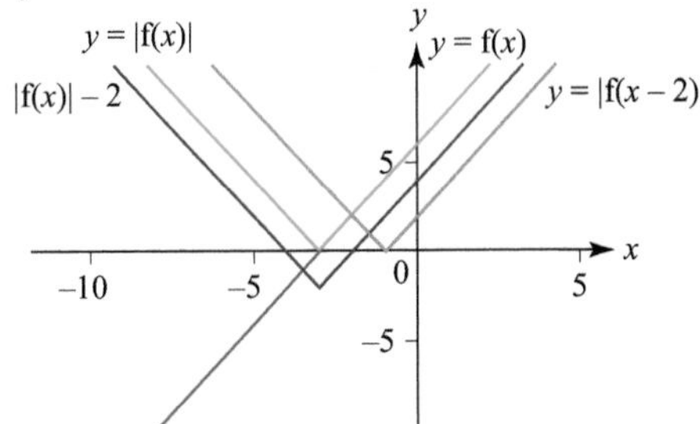

b $y = -x - 5$

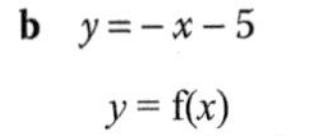

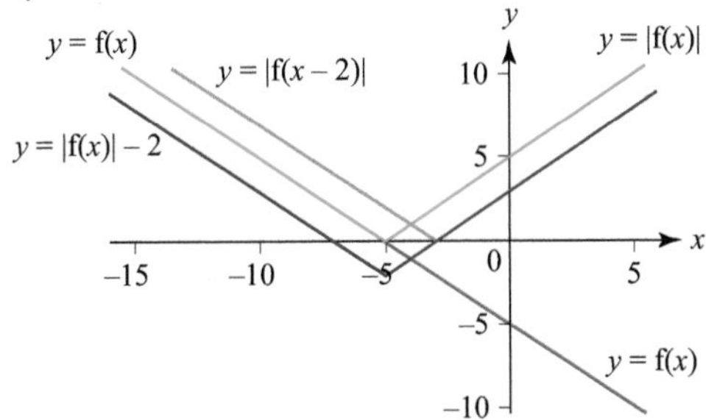

10 a i $0 \leq f(x) < 64$
ii $f^{-1}(x) = \sqrt{x}, 0 \leq x < 64$
iii

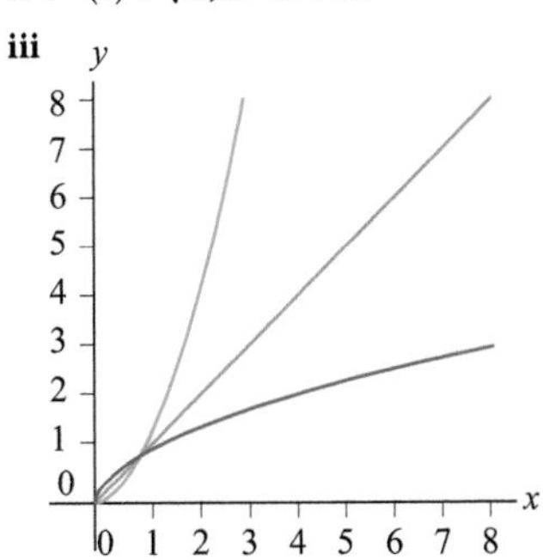

iv $0 \leq f^{-1}(x) < 8$
b i $0 < f(x) < 512$
ii $f^{-1}(x) = 2 + \sqrt[3]{x}, 0 < x < 512$
iii

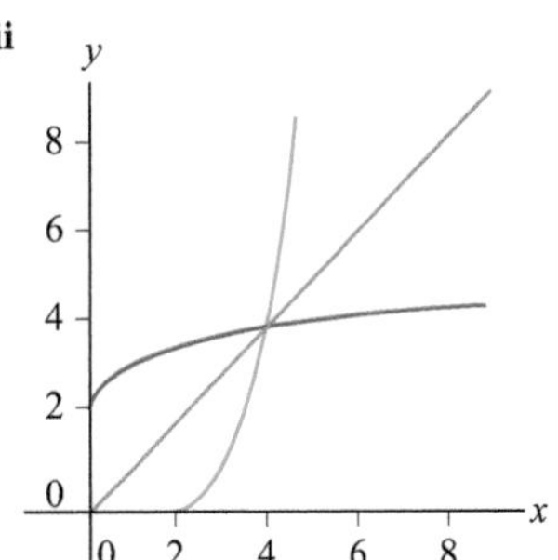

iv $f^{-1}(x) \in (2, 10)$
c i $0 < f(x) < 1$
ii $f^{-1}(x) = \log_2 x, 0 < x < 1$
iii

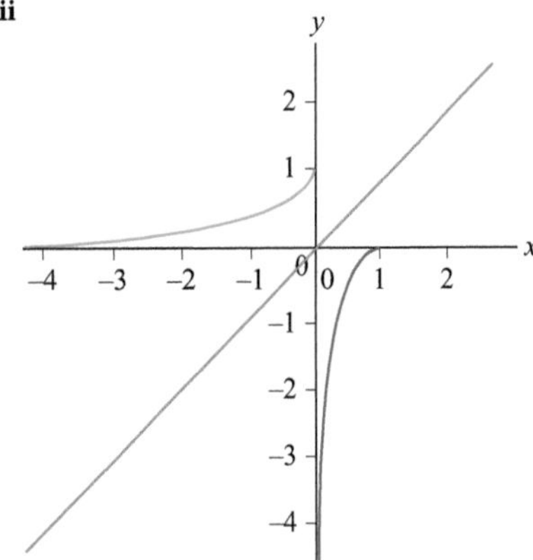

iv $f^{-1}(x) < 0$
11 a $f(x) = x^2, g(x) = 2x$
i $fg(x) = (2x)^2 = 4x^2$
ii This transformation is a vertical stretch with sf 4 (or horizontal stretch with sf ½).

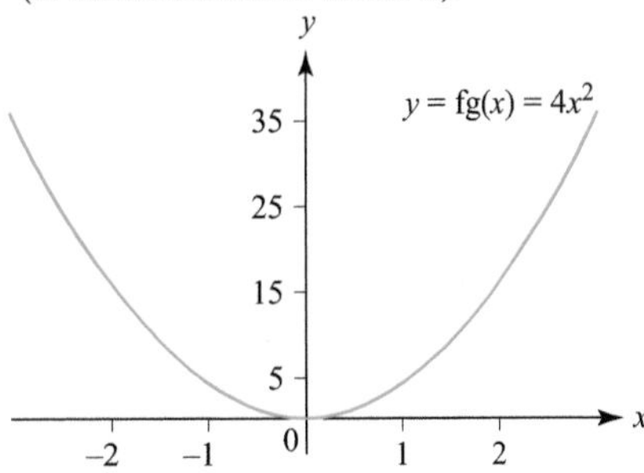

b $f(x) = 4x^2, h(x) = x + 3$
i $fh(x) = 4(h(x))^2$
$= 4(x + 3)^2$
$= 4(x^2 + 6x + 9)$
$= 4x^2 + 24x + 36$
ii This transformation is a translation through vector $\begin{pmatrix} -3 \\ 0 \end{pmatrix}$

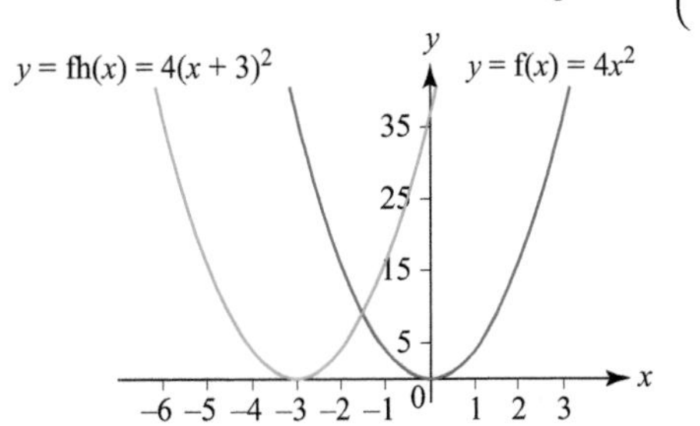

12 a $f(x) = (x + 3)^2, x \in \mathbb{R}$
b $g(x) = 4x, x \in \mathbb{R}$

c $\text{gf}(x) = 4(x+3)^2, x \in \mathbb{R}$

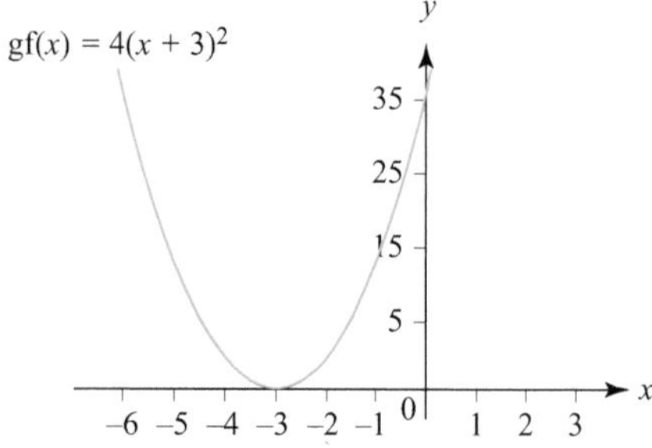

13 a i $x^2 - 6x + 13 \equiv (x-3)^2 + 4$

Translate $y = x^2$ by the vector $\begin{pmatrix} 3 \\ 4 \end{pmatrix}$

ii $4x^2 + 12x + 8 \equiv (2x+3)^2 - 1$

First, transform $y = x^2$ into $y = (2x)^2$ by a stretch sf $\frac{1}{2}$ parallel to the x-axis.

Then transform to $y = (2(x+1.5))^2 = (2x+3)^2$ by a translation by the vector $\begin{pmatrix} -1.5 \\ 0 \end{pmatrix}$

Then transform to $y = (2x+3)^2 - 1$ by a translation by the vector $\begin{pmatrix} 0 \\ -1 \end{pmatrix}$

b i $(x+2)^3 - 7$

Start with $y = x^3$

Use translation $\begin{pmatrix} -2 \\ 0 \end{pmatrix}$ to transform to $(x+2)^3$

Use translation $\begin{pmatrix} 0 \\ -7 \end{pmatrix}$ to transform to $(x+2)^3 - 7$

ii $(3x-5)^3 + 6$

Start with $y = x^3$

Transform to $(3x)^3$. This is a stretch sf $\frac{1}{3}$ parallel to the x-axis.

Then transform to $(3x-5)^3 = \left(3\left(x - \frac{5}{3}\right)\right)^3$. This is a translation $\begin{pmatrix} \frac{5}{3} \\ 0 \end{pmatrix}$

Then transform to $(3x-5)^3 + 6$. This is a translation $\begin{pmatrix} 0 \\ 6 \end{pmatrix}$

OR

Transform to $27x^3$. This is a stretch sf 27 parallel to the y-axis.

Then transform to $27\left(x - \frac{5}{3}\right)^3$. This is a translation $\begin{pmatrix} \frac{5}{3} \\ 0 \end{pmatrix}$

Then transform to $27\left(x - \frac{5}{3}\right)^3 + 6 = (3x-5)^3 + 6$. This is a translation $\begin{pmatrix} 0 \\ 6 \end{pmatrix}$

14 a $f(x) = e^x$ **b** $f(x) = \log(x-1)^3$

15 a $\text{gf}(x) = (-x)^2, x \in \mathbb{R}$

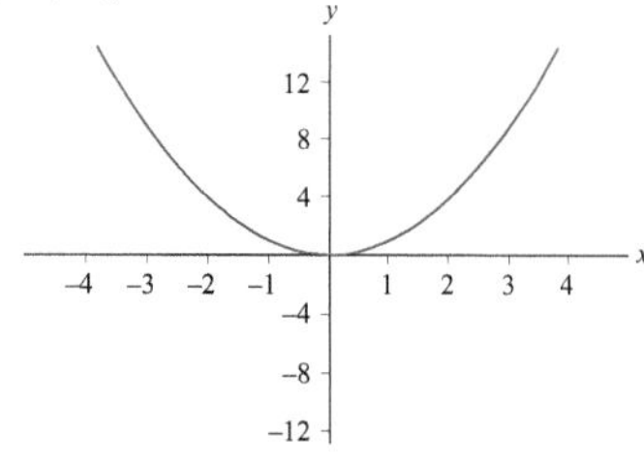

b $\text{gf}(x) = x^3, x \in \mathbb{R}$

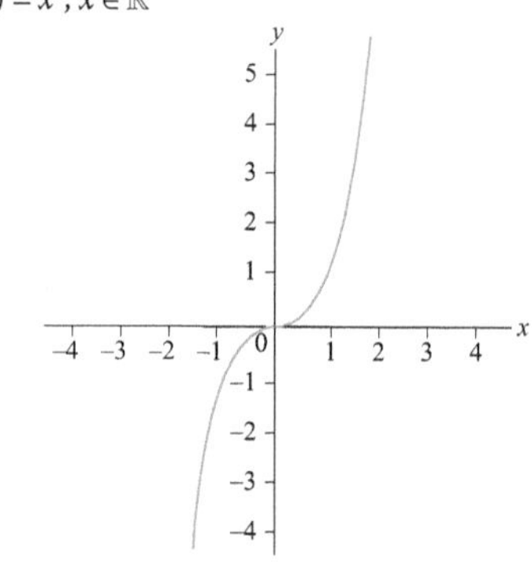

Exercise 12.2B Reasoning and problem-solving

1 a

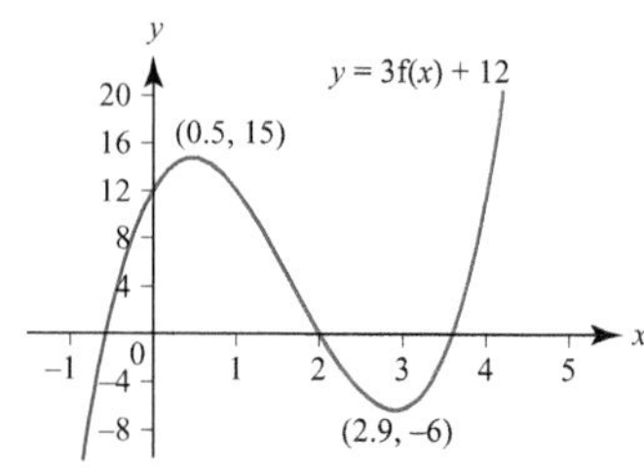

b Simple ways in which the domain could be restricted include $0.5 < x < 2.9$ *or* $x < 0.5$ *or* $x > 2.9$

2 a Transform w to $(w - 15)$ by the translation $\begin{pmatrix} 15 \\ 0 \end{pmatrix}$

Transform $(w-15)$ to $|(w-15)|$ by a reflection in the w-axis of the portion of the graph below the w-axis.

Transform $|(w-15)|$ to $2|(w-15)|$ by a stretch parallel to the S-axis, sf 2

Transform $2|(w-15)|$ to $-2|(w-15)|$ by a reflection in the w-axis

Transform $-2|(w-15)|$to $-2|(w-15)| + 30$ by the translation $\begin{pmatrix} 0 \\ 30 \end{pmatrix}$

b

w	0	2	4	6	8	10	12	14	16	18	20	22	24	26	28	30
S	0	4	8	12	16	20	24	28	28	24	20	16	12	8	4	0

c

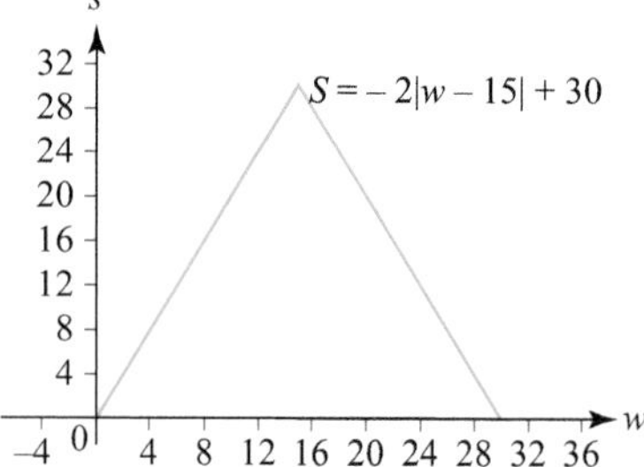

d £30 000

3 $y = |5x - 36|$

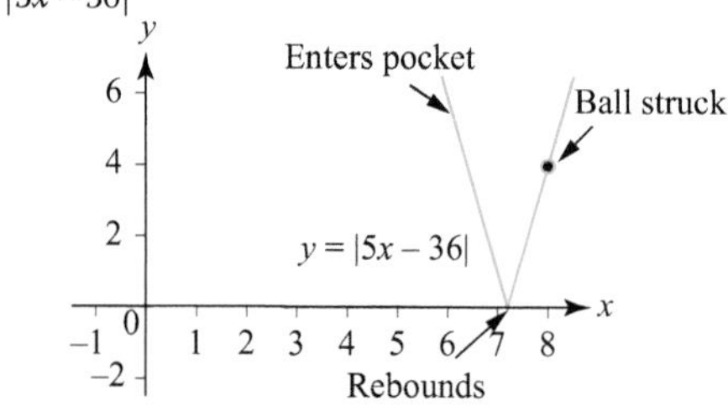

4 $f(x) = \frac{-4}{3}|x - 16.5| + 22$

a

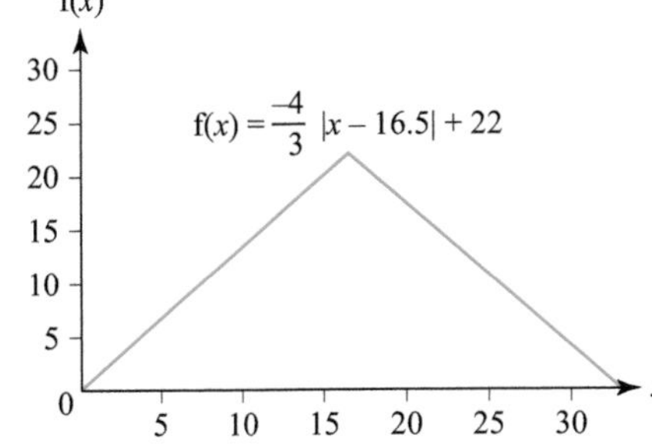

b Domain is $\{x \in R : 0 \le x \le 33\}$, Range is $\{y \in R : 0 \le y \le 22\}$

c 22 m

d 33 m

5 **a** There is no inverse function for f(x) because f(x) is MANY-ONE.

b There is no inverse function for g(x) because g(x) is MANY-ONE.

c The inverse function is $h^{-1}(x) = \sqrt{x} - 2, x \ge 0$

6 **a** Domain is $\{x \in R; x \ne 2\}$

Range is $\{y \in R; y \ne 0\}$

b $f^{-1}(x) = \frac{1}{x} + 2$

Domain of $f^{-1}(x)$ is $\{x \in \mathbb{R}; x \ne 0\}$; Range of $f^{-1}(x)$ is $\{y \in R; y \ne 2\}$

c The domain of f(x) is the same as the range of $f^{-1}(x)$

The range of f(x) is the same as the domain of $f^{-1}(x)$

d

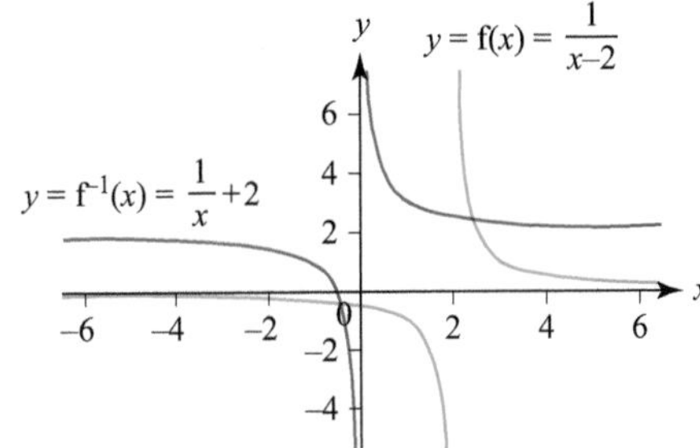

7 **a, b, c** and **d**

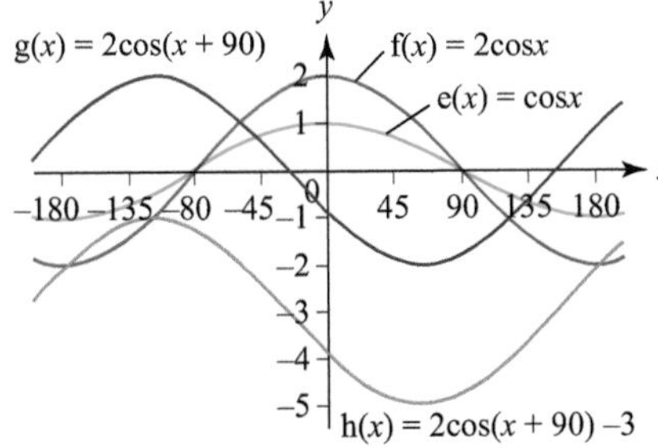

e f(x) is a vertical stretch of e(x) by sf 2

g(x) is a translation of f(x) by $\begin{pmatrix} -90 \\ 0 \end{pmatrix}$

h(x) is a translation of g(x) by $\begin{pmatrix} 0 \\ -3 \end{pmatrix}$

8 Reflection in the x-axis;

Translation by $\begin{pmatrix} 30 \\ 0 \end{pmatrix}$

9 **a**

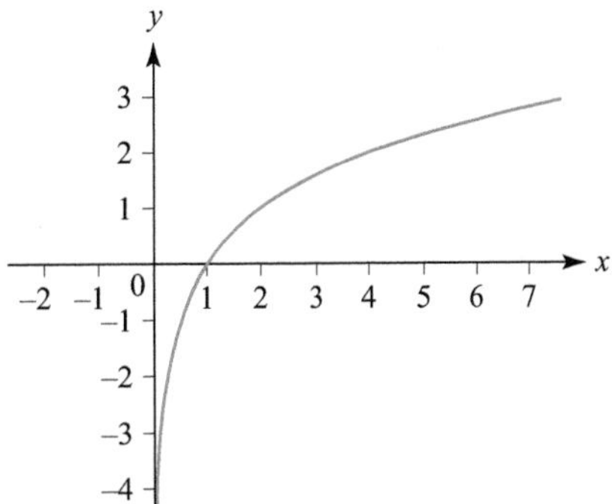

b

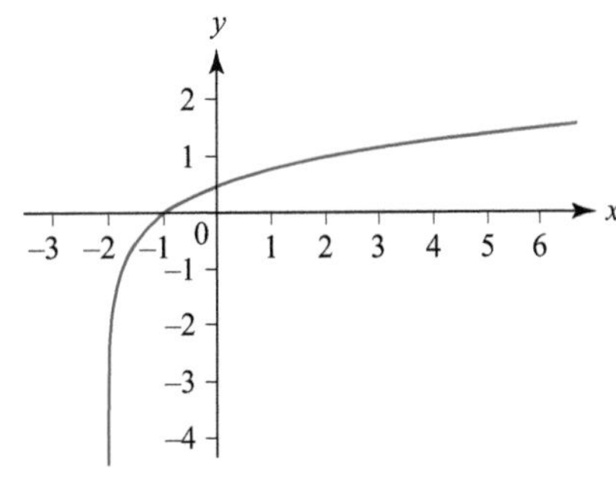

c $y = \frac{1}{2}\log_2(x+2)$

10 **a** **i** Stretch, parallel to y-axis, sf 3; translation $\begin{pmatrix} 0 \\ -12 \end{pmatrix}$

ii $fg(e^x) = 3e^x - 12$

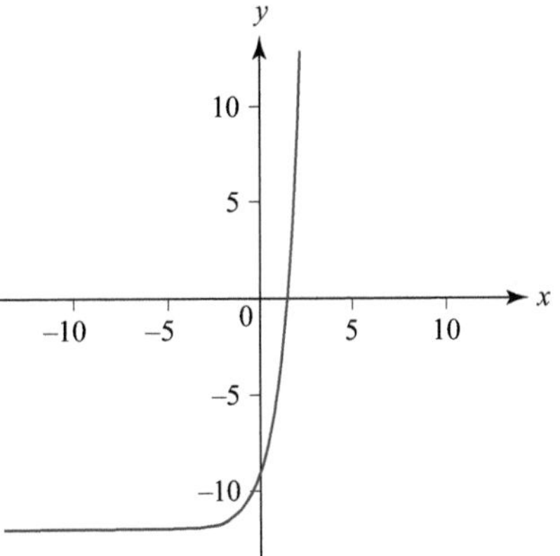

b **i** Stretch, parallel to y-axis, sf 3; translation $\begin{pmatrix} 0 \\ -4 \end{pmatrix}$

ii $gf(e^x) = 3e^x - 4$

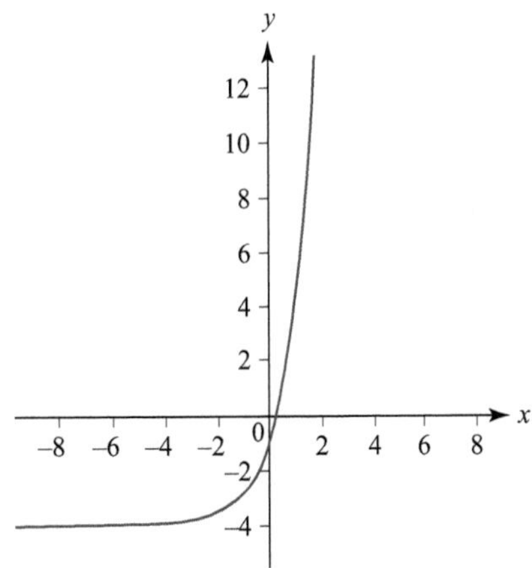

11 Her mistake is that $|4-x| \neq 4+x$

$|f(x)| = \frac{1}{2}x \Rightarrow$

$4-x = \frac{1}{2}x$ or $4-x = -\frac{1}{2}x$

$4 = \frac{3}{2}x$ or $4 = \frac{1}{2}x$

$x = \frac{8}{3}$ or $x = 8$

12 a Domain $\{x \in \mathbb{R} : x > 0\}$
Range $\{y \in \mathbb{R}\}$

b Saqib is not correct.
$fg(x) = (\ln x)^3$ is a function with domain $\{x \in \mathbb{R} : x > 0\}$
$gf(x) = \ln x^3$ is also a function if you restrict the domain to $\{x \in \mathbb{R} : x^3 > 0\}$, that is, $\{x \in \mathbb{R} : x > 0\}$

Exercise 12.3A Fluency and skills

1 a $t = 5 \to \left(5, \frac{4}{5}\right)$

$t = 2 \to (2, 2)$

$t = -3 \to \left(-3, \frac{-4}{3}\right)$

b $t = 5 \to \left(\frac{3}{25}, -10\right)$

$t = 2 \to \left(\frac{3}{4}, -4\right)$

$t = -3 \to \left(\frac{1}{3}, 6\right)$

c $t = 5 \to \left(-\frac{3}{2}, -\frac{3}{7}\right)$

$t = 2 \to (-3, 0)$

$t = -3 \to \left(\frac{-1}{2}, -5\right)$

d $t = 5 \to \left(\frac{15}{6}, 43\right)$

$t = 2 \to (3, 4)$

$t = -3 \to \left(\frac{1}{2}, \frac{-23}{3}\right)$

2 a $(y-1)^2 = 4x - 8$

b $y = -\frac{x^3}{2}$

c $y = \frac{3}{2x}$

d $y = \frac{48}{x^4}$

e $y = \frac{2(x-1)}{1+x}$

f $x^2 + y^2 = 4$

3 a

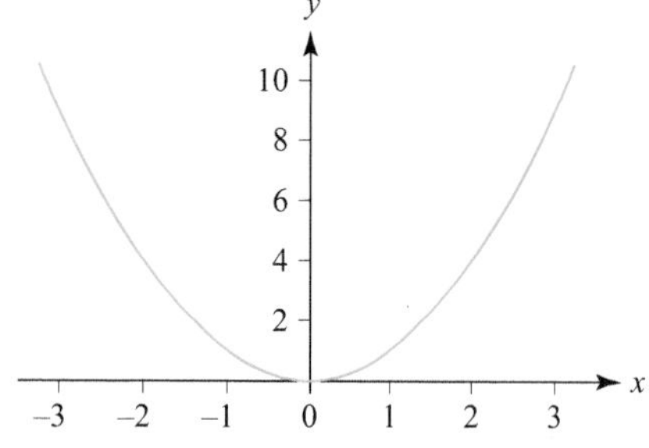

b

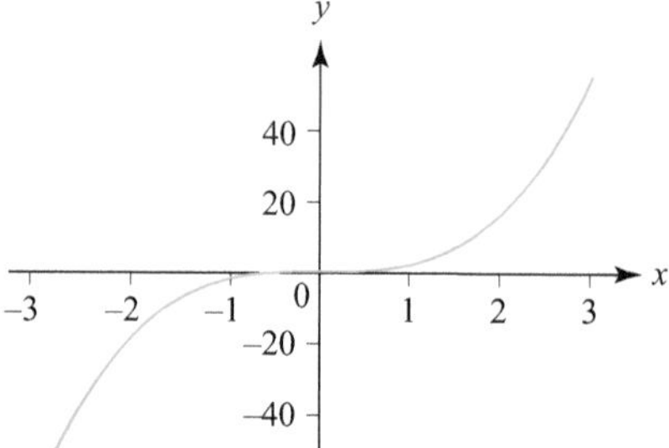

c

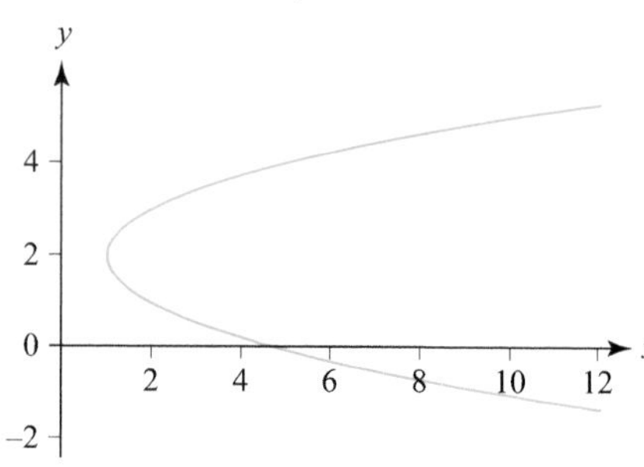

d

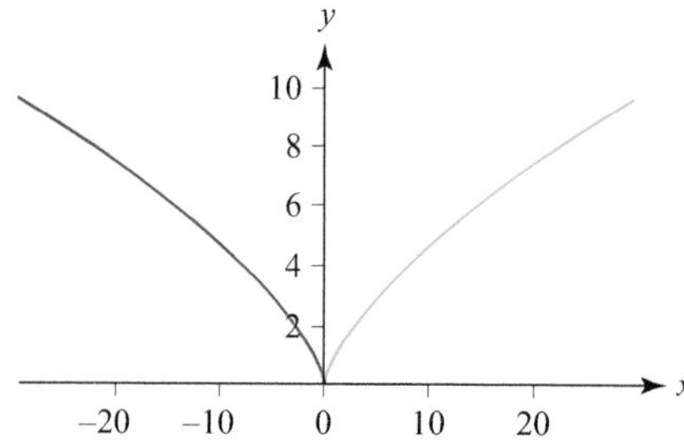

e

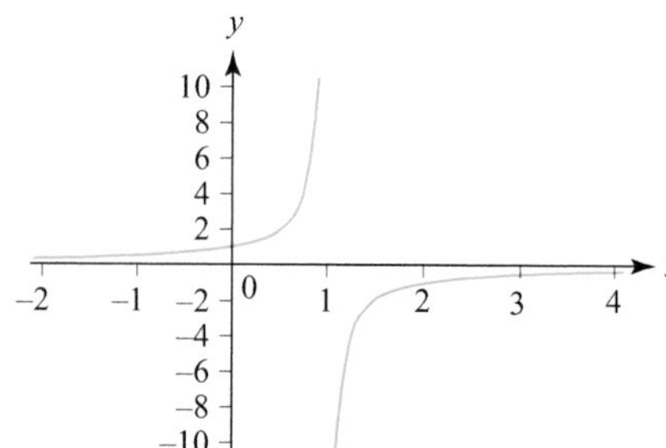

f

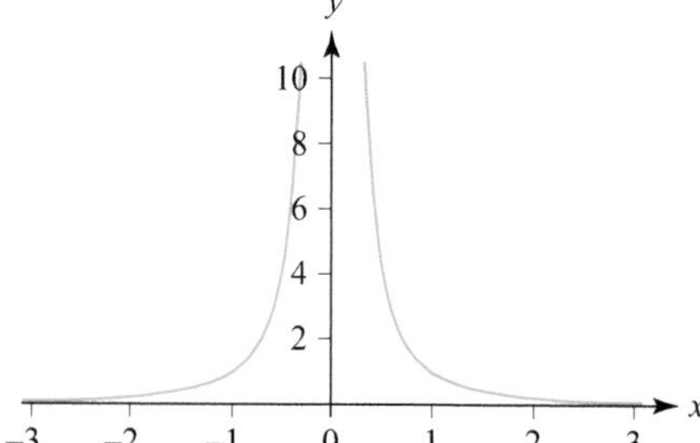

4 a Akeem is correct since, if $t = 2$ then the expression for x would be undefined.

b $y^2 = \frac{4}{x}$

$t > 2 \Rightarrow x > 0$ so domain is $\{x \in \mathbb{R} : x > 0\}$

$\Rightarrow y > 0$ so range is $\{y \in \mathbb{R} : y > 0\}$

5 $y^2 = (2at)^2 = 4a^2t^2$

$4ax = 4a(at^2) = 4a^2t^2$

So $y^2 = 4ax$

6 $25 = xy^2$ domain $x > 0$; range $y > 0$

7 $\sin^2\theta + \cos^2\theta = 1$

$x = 5\sin\theta;\ y = 5\cos\theta \Rightarrow \frac{x^2}{25} + \frac{y^2}{25} = 1$

So $x^2 + y^2 = 25$, which is a circle, centre O, radius 5

8 a $(x-8)^2 + (y-6)^2 = 49$ b $(x-3)^2 + (y+1)^2 = 25$

c $(x+4)^2 + (y-1)^2 = \frac{1}{4}$ d $(x+4)^2 + (y+3)^2 = 2$

9 a $\left(\frac{1}{2}, \frac{-3}{2}\right)$ b $\left(\frac{-19}{7}, \frac{-6}{25}\right)$

10 a $x = t^3;\ y = t^4$

b $x = t + 3;\ y = t^2 + 3t$

c $x = \frac{5}{1-t^2};\ y = \frac{5t}{1-t^2}$

d $x = \frac{1 \pm \sqrt{1+4t^4}}{2t^4};\ y = \frac{1 \pm \sqrt{1+4t^4}}{2t^3}$

e $x = \frac{-3t \pm \sqrt{9t^2+4t}}{2};\ y = \frac{-3t^2 \pm t\sqrt{9t^2+4t}}{2}$

Exercise 12.3B Reasoning and problem-solving

1 a When $t = 3.16$ s

b 2371.71 m

c $750\sqrt{5}$ m

2 $x = -10\sin\theta;\ y = -10\cos\theta$

When $\theta = 90°$, child is at (–10, 0)

When $\theta = 135°$, child is at $(-5\sqrt{2}, 5\sqrt{2})$

When $\theta = 180°$, child is at (0, 10)

When $\theta = 270°$, child is at (10, 0)

3

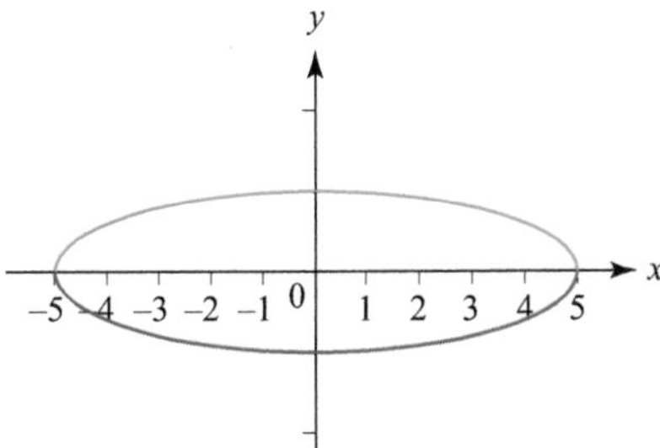

4 Yes he does succeed

5 a 125 m b $t = 7$ s c 35 m

6 $y^2 = 4x^2(1 - x^2)$

7 a $x = \sin\theta + 6;\ y = \frac{3}{2}\cos\theta$

b $(x-6)^2 + \frac{4}{9}y^2 = 1$

8 a Curve A:

$y = 3 \times 3^{-2t}$

$\Rightarrow y = 3 \times \frac{5}{x}$

$\Rightarrow y = \frac{15}{x}$

Curve B:

$t = \frac{3}{x}$

$\Rightarrow y = 5 \times \frac{3}{x}$

$\Rightarrow y = \frac{15}{x}$

Hence curves A and B are the same.

b $y = \frac{15}{x};\ x > 0$

9 $(2\sqrt{5}+7,\ 2\sqrt{5}+5)$ and $(7-2\sqrt{5},\ 5-2\sqrt{5})$

10 (9, 6) and (9, –6)

11 (5, –4) and (29, –20)

12 a $x = 1 + 5\cos\theta$, $y = 3 + 5\sin\theta$ b $x = 1 - 5\sin\alpha$, $y = 3 + 5\cos\alpha$

13 Circle is $x^2 + (y-4)^2 = 9$

For the parabola, $t = -\frac{x}{5}$

$\Rightarrow y = \frac{2x^2}{25}$

$\Rightarrow \frac{25y}{2} = x^2$

Parabola and circle intersect when

$\frac{25y}{2} + (y-4)^2 = 9$

$\Rightarrow y^2 + \frac{9}{2}y + 7 = 0$

Discriminant $= -\frac{31}{4} < 0 \Rightarrow$ no solutions

Hence the circle and parabola do not intersect.

Exercise 12.4A Fluency and skills

1 a $\frac{x(3x-11)}{(x-3)(x-5)}$ b $\frac{-2y(y+18)}{(y+8)(y-2)}$

c $\frac{x(7y-23)}{(y-3)(2y-7)}$ d $\frac{-z(z+57)}{(2z+9)(3z-4)}$

2 a i $\frac{2x-1}{x+2}$ ii $\frac{x+2}{2x-1}$

b 1

3 a $\frac{25(z+9)}{96(z+3)}$ b $\frac{3}{4w^2}$

c $\frac{3n-7}{2n+3}$ d $\frac{2(m-2)}{3m+1}$

4 a $8x + 14$ b $3x + 21$

5 a $x^2 + 5x - 1$

b $x^2 + 2x + \frac{5}{2}$ remainder $= -\frac{111}{2}$

c $x^2 + 5x - 7$ remainder $= -19$

d $2x^2 - \frac{1}{3}x - \frac{56}{9}$ remainder $= -\frac{52}{9}$

e $x^2 - 3x + 9$

6 a $\frac{107}{4}$ b 13

c $\frac{89}{9}$ d $\frac{27\,609}{8}$

7 $(x-1)(x+1)(x^2-3)$

8 a Remainder $= 6\left(\frac{4}{3}\right)^3 - 5\left(\frac{4}{3}\right)^2 - 16\left(\frac{4}{3}\right) + 16 = 0$

Hence, $(3x-4)$ is a factor

b Remainder $= 2\left(-\frac{1}{2}\right)^3 - \left(-\frac{1}{2}\right)^2 + 7\left(-\frac{1}{2}\right) + 4 = 0$

Hence, $(2x+1)$ is a factor

c Remainder $= 8(4)^3 - 36(4)^2 + 18(4) - 8 = 0$

Hence, $(2x-8)$ is a factor

9 a Remainder $= 6\left(-\frac{1}{3}\right)^3 + 5\left(-\frac{1}{3}\right)^2 + 13\left(-\frac{1}{3}\right) + 4 = 0$

Hence, $(3x+1)$ is a factor

b $(3x+1)(2x^2+x+4)$

10 a Remainder $= 3\left(\frac{2}{3}\right)^3 + 10\left(\frac{2}{3}\right)^2 - 23\left(\frac{2}{3}\right) + 10 = 0$

Hence, $(3x-2)$ is a factor

b $(3x-2)(x+5)(x-1)$

c $\frac{(x+5)(x-1)}{(x+2)}$

11 a $(x+2)(2x-1)(3x-2)$

b $(4x+3)(x-1)(x-2)$

c $x(6x-5)(3x^2-2x-2)$

Exercise 12.4B Reasoning and problem-solving

1 $12x^3 + 2x^2 - 54x + 40$

2 $q = -4, p = -20$

3 $$\frac{x}{x-1} + \frac{1}{(x-1)(x-2)} = \frac{x(x-1)(x-2)+(x-1)}{(x-1)^2(x-2)} = \frac{x(x-2)+1}{(x-1)(x-2)} = \frac{x^2-2x+1}{(x-1)(x-2)} = \frac{(x-1)^2}{(x-1)(x-2)} = \frac{x-1}{x-2}$$

So the only vertical asymptote is $x = 2$

4 $x = a$ or $x = 2 - a$

5 $(x^2-9) = (x+3)(x-3)$ so we need to show that both $f(3) = 0$ and $f(-3) = 0$

$f(3) = (3)^4 + 6(3)^3 - 4(3)^2 - 54(3) - 45 = 0$ so $(x-3)$ is a factor

$f(-3) = (-3)^4 + 6(-3)^3 - 4(-3)^2 - 54(-3) - 45 = 0$ so $(x+3)$ is a factor

Hence, (x^2-9) is a factor of $x^4 + 6x^3 - 4x^2 - 54x - 45$

6 $(x-2)(x+3) = x^2 + x - 6$

7 $(x-1)(x+4)$

8 $f(2) = (2)^4 - 10(2)^3 + 37(2)^2 - 60(2) + 36$

$= 16 - 80 + 148 - 120 + 36 = 0$

$x^4 - 10x^3 + 37x^2 - 60x + 36 \equiv (x-2)(x^3 - 8x^2 + 21x - 18)$

$(2)^3 - 8(2)^2 + 21(2) - 18 = 0$

So $(x-2)(x-2)$ is a factor.

$x^4 - 10x^3 + 37x^2 - 60x + 36 \equiv (x-2)^2(x-3)^2$

9 a $30\left(-\frac{2}{5}\right)^3 + 7\left(-\frac{2}{5}\right)^2 - 12\left(-\frac{2}{5}\right) - 4 = 0$

$\Rightarrow (5x+2)$ is a factor

b $\frac{(5x+2)(3x-2)}{x}$

c i $x \in \mathbb{R}, x \neq 0; f(x) \in \mathbb{R}$

ii $f'(x) = \frac{15x^2+4}{x^2}$

$15x^2 + 4 > 4 > 0$ for all $x \neq 0$

$x^2 > 0$ for all $x \neq 0$

Positive ÷ positive = positive

$\therefore f'(x) > 0$

10 a $(2x-5)$ is a factor of $f(x) \Rightarrow f\left(\frac{5}{2}\right) = 0$

$a\left(\frac{5}{2}\right)^4 + b\left(\frac{5}{2}\right)^2 - 75 = 0$

$\Rightarrow \frac{625}{16}a + \frac{25}{4}b - 75 = 0$

Let $x = -\frac{5}{2}$

$$f\left(-\frac{5}{2}\right) = a\left(-\frac{5}{2}\right)^4 + b\left(-\frac{5}{2}\right)^2 - 75 = \frac{625}{16}a + \frac{25}{4}b - 75 = 0$$

$\Rightarrow (2x+5)$ is a factor of $f(x)$

b $a = 4, b = -13, c = 3$

11 The remainder is not a constant term that can be added on at the end of the quotient.

$(4x^4 + 6x^3 + x - 8) \div (2x+1) = 2x^3 + 2x^2 - 0.5 - \frac{7.5}{2x+1}$

12 $b = \frac{1}{2}$ or -4

13 a If $(x-a)$ is a factor of $f(x)$ then $f(x) = (x-a)p(x)$

If $(x-a)$ is a factor of $g(x)$ then $g(x) = (x-a)q(x)$

So $[f(x) - g(x)] \equiv (x-a)p(x) - (x-a)q(x)$

$\equiv (x-a)[p(x) - q(x)]$

$\equiv (x-a)r(x)$

So $(x-a)$ is a common factor of $f(x) - g(x)$

b $(x-a)$ is a common factor of

$kx^3 + 3x^2 + x + 4 - (kx^3 + 2x^2 + 9x - 8) = 0$

or $x^2 - 8x + 12 = 0$ or $(x-2)(x-6) = 0$

So for these polynomials to have a common factor then $x = 2$ or 6

When $x = 2$, $k(2)^3 + 3(2)^2 + 2 + 4 = 0$ and

$k(2)^3 + 2(2)^2 + 9(2) - 8 = 0 \to k = \frac{-9}{4}$

When $x = 6$, $k(6)^3 + 3(6)^2 + 6 + 4 = 0$ and

$k(6)^3 + 2(6)^2 + 9(6) - 8 = 0 \to k = \frac{-59}{108}$

Exercise 12.5A Fluency and skills

1 $\frac{7}{(x-4)} - \frac{3}{(3x+2)}$

2 a B = 4 and A = −4

b C = 2, D = 4, E = −5

c F = 3, H = 5 and G = 4

3 a B = 4, A = 3

b E = 17, C = −9 and D = 4

c G = 3, F = 2, H = −5

4 A = 3, C = −1, B = 2

5 $\frac{1}{(1-x)^2} = \frac{1}{(1-x)^2}$ so there are no partial fractions.

$\frac{x}{(1-x)^2} = \frac{-1}{(1-x)} + \frac{1}{(1-x)^2}$ so there are partial fractions.

6 a $\frac{4}{x+3} - \frac{3}{x-2}$

b $\frac{2}{x-1} - \frac{2}{x+1} + \frac{6}{(x+1)^2}$

c $\frac{5}{2x+5} - \frac{5}{2x-5} + \frac{40}{(2x-5)^2}$

7 a $\frac{2}{x} - \frac{2}{x+4}$ **b** $\frac{-1}{x+1} + \frac{1}{x-3}$

c $\frac{23}{18(x+5)} - \frac{23}{18(x-7)} - \frac{62}{3(x-7)^2}$

d $\frac{1}{x} + \frac{20}{11(2x-3)} - \frac{21}{11(x+4)}$

8 a $3 - \frac{2}{x-1} + \frac{4}{x-3}$ **b** $5 + \frac{1}{(x+1)} - \frac{4}{x+5}$

9 a $\frac{3}{x-\sqrt{3}} + \frac{1}{x+\sqrt{3}}$ **b** $\frac{5}{\sqrt{6}x-\sqrt{5}} - \frac{4}{\sqrt{6}x+\sqrt{5}}$

10 $\dfrac{a}{(a-b)(x-a)}+\dfrac{b}{(b-a)(x-b)}$

11 $\mathrm{Q}=b-ac;\ \mathrm{P}=a$

Exercise 12.5B Reasoning and problem-solving

1 $\dfrac{5t-27}{(t-3)(t-7)}$ minutes

2 $\dfrac{1}{2x-1}$

3 $\mathrm{A}=3,\ \mathrm{B}=5$

4 a Let $a=b=c=d=1$

Then $\dfrac{a}{x+c}+\dfrac{b}{x+d}\equiv\dfrac{1}{x+1}+\dfrac{1}{x+1}\equiv\dfrac{2}{x+1}$

Now $\dfrac{a+b}{(x+c)(x+d)}\equiv\dfrac{2}{(x+1)^2}$

Since $\dfrac{2}{(x+1)}\neq\dfrac{2}{(x+1)^2}$ for all x, then

$\dfrac{a+b}{(x+c)(x+d)}\neq\dfrac{a}{x+c}+\dfrac{b}{x+d}$ for all real a,b,c,d

b $a=\dfrac{a+b}{d-c}$ and $b=\dfrac{a+b}{c-d}$

5 Need three partial fractions in order to cope with the repeated $x+1$ term in the denominator

$$\frac{14}{(x-5)(x+1)^2}\equiv\frac{A}{(x-5)}+\frac{B}{(x+1)}+\frac{C}{(x+1)^2}$$

$$14=A(x+1)^2+B(x-5)(x+1)+C(x-5)$$

$$x=5\Rightarrow A=\frac{7}{18}$$

$$x=-1\Rightarrow C=-\frac{7}{3}$$

$$x^2:\ 0=\frac{7}{18}+B\Rightarrow B=-\frac{7}{18}$$

$$\frac{14}{(x-5)(x+1)^2}\equiv\frac{7}{18(x-5)}-\frac{7}{18(x+1)}-\frac{7}{3(x+1)^2}$$

6 a $\dfrac{1}{r+1}-\dfrac{1}{r+2}$

b $\dfrac{1}{1(2)}+\dfrac{1}{2(3)}+\dfrac{1}{3(4)}+\dfrac{1}{4(5)}+\ldots+\dfrac{1}{(n+1)(n+2)}$

$$=\left(1-\frac{1}{2}\right)+\left(\frac{1}{2}-\frac{1}{3}\right)+\left(\frac{1}{3}-\frac{1}{4}\right)+\left(\frac{1}{4}-\frac{1}{5}\right)+\ldots+\left(\frac{1}{n+1}-\frac{1}{n+2}\right)$$

$$=1+\left(\frac{1}{2}-\frac{1}{2}\right)+\left(\frac{1}{3}-\frac{1}{3}\right)+\left(\frac{1}{4}-\frac{1}{4}\right)+\ldots+\left(\frac{1}{n+1}-\frac{1}{n+1}\right)-\frac{1}{n+2}$$

$$=1-\frac{1}{n+2}$$

c As $n\to\infty,\ \dfrac{1}{n+2}\to 0$ so the sum tends to 1

Review exercise 12

1 Let n be any positive integer such that $n\geq 2$

Then $n!=n\times(n-1)\times(n-2)\times\ldots\times 2\times 1$

which has a factor of 2, so is even.

2 $385=5\times 77=5\times 7\times 11$

$385=7\times 55=7\times 5\times 11$

$385=11\times 35=11\times 5\times 7$

So, whichever way you factorise 385 you end up with the same factors.

3 Assume x is rational, y is irrational, but $x-y$ is rational.

Then $x=\dfrac{p}{q}$, where $p,q\in\mathbb{Z}$, and $q\neq 0$.

Since $x-y$ is rational, $x-y=\dfrac{r}{s}$, for some $r,s\in\mathbb{Z}, s\neq 0$.

$y=x-\dfrac{r}{s},=\dfrac{p}{q}-\dfrac{r}{s}=\dfrac{ps-qr}{qs}$

Since $ps-qr$ and qs are integers, and $qs\neq 0$, since $q\neq 0$ and $s\neq 0$, y is rational.

This contradicts our original statement.

Therefore $x-y$ is irrational.

4 a 14 b -125

c 26

5 a Transform $y=x$ into $y=5x$ by a vertical stretch sf 5

Transform $y=5x$ into $y=5x-4$ by a translation $\begin{pmatrix}0\\-4\end{pmatrix}$

b

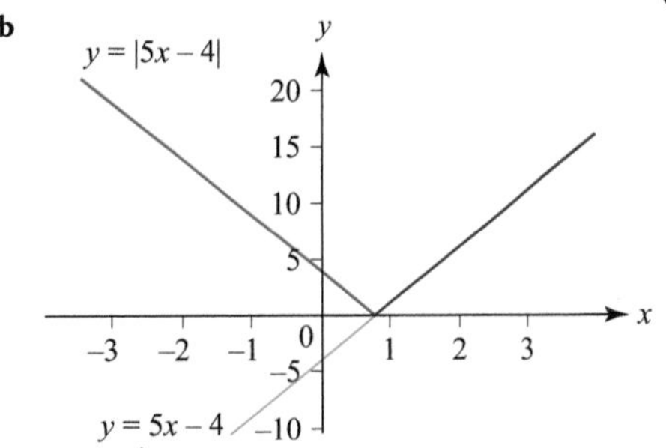

c Either 11 or 19

The sketch of $y=|5x-4|$ shows these values when $x=3$ or -3

6 a $\mathrm{f}^{-1}(x)=\sqrt{\dfrac{5}{x}},\ x>0$ b $\mathrm{f}^{-1}(x)=\dfrac{4x-7}{3},\ x\in\mathbb{R}$

c $\mathrm{f}^{-1}(x)=\dfrac{4x-2}{x-1},\ x\neq 1$

7 a $\mathrm{f}(x)\geq -2$

b $\mathrm{f}^{-1}(x)=\sqrt{x+2}$ which has domain $x\geq -2$

c

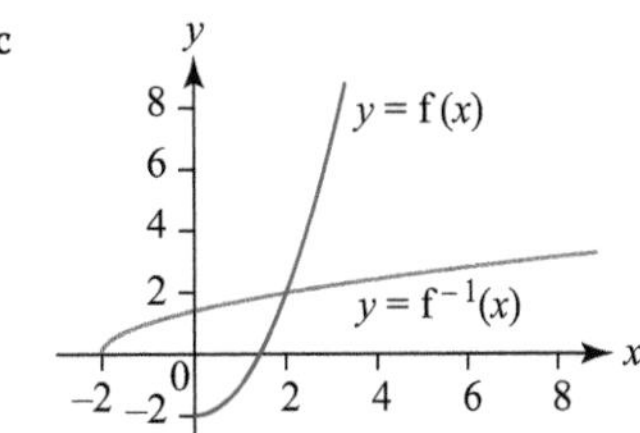

d $\mathrm{f}(x)=\mathrm{f}^{-1}(x)$ when $x=2$

8 a $\mathrm{g}(x)\geq 5$ b $x=3$ or $x=5$

9 a

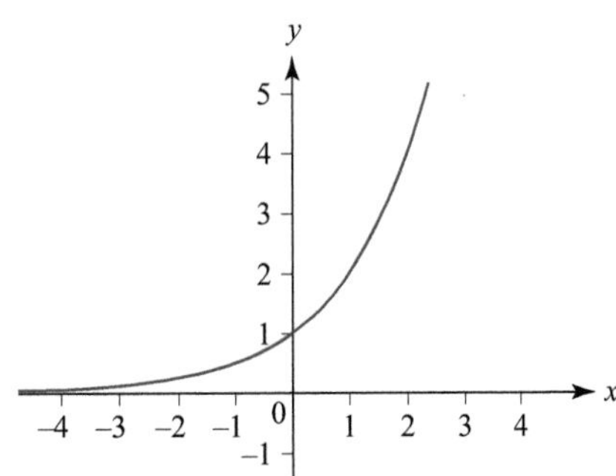

b i

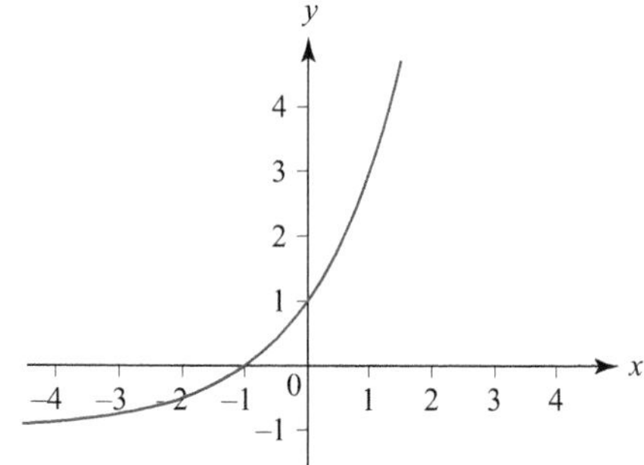

ii $y = 2^{x+1} - 1$

10 $xy + 20 = 0$

11 (5,–1) and (–7,–13)

12 $x^3 - x^2 - x - 1$

13 $\dfrac{n^2 - 4n - 5}{8n^2 + 4n + 3}$

14 $(x - 3)(2x - 1)(2x + 1)$

15 $3 \times \left(\frac{2}{3}\right)^4 - 2 \times \left(\frac{2}{3}\right)^3 + 12 \times \left(\frac{2}{3}\right)^2 - 11 \times \frac{2}{3} + 2 = 0$

16 $\dfrac{2}{x+1} - \dfrac{2}{x-5} + \dfrac{12}{(x-5)^2}$

17 a i $\dfrac{1}{(x+1)} + \dfrac{3}{(x-2)}$

ii $\dfrac{-6}{(x-1)} - \dfrac{3}{(x-2)}$

b $x = -2$ or $x = -3$

Assessment 12

1 a A

b C

2 a B

b C

3 $A = 3, B = -5, C = 2, D = -1, E = 3$

4 a $f^{-1}(x) = \dfrac{3x}{x-1}, x \neq 1$

b $gf(x) = \dfrac{3x+6}{x}, x \neq 0$

c $\dfrac{3x}{x-1} = \dfrac{3x+6}{x}$

$3x^2 = 3x^2 + 3x - 6$

$x = 2$

5 a

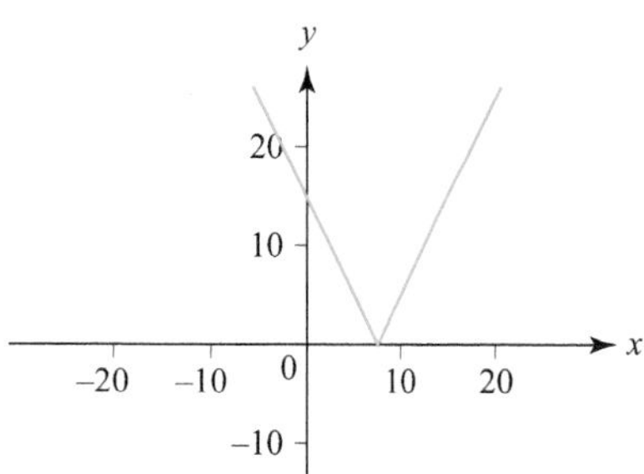

b $x = 9$ or $x = 6$

c $6 \le x \le 9$

6 **A** is one-to-one

B is many-to-one

C is one-to-one

D is many-to-one

7 Assume, for a contradiction, that n is even.

Then $n = 2N$ where N is an integer.

So $n^n = (2N)^{2N} = 2^{2N} \times N^{2N}$

So n^n is even, which contradicts our original assumption.

So our original assumption that n is even, must be wrong, and n must be odd.

8 a $\dfrac{3x-7}{(x-4)(x+1)} - \dfrac{1}{(x-4)} \times \dfrac{(x+1)}{(x+1)}$

$f(x) = \dfrac{3x-7-x-1}{(x-4)(x+1)}$

$= \dfrac{2(x-4)}{(x-4)(x+1)}$

$= \dfrac{2}{x+1}$

b $\{x \in \mathbb{R} : x \neq -1\}$

c $f^{-1}(x) = \dfrac{2-x}{x}, x \in \mathbb{R}, x \neq 0$

d $x = 1$ or -2

9 **A** True

Assume for a contradiction, there exists $x \neq -1$ such that $\dfrac{4x}{(x+1)^2} > 1$

$4x > (x+1)^2$

$x^2 - 2x + 1 < 0$

$(x-1)^2 < 0$

But a square number cannot be negative, so this gives a contradiction, therefore $\dfrac{4x}{(x+1)^2} \le 1$ for all $x \neq -1$

B False

$4! + 1 = 25$, which is not prime

C False

$7 \times 9 \times 11 = 693$, which is not a multiple of 15

D True

$n^3 - n = (n-1)n(n+1)$

In three consecutive numbers, at least one is bound to be even.

In three consecutive numbers, at least one is bound to be a multiple of 3

$2 \times 3 = 6$, hence $n^3 - n$ is divisible by 6

10 $3(x+1) - 4(x-2) = 2(x+1)(x-2)$

$0 = 2x^2 - x - 15$

$0 = (2x+5)(x-3)$

$x = -\dfrac{5}{2}$ or 3

11 a

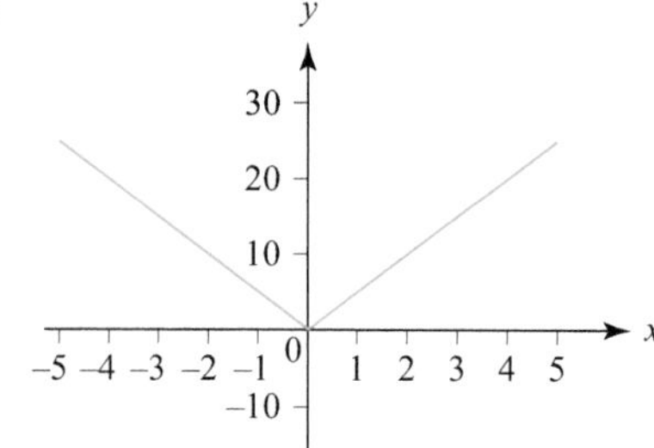

b

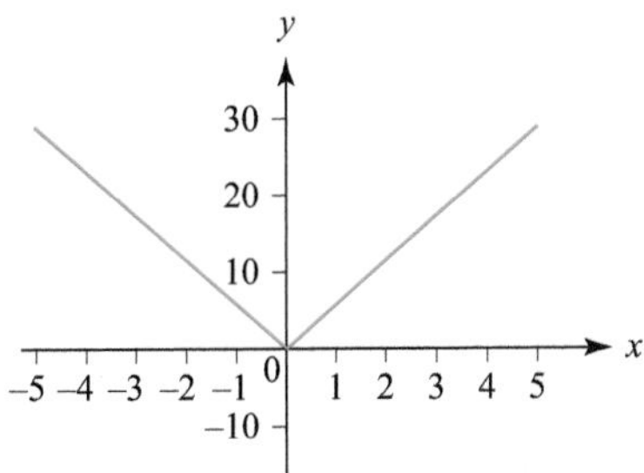

c Stretch, sf 2 in y-direction followed by translation $\begin{pmatrix} 0 \\ -1 \end{pmatrix}$

12 $A=3, B=-2, C=1, D=-3, E=10$

13 a $x^4=\sin^2 t$

$y^2=9\sin^2 t\cos^2 t$

$y^2=9\sin^2 t(1-\sin^2 t)$

$y^2=9x^4(1-x^4)$

$y=3x^2\sqrt{(1-x^4)}$

b Let $y=2$

$3\sin t\cos t=2$

$\sin t\cos t=\frac{2}{3}$

$\sin^2 t\cos^2 t=\frac{4}{9}$

$\sin^2 t(1-\sin^2 t)=\frac{4}{9}$

$\sin^4 t-\sin^2 t+\frac{4}{9}=0$

$(\sin^2 t)^2-(\sin^2 t)+\frac{4}{9}=0$

$b^2-4ac=(-1)^2-4\times1\times\frac{4}{9}$

$=-\frac{7}{9}$

$b^2-4ac<0\Rightarrow$ no solutions for t

14 a $\text{fg}(x)=(3x-2)^2$

b $x=1$

c $\text{f}^{-1}(x)=\frac{1}{2}\ln x, x>0$

d $x=\frac{1}{2}\ln 5$

15 a $1=\text{A}$

$-2=-\text{B}$

b For $x>2$, both $\text{f}(x)=\frac{1}{x-2}$ and $\text{F}(x)=\frac{1}{x-1}$ are decreasing functions

Hence f(x) + 2F(x) is also a decreasing function.

16 a Let a be an integer

By the principle of prime factorisation, there exist primes $p_1, p_2, ..., p_n$ such that $a=p_1\times p_2\times...\times p_n$

Then $a^2=(p_1\times p_1)\times(p_2\times p_2)\times...\times(p_n\times p_n)$

Suppose 3 divides a^2

Since 3 is prime, at least one of $p_1, p_2, ..., p_n$ must be equal to 3

Since $a=p_1\times p_2\times...\times p_n$ and one of $p_1, p_2, ..., p_n$ must be equal to 3, then it follows that 3 also divides a

b Suppose $\sqrt{3}$ is rational.

Let $\left(\frac{m}{n}\right)^2=3$ where m and n are integers which share no common factors.

Thus $m^2=3n^2$ and 3 divides m^2

Since 3 is prime, it must also divide m

So $m=3k$ for some integer k

Hence, $9k^2=3n^2$

$\Rightarrow 3k^2=n^2$ and 3 divides n^2

Since 3 is prime, it must also divide n

So $n=3j$ for some integer j

Hence, $\frac{m}{n}=\frac{3k}{3j}$ which implies m and n share a common factor, 3

This contradicts our original assumption that m and n share no common factors.

Hence $\sqrt{3}$ is irrational.

Chapter 13

Exercise 13.1A Fluency and skills

1 a $1-3x+6x^2-10x^3+...$ **b** $1+\frac{1}{2}x-\frac{1}{8}x^2+\frac{1}{16}x^3+...$

c $1+\frac{2}{3}x-\frac{1}{9}x^2+\frac{4}{81}x^3+...$ **d** $1-4x+16x^2-64x^3+...$

e $1+6x+27x^2+108x^3+...$ **f** $1+\frac{1}{6}x-\frac{1}{36}x^2+\frac{5}{648}x^3+...$

2 a $\frac{1}{16}-\frac{1}{8}x+\frac{5}{32}x^2+...$ **b** $\frac{1}{3}-\frac{2}{9}x+\frac{4}{27}x^2+...$

c $8-9x+\frac{27}{16}x^2+...$ **d** $\frac{1}{3}-\frac{2}{9}x+\frac{1}{9}x^2+...$

e $3+\frac{1}{6}x-\frac{1}{216}x^2+...$ **f** $2+\frac{8}{3}x+\frac{32}{9}x^2+...$

g $4-\frac{9}{2}x+\frac{243}{32}x^2+...$ **h** $\frac{3}{2}+2x+2x^2+...$

3 a $1-3kx+6k^2x^2+...$ **b** $2+\frac{k}{4}x-\frac{k^2}{64}x^2+...$

c $1-\frac{2}{k}x+\frac{3}{k^2}x^2+...$ **d** $\frac{1}{k}-\frac{2}{k^{\frac{3}{2}}}x+\frac{3}{k^2}x^2+...$

4 a $-\frac{1}{4}<x<\frac{1}{4}$, or $|x|<\frac{1}{4}$ **b** $-3<x<3$, or $|x|<3$

c $-\frac{4}{5}<x<\frac{4}{5}$, or $|x|<\frac{4}{5}$ **d** $-\frac{3}{2}<x<\frac{3}{2}$, or $|x|<\frac{3}{2}$

5 a $x-2x^2+...$ **b** $2-3x+4x^2+...$

c $3-8x+13x^2+...$ **d** $1-2x+...$

6 a $\frac{1}{4}-\frac{1}{4}x+\frac{3}{16}x^2+...$ **b** $1-\frac{3}{2}x+\frac{9}{4}x^2+...$

c $3-6x+12x^2+...$ **d** $3+\frac{3}{4}x+\frac{3}{16}x^2+...$

7 a 2 **b** −10

Exercise 13.1B Reasoning and problem-solving

1 a $1+\frac{5}{2}x-\frac{25}{8}x^2+...$

b Let $x=0.05$

$\sqrt{1+5(0.05)}\approx1+\frac{5}{2}(0.05)-\frac{25}{8}(0.05)^2$

$\text{LHS}=\sqrt{\frac{5}{4}}=\frac{\sqrt{5}}{2}$ $\text{RHS}=\frac{143}{128}$

$\therefore\frac{\sqrt{5}}{2}\approx\frac{143}{128}$

so $\sqrt{5}\approx\frac{143}{64}$

c Although when $x = 0.8$, $\sqrt{1+5x} = \sqrt{1+5(0.8)}$
$= \sqrt{5}$

the expansion of $\sqrt{1+5x}$ is only valid for $|x| < \frac{1}{5} = 0.2$

2 a $2 + \frac{1}{4}x - \frac{1}{64}x^2 + \dots$

The full expansion is valid for $-4 < x < 4$

b i Let $x = 0.25$

$\sqrt{4+0.25} \approx 2 + \frac{1}{4}(0.25) - \frac{1}{64}(0.25)^2$

LHS $= \sqrt{\frac{17}{4}} = \frac{\sqrt{17}}{2}$ RHS $= \frac{2111}{1024}$

$\sqrt{17} \approx \frac{2111}{512}$

ii $\frac{1983}{512}$

c $\sqrt{17} - \sqrt{15} \approx \frac{2111}{512} - \frac{1983}{512}$

$= \frac{128}{512}$

$= \frac{1}{4}$

3 a $n = \frac{2}{3}, k = 2$

b 1.19

4 $(1+px)^n = 1 + npx + \frac{n(n-1)}{2!}p^2x^2 + \dots$

$1 + npx + \frac{n(n-1)}{2!}p^2x^2 + \dots \equiv 1 + 12x + 24x^2 + \dots$

Equate x coefficients: $np = 12$ (1)

Equate x^2 coefficients: $\frac{n(n-1)}{2}p^2 = 24$ (2)

Solve (1) and (2) simultaneously to give $p = 8$, $n = \frac{3}{2}$

$(1+8x)^{\frac{3}{2}} \approx 1 + 12x + 24x^2$

Let $x = 0.01$

$(1+8(0.01))^{\frac{3}{2}} \approx 1 + 12(0.01) + 24(0.01)^2$

LHS $= \left(\frac{27}{25}\right)^{\frac{3}{2}}$ RHS $= \frac{1403}{1250}$

$= \frac{81}{125}\sqrt{3}$

$\therefore \frac{81}{125}\sqrt{3} = \frac{1403}{1250}$

so $\sqrt{3} \approx \frac{1403}{810}$

5 a $3 - 5x + 11x^2 - 29x^3 + \dots$

b $|x| < \frac{1}{3}$

6 a $1 + x^2 + \dots$

b

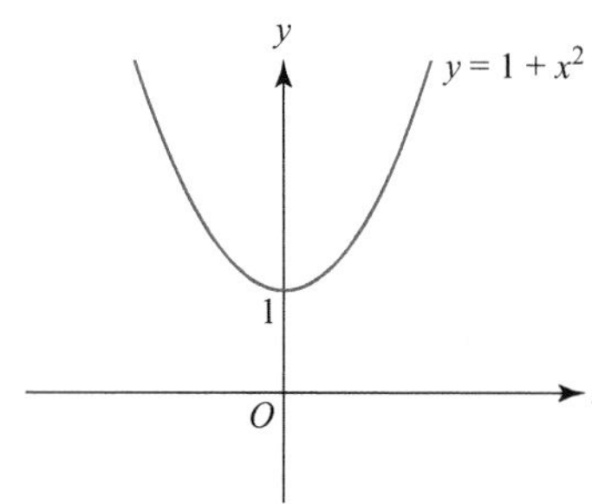

7 a 1.2

b $1 - \frac{7}{4}x + \frac{9}{4}x^2 + \dots$

c $f(x) \approx 1 - \frac{7}{4}x + \frac{9}{4}x^2 + \dots$

$\therefore f'(x) \approx -\frac{7}{4} + \frac{9}{2}x + \dots$

Gradient of tangent at $P = f'(-0.1)$

$\approx -\frac{7}{4} + \frac{9}{2}(-0.1)$

$= -2.2$

Tangent at P is approximated by the line with equation

$y - 1.2 = -2.2(x + 0.1)$

leading to $y = 0.98 - 2.2x$

8 a $\frac{1}{\cos^2 x} = \frac{1}{1 - \sin^2 x}$

$= (1 - \sin^2 x)^{-1}$

$= 1 + (-1)(-\sin^2 x) + \frac{(-1)(-2)}{2!}(-\sin^2 x)^2$

$+ \frac{(-1)(-2)(-3)}{3!}(-\sin^2 x)^3 + \dots$

$= 1 + \sin^2 x + \sin^4 x + \sin^6 x + \dots$

The expansion is valid provided $-1 < \sin x < 1$ which holds provided $-90° < x < 90°$

b The approximation $\frac{1}{\cos^2 x} \approx 1 + \sin^2 x$ is good only if $\sin x$ is close to 0

The approximation $\frac{1}{\sin^2 x} \approx 1 + \cos^2 x$ is good only if $\cos x$ is close to 0

Since $\sin^2 x + \cos^2 x \equiv 1$, it is not possible for both $\sin x$ *and* $\cos x$ to be close to 0

for a given value of x (or, considering their graphs for $0° < x < 90°$, $\sin x \approx 0$ when $x \approx 0$ and $\cos x \approx 0$ when $x \approx 90°$)

Exercise 13.2A Fluency and skills

1 a $u_1 = 8, u_2 = 13, u_3 = 18, u_4 = 23$

b $u_1 = -2, u_2 = 1, u_3 = 6, u_4 = 13$

c $u_1 = 0, u_2 = 8, u_3 = 36, u_4 = 96$

d $u_1 = -\frac{1}{5}, u_2 = -1, u_3 = 1, u_4 = \frac{2}{5}$

2 a $u_2 = 7, u_3 = 19, u_4 = 43$

b $u_2 = 3, u_3 = 11, u_4 = 123$

c $u_2 = \frac{1}{4}, u_3 = \frac{4}{5}, u_4 = \frac{5}{9}$

d $u_2 = 4, u_3 = 1, u_4 = 4$

3 a Decreasing **b** Increasing

c Neither **d** Neither

e Decreasing **f** Neither

4 a 3 **b** 3 **c** 4 **d** 4

5 a 77 **b** $n > 334$

6 a 13 **b** $u_4 = 2$

7 a $L = 5$ $[u_9 = 5.00\ (3\text{ sf})]$

b $L = -7.5$ $[u_{16} = -7.50\ (3\text{ sf})]$

c $L = 2.4$ $[u_6 = 2.40\ (3\text{ sf})]$

d $L = -\frac{5}{4}$ $[u_{13} = -1.25\ (3\text{ sf})]$

e $L=\frac{2}{3}$ $[u_{18}=0.667\text{ (3 sf)}]$

f $L=3+\sqrt{5}$ $[u_{30}=5.24\text{ (3 sf)}]$

g $L=1+\sqrt{3}$ $[u_6=2.73\text{ (3 sf)}]$

h $L=2$ $[u_{792}=2.00\text{ (3 sf)}]$

Exercise 13.2B Reasoning and problem-solving

1 a 127 b 166

2 a $u_2=au_1+7$
$=a(-1)+7$
$=7-a$

$u_3=au_2+7$
$=a(7-a)+7$
$=-a^2+7a+7$

$-a^2+7a+7=19$

$a^2-7a+12=0$

b 64 or 83

3 a $a=2, b=-6$ b 582

4 a i $u_4=a-\frac{2a}{u_3}$
$=a-\frac{2a}{(-2)}$
$=2a$

So $u_1=u_4$ (as the sequence has order 3)
$=2a$

$u_2=a-\frac{2a}{u_1}$
$=a-\frac{2a}{2a}$
$=a-1$

ii $u_3=a-\frac{2a}{u_2}$
$=a-\frac{2a}{(a-1)}$

and $u_3=-2$

So $a-\frac{2a}{(a-1)}=-2$

$a(a-1)-2a=-2(a-1)$

leading to $a^2-a-2=0$

b 300

5 a $p_1=56$

$p_2=0.75(p_1+8)$
$=0.75(56+8)$
$=48$

$p_3=0.75(p_2+8)$
$=0.75(48+8)$
$=42$

Decrease from 2nd to 3rd observations $=48-42=6$ otters

% decrease $=\frac{6}{48}\times 100=12.5\%$

b $p_n\geq 24$ for some n

$p_{n+1}=0.75(p_n+8)$
$\geq 0.75(24+8)$
$=24$

So at least 24 otters were seen in the $(n+1)$th observation

c Using the same argument as in part **b**,
$p_{n+1}\geq 24$ implies $p_{n+2}\geq 24$,
which, in turn, implies $p_{n+3}\geq 24$ and so on
The number of otters seen from the nth observation onwards is always at least 24 and hence the population will never fall to zero.

6 a $m_2=4$

$m_3=a(m_2-1)$
$=a(3)$
$=3a$

$m_4=a(m_3-1)$
$=a(3a-1)$

$m_4=10$

So $a(3a-1)=10$

leading to $3a^2-a-10=0$

b 3 minutes

c 7 rounds

7 a 6 miles

b $e_2=\frac{2}{3}e_1+2$
$=\frac{2}{3}(1.5)+2$
$=3$

$s_2=\frac{1}{2}s_1+k$
$=\frac{1}{2}(2)+k$
$=1+k$

Hypotenuse $=\sqrt{3^2+(1+k)^2}$
$=\sqrt{10+2k+k^2}$

Total distance run $=3+(1+k)+\sqrt{10+2k+k^2}$

So $4+k+\sqrt{10+2k+k^2}=12$

$\sqrt{10+2k+k^2}=8-k$, as required

$k=3$

c 20 miles

8 a $u_2=5p+6$

$\Rightarrow u_3=p(5p+6)+6$
$=5p^2+6p+6$

$u_3=9.2\ \therefore\ 5p^2+6p+6=9.2$

$\Rightarrow 5p^2+6p-3.2=0$

b $L=10$

9 a $p=0.8, q=0.4$ b $L=\frac{4}{3}$

Exercise 13.3A Fluency and skills

1 a 5, 7, 9, 11 b 9, 6, 3, 0

c 3, 10, 17, 24 d $4, \frac{13}{2}, 9, \frac{23}{2}$

2 a $u_n=4n+3$

b $u_n=5-3n$

c $u_n=\frac{3}{2}n+\frac{5}{2}$ or $u_n=\frac{1}{2}(3n+5)$

d $u_n=\frac{2}{3}n+\frac{4}{3}$ or $u_n=\frac{2}{3}(n+2)$

e $u_n=\frac{2}{5}n+\frac{1}{5}$ or $u_n=\frac{1}{5}(2n+1)$

f $u_n=\sqrt{2}n$

3 a 150 b −139

c $\frac{125}{2}$ d −102.4

4 a 585 b −144

c 2623.5 d $100\sqrt{3}$

5 a 50 b 10

c 100 d 10

6 a 480 b 306

c 119 d 1080

Exercise 13.3B Reasoning and problem-solving

1 a $d=5, a=6$ b 61

2 a Find the first term a and the common difference d

$u_n = a+(n-1)d$

$u_5 = 13$ so $a+4d=13$ ...[1]

$u_{10} = 7u_2$ so $a+9d=7(a+d)$

leading to $d=3a$ ...[2]

Solve [1] and [2] to give $a=1, d=3$

$u_n = 1+(n-1)\times 3$
$= 3n-2$

$u_{18} = 3\times 18-2$
$= 52$

b 4: $u_1(=1), u_2(=4), u_5(=13)$ and $u_{18}(=52)$

3 475

4 a $u_{16}=2$ b 392

5 a $S_n = \frac{1}{2}n[2\times 5+(n-1)\times 4]$

$= \frac{1}{2}n(6+4n)$

$= n(2n+3)$, as required

b 77

6 a 17 b 790

7 Let c_n = number of complaints received in the nth month.

Let the first term be a and the common difference d

$a = 152(=c_1)$

$d = c_2 - c_1$
$= 140-152$
$= -12$

a $S_{12} = \frac{1}{2}(12)[2\times 152+11\times(-12)]$
$= 1032$

The total number of complaints received in the first year was 1032, as required.

b $S_6 = \frac{1}{2}(6)[2\times 152+5\times(-12)]$
$= 732$

Number of complaints in 2nd half of the year $= S_{12} - S_6$
$= 1032 - 732$
$= 300$

Reduction in the number of complaints from 1st to 2nd half of the year $= 732-300$
$= 432$

% reduction $= \frac{432}{732}\times 100\%$
$= 59.01\%$

The claim is accurate.

c In the following year,

$n=13$ and $c_{13} = 152+12\times(-12)$
$= 8$ (very low number of complaints, but still possible)

$n=14$, $c_{14} = -4$ which is meaningless

So the model is not suitable for use in the following year.

8 a 3300 b £50 000

9 a i £142 500 ii £450 000

b 13

c The amount Jim earns each year is not likely to increase by the same amount each year forever. Sales made may increase or decrease over time.

10 $1^3+2^3+3^3+...+n^3 \equiv (1+2+3+...+n)^2$; $n=24$

Exercise 13.4A Fluency and skills

1 a 4, 12, 36, 108 b 3, 12, 48, 192

c 4, 2, 1, $\frac{1}{2}$ d $2, \frac{2}{3}, \frac{2}{9}, \frac{2}{27}$

e $\frac{5}{4}, \frac{5}{8}, \frac{5}{16}, \frac{5}{32}$ f $\frac{1}{2}$, 1, 2, 4

2 a $u_n = 5\times 2^{n-1}$ b $u_n = 36\times\left(\frac{2}{3}\right)^{n-1}$

c $u_n = 2\times(-3)^{n-1}$ d $u_n = (-8)\times\left(-\frac{3}{4}\right)^{n-1}$

e $u_n = 7\times\left(-\frac{7}{2}\right)^{n-1}$ f $u_n = (1)\times\left(\frac{1}{6}\right)^{n-1}$

3 a 6138 b 2059

c 241 d −5320

4 a 1594 320 b 189

c 10 200 d 32 784

Exercise 13.4B Reasoning and problem-solving

1 a $r=4, a=\frac{1}{8}$

b 2048

c $u_n = 2^{2n-5}$ so $p=2, q=-5$

2 a 1820 b 1328 600

3 a 2 b 49 146

4 a 9310 b $\frac{3}{16}$ or 0.1875

5 a 125

b i 10 ii 1.260

6 a 1.02

b £2653 (nearest £)

c £56 (nearest £)

7 a 0.75

b v_n = value of car (in £) n years after its purchase

First term $a = v_1$
$= 24\,000\times 0.75$
$= 18\,000$

$v_n = 18\,000\times 0.75^{n-1}$

so $v_3 = 18\,000\times 0.75^2$
$= 10\,125$

3 years after purchase the value of the car was £10 125, as required.

c 8 whole years

8 a i 1600 miles

ii 11 529 miles

iii The total distance driven in Tom's lifetime must be less than the sum to infinity of the series $3125+2500+....$

$S_\infty = \frac{3125}{1-0.8}$
$= 15\,625$

Tom will drive less than 15 625 miles in his lifetime.

b The assumption that the distance driven decreases by 20% every year is unrealistic.

9 a 6.4 mm

b 160 cm^3

c 439 800 km, which is over 55 000 km more than the distance from the Earth to the Moon

10 b_n = amount due (in £) at end of nth month

A geometric model is appropriate since the amount due increased by 5% each month

So the common ratio $r = 1.05$

First term $a = 450$

$b_n = 450 \times 1.05^{n-1}$

and $S_n = \frac{450(1.05^n - 1)}{1.05 - 1}$

$S_{12} = \frac{450(1.05^{12} - 1)}{1.05 - 1}$

$= 7162.706...$

$\frac{S_{12}}{12} = \frac{7162.706...}{12}$

$= 596.892...$

Mean monthly bill in 1st year was £600 (nearest £10)
The bills are increasing in value so median bill = mean of the 6th and 7th terms

$b_6 = 450 \times 1.05^5$

$= 574.326...$

$b_7 = 450 \times 1.05^6$

$= 603.043...$

Mean of 6th and 7th terms $= \frac{574.326... + 603.043...}{2}$

$= 588.6848...$

Median bill for 1st year = £590 (nearest £10)
So the statement used only the mean.

11 21 hours

Review exercise 13

1 **a** $1 + \frac{4}{3}x + \frac{2}{9}x^2 - \frac{4}{81}x^3 + ..., -1 < x < 1$

b $1 - 6x + 24x^2 - 80x^3 + ..., -\frac{1}{2} < x < \frac{1}{2}$

c $27 - 18x + 2x^2 + \frac{4}{27}x^3 + ..., -\frac{9}{4} < x < \frac{9}{4}$

d $1 - \frac{3}{8}x + \frac{15}{128}x^2 - \frac{35}{1024}x^3 + ..., -4 < x < 4$

2 **a** $\sqrt{3} \approx \frac{433}{250} (= 1.732)$

b $\sqrt{3} \approx \frac{111}{64} (= 1.734375)$

3 **a** Increasing **b** Decreasing
c Increasing **d** Decreasing
e Neither **f** Neither

4 **a** 2 **b** 4
c 3 **d** 4

5 **a** nth term $= 6n + 1$; 480

b nth term $= 19 - 4n$; -825

c nth term $= 5n + 4$; 2445

d nth term $= \frac{1}{2}n$; 637.5

e nth term $= 5.5n + 13$; 432.5

f nth term $= \frac{2}{3} - \frac{1}{6}n$; -19

6 **a** nth term $= 3 \times 6^{n-1}$; 1007769

b nth term $= 100 \times \left(\frac{1}{10}\right)^{n-1}$; 111

c nth term $= 200 \times (-0.95)^{n-1}$; 150

d nth term $= \log_2(3) \times 2^{n-1}$; 810

7 **a** 2415 **b** 6140
c 40 **d** 375

Assessment 13

1 C
2 A
3 D
4 A

5 **a** **i** $\frac{1}{2}$ **ii** -1 **iii** 2

b $\frac{1}{2}$

6 **a** Periodic **b** Decreasing **c** Increasing

7 $1 + 4x + 12x^2 + 32x^3 + ...$, valid for $|x| < \frac{1}{2}$

8 **a** £24000 **b** £12288 **c** Year 9

9 **a** $u_2 = 4 - 3k$ **b** $u_3 = 4 - 4k + 3k^2$ **c** $k = 2$ or $\frac{1}{3}$

10 **a** 0.8 **b** 2.56

11 **a** $800 + 2 \times 100 = 1000$ **b** £35000 **c** 120

12 **a** $n = -\frac{2}{3}, a = 6$ **b** $-\frac{320}{3}$

13 **a** 0.5 **b** 240 **c** 480

14 **a** 2500 **b** 80

15 **a** $1 - x - \frac{1}{2}x^2 + ...$

b $1 - \frac{1}{2}x + \frac{3}{8}x^2 + ...$

c $\left(1 - x - \frac{1}{2}x^2\right)\left(1 - \frac{1}{2}x + \frac{3}{8}x^2\right)$

$1 - \frac{3}{2}x + \frac{3}{8}x^2$

16 **a** -7 **b** 6 **c** 15

17 **a** $p = 1$ or $p = -1.8$ **b** -25.6 **c** 2

18 **a** $1 - \frac{1}{2}x - \frac{1}{8}x^2 - \frac{1}{16}x^3 + ...$ **b** $\frac{887}{512}$

19 **a** **i** £2.40 **ii** £12 **iii** £36

b £1188

20 **a** $\frac{a}{1 - r} = 48$

$a + ar = 45$

$\frac{45}{1 + r} = 48(1 - r)$

$1 - 16r^2 = 0$

b 47.8125

21 **a** 15 km **b** 120 km
c 50th day **d** 31st day

22 **a** $2 - \frac{x}{4} - \frac{x^2}{64} - ...$, valid for $|x| < 4$

b 19.975

23 **a** £1000 + 1.03 × £1000
= £2030

b £3090.90

c Year 31

24 **a** 8 **b** -3 **c** 18

25 **a** $A = -2, B = 1$

b $\frac{5}{2}x + \frac{15}{4}x^2 + \frac{65}{8}x^3 + ...$

c $|x| < \frac{1}{2}$

26 **a** $-\frac{33}{5} + -\frac{527}{25}x + -\frac{7873}{125}x^2 + \frac{118127}{625}x^3 + ...$

b $|x| < \frac{1}{3}$

27 a 1.5 m b 5 m c 14 m

28 $x=-6$ or $x=5$
$y=36$ or $y=25$

29 a 5 b 7.5 c −6
d $4+2\sqrt{2}$ e $2+2\sqrt{2}$ f 1

Chapter 14

Exercise 14.1A Fluency and skills

1 a 114.6° b 171.9° c 28.6°

2 a $\frac{\pi}{2}=1.57$ radians b $\frac{\pi}{3}=1.05$ radians
c $\frac{3\pi}{2}=4.71$ radians

3 8 cm, 16 cm²

4

θ		sin θ	cos θ	tan θ
deg	rad			
45	$\frac{\pi}{4}$	$\frac{1}{\sqrt{2}}$	$\frac{1}{\sqrt{2}}$	1
120	$\frac{2\pi}{3}$	$\frac{\sqrt{3}}{2}$	$-\frac{1}{2}$	$-\sqrt{3}$
135	$\frac{3\pi}{4}$	$\frac{1}{\sqrt{2}}$	$-\frac{1}{\sqrt{2}}$	−1
270	$\frac{3\pi}{2}$	−1	0	∞
360	2π	0	1	0

5 $DE=51.8$ cm (3 sf)
Angle $D=1.73$ radians
Angle $E=0.630$ radians

6 $\frac{9}{2}\pi$ cm, $\frac{27}{2}\pi$ cm²

7 a

θ	θ (radians)	sin θ	tan θ
10°	0.1745	0.1736	0.1763
5°	0.0873	0.0872	0.0875
2°	0.0349	0.0349	0.0349

b 2°

8 1.5%

9 43.6 m

10 a 1 b $\frac{1}{4}$ c $\frac{1}{2}$

11 a $\theta=\frac{\pi}{6}$ or $\theta=\pi-\frac{\pi}{6}=\frac{5\pi}{6}$
b $\theta=-\frac{5\pi}{6}$ or $\theta=\frac{5\pi}{6}$
c $\theta=\frac{\pi}{4}$ or $\theta=-\pi+\frac{\pi}{4}=-\frac{3\pi}{4}$

Exercise 14.1B Reasoning and problem-solving

1 $\frac{2}{3}\pi$

2

r cm	θ radians	s cm	A cm²
4	3	12	24
5	0.4	2	5
4	1.25	5	10
5	2.4	12	30

3 175 m

4 30.6 km

5 6980 km

6 π rad

7 0.25 rad

8 $4+2\pi$ cm, 2π cm²

9 a $\theta=\frac{\pi}{6}$ or $\frac{\pi}{3}$ b $\theta=\frac{\pi}{4}-\frac{\pi}{6}=\frac{\pi}{12}$
c $\theta=\pm\frac{\pi}{2}$ or $\frac{\pi}{4}$ d $\theta=\frac{\pi}{4}$ or −0.464 rad

10 1.1 cm²

11 27.0 cm²

12 a $\left(\frac{7}{3}\pi+4\right)$cm b $\frac{5}{3}\pi$ cm²

13 $\left(\sqrt{3}-\frac{\pi}{2}\right)r^2$

Exercise 14.2A Fluency and skills

1 The graphs have asymptotes.
The values are also ±1
The graphs are also positive (or negative).

2 a −1.06 b −0.839 c −2.92
d −3.24 e −1.73 f 1.56

3

x°	x rad	cosec x	sec x	cot x
30°	$\frac{\pi}{6}$	2	$\frac{2}{\sqrt{3}}$	$\sqrt{3}$
60°	$\frac{\pi}{3}$	$\frac{2}{\sqrt{3}}$	2	$\frac{1}{\sqrt{3}}$
90°	$\frac{\pi}{2}$	1	∞	0

4 a −1 b −2 c −2
d $\frac{1}{\sqrt{3}}$ e $\sqrt{2}$ f −1

5 a $x=30°$, $x=150°$ b $x=75.5°$, $x=-75.5°$
c $x=135°$, $x=-45°$ d $x=80.5°$, $-60.5°$
e $x=-24.5°$, $-175.5°$ f $x=-56.6°$, $123.4°$

6 a 1 b sin θ
c cot θ d tan θ

7 a 71.6° b 75.5° c 48.6°
d 113.6° e −63.4° f −14.5

8 a $\frac{\pi}{4}$ b $\frac{\pi}{3}$ c 0
d $-\frac{\pi}{4}$ e $\frac{\pi}{6}$ f 0
g $\frac{3\pi}{4}$ h $-\frac{\pi}{6}$

9 a $x=\frac{\pi}{6}, y=1+\frac{2}{\sqrt{3}}, x=\frac{5\pi}{6}, y=1-\frac{2}{\sqrt{3}}$
b $\cos x=0$, $x=0$

10 a 30°, 150° b 90°, 150° c 66.4°
d 35.6° e 45° f 5°
g 0°, 90° h 105°

11 a 0° b 90°

12 a i Translation of $\begin{pmatrix}45°\\0\end{pmatrix}$
ii Stretch (s.f. 2) parallel to y-axis and stretch (s.f. 3) parallel to x-axis
b i Translation of $\begin{pmatrix}-45°\\0\end{pmatrix}$ and stretch (s.f. 2) parallel to y-axis
ii Stretch (s.f. 2) parallel to x-axis
c i Reflection in x-axis and stretch $\left(\text{s.f.}\frac{1}{4}\right)$ parallel to the y-axis
ii Translation of $\begin{pmatrix}0\\2\end{pmatrix}$

13 a

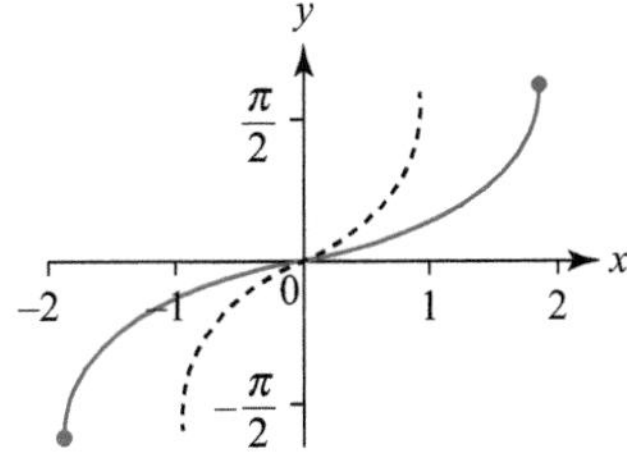

End points $(\pm 2), \left(\pm\frac{\pi}{2}\right)$

b

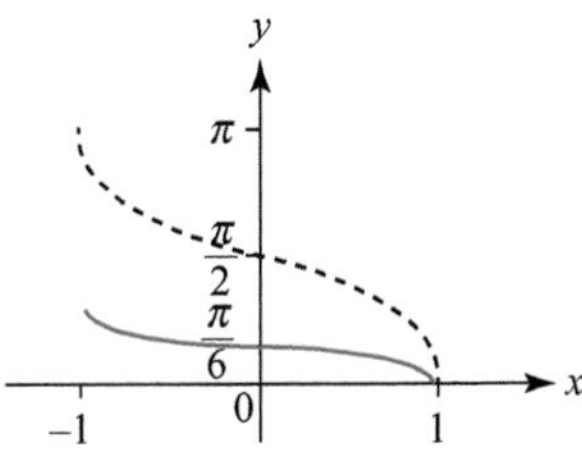

End points $\left(-1, \frac{\pi}{3}\right), (1, 0)$

c

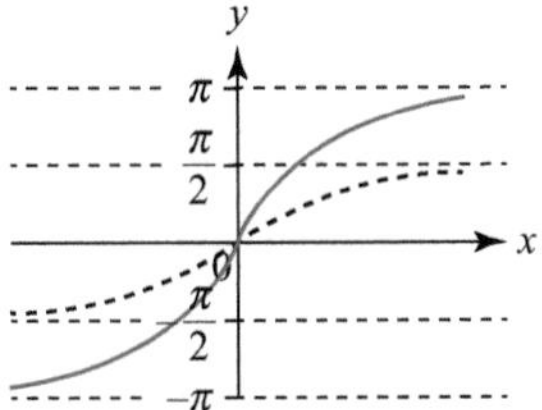

End points $(\pm\infty), (\pm\pi)$

d

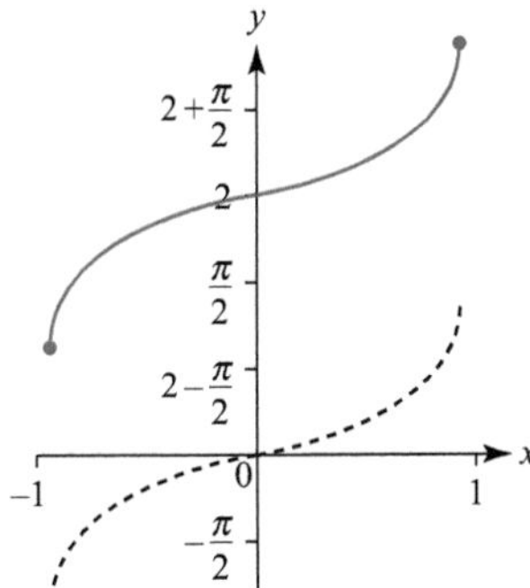

End points $\left(-1, 2-\frac{\pi}{2}\right), \left(1, 2+\frac{\pi}{2}\right)$

Exercise 14.2B Reasoning and problem-solving

1 a $\theta = \pm\frac{\pi}{4}, \pm\frac{3\pi}{4}$ **b** $\theta = \pm\frac{\pi}{6}, \pm\frac{5\pi}{6}$

c $\theta = \pm\frac{\pi}{3}, \pm\frac{2\pi}{3}$ **d** $\theta = 0, \pm\frac{\pi}{3}, \pm\frac{2\pi}{3}$

e $x = \pm 0.955^c, \pm 2.19^c$ **f** $x = \pm 0.857^c, \pm 2.28^c$

g $\theta = \pm\frac{\pi}{8}, \pm\frac{3\pi}{8}, \pm\frac{5\pi}{8}, \pm\frac{7\pi}{8}$

h $\theta = \frac{\pi}{12}, \frac{5\pi}{12}, -\frac{\pi}{4}, -\frac{7\pi}{12}, -\frac{11\pi}{12}$

2 a $\theta = 45°, 71.6°, 225°, 251.6°$

b $\theta = 45°, 116.6°, 296.6°, 225°$

c $\theta = 90°, 221.8°, 318.2°$

d $\theta = 60°, 300°$

3 a $4y^2 + 16 = x^2$ **b** $9y^2 + 36 = 4x^2$

c $x^2(y^2+9) = 144$ **d** $1 + (1-x)^2 = (y-1)^2$

4 a i $\frac{5}{4}$ **ii** $\frac{3}{4}$

b i $-\frac{17}{15}$ **ii** $-\frac{15}{8}$

c i $-\frac{41}{40}$ **ii** $-\frac{9}{40}$

5 a $\cot^2\theta$ **b** $\sec^3\theta$ **c** $\mathrm{cosec}^2\theta$

6 a $\frac{2}{3}$ **b** 4

7 a $\text{LHS} = \frac{\sin^2 x}{\cos x} + \cos x$

$= \frac{\sin^2 x + \cos^2 x}{\cos x}$

$= \sec x$

$= \text{RHS}$

b $\text{LHS} = \cos\theta \times \frac{\sin\theta}{\cos\theta} \times \sin^2\theta$

$= \sin^3\theta$

$\text{RHS} = \cos^3\theta \times \frac{\sin^3\theta}{\cos^3\theta}$

$= \sin^3\theta$

$\therefore \text{LHS} = \text{RHS}$

c $\tan\theta + \cot\theta$

$\equiv \frac{\sin\theta}{\cos\theta} + \frac{\cos\theta}{\sin\theta}$

$\equiv \frac{\sin^2\theta + \cos^2\theta}{\cos\theta\sin\theta}$

$\equiv \sec\theta \times \mathrm{cosec}\,\theta$

d $(\sec\theta - \tan\theta)(\sec\theta + \tan\theta)$

$\equiv \sec^2\theta - \tan^2\theta$

$\equiv (1 + \tan^2\theta) - \tan^2\theta$

$\equiv 1$

Dividing by $\sec\theta + \tan\theta$ gives

$\sec\theta - \tan\theta \equiv \frac{1}{\sec\theta + \tan\theta}$

e $(1 + \sec\theta)(1 - \cos\theta)$

$\equiv 1 - \cos\theta + \sec\theta - 1$

$\equiv -\cos\theta + \frac{1}{\cos\theta}$

$\equiv \frac{1 - \cos^2\theta}{\cos\theta}$

$\equiv \frac{\sin^2\theta}{\cos\theta}$

$\equiv \tan\theta\sin\theta$

8 0.78 s

9 $\tan\theta = \frac{x}{\sqrt{1-x^2}}$

10 $\mathrm{arccot}\, x = 90° - \theta$

11 $f(\theta) = \frac{1 - \sin\theta + 1 + \sin\theta}{(1+\sin\theta)(1-\sin\theta)}$

$= \frac{2}{1 - \sin^2\theta}$

$= 2\sec^2\theta$

$x = \frac{\pi}{2}, \frac{5\pi}{6}$

12 a 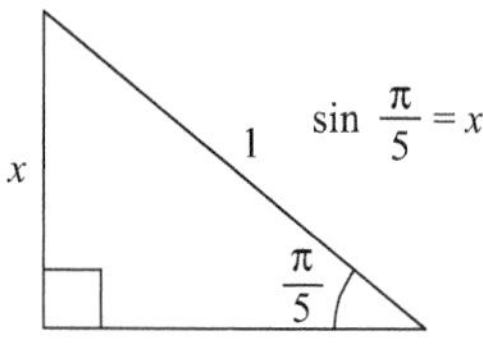

Let $\arccos x=\alpha$

$x=\cos\alpha$

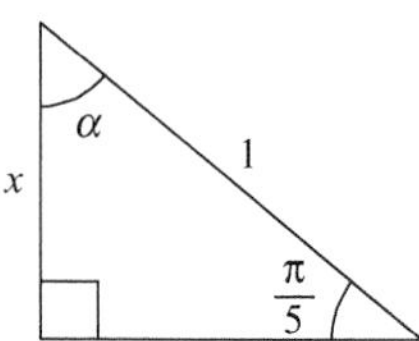

$\alpha+\frac{\pi}{5}=\frac{\pi}{2}$

$\alpha=\frac{\pi}{2}-\frac{\pi}{5}=\frac{3\pi}{10}$

$\arccos x=\frac{3\pi}{10}$

b $x=\frac{1}{2}$

13 a $\text{LHS}=\frac{\sin^2\theta+(1+\cos\theta)^2}{\sin\theta(1+\cos\theta)}$

$=\frac{\sin^2\theta+1+2\cos\theta+\cos^2\theta}{\sin\theta(1+\cos\theta)}$

$=\frac{2(1+\cos\theta)}{\sin\theta(1+\cos\theta)}$

$=2\operatorname{cosec}\theta$

$=\text{RHS}$

b $\text{LHS}=\frac{1}{\frac{\cos\theta}{\sin\theta}+\frac{1}{\sin\theta}}$

$=\frac{\sin\theta}{1+\cos\theta}\times\frac{1-\cos\theta}{1-\cos\theta}$

$=\frac{\sin\theta(1-\cos\theta)}{1-\cos^2\theta}$

$=\frac{1-\cos\theta}{\sin\theta}$

$=\text{RHS}$

c Consider $(\sec\theta-\sin\theta)(\sec\theta+\sin\theta)$

$\sec^2\theta+\sec\theta\sin\theta-\sec\theta\sin\theta-\sin^2\theta$

$1+\tan^2\theta-(1-\cos^2\theta)$

$\tan^2\theta-\cos^2\theta$

Divide by $\sec\theta+\sin\theta$ $(\neq 0)$

$\sec\theta-\sin\theta\equiv\frac{\tan^2\theta-\cos^2\theta}{\sec\theta+\sin\theta}$

14 $\frac{4}{\cos^2\theta}-\frac{\sin^2\theta}{\cos^2\theta}=k$

$4-\sin^2\theta=k(1-\sin^2\theta)$

$(k-1)\sin^2\theta=k-4$

$\operatorname{cosec}^2\theta=\frac{k-1}{k-4}$

$x=\frac{\pi}{2}$

Exercise 14.3A Fluency and skills

1 a $\frac{1}{4}\sqrt{2}(\sqrt{3}-1)$

b i $\frac{1}{4}\sqrt{2}(\sqrt{3}-1)$

ii $\frac{\sqrt{3}+1}{\sqrt{3}-1}$

iii $-\frac{\sqrt{3}+1}{\sqrt{3}-1}$

2 a $\sin(35+10)=\sin 45°$

$=\frac{1}{\sqrt{2}}$

b $\sin(70-10)=\sin 60°$

$=\frac{\sqrt{3}}{2}$

c $\cos(40-10)=\cos 30°$

$=\frac{\sqrt{3}}{2}$

d $\tan(74-45)=\tan 30°$

$=\frac{1}{\sqrt{3}}$

3 a $\sin(2\theta+\theta)=\sin 3\theta$ **b** $\cos(3\theta+\theta)=\cos 4\theta$

c $\tan(2x+x)=\tan 3x$ **d** $\frac{1}{\tan(3x-x)}=\frac{1}{\tan 2x}=\cot 2x$

4 a $\sin(60°+x)$ or $\cos(30°-x)$ **b** $\tan(60°+x)$

c $\sin(45°-x)$ or $\cos(45°+x)$ **d** $\tan(45°+x)$

e $\cot(x+60°)$

5 a $\sin(90-A)$

$=\sin 90\cos A-\cos 90\sin A$

$=1\times\cos A-0\times\sin A$

$=\cos A$

b $\sin(180-A)$

$=\sin 180\cos A-\cos 180\sin A$

$=0\times\cos A-(-1)\times\sin A$

$=\sin A$

c $\cos(90-A)$

$=\cos 90\cos A+\sin 90\sin A$

$=0\times\cos A+1\times\sin A$

$=\sin A$

d $\cos(180-A)$

$=\cos 180\cos A+\sin 180\sin A$

$=-1\times\cos A+0\times\sin A$

$=-\cos A$

6 a 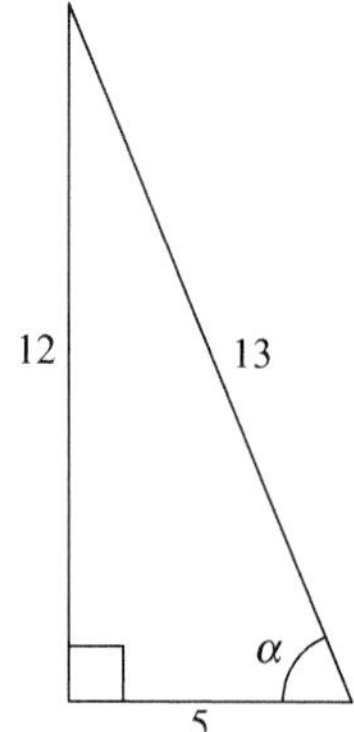

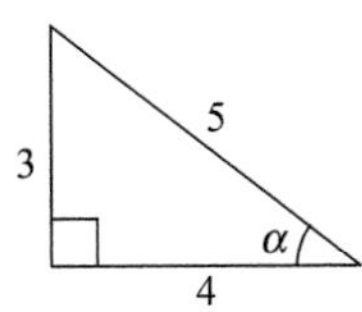

$\sin\alpha\cos\beta+\cos\alpha\sin\beta$

$=\frac{12}{13}\bullet\frac{4}{5}+\frac{5}{13}\bullet\frac{3}{5}$

$=\frac{63}{65}$

b $\frac{\tan\alpha-\tan\beta}{1+\tan\alpha\tan\beta}$

$=\frac{\frac{12}{5}-\frac{3}{4}}{1+\frac{12}{5}\bullet\frac{3}{4}}$

$=\frac{33}{20}\times\left(\frac{5}{14}\right)$

$=\frac{33}{56}$

7 a $\sin 46°$ **b** $\cos 84°$ **c** $\tan 140°$

d $\cos 100°$ **e** $\sin 6\theta$ **f** $\frac{1}{2}\sin 2\theta$

g $\cos^2 20°$ **h** $\cos 8\theta$ **i** $2\cos^2\theta$

j $\sin^2 25°$ **k** $\cos\frac{2\pi}{5}$ **l** $2\operatorname{cosec} 2\theta$

m $2\operatorname{cosec} 2\theta$ **n** $\cot 8\theta$

8 a $\sin\left(2\times\frac{\pi}{12}\right)=\sin\left(\frac{\pi}{6}\right)$

$=\frac{1}{2}$

b $\cos\left(2\times\frac{\pi}{8}\right)=\cos\left(\frac{\pi}{4}\right)$

$=\frac{1}{\sqrt{2}}$

c $\cos\left(2\times\frac{\pi}{4}\right)=\cos\left(\frac{\pi}{2}\right)$

$=0$

d $\frac{1}{\tan\left(2\times 22\frac{1}{2}\right)}=\frac{1}{\tan 45°}$

$=1$

9 a $\frac{24}{25},\frac{1}{\sqrt{5}}$ **b** $\frac{120}{169},\frac{1}{\sqrt{26}}$

10 a $\sin\frac{x}{2}=\frac{2}{3}$

$\cos\frac{x}{2}=\frac{\sqrt{5}}{3}$

$\tan\frac{x}{2}=\frac{2}{\sqrt{5}}$

b $\sin\frac{x}{2}=\frac{1}{\sqrt{10}}$

$\cos\frac{x}{2}=\frac{3}{\sqrt{10}}$

$\tan\frac{x}{2}=\frac{1}{3}$

11 a $\frac{\sqrt{5}}{7}$ **b** $\frac{3}{\sqrt{10}}$

12 a $\frac{1}{13}$ **b** $\frac{1}{2}$ **c** $\frac{5+\tan\beta}{1-5\tan\beta}$

Exercise 14.3B Reasoning and problem-solving

1 a $\theta=17.1°$ **b** $\theta=113.8°$ **c** $\theta=160.9°$

2 a 12 cm, 13 cm

b $\sin A\hat{O}C=0.862$ to 3 sf

$\cos A\hat{O}C=0.508$ to 3 sf

$\tan A\hat{O}C=1.70$ to 3 sf

3 a $\text{LHS}=\frac{\sin A}{\cos A}+\frac{\cos A}{\sin A}$

$=\frac{\sin^2 A+\cos^2 A}{\cos A\sin A}$

$=\frac{2}{2\cos A\sin A}$

$=\frac{2}{\sin 2A}$

$=2\operatorname{cosec}2A$

$=\text{RHS}$

b $\text{LHS}=\frac{1}{\sin 2A}+\frac{\cos 2A}{\sin 2A}$

$=\frac{2\cos^2 A}{2\cos A\sin A}$

$=\cot A$

$=\text{RHS}$

$\text{LHS}=\frac{1}{\sin 2A}+\frac{\cos 2A}{\sin 2A}$

$=\frac{2\cos^2 A}{2\cos A\sin A}$

$=\cot A$

$=\text{RHS}$

$=\frac{2}{\sin 2A}$

$=2\operatorname{cosec}2A$

$=\text{RHS}$

c $\text{RHS}=\frac{1}{\cos A\sin A}$

$=\frac{2}{2\cos A\sin A}$

$=2\operatorname{cosec}2A$

$=\text{LHS}$

d $\text{LHS}=\frac{1}{\sin A}-\frac{\cos A}{\sin A}$

$=\frac{1-\cos A}{\sin A}$

$=\frac{2\sin^2\frac{A}{2}}{2\sin\frac{A}{2}\cos\frac{A}{2}}$

$=\tan\frac{A}{2}$

$=\text{RHS}$

4 $\sin\theta\cos\phi+\cos\theta\sin\phi=\cos\phi$

$\tan\theta\cos\phi+\sin\phi=\frac{\cos\phi}{\cos\theta}$

$\tan\theta+\tan\phi=\frac{1}{\cos\theta}$

$=\sec\theta$

5 Let $\arcsin\frac{1}{2}=\alpha$

So $\sin\alpha=\frac{1}{2}$

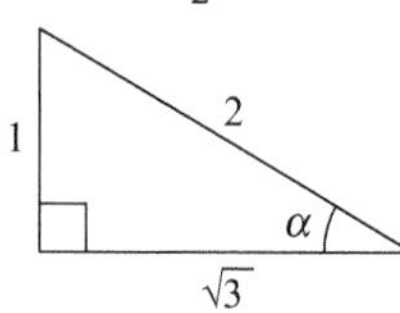

Let $\arcsin\frac{1}{3}=\beta$

So $\sin\beta=\frac{1}{3}$

1 3 β √8

$$\sin(\alpha+\beta)=\frac{1}{2}\times\frac{\sqrt{8}}{3}+\frac{\sqrt{3}}{2}\times\frac{1}{3}$$

$$=\frac{\sqrt{8}+\sqrt{3}}{6}$$

$$\alpha+\beta=\arcsin\left(\frac{2\sqrt{2}+\sqrt{3}}{6}\right)$$

6 a i $12+5\sqrt{5}$ **ii** $\frac{2-\sqrt{5}}{\sqrt{30}}$

b $\frac{\tan\theta+\sqrt{3}}{1-\sqrt{3}\tan\theta}=\frac{1}{3}$

$3\tan\theta+3\sqrt{3}=1-\sqrt{3}\tan\theta$

$(3+\sqrt{3})\tan\theta=1-3\sqrt{3}$

$\tan\theta=\frac{1-3\sqrt{3}}{3+\sqrt{3}}\times\frac{3-\sqrt{3}}{3-\sqrt{3}}$

$=\frac{12-10\sqrt{3}}{6}$

$=2-\frac{5}{3}\sqrt{3}$

7 a $\theta=0,\pi,2\pi,\frac{\pi}{6},\frac{5\pi}{6},\frac{7\pi}{6},\frac{11\pi}{6}$

b $\theta=\frac{7\pi}{6}$ or $\frac{11\pi}{6}$

c $\theta=\frac{\pi}{2},\frac{3\pi}{2},\frac{\pi}{4},\frac{5\pi}{4}$

d $\theta=0,\pi,2\pi,\frac{\pi}{3},\frac{5\pi}{3}$

e $\theta=0.332,1.239,3.473,4.381$ radians

f $\theta=0,\pi,2\pi,\ldots$ or $\theta=\frac{\pi}{3},\frac{2\pi}{3},\frac{4\pi}{3},\frac{5\pi}{3}$

8 a $\theta=0°,360°,120°$

b $\theta=0°,360°,78.4°,281.6°$

c $\theta=120°$ or $0°$

d $\theta=180°,90°,270°$

9 $4y=x^2-8$

10 $36+y^2=4x^2$ or $y^2=4(x^2-9)$

11 a $\tan(A+B)=\frac{\sin A\cos B+\cos A\sin B}{\cos A\cos B-\sin A\sin B}$

Divide by $\cos A\cos B$ top and bottom

$=\frac{\tan A-\tan B}{1-\tan A\tan B}$

b $\tan(A-B)=\frac{\tan A-\tan B}{1+\tan A\tan B}$

12 a $\frac{\sin\theta\cos\phi}{\cos\theta\cos\phi}+\frac{\cos\theta\sin\phi}{\cos\theta\cos\phi}\equiv\tan\theta+\tan\phi$

b $\frac{2\sin\theta\cos\theta+\cos\theta}{1+\sin\theta-(1-2\sin^2\theta)}$

$\frac{\cos\theta(2\sin\theta+1)}{\sin\theta(1+2\sin\theta)}\equiv\cot\theta$

c $\text{RHS}=\frac{1}{\cos2\theta}+\frac{\sin2\theta}{\cos2\theta}$

$=\frac{1+2\sin\theta\cos\theta}{\cos^2\theta-\sin^2\theta}$

$=\frac{\cos^2\theta+\sin^2\theta+2\sin\theta\cos\theta}{\cos^2\theta-\sin^2\theta}$

$=\frac{(\cos\theta+\sin\theta)^2}{(\cos\theta+\sin\theta)(\cos\theta-\sin\theta)}$

$=\frac{\cos\theta+\sin\theta}{\cos\theta-\sin\theta}$

$=\text{LHS}$

d $\text{RHS}=\left(\frac{\tan\theta+\tan45°}{1-\tan\theta\tan45°}\right)^2$

$=\left(\frac{1+\tan\theta}{1-\tan\theta}\right)^2$

$\text{LHS}=\frac{1+2\sin\theta\cos\theta}{1-2\sin\theta\cos\theta}$

Divide by $\cos^2\theta$

$=\frac{\sec^2\theta+2\tan\theta}{\sec^2\theta-2\tan\theta}$

$=\frac{1+\tan^2\theta+2\tan\theta}{1+\tan^2\theta-2\tan\theta}$

$=\left(\frac{1+\tan\theta}{1-\tan\theta}\right)^2$

$\equiv\text{RHS}$

e $\frac{\cos^3\theta+\sin^3\theta}{\cos\theta+\sin\theta}=\frac{(\cos\theta+\sin\theta)(\cos^2\theta-\sin\theta\cos\theta+\sin^2\theta)}{\cos\theta+\sin\theta}$

$=\cos^2\theta-\sin\theta\cos\theta+\sin^2\theta$

$=1-\sin\theta\cos\theta$

$=1-\frac{1}{2}\sin2\theta$

f $\frac{2\sin\theta\cos\theta}{2\sin^2\theta}\equiv\frac{\cos\theta}{\sin\theta}$

$\equiv\cot\theta$

g $\frac{2\tan\theta}{1+\tan^2\theta}\equiv\frac{2\sin\theta}{\cos\theta}\bullet\frac{1}{\sec^2\theta}$

$\equiv2\sin\theta\bullet\cos\theta$

$\equiv\sin2\theta$

h $\frac{2\sin\theta\cos\theta+\sin\theta}{2\cos^2\theta+\cos\theta}$

$\equiv\frac{\sin\theta(2\cos\theta+1)}{\cos\theta(2\cos\theta+1)}$

$\equiv\tan\theta$

13 $\cos^4\theta\equiv(1-\sin^2\theta)^2$

$\equiv1-2\sin^2\theta+\sin^4\theta$

So $\cos^4\theta+2\sin^2\theta-\sin^4\theta\equiv1$

14 $\tan45°=1$

So $\frac{\tan45-\tan15}{1+\tan45\times\tan15}$

$=\tan(45-15)$

$= \tan 30°$

$= \frac{1}{\sqrt{3}}$

15 $\sin\theta \cdot \frac{\sin\theta}{\cos\theta} = 2 - \cos\theta$

$1 - \cos^2\theta = 2\cos\theta - \cos^2\theta$

$2\cos\theta = 1$

$\cos\theta = \frac{1}{2}$

$\varphi = \frac{\pi}{3}$

16 $3\frac{\cos\theta}{\sin\theta} = \frac{8}{\cos\theta}$

$3\cos^2\theta = 8\sin\theta$

$3(1 - \sin^2\theta) = 8\sin\theta$

$3\sin^2\theta + 8\sin\theta - 3 = 0$

$\theta = 19.5°$ or $160.5°$

17 a $\cos^2 x + 2\cos x\cos y + \cos^2 y + \sin^2 x + 2\sin x\sin y + \sin^2 y$

$\equiv 2 + 2(\cos x\cos y + \sin x\sin y)$

$\equiv 2 + 2\cos(x - y)$

$\equiv 2(1 + \cos(x - y))$

$\equiv 2\times 2\cos^2\left(\frac{x-y}{2}\right)$

$\equiv 4\cos^2\left(\frac{x-y}{2}\right)$

b Let $x = 120°$, $y = 90°$

So $\frac{x-y}{2} = 15°$

$4\cos^2 15° = (\cos 120 + \cos 90)^2 + (\sin 120 + \sin 90)^2$

$= \left(-\frac{1}{2} + 0\right)^2 + \left(\frac{\sqrt{3}}{2} + 1\right)^2$

$= \frac{1}{4} + \frac{3}{4} + \sqrt{3} + 1$

$= 2 + \sqrt{3}$

But $4\left(\frac{\sqrt{2}+\sqrt{6}}{4}\right)^2$

$= \frac{1}{4}(2 + 2\sqrt{12} + 6)$

$= 2 + \sqrt{3}$

So $\cos 15° = \frac{\sqrt{2}+\sqrt{6}}{4}$

18 a $\sin A + \sin B = 2\sin\left(\frac{A+B}{2}\right)\cos\left(\frac{A-B}{2}\right)$

$\frac{1}{2}[\sin(A+B) + \sin(A-B)] = \sin A\cos B$

$\sin(A+B) + \sin(A-B) = 2\sin A\cos B$

$A + B = x$

$A - B = y$

$\text{LHS} = \sin x + \sin y$

$2A = x + y \Rightarrow A = \frac{1}{2}(x+y)$

$2B = x - y \Rightarrow B = \frac{1}{2}(x-y)$

$\text{RHS} = 2\sin\frac{1}{2}(x+y)\cos\frac{1}{2}(x-y)$

as required.

b $x = 0, y = 0, \frac{\pi}{2}$

19 $\theta = 0.535\,\text{rad}, 3.68\,\text{rad}$

20 a $\frac{1 + \sin x + 1 - \sin x}{(1 - \sin x)(1 + \sin x)} = 8$

$\frac{2}{1 - \sin^2 x} = 8$

$2\sec^2 x = 8$

$\sec^2 x = 4$

b $\theta = 25°, 55°, 115°, 145°$

21 $k = 2$

22 $y = 4\cot t$

23 a $3(1 + \cot^2 x) - \cot^2 x = n$

$2\cot^2 x = n - 3$

$\tan^2 x = \frac{2}{n-3}$

$\sec^2 x = 1 + \frac{2}{n-3}$

$= \frac{n-3+2}{n-3}$

$= \frac{n-1}{n-3}$

($n \neq 3$ for $\sec^2 x$ to exist)

b $\theta = 15°, 120°, 210°, 300°, \ldots$

24 The tides do not clash at 05:13 and 11:13

25 a i $\sin 3A = \sin(2A + A)$

$= \sin 2A\cos A + \cos 2A\sin A$

$= 2\sin A(1 - \sin^2 A) + (1 - 2\sin^2 A)\sin A$

$= 3\sin A - 4\sin^3 A$

ii $\cos 3A = \cos(2A + A)$

$= \cos 2A\cos A - \sin 2A\sin A$

$= (2\cos^2 A - 1)\cos A - 2(1 - \cos^2 A)\cos A$

$= 4\cos^3 A - 3\cos A$

b $\tan 3A = \frac{3\tan A - \tan^3 A}{1 - 3\tan^2 A}$

c $\sin 4A = 4\cos^3 A\sin A - 4\sin^3 A\cos A$

$\cos 4A = \cos^4 A - 6\cos^2 A\sin^2 A + \sin^4 A$

26 a Let $\arctan 3 = \alpha$ so $\tan\alpha = 3$

and $\arctan x = \beta$ so $\cot\beta = x$

So $2\alpha = \beta$

$\tan 2\alpha = \tan\beta$

$\frac{2\tan\alpha}{1 - \tan^2\alpha} = \tan\beta$

$\frac{2\times 3}{1 - 3^2} = \tan\beta$

$-\frac{3}{4} = \tan\beta$

$-\frac{3}{4} = \cot\beta$

But $\cot\beta = x$ so $x = -\frac{4}{3}$

b

Let $\arcsin\frac{4}{5} = \alpha$ so $\sin\alpha = \frac{4}{5}$

and $\arccos\frac{12}{13} = \beta$ so $\cos\beta = \frac{12}{13}$

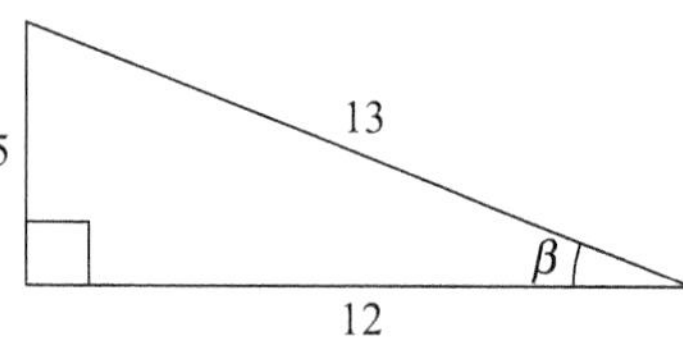

So

$$\sin(\alpha - \beta) = \frac{4}{5} \times \frac{12}{13} - \frac{3}{5} \times \frac{5}{13}$$

$$= \frac{33}{65}$$

$$\alpha - \beta = \arcsin\left(\frac{33}{65}\right)$$

So $x = \frac{33}{65}$

c $x = 90°\left(\text{or } \frac{\pi}{2}\right)$

Exercise 14.4A Fluency and skills

1 **a** $r = 13, \alpha = 67.4°$ **b** $r = 5, \alpha = 36.9°$

c $r = \sqrt{5}, \alpha = 63.4°$ **d** $r = 2\sqrt{5}, \alpha = 26.6°$

e $r = 5, \alpha = -53.1°$ **f** $r = 17, \alpha = -61.9°$

g $r = \sqrt{29}, \alpha = 21.8°$

2 **a** $r = \sqrt{2}$ **b** $r = \sqrt{2}$

3 **a** $\alpha = \frac{\pi}{3}$ (60°) **b** $\alpha = \frac{\pi}{3}$ (60°)

4 **a**

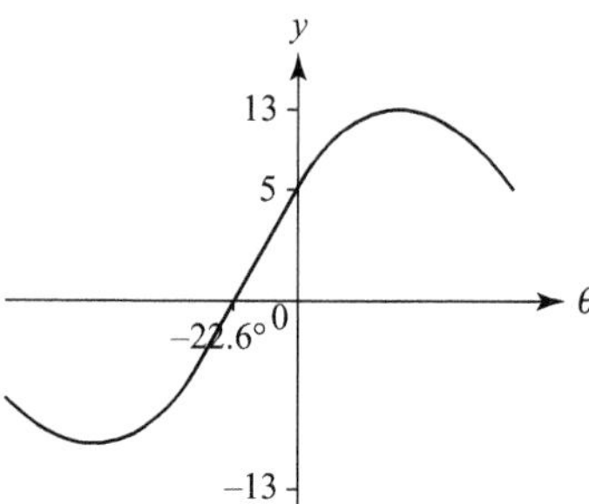

$r = 13, \alpha = 22.6°$

b $r = 6.4, \alpha = -51.3°$

5 The transformations are: a stretch (s.f. = 13) parallel to the y-axis, followed by a translation of $\begin{pmatrix} 22.6° \\ 0 \end{pmatrix}$.

Exercise 14.4B Reasoning and problem-solving

1 **a** $\theta = 4.9°, 129.9°$ **b** $\theta = 17.6°, 229.8°$

c $\theta = 82.0°, 334.1°$ **d** $\theta = 102.3°, 195.7°$

e $\theta = 60°, 240°$ **f** $\theta = 122.0°$ or $302.0°$

2 **a** $\theta = \frac{\pi}{3}$ **b** $\theta = \frac{7\pi}{12}$ or $-\frac{\pi}{12}$

c $\theta = \frac{\pi}{2}$ or $-\frac{\pi}{6}$ **d** $\theta = \frac{\pi}{12}$ or $\frac{-5\pi}{12}$

3 **a** $\theta = 0, 63.4°, 180°$ **b** $\theta = 83.6°, 108.7°$

c $\theta = 54.6°, 157.9°$ **d** $\theta = 180°$

4 **a** Points of intersection are $\left(-15°, \frac{1}{\sqrt{2}}\right)$ and $\left(165°, -\frac{1}{\sqrt{2}}\right)$

b $\sqrt{2}\sin(\theta + 45°) = \cos(\theta - 30°)$

$$\sin\theta + \cos\theta = \cos\theta \times \frac{\sqrt{3}}{2} + \sin\theta \times \frac{1}{2}$$

$$\frac{1}{2}\sin\theta = \left(\frac{\sqrt{3} - 2}{2}\right)\cos\theta$$

$$\tan\theta = \sqrt{3} - 2$$

$$\theta = -15° \text{ or } 165°$$

$$y = \frac{1}{\sqrt{2}} \text{ or } -\frac{1}{\sqrt{2}}$$

5

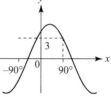

Let $3\sin x + 4\cos x = r\sin(x + \alpha)$

$= r\sin x\cos\alpha + r\cos x\sin\alpha$

Equate coefficients

$r\cos\alpha = 3$

$r\sin\alpha = 4$

$r = 5$

$\tan\alpha = \frac{4}{3}$

$\alpha = 53.1°$

So $y = 5\sin(x + 53.1°)$

$= 3$ when $\sin(x + 53.1°) = \frac{3}{5}$

$\sin(x + 53.1°) = 0.6$

$x + 53.1° = 36.9°$

or

$x + 53.1° = (180 - 36.9)°$

$= 36.9°$

or

$143.1°$

$x = -16.2°$

or

$90°$

For $-90° \le x \le 90°$

From graph, $y \ge 3$ for $-16.2° \le x \le 90°$

6

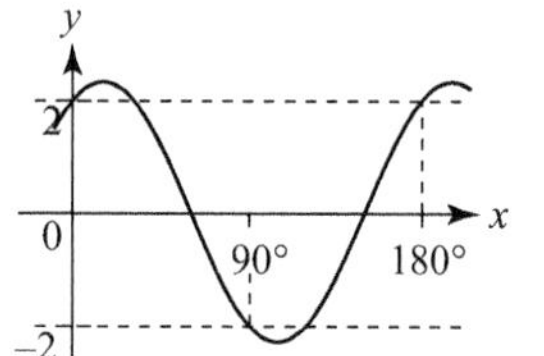

a For $y = 2$

$\sqrt{5}\sin(2x + 63.4°) = 2$

$2x + 63.4° = 63.4°$

$= 180 - 63.4°$

$= 360 + 63.4°$

$2x = 0, 53.2°, 360°, ...$

$x = 0, 26.6°, 180°$

for $0 < x < 180°$

From graph, $y > 2$ for $0 < x < 26.6°$

b For $y > -2$

$\sin(2x+63.4°) = -\frac{2}{\sqrt{5}}$

$2x+63.4° = 180+63.4°$

$= 360-63.4°$

$= 540+63.4°, ...$

$2x = 180°, 233.2°, ...$

$x = 90°, 116.6°$

for $0 < x < 180°$

From graph, $y > -2$ for $90° < x < 116.6°$

7 a 12 amps

b 0.26 seconds

8 a $\sqrt{2}\cos\theta + \sqrt{3}\sin\theta = r\cos\theta\cos\alpha + r\sin\theta\sin\alpha$

$r\cos\alpha = \sqrt{2}$

$r\sin\alpha = \sqrt{3}$

$r^2 = 5$

$r = \sqrt{5}$

$\tan\alpha = \sqrt{\frac{3}{2}}$

$\alpha = 50.8°$

$r\cos(\theta-\alpha) = \sqrt{5}\cos(\theta-50.8°)$ has a max of $\sqrt{5}$ when $\theta = 50.8°$

b Min value $= \frac{1}{\sqrt{5}}$ when $\theta = 50.8°$

9 a Max of 2 when $\theta = 300°$
Min of -2 when $\theta = 120°$

b Max of 25 when $\theta = 106.3°$
Min of -25 when $\theta = 286.3°$

c Max of $\sqrt{13}$ when $\theta = 123.7°$
Min of $-\sqrt{13}$ when $\theta = 303.7°$

d Max of 10 when $\theta = 161.6°$
Min of -10 when $\theta = 71.6°$

10 a $r = 25, \alpha = -73.7°$

$7\cos\theta - 24\sin\theta + 3 = 25\cos(\theta+73.7°)+3$

has a max value of $(25 \times 1) + 3 = 28$

and a min value of $(25 \times (-1)) + 3 = -22$

So $-22 \le 7\cos\theta - 24\sin\theta + 3 \le 28$

b A stretch of scale factor $\frac{1}{25}$ parallel to the y-axis, followed by a translation of $-73.7°$ parallel to the θ-axis

11 a i Max = 33 when $\theta = 157.4°$

ii Max = 33 when $\theta = 337.4°$

b Min = 5 when $\theta = 157.4°$

12 a Max $I = 2\sqrt{5}$ when $t = 74.1$ sec

b 30 seconds

Review exercise 14

1 a i 57.3° **ii** 143.2° **iii** 200.5°

b i $\frac{\pi}{12}$ rad **ii** $\frac{2\pi}{5}$ rad **iii** $\frac{7\pi}{6}$ rad

2 a 1 **b** $\frac{1}{2}$ **c** 2θ

3 a 2 **b** $\sqrt{3}$ **c** $\sqrt{2}$

d -1 **e** $\frac{\pi}{4}$ **f** $\frac{\pi}{6}$

g $\frac{\pi}{2}$ **h** $-\frac{\pi}{4}$

4 a $\cos(40+10) = \cos 50°$

b $\frac{1}{\tan(100+35)} = \cot 135°$

5 a $x = 31.7°, 121.7°$

b $x = 10°, 130°$

c $x = 90°$ or $x = 19.5°, 160.5°$

d $x = 60°$

e $x = 15°$ or $75°$

f $x = 26.6°$

g $x = 45°$

6 a $2\pi\,\text{cm}^2$ **b** $2\pi - 4\sqrt{2}\,\text{cm}^2$

7 a 1 **b** 1

8 $\tan 15° = 2 - \sqrt{3}$

$\sec 75° = \sqrt{2}\left(\sqrt{3}+1\right)$

9

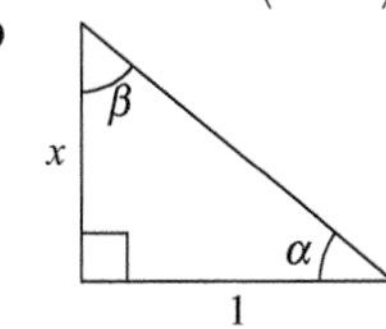

In the triangle

$\tan\alpha = x$

$\cot\beta = x$

and

$\alpha + \beta = \frac{\pi}{2}$

So

$\arctan x + \text{arccot}\, x = \frac{\pi}{2}$

$\arctan x = \frac{\pi}{2} - \text{arccot}\, x$

10 a $\theta = 106.1°$ **b** $\theta = 0°, 180°, 82.8°$

c $\theta = 0, 180°, 60°$ **d** $\theta = 30°, 150°$

11 a $7\sin\theta - 24\cos\theta = 25\sin(\theta - 73.7°)$

b $\theta = 110.6°, 216.8°$

12 a $7\cos\theta + 6\sin\theta = \sqrt{85}\sin(\theta + 49.4°)$

b Max is $\sqrt{85}$ when $\theta = 40.6°$

13 a LHS $= 2\sin^2 A$

RHS $= \frac{\sin A}{\cos A} \bullet 2\sin A\cos A = 2\sin^2 A$

LHS = RHS

b LHS $= 2\sin A\cos A \bullet \sec^2 A$

$= 2\sin A\cos A\frac{1}{\cos^2 A}$

$= 2\tan A$

$=$ RHS

c RHS $= 2 \times \frac{1-\tan^2 A}{2\tan A}$

$= \frac{1-\tan^2 A}{\tan A}$

$= \frac{1}{\tan A} - \tan A$

$= \cot A - \tan A$

$=$ LHS

14

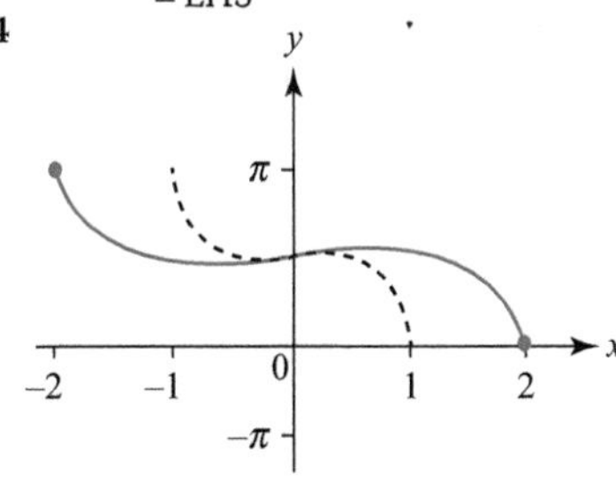

End points $(-2, \pi), (2, 0)$

15 $\frac{(1-\sin x)^2 + \cos^2 x}{\cos x(1-\sin x)} = \frac{1-2\sin x+1}{\cos x(1-\sin x)}$

$= \frac{2(1-\sin x)}{\cos x(1-\sin x)}$

$= 2\sec x$

So $f(x) = 2\sec x$

$x = 180°, 70.5°, 289.5°$

Assessment 14

1 D

2 B

3 **a** $x = 60°, -120°$

b

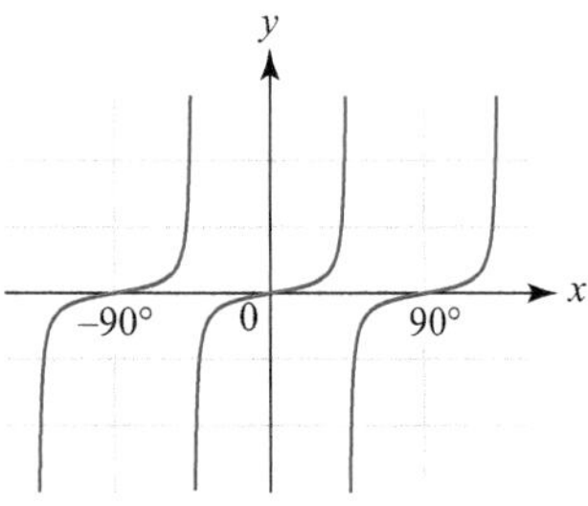

$x = \pm 45$

4 **a** $\frac{25\pi}{3}$ **b** $\frac{10\pi}{3}$

5 **a** **i** $\sqrt{2}$ **ii** $-\sqrt{3}$

b

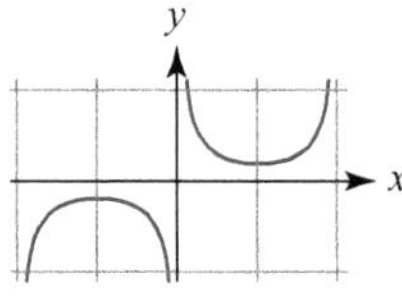

$x = \pi, x = -\pi, x = 0$

c $y \in \mathbb{R}, y \geq 1, y \leq -1$

6 **a** $x = 30°, 150°$ **b** $x = 135°, 225°$

7 **a** 5 m s^{-1} **b** $0, \frac{2\pi}{3}(2.09)$ s, $\frac{14\pi}{3}$ (14.66) s, $\frac{26\pi}{3}$ (27.2) s

8 **a**

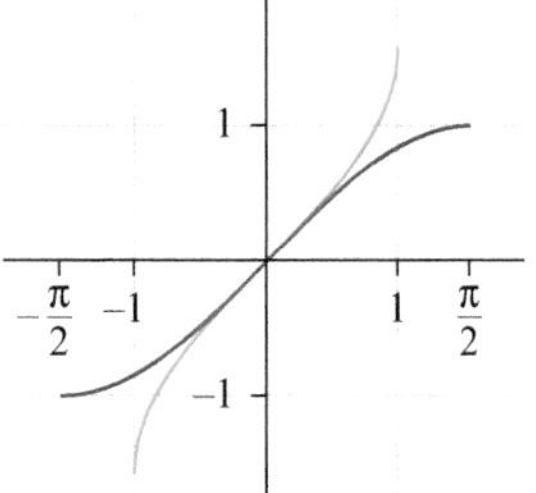

b Reflection in line $y = x$

c **i** $-\frac{\pi}{3}$ **ii** $\frac{\pi}{6}$

9 **a** $\frac{\sin^2\theta}{\cos^2\theta} + \frac{\cos^2\theta}{\cos^2\theta} \equiv \frac{1}{\cos^2\theta}$

$\tan^2\theta + 1 \equiv \sec^2\theta$

$\sec^2\theta - \tan^2\theta \equiv 1$

b **i** $\pm\sqrt{6}$ **ii** $\pm\frac{1}{\sqrt{6}}$

10 **a** $\frac{\sqrt{6}}{4} - \frac{\sqrt{2}}{4}$ **b** $\frac{\sqrt{6}}{4} + \frac{\sqrt{2}}{4}$

11 Let $A = B = x$

Then $\sin 2x = \sin x \cos x + \sin x \cos x$

$= 2\sin x \cos x$

12 $10\sin(\theta + 53.1)$

13 **a** 0.05 **b** 1 **c** 0.05

14 **a** **i** 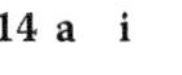

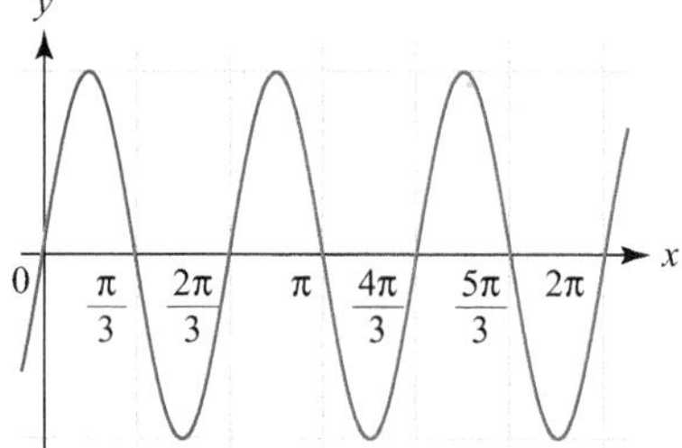

ii

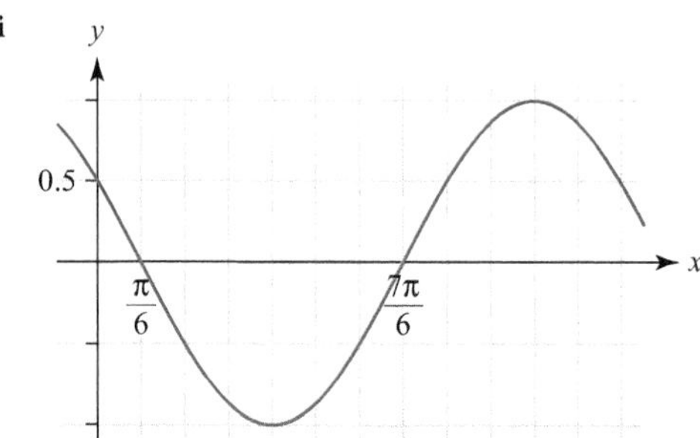

b $x = \frac{\pi}{12}, \frac{\pi}{4}, \frac{3\pi}{4}, \frac{11\pi}{12}$

15 50.4 cm

16 **a** 6500 **b** 10 600

c 3 weeks 3 days or 24 days

17 **a** **i** 4.9 cm^2 **ii** 9.8 cm

b **i** 0.51 cm^2 **ii** 5.53 cm

18 **a** **i**

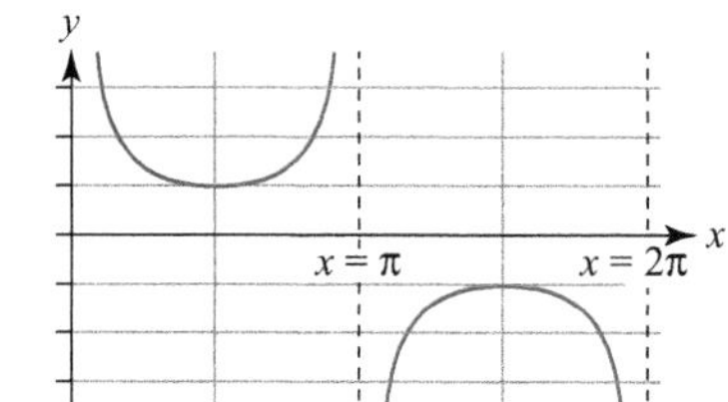

ii

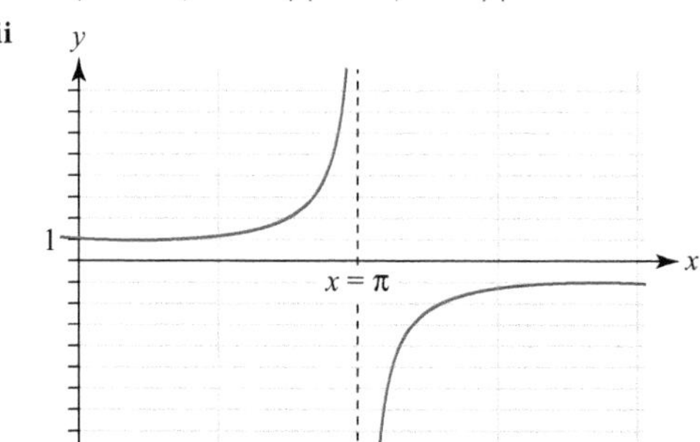

b **i** $y \in \mathbb{R}, y \leq -2, y \geq 2$

ii $y \in \mathbb{R}, y \leq -1, y \geq 1$

19 **a**

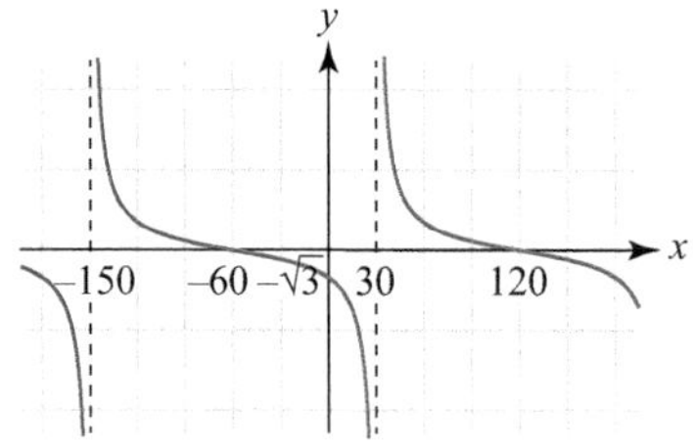

$x = 30, x = -150$

b $x = 108.7°, -71.3°$

20 i $x = \frac{2\pi}{3}, \frac{4\pi}{3}$ **ii** $x = \frac{3\pi}{8}, \frac{7\pi}{8}$

21 a

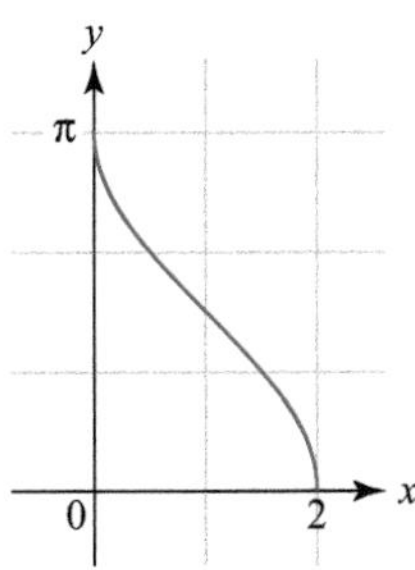

b $y \in \mathbb{R}, 0 \le y \le \pi$

c $f(x) = 1 + \cos x$

Domain is $0 \le x \le \pi$

Range is $0 \le y \le 2$

22 a $\frac{1+\sqrt{3}}{2}$ **b** $5 - \frac{\sqrt{3}}{3}$

23 a $2(\sec^2 x - 1) = \sec x - 1$

$2\sec^2 x - 2 = \sec x - 1$

$2\sec^2 x - \sec x - 1 = 0$

b $x = 0, 2\pi$

24 a $\frac{\sin^2\theta}{\sin^2\theta} + \frac{\cos^2\theta}{\sin^2\theta} \equiv \frac{1}{\sin^2\theta}$

$1 + \cot^2\theta \equiv \operatorname{cosec}^2\theta$

$\operatorname{cosec}^2\theta - \cot^2\theta \equiv 1$

b $\theta = 18.4°, 198.4°$

25 a $\sec^4 x - \tan^4 x$

$= (\sec^2 x - \tan^2 x)(\sec^2 x + \tan^2 x)$

$= \sec^2 x + \tan^2 x$

b $x = \pm 1.11$ rad, ± 2.03 rad

26 a $\cos 3x = \cos(2x + x)$

$= \cos 2x \cos x - \sin x \sin 2x$

$= (2\cos^2 x - 1)\cos x - 2\sin x(2\sin x \cos x)$

$= 2\cos^3 x - \cos x - 2\sin^2 x \cos x$

$= 2\cos^3 x - \cos x - 2(1 - \cos^2 x)\cos x$

$= 2\cos^3 x - \cos x - 2\cos x + 2\cos^3 x$

$= 4\cos^3 x - 3\cos x$

b $x = \frac{\pi}{18}, \frac{11\pi}{18}, \frac{13\pi}{18}, \frac{23\pi}{18}, \frac{25\pi}{18}, \frac{35\pi}{18}$

27 a $\cos 2x = \cos x \cos x - \sin x \sin x$

$= \cos^2 x - \sin^2 x$

$= \cos^2 x - (1 - \cos^2 x)$

$= 2\cos^2 x - 1$

b $x = 120°, 180°, 240°$

28 e.g. $A = 30°, B = 60°$

$\cos(30 + 60) = \cos 90 = 0$

$\cos 30 + \cos 60 = \frac{\sqrt{3}}{2} + \frac{1}{2} \ne 0$

Therefore, in general,

$\cos(A + B) \ne \cos A + \cos B$

29 a $f(x) = 4\sqrt{5}\cos(x - 0.464)$

b $x = 1.78$ rad, 5.43 rad

30 a

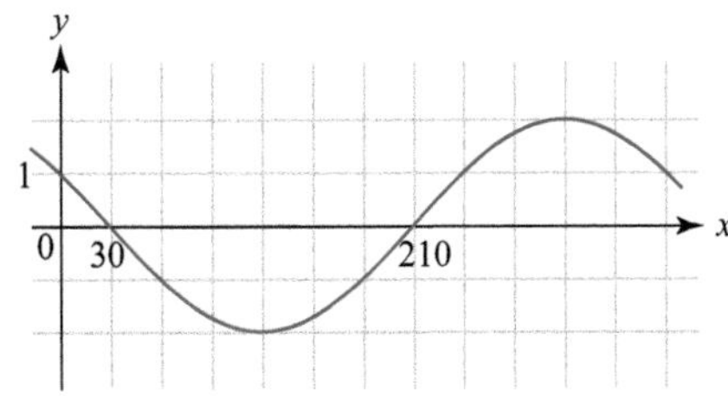

b

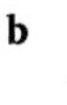

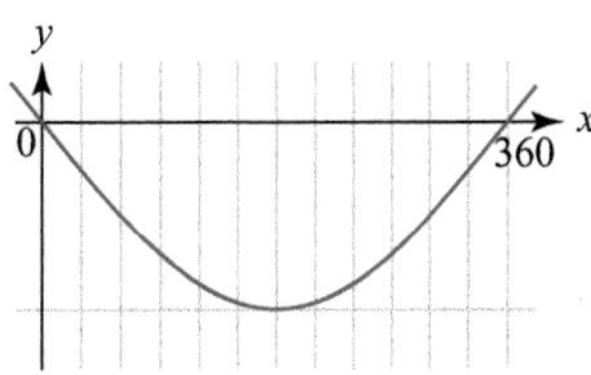

31 a

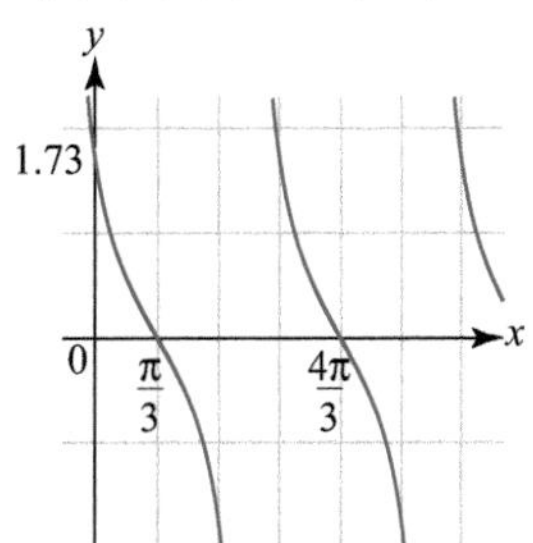

b $x = \frac{5\pi}{6}, x = \frac{11\pi}{6}$

$x = \frac{\pi}{6}, \frac{7\pi}{6}$

32 $16.8\,\text{cm}^2$

33 a 10.7 cm

b i $59.2\,\text{cm}^2$ **ii** 33.7 cm

34 a i

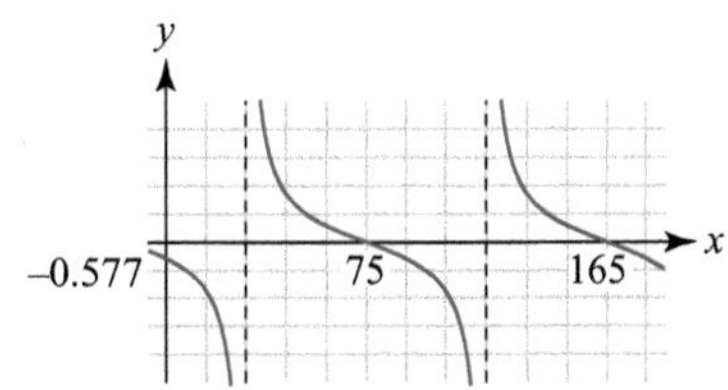

$x = 30, x = 120$

ii

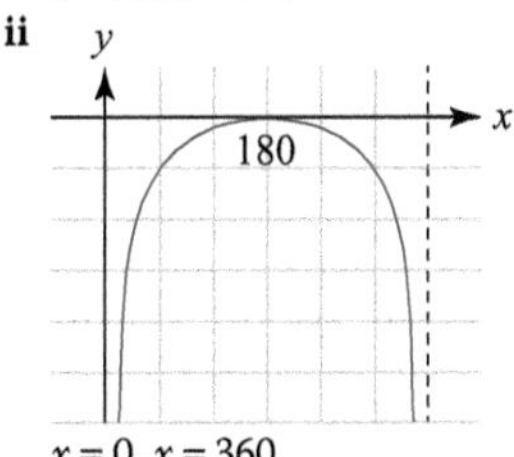

$x = 0, x = 360$

b $x = 8.3°, 98.3°, 128.3°$

35 a $\theta = 0.421, 3.56, 2.72, 5.86$ rad

b $\theta = \frac{\pi}{12}, \frac{19\pi}{12}, \frac{7\pi}{12}, \frac{13\pi}{12}$

36 $x = 90°, 270°, y = 1, -1$

37 a

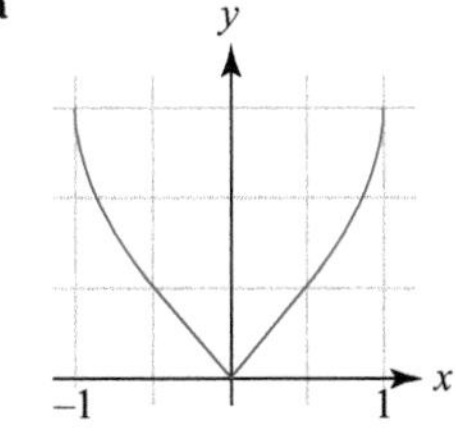

b

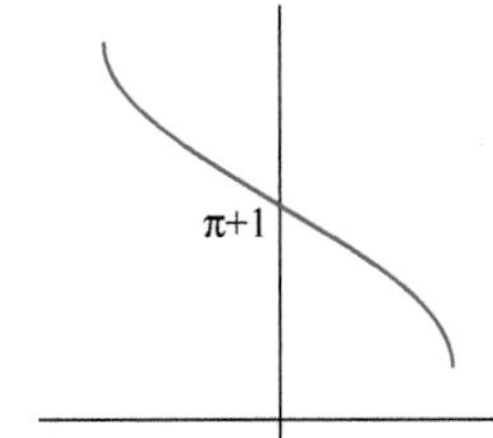

38 $\frac{5^2\sec^2 t}{25} - \frac{3^2\tan^2 t}{9} = \sec^2 t - \tan^2 t$
$= 1$

39 $x = 36.9°, 216.9°, 135°, 315°$

40 $\theta = 0.421, 2.721, \frac{\pi}{3}, \frac{2\pi}{3}$ rad

41 LHS $= \frac{\cos^2\theta}{\sin\theta} - \sin\theta$
$= \frac{\cos^2\theta - \sin^2\theta}{\sin\theta}$
$= \frac{1-2\sin^2\theta}{\sin\theta}$
$= \operatorname{cosec}\theta - 2\sin\theta$

42 $x < \frac{\pi}{6}, x > \frac{5\pi}{6}$

43 a i LHS $= \frac{\cos x(1-\cos x) - \sin x\sin x}{\sin x(1-\cos x)}$
$= \frac{\cos x - \cos^2 x - \sin^2 x}{\sin x(1-\cos x)}$
$= \frac{\cos x - 1}{\sin x(1-\cos x)}$
$= \frac{-1}{\sin x}$
$= -\operatorname{cosec} x$

ii $x \neq 2n\pi$

b $x = 3.48, 5.94$ rad

44 a $\cos x = \cos\left(\frac{x}{2}\right)\cos\left(\frac{x}{2}\right) - \sin\left(\frac{x}{2}\right)\sin\left(\frac{x}{2}\right)$
$= \cos^2\left(\frac{x}{2}\right) - \sin^2\left(\frac{x}{2}\right)$
$= 1 - \sin^2\left(\frac{x}{2}\right) - \sin^2\left(\frac{x}{2}\right)$
$= 1 - 2\sin^2\left(\frac{x}{2}\right)$ as required

b $x = 250.5°, 109.5°$

45 Earliest time 15:04
Latest time 20:32

46 $x = 120°, 180°, 240°$

47 a $r = \sqrt{5}$ or 2.24
$\alpha = 63.4$

b Minimum is $\frac{1}{5}$
$\theta = 26.6°$

c Maximum is $\frac{1}{3-\sqrt{5}}$
$\theta = 206.6°$

Chapter 15

Exercise 15.1A Fluency and skills

1 a Concave for $x < 0$
Convex for $x > 0$

b Convex for $x < -2$
Concave for $-2 < x < -1$
Convex for $x > -1$

2 i a 1

b Horizontal point of inflection ($x = 0$) as curve goes from convex to concave.

ii a 1

b Point of inflection on a decreasing section of the curve ($x \approx -0.5$), as curve goes from concave to convex.

iii a 2

b Point of inflection on a decreasing section of the curve ($x \approx -2$), as curve goes from concave to convex; horizontal point of inflection ($x = 0$), as curve goes from convex to concave.

iv a 3

b Point of inflection on a decreasing section of the curve ($x \approx -1.5$) as curve goes from concave to convex; point of inflection on an increasing section of the curve ($x \approx 1$) as curve goes from convex to concave; horizontal point of inflection ($x \approx 2$) as curve goes from concave to convex.

3 a Convex at (1, 2)

b Concave at (−1, −1)

c Convex at (1, 2)

d Concave at (1, 0)

4 a −2

b When $x = -2$, $\frac{dy}{dx} = (-2)^2 + 4\times(-2) + 4 = 0$
$\Rightarrow$ gradient is zero.

c i Concave

ii Convex

5 (1, −1), on a decreasing section of the curve.
(0, 0), horizontal point of inflection.

6 (−2, 0), on decreasing section of the curve.

7 (0, 0), horizontal point of inflection. (1, −7) on a decreasing section of the curve.
(−1, 7) on a decreasing section of the curve.

8 a $\frac{dy}{dx} = 2 - 2x$
$\Rightarrow \frac{d^2y}{dx^2} = -2$
Since it is a constant and negative, the curve is always concave.

b $\frac{dy}{dx} = -\frac{1}{2\sqrt{x}}$
$\Rightarrow \frac{d^2y}{dx^2} = \frac{1}{4(\sqrt{x})^3}$
Since x is positive, the second derivative exists and is positive.

Exercise 15.1B Reasoning and problem-solving

1 a

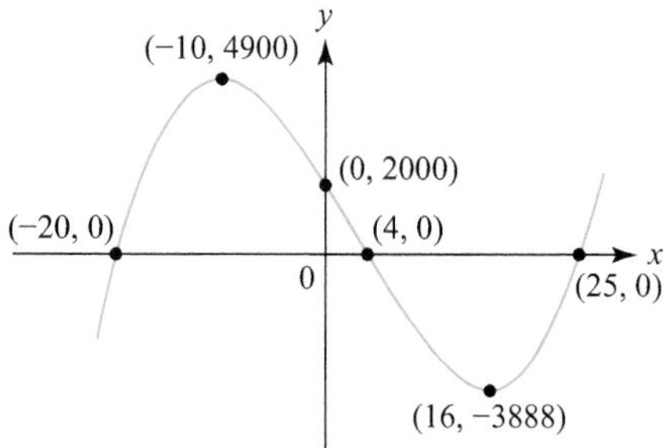

b

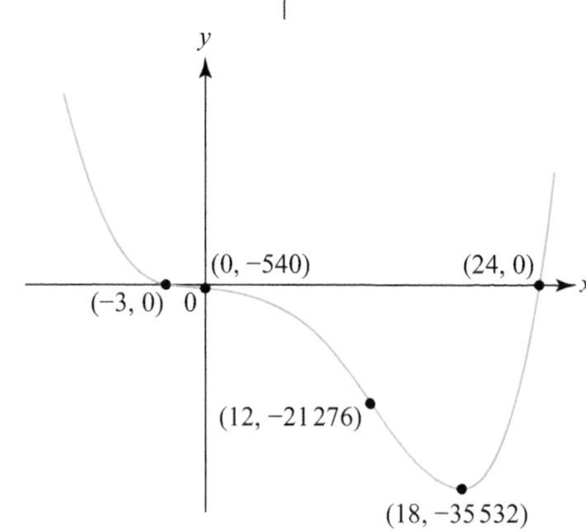

c

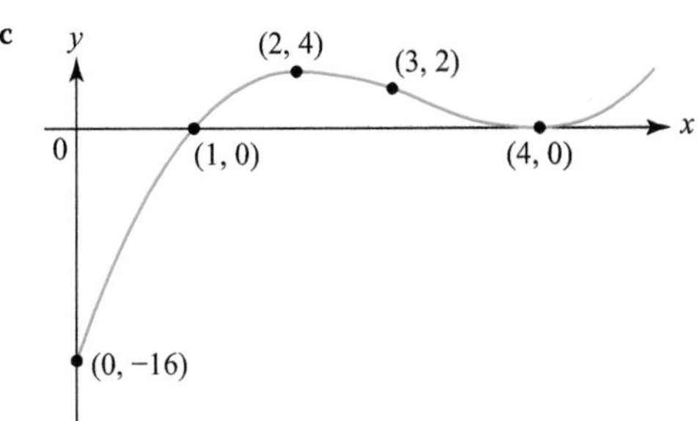

2 $y = ax^2 + bx + c$

$\Rightarrow \frac{dy}{dx} = 2ax + b$

$\Rightarrow \frac{d^2y}{dx^2} = 2a$

a If $a > 0$ then $\frac{d^2y}{dx^2} > 0$ so curve is convex.

b If $a < 0$ then $\frac{d^2y}{dx^2} < 0$ so curve is concave.

3 a $y = x^4 + 8x^3 + 25x^2 - 5x + 10$

$\Rightarrow \frac{dy}{dx} = 4x^3 + 24x^2 + 50x - 5$

$\Rightarrow \frac{d^2y}{dx^2} = 12x^2 + 48x + 50$

$= 12(x^2 + 4x) + 50$

$= 12[(x+2)^2 - 2^2] + 50$

$\Rightarrow \frac{d^2y}{dx^2} = 12(x+2)^2 + 2$ which is greater than zero for all x

So y is convex for all x

b $y = 1 - 2x - 10x^2 + 4x^3 - x^4$

$\Rightarrow \frac{dy}{dx} = -2 - 20x + 12x^2 - 4x^3$

$\Rightarrow \frac{d^2y}{dx^2} = -20 + 24x - 12x^2$

$= -12[x^2 - 2x] - 20$

$= -12[(x-1)^2 - 1^2] - 20$

$\Rightarrow \frac{d^2y}{dx^2} = -12(x-1)^2 - 8$ which is less than zero for all x

So y is concave for all x

c $y = 1 - 2x - 6x^2 + 4x^3 - x^4$

$\Rightarrow \frac{dy}{dx} = -2 - 12x + 12x^2 - 4x^3$

$\Rightarrow \frac{d^2y}{dx^2} = -12 + 24x - 12x^2$

$= -12[x^2 - 2x] - 12$

$= -12[(x-1)^2 - 1^2] - 12$

$\Rightarrow \frac{d^2y}{dx^2} = -12(x-1)^2$ which is less than **or equal to** zero for all x

So it is never greater than zero. So y is never convex for any value of x

4 $y = k(x-a)(x-b)(x-c)$

$= k(x^3 - (a+b+c)x^2 + (ab+ac+bc)x - abc)$

$\Rightarrow \frac{dy}{dx} = k(3x^2 - 2(a+b+c)x + (ab+ac+bc))$

$\Rightarrow \frac{d^2y}{dx^2} = k(6x - 2(a+b+c))$

At point of inflection, $\frac{d^2y}{dx^2} = 0$

$\Rightarrow k(6x - 2(a+b+c)) = 0$

$\Rightarrow x = \frac{a+b+c}{3}$

You know from the shape of cubic graphs that there is always a point of inflection, so the point where the second derivative is 0 must be that point.

5 Roots: $x = 0$ or $\pm\sqrt{\frac{5}{3}}$

Stationary points:

Horizontal point of inflection at (0, 0)

Maximum turning point at (−1, 2)

Minimum turning point at (1, −2)

Non-horizontal points of inflection, both on decreasing sections of curve at $\left(\frac{1}{\sqrt{2}}, -\frac{7\sqrt{2}}{8}\right)$, and $\left(\frac{-1}{\sqrt{2}}, \frac{7\sqrt{2}}{8}\right)$

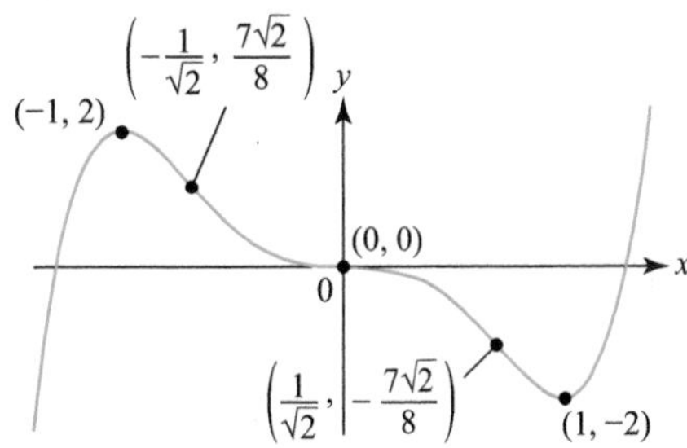

6 $y = -13x - 22$

7 When $x > \frac{a}{3}$

8 a i When $b^2 - 3ac < 0$

ii When $b^2 - 3ac = 0$

iii When $b^2 - 3ac > 0$

b i When $3ax + b < 0$ or $x < -\frac{b}{3a}$

ii When $3ax + b > 0$ or $x > -\frac{b}{3a}$

iii When $3ax + b = 0$ or $x = -\frac{b}{3a}$

9 $12ax^2 + 6bx + 2c = 0$ must have two roots.

The discriminant must be greater than zero.

So, $(6b)^2 - 4.12a.2c > 0$

giving $3b^2 - 8ac > 0$

Exercise 15.2A Fluency and skills

1 a $3\cos x$ **b** $-\frac{\sin x}{3}$ **c** $\sin x$ **d** $1 + \cos x$

e $2x - \sin x$ **f** $\cos x - \sin x$ **g** $3\cos x + 4\sin x$

2 a $\frac{dy}{dx} = 2\cos x$

$\frac{d^2y}{dx^2} = -2\sin x$

b $\frac{dy}{dx} = 4x + \sin x$

$\frac{d^2y}{dx^2} = 4 + \cos x$

c $\frac{dy}{dx} = 2\cos x + 3\sin x$;

$\frac{d^2y}{dx^2} = -2\sin x + 3\cos x$

d $\frac{dy}{dx} = \cos x - \sin x + 3x^2$

$\frac{d^2y}{dx^2} = -\sin x - \cos x + 6x$

e $\frac{dy}{dx} = \frac{-\sin x - \cos x}{2}$

$\frac{d^2y}{dx^2} = \frac{-\cos x + \sin x}{2}$

3 a i $\frac{3}{2}$ **ii** $-\frac{3\sqrt{3}}{2}$

b i -1 **ii** $-\sqrt{3}$

c i 0 **ii** $-\sqrt{2}$

d i 1 **ii** 0

e i 3 **ii** -4

4 a $\left(\frac{\pi}{2}, 2\right)$ and $\left(\frac{3\pi}{2}, 0\right)$

b $\left(\frac{\pi}{2}, 1\right)$ and $\left(\frac{3\pi}{2}, 1\right)$

5 4

6 Concave

7 a $2\cos x$ **b** $\sin x$ **c** $1 - \cos x$

Exercise 15.2B Reasoning and problem-solving

1 a $\cos x$ **b** $-\sin x$

c $-\sin x$ **d** $\cos x$

2 a $3x - 6y = \pi - 3\sqrt{3}$

b $\frac{d^2y}{dx^2} = -\sin x$

When $x = \frac{\pi}{3}, \frac{d^2y}{dx^2} = -\sin\frac{\pi}{3}$

$= -\frac{\sqrt{3}}{2}$

Since the result is negative, the curve is concave.

3 a $\frac{1}{2} - \frac{1}{2}\cos x$

b $-12x^{-4} + \sin x$

c $\cos x - \sin x$

d $-\sin x + \cos x$

4 a i -2 **ii** $\frac{1}{2}$ **iii** 3

iv $\frac{3}{2}$ **v** 0

b $-1 \le \cos x \le 1$

$\Rightarrow -2 \le -2\cos x \le 2$

$\Rightarrow \frac{x}{\pi} - 2 \le \frac{x}{\pi} - 2\cos x \le \frac{x}{\pi} + 2$

$\Rightarrow \frac{x}{\pi} - 2 \le \frac{dy}{dx} \le \frac{x}{\pi} + 2$

When $x > 2\pi$, $0 < \frac{dy}{dx}$

Since the derivative is always positive when $x > 2\pi$ then the function is always increasing.

c $a > 2\pi$

5 a 11 m **b** 10 m s^{-1} **c** 0 m s^{-2}

6 a 2.40 hours per unit (3 sf)

b When $x = 0, \pi, 2\pi$ units

c $\frac{\pi}{2}, \frac{3\pi}{2}$

7 a $2\cos\left(x + \frac{\pi}{3}\right)$

b i 0.657(3 sf) **ii** -0.459 (3 sf)

c Rising

8 Remember the limits: $\lim_{\theta \to 0}\left(\frac{\sin\theta}{\theta}\right) = 1$ and $\lim_{\theta \to 0}\left(\frac{\cos\theta - 1}{\theta}\right) = 0$

a By definition $\frac{d(\sin 2x)}{dx} = \lim_{h \to 0}\left(\frac{\sin(2(x+h)) - \sin 2x}{h}\right)$

$= \lim_{h \to 0}\left(\frac{\sin 2x\cos 2h + \cos 2x\sin 2h - \sin 2x}{h}\right)$

$= \lim_{h \to 0}\left(\frac{\sin 2x(\cos 2h - 1) + \cos 2x\sin 2h}{h}\right)$

$= \lim_{h \to 0}\left(\frac{2\sin 2x(\cos 2h - 1)}{2h}\right) + \lim_{h \to 0}\left(\frac{2\cos 2x\sin 2h}{2h}\right)$

Note that as h tends to zero then so does $2h$

$= \lim_{h \to 0}\left(2\sin 2x\frac{(\cos 2h - 1)}{2h}\right) + \lim_{h \to 0}\left(2\cos 2x\frac{\sin 2h}{2h}\right)$

Treating $2h$ as θ

$= 2\sin 2x \times 0 + 2\cos 2x \times 1$

$= 2\cos 2x$

b By definition $\frac{d(\sin ax)}{dx} = \lim_{h \to 0}\left(\frac{\sin(a(x+h)) - \sin ax}{h}\right)$

$= \lim_{h \to 0}\left(\frac{\sin ax\cos ah + \cos ax\sin ah - \sin ax}{h}\right)$

$= \lim_{h \to 0}\left(\frac{\sin ax(\cos ah - 1) + \cos ax\sin ah}{h}\right)$

$= \lim_{h \to 0}\left(\frac{a\sin ax(\cos ah - 1)}{ah}\right) + \lim_{h \to 0}\left(\frac{a\cos ax\sin ah}{ah}\right)$

Note that as h tends to zero then so does ah

$= \lim_{h \to 0}\left(a\sin ax\frac{(\cos ah - 1)}{ah}\right) + \lim_{h \to 0}\left(a\cos ax\frac{\sin ah}{ah}\right)$

Treating ah as θ

$= a\sin ax \times 0 + a\cos ax \times 1$

$= a\cos ax$

Exercise 15.3A Fluency and skills

1 a $4e^x$ **b** $1 - 2e^x$ **c** $5e^x - 6x$ **d** $-x^{-2} + e^x$

e $-\frac{1}{x^2} + e^x$ **f** $e^x + \frac{2}{x^3}$ **g** $3x^2 + e^x + \sin x$ **h** $-\frac{e^x}{4}$

2 a $\frac{dy}{dx} = 3e^{3x}$, $\frac{d^2y}{dx^2} = 9e^{3x}$

b $\frac{dy}{dx} = -8e^{-4x}$, $\frac{d^2y}{dx^2} = 32e^{-4x}$

c $\frac{dy}{dx} = e^x - 2e^{2x}$, $\frac{d^2y}{dx^2} = e^x - 4e^{2x}$

d $\frac{dy}{dx} = 3e^{0.5x}$, $\frac{d^2y}{dx^2} = 1.5e^{0.5x}$

e $\frac{dy}{dx} = e^x - e^{-x}$, $\frac{d^2y}{dx^2} = e^x + e^{-x}$

f $\frac{dy}{dx} = e^x + e^{-x}$, $\frac{d^2y}{dx^2} = e^x - e^{-x}$

g $\frac{dy}{dx} = 1.25e^{1.25x} + 0.5e^{0.5x}$, $\frac{d^2y}{dx^2} = 1.5625e^{1.25x} + 0.25e^{0.5x}$

h $\frac{dy}{dx} = e^x - 2e^{-2x}$, $\frac{d^2y}{dx^2} = e^x + 4e^{-2x}$

3 a $f'(x) = \frac{3}{x}$, $f''(x) = -\frac{3}{x^2}$

b $f'(x) = -\frac{2}{x}$, $f''(x) = \frac{2}{x^2}$

c $f'(x) = \frac{1}{x}$, $f''(x) = -\frac{1}{x^2}$

d $f'(x) = \frac{3}{x} + \frac{1}{x} = \frac{4}{x}$, $f''(x) = -\frac{4}{x^2}$

e $f'(x) = \frac{5}{x}$, $f''(x) = -\frac{5}{x^2}$

f $f'(x) = -\frac{1}{x}$, $f''(x) = \frac{1}{x^2}$

g $f'(x)=\frac{2}{x}$

$f''(x)=-\frac{2}{x^2}$

h $f'(x)=-\frac{1}{x}$

$f''(x)=\frac{1}{x^2}$

i $f'(x)=\frac{1}{2x}$

$f''(x)=-\frac{1}{2x^2}$

4 a $5^x\ln 5$ **b** $2^x\ln 2$ **c** $6^x\ln 6$

d $\frac{3^x\ln 3}{5}$ **e** $8^x\ln 8-7^x\ln 7$ **f** $5-5^x\ln 5$

g $3\times 4^x\ln 4$

5 a 3 **b** $1-2e$ **c** $1+6e$

d 2 **e** 8 **f** $\frac{1}{2}$

6 a 3 **b** 5e **c** ln4

d 5 ln5 **e** $\frac{1}{2}$ ln2 **f** $1+36\ln 6$

g 3 **h** $1-\frac{2}{e}$

Exercise 15.3B Reasoning and problem-solving

1 a 4.27 years (3 sf)

b 14.25 pheasant pairs per year

c $\frac{d^2P}{dt^2}=0.9025e^{0.095t}>0$ for all t

Curve is always convex.

2 a $y=230e^{-0.134\times 0}-60=230-60$

$=170$

b Machine is losing £15.77 per year.

c Year 7

d i Year 10

ii Around £8 per year

e $\frac{dy}{dt}=-30.82e^{-0.134t}$

$\Rightarrow\frac{d^2y}{dt^2}=-0.134\times-30.82e^{-0.134t}$

$=4.13e^{-0.134t}$

which is positive for $0\le t\le 10$. So the curve is always convex during its lifetime.

3 a $A=82$

$k=0.093$

b i 7.6°C per minute

ii 3.0°C per minute

c 21.82 minutes

d 0.4°C per minute per minute

4 a 1085 m

b 11.7 metres per unit

c 12.1 metres per unit

5 a $c=250$, $k=-77.1$

b Decreasing by 19.3 tics per day

c i 25.3

ii Dropping by 3.05 tics per day

6 a $c=80$, $k=285.2$

b $w=285.2\ln t+80\Rightarrow\frac{dw}{dt}=\frac{285.2}{t}$

i 40.7 g/day

ii 20.4 g/day

c 11.3 g/day

d $\frac{dw}{dt}=\frac{285.2}{t}>0$ for $1\le t\le 30$, so the function is increasing in this domain.

$\frac{d^2w}{dt^2}=-\frac{285.2}{t^2}<0$ for $1\le t\le 30$, so the rate of growth is slowing down.

Exercise 15.4A Fluency and skills

1 a $\sin x+x\cos x$

b $e^x(1+x)$

c $e^x(\cos x-\sin x)$

d $3(1+\ln x)$

e $\cos x\bullet\ln x+\frac{\sin x}{x}$

f $e^{3x}(3x^2+2x+1)$

g $\frac{\ln 3x+2}{2\sqrt{x}}$

h $\frac{-\cos x+x\sin x}{x^2}$

i $2\left(x+\frac{1}{x^3}\right)$

j $e^{3x}\left(3\ln x+\frac{1}{x}\right)$

k $2x(1-3x+4x^2-5x^3)$

l $2x(1+2\ln x)$

m $-3\sqrt{x}\left(2+\frac{1}{2x^2}\right)$

n $\frac{\ln 3x+2}{2\sqrt{x}}$

o $-2\sin x\cos x$

p $4e^{2x+1}(x+1)$

2 a $\frac{d(\sin^2 x)}{dx}=\frac{d(\sin x\sin x)}{dx}$

Let $u=\sin x$ and $v=\sin x$

So $\frac{du}{dx}=\cos x$ and $\frac{dv}{dx}=\cos x$

Therefore, $\frac{dy}{dx}=\sin x\cos x+\sin x\cos x=2\sin x\cos x$
$=\sin 2x$

b $\frac{d(\cos^2 x)}{dx}=\frac{d(\cos x\cos x)}{dx}$

Let $u=\cos x$ and $v=\cos x$

So $\frac{du}{dx}=-\sin x$ and $\frac{dv}{dx}=-\sin x$

Therefore, $\frac{dy}{dx}=(\cos x)\cdot(-\sin x)+(\cos x)(-\sin x)$

$=-2\sin x\cos x\ =-\sin 2x$

c $\frac{d(\sin 2x)}{dx}=\frac{d(2\sin x\cos x)}{dx}$

Let $u=2\sin x$ and $v=\cos x$

So $\frac{du}{dx}=2\cos x$ and $\frac{dv}{dx}=-\sin x$

Therefore, $\frac{dy}{dx}=(\cos x).(2\cos x)+(2\sin x)(-\sin x)$

$=2(\cos^2 x-\sin^2 x)\ =2\cos 2x$

d $\frac{d(\cos 2x)}{dx}=\frac{d(\cos^2 x-\sin^2 x)}{dx}$

$=-2\sin x\cos x-2\sin x\cos x$

$=-4\sin x\cos x=-2\sin 2x$

3 a i 1 **ii** $\frac{6+\pi\sqrt{3}}{12}$

b i 0 **ii** 4e

c i −1 **ii** 0

4 a $-\frac{2}{(x-1)^2}$ **b** $\frac{x\cos x-\sin x}{x^2}$

c $\frac{\cos x-\sin x}{e^x}$ **d** $\frac{x^2+2x-1}{(x+1)^2}$

e $-\frac{1}{\sin^2 x}$ **f** $\frac{1-\ln x}{x^2}$

g $-\frac{3\cos x}{\sin^2 x}$ **h** $\frac{\cos x+x\sin x}{\cos^2 x}$

i $\frac{x(4-3\sqrt{x})}{2(1-\sqrt{x})^2}$ **j** $\frac{3x^2-3x-1}{2x\sqrt{x}}$

k $\frac{e^{2x+3}(2x-3)}{x^4}$ **l** $\frac{1+\sin x}{\cos^2 x}$

5 a i −1 **ii** 1

b i $\frac{1-e}{e^{e+1}}$ **ii** e^{-1}

c i 0 **ii** $\frac{-4}{9}$ **iii** 0

Exercise 15.4B Reasoning and problem-solving

1 −2

2 $y=x+1$

3 $1-\frac{\pi}{2}$

4 a At $x=1$

b Let $u=\ln x$ and $v=e^x$

So $\frac{du}{dx}=\frac{1}{x}$ and $\frac{dv}{dx}=e^x$

Therefore, $f'(x)=\frac{e^x\frac{1}{x}-e^x\ln x}{e^{2x}}$

$=\frac{1-x\ln x}{xe^x}$

At stationary point $f'(x)=0$

$\Rightarrow \frac{1-x\ln x}{xe^x}=0$

$\Rightarrow x\ln x=1$

$\Rightarrow \ln x^x=1$

$\Rightarrow x^x=e$

Substituting $x=1.763$ gives $1.763^{1.763}$ $=2.717$...which is e correct to 3 sf

c $f'(x)=\frac{1-x\ln x}{xe^x}$

$\Rightarrow f''(x)=\frac{xe^x\frac{d(1-x\ln x)}{dx}-(1-x\ln x)\frac{d(xe^x)}{dx}}{x^2e^{2x}}$

$=\frac{xe^x(-1-\ln x)-(1-x\ln x)(xe^x+e^x)}{x^2e^{2x}}$

Points of inflection occur when

$\frac{xe^x(-1-\ln x)-(1-x\ln x)(xe^x+e^x)}{x^2e^{2x}}=0$

$\Rightarrow xe^x(-1-\ln x)-(1-x\ln x)(xe^x+e^x)=0$

$\Rightarrow x^2\ln x-2x-1=0$

$\Rightarrow \ln x=\frac{2x+1}{x^2}$

$\Rightarrow x=e^{\frac{2x+1}{x^2}}$

$\Rightarrow e=x^{\frac{x^2}{2x+1}}$

Substituting 2.55245 for x:

$2.55245^{\frac{2.55245^2}{2\times 2.55245+1}}=2.718284357$... which is e to 6 sf

5 a $\frac{d\sec x}{dx}=\frac{d\left(\frac{1}{\cos x}\right)}{dx}$

Let $u=1$ and $v=\cos x$

So $\frac{du}{dx}=0$ and $\frac{dv}{dx}=-\sin x$

Therefore, $\frac{dy}{dx}=\frac{\cos x\cdot 0+1\cdot\sin x}{\cos^2 x}$

$=\frac{\sin x}{\cos x\cos x}$

$=\sec x\tan x$

b $\frac{d\,\mathrm{cosec}\,x}{dx}=\frac{d\left(\frac{1}{\sin x}\right)}{dx}$

Let $u=1$ and $v=\sin x$

So $\frac{du}{dx}=0$ and $\frac{dv}{dx}=\cos x$

Therefore, $\frac{dy}{dx}=\frac{\sin x\cdot 0-1\cdot\cos x}{\sin^2 x}$

$=\frac{-\cos x}{\sin x\sin x}$

$=-\mathrm{cosec}\,x\cot x$

c $\frac{d\cot x}{dx}=\frac{d\left(\frac{1}{\tan x}\right)}{dx}$

Let $u=1$ and $v=\tan x$

So $\frac{du}{dx}=0$ and $\frac{dv}{dx}=\sec^2 x$

Therefore, $\frac{dy}{dx}=\frac{\tan x\cdot 0+1\cdot\sec^2 x}{\tan^2 x}$

$=\frac{-\cos x\cos x}{\cos^2 x\sin^2 x}$

$=-\mathrm{cosec}^2 x$

6 a $3\sin^2 x\cos x$

b $xe^x\cos x+(x+1)e^x\sin x$

7 a $T(x)=\frac{20}{100-x^2}$

You can see that $100-x^2$ is decreasing for $0\le x<10$

So $\frac{20}{100-x^2}$ must be increasing in this range.

b 128 secs per mph

c i 0.2 **ii** 0

d As x approaches 10, the time taken to make the journey approaches infinity. This means that the closer the speed of the river is to 10 mph, the longer it takes, and if the speed of the river were to be 10mph, you would never complete the journey.

Exercise 15.5A Fluency and skills

1 a $y = u^2, u = \sin x$ b $y = \tan u, u = 2x$

c $y = u^{\frac{1}{2}}, u = 7x$ d $y = \ln u, u = \cos x$

e $y = e^u, u = 7x$ f $y = u^4, u = 2x + 1$

g $y = u^3, u = 3x - 2$ h $y = u^{-\frac{1}{2}}, u = \sin x$

i $y = 2u^{-\frac{1}{2}}, u = x^3$ j $y = u^{\frac{1}{2}}, u = x + \ln x$

k $y = u^5, u = \sin x + \ln x$ l $y = u^{\frac{1}{3}}, u = \cos x - \ln x$

2 a $15(3x+4)^4$ b $14(2x-1)^6$

c $12x(x^2+1)^5$ d $-6(1-2x-3x^2)^2(1+3x)$

e $\frac{1}{\sqrt{2x+1}}$ f $-\frac{5}{2\sqrt{3-5x}}$

g $\frac{3x}{\sqrt{3x^2+4}}$ h $-\frac{2}{3\sqrt[3]{(1-2x)^2}}$

i $\frac{2x+3}{2\sqrt{x^2+3x+4}}$ j $-\frac{4}{(2x+3)^3}$

k $\frac{3}{2}(1-3x)^{-\frac{3}{2}}$ l $\frac{6(1-x)}{(x^2-2x+5)^2}$

m $-2\cos x \sin x$ n $-\frac{\sin x}{2\sqrt{\cos x}}$

o $3\cos(3x+2)$ p $5\sec^2(5x-1)$

q $-\frac{\sin\sqrt{x+1}}{2\sqrt{x+1}}$ r $-\cos(\cos x)\sin x$

s $e^{\sin x}\cos x$ t $\frac{e^{\sqrt{2x-1}}}{\sqrt{2x-1}}$

u $e^{(e^x)} \cdot e^x$ v $\cot x$

w $\frac{1}{2x+3}$ x $\frac{1}{x\ln x}$

y $\frac{\cos(\ln x)}{x}$ z $-\frac{1}{x(\ln x)^2}$

3 a 6 b $\frac{9}{8}$ c $\frac{3}{8}$

d -4 e $\frac{-3e^2}{4}$ f -1

g -12 h 3

4 a $2^{\sin x}\cos x \ln 2$

b $\frac{2}{(2x+1)\ln 10}$

c $2(x+1)3^{x^2+2x-1}\ln 3$

5 a $y = -5x + 4$ b $y = 4ex - 3e$

c $8y - x = 15$ d $y = 2(x-1)$

e $y = 4x - 3$ f $y = 2x - 2 + e$

6 a $-\frac{3}{4}$ b $4y - 3x = 13$

c $\frac{4}{3}$ d $\frac{-25}{27}$

Exercise 15.5B Reasoning and problem-solving

1 a $-2\sin 2x\cos(\cos 2x)$

b $3\cos 3x\, e^{\sin 3x}$

c $5(\sin x + \cos 2x)^4(\cos x - 2\sin 2x)$

d $\frac{2\cos(4x+1)}{\sqrt{\sin(4x+1)}}$

e $\frac{2x+1}{2\sqrt{(x-1)(x+2)}}$

f $e^x(x+1)\cos(xe^x)$

g $2(3x+2)e^{x(3x+4)}$

h $-x(x\cos x + 2\sin x)\sin(x^2\sin x)$

i $-\frac{3}{(x-1)^2}\left(\frac{x+1}{x-1}\right)^{\frac{1}{2}}$

j $\frac{1}{(x+1)^2}\cos\left(\frac{x}{x+1}\right)$

k $\frac{1}{x(1-x)}$

l $\frac{x+1}{x}$

2 a $-\frac{\pi}{72}\sin\left(\frac{\pi D}{180}\right)$

b Minimum T of 16:30 pm occurs 180 days after the longest day.

c $\frac{d^2T}{dD^2} = \frac{\pi^2}{12\,960} > 0$

$\Rightarrow$ a minimum turning point

3 40 $cm^2 s^{-1}$

4 a $\frac{b}{\sqrt{b^2-1}} - 1$

b $b^2 > b^2 - 1 > 0$

$\Rightarrow b > \sqrt{b^2-1}$

$\Rightarrow \frac{b}{\sqrt{b^2-1}} > 1$

$\Rightarrow \frac{b}{\sqrt{b^2-1}} - 1 > 0$

$\Rightarrow \frac{dx}{db} > 0$

So x increases as b increases.

c $-\frac{1}{(b^2-1)\sqrt{b^2-1}} < 0$

So graph of function is concave.

5 a $e^{0.001y}$

b 0.5% per annum

c $0.001\,e^{0.001y}$

6 a $3.3e^{-0.22t}$

b $h = 15(1 - e^{-0.22t})$

$\Rightarrow e^{-0.22t} = 1 - \frac{h}{15}$

So $\frac{dh}{dt} = 0.22 \times 15e^{-0.22t}$

$= 0.22 \times 15\left(1 - \frac{h}{15}\right)$

$= 0.22(15 - h)$

7 a $\frac{2660e^{-0.7t}}{(19e^{-0.7t}+1)^2}$

b $\frac{200-N}{N}$

c $\frac{dN}{dt} = \frac{2660e^{-0.7t}}{(19e^{-0.7t}+1)^2}$

$= 140\frac{19e^{-0.7t}}{(19e^{-0.7t}+1)^2}$

$$=140\frac{\left(\frac{200}{N}-1\right)}{\left(\frac{200}{N}\right)^2}$$

$$=140\frac{200N-N^2}{40000}$$

$$=\frac{7}{2000}N(200-N)$$

8 **a** $3\tan^2 x\sec^2 x$

b $\frac{\cos^2 x-\sin^2 x}{2\sqrt{\sin x\cos x}}=\frac{1}{\sqrt{2}}\cot 2x$

c $\frac{2\sin x\cos x\cos 7x-7\sin^2 x\sin 7x}{2\sqrt{\sin^2 x\cos 7x}}$

9 When $x=8$, $\frac{dV}{dt}=288\text{ m}^3\text{ minute}^{-1}$

Exercise 15.6A Fluency and skills

1 **a** **i** $f'(x)=6x^5$

$f^{-1}(x)=\sqrt[6]{x}$

ii $\frac{1}{6\left(\sqrt[6]{x}\right)^5}$

b **i** $f'(x)=\frac{2}{3\sqrt[3]{x}}$

$f^{-1}(x)=x^{\frac{3}{2}}$

ii $\frac{3\sqrt{x}}{2}$

c **i** $f'(x)=2x$

$f^{-1}(x)=\sqrt{x-2}$

ii $\frac{1}{2\sqrt{x-2}}$

d **i** $f'(x)=2(x+4)$

$f^{-1}(x)=\sqrt{x-1}-4$

ii $\frac{1}{2\sqrt{x-1}}$

e **i** $f'(x)=\frac{3}{2\sqrt{3x+2}}$

$f^{-1}(x)=\frac{x^2-2}{3}$

ii $\frac{2x}{3}$

f **i** $f'(x)=\frac{2}{3(2x+1)^{\frac{2}{3}}}$

$f^{-1}(x)=\frac{x^3-1}{2}$

ii $\frac{3x^2}{2}$

2 **a** $y=\cos^{-1}x\Rightarrow x=\cos y$

$\Rightarrow\frac{dx}{dy}=-\sin y$

$0\le\cos^{-1}x\le\pi$

$\Rightarrow 0\le y\le\pi$

$\Rightarrow\sin y\ge 0$

$\Rightarrow\sin y=\sqrt{1-\cos^2 y}$

$=\sqrt{1-x^2}$

$\Rightarrow\frac{dy}{dx}=-\frac{1}{\sin y}$

$=-\frac{1}{\sqrt{1-x^2}}$

b $y=\tan^{-1}x\Rightarrow x=\tan y$

$\Rightarrow\frac{dx}{dy}=1+\tan^2 y$

$=1+x^2$

$\Rightarrow\frac{dy}{dx}=\frac{1}{1+x^2}$

c $y=\sec^{-1}x\Rightarrow x=\sec y=\frac{1}{\cos y}$

$\Rightarrow\frac{dx}{dy}=\frac{\sin y}{\cos^2 y}$

$=\sin y\sec^2 y$

$0\le\sec^{-1}x\le\pi$

$\Rightarrow 0\le y\le\pi$

$\Rightarrow\sin y\ge 0$

$\Rightarrow\sin y=\sqrt{1-\cos^2 y}=\sqrt{1-\frac{1}{x^2}}$

$\frac{dx}{dy}=\sin y\sec^2 y$

$=x^2\sqrt{1-\frac{1}{x^2}}$

$=x\sqrt{x^2-1}$

$\Rightarrow\frac{dy}{dx}=\frac{1}{x\sqrt{x^2-1}}$

d $y=\cot^{-1}x\Rightarrow x=\cot y$

$\Rightarrow\frac{dx}{dy}=-\text{cosec}^2 y=-(1+\cot^2 y)$

$=-(1+x^2)$

$\Rightarrow\frac{dy}{dx}=-\frac{1}{1+x^2}$

3 **a** **i** 3 **ii** $\frac{x-1}{3}$ **iii** $\frac{1}{3}$

b **i** -5 **ii** $\frac{1-x}{5}$ **iii** $-\frac{1}{5}$

c **i** $6x$ **ii** $\sqrt{\frac{x}{3}}$ **iii** $\frac{1}{2\sqrt{3x}}$

d **i** $\frac{1}{2}$ **ii** $2x-1$ **iii** 2

e **i** $-\frac{1}{(x+1)^2}$ **ii** $\frac{1}{x}-1$ **iii** $-\frac{1}{x^2}$

f **i** $\frac{1}{(x+1)^2}$ **ii** $\frac{x}{1-x}$ **iii** $\frac{1}{(1-x)^2}$

g **i** $-\frac{2}{x^3}$ **ii** $\frac{1}{\sqrt{x}}$ **iii** $-\frac{1}{2x\sqrt{x}}$

h **i** $\frac{1}{2\sqrt{x}}$ **ii** x^2 **iii** $2x$

i **i** $\frac{1}{2\sqrt{x}}$ **ii** $(x+1)^2$ **iii** $2(x+1)$

j **i** $\frac{1}{3}x^{-\frac{2}{3}}$ **ii** x^3 **iii** $3x^2$

k **i** $2e^{2x+1}$ **ii** $\frac{\ln x - 1}{2}$ **iii** $\frac{1}{2x}$

l **i** $\frac{1}{x}$ **ii** $\frac{e^x}{4}$ **iii** $\frac{e^x}{4}$

Exercise 15.6B Reasoning and problem-solving

1 $y = 3x - 5$

2 $6\sqrt{3}y - 6x = \pi\sqrt{3} - 3$

3 **a** $6\sqrt{3}y - 12x = \pi\sqrt{3} + 6$

b The gradient is undefined, so the tangent is parallel to the y-axis. Equation: $x = -2$

4 **a** $A = \frac{1}{2}r^2\theta$

$= \frac{1.5^2}{2}\theta$

$= \frac{9}{8}\theta$

$h = 1.5\sin\theta$

$\Rightarrow \theta = \sin^{-1}\frac{2h}{3}$

$A = \frac{9}{8}\sin^{-1}\frac{2h}{3}$

b $\frac{9}{4\sqrt{9-4h^2}}$

c $\frac{5}{4}$ m² per m

5 **a** $\frac{1.2}{\pi\sqrt{1-(2-0.2D)^2}}$

b 0.382

Exercise 15.7A Fluency and skills

1 **a** $\frac{2x}{3y^2}$ **b** $-\frac{x+1}{y-3}$

c $-\frac{2x}{y}$ **d** $-\frac{y^2}{x^2}$

e $-\frac{2(x+y)}{2x+3}$ **f** $\frac{2xy^2}{y^2-1}$

g -1 **h** $6e^x\sqrt{y}$

i $-\frac{\cos x}{2\sin 2y}$ **j** $2x(y+1)$

k $\frac{\cos y - \sec^2 x}{x\sin y}$ **l** $-\frac{y^{\frac{1}{4}}}{x^{\frac{1}{4}}}$

m $\frac{\sqrt{1-y^2}(\sqrt{1-x^2}+1)}{\sqrt{1-x^2}(\sqrt{1-y^2}-1)}$ **n** $\frac{(x+y)(1+y)-1}{(1-x(x+y))}$

o $\frac{2(x+y)}{e^{y+1}-2x}$ **p** $\frac{y(1+y^2)}{1+x(1+y^2)}$

q $-\frac{y^3}{x^3}$

2 $xy + 2x^2y^2 = 1$

$\Rightarrow x\frac{dy}{dx} + y + 4x^2y\frac{dy}{dx} + 4xy^2 = 0$

$\Rightarrow \frac{dy}{dx} = -\frac{4xy^2 + y}{x + 4x^2y}$

When $x = 1$ and $y = 0.5$

$\Rightarrow \frac{dy}{dx} = -\frac{4\times1\times0.25+0.5}{1+4\times1\times0.5}$

$= -\frac{1}{2}$

3 **a** $-\frac{x}{y}$

b When $x = 3$

$\Rightarrow y = 4$ or -4

At point (3, 4) the gradient $\frac{dy}{dx} = -\frac{3}{4}$

and at point (3, –4) the gradient is $\frac{dy}{dx} = \frac{3}{4}$

c When $x = 0$, $y = 5$

$\Rightarrow \frac{dy}{dx} = -\frac{0}{5}$

$= 0$

4 **a** $(10, \pm2\sqrt{5})$

b $\left(2, -\frac{5}{2}\right)$ and $(2, -1)$

c (0, 3)

5 **a** $e^x + e^{2y} + 3x = 2 - 4y$

When $x = 0$

$1 + e^{2y} = 2 - 4y$

$\Rightarrow e^{2y} = 1 - 4y$

$\Rightarrow y = 0$

Therefore (0, 0) lies on the curve.

b $-\frac{3+e^x}{2(e^{2y}+2)}$

c $-\frac{2}{3}$

6 **a** $(1 + \ln x)(y - 1)$

b $\ln(y - 1) = 0$

$\Rightarrow y - 1 = 1$

$\Rightarrow y = 2$

c 1

7 **a** $\frac{1-x}{y-2}$; $y \neq 2$

b $\frac{4}{3}$

8 **a** $\frac{2x[3(x^2+y^2-1)^2 - y^3]}{3y[x^2y - 2(x^2+y^2-1)^2]}$

b $-\frac{4}{3}$

Exercise 15.7B Reasoning and problem-solving

1 $y = -2x + 10$

2 At points (1, 6) and (7, –2)

3 **a** Ellipse cuts x-axis when $y = 0$

$\frac{x^2}{a^2} + \frac{0^2}{b^2} = 1$

$\Rightarrow x = \pm a$

Hence result.

b $-\frac{b^2x}{a^2y}$

c $(0, b)$ and $(0, -b)$

d y is directly proportional to x

4 **a** $\frac{4x+7}{2(y+1)}$

b $2y - 3x = 2$

or $2y + 3x = -6$

c $\left(-\frac{4}{3}, -1\right)$

5 **a** (3, –3) and (–3, 3)

b Maximum: $2\sqrt{3}$; minimum: $-2\sqrt{3}$

c $x = \mp2\sqrt{3}$

6 a i $3y-4x=11$

ii $4y-3x=17$

b (1, 5); the centre of the circle

7 a $11y-x=96$

b $\left(-\frac{12\sqrt{7}}{7}, -\frac{24\sqrt{7}}{7}\right)$

$\left(\frac{12\sqrt{7}}{7}, \frac{24\sqrt{7}}{7}\right)$

Exercise 15.8A Fluency and skills

1 a $\frac{1}{2}$ b 2

c $\frac{1}{2t}$ d $-\tan\theta$

e $-\frac{\sin(\theta+2)}{\cos(\theta+3)}$ f $-\tan\theta$

g $-\tan(2\theta+1)$ h $\frac{e^t+e^{-t}}{e^t-e^{-t}}$

i $-te^{-t}$ j $\frac{1}{t}$

k $\frac{2t}{\cos t}$

2 a $\frac{\cos 3\theta-\cos\theta}{\sin 3\theta+\sin\theta}$ b $-\frac{4}{3}\tan 2\theta$

c -1 d $\frac{\sqrt{1-\theta^2}}{1+\theta^2}$

e -1 f $-\frac{1}{t\sqrt{3}}$

g $-\tan t\sqrt{\tan t}$ h $-\frac{1}{\sin\theta}$

i $-\frac{e^{\cos t}}{e^{\sin t}}\tan t$ j $\sec t$

k $2t-4$

3 a $(-\pi, 0)$ b π

4 a $x=8, y=2$

gradient $=\frac{1}{12}$

c $x=1, y=2$

gradient $=-2$

e $x=6, y=4\sqrt{3}$

gradient $=\frac{8}{3\sqrt{3}}$

Exercise 15.8B Reasoning and problem-solving

1 a 768

b $16x^3-4x^2+1$

2 a $x^2+\frac{y^2}{25}=1$

b $y=\pm 5$

3 a $\frac{1-9t^2}{8t}$ b $\pm\frac{1}{3}$

c $\left(\frac{14}{9}, \frac{7}{9}\right)$; minimum

$\left(\frac{14}{9}, \frac{11}{9}\right)$; maximum

d i $t=0, \pm\frac{1}{\sqrt{3}}$

ii $t=0$ corresponds to the point (2, 1) where the loop occurs (tangent vertical); $t=\pm\frac{1}{\sqrt{3}}$ corresponds to $\left(\frac{2}{3}, 1\right)$, the point where the curve intersects itself.

$y=1$ is an axis of reflective symmetry.

4 a $-\cot\theta$

b $\sqrt{3}$

c $x=1+5\cos\theta$

$\Rightarrow \cos\theta=\frac{x-1}{5}$

$y=2+5\sin\theta$

$\Rightarrow \sin\theta=\frac{y-2}{5}$

Squaring

$\cos^2\theta=\frac{(x-1)^2}{25}$

$\sin^2\theta=\frac{(y-2)^2}{25}$

Adding gives

$\frac{(x-1)^2}{25}+\frac{(y-2)^2}{25}=\cos^2\theta+\sin^2\theta$

$=1$

Hence $(x-1)^2+(y-2)^2=25$

5 a $1-t$

b i $t=0, 2$

ii $x=0$ and $x=20$

c $x=10t$

$\Rightarrow t=\frac{x}{10}$

$\Rightarrow y=10\left(\frac{x}{10}\right)-5\left(\frac{x}{10}\right)^2$

$=x-\frac{1}{20}x^2$

which is of the required form with $a=-\frac{1}{20}, b=1, c=0$

6 $(8\sqrt{2}, 0)$

Review exercise 15

1 Curve is convex at $\left(1, -\frac{2}{3}\right)$ a minimum turning point.

Curve is concave at $(-3, 10)$ a maximum turning point.

$\left(-1, 4\frac{2}{3}\right)$ is a point of inflection on a decreasing section of the curve.

2 a (0, 1) and (–3, 28)

b (0, 1) is a minimum turning point

(–3, 28) is a point of inflection on a decreasing section of the curve.

3 a $f(x)=\sin 2x$

$\Rightarrow f'(x)=\lim_{h\to 0}\frac{\sin 2(x+h)-\sin 2x}{h}$

$\Rightarrow f'(x)=\lim_{h\to 0}\frac{\sin 2x\cos 2h+\cos 2x\sin 2h-\sin 2x}{h}$

$=2\lim_{h\to 0}\frac{\sin 2x(\cos 2h-1)+\cos 2x\sin 2h}{2h}$

$=2\lim_{h\to 0}\sin 2x\frac{(\cos 2h-1)}{2h}+2\lim_{h\to 0}\cos 2x\frac{\sin 2h}{2h}$

$$= 2\sin 2x \lim_{h\to 0}\frac{(\cos 2h - 1)}{2h} + 2\cos 2x \lim_{h\to 0}\frac{\sin 2h}{2h}$$

As h tends to zero so does $2h$

$\Rightarrow f'(x) = 2\sin 2x \times 0 + 2\cos 2x \times 1$

$\Rightarrow f'(x) = 2\cos 2x$

b **i** $-\frac{1}{\sqrt{2}}$

ii $-\frac{5}{\sqrt{2}}$

4 **a** 9.69 cm s^{-1}

b Falling

c After $\frac{\pi}{4}$ seconds

5 **a** $8e^{2x+1}$ **b** $\frac{3}{3x+1}$

c $2\times 3^x \ln 3$ **d** $-2^{-x}\ln 2$

6 **a** $e^x \sin x(\sin x + 2\cos x)$

b $\frac{3(1-\ln x)}{2x^2}$

c $\sqrt{x}\sec^2 x + \frac{\tan x}{2\sqrt{x}}$

d $\frac{(3x^2+6)\cos x + (x^3+6x+11)\sin x}{\cos^2 x}$

7 **a** $\frac{e^x[(1+x)\ln x - 1]}{(\ln x)^2}$

b $\frac{-x-\cos x \sin x}{x^2 \sin^2 x}$

c $\frac{-2x^2 \sin x - x\sin x + \cos x}{(2x+1)^2}$

8 **a** $20x(2x^2-3)^4$ **b** $x > 0$

9 **a** $\frac{1}{4}$ **b** 0

c -6 **d** $\frac{e^3}{3}$

10 **a** $-3^{\cos x}\ln 3 \sin x$

b $\frac{4^{\ln x}\ln 4}{x}$

11 **a** $-3x^{-3}y^3$ **b** $\tan x \tan y$

c $\frac{1-e^y-ye^x}{xe^y+e^x}$ **d** $\frac{x-e^y}{x(e^y\ln x - 2)}$

12 **a** $\frac{3}{6-4y}$ **b** $\frac{1}{6}$

13 $-\frac{1}{3}$

14 **a** $\frac{1}{e^y+2}$ **b** $\frac{1}{3}$

15 **a** $\frac{-2x}{\sqrt{1-x^4}}$ **b** $-\frac{1}{10\sqrt{2e^{\frac{x}{10}}-1}}$

c $\frac{1}{x\sqrt{x^2-1}}$

16 **a** 2 **b** $2t^2$

c $-\frac{4}{3}\tan\theta$ **d** 2

17 **a** $-\frac{3}{t^3}$ **b** $-\frac{3}{64}$

18 **a** At $x = -0.3779645\ldots\left(=-\frac{1}{\sqrt{7}}\right)$ and $x = 0.3779645\left(=\frac{1}{\sqrt{7}}\right)$

b $-\frac{7}{3\sqrt{3}}$

Assessment 15

1 **a** B $-10\sin 2x$

b A $x^2(3\ln x + 1)$

c D $\frac{e^x(x-1)}{4x^2}$

2 B $\frac{\sqrt{3}}{2}$

3 **a** $6x + x^{-\frac{3}{2}}$

b $\frac{193}{8}$ or 24.125

c $y - 1 = -\frac{1}{7}(x-1)$

4 2 s

5 $x + 2y - 1 = 0$

6 **a** **i** $e^x + xe^x$

ii $2e^x + xe^x$

b $ke^x + xe^x$

7 **a** $3\sin x + 3x\cos x$ **b** $y = -3\pi(x-\pi)$

8 **a** **i** $\frac{2}{(x+2)^2}$

ii $\frac{6x\cos x + 3x^2 \sin x}{\cos^2 x}$

iii $9x^2e^x + (3x^3+5)e^x$

b $\frac{(2x+3)(x-5) - 1(x^2+3x)}{(x-5)^2}$

$= \frac{x^2-10x-15}{(x-5)^2}$

9 **a** $-3\sin 3x$ **b** $6e^{3x}$

c $2\cos(2x-5)$

10 **a** **i** $4y - \frac{4}{\sqrt{y}}$ **ii** $\frac{1}{4y-\frac{4}{\sqrt{y}}}$

b $y - 4 = -14(x-16)$

11 $\frac{5y}{3y^2-5x}$

12 **a** 1

b $y - 12 = \frac{3}{2}(x-4)$

13 **a** $x = 0, y = 0$ or $x = -2, y = 8e^{-2}$

b $(0, 0)$ is a minimum, $(-2, 8e^{-2})$ is a maximum

14 **a** $x = 0$, there is a point of inflection on a decreasing part of the curve. At the point of inflection, the gradient of the curve is -4

b

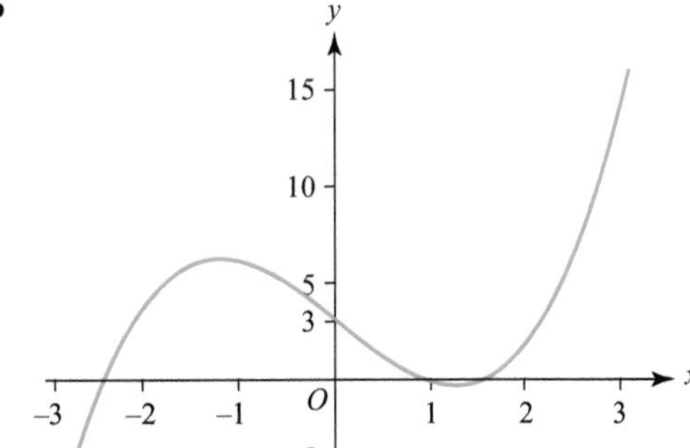

c $x < 0$

15 **a** Stationary point at $(-1, 2e^{-1})$

Either side of the stationary point $\frac{dy}{dx} > 0$

So a point of inflection.

b $\frac{d^2y}{dx^2} = e^x(x^2+4x+3)$

$= e^x(x+3)(x+1)$

$\frac{d^2y}{dx^2} = 0$ when $x = -3$ and when $x = -1$

When $x = -3$, $\frac{dy}{dx} = 4e^{-3} > 0$

At $(-3, 10e^{-3})$ there is a point of inflection on an increasing part of the curve.

16 a $V = 2x^2h$ where h is the height of the cuboid.

$2x^2h = 192$ gives $h = \frac{96}{x^2}$

$S = 2(2x^2 + xh + 2xh)$

$S = 4x^2 + 6x\left(\frac{96}{x^2}\right)$

$S = 4x^2 + \frac{576}{x} \quad (k = 576)$

b 208 cm²

c $\frac{d^2S}{dx^2} = 8 + \frac{1152}{x^3}$

which is always positive as $x > 0$

Therefore a minimum.

17 $(-1, -6)$ is a point of inflection.

18 a $\frac{dy}{dx} = \lim_{h\to 0}\left(\frac{\sin(x+h) - \sin x}{x+h-x}\right)$

$= \lim_{h\to 0}\left(\frac{\sin x\cos h + \sin h\cos x - \sin x}{h}\right)$

$= \sin x \lim_{h\to 0}\left(\frac{\cos h - 1}{h}\right) + \cos x \lim_{h\to 0}\left(\frac{\sin h}{h}\right)$

$= \cos x$

Since $\lim_{h\to 0}\left(\frac{\cos h - 1}{h}\right) = 0$ and $\lim_{h\to 0}\left(\frac{\sin h}{h}\right) = 1$

b $\frac{dy}{dx} = \lim_{h\to 0}\left(\frac{\cos(2x+h) - \cos 2x}{2x+h-2x}\right)$

$= \lim_{h\to 0}\left(\frac{\cos 2x\cos h - \sin h\sin 2x - \cos 2x}{h}\right)$

$= \cos 2x \lim_{h\to 0}\left(\frac{\cos h - 1}{h}\right) - \sin 2x \lim_{h\to 0}\left(\frac{\sin h}{h}\right)$

$= -\sin 2x$

Since $\lim_{h\to 0}\left(\frac{\cos h - 1}{h}\right) = 0$ and $\lim_{h\to 0}\left(\frac{\sin h}{h}\right) = 1$

19 a $\sec^2 x$

b $\sec^2 x = (\cos x)^{-2}$

$f''(x) = -2(\cos x)^{-3}(-\sin x)$

$= 2\frac{\sin x}{\cos x} \cdot \frac{1}{\cos^2 x}$

$= 2\tan x \sec^2 x$ as required

20 a $4^x \ln 4$

b $x = \frac{-1 - \ln 4}{\ln 4}$

21 a i $e^x \sin 5x + 5e^x \cos 5x$

ii $10e^x \cos 5x - 24e^x \sin 5x$

b $x = 0.354$

$e^{0.354}[10\cos(5\times 0.354) - 24\sin(5\times 0.354)]$

$= -36.3 < 0$ so a maximum.

22 a i $\frac{3e^{3t}(t^2+1) - 2te^{3t}}{(t^2+1)^2}$

ii $3\ln t + 3$

iii $-e^{-t}\sin 4t + 4e^{-t}\cos 4t$

b $2^x \ln 2 \tan x + 2^x \sec^2 x$

23 a $y = (\cos x)^{-1}$

$\frac{dy}{dx} = -(\cos x)^{-2}(-\sin x)$

$\frac{dy}{dx} = \frac{\sin x}{\cos^2 x} = \frac{1}{\cos x} \cdot \frac{\sin x}{\cos x}$

$\frac{dy}{dx} = \sec x \tan x$ as required

b $2\sec^3 x - \sec x$

24 a £6749

b £174 per year

25 a $\frac{1}{1 + \ln 2}$

b $y = 2 + 2\ln 2 + 2(\ln 2)^2$

26 $\frac{dx}{dy} = -\sin y$

$\frac{dy}{dx} = -\frac{1}{\sin y}$

$\sin y = \sqrt{1 - \cos^2 y}$

$= \sqrt{1 - x^2}$

(because $\sin y > 0$)

Therefore $\frac{dy}{dx} = -\frac{1}{\sqrt{1-x^2}}$ as required

27 a $\frac{1 - xe^y}{x^2e^y}$

b $\frac{4xy - \sin y}{x\cos y - 2x^2}$

28 $-\frac{1}{4}$

29 a -1

b $y = \frac{2x - 5}{x - 7}$

30 a $-\frac{3\sin t}{2\cos t}$

b $3\sqrt{3}\,x - 2y - 12 = 0$

31 $(0, -3)$ is a minimum.

32 a Convex

b Convex

33 $x < -1$ or $x > 1$

34 a Arc length is $r\theta = 6$

So $r = \frac{6}{\theta}$

Area of triangle $= \frac{1}{2}r^2\sin\theta$

$A = \frac{1}{2}\left(\frac{6}{\theta}\right)^2 \sin\theta$

$= \frac{18}{\theta^2}\sin\theta$

b $\frac{dA}{d\theta} = -\frac{36}{\theta^3}\sin\theta + \frac{18}{\theta^2}\cos\theta$

$-36\sin\theta + 18\theta\cos\theta = 0$

$36\sin\theta = 18\theta\cos\theta$

$\frac{\sin\theta}{\cos\theta} = \frac{18\theta}{36}$

$\tan\theta = \frac{\theta}{2}$ as required

35 $-6 + \ln 27$

36 a $y = (\sin 3x)^{-1}$

$\frac{dy}{dx} = -(\sin 3x)^{-2}(3\cos 3x)$

$\frac{dy}{dx} = -\frac{3\cos 3x}{\sin^2 3x} = -3\frac{\cos 3x}{\sin 3x} \cdot \frac{1}{\sin 3x}$

$\frac{dy}{dx} = -3\cot 3x \operatorname{cosec} 3x$ as required

b $9\cot^2 3x \operatorname{cosec} 3x + 9\operatorname{cosec}^3 3x$

37 a $\frac{1}{1+\tan^2 y}=\frac{1}{1+x^2}$

b $y-\frac{\pi}{4}=\frac{1}{2}(x-1)$

c At B, $x=0$

$y-\frac{\pi}{4}=\frac{1}{2}(-1)$

$y=\frac{\pi}{4}-\frac{1}{2}$

At A, $y=0$

$-\frac{\pi}{4}=\frac{1}{2}(x-1)$

$x=1-\frac{\pi}{2}$ (negative)

$\text{Area}=\frac{1}{2}\left(\frac{\pi}{2}-1\right)\left(\frac{\pi}{4}-\frac{1}{2}\right)$

$=\frac{\pi^2}{16}-\frac{\pi}{8}-\frac{\pi}{8}+\frac{1}{4}$

$=\frac{1}{16}(\pi^2-4\pi+4)$ as required

38 $\frac{1}{9}$

39 $\frac{1}{6\pi}$

40 Let $y=\sin^{-1}2x$

Then $\sin y=2x$

$\cos y\frac{dy}{dx}=2$

$\frac{dy}{dx}=\frac{2}{\cos y}$

$=\frac{2}{\sqrt{1-\sin^2 y}}$

(because $\cos y\geq 0$)

$=\frac{2}{\sqrt{1-(2x)^2}}$ or $=\frac{2}{\sqrt{1-4x^2}}$

41 a $\frac{6x-y^2}{2xy+2}$

b $\frac{5}{4}$ or $-\frac{15}{2}$

42 Let $y=a^x$

Then $\ln y=\ln a^x$

$=x\ln a$

Differentiate both sides with respect to x:

$\frac{1}{y}\frac{dy}{dx}=\ln a$

$\frac{dy}{dx}=y\ln a$

$=a^x\ln a$ as required

43 a $\frac{dx}{dt}=\frac{2}{(3-t)^2}$

$\frac{dy}{dt}=\frac{2t}{3-t}+\frac{t^2}{(3-t)^2}$

or $\frac{2t(3-t)+t^2}{(3-t)^2}\left(=\frac{6t-t^2}{(3-t)^2}\right)$

$\frac{dy}{dx}=\frac{6t-t^2}{(3-t)^2}\cdot\frac{(3-t)^2}{2}$

$=\frac{6t-t^2}{2}$

b $y=\frac{9}{2}x+\frac{2}{x}-6$

44 a $\frac{dy}{d\theta}=\sec^2(\theta-4)$

$\frac{dx}{d\theta}=\sec(\theta-4)\tan(\theta-4)$

$\frac{dy}{dx}=\frac{\sec^2(\theta-4)}{\sec(\theta-4)\tan(\theta-4)}$

$\frac{dy}{dx}=\frac{\sec(\theta-4)}{\tan(\theta-4)}$

$\frac{dy}{dx}=\frac{1}{\cos(\theta-4)}\cdot\frac{\cos(\theta-4)}{\sin(\theta-4)}$

$\frac{dy}{dx}=\text{cosec}(\theta-4)$ as required

b $x^2-y^2=1$

c When $x=3$,

$9-y^2=1$ gives $y=\sqrt{8}=2\sqrt{2}$

Either differentiate the Cartesian equation to get

$2x-2y\frac{dy}{dx}=0$

$\frac{dy}{dx}=\frac{x}{y}$

So when $x=3$ and $y=2\sqrt{2}$,

$\frac{dy}{dx}=\frac{3}{2\sqrt{2}}$

Or use $3=\sec(\theta-4)$

Which gives $\frac{1}{3}=\cos(\theta-4)$

So $\sin(\theta-4)=\sqrt{1-\left(\frac{1}{3}\right)^2}=\frac{2\sqrt{2}}{3}$

Therefore $\frac{dy}{dx}=\frac{1}{\frac{2\sqrt{2}}{3}}=\frac{3}{2\sqrt{2}}$

$y-2\sqrt{2}=\frac{3}{2\sqrt{2}}(x-3)$

$3x-2y\sqrt{2}=1$

Chapter 16

Exercise 16.1A Fluency and skills

1 a $4x^6+c$ b $-7\cos x+c$ c $4\sin x+c$

d $\frac{2}{3}\tan x+c$ e $5e^x+c$ f $x^2+\frac{2}{3}x^{\frac{3}{2}}+c$

g $-\frac{3}{x}+c$ h $3\ln|x|+c$ i $-\cos x+\tan x+c$

j $4\sin x+3\cos x+c$

k $3\tan x+2\sin x+c$

l $5e^x+c$

2 a $3x-4e^x+c$ b $\frac{1}{2}\ln|x|+c$ c $\frac{11}{15}\ln|x|+c$

d $x-3e^x+c$ e $7x+6\cos x+c$ f $x^4-5\ln|x|+c$

g $x-2\sin x+c$ h $\frac{x^2}{2}+\frac{1}{2x^2}+c$ i $\frac{x}{e}+\frac{e^x}{3}+c$

j $\frac{3x}{4}-\frac{1}{x^2}+\frac{\cos x}{3}+c$

k $-\frac{4}{x}-\frac{10}{3}x^{\frac{3}{2}}+c$

l $x+\frac{3}{5}x^{\frac{5}{3}}+c$

3 **a** 2 **b** 1 **c** $\frac{\sqrt{3}-1}{2}$

d 2 **e** 1.718 **f** 2

g 8 **h** −12

4 **a** $71\frac{2}{3}$ **b** $\frac{\pi}{2}+1$ **c** $\sqrt{3}-1+\frac{\pi}{2}$

d 2 **e** $e-\frac{1}{2}$ **f** $\frac{4}{3}$

5 **a** **i** $\frac{13}{4}$ **ii** $\int_1^2 6-2x\,dx$ **iii** 3

b **i** $\frac{165}{32}$ **ii** $\int_{-1}^2 x^2+1\,dx$ **iii** 6

6 **a** **i** 4.41 (to 3 sf) **ii** $\int_1^4 -\sqrt{x}\,dx$ **iii** $\frac{14}{3}$

b **i** $\frac{197}{27}$ **ii** $\int_1^3 4x-x^2\,dx$ **iii** $\frac{22}{3}$

7 $\frac{2\pi}{3}+3$

8 $\frac{1}{2}$

Exercise 16.1B Reasoning and problem-solving

1 **a** $\frac{2}{5}x^{\frac{5}{2}}+c$ **b** $\frac{1}{3}x+\frac{1}{3}\ln x+c$

c $\frac{1}{2}x^4+x^3+c$ **d** $\frac{2}{5}x^{\frac{5}{2}}+x^2+c$

e $\frac{1}{2}x^2+2x+\ln x+c$ **f** $\frac{2}{3}x^{\frac{3}{2}}+x+c$

g $\tan x+\sin x+c$ **h** $\tan x-\cos x+c$

i $\frac{1}{2}e^{2x}+e^x+c$ **j** $\frac{1}{2}e^{2x}+3e^x-4x+c$

k $\frac{1}{2}e^{2x}+\frac{1}{2}e^{-2x}+c$ **l** $-\cos x+c$

m $x+c$

2 **a** $4\frac{1}{2}$ **b** $204\frac{17}{24}$

c 30.375 **d** $\frac{1}{3}$

3 **a** $\left(\frac{\pi}{4},\frac{1}{\sqrt{2}}\right)$ and $\left(\frac{5\pi}{4},-\frac{1}{\sqrt{2}}\right)$

b $2\sqrt{2}$

c $\sqrt{2}-1$

4 **a** $3\frac{1}{12}$ **b** $3\frac{1}{12}$

c $19\frac{1}{5}$ **d** 0.274 (to 3 sf)

5 512 m^3

6 **a** When $t=4$,

$v_A=4\sqrt{4}=8\,\text{m s}^{-1}$

$v_B=4^2-\left(\frac{4}{2}\right)^3$

$=16-8$

$=8\,\text{m s}^{-1}\ (=v_A)$

b Distance $=\left|\int_0^4 4\sqrt{t}-\left(t^2-\left(\frac{t}{2}\right)^3\right)dx\right|$

$=\left|\int_0^4 4t^{\frac{1}{2}}-t^2+\frac{t^3}{8}dx\right|$

$=\left|\left[\frac{8}{3}t^{\frac{3}{2}}-\frac{t^3}{3}+\frac{t^4}{32}\right]_0^4\right|$

$=\left|(21.3...-21.3...+8)-(0-0+0)\right|$

$=8$ m

7 **a** **i** x^2+2x+1

ii x^3+3x^2+3x+1

iii $x^4+4x^3+6x^2+4x+1$

iv $x^5+5x^4+10x^3+10x^2+5x+1$

b **i** $\frac{x^3}{3}+x^2+x+c$

ii $\frac{x^3}{3}+x^2+x+\frac{1}{3}+c_1$

iii $\frac{1}{3}(x^3+3x^2+3x+1)+c_1$

iv $\frac{1}{3}(x+1)^3+c_1$

c **i** $\frac{x^4}{4}+x^3+\frac{3}{2}x^2+x+c$

ii $\frac{x^4}{4}+x^3+\frac{3}{2}x^2+x+\frac{1}{4}+c_1$

iii $\frac{1}{4}(x^4+4x^3+6x^2+4x+1)+c_1$

iv $\frac{1}{4}(x+1)^4+c_1$

d $\int(x+1)^n\,dx=\frac{1}{n+1}(x+1)^{n+1}+c$

e $\int(ax+1)^n\,dx=\frac{1}{a(n+1)}(ax+1)^{n+1}+c$

Exercise 16.2A Fluency and skills

1 **a** **i** $\sqrt{2x}+c$ **ii** $\sqrt{4x-1}+c$

b **i** $-\frac{1}{4}\cos 4x+c$ **ii** $-\frac{1}{5}\cos(5x-2)+c$

c **i** $\frac{1}{3}\sin 3x+c$ **ii** $-\sin(2-3x)+c$

d **i** $\frac{1}{5}\tan 5x+c$ **ii** $\frac{1}{3}\tan(4+3x)+c$

e **i** $\frac{1}{7}e^{7x}+c$ **ii** $-2e^{-2x}+c$

f **i** $\frac{1}{2}\ln|x|+c$ **ii** $\frac{1}{2}\ln|4x+5|+c$

2 **a** $\frac{1}{33}(3x+2)^{11}+c$ **b** $\frac{1}{45}(5x-1)^9+c$

c $\frac{1}{707}(7x-3)^{101}+c$ **d** $-\frac{1}{21}(3x-8)^{-7}+c$

e $-\frac{1}{30}(1-3x)^{10}+c$ **f** $-\frac{1}{8}(6-x)^8+c$

g $-\frac{3}{8}(2x-1)^{-4}+c$ **h** $\frac{1}{4}(10-x)^{-4}+c$

i $\frac{1}{5}\cos(3-5x)+c$ **j** $\frac{1}{2}\sin(4x-1)+c$

k $\frac{1}{2}\sin(2x)+c$ l $\frac{1}{4}\tan(4x+3)+c$

m $\frac{3}{2}\tan(2x+1)+c$ n $\frac{1}{5}e^{5x+2}+c$

o $\frac{7}{3}\ln|3x+9|+c$ p $-4\ln|8-x|+c$

3 a $\frac{1}{5}(x^2+3)^5+c$ b $\frac{1}{4}(x^2+x-1)^4+c$

c $\frac{2}{3}(2x^3-1)^{\frac{3}{2}}+c$ d $\frac{1}{16}(2x^2-5)^4+c$

e $-\frac{1}{6(x^2-7)^3}+c$ f $-\frac{1}{x^2+3x-1}+c$

g $-3\sqrt{1-x^2}+c$ h $\frac{3}{2}(x+4)^{\frac{2}{3}}+c$

i $-\frac{1}{6}\cos(3x^2+1)+c$ j $\frac{1}{4}\tan^4 x+c$

k $\ln|\sin x-1|+c$ l $\frac{1}{2}\ln|3-2\cos x|+c$

m $\frac{1}{4}e^{2x^2+1}+c$ n $2e^{\sqrt{x}}+c$

o $-e^{\cos x}+c$ p $\frac{1}{2}(\ln x)^2+c$

4 a $\frac{1}{16}(2x^2-1)^4+c$ b $\frac{2}{3}\left(\frac{x^3}{3}+\frac{x^2}{2}\right)^{\frac{3}{2}}+c$

c $\frac{2}{3}\sqrt{x^3-3x^2}+c$ d $-\frac{1}{2}\cos(x^2-2x+3)+c$

e $\frac{1}{3}\ln|x^3+3x^2+9x+1|+c$ f $-2\cos\sqrt{x}+c$

g $2\left(\frac{\left(\sqrt{x+1}\right)^3}{3}-\sqrt{x+1}+\sin\sqrt{x+1}\right)+c$

h $\frac{3}{4}(2x-1)^{\frac{2}{3}}+c$ i $-\cos x+\frac{\cos^4 x}{4}+c$

j $\tan x+\frac{\tan^3 x}{3}+c$ k $2\sqrt{1+\sin x}+c$

l $\ln|1+e^x|+c$ m $-\frac{1}{2}e^{\cos 2x}+c$

n $\frac{1}{3}(\ln(2x+3))^{\frac{3}{2}}+c$

5 a $\frac{896}{3}$ b $\frac{1}{3}$ c $\frac{15}{8}$

d $\ln\frac{5}{2}$ e 8 f $\frac{1}{2}\ln\frac{3}{2}$

g $\frac{2}{\sqrt{3}}-1$ h $2-(12)^{\frac{1}{4}}$ i $\frac{38}{3}$

Exercise 16.2B Reasoning and problem-solving

1 a i $\frac{1}{2}x-\frac{1}{4}\sin 2x+c$ ii $\frac{1}{4}\sin 2x+\frac{1}{2}x+c$

b i $\frac{1}{2}x-\frac{1}{8}\sin 4x+c$ ii $\frac{1}{8}\sin 4x+\frac{1}{2}x+c$

iii $\frac{1}{2}x-\frac{1}{8}\sin(4x-2)+c$ iv $\frac{1}{8}\sin(4x+6)+\frac{1}{2}x+c$

v $\frac{1}{2}x+\frac{1}{4}\sin(2-2x)+c$ vi $-\frac{1}{8}\sin(2-4x)+\frac{1}{2}x+c$

2 a $-\frac{1}{2}\cos 2x+c$ b $-\frac{1}{2}\cos 2x-2\cos x+c$

c $\frac{1}{2}\sin 2x+c$

3 a $\frac{1}{2}\int\frac{\sec^2 u}{1+\tan^2 u}\,du$ b $\frac{1}{2}\int du$

c $\frac{1}{2}u+c$ d $\frac{1}{2}\tan^{-1}\left(\frac{x}{2}\right)+c$

e i $\frac{1}{3}\tan^{-1}\frac{x}{3}+c$ ii $\frac{1}{2}\tan^{-1}2x+c$

4 a $\int u\,du$

b $\frac{1}{2}u^2+c$

c $\frac{1}{2}[f(x)]^2+c$

d i $\frac{1}{2}[x^2+x+5]^2+c$ ii $\frac{1}{2}[\sin x]^2+c$

iii $\frac{1}{2}[\ln x]^2+c$ iv $\frac{1}{2}[e^{2x}+1]^2+c$

e $u=f(x)\Rightarrow du=f'(x)dx$

$$\int\frac{f'(x)}{f(x)}\,dx=\int\frac{1}{u}du$$
$$=\ln|u|+c$$
$$=\ln|f(x)|+c$$

f i $\ln|\sin x|+c$ ii $\ln|\tan x|+c$

iii $\ln|e^x+3|+c$ v $\ln|\ln x|+c$

5 a $\int e^{x\ln(a)}\,dx$

b $dx=\frac{du}{\ln a}$

c $\frac{1}{\ln a}a^x+c$

d i $\frac{1}{\ln 4}4^x+c$

ii $\frac{1}{2\ln 5}5^{2x}+c$

iii $\frac{1}{\ln 3}3^{x+1}+c$

Exercise 16.3A Fluency and skills

1 a $3x\sin x+3\cos x+c$

b $-2x\cos x+2\sin x+c$

c $\frac{1}{2}x\sin x+\frac{1}{2}\cos x+c$

d $-\frac{3}{2}x\cos 2x+\frac{3}{4}\sin 2x+c$

e $\frac{1}{3}x\sin(3x+1)+\frac{1}{9}\cos(3x+1)+c$

f $-\frac{x}{20}\sin(1-4x)+\frac{1}{80}\cos(1-4x)+c$

g $-(2x+3)\cos x+2\sin x+c$

h $(2x+1)\sin(x+1)+2\cos(x+1)+c$

i $\frac{1}{4}(1-3x)\sin(4x)-\frac{3}{16}\cos(4x)+c$

2 a $3xe^x-3e^x+c$

b $xe^{2x}-\frac{1}{2}e^{2x}+c$

c $\frac{1}{4}xe^{2x+1}-\frac{1}{8}e^{2x+1}+c$

d $\frac{1}{2}(x+2)e^{2x}-\frac{1}{4}e^{2x}+c$

e $\frac{1}{2}(3-5x)e^{1+2x}+\frac{5}{4}e^{1+2x}+c$

f $-\frac{1}{3}(3x-2)e^{1-3x}-\frac{1}{3}e^{1-3x}+c$

3 a $x^2\ln x-\frac{x^2}{2}+c$

b $\frac{5}{2}x^2\ln x-\frac{5}{4}x^2+c$

c $\left(\frac{x^2}{2}+x\right)\ln x-\left(\frac{x^2}{4}+x\right)+c$

d $(x^2-x)\ln 2x-\left(\frac{x^2}{2}-x\right)+c$

e $-\frac{\ln(3x)}{x}+\int\frac{1}{x^2}\mathrm{d}x=-\frac{\ln(3x)}{x}-\frac{1}{x}+c$

f $\frac{1}{3}(x+1)^3\ln x-\frac{1}{3}\left(\frac{x^3}{3}+\frac{3x^2}{2}+3x+\ln x\right)+c$

g $2\sqrt{x}\ln 2x-4\sqrt{x}+c$

h $\frac{2}{5}x^{\frac{5}{2}}+c$

i $\frac{4}{3}x(x+1)^{\frac{3}{2}}-\frac{8}{15}(x+1)^{\frac{5}{2}}+c$

j $x\tan x+\ln|\cos x|+c$

4 a 1 **b** $\frac{10}{9}$

c $\frac{1}{\sqrt{2}}\left(\frac{\pi}{4}+2\right)-1$ **d** $\frac{4}{5}-\frac{\ln 5}{5}$

e $5-10e^{-1}$ **f** 0

g $\frac{1076}{15}$ **h** $\frac{\pi}{2}-1$

5 a $0\sin 0=\pi\sin\pi=2\pi\sin 2\pi=0$

b i π **ii** 3π

c 4π

Exercise 16.3B Reasoning and problem-solving

1 a $x\ln 4x-x+c$

b $x\ln(3x+1)-x+\frac{1}{3}\ln|3x+1|+c$

c $\frac{1}{8}x^8\ln(x^3)-\frac{3}{64}x^8+c$

d $-\cos x\ln(\cos x)+\cos x+c$

e $x\ln(1-5x)-x-\frac{1}{5}\ln|1-5x|+c$

f $-x\ln x+x+c$

g $x(\ln x)^2-2(x\ln x-x)+c$

h $x(\ln x^2)-2x+c$

i $\frac{1}{4}x^2-\frac{1}{4}x\sin 2x-\frac{1}{8}\cos 2x+c$

2 a $x^2e^x-2xe^x+2e^x+c$

b $-x^2\cos x+2x\sin x+2\cos x+c$

c $x^2\sin x+2x\cos x-2\sin x+c$

d $-(x+1)^2\cos x+2(x+1)\sin x+2\cos x+c$

e $(x^2+2x)\sin x+(2x+2)\cos x-2\sin x+c$

f $(1-3x)^2e^x+6(1-3x)e^x+18e^x+c$

g $-(x^2+x+1)e^{-x}-(2x+1)e^{-x}-2e^{-x}+c$

h $\frac{1}{8}x^2(x+1)^8-\frac{1}{36}x(x+1)^9+\frac{1}{360}(x+1)^{10}+c$

i $\frac{-x^2}{x+1}+2x-2\ln(x+1)+c$

j $\frac{1}{6}(x+1)^2(x+3)^6-\frac{1}{21}(x+1)(x+3)^7+\frac{1}{168}(x+3)^8+c$

k $-(x^2+x)\cos x+(2x+1)\sin x+2\cos x+c$

l $-\frac{1}{2}x^2\cos 2x+\frac{1}{2}x\sin 2x+\frac{1}{4}\cos 2x+c$

m $\frac{1}{3}x^2\sin 3x+\frac{2}{9}x\cos 3x-\frac{2}{27}\sin 3x+c$

3 a $54e^{\frac{1}{3}}-69$ **b** $\frac{\pi^2}{32}-\frac{1}{4}$ **c** 2π

4 a $x^2e^x-2xe^x+2e^x+c$

b $y=x^2e^x-2xe^x+2e^x=e^x(x^2-2x+2)$

c The discriminant of x^2-2x+2 is $(-2)^2-4.1.2=-12$ so x^2-2x+2 has no real roots and so the curve is always above the x-axis and $x^2-2x+2>0$ for all x.
Thus $y>0$ for all x.

5 a $\frac{x^2}{2}\ln x-\frac{x^2}{4}+c$

b $2x^2(\ln x)^2-2x^2\ln x+x^2+c$

6 a $(\pi-0)^2\sin 0=(\pi-\pi)^2\sin\pi=(\pi-2\pi)^2\sin 2\pi=0$

b $2(\pi^2-4)$

7 $\frac{178}{60}$

8 a $P=-e^x\cos x+\int e^x\cos x\,\mathrm{d}x$

b $P=-e^x\cos x+\left(e^x\sin x-\int e^x\sin x\,\mathrm{d}x\right)$

c $P=-e^x\cos x+e^x\sin x-P$

d $\frac{e^x}{2}(\sin x-\cos x)+c$

Exercise 16.4A Fluency and skills

In each answer the form ln A has been used as the constant of integration.

1 a $\ln\left|A\frac{x}{x+1}\right|$ **b** $\ln\left|A\frac{x^2}{(x+3)^2}\right|$

c $\ln\left|A\frac{x}{(3x+1)}\right|$ **d** $\ln\left|A\sqrt{x(x+2)}\right|$

e $\ln\left|A\frac{(x-2)^2}{(x+2)^2}\right|$ **f** $\ln\left|A\frac{(x-4)}{(x+1)}\right|$

g $\ln\left|A\frac{(x+1)^2}{(x+2)^2}\right|$ **h** $\ln\left|A\frac{(x-4)^3}{(x+3)^3}\right|$

2 a $\ln A\left|\frac{(2x-3)}{(2x+1)}\right|^{\frac{1}{2}}$ **b** $\ln\left|A\frac{(3x+1)}{(x+2)}\right|$

c $\ln A\left|\frac{(3x+1)}{(x+5)}\right|^{\frac{1}{2}}$ **d** $\ln\left|A\sqrt{(x-1)(x+3)}\right|$

e $\ln\left|A(x+4)^3(x-2)\right|$ **f** $\ln\left|A\frac{(3x-1)}{(2x+1)}\right|$

g $\ln\left|A(x-2)^4(x+1)^2\right|$ **h** $\ln\left(A|(2x+3)|^{\frac{2}{5}}|(x-1)|^{\frac{3}{5}}\right)$

3 a $\ln\left|A(x-1)(2x+1)\right|$ **b** $\ln\left(A|(2x+1)(2x-1)|^{\frac{3}{8}}\right)$

c $\ln\left(A\frac{|(3x-1)|^{\frac{1}{3}}}{|(2x+1)|^{\frac{1}{2}}}\right)$ **d** $\ln\left|A\frac{x}{1-x}\right|$

e $\ln\left(A\frac{|x|^{\frac{1}{4}}}{|4-x|^{\frac{5}{4}}}\right)$ **f** $\ln\left(A\left|\frac{x}{10-x}\right|^{\frac{1}{20}}\right)$

4 a $\ln\left(\frac{A\sqrt{|x-1||x+1|}}{|x|}\right)$ **b** $\ln\left(A\frac{|3x+2|^3|x-1|^2}{|x|^5}\right)$

c $\ln\left(A\frac{|x||x-1|^4}{|4x+3|^5}\right)$ **d** $\ln\left(A\frac{|x-2|^2|x+1|}{|x-1|^3}\right)$

e $\ln\left(A\frac{|2x-1||x-2|^2}{|x-1|^3}\right)$ **f** $\ln\left(A\frac{|x||x-2|^3}{|x-1|^3}\right)$

g $\ln\left(A\frac{|x+4|^{64}|x-1|}{|x+3|^{45}}\right)$ **h** $\ln\left(A\frac{|x-3|^{\frac{1}{2}}|2x-1|^{\frac{1}{6}}}{|x+1|^{\frac{1}{6}}}\right)$

i $\ln\left(A\dfrac{|x-1||x+2|^{\frac{1}{2}}}{|x|^{\frac{1}{2}}}\right)$ **j** $\ln\left(A\dfrac{|x|^{\frac{1}{2}}}{|1-x||x+2|^{\frac{1}{2}}}\right)$

k $\ln\left(A|x||1-2x|^{-\frac{11}{14}}|x+3|^{\frac{2}{7}}\right)$

5 **a** $-\ln 6$ **b** $\ln 4$ **c** $\ln\left(\dfrac{36}{25}\right)$

d $\ln\sqrt[4]{3}$ **e** $\ln\dfrac{3}{2}$ **f** $-\ln 4$

g $-\ln 2$

Exercise 16.4B Reasoning and problem-solving

1 **a** $\ln\left(A\left|\dfrac{x}{x+2}\right|^{\frac{1}{2}}\right)$ **b** $\ln\left(A\left|\dfrac{x}{x+5}\right|^{\frac{2}{5}}\right)$

c $\ln\left(A\left|\dfrac{x}{2x+1}\right|\right)$ **d** $\ln\left(A\dfrac{x^2}{|x+1|}\right)$

e $\ln\left(A\left|\dfrac{x-1}{x+1}\right|^{\frac{1}{2}}\right)$ **f** $\ln\left(A\left|\dfrac{x-3}{x+3}\right|^{2}\right)$

2 **a** $\ln\left(A\left|\dfrac{2x-1}{2x+1}\right|^{\frac{1}{2}}\right)$ **b** $\ln\left(A|(3x-2)(3x+2)|^{\frac{1}{6}}\right)$

c $\ln\left(A\left|\dfrac{x-3}{x-1}\right|^{\frac{1}{2}}\right)$ **d** $\ln\left(A\left|\dfrac{x-4}{x-1}\right|\right)$

e $\ln\left(A|x+2|^{\frac{2}{5}}|2x-1|^{\frac{1}{10}}\right)$ **f** $\ln\left(A\dfrac{|x-5|^{\frac{10}{3}}}{|x-2|^{\frac{4}{3}}}\right)$

g $\ln\left(A|x-2|^{\frac{3}{7}}|x+5|^{\frac{4}{7}}\right)$ **h** $\ln\left(A|2x-1|^{\frac{1}{18}}|x+4|^{\frac{13}{9}}\right)$

3 **a** $x+\ln\left(A\left|\dfrac{x-1}{x+2}\right|^{\frac{1}{3}}\right)$ **b** $x+\ln\left(A\left|\dfrac{x+1}{x+2}\right|^{4}\right)$

c $x+\ln\left(A\left|\dfrac{x-1}{x+3}\right|^{\frac{3}{4}}\right)$ **d** $2x+\ln\left(A\left|\dfrac{x-2}{x+2}\right|^{\frac{3}{4}}\right)$

e $2x+\ln\left(A\left|\dfrac{x+1}{x+3}\right|^{\frac{1}{2}}\right)$ **f** $3x+\ln\left(A\dfrac{|x-2|^2}{|x-1|}\right)$

4 **a** $\ln 4$ **b** $\ln 2$ **c** $\ln 16$

d $\ln 9$ **e** $\ln 27$ **f** $\ln 81$

5 **a** $\ln\dfrac{8}{3}$ **b** $\ln\dfrac{8}{3}$

c The interval contains $x=1$ for which the function is undefined. The area is unbounded.

6 **a** $y=\dfrac{-8}{x(x-4)}$: lower curve

$y=\dfrac{-9}{x(x-3)}$: upper curve

b **i** $\ln 9$ **ii** $\ln 64$

c $6\ln 2-2\ln 3$

7 **a** $4\ln 2-\ln 3$

b Any interval containing the x-values 1, 2 or 5

Exercise 16.5A Fluency and skills

1 **a** $y=x^2-x+c$ **b** $y=x-\dfrac{x^3}{3}+c$

c $y=\dfrac{(\sqrt{2x+1})^3}{3}+c$ **d** $y=-\dfrac{3}{x}+c$

e $y=\ln Ax^2$ **f** $y=Ae^{3x}$

g $y=Ae^{\frac{5}{2}x^2}$ **h** $y=\pm\sqrt{x^2+2c}$

i $y=\pm\sqrt{(x+1)^2+2c}+3$ **j** $y=Ae^{\frac{x^2}{2}}-1$

k $y=5-Ae^{-x}$ **l** $y=\dfrac{Ae^{x^2+4x}-1}{2}$

m $y^2=x^3+c$ **n** $y=\sin^{-1}(-\cos x+c)$

o $y=\tan^{-1}\left(\dfrac{1}{2}e^{2x}+c\right)$ **p** $y=\ln(e^x+c)$

q $y=\dfrac{1}{3}\ln\left(\dfrac{3}{2}e^{2x+1}+c\right)$

2 **a** $y=2x^2-x+3$ **b** $y=2x-3x^3+10$

c $y=\ln|x|+3$ **d** $y=e^{6x-2}$

e $y=e^{\frac{1}{2}x^2-2}-3$ **f** $y=e^{\frac{1}{2}x^2+5x+1}$

g $y^2-y=\dfrac{1}{2}x^2+2x-6$ **h** $y=\cos^{-1}(1-\sin x)$

i $y=\ln\left(\dfrac{1}{2}e^{2x}+1-\dfrac{1}{2}e\right)$ **j** $y=2\sqrt{x}$

3 **a** **i** $-\dfrac{1}{8(y^2+1)^4}=x+c$

ii $y=\sin^{-1}\left(e^{\frac{1}{2}x^2+c}\right)$

iii $y=\left(\ln|x+c|\right)^2$

b **i** $-y\cos y+\sin y=-\cos x+c$

ii $y=\sqrt{2(xe^x-e^x+c)}$

iii $\dfrac{1}{3}y^3\ln y-\dfrac{1}{9}y^3=9x+c$

c **i** $y=A\left(\dfrac{2x-3}{x}\right)^2$

ii $y=\dfrac{1}{\ln\left(A^{-2}\left|\dfrac{x+1}{x-1}\right|^2\right)}$

iii $y=\left[\dfrac{1}{2}\ln\left(A\left|\dfrac{x-3}{x-2}\right|\right)\right]^2$

4 $y=\ln(\sin x+1)$

5 $y=\dfrac{1}{4}\cos^2 x-\sqrt{7}\cos x+2$

Exercise 16.5B Reasoning and problem-solving

1 **a** $\dfrac{dC}{dw}=k$ **b** $\dfrac{dS}{dt}=kt$

c $\dfrac{dA}{dT}=kA$ **d** $\dfrac{dT}{dt}=k(T-20)$

e $\dfrac{dI}{dR}=k10^R$ **f** $\dfrac{dy}{dx}=\dfrac{k}{\sqrt{y}}$

g $\dfrac{dh}{dt}=\dfrac{k}{\sec^2 h}$

2 **a** $V=2t+0.3t^2+c$

b $V=2t+0.3t^2$

c $t=10$

3 **a** $P=50\sin\left(\dfrac{\pi t}{14}\right)+c$ **b** $P=50\sin\left(\dfrac{\pi t}{14}\right)+50$

c 89% **d** Day 7

4 **a** $y^2=kx^2+2c$

b 12.5

c $y^2=-x^2+25$ or $x^2+y^2=25$

This is the equation of a circle centre the origin and radius 5. Here the origin is the meeting of the wall and the ground.

5 a $\frac{dN}{dt} = kN$

b $N = e^{kt+c}$

c $c = \ln 500$

d $k = \ln 2$

$N = 500 \times 2^t$

e On the 6th day

f Space and nutrients etc will prevent infinite growth.

6 a $\frac{dw}{dt} = k(2000 - w)$

b $-7.58\,(3\,\text{sf})$

c $0.0879\,(3\,\text{sf})$

d $w = 2000 - e^{-0.0879t + 7.58}$

e On the 15th day

f Will reach a point when weight doesn't increase further.

7 a $P = \frac{500e^{500(kt+c)}}{(1+e^{500(kt+c)})}$ b $P = \frac{500e^{1.69t-3.89}}{(1+e^{1.69t-3.89})}$

8 Students' own investigations.

Review exercise 16

1 a $\frac{4}{7}x^7 + c$ b $-\frac{2}{3}\cos(3x+1) + c$

c $\frac{1}{2}\sin 2x + c$ d $\frac{1}{2}x^2 + \tan x + c$

e $e^x - e^{-x} + c$

2 a 4 b $\frac{\sqrt{3}}{2}$ c 1

d $\frac{\pi}{4} - 1$ e 3 f 6

3 a $\frac{9}{2}$ b $14\frac{17}{24}$

4 a $\frac{1}{12}(x^2+4)^6 + c$ b $\frac{1}{2}(x^4-2)^4 + c$

c $\frac{1}{4}(x^2-4x+1)^2 + c$ d $\frac{2}{3}(\sin x)^{\frac{3}{2}} + c$

5 a 1 b $\frac{1}{2}$

6 a $-2x\cos x + 2\sin x + c$

b $\frac{1}{2}(3x+1)e^{2x} - \frac{3}{4}e^{2x} + c$

c $\frac{1}{3}x^2\sin 3x + \frac{2}{9}x\cos 3x - \frac{2}{27}\sin 3x + c$

7 a $4e^4 - e^4 - 3e^3 + e^3$

b $2e^3 + 1$

8 a $(x^2-1)e^x + c$

b $x^2(\ln x)^2 - x^2\ln x + \frac{1}{2}x^2 + c$

9 a $\ln\left(\frac{x-2}{x}\right)^4 + c$ b $\ln\left(\frac{x+2}{x+3}\right)^2 + c$

10 $\ln 18$

11 a $y = e^{\frac{x^2}{2} + x - \frac{3}{2}}$ b $y = \sin^{-1}\left(x^2 - \frac{1}{2}\right)$

12 a $\frac{dV}{dT} = kV$

b $c \approx 9.90$

$k \approx -0.105$

c $V = 20000e^{-0.105T}$

Assessment 16

1 D

2 C

3 a $\ln|x|\,(+c)$ b $-\cos(x-3)(+c)$

c $\frac{1}{2}e^{2x}(+c)$ d $\frac{1}{2}\sin 2x\,(+c)$

4 a i $\ln|x+3|\,(+c)$ ii $-\frac{1}{3}\cos^3 x\,(+c)$ iii $\frac{1}{4}(x^2+4)^4\,(+c)$

b $\frac{1}{2}\ln 2$

5 a i $xe^x - e^x\,(+c)$ ii $-x\cos x + \sin x\,(+c)$

b $-\frac{1}{2}x\cos 2x + \frac{1}{2}\int \cos 2x\,dx$

$\left[-\frac{1}{2}x\cos 2x + \frac{1}{4}\sin 2x\right]_0^{\frac{\pi}{6}}$

$\left(-\frac{1}{2}\frac{\pi}{6}\cos\frac{\pi}{3} + \frac{1}{4}\sin\frac{\pi}{3}\right) - (0+0)$

$= -\frac{\pi}{12}\cdot\frac{1}{2} + \frac{1}{4}\frac{\sqrt{3}}{2}$

$= \frac{1}{8}\sqrt{3} - \frac{1}{24}\pi$

6 a $y = A(x+1)$ b $y = 4(x+1)$

7 a $A = 7$

$B = 1$

b $\frac{15}{2}\ln 3 - 7\ln 2$

8 a $\frac{(1+2x)}{2} - \frac{1}{2}\ln|1+2x| + c$ b $1 - \frac{1}{2}\ln 3$

9 $\ln 4$

10 a $A = 2$

$B = \frac{7}{5}$

$C = -\frac{6}{5}$

b $y = x^2 + \frac{7}{5}\ln|x+2| - \frac{6}{15}\ln|3x+1| - \frac{7}{5}\ln 2$

11 $y = \frac{A(x+2)}{1-2x}$

12 a $\frac{dN}{dt} = -kN$ b $N = \frac{N_0}{2^{\frac{t}{T}}}$

13 a

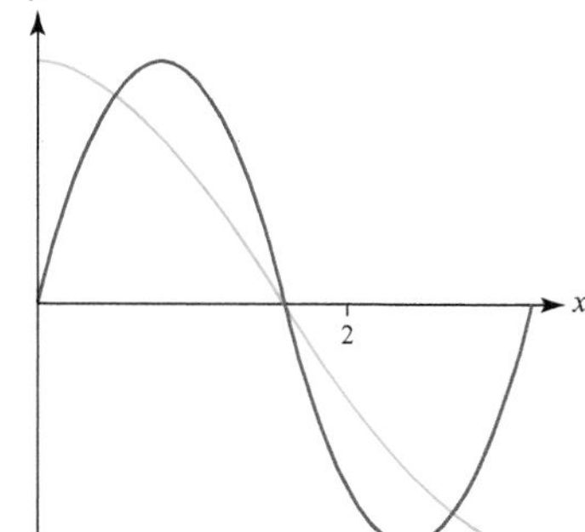

b $x = \frac{\pi}{2}$

or $x = \frac{\pi}{6}$ or $\frac{5\pi}{6}$

c $\frac{1}{2}$

14 a $y - 3e = 2e(x-3)$

b $3xe^{\frac{x}{3}} - 9e^{\frac{x}{3}} + c$

c $9 - \frac{9e}{4}$

15 $2\ln\left|\frac{A(\sqrt{x}-2)}{\sqrt{x}+2}\right|$

or $\ln\left(\frac{A(\sqrt{x}-2)}{\sqrt{x}+2}\right)^2$

16 $\frac{1}{24}(2x-5)^6+\frac{1}{4}(2x-5)^5+c$

17 a $=-x^2\cos x+2x\sin x+2\cos x\ (+c)$

b $6\pi^2-8$

18 a You are given $\frac{dV}{dt}=-c\sqrt{V}$

$V=\pi r^2 h$ so $\frac{dV}{dh}=\pi r^2$

$\frac{dh}{dt}=\frac{dh}{dV}\cdot\frac{dV}{dt}$

$\frac{dh}{dt}=\frac{1}{\pi r^2}\cdot -c\sqrt{V}$

$=\frac{-c\sqrt{\pi r^2 h}}{\pi r^2}$

$=-k\sqrt{h}$ where $k=\frac{c}{\sqrt{\pi r^2}}$

b $h=\left(\frac{A-kt}{2}\right)$

c When $t=0$, $h=2$ so $2=\left(\frac{A}{2}\right)^2$

$\Rightarrow A=2\sqrt{2}$

When $t=2$, $h=0$ so $0=\left(\frac{2\sqrt{2}-2k}{2}\right)^2$

$\Rightarrow k=\sqrt{2}$

$h=\left(\frac{2\sqrt{2}-\sqrt{2}t}{2}\right)^2$

$h=2\left(1-\frac{1}{2}t\right)^2$ as required

Chapter 17

Exercise 17.1A Fluency and skills

1 a $f(x)=7-3x-x^3$

$f(1)=3>0, f(2)=-7<0$

The continuous function $f(x)$ changes sign between $x=1$ and $x=2$

By the change of sign method, the equation $f(x)=0$ has a root between 1 and 2

b $f(x)=x^2-\frac{1}{x}-4$

$f(2.1)=-0.066..<0, f(2.2)=0.385..>0$

The continuous function $f(x)$ changes sign between $x=2.1$ and $x=2.2$

By the change of sign method, the equation $f(x)=0$ has a root between 2.1 and 2.2

c $f(x)=\sin(2x^c)-x^2+3$

$f\left(\frac{1}{2}\pi\right)=0.532..>0, f\left(\frac{2}{3}\pi\right)=-2.252...<0$

The continuous function $f(x)$ changes sign between $x=\frac{1}{2}\pi$ and $x=\frac{2}{3}\pi$

By the change of sign method, the equation $f(x)=0$ has a root between $\frac{1}{2}\pi$ and $\frac{2}{3}\pi$

d $f(x)=e^x\ln x-x^2$

$f(1.69)=-0.012...<0, f(1.71)=0.042...>0$

The continuous function $f(x)$ changes sign between $x=1.69$ and $x=1.71$

By the change of sign method, the equation $f(x)=0$ has a root between 1.69 and 1.71

2 a $f(x)=e^x-x^3$

i $f(1.85)=0.028...>0, f(1.95)=-0.386...<0$

The continuous function $f(x)$ changes sign between $x=1.85$ and $x=1.95$

By the change of sign method, the equation $f(x)=0$ has a root, α, between 1.85 and 1.95

ii $f(4.535)=-0.044...<0, f(4.545)=0.274...>0$

The continuous function $f(x)$ changes sign between $x=4.535$ and $x=4.545$

By the change of sign method, the equation $f(x)=0$ has a root between 4.535 and 4.545

b $\alpha=1.9$ (1 dp), $\beta=4.54$ (2 dp)

3 a i $f(x)=\frac{2}{x^3}-\frac{1}{x}-2$

$f(0.85)=0.08...>0, f(0.95)=-0.71...<0$

so by the change of sign method, the equation $f(x)=0$ has a root between $x=0.85$ and $x=0.95$

ii 0.9 (1 dp)

b i $f(x)=e^{-x}+2x-1$

$f(-1.35)=0.15...>0, f(-1.25)=-0.009...<0$

so by the change of sign method, the equation $f(x)=0$ has a root between $x=-1.35$ and $x=-1.25$

ii −1.3 (1 dp)

c i $f(x)=x^2\sin x-0.5$

$f(3.05)=0.35...>0, f(\pi)=-0.5<0$

so by the change of sign method, the equation $f(x)=0$ has a root between $x=3.05$ and $x=\pi$

ii 3.1 (1 dp)

4 a i $f(x)=x^4-3x^3+1$

$f(2.955)=-0.16...<0, f(2.965)=0.08...>0$

so by the change of sign method, the equation $f(x)=0$ has a root between $x=2.955$ and $x=2.965$

ii 2.96 (2 dp)

b i $f(x)=e^{\frac{1}{x}}-x^2$

$f(1.414)=0.028...>0, f(1.424)=-0.009...<0$

so by the change of sign method, the equation $f(x)=0$ has a root between $x=1.414$ and $x=1.424$

ii 1.4 (1 dp)

c i $f(x)=x^2-\sqrt{x}-2$

$f(1.8305)=-0.002...<0, f(1.8315)=0.001...>0$

so by the change of sign method, the equation $f(x)=0$ has a root between $x=1.8305$ and $x=1.8315$

ii 1.831 (3 dp)

d i $f(x)=2\ln x-\sec x$

$f\left(\frac{8}{5}\pi\right)=-0.007<0, f(5.02725)\approx 6\times10^{-4}>0$

so by the change of sign method, the equation $f(x)=0$ has a root between $x=\frac{8}{5}\pi$ and $x=5.02725$

ii Root is between 5.02654… and 5.02725, so the root is 5.027 (3 dp)

e i $f(x)=e^{\cos x}-\cos(e^x)$

$f\left(-\frac{3}{5}e\right)=-0.03...<0, f\left(-\frac{1}{2}\pi\right)=0.021...>0$

so by the change of sign method, the equation $f(x)=0$ has a root between $x=-\frac{3}{5}e$ and $x=-\frac{1}{2}\pi$

ii −1.6 (1 dp)

5 i Equations **b** and **c**

ii Equation **c**

6 a Equation i $\leftrightarrow$ C, Equation ii $\leftrightarrow$ B, Equation iii $\leftrightarrow$ A

b i 1.6 (1 dp) ii 2.2 (1 dp) iii 2.21 (2 dp)

7 a $f(x)=x^2+2x-2a$

$f(0)=-2a<0,\ f(a)=a^2+2a-2a \Rightarrow f(a)=a^2>0$

Change of sign $\Rightarrow$ equation $f(x)=0$ has a root between 0 and a

b $f(x)=ax^2+x-a^3$

$f(0)=-a^3<0,\ f(a)=a(a)^2+a-a^3 \Rightarrow f(a)=a>0$

Change of sign $\Rightarrow$ equation $f(x)=0$ has a root between 0 and a

c $f(x)=\cos\left(\frac{\pi}{a}x\right)-\frac{a}{\pi}x$

$f(0)=1>0,\ f(a)=\cos\pi-\frac{a^2}{\pi} \Rightarrow f(a)=-1-\frac{a^2}{\pi}<0$

Change of sign $\Rightarrow$ equation $f(x)=0$ has a root between 0 and a

d $f(x)=x^3+(a+1)x^2-2a^3$

$f(0)=-2a^3<0,\ f(a)=a^3+(a+1)a^2-2a^3 \Rightarrow f(a)=a^2>0$

Change of sign $\Rightarrow$ equation $f(x)=0$ has a root between 0 and a

Exercise 17.1B Reasoning and problem-solving

1 a $f(x)=x^3-x^2-1$

$f(1.4)=-0.216<0,\ f(1.5)=0.125>0$

The continuous function $f(x)$ changes sign between $x=1.4$ and $x=1.5$ so the equation $f(x)=0$ has a solution, α, in the interval (1.4, 1.5)

b The curve $y=x^3-x^2$ and line $y=1$ intersect exactly once, so there is exactly one real solution, α, to the equation $x^3-x^2=1$

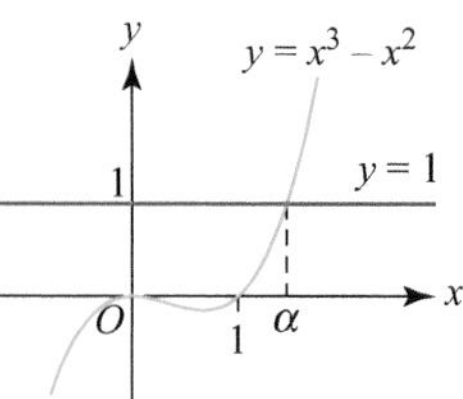

c $\beta=0.7$ (1 dp)

2 a (1, 1)

b

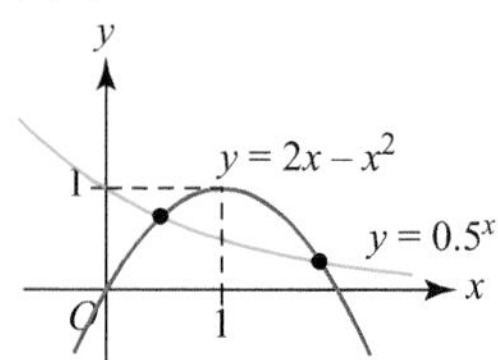

The curves $y=2x-x^2$ and $y=0.5^x$ intersect at exactly two points, so the equation $2x-x^2=0.5^x$ has exactly two solutions.

c $f(x)=2x-x^2-0.5^x$

i $f(0.44)=-0.0507...<0,\ f(0.48)=0.0126...>0$

The continuous function $f(x)$ changes sign between $x=0.44$ and $x=0.48$ so the equation $f(x)=0$ has a root in the interval (0.44, 0.48)

Since $\beta>0.5$, it follows that α lies in the interval (0.44, 0.48)

ii $f(1.84)=0.015...>0,\ f(1.88)=-0.046...<0$

The continuous function $f(x)$ changes sign between $x=1.84$ and $x=1.88$ so the equation $f(x)=0$ has a root in the interval (1.84, 1.88)

Since $\beta>0.5$ and $\alpha<0.48$, it follows that β lies in the interval (1.84, 1.88)

d $\beta-\alpha=1.4$ (1 dp)

3 a $x=1.8$ radians (1 dp)

b $f(2)=0.557..>0,\ f(3)=-3.185...<0$

So $f(x)$ changes sign across the interval (2, 3)

c If $f(x)$ was continuous on the interval (2,3) then, by the change of sign method, the equation $f(x)=0$ would have a root β in this interval.

But then $f(x)=0$ would have two roots, α and β, in the interval $(0, \pi)$

This would contradict the result of part **a**, which stated that $f(x)=0$ had only one root, α, in the interval $(0, \pi)$

Hence $f(x)$ cannot be continuous on the interval (2, 3)

d $x=2.6$ radians (1 dp)

4 a Line to be drawn is $y=3-\frac{1}{2}x$

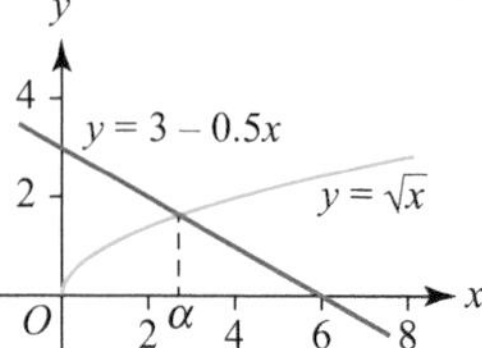

The line $y=3-\frac{1}{2}x$ and curve $y=\sqrt{x}$ intersect at exactly one point, so the equation $2\sqrt{x}=6-x$ has exactly one real solution, α

b $f(x)=2\sqrt{x}-6+x$

$f(2.65)=-0.094...<0,\ f(2.75)=0.066...>0$

The continuous function $f(x)$ changes sign between $x=2.65$ and $x=2.75$ so the equation $f(x)=0$ has a root in the interval (2.65, 2.75)

Since α is the only root of this equation, $2.65<\alpha<2.75$ and therefore $\alpha=2.7$ (1 dp)

c i $x=8-2\sqrt{7}$, $a=8$ and $b=-2$

ii Combining the results of parts **b** and **c i** gives

$8-2\sqrt{7}\approx 2.7$

$\Rightarrow \sqrt{7}\approx\frac{8-2.7}{2}=\frac{53}{20}$

$\Rightarrow \sqrt{7}\approx\frac{53}{20}$

5 a $f(0)=-1<0$ and $f(1)=-7<0$

Since $f(0)$ and $f(1)$ have the same sign, $f(x)$ does not change sign across the interval (0, 1)

b $f\left(\frac{1}{6}\right)=0$

So $x=\frac{1}{6}$ is a root of the equation $f(x)=0$

$f(x)$ does not change sign across the interval (0, 1), but $\frac{1}{6}$ lies in (0, 1) and $f\left(\frac{1}{6}\right)=0$

Hence the equation $f(x)=0$ must have an even number of roots between 0 and 1

Therefore there must be at least two roots of $f(x)=0$ in the interval (0, 1)

c $y=4\sin(\pi x)$ and $y=6x+1$ intersect twice only, so $x=\frac{1}{6}$ and $x=\frac{1}{2}$ are the only roots of $f(x)=0$

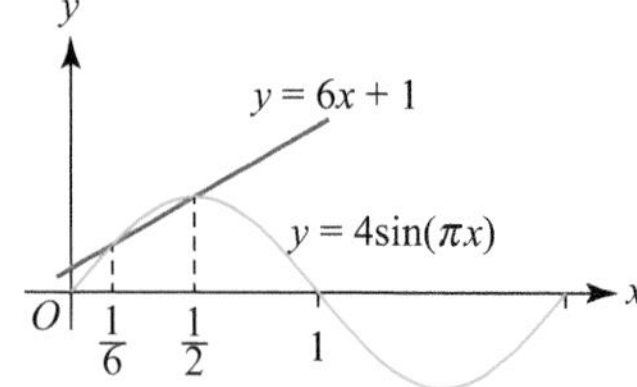

6 a True.

$f(a)$ has opposite sign to $f(b)$, and $f(c)$ has opposite sign to $f(b)$

Therefore $f(a)$ and $f(c)$ have the same sign.

b False.
$f(x)$ is continuous and changes sign across the interval (a, b), so $f(x)=0$ has a root in the interval (a, b)
Since (a, b) is contained in (a, c), therefore $f(x)=0$ has a root in the interval (a, c)

c True.
The equation $f(x)=0$ has a root α in the interval (a, b) and a root β in the interval (b, c)
At some point between $x=\alpha$ and $x=\beta$ the graph of $y=f(x)$ must have a stationary point.
The x-coordinate of this point is a root of the equation $f'(x)=0$

7 a $f(1)=-2,\ f(2)=3$
The curve $y=f(x)$ passes through $A(1,-2)$ and $B(2, 3)$ so the gradient of the line AB is $\frac{3-(-2)}{2-1}=5$
Hence, the equation of the line through A and B is $y-3=5(x-2)$
The x-intercept of this line is an estimate for α:
When $y=0$, $0-3=5(x-2)$
$\Rightarrow x=-\frac{3}{5}+2$
$\Rightarrow x=1.4$, as required.

b $x=1.6$ (1 dp) is another approximation for α
Use a change of sign across a suitable interval to show $\alpha=1.6$ to 1 decimal place:
$f(1.55)=-0.37...<0$, $f(1.65)=0.19...>0$
The continuous function $f(x)$ changes sign between $x=1.55$ and $x=1.65$ so $x=1.6$ (1 dp) is a root of the equation $f(x)=0$
Since α is the only root of this equation in the interval (1, 2), therefore $\alpha=1.6$ (1 dp), which agrees with the second estimate found.

c $\alpha=\frac{1+\sqrt{5}}{2}$

Exercise 17.2A Fluency and skills

1 a i $x_2=0.9$, $x_3=0.927$, $x_4=0.920$ (all to 3 dp where appropriate)
ii $x_5=0.922$ (3 dp)
b i $x_2=3$, $x_3=3.442$, $x_4=3.510$ (all to 3 dp where appropriate)
ii $x_6=3.521$ (3 dp)
c i $x_2=-1$, $x_3=-0.5$, $x_4=-0.64$
ii $x_7=-0.618$ (3 dp)
d i $x_2=1.6$, $x_3=1.425$, $x_4=1.511$ (all to 3 dp where appropriate)
ii $x_{10}=1.485$ (3 dp)

2 a i $\alpha=2.12$ (2 dp) **ii** $(2.115, 2.125)$
b i $\alpha=3.69$ (2 dp) **ii** $(3.685, 3.695)$
c i $\alpha=2.36$ (2 dp) **ii** $(2.355, 2.365)$
d i $\alpha=3.94$ (2 dp) **ii** $(3.935, 3.945)$
e i $\alpha=2.20$ (2 dp) **ii** $(2.195, 2.205)$
f i $\alpha=1.24$ (2 dp) **ii** $(1.235, 1.245)$

3 a $N=4$ iterations are required.
b $x_1=100$
To 3 dp: $x_2=0.885$, $x_3=0.877$, $x_4=0.878$ (stop)
Only 3 iterations are required.

4 a 

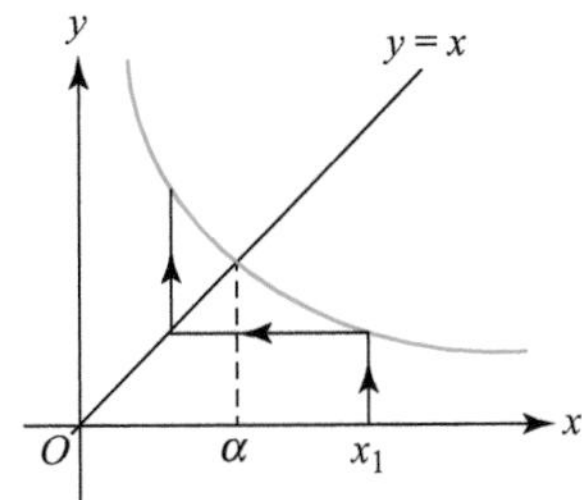

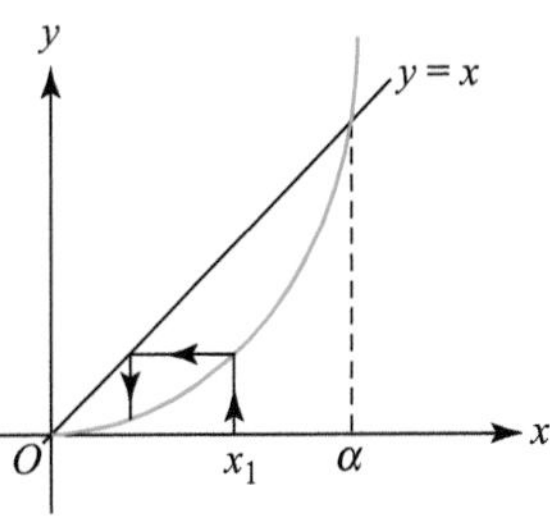

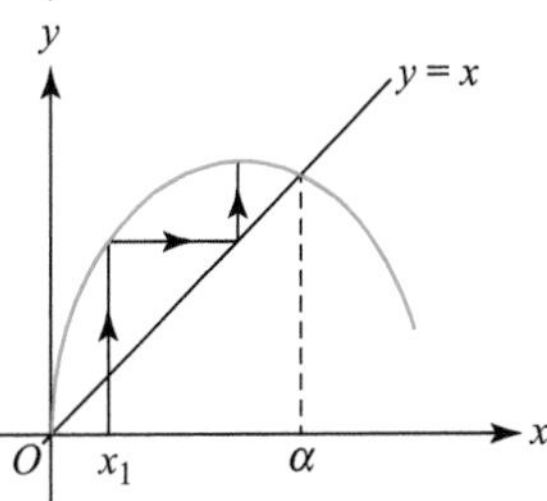

b Fig 2, if continued, would definitely not illustrate convergence to α

5 a 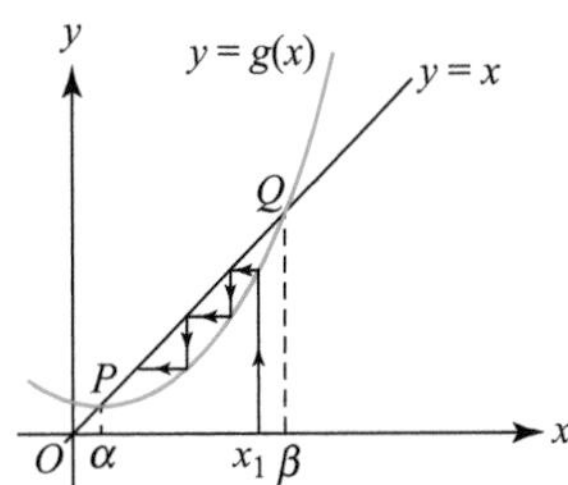

The staircase diagram shows the iterates converging to α even though the starting value is close to β

b 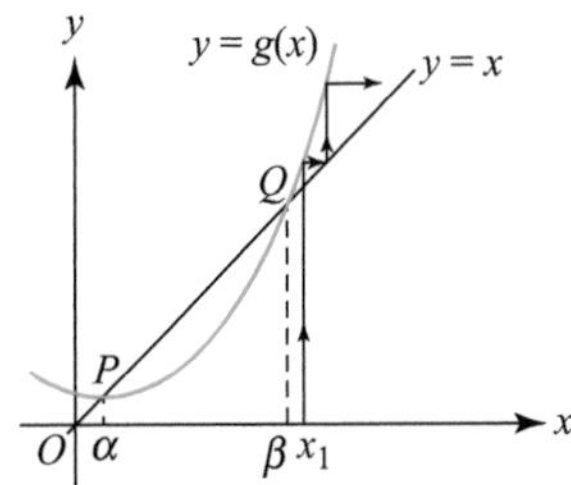

The staircase diagram shows the iterates diverging from β even though the starting value is close to this root.

Exercise 17.2B Reasoning and problem-solving

1 a $x^4-2x-1=0 \Rightarrow x^4=2x+1$
$\Rightarrow x^2=\frac{2x+1}{x^2}$
$\Rightarrow x=\sqrt{\frac{2x+1}{x^2}}=\frac{\sqrt{2x+1}}{x}$, as required

b The iteration $x_{n+1}=\frac{\sqrt{2x_n+1}}{x_n}$ converges to $1.3958...=1.4$ (1 dp)
Now $f(1.35)=-0.37...<0$ and $f(1.45)=0.52...>0$
Since $f(x)$ is continuous on (1.35, 1.45), change of sign $\Rightarrow$ 1.4 (1 dp) is a root of the equation $f(x)=0$
$\therefore \alpha=1.4$ (1 dp)

c $x_1=-0.5$, $x_2=0$, $x_3=$ cannot be found (division by zero has been attempted)

2 a $f(3.3545)=-0.01..<0$, $f(3.3555)=0.002...>0$
Since $f(x)$ is continuous on (3.3545, 3.3555), change of sign $\Rightarrow$ 3.355 (3dp) is a root of the equation $f(x)=0$
$\therefore \alpha=3.355$ (3 dp)

b 12 iterations are needed to produce an estimate for α (correct to 3 dp) using (I)

7 iterations are needed to produce an estimate for α (correct to 3 dp) using (II)

Hence formula (II) converges to α roughly twice as quickly as formula (II)

c **i** $x^3-3x^2-4=0 \Rightarrow x^3-3x^2=4$

$\Rightarrow x^2(x-3)=4$

$\Rightarrow x^2=\dfrac{4}{x-3}$

$\Rightarrow x=\sqrt{\dfrac{4}{x-3}}$

$\Rightarrow x=\dfrac{2}{\sqrt{x-3}}$, as required

ii $g'(x)=-\dfrac{1}{(x-3)^{\frac{3}{2}}}$

$g'(x)<-1$ for values near the root α

Any starting value used in this iterative formula will lead to failure e.g. $x_1=4 \Rightarrow x_2=2$, $x_3=$ cannot be found (square root of a negative number)

Hence this is not a suitable rearrangement for estimating α

3 **a** $T_A=T_B \Rightarrow 10e^{0.1t}=16e^{-0.2t}+25$

$\Rightarrow e^{0.1t}=1.6e^{-0.2t}+2.5$

$\Rightarrow 0.1t=\ln(1.6e^{-0.2t}+2.5)$

$\Rightarrow t=10\ln(1.6e^{-0.2t}+2.5)$, as required

b $t=10$ minutes (nearest minute)

c Use any suitable arrangement

e.g. $x_{n+1}=\sqrt[3]{2.5x_n^2+1.6}$, $x_1=0$

which gives $t\approx 10$ minutes

4 **a**

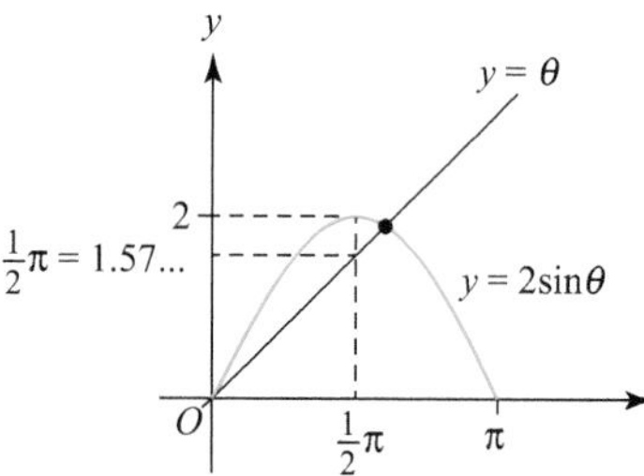

The line $y=\theta$ intersects the curve with equation $y=2\sin\theta$ exactly once between $\theta=\frac{1}{2}\pi$ and $\theta=\pi$

So there is exactly one solution to $\theta=2\sin\theta$ between $\theta=\frac{1}{2}\pi$ and $\theta=\pi$

b Area of ΔOAB = area of shaded segment

$\Rightarrow$ Area of $\Delta OAB=\frac{1}{2}\times$ area of sector OAB

$\Rightarrow \frac{1}{2}\times 4\times 4\times\sin\alpha=\frac{1}{2}\times\frac{1}{2}\times 4^2\alpha$

$\Rightarrow \sin\alpha=\frac{1}{2}\alpha$

$\Rightarrow \alpha=2\sin\alpha$, as required

c The iterates converge to $1.8954...=1.90$ (2 dp)

Now $f(\theta)=\theta-2\sin\theta$

$f(1.895)\approx -8\times10^{-4}<0$, $f(1.905)=0.015...>0$

Change of sign $\Rightarrow$ 1.90 (2 dp) is a root of the equation $f(\theta)=0$

Since there is exactly one solution to $\theta=2\sin\theta$ between $\theta=\frac{1}{2}\pi$ and $\theta=\pi$

$\Rightarrow \alpha=1.90$ to 2 dp

5 **a** Arc length $AB=r\alpha$

Cosine rule: Chord length $AB^2=r^2+r^2-2(r)(r)\cos\alpha$

$\Rightarrow$ Chord length $AB=\sqrt{2r^2-2r^2\cos\alpha}$

(Arc length AB) = 2 × (chord length AB)

$\Rightarrow r\alpha=2\sqrt{2r^2-2r^2\cos\alpha}$

$\Rightarrow r\alpha=r\sqrt{4}\sqrt{2-2\cos\alpha}$

$\Rightarrow \alpha=\sqrt{8-8\cos\alpha}$, as required

b The iterates converge to $3.7909...=3.8$ (1 dp)

Let $f(\theta)=\theta-\sqrt{8-8\cos\theta}$

$f(3.75)=-0.06...<0$, $f(3.85)=0.09...>0$

Change of sign $\Rightarrow$ 3.8 (1 dp) is a root of the equation $f(\theta)=0$

Since there is only one positive solution to $f(\theta)=0$, so $\alpha=3.8$ (1 dp)

c 1.2 cm² (1 dp)

d **i** Arc length PQ = chord length PQ

ii The student who claims the sequence converges to 0 is correct.

6 **a** The sequence is constant for any given starting value.

b The result of part **a** suggests that *any* positive number x is a root of the equation $x^{\ln 2}-2^{\ln x}=0$

This in turn suggests that $x^{\ln 2}-2^{\ln x}\equiv 0$

To justify this identity, notice that

$\ln(x^{\ln 2})\equiv \ln 2\times\ln x$ (by the power rule for logs)

$\equiv \ln x\times\ln 2$

$\equiv \ln(2^{\ln x})$ (again, by the power rule for logs)

$\therefore x^{\ln 2}\equiv 2^{\ln x}$

$\Rightarrow x^{\ln 2}-2^{\ln x}\equiv 0$

Also, note that $2=e^{\ln 2}$ and so

$(2^{\ln x})^{\frac{1}{\ln 2}}=\left((e^{\ln 2})^{\ln x}\right)^{\frac{1}{\ln 2}}$

$=(e^{\ln x\times\ln 2})^{\frac{1}{\ln 2}}$

$=e^{\ln x}$

$=x$

which means that the iteration scheme reduces to $x_{n+1}=x_n$

Exercise 17.3A Fluency and skills

1 **a** $x_2=2.2$, $x_3=2.18$ (2 dp)

b $x_2=2.1$, $x_3=2.09$ (2 dp)

c $x_2=-0.8$, $x_3=-0.73$ (2 dp)

d $x_2=3.45$, $x_3=3.42$ (2 dp)

e $x_2=1.64$, $x_3=1.51$ (2 dp)

f $x_2=4.54$, $x_3=4.56$ (2 dp)

2 **a** The continuous function $f(x)=e^x+3x-4$ changes sign across the interval (0.6765, 0.6775), and hence 0.677 (3 dp) is a root of the equation $f(x)=0$

Since α is the only root of this equation, it follows that $\alpha=0.677$ (3 dp)

b The continuous function $f(x)=x^2-3e^{2x}$ changes sign across the interval (−0.7885, −0.7875), and hence −0.788 (3 dp) is a root of the equation $f(x)=0$

Since α is the only root of this equation, it follows that $\alpha=-0.788$ (3 dp)

c The continuous function $f(x)=x^2+3\ln x$ changes sign across the interval (0.8055, 0.8065), and hence 0.806 (3 dp) is a root of the equation $f(x)=0$

Since α is the only root of this equation, it follows that $\alpha=0.806$ (3 dp)

d The continuous function $f(x) = \sin x + x - 3$ changes sign across the interval (2.1795, 2.1805), and hence 2.180 (3 dp) is a root of the equation $f(x) = 0$
Since α is the only root of this equation, it follows that $\alpha = 2.180$ (3 dp)

e The continuous function $f(x) = x - \cos^2 x - 3$ changes sign across the interval (3.7095, 3.7105), and hence 3.710 (3 dp) is a root of the equation $f(x) = 0$
Since α is the only root of this equation, it follows that $\alpha = 3.710$ (3 dp)

f The continuous function $f(x) = x^2 \ln x - 2$ changes sign across the interval (1.8235, 1.8245), and hence 1.824 (3 dp) is a root of the equation $f(x) = 0$
Since α is the only root of this equation, it follows that $\alpha = 1.824$ (3 dp)

3 a $f(x) = x^3 - 2x^2 - 7$
$f'(x) = 3x^2 - 4x$
$x_1 = 3 \Rightarrow x_2 = 2.8666...$
$x_3 = 2.8574... = 2.857$ (3 dp)
$f(2.8565) = -0.011... < 0,\ f(2.8575) = 0.001... > 0$
The continuous function $f(x)$ changes sign across the interval (2.8565, 2.8575), and hence 2.857 (3 dp) is a root of the equation $f(x) = 0$
Since α is the only real root of this equation, it follows that $\alpha = 2.857$ (3 dp)
Hence two iterations are sufficient to locate α to 3 decimal places.

b Yes: the iterates converge to $\alpha = 2.857$ (3 dp)
{11 iterations required}

4 a i $f(x) = (x + \sin x)^2 - 1$
$f'(x) = 2(x + \sin x)(1 + \cos x)$
$x_1 = 1 \Rightarrow x_2 = 0.5785...$
$\Rightarrow x_3 = 0.5141... = 0.51$ (2 dp)
$f(0.505) = -0.02... < 0,\ f(0.515) = 0.01... > 0$
The continuous function $f(x)$ changes sign across the interval (0.505, 0.515), and hence 0.51 (2 dp) is a root of the equation $f(x) = 0$
Since α is the only positive root of this equation, it follows that $\alpha = 0.51$ (2 dp)
Hence two iterations are sufficient to locate α to 2 dp.

ii $x_3 = 0.514$ (3 dp)
$f(0.5135) = 0.009... > 0,\ f(0.5145) = 0.01... > 0$
The continuous function $f(x)$ does *not* change sign across the interval (0.5135, 0.5145) so α does not lie in this interval.
$\therefore$ as an approximation to α, x_3 is not accurate to 3 decimal places.

b No: Newton-Raphson method with starting value $x_1 = 2$ does not produce a reliable estimate for α (since this iteration converges to a different root which is, in fact, $-\alpha$).

5 a $f(x) = \sin x + e^{-x}$
$f(3.1825) \approx 5.9 \times 10^{-4} > 0,\ f(3.1835) \approx -4.5 \times 10^{-4} < 0$
This change of sign means 3.183 (3 dp) is a root of the equation $f(x) = 0$

b $h'(x) = -\sin x + e^{-x}$
$h(1.2925) \approx 1.3 \times 10^{-4} > 0,\ h(1.2935) \approx -5.5 \times 10^{-4} < 0$
This change of sign means 1.293 (3 dp) is a root of the equation $f'(x) = 0$

6 a i $f(x) = 2\sqrt{x} - \dfrac{1}{x} + 1$
$f'(x) = x^{-\frac{1}{2}} + x^{-2}$
$x_1 = 1 \Rightarrow x_2 = 0,\ x_3 = 0 - \dfrac{f(0)}{f'(0)}$
Neither $f(0)$ nor $f'(0)$ can be evaluated because this would involve division by zero.
Hence x_3 cannot be calculated.

ii $f(0.4315) = -0.003... < 0,\ f(0.4325) = 0.003... > 0$
This change of sign means 0.432 (to 3 dp) is a root of the equation $f(x) = 0$

b i $f(x) = x^3 - 8\sqrt{x} + 2$
$f'(x) = 3x^2 - 4x^{-\frac{1}{2}}$
$x_1 = 1 \Rightarrow x_2 = -4,\ x_3 = -4 - \dfrac{f(-4)}{f'(-4)}$
Neither $f(-4)$ nor $f'(-4)$ can be evaluated because they include $\sqrt{-4}$, which is not a real number.
Hence x_3 cannot be calculated.

ii $f(2.1305) = -0.006... < 0,\ f(2.1315) = 0.004... > 0$
Also $f(0.0625) \approx 2 \times 10^{-4} > 0,\ f(0.0635) = -0.01... < 0$
These changes of sign means 2.131 and 0.063 (to 3 dp) are roots of the equation $f(x) = 0$

Exercise 17.3B Reasoning and problem-solving

1 a $f(x) = 4x^3 - 12x^2 + 9x - 1$
$f(0.125) = -0.054... < 0\ f(0.135) = 0.006... > 0$
This change of sign means 0.13 (2 decimal places) is a root of the equation $f(x) = 0$
Since α is the only root of this equation, $\alpha = 0.13$ (2 dp)

b i $x_1 = 0.5$
$f'(0.5) = 12(0.5)^2 - 24(0.5) + 9$
$= 0$
Hence $x_2 = 0.5 - \dfrac{f(0.5)}{0}$ which cannot be evaluated.
So the Newton-Raphson method fails when $x_1 = 0.5$

ii $x_1 = 1 \Rightarrow x_2 = 1,\ x_3 = 1, ...$
The sequence is constant because we are already at a root and therefore it does not converge to α

c $\alpha = \dfrac{2 - \sqrt{3}}{2} = 0.1339745962...$
Only four iterations are required to find α to 8 dp

2 a $x = 0$

b i $x = 1.1$ (1 dp)

ii (0.6, −2.0) (both to 1 dp)

c

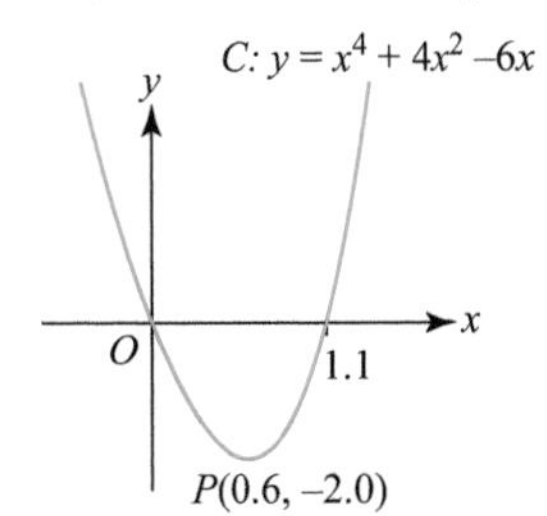

3 a $y = x\sin x + 2\cos x \Rightarrow \dfrac{dy}{dx} = x\cos x + \sin x - 2\sin x$
$= x\cos x - \sin x$

At a stationary point $\dfrac{dy}{dx} = 0 \Rightarrow x\cos x - \sin x = 0$
$\Rightarrow x = \dfrac{\sin x}{\cos x}$
$\Rightarrow \tan x - x = 0$
{as $\dfrac{\sin x}{\cos x} \equiv \tan x$}
$\Rightarrow \tan\beta - \beta = 0$ {as P is the stationary point of C}
$\therefore \beta$ is a root of the equation $g(x) = 0$, where $g(x) = \tan x - x$

b $g(x) = \tan x - x$
$g'(x) = \sec^2 x - 1$
$= \tan^2 x$

$\therefore\ x_{n+1} = x_n - \dfrac{g(x_n)}{g'(x_n)}$ can be written as

$x_{n+1} = x_n - \left(\dfrac{\tan x_n - x_n}{\tan^2 x_n}\right)$

c $(4.49, -4.82)$ (each to 2 dp)

d Using 2nd derivative: $\dfrac{dy}{dx} = x\cos x - \sin x$

$$\Rightarrow \frac{d^2y}{dx^2} = x(-\sin x) + \cos x - \cos x$$
$$= -x\sin x$$

Across the interval $(\pi, 2\pi)$, $x > 0$ and $\sin x < 0$

$\therefore\ \dfrac{d^2y}{dx^2} > 0$ for all values $\pi < x < 2\pi$ and, in particular, when $x = \beta$

Hence P is a minimum point.

OR Using a logical approach:

P is the only stationary point in the interval $(\pi, 2\pi)$

$f(\pi) = -2$ and $f(2\pi) = 2$

Since $f(\beta) \approx -4.8$, P must be a minimum point

4 a i 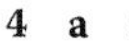

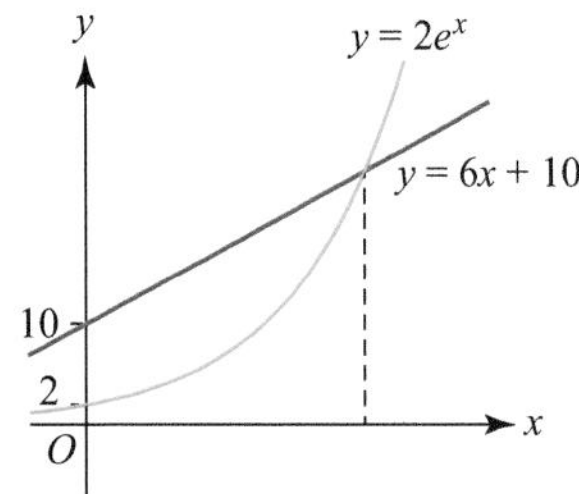

For $x \geq 0$, the line with equation $y = 6x + 40$ and curve with equation $y = 2e^x$ intersect exactly once.

Hence the equation $2e^x = 6x + 40$ has exactly one positive solution.

ii $f(3.35) = -3.09... < 0$; $f(3.45) = 2.30... > 0$

The continuous function $f(x) = 2e^x - 6x - 40$ changes sign between $x = 3.35$ and $x = 3.45$, and hence $x = 3.4$ (1 dp) is the only positive solution of the equation $f(x) = 0$

b 2017

c £127 million

5 a Converges to $\sqrt{k}$

b Use the iteration $x_{n+1} = \dfrac{1}{2}\left(x_n + \dfrac{54321}{x_n}\right)$, $x_1 = 1$

c Let $f(x) = x^2 - k$

$f'(x) = 2x$

Newton-Raphson formula: $x_{n+1} = x_n - \left(\dfrac{x_n^2 - k}{2x_n}\right)$

$$= \frac{1}{2}\left(2x_n - \left(\frac{x_n^2}{x_n} - \frac{k}{x_n}\right)\right)$$
$$= \frac{1}{2}\left(x_n + \frac{k}{x_n}\right)$$

This sequence converges to the root of the equation $x^2 - k = 0$ that is, to $\sqrt{k}$

Exercise 17.4A Fluency and skills

1 a 23.19 (2 dp) **b** 7.06 (2 dp) **c** 4.14 (2 dp)
d 0.77 (2 dp) **e** 0.52 (2 dp) **f** 2.07 (2 dp)

2 a i 0.697 (3 dp) **ii** underestimate
b i 3.268 (3 dp) **ii** overestimate
c i 7.402 (3 dp) **ii** overestimate
d i 3.715 (3 dp) **ii** underestimate

3 a

x_i	1	1.1	1.2	1.3	1.4	1.5	1.6	1.7	1.8	1.9	2
y_i	1	1.26	1.62	**2.13**	2.87	**3.95**	5.55	7.98	**11.72**	17.59	27

b 6.9 (1 dp)

4 a i 12.211 (3 dp) **ii** 11.214 (3 dp)

b Graph 1 pairs with the equation $y = 6 - e^{0.1x^2}$ and Graph II pairs with the equation $y = \dfrac{e^x}{x} - 2$

Graph 1 is concave. Each trapezium lies entirely under the curve, so the answer 11.213 is an underestimate, that is $\displaystyle\int_1^4 6 - e^{0.1x^2}\,dx > 11.213$

Graph 2 is convex. Each trapezium lies partly above the curve, so the answer 12.211 is an overestimate, that is $\displaystyle\int_1^4 \frac{e^x}{x} - 2\,dx < 12.211$

5 a i

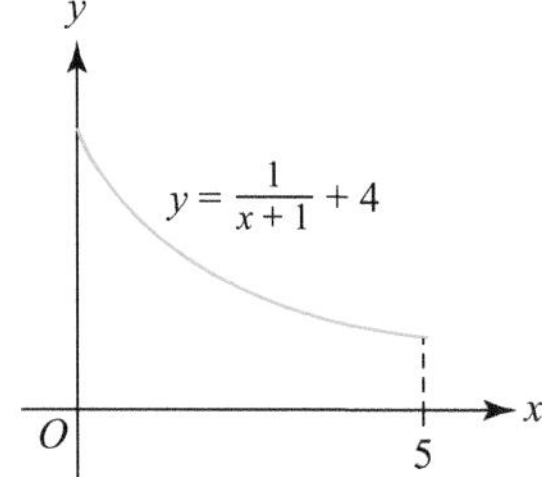

ii 21.91 (2 dp)

iii The answer is an overestimate, due to the concave upwards shape of the curve.

When drawn, each trapezium lies partly above the curve.

iv $\displaystyle\int_0^5 \frac{1}{x+1} + 4\,dx = \ln 6 + 20$ $(< 21.91$, as required)

b i

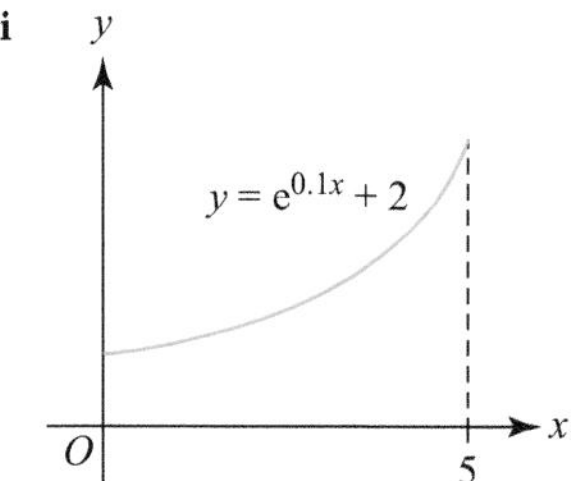

ii 16.49 (2 dp)

iii The answer is an overestimate, due to the concave upwards shape of the curve.

When drawn, each trapezium lies partly above the curve.

iv $\displaystyle\int_0^5 e^{0.1x} + 2\,dx = 10\sqrt{e}$ $(< 16.49$, as required)

Exercise 17.4B Reasoning and problem-solving

1 a 0.52 (2 sf)

b i 4.44 km^2 (3 sf)

ii Underestimate for the actual area of the field because 0.52110420 is an overestimate for $\displaystyle\int_0^2 1 - \sqrt{\cos(0.25\pi x)}\,dx$ (due to convex curve).

2 a 4.04 (2 dp)

b 3.56 (2 dp)

c 3.5

d Using 5 intervals: $I \approx 4.04 \Rightarrow$ error $= 4.04 - 3.5 = 0.54$
Using 20 intervals: $I \approx 3.56 \Rightarrow$ error $= 3.56 - 3.5 = 0.06$
$\frac{0.54}{0.06} = 9$ so the answer (to 2 dp) obtained using 20 intervals is nine times more accurate than that obtained using five intervals.

3 a 6 metres

b 11.8 m^2

c 6000 litres per second (1 significant figure)

4 a i 17.4 (1 dp)

ii

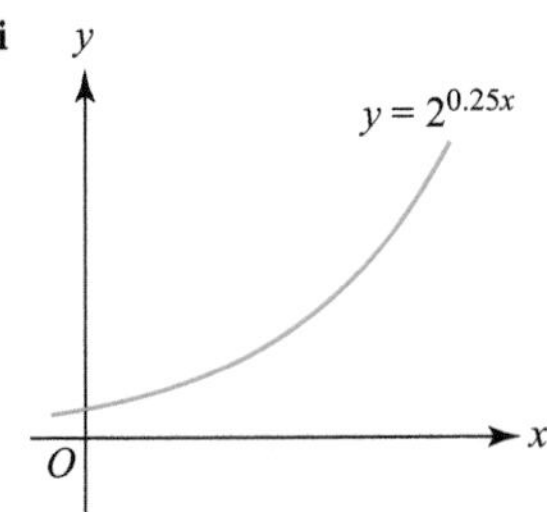

Due to the shape of the curve (convex), the answer found is an overestimate for the integral.

b $k = 1$

c A's distance from start after 8 seconds $= \int_0^8 1 + 2^{0.25t}\, dt$
$\approx [t]_0^8 + 17.4$
$= 25.4$ m (an overestimate)

B's distance from start after
8 seconds $= 10\text{m} + \int_0^8 0.25t + 1\, dt$
$= 10\text{m} + [0.125t^2 + t]_0^8$
$= 26$ m

Hence B wins the race

5 a $S_n = \frac{a(r^n - 1)}{r - 1}$

b $h = \frac{1-0}{n} = \frac{1}{n}$

x_i	0	$\frac{1}{n}$	$\frac{2}{n}$	...	$\frac{n-1}{n}$	1
y_i	1	2	2^2	....	2^{n-1}	2^n

$$\int_0^1 2^{nx}\, dx \approx \frac{1}{2}\left(\frac{1}{n}\right)[1 + 2^n + 2(2 + 2^2 + ... + 2^{n-1})]$$

$2 + 2^2 + ... + 2^{n-1}$ is the sum of the first $(n-1)$ terms of a geometric series, first term and common ratio 2

Hence $2 + 2^2 + ... + 2^{n-1} = \frac{2(2^{n-1} - 1)}{2 - 1}$
$= 2^n - 2$

$$\therefore \int_0^1 2^{nx}\, dx \approx \frac{1}{2}\left(\frac{1}{n}\right)[1 + 2^n + 2(2^n - 2)]$$
$$= \frac{1}{2n}[1 + 2^n + 2^{n+1} - 4]$$
$$= \frac{1}{2n}[2^n(1+2) - 3]$$
$$= \frac{3}{2n}(2^n - 1), \text{ as required}$$

c i $n = 16$
To justify, Let $f(x) = 2^x - 1 - 4096x$
$f(15.5) = -17148.04.. < 0$, $f(16.5) = 25096.9... > 0$
Change of sign $\Rightarrow f(x) = 0$ has a root between 15.5 and 16.5
By sketching the graphs of $y = 2^x - 1$ and $y = 4096x$, it is clear that they intersect only once for positive x
Hence, 16 is a root of the equation $2^x - 1 = 4096x$ (to the nearest integer)

ii 47.8 (1 dp)

6 a 0.299 (3 dp)

b Letting $u = -\frac{3}{x} \Rightarrow \frac{dx}{du} = \frac{3}{u^2}$
Change limits: $x = 3 \Rightarrow u = -1$
$x = 2 \Rightarrow u = -1.5$

$$\int_2^3 e^{-\frac{3}{x}}\, dx = \int_{-1.5}^{-1} \frac{3}{u^2} e^u\, du$$

c 0.301 (3 dp)

d $I = 0.30$ (2 dp)

7 a i $I \approx 3.1312$ **ii** $I \approx 3.1349$ **iii** $I \approx 3.1390$

b As n increases, the estimates for I seem to approach π

c By using the substitution, $I = \pi$

Review exercise 17

1 a i $f(x) = x^3 + 3 - 5x^2$
$f(0) = 3 > 0$, $f(1) = -1 < 0$
The continuous function $f(x)$ changes sign across the interval $(0, 1)$
$\therefore$ the equation $f(x) = 0$ has a root between $x = 0$ and $x = 1$
$\therefore$ the equation $x^3 + 3 = 5x^2$ has a solution between $x = 0$ and $x = 1$

ii $f(x) = x^3 + 3 - 5x^2$
$f(-0.8) = -0.712 < 0$, $f(-0.7) = 0.207 > 0$
The change of sign shows that the equation $x^3 + 3 = 5x^2$ has a solution between $x = -0.8$ and $x = -0.7$

iii $f(x) = 2^x - e^{\sqrt{x}-1} - 3$
$f(2.15) = -0.15... < 0$, $f(2.25) = 0.108... > 0$
The change of sign shows that the equation $2^x - e^{\sqrt{x}-1} = 3$ has a solution between $x = 2.15$ and $x = 2.25$

iv $f(x) = x\sin x - \cos(\pi x) - 1$
$f(3) = 0.42... > 0$, $f(\pi) = -0.09... < 0$
The change of sign shows that the equation $x\sin x = \cos(\pi x) + 1$ has a solution between $x = 3$ and $x = \pi$

b The solution to the equation $2^x - e^{\sqrt{x}-1} = 3$ (part **iii**)

2 a $x = 1 \pm \sqrt{2}$

b $f(1.5) = -3.5 < 0$, $f(2.4) = 0.1 > 0$

c The equation $f(x) = 0$ does not have a root between $x = 1.5$ and $x = 2.4$ since
$1 - \sqrt{2} = -0.41... < 1.5$ and $1 + \sqrt{2} = 2.414... > 2.4$
However, $f(x)$ changes sign between $x = 1.5$ and $x = 2.4$
Therefore, $f(x)$ cannot be continuous on the interval $(1.5, 2.4)$

3 a

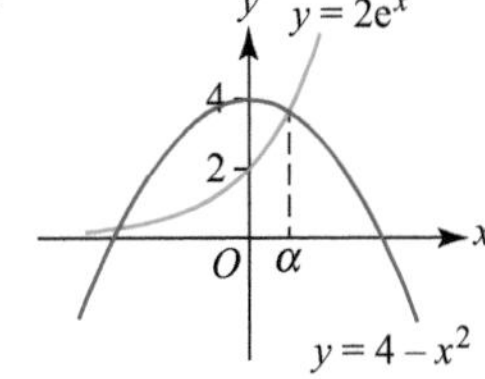

The two curves $y = 2e^x$ and $y = 4 - x^2$ intersect exactly once for positive x

$\therefore$ the equation $2e^x = 4 - x^2$ has exactly one positive solution

b $2e^x = 4 - x^2 \Rightarrow e^x = 2 - 0.5x^2$

$\Rightarrow x = \ln(2 - 0.5x^2)$ as required

c Let $f(x) = 2e^x - 4 + x^2$

$f(0.5985) = -0.003... < 0, f(0.5995) = 0.001... > 0$

This change of sign $\Rightarrow$ 0.60 (2 dp) is a root of the equation $f(x) = 0$

Since the equation $f(x) = 0$ has exactly one positive solution, $\alpha = 0.599$ (3 dp)

4 a $x = 0.481$ (3 dp) **b** $x = 5.236$ (3 dp)

c $x = 1.073$ (3 dp) **d** $x = 2.308$ (3 dp)

5 a $f(2.7655) = -0.003... < 0,\ f(2.7665) \approx 4 \times 10^{-4} > 0$

This change of sign $\Rightarrow$ 2.766 (3 dp) is a root of the equation $f(x) = 0$

b $f'(1) = 2(1) - 2(1)^{-\frac{1}{2}}$

$= 0$

The starting value is the x-coordinate of a stationary point of the curve with equation $y = f(x)$

This means division by zero occurs when using the Newton-Raphson formula to calculate x_2

(or, graphically, the tangent to the graph at the first approximation never intersects the x-axis, which is required for the Newton-Raphson method to work).

6 a $n = 5,\ h = \frac{2-0}{5} = 0.4$

x_i	0	0.4	0.8	1.2	1.6	2
y_i (3 dp where appropriate)	1	1.741	3.031	5.278	9.190	16

$$\int_0^2 4^x \, dx \approx \frac{1}{2}(0.4)\{1 + 16 + 2(1.741 + ... + 9.190)\}$$

$$= \frac{1}{2}(0.4)\{55.48\}$$

$$= 11.096$$

$$= 11.10 \text{ (2 dp)}$$

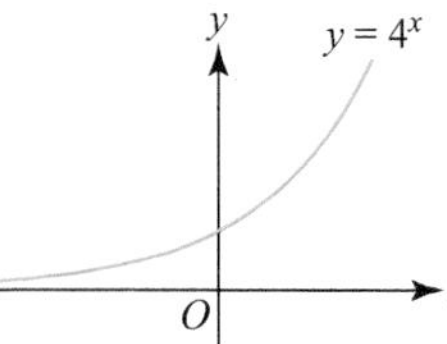

This answer is an overestimate for $\int_0^2 4^x \, dx$ because of the concave upwards shape of the graph with equation $y = 4^x$

b $2^{2x+3} = 2^{2x} \times 2^3$

$= (2^2)^x \times 8$

$= 4^x \times 8$

$$\therefore \int_0^2 2^{2x+3} \, dx = 8 \times \int_0^2 4^x \, dx$$

$$\approx 8 \times 11.096$$

$$= 88.768$$

$$= 88.8 \text{ (3 sf)}$$

Assessment 17

1 C

2 A

3 $1^3 + 2(1)^2 - 3(1) - 2 = -2,\ 2^3 + 2(2)^2 - 3(2) - 2 = 8$

Change of sign, hence root in interval.

4 $f(3) = 0.42,\ f(3.5) = -1.23$

Change of sign, hence root in interval.

5 a $2^3 - 4(2) - 1 = -1, 2.5^3 - 4(2.5) - 1 = 4.6$

Change of sign, hence root in interval.

b $x_2 = \sqrt{4 + \frac{1}{2}} = 2.12,\ x_3 = \sqrt{4 + \frac{1}{2.12}} = 2.11$

6 a 1.31

b $e^{1.305} + 1.305 - 5 = -0.007,\ e^{1.315} + 1.315 - 5 = 0.04$

Change of sign, hence must be correct to 2 decimal places.

7 a $1^3 - 3(1) + 1 = -1,\ 2^3 - 3(2) + 1 = 3$

Change of sign, hence root in interval.

b 1.55

8 a -0.91

b $f(-0.915) = 0.04,\ f(-0.905) = -0.055$

Change of sign, hence must be correct to 2 decimal places.

9 1.13

10 -6.66

11 a $f(0) = 5(0) - e^0 = -1,\ f(0.5) = 5(0.5) - e^{0.5} = 0.85$

Change of sign, hence root in (0, 0.5)

b Because the curve intersects the x-axis between 2 and 3.

c $f(2.45) = 5(2.45) - e^{2.45} = 0.66,$

$f(2.55) = 5(2.55) - e^{2.55} = -0.06$

Change of sign, hence must be correct to 1 decimal place.

12 a

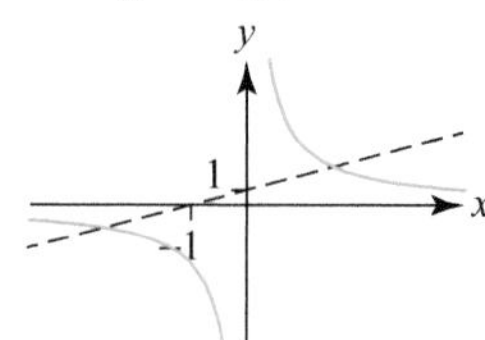

b Two as they intersect twice.

c Let $f(x) = x + 1 - \frac{4}{x},\ f(1.5) = 1.5 + 1 - \frac{4}{1.5} = -0.17,\ f(1.6) = 0.1$

Change of sign so root in interval.

d $x = 1.56, -2.56$

13 a Three as the curves intersect 3 times.

b Let $f(x) = e^x \sin x - x - 2,\ f(1.2) = e^{1.2} \sin 1.2 - 1.2 - 2 = -0.106,$ $f(1.3) = 0.236$

Change of sign so root in interval

c Root in interval (2.8, 2.9)

14 a $x^3 = 5 + 3x^2,\ x^2 = \frac{5}{x} + 3x,\ x = \sqrt{\frac{5}{x} + 3x}$

b $x_5 = 3.42$

15 a $x_2 = 0.85,\ x_3 = 0.77,\ x_4 = 0.89,\ x_5 = 0.71$

b The sequence is diverging from the root.

c $x_2 = 0.83,\ x_3 = 0.81,\ x_4 = 0.83,\ x_5 = 0.82$

16 a

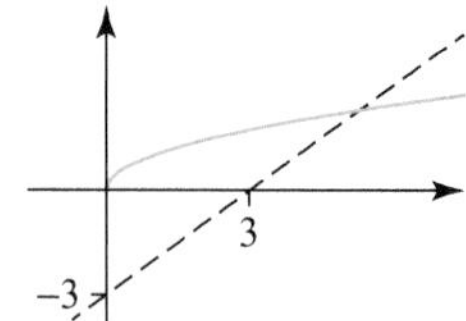

b 5.30

c i 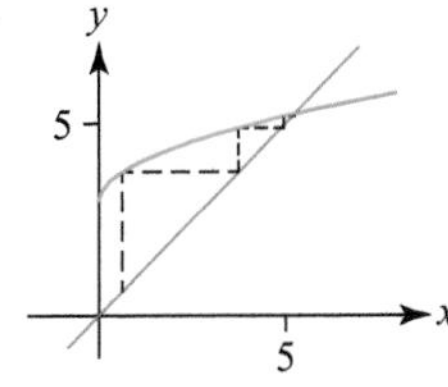

ii Staircase diagram.

17 a 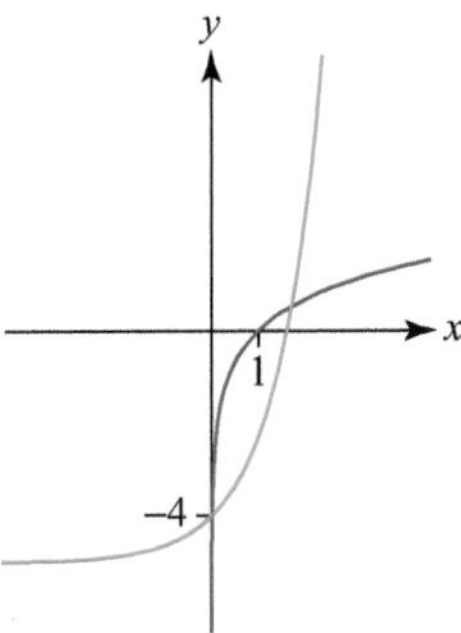

b Two as curves intersect twice.

c Let $f(x) = \ln x - e^x + 5$, $f(1.6) = \ln 1.6 - e^{1.6} + 5 = 0.517$, $f(1.8) = -0.462$
Change of sign so root in interval.

d 1.71

18 a $x^3 + 4x - 3 = 0$

$4x = 3 - x^3$

$x = \dfrac{3 - x^3}{4}$

$a = \dfrac{1}{4}, b = 3$

b $x_5 = 0.67$

c i 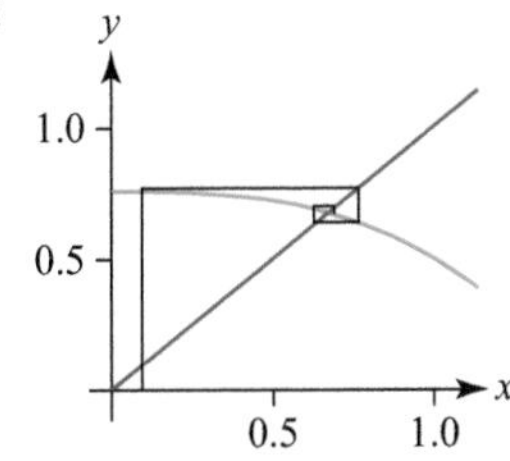

ii Cobweb diagram.

19 2.33

20 a 2.168

b Overestimate as curve is convex.

c i 2.097

ii 3.4%

21 a Three as the curve intersects the x-axis three times.

b $x^3 = 5x + 3$, $x^2 = 5 + \dfrac{3}{x}$, $x = \pm\sqrt{5 + \dfrac{3}{x}}$

c 2.49

d –0.66

e $f(-1.835) = (-1.835)^3 - 5(-1.835) - 3 = -0.004$, $f(-1.825) = 0.047$
Change of sign, hence must be correct to 3 significant figures.

22 a There is an asymptote $x = 3$ so although there is a change of sign in the interval, there is no root.

b f(0) and f(1) are both negative because there are two roots $\left(\dfrac{1}{2} \text{and } \dfrac{2}{3}\right)$ between them.

23 a 0.76 b 2.91

24 a 0.568

b Underestimate as curve is concave. Hence, actual area will be significantly higher than 0.5

25 a $f(1) = -1$, $f(2) = 0.39$, $f(1.5) = -0.39$, $f(1.7) = -0.10$, $f(1.9) = 0.22$
So solution is in interval (1.7, 1.9)

b 1.763

Assessment chapters 12–17: Pure

1 D $6 - 2\sqrt{5}$

2 B $x = 1, 0.585$

3 a $k > -1 + \sqrt{8}, k < -1 - \sqrt{8}$

b $x = \pm 1, \pm\sqrt{2}$

4 (5, –1) and (–2, 6)

5 $-\dfrac{1}{3} - \dfrac{\sqrt{7}}{3} \le x \le -\dfrac{1}{3} + \dfrac{\sqrt{7}}{3}$

6 $\theta = 30°, 150°$ or $\theta = 199.5°, 340.5°$

7 $y = \ln 27, x = \ln\left(\dfrac{1}{9}\right)$

8 a $\left(-\dfrac{3}{2}, 0\right), \left(\dfrac{5}{3}, 0\right)$

b 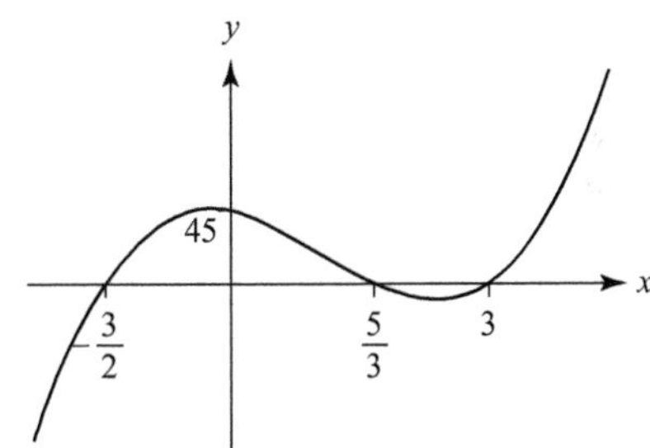

c 101.7 square units

9 $\left(-1, -\dfrac{17}{2}\right)$, a minimum

$\left(3, \dfrac{47}{2}\right)$, a minimum

$\left(2.5, \dfrac{757}{32}\right)$, a maximum

10 a $A = 2, k = \dfrac{1}{3}\ln 2$

b 23.9 days

c 10.7 days

d 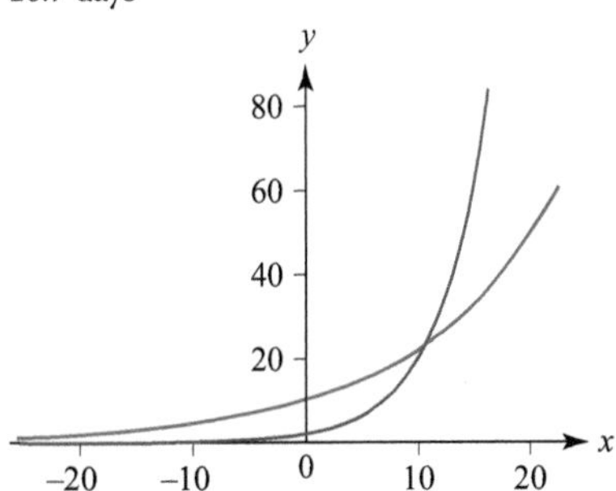

e Only realistic for small values of t.
Justification, e.g. Students who catch one infection may be off school so not catch other infection/may have weakened immune system so more likely to catch other infection.
Numbers of cases cannot increase indefinitely.
There is only a finite population/some students may have a natural immunity and not catch an infection.

11 a $f\left(-\frac{1}{2}\right)$

$= 6\left(-\frac{1}{2}\right)^3 - 19\left(-\frac{1}{2}\right)^2 - 51\left(-\frac{1}{2}\right) - 20$

$= 0$

Therefore $2x+1$ is a factor of $f(x)$

b $x = \left(-\frac{1}{2}\right), 5, -\frac{4}{3}$

12 a $\frac{3x-1}{(x+3)(2x+1)} = \frac{2}{x+3} - \frac{1}{2x+1}$

b $\frac{3x-5}{x^2-25} = \frac{2}{x+5} + \frac{1}{x-5}$

13 a $f(x) = \frac{2x-14}{(x-3)(x+1)} + \frac{2(x+1)}{(x-3)(x+1)}$

$= \frac{4x-12}{(x-3)(x+1)}$

$= \frac{4(x-3)}{(x-3)(x+1)}$

$= \frac{4}{x+1}$

b i $x > 3$ **ii** $0 < f(x) < 1$

c $f^{-1}(x) = \frac{4}{x} - 1$

Domain is $0 < x < 1$

14 a $\frac{3}{5}$ or 0.6

b $x = 3, 1$

c Range of gh(x) is gh(x) $\neq 3$

Range of hg(x) is hg(x) $\neq \pm 1$

So not the same (can use graphs to illustrate).

15 $1 + \frac{1}{2}x - \frac{3}{8}x^2 + \frac{7}{16}x^3 + \ldots$

16 a –3 **b** –67

17 a 5.18 cm² **b** 13.4 cm

18 a i

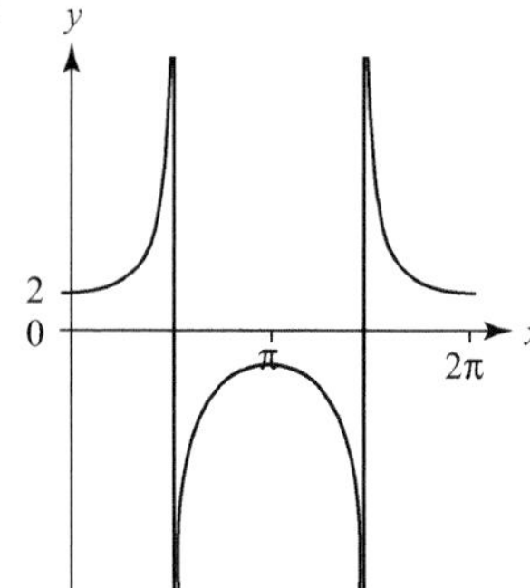

ii

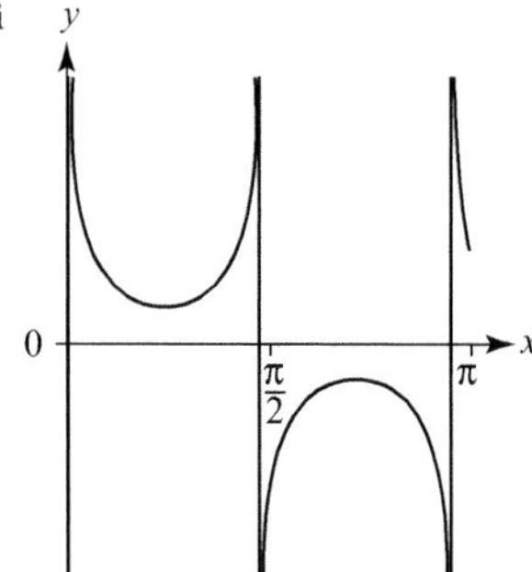

b $x = 56.3°, 236.3°$

19 a $x = \frac{\sqrt{2}}{2}$ **b** $x = \frac{\sqrt{3}}{4}$

20 a i $3\cos x$ **ii** $\ln x + 1$

b $\sqrt{3} - \frac{\pi}{6} - \frac{1}{2}$

21 a $x = -\frac{1}{2}, y = -\frac{1}{2}e^{-1}$

b $\frac{d^2y}{dx^2} = 2e^{2x} + 2e^{2x} + 4xe^{2x}$

When $x = -\frac{1}{2}$,

$\frac{d^2y}{dx^2} = 4e^{-1} - 2e^{-1}$

$= 2e^{-1}$

> 0 so a minimum

22 a i $-\cos x\ (+c)$ **ii** $3\ln|x|\ (+c)$

b $2\ln 5$ or $\ln 25$

23 2

24 a $3^3 - 6(3) - 12 = -3$

$3.5^3 - 6(3.5) - 12 = 9.875$

Change of sign so root in interval (3, 3.5)

b $x_2 = 3.16, x_3 = 3.13$

c $3.125^3 - 6(3.125) - 12 = -0.23$

$3.135^3 - 6(3.135) - 12 = 0.0015$

Change of sign so 3.13 is correct to 2 decimal places

25 0.527

26 Assume there exists an even integer n such that n^2 is odd.

Then $n = 2k$ for some integer k.

So $n^2 = (2k)^2$

$= 4k^2$

$= 2(2k^2)$ hence it is even

Which is a contradiction so we have proved the statement.

27 $\frac{2x^2+4x+3}{2x^2-x-1}$

$= \frac{A(x-1)(2x+1) + B(2x+1) + C(x-1)}{(2x+1)(x-1)}$

$2x^2 + 4x + 3 = A(x-1)(2x+1) + B(2x+1) + C(x-1)$

$A = 1, B = 3, C = -1$

28 a

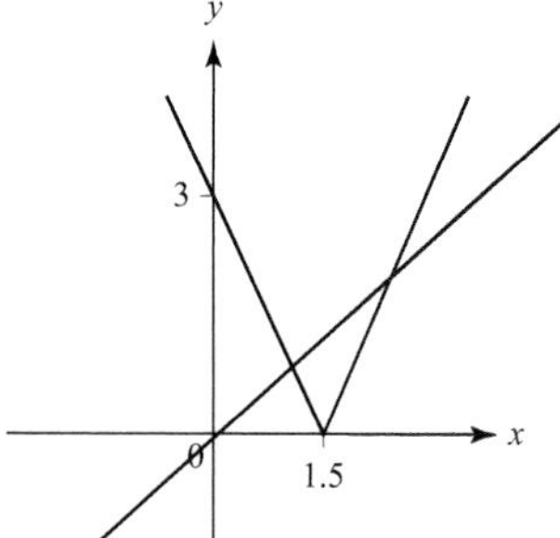

b 2 as they will intersect twice – either sketch to demonstrate or explain that gradient of $y = x$ is less than gradient of $y = -(3 - 2x)$.

c $x \leq 1, x \geq 3$

29 a $f^{-1}(x)=\frac{1}{3}(e^x-1)$

b

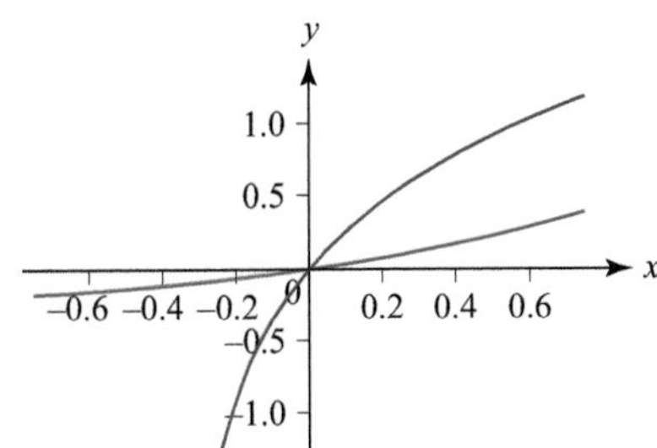

c Domain is $x \in \mathbb{R}$

Range is $y > -\frac{1}{3}$

30 a $\frac{1}{x-1}+\frac{-1}{x+3}+\frac{2}{(x+3)^2}$

b $\ln|x-1|-\ln|x+3|-\frac{2}{x+3}(+c)$

31 a $x_2=-1, x_3=3$ **b** $3n-6$

32 a 8 **b** $\frac{2}{9}$

33 a $S_n=a+(a+d)+(a+2d)+\ldots+(a+(n-2)d)+(a+(n-1)d)$

Also,

$S_n=(a+(n-1)d)+(a+(n-2)d)+\ldots$
$+(a+2d)+(a+d)+a$

So $2S_n=(2a+(n-1)d)+(2a+(n-1)d)+\ldots$
$+(2a+(n-1)d)+(2a+(n-1)d)$

$\Rightarrow 2S_n=n(2a+(n-1)d)$

$\Rightarrow S_n=\frac{n}{2}(2a+(n-1)d)$ as required

b 12

34 a i

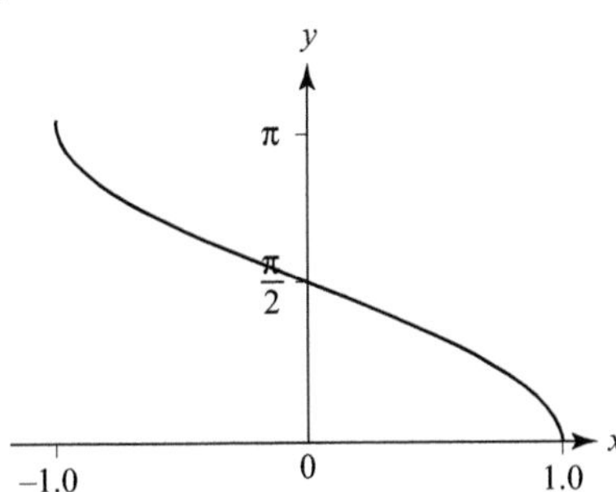

ii

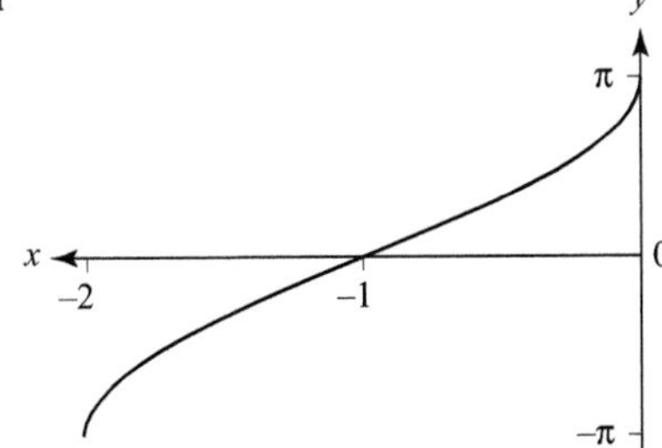

b i $0 \le y \le \pi$

ii $-\pi \le y \le \pi$

c $f^{-1}(x)=\sin\left(\frac{x}{2}\right)-1$

Domain is $-\pi \le x \le \pi$

35 a

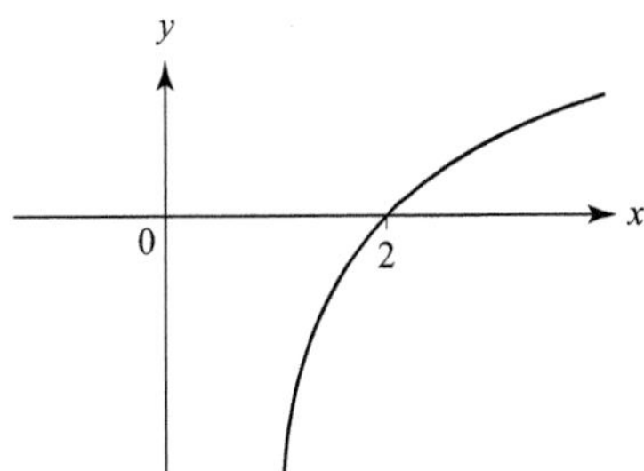

Asymptote at $x=1$

b $\frac{3}{2}$

36 a $\sec 2x=\frac{1}{\cos 2x}$

$$\approx\frac{1}{1-\frac{(2x)^2}{2}}$$

$$=\frac{2}{2-(2x)^2}$$

$$=\frac{2}{2-4x^2}$$

$$=\frac{1}{1-2x^2}$$

b $1+2x^2+4x^4$

c 1.0204

37 a $x=0.421, 2.72, 3.56, 5.86$

b $x=1.85$

38 a $\text{LHS} \equiv 3(1+\tan^2\theta)-7\tan^2\theta$
$\equiv 3+3\tan^2\theta-7\tan^2\theta$
$\equiv 3-4\tan^2\theta$
$\equiv \text{RHS}$

b $\theta=0.669, 3.81, 2.47$

39 a $\cos 2x \equiv \cos x\cos x-\sin x\sin x$
$\equiv \cos^2 x-\sin^2 x$

$\equiv 1-2\sin^2 x$ as required

b $x=90°, -30°, -150°$

40 a $\sqrt{17}\sin(\theta+0.245)$

b Max is $\sqrt{17}$

Occurs when $\theta=1.33°$

41 $-3<x<1$

42 $y=2x+1$

43 a $\frac{(x+2)\cos x-\sin x}{(x+2)^2}$

b $\frac{(x+2)\cos x-\sin x}{(x+2)^2}=0$

$\Rightarrow (x+2)\cos x-\sin x=0$

$\Rightarrow (x+2)=\tan x$

$\Rightarrow \arctan(x+2)=x$ as required

c $x=1.27$

44 a $\frac{1}{t}$

b $y-5=-2(x-1)$

c $x=\left(\frac{y-1}{2}\right)^2-3$

45 When $y=0$, $\frac{dy}{dx}=\frac{2}{3}$

When $y=\frac{3}{2}$, $\frac{dy}{dx}=\frac{5}{6}$

46 $\cos(x+h)-\cos x$

$= \cos x\cos h - \sin x \sin h - \cos x$

when h small

$\approx \cos x\left(1-\frac{h^2}{2}\right) - h\sin x - \cos x$

$= -\frac{h^2}{2}\cos x - h\sin x$

$\frac{dy}{dx} = \lim_{h\to 0}\left(\frac{\cos(x+h)-\cos x}{(x+h)-x}\right)$

$= \lim_{h\to 0}\left(\frac{-\frac{h^2}{2}\cos x - h\sin x}{h}\right)$

$= \lim_{h\to 0}\left(-\frac{h}{2}\cos x - \sin x\right)$

$= -\sin x$

47 a **i** $-x\cos x + \sin x\ (+c)$

ii $x\ln x - x\ (+c)$

b $-\frac{1}{4}(x^2+1)^{-2}(+c)$

48 $\ln\left(\frac{27}{4}\right)$

49 $y = e^{\sin x - \frac{1}{2}}$

50 a

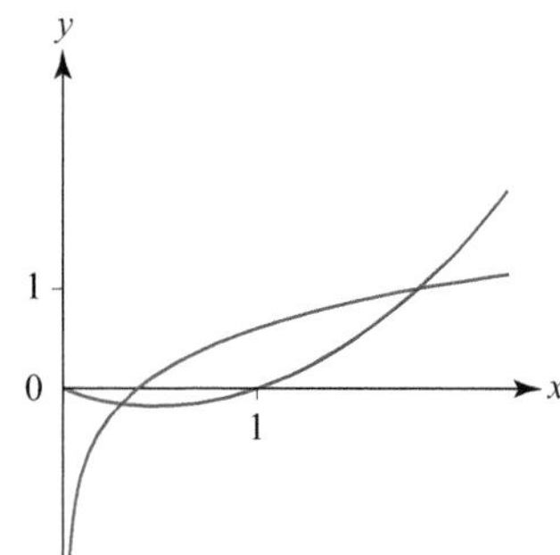

b $x = 0.298$ (or $x = 1.87$)

c e.g.

$0.2975^2 - 0.2975 - \ln 0.2975 - 1 = 0.003$

$0.2985^2 - 0.2985 - \ln 0.2985 - 1 = -0.0004$

Change of sign so root in interval, hence solution is correct to 3 significant figures.

51 a $x = -1.343$

b 1.9 is near a stationary point so process could be unstable/next approximation is 19.6 which is nowhere near a root.

52 a One solution as the graphs intersect once

b Let $f(x) = \sqrt{x} - e^{-x}$

Then $f(0.4) = \sqrt{0.4} - e^{-0.4}\ (= -0.04)$

$f(0.5) = \sqrt{0.5} - e^{-0.5} (= 0.1)$

Change of sign so solution is between 0.4 and 0.5

c $x = 0.43$

$f(0.425) = \sqrt{0.425} - e^{-0.425}\ (= -0.002)$

$f(0.435) = \sqrt{0.435} - e^{-0.435}\ (= 0.012)$

Change of sign so 0.43 is correct to 2 dp

53 a

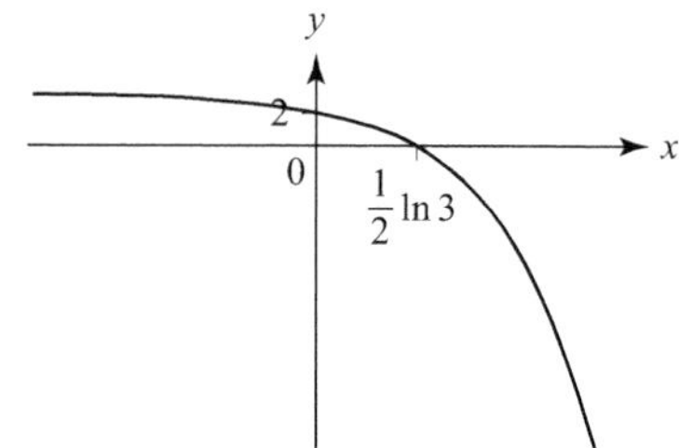

Asymptote at $y = 3$

b $f^{-1}(x) = \frac{1}{2}\ln(3-x)$

Domain is $x < 3$

Range is $f^{-1}(x) \in \mathbb{R}$

54 $d = 2.5,\ a = 1$

55 $\frac{12}{5}$ or 2.4

56 a **i** Geometric, $a = \frac{2}{3}, r = \frac{1}{3}$

ii 1

b Arithmetic, $\frac{n}{2}(n+1)(\ln 3)$

57 a $2 - \frac{5}{12}x - \frac{25}{288}x^2 + \ldots$

b 19.9582

58 Let $\frac{A+B}{2} = P$ and $\frac{A-B}{2} = Q$

Then $P+Q = A$ and $P-Q = B$

LHS $= \cos(P+Q) - \cos(P-Q)$

$\equiv (\cos P\cos Q - \sin P \sin Q)$

$-(\cos P\cos Q + \sin P\sin Q)$

$\equiv -2\sin P\sin Q$

So $K = -2$

59 $0.554 \le x \le 2.12$

60 $x = 214.7°, 325.3°$

61 $\left(53.1°, \frac{1}{5}\right)$

62 a LHS

$\equiv (\sin^2 x + \cos^2 x)^2 - 2\sin^2 x\cos^2 x$

$\equiv 1 - 2(\sin x\cos x)^2$

$\equiv 1 - 2\left(\frac{1}{2}\sin 2x\right)^2$

$\equiv 1 - 2\left(\frac{1}{4}\sin^2 2x\right)$

$\equiv 1 - \frac{1}{2}\sin^2 2x$

$\equiv 1 - \frac{1}{2}\left(\frac{1-\cos 4x}{2}\right)$

$\equiv 1 - \frac{1}{4}(1-\cos 4x)$

$\equiv \frac{1}{4}(4 - 1 + \cos 4x)$

$\equiv \frac{1}{4}(3 + \cos 4x) \equiv$ RHS

b $x = \pm 0.48, \pm 1.09, \pm 2.05, \pm 2.66$

63 a i

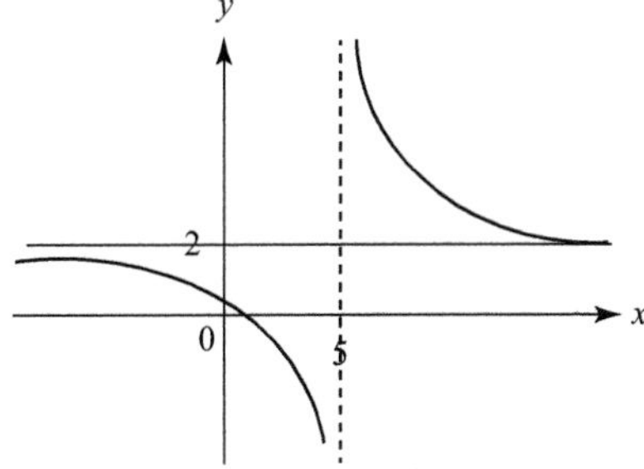

$x = 5,\ y = 2$

ii

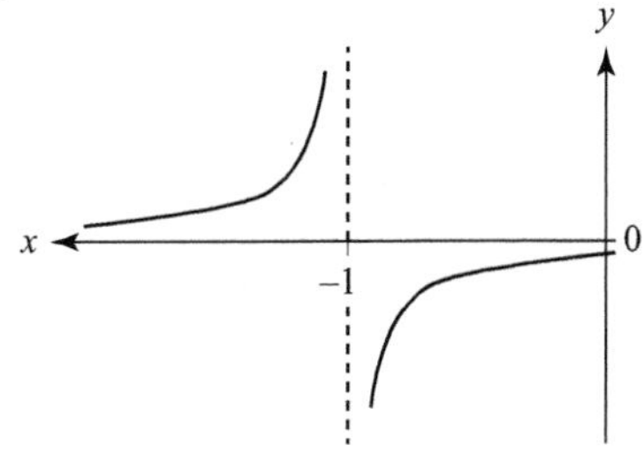

$x = -1,\ y = 0$

b $b = -4\ ,\ a = 8$

c i $y = \frac{2x-2}{x-5}$

ii $y = -\frac{8}{1+x}$

64 a $\frac{2}{27}$

b $y = \frac{1}{2}\sqrt{x} + 10 + \frac{50}{\sqrt{x}}$

65 $\frac{5}{2}, -\frac{20}{23}$

66 149 cm^2, $r = 9.73$ cm

67 $-\ln 2$

68 Let $y = \arccos 3x$

Then $\cos y = 3x$

$$-\sin y \frac{dy}{dx} = 3$$

$$\Rightarrow \frac{dy}{dx} = -\frac{3}{\sin y}$$

$$= -\frac{3}{\sqrt{1-\cos^2 y}}$$

$$= -\frac{3}{\sqrt{1-(3x)^2}}$$

$$= -\frac{3}{\sqrt{1-9x^2}} \text{ as required}$$

69 $f(\theta) = \int \frac{\sin\theta}{\cos\theta}\,d\theta$

$$= -\ln|\cos\theta| + c$$

$$= \ln|\cos\theta|^{-1} + c$$

$$= \ln|\sec\theta| + c$$

$$f\left(\frac{\pi}{3}\right) = \ln 6$$

$$\Rightarrow \ln 2 + c = \ln 6$$

$$\Rightarrow c = \ln 3$$

$$\therefore f(\theta) = \ln|\sec\theta| + \ln 3$$

70 a $ye^y - e^y = \frac{1}{2}e^{2x} + c$

b $x = \frac{1}{2}\ln(2ye^y - 2e^y + 1)$

71 $\frac{1}{216}(3x-1)^8 + \frac{2}{189}(3x-1)^7 + \frac{1}{162}(3x-1)^6 (+c)$

72 4

73 a 3.955

b Not true, estimation is an over-estimate since curve is convex.

74 13.5 square units

75 $\frac{\pi}{4} + \frac{\sqrt{3}}{4}$

Chapter 18 Motion in two dimensions

Exercise 18.1A

1 a $(13\mathbf{i} - 3\mathbf{j})\,\text{m s}^{-1}$ **b** $(18\mathbf{i} + 1.5\mathbf{j})\,\text{m}$

2 a $(13\mathbf{i} + 6\mathbf{j})\,\text{m s}^{-1}$ **b** $(-\mathbf{i} + 2\mathbf{j})\,\text{m s}^{-2}$ **c** $16.5\mathbf{i}\,\text{m}$

3 $(24\mathbf{i} + 4\mathbf{j})\,\text{m}$

4 $(6\mathbf{i} + 3\mathbf{j})\,\text{m s}^{-1}$

5 a $\mathbf{s} = (36\mathbf{i} - 6\mathbf{j})$ **b** $\mathbf{s} = (80\mathbf{i} - 104\mathbf{j})$

c $\mathbf{a} = (-2\mathbf{i} + 3\mathbf{j})$ **d** $\mathbf{v} = (3\mathbf{i} + 11\mathbf{j})$

6 $(13.5\mathbf{i} + 4\mathbf{j})\,\text{m}$

7 Speed = $11.4\,\text{m s}^{-1}$

Direction = 37.9° to the x-direction

8 $(25\mathbf{i} + 13\mathbf{j})\,\text{m}$

Exercise 18.1B

1 a Speed = $38.5\,\text{m s}^{-1}$

Direction = 27.9° to the x-axis

b 152 m

2 $(41\mathbf{i} + 8\mathbf{j})\,\text{m}$

3 12.2 m

4 $\mathbf{u} = (6\mathbf{i} + 2\mathbf{j})$

$\mathbf{v} = (6\mathbf{i} + 2\mathbf{j}) + (-\mathbf{i} - 7\mathbf{j}) \times 2$

$= (4\mathbf{i} - 12\mathbf{j})$

Initial speed $u = \sqrt{6^2 + 2^2}$

$= 2\sqrt{10}$

$= \sqrt{4^2 + (-12)^2}$

Final speed $= 4\sqrt{10}$

$= 2u$

Initial direction along a line with gradient $m_1 = \frac{2}{6}$

$= \frac{1}{3}$

Final direction along a line with gradient $m_2 = \frac{-12}{4}$

$= -3$

$m_1 m_2 = -1$, so directions are perpendicular.

5 $\mathbf{v} = (10\mathbf{i} - 3\mathbf{j})\,\text{m s}^{-1}$

$\mathbf{s} = (28\mathbf{i} + 4\mathbf{j})\,\text{m}$

6 Speed = $60.2\,\text{m s}^{-1}$

$\mathbf{s} = (3\mathbf{i} + \mathbf{j}) \times 2 + \frac{1}{2} \times (16\mathbf{i} + 24\mathbf{j}) \times 4$

$= (38\mathbf{i} + 50\mathbf{j})$

Final position $\mathbf{r} = (2\mathbf{i} + 6\mathbf{j}) + (38\mathbf{i} + 50\mathbf{j})$

$= (40\mathbf{i} + 56\mathbf{j})$

$\mathbf{r} = \frac{8}{7}\mathbf{v}$ so $\mathbf{v}$ has same direction as $\mathbf{r}$.

Particle is moving directly away from O.

7 $\mathbf{u} = (-3\mathbf{i} + 2\mathbf{j})$, $\mathbf{a} = (2\mathbf{i} - 4\mathbf{j})$

8 a $(8\mathbf{i}+12\mathbf{j})$ m
b $(8\mathbf{i}-15\mathbf{j})$ m s^{-1}
c $(-2\mathbf{i}+3.75\mathbf{j})$ m s^{-2}
d $(54\mathbf{i}-54\mathbf{j})$ m
9 a $\overline{OP}=(120t-2t^2)\mathbf{i}+(400-4t^2)\mathbf{j}$ m
b 10 s
c 1000 m

Exercise 18.2A

1 a $\mathbf{v}=4\mathbf{i}+4t\mathbf{j}$
b $\mathbf{v}=(2t-4)\mathbf{i}+(3t^2-4t)\mathbf{j}$
c $\mathbf{v}=-2\sin t\mathbf{i}+4\cos t\mathbf{j}$
d $\mathbf{v}=e^t\mathbf{i}+\frac{1}{t+1}\mathbf{j}$
2 a $\mathbf{v}=(3t-t^2+3)\mathbf{i}+(t^2-1.5t^4)\mathbf{j}$
b $\mathbf{v}=(2+2\sin 2t)\mathbf{i}+(5-4\cos 2t)\mathbf{j}$
3 a $\mathbf{a}=-10\mathbf{j}$
$\mathbf{s}=(15t+4)\mathbf{i}+(20t-5t^2-3)\mathbf{j}$
b $\mathbf{a}=-8\cos t\,\mathbf{i}+8e^{2t}\,\mathbf{j}$
$\mathbf{r}=(8\cos t-7)\mathbf{i}+(2e^{2t}+1)\mathbf{j}$
c $\mathbf{a}=(2-12t)\mathbf{i}+(6t-12t^2)\mathbf{j}$
$\mathbf{r}=(t^2-2t^3+2)\mathbf{i}+(t^3-t^4)\mathbf{j}$
4 $(5\mathbf{i}+9\mathbf{j})$ m s^{-1}
5 $-\pi\mathbf{i}$ m s^{-1}
6 $(6\mathbf{i}+11\mathbf{j})$ m
7 a $(3(t^2+1)\mathbf{i}+2(t+2)\mathbf{j})$ m s^{-1}
b 17 m s^{-1}
c 28.1° to the x-direction
8 $\mathbf{v}=\left(12\mathbf{i}+\frac{8}{3}\mathbf{j}\right)$ m s^{-1}, $\mathbf{a}=\left(12\mathbf{i}-\frac{8}{9}\mathbf{j}\right)$ m s^{-2}
9 a $-18\sin 3t\mathbf{i}+16\cos 2t\mathbf{j}$
b $(2\sin 3t+2)\mathbf{i}-(4\cos 2t+1)\mathbf{j}$

Exercise 18.2B

1 a 3.16 m s^{-1} b $8\mathbf{i}+\frac{34}{3}\mathbf{j}$
2 $\sqrt{\frac{656}{5}}$ m
3 a $\mathbf{v}=(3\mathbf{i}+(4-4t)\mathbf{j})$ m s^{-1}
$\mathbf{a}=-4\mathbf{j}$ m s^{-2}
b 5 m s^{-1}
c 1 s
d No; to be stationary **v** must be $0\mathbf{i}+0\mathbf{j}$, but x-component is always 3.
4 a $\frac{d\mathbf{v}}{dt}=\mathbf{a} \Rightarrow \mathbf{v}=\int \mathbf{a}\,dt=\mathbf{a}t+\mathbf{c}$

When $t=0$, $\mathbf{v}=\mathbf{u}$, so $\mathbf{c}=\mathbf{u}$
giving $\mathbf{v}=\mathbf{u}+\mathbf{a}t$
$\mathbf{s}=\int \mathbf{v}\,dt$

b $=\int \mathbf{u}+\mathbf{a}t\,dt$
$=\mathbf{u}t+\frac{1}{2}\mathbf{a}t^2+\mathbf{c}$
When $t=0$, $\mathbf{s}=\mathbf{0}$, so $\mathbf{c}=\mathbf{0}$
giving $\mathbf{s}=\mathbf{u}t+\frac{1}{2}\mathbf{a}t^2$
5 $\mathbf{v}=(70\mathbf{i}-6\mathbf{j})$ m s^{-1}
$\mathbf{r}=(134\mathbf{i}+21\mathbf{j})$ m
6 a $t=\frac{\pi}{2}$
b $\mathbf{v}=(\mathbf{i}-2\mathbf{j})$, $\mathbf{a}=\mathbf{0}$

7 a 24.1 m s^{-1}
b 30.0 m
8 a $\mathbf{v}=\int \mathbf{a}\,dt$
$=-2\sin 2t\mathbf{i}+2\cos 2t\mathbf{j}+\mathbf{c}$
When $t=0$, $\mathbf{v}=2\mathbf{j}$, so $\mathbf{c}=\mathbf{0}$
$\mathbf{v}=-2\sin 2t\mathbf{i}+2\cos 2t\mathbf{j}$
giving speed $=\sqrt{4(\sin^2 2t+\cos^2 2t)}$
$=2$ m s^{-1}
Therefore constant speed.
b Distance = 1 m
Path is a circle, radius 1 m, about the origin.
9 a $t=4.20$ s
b $t=2.11$ s
10 a $t=2$
b $\mathbf{r}=\frac{8}{7}\mathbf{v}$
v and **r** point the same way, so particle is moving away from O.

Exercise 18.3A

1 a 1.76 s b 21.6 m c 0.878 s d 3.78 m
2 a 1.77 s b 8.83 m c 0.883 s d 3.82 m
3 a 17.2 m
b Speed = 15.4 m s^{-1}
Direction = −2.57°, that is 2.57° below the horizontal
4 a i $(30\mathbf{i}+35.1\mathbf{j})$ m ii $(60\mathbf{i}+60.4\mathbf{j})$ m
b 8.16 s
c 245 m
d $y=\frac{4x}{3}-\frac{49x^2}{9000}$
5 $\alpha=30.0°$
6 31.6 m s^{-1}

Exercise 18.3B

1 a 64.8 m
b Assumes stone is a particle, no air resistance, beach is horizontal.
2 a 22.6°
b The available headroom would be reduced by the radius of the ball, so the maximum angle would be less.
3 a $\frac{50V^2\sin^2 40°}{981}$ b 21.8 m s^{-1} c 47.7 m
4 2.67 s
5 28.5 m
6 a Speed = 54.3 m s^{-1}
Direction = 46.3° above horizontal
b Speed = 42.3 m s^{-1}
Direction = 27.6°
c 5.61 s
7 $v_y=u\sin\alpha-gt=0$ for max height $\Rightarrow t=\frac{u\sin\alpha}{g}$
$y=ut\sin\alpha-\frac{1}{2}gt^2 \Rightarrow h=u\left(\frac{u\sin\alpha}{g}\right)-\frac{1}{2}g\left(\frac{u\sin\alpha}{g}\right)^2$
$=\frac{u^2\sin^2\alpha}{2g}$
8 a $u=60$ m s^{-1}, $v=9.20$ m s^{-1}
b 0.33 m

9 a $x = ut\cos\alpha \Rightarrow t = \frac{x}{u\cos\alpha}$

$y = ut\sin\alpha - \frac{1}{2}gt^2$

$= u\left(\frac{x}{u\cos\alpha}\right)\sin\alpha - \frac{1}{2}g\left(\frac{x}{u\cos\alpha}\right)^2$

$= x\tan\alpha - \left(\frac{g\sec^2\alpha}{2u^2}\right)x^2$

b Let range be R. Projectile passes through (a, b), (b, a) and $(0, R)$.

$b = a\tan\alpha - \left(\frac{g\sec^2\alpha}{2u^2}\right)a^2$ ...[1]

$a = b\tan\alpha - \left(\frac{g\sec^2\alpha}{2u^2}\right)b^2$ ...[2]

$0 = R\tan\alpha - \left(\frac{g\sec^2\alpha}{2u^2}\right)R^2$ ...[3]

Substitute tan α from [3] into [1] and [2]:

$b = aR\left(\frac{g\sec^2\alpha}{2u^2}\right) - \left(\frac{g\sec^2\alpha}{2u^2}\right)a^2$ ...[4]

$a = bR\left(\frac{g\sec^2\alpha}{2u^2}\right) - \left(\frac{g\sec^2\alpha}{2u^2}\right)b^2$ ...[5]

Divide [5] by [4]:

$\frac{a}{b} = \frac{bR - b^2}{aR - a^2} \Rightarrow R = \frac{a^3 - b^3}{a^2 - b^2} = \frac{(a-b)(a^2+ab+b^2)}{(a-b)(a+b)}$

$= \frac{a^2+ab+b^2}{a+b}$

Exercise 18.4A

1 a $7.21\,\text{m s}^{-2}$
b $33.7°$

2 a $4.47\,\text{m s}^{-2}$
b $26.6°$

3 $P = 32.5\,\text{N}$, $Q = 19.7\,\text{N}$

4 a $17.2\,\text{N}$
b $11.1\,\text{m}$

5 a $68.4\,\text{N}$
b $1.40\,\text{m s}^{-2}$

6 a $38.0°$
b $0.959\,\text{m s}^{-2}$
c $4.39\,\text{m s}^{-1}$

7 a $2.80\,\text{m s}^{-2}$
b $21.3\,\text{N}$

8 Distance = 77.5 m
Bearing = 025° (to nearest degree)

9 Magnitude = $9.97\,\text{m s}^{-2}$
Bearing = 108° (to nearest degree)

10 a $37.6\,\text{N}$ b $4.46\,\text{m s}^{-2}$

Exercise 18.4B

1 a $62.5\,\text{N}$ b $2.02\,\text{m s}^{-2}$

2 $7.70\,\text{m s}^{-2}$

3 $T_1 = 4098\,\text{N}$, $T_2 = 2698\,\text{N}$

4 a Resolve vertically:

$R - mg + mg\sin\theta = 0$

$R = mg(1 - \sin\theta)$ ...[1]

Resolve horizontally:

$mg\cos\theta - 0.75R = ma$ ...[2]

Substitute from [1] into [2]:

$ma = mg\cos\theta - 0.75mg(1 - \sin\theta)$

$a = \frac{g(3\sin\theta + 4\cos\theta - 3)}{4}$

b $\frac{da}{d\theta} = \frac{g}{4}(3\cos\theta - 4\sin\theta)$

$= 0$ for maximum

$3\cos\theta = 4\sin\theta \Rightarrow \tan\theta = \frac{3}{4} \Rightarrow \theta = 36.9°$

To prove a maximum, $\frac{d^2a}{d\theta^2} = \frac{g}{4}(-3\sin\theta - 4\cos\theta)$ which is negative for $\theta = 36.9°$. Hence, a maximum.

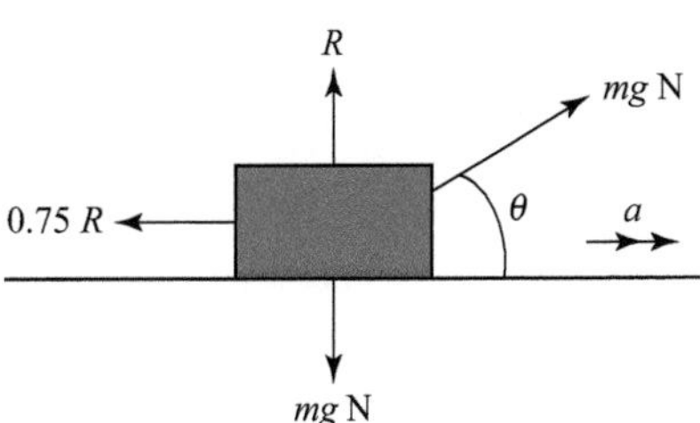

5 $2.84\,\text{m s}^{-2}$

6 18.2 m

7 24.7 m

8 $T_1 = 508\,\text{N}$, $T_2 = 584\,\text{N}$

Review exercise

1 a $\mathbf{s} = 25\mathbf{i} + 10\mathbf{j}$ b $\mathbf{s} = 30\mathbf{i} - 54\mathbf{j}$
c $\mathbf{a} = -2\mathbf{i} + 3\mathbf{j}$ d $\mathbf{v} = 4.2\mathbf{i} - 5.2\mathbf{j}$

2 $(21\mathbf{i} + 43\mathbf{j})$ m

3 $\mathbf{v} = 10\mathbf{i} + 20\mathbf{j}$
$\mathbf{a} = 6\mathbf{i} + 22\mathbf{j}$

4 $-13\mathbf{i} + 10\mathbf{j}$

5 $\mathbf{a} = 4\sin 4t\,\mathbf{i} + 4\cos 2t\,\mathbf{j}$

When $t = \frac{\pi}{4}$, $\mathbf{v} = \mathbf{i} + 2\mathbf{j}$ and $\mathbf{r} = 3\mathbf{i} + \mathbf{j}$

$\mathbf{v} = \int \mathbf{a}\,dt$

$= -\cos 4t\mathbf{i} + 2\sin 2t\mathbf{j} + \mathbf{c}$

When $t = \frac{\pi}{4}$, $\mathbf{v} = \mathbf{i} + 2\mathbf{j}$

$\mathbf{i} + 2\mathbf{j} = -\cos\pi\mathbf{i} + 2\sin\frac{\pi}{2}\mathbf{j} + \mathbf{c}$

$\mathbf{i} + 2\mathbf{j} = \mathbf{i} + 2\mathbf{j} + \mathbf{c}$

$\mathbf{c} = 0$

$\mathbf{v} = -\cos 4t\mathbf{i} + 2\sin 2t\mathbf{j}$

$\mathbf{r} = \int \mathbf{v}\,dt$

$= -\frac{1}{4}\sin 4t\mathbf{i} - \cos 2t\mathbf{j} + \mathbf{c}$

When $t = \frac{\pi}{4}$, $\mathbf{r} = 3\mathbf{i} + \mathbf{j}$

$3\mathbf{i} + \mathbf{j} = -\frac{1}{4}\sin\pi\mathbf{i} - \cos\frac{\pi}{2}\mathbf{j} + \mathbf{c}$

$3\mathbf{i} + \mathbf{j} = \mathbf{c}$

$\mathbf{r} = \left(-\frac{1}{4}\sin 4t + 3\right)\mathbf{i} - (\cos 2t + 1)\mathbf{j}$

When $t = \frac{\pi}{2}$

$\mathbf{r} = \left(-\frac{1}{4}\sin 2\pi + 3\right)\mathbf{i} - (\cos\pi - 1)\mathbf{j}$

$\mathbf{r} = 3\mathbf{i} + 2\mathbf{j}$

Distance from origin is $|\mathbf{r}| = \sqrt{3^2 + 2^2}$

$= \sqrt{13}$

6 a $1.03\,s$
b $11.3\,m$
c $0.517\,s$
d $1.31\,m$

7 a $91.8\,m$
b $y = x - \frac{9.8x^2}{900}$

8 a $P = 29.3\,N$
b $54.4\,m$

Assessment

1 a D $10\mathbf{i} + 24\mathbf{j}\,m\,s^{-1}$ b A $26\,m\,s^{-1}$

2 $2\sqrt{3}\,g = 33.9\,N$ (to 3 sf)

3 a $\mathbf{i} - 2\mathbf{j}$ b $6\sqrt{2}\,m = 8.49\,m$ (to 3 sf)

4 a $44.1\,m$ b $305\,m$

5 a $122.5\,m$ b $100\,m$
c $67.8°$ below horizontal

6 a $\mathbf{i} - 2\sqrt{3}\mathbf{j}\,m\,s^{-1}$ b $-2\sqrt{3}\,\mathbf{i} + 8\mathbf{j}\,m\,s^{-2}$

7 a $-4\mathbf{i} + 9\mathbf{j}\,m\,s^{-1}$ b $-\mathbf{i} + 2\mathbf{j}\,m\,s^{-2}$
c $1.5\,s$

8 a $3g\,N$ b $3\sqrt{3}\,g\,N$

9 a $\begin{pmatrix} 4.8 \\ 5.6 \end{pmatrix} N$ b $30\,m$

10 a $12.1\,m$ b $7.00\,m\,s^{-1}$
c $31.0°$ below horizontal

11 a $\begin{pmatrix} 16 \\ 1 \end{pmatrix} m\,s^{-2}$
b $\begin{pmatrix} 0 \\ -\frac{1}{\sqrt{2}} \end{pmatrix} m\,s^{-1}$
c $\begin{pmatrix} 2 \\ 0 \end{pmatrix} m$

12 a $\mathbf{S}_P = (4 + 3t)\,\mathbf{i} + (11 - 8t)\,\mathbf{j}$
b $\mathbf{S}_Q = (9 - 7t)\,\mathbf{i} + (3.5 + 12t)\,\mathbf{j}$
c $\mathbf{S}_P - \mathbf{S}_Q = (-5 + 10t)\,\mathbf{i} + (7.5 - 20t)\,\mathbf{j}$

$$\left|\mathbf{S}_P - \mathbf{S}_Q\right|^2 = (-5+10t)^2 + (7.5-20t)^2$$

d 12.24 pm
e $\frac{\sqrt{5}}{2}$ km

13 a $x = u\cos\alpha \times t$

$$t = \frac{x}{u\cos\alpha}$$

$$y = u\sin\alpha \times t - \frac{1}{2}gt^2$$

$$y = u\sin\alpha \times \left(\frac{x}{u\cos\alpha}\right) - \frac{1}{2}g\left(\frac{x}{u\cos\alpha}\right)^2$$

$$y = x\tan\alpha - \frac{gx^2}{2u^2}\sec^2\alpha$$

b $76.0°$ or $63.4°$

Chapter 19 Forces

Exercise 19.1A

1 a i $-3\mathbf{i} + 2\mathbf{j} + 4\mathbf{k}$ ii $11\mathbf{i} + 3\mathbf{j} - 11\mathbf{k}$
iii 7 iv $\sqrt{33}$ v $\frac{1}{7}(6\mathbf{i} - 3\mathbf{j} - 2\mathbf{k})$
vi $31.0°, 115.4°, 106.6°$
b $24\mathbf{i} - 12\mathbf{j} - 8\mathbf{k}$
c $p = 2, q = -3, r = 13$

2 $\mathbf{r} = 7.13\mathbf{i} + 0.697\mathbf{j} + 3.57\mathbf{k}$

3 a $27.1°$ or $152.9°$
b $\pm 10.7\mathbf{i} + 4.50\mathbf{j} + 3.11\mathbf{k}$

4 a $AB = 3\sqrt{2},\ BC = 3\sqrt{2},\ AC = \sqrt{36} = 6$
b
$$\begin{aligned} AB^2 + BC^2 &= 18 + 18 \\ &= 36 \\ &= AC^2 \end{aligned}$$
Triangle ABC satisfies Pythagoras' theorem
Hence, is right-angled at B
c Triangle is isosceles since $AB = BC$

5 a $6\mathbf{i} - 2\mathbf{j} + 3\mathbf{k}$
b 7 units
c $\frac{6}{7}\mathbf{i} - \frac{2}{7}\mathbf{j} + \frac{3}{7}\mathbf{k}$

6 $4.8\mathbf{i} + 4.1\mathbf{j} - 0.5\mathbf{k}$

7 a $\begin{pmatrix} 7 \\ 0 \\ -8 \end{pmatrix}$
b 3 units
c $\begin{pmatrix} -10 \\ 10 \\ 5 \end{pmatrix}$
d $132°$
e $\lambda = 2, \mu = -3$

Exercise 19.1B

1
$$\begin{aligned} AB &= |\mathbf{b} - \mathbf{a}| \\ &= |-\mathbf{i} - 7\mathbf{j} + 5\mathbf{k}| \\ &= \sqrt{75} \end{aligned}$$
$BC = \sqrt{6}$
$AC = \sqrt{81}$
$AB^2 + BC^2 = AC^2$, hence ABC satisfies Pythagoras' theorem, and is right-angled at B

2 $\mathbf{d}$ could be at $4\mathbf{i} - 2\mathbf{j} - \mathbf{k}$ or $2\mathbf{i} + 3\mathbf{k}$ or $4\mathbf{j} - 3\mathbf{k}$

3 $\mathbf{V} = \pm 2\sqrt{3}(\mathbf{i} + \mathbf{j} + \mathbf{k})$

4 a $26.6°$ b $50\,m$

5 a $\mathbf{p} = \frac{1}{2}(\mathbf{a} + \mathbf{b}),\ \mathbf{q} = \frac{1}{2}(\mathbf{a} + \mathbf{d}),\ \mathbf{r} = \frac{1}{2}(\mathbf{b} + \mathbf{c})$

$$\mathbf{s} = \frac{2\mathbf{p} + \mathbf{c}}{3} = \frac{1}{3}(\mathbf{a} + \mathbf{b} + \mathbf{c})$$

$$\mathbf{t} = \frac{1}{2}(\mathbf{q} + \mathbf{r}) = \frac{1}{4}(\mathbf{a} + \mathbf{b} + \mathbf{c} + \mathbf{d})$$

Now $\overrightarrow{DT} = \mathbf{t} - \mathbf{d} = \frac{1}{4}(\mathbf{a} + \mathbf{b} + \mathbf{c} - 3\mathbf{d})$
and $\overrightarrow{DS} = \mathbf{s} - \mathbf{d} = \frac{1}{3}(\mathbf{a} + \mathbf{b} + \mathbf{c} - 3\mathbf{d})$
$\overrightarrow{DT}$ and $\overrightarrow{DS}$ are parallel and both pass through D, so D, T and S are collinear.
b $3:1$

6 a 19.3
b $\mathbf{c} = \frac{\mu\mathbf{a} + \lambda\mathbf{b}}{\lambda + \mu},\ \mathbf{d} = \frac{-\mu\mathbf{a} + \lambda\mathbf{b}}{\lambda - \mu}$

$$\begin{aligned} CD &= |\mathbf{d} - \mathbf{c}| \\ &= \left|\frac{(\lambda + \mu)(-\mu\mathbf{a} + \lambda\mathbf{b}) - (\lambda - \mu)(\mu\mathbf{a} + \lambda\mathbf{b})}{\lambda^2 - \mu^2}\right| \\ &= \frac{2\lambda\mu}{\lambda^2 - \mu^2}|\mathbf{b} - \mathbf{a}| \end{aligned}$$

So $CD : AB = 2\lambda\mu : \left(\lambda^2 - \mu^2\right)$

Exercise 19.2A

1 **a** **i** $X = -1.83\text{ N}$ (to 3 sf)
$Y = 3.17\text{ N}$ (to 3 sf)
ii $R = 3.7\text{ N}$ (to 2 sf)
$\theta = 30°$ to the nearest degree
b **i** $X = 42.8\text{ N}$ (to 3 sf)
$Y = -1.67\text{ N}$ (to 3 sf)
ii $R = 43\text{ N}$ (to 2 sf)
$\theta = 92°$ to the nearest degree

2 $R = 625\text{ N}$
$\mu = 0.485$ (to 3 sf)

3 $R = 570\text{ N}$ (to 2 sf)
$W = 990\text{ N}$ (to 2 sf)

4
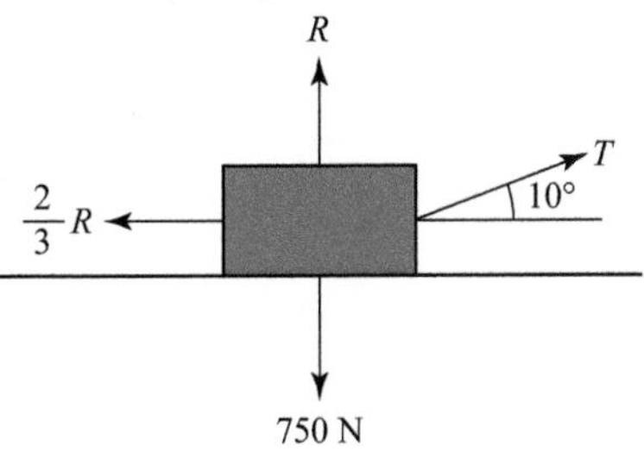

$(\rightarrow) \quad T\cos 10° - \frac{2}{3}R = 0 \quad (1)$

$(\uparrow) \quad R + T\sin 10° - 750 = 0 \quad (2)$

From (2) $R = 750 - T\sin 10°$
and so (1) gives $T\cos 10° - \frac{2}{3}(750 - T\sin 10°) = 0$
So

$$T = \frac{750 \times \frac{2}{3}}{\left(\cos 10° + \frac{2}{3}\sin 10°\right)}$$

$= 450\text{ N}$ (to 2 sf)
$R = 670\text{ N}$ (to 2 sf)

5 Magnitude $= 113\text{ N}$ (to 3 sf)
Bearing $= 246°$ (to 3 sf)

6 $T = 156\text{ N}$ (to 3 sf)
$\mu = 0.397$ (to 3 sf)

7 $W = 36.2\text{ N}$ (to 3 sf)
$\mu = 0.567$ (to 3 sf)

Exercise 19.2B

1 $S = 58.7\text{ N}$ (to 3 sf)
$R = 40.2\text{ N}$ (to 3 sf)

2 $W = 116\text{ N}$ (to 3 sf)

3 **a** 102 N (to 3 sf) **b** 184 N (to 3 sf)

4 $\mu = \frac{2}{3}$

5 $k = \sqrt{3} + 3$

6 $S = 40.0\text{ N}$, $T = 56.6\text{ N}$ (to 3 sf)

7 $\mu = \frac{1}{k}$

8 $\mu = 0.331$

9 **a** $m = 2$ **b** $m = 8$

10 1320 N

Exercise 19.3A

1 $X = 11.5\text{ N}$ (to 3 sf)
$a = 11.5\text{ m s}^{-2}$ (to 3 sf)

2 146 m s^{-2} (to 3 sf), 4°

3 2.95 m s^{-2} (to 3 sf)

4 0.743 (to 3 sf)

5 2.84 m s^{-2} (to 3 sf) at 13.0° above the 55 N force

6 0.89 m s^{-2} (to 2 sf)

7 0.28 (to 2 sf)

Exercise 19.3B

1 0.72 m s^{-2} (to 2 sf)

2 0.581 (to 3 sf)

3 950 N (to 2 sf)

4
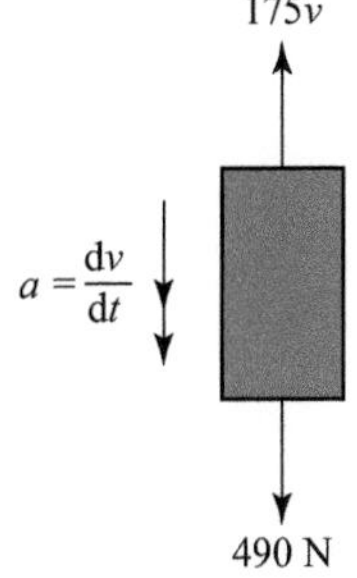

$(\downarrow) \quad 50\frac{dv}{dt} = 490 - 175v$

$\frac{dv}{dt} = 9.8 - 3.5v$
$= -3.5(v - 2.8)$

$v = 10e^{-3.5t} + 2.8$

5
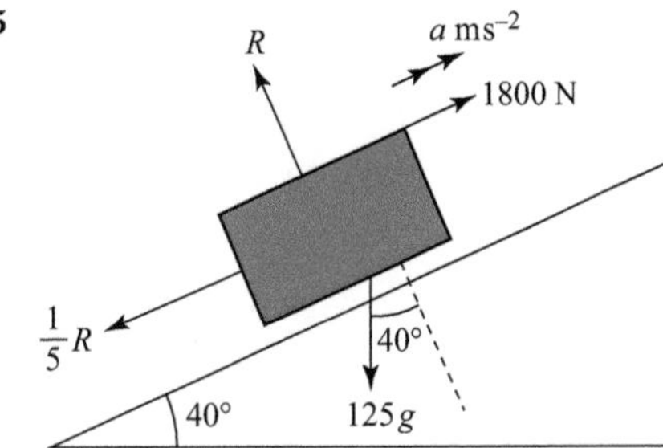

$(\nwarrow)\ R - 125g\cos 40° = 0$
So $R = 125g\cos 40°$
$(\nearrow)\ 1800 - \frac{1}{5}R - 125g\sin 40° = 125a$

$a = \frac{825.904\ldots}{125}$
$= 6.6\text{ m s}^{-2}$ (to 2 sf)
$t = 1.7\text{ s}$ (to 2 sf)

6 $X = 411\text{ N}$ (to 3 sf)
$a = 8.71\text{ m s}^{-2}$ (to 3 sf)

7
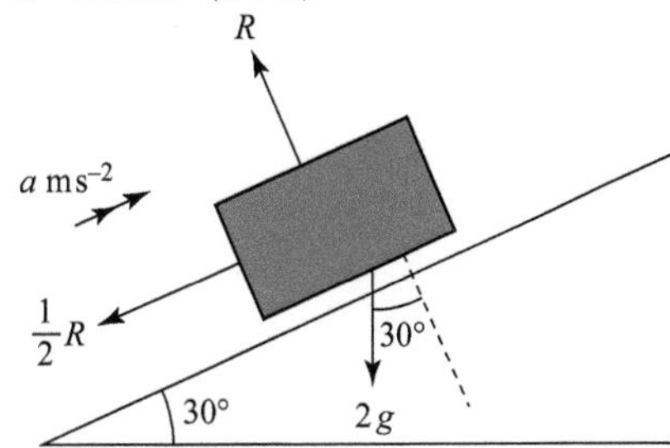

a $(\nwarrow)\ R-2g\cos 30°=0$

So $R=2g\cos 30°$

$(\swarrow)\ -\frac{1}{2}R-2g\sin 30°=2a$

So

$a=-\frac{18.287\ldots}{2}$

$=-9.1\,\text{m s}^{-2}$ (to 2 sf)

So deceleration is $9.1\,\text{m s}^{-2}$

b 0.49 m (to 2 sf)

c $0.66\,\text{m s}^{-2}$ (to 2 sf)

d $s=0.492\ldots \quad u=0 \quad a=0.656\ldots$

Using $v^2=u^2+2as$ gives

$v=\sqrt{0^2+2\times 0.656\ldots\times 0.492\ldots}$

$=0.8\,\text{m s}^{-2}$ (to 1 sf)

8 $v=30(1-e^{-0.15t})$

As $t\to\infty$, $e^{-0.15t}\to 0$ and so $v\to 30\,\text{m s}^{-1}$

9 a

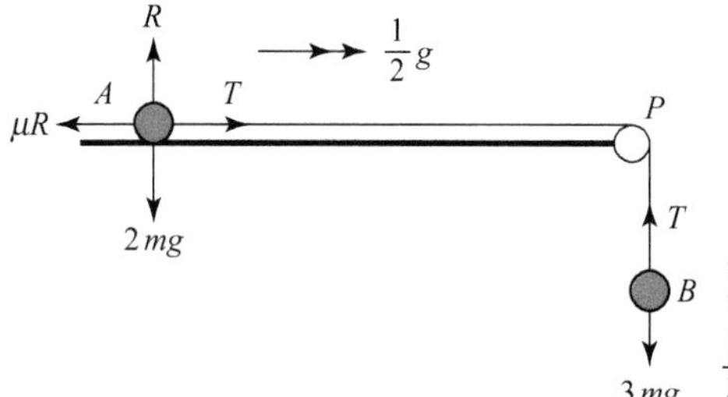

$3m$	$(\downarrow)$	$3mg-T=3m\times\frac{1}{2}g$	(1)
$2m$	$(\uparrow)$	$R-2mg=0$	(2)
	$(\rightarrow)$	$T-\mu R=2m\times\frac{1}{2}g$	(3)

(2) in (3) gives $(\rightarrow)$ $T-2mg\mu=mg$

(1) $3mg-T=\frac{3mg}{2}$

Adding these up gives

$3mg-2mg\mu=\frac{3mg}{2}+mg$

$=\frac{5mg}{2}$

and so $\mu=\frac{1}{4}$

$T=\frac{3mg}{2}$

b $3h$ m

Exercise 19.4A

1 a -16 N m **b** -30 N m
c -1 N m **d** -39 N m

2 2 N m

3 a The object will not turn.
b The object will turn clockwise.

4 80 N

5 10 N m anticlockwise

6 350 N

7 $p=-3$ or $p=2$

8 $x=-6$

Exercise 19.4B

1 a At the edge of the table
b 50 N
c 100 N

2 0.192 m (19.2 cm)

3 32 N, 48 N

4 $W=\frac{90}{7}$ N, $R=\frac{650}{7}$ N

5 a 1 m
b 10 kg

6 a 1220 N
b 50 N

7 a $80g$ at C, $70g$ at D
b i 70 kg **ii** 85.7 kg (to 3 sf)

8 a $(\mathbf{i}-3\mathbf{j})$
b $(\mathbf{i}+3\mathbf{j})$

Review exercise

1 a $3\mathbf{i}-3\mathbf{j}+6\mathbf{k}$
b 7.35
c 35.3°
d $4\mathbf{i}-\mathbf{j}+\mathbf{k}$

2 0.08

3 a $20g$ at C, $15g$ at D
b 3.18 kg (to 3 sf)

4 a $1.34\,\text{m s}^{-1}$ (to 3 sf)
b $1.25\,\text{m s}^{-1}$

5 a $\mu<\frac{2}{5}$
b 2 s

Assessment

1 a A
b C

2 D

3 a $F=10g$
b $10\sqrt{3}g$

4 a 122 N
b 0.723

5 Magnitude $=4\sqrt{5}=8.94$
Direction is 63.4° below **i**

6 a 45 N
b 15 N

7 a $\frac{5g\sqrt{3}}{2}$ **b** $g\le P\le 4g$

8 a

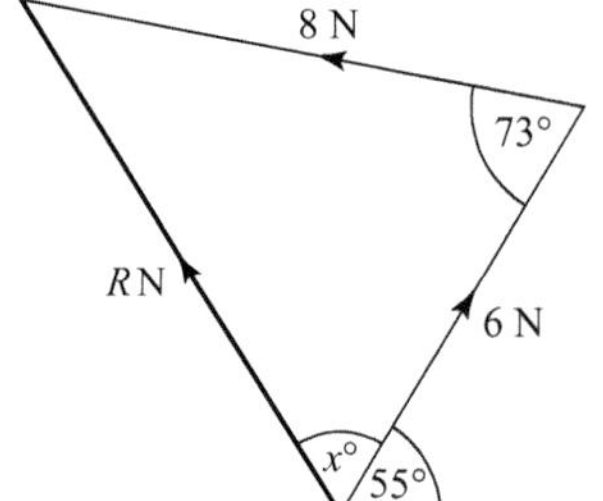

8.48

b 330.6°

9 a $\frac{5\sqrt{2}}{(\sqrt{3}+1)}g$ or equivalent exact form

b $\frac{10}{(\sqrt{3}+1)}g$ or equivalent exact form

10 a $1.8\,\text{m s}^{-2}$
b $5.7\,\text{m s}^{-1}$

11 a 80

b 4.56 m

12 a 27°

b 110 N

13 11.4

14 a $25g$

b $F = T\cos x$

$F = \frac{4}{5} \times T$

$F = 20g$

$2R = 30g + 20g + 20g$

$R = 35g$

$20g = \mu \times 35g$

$\mu = \frac{4}{7}$

15 a $10\sqrt{3}g$ b $8\sqrt{3}g$ c $12g$

16 a $-3.48\,\text{m s}^{-2}$

b 2.30 m

c $mg\sin 15° > \frac{mg\cos 15°}{10}$

Component of weight down the slope exceeds F_{max} so the particle does start to move back down the slope.

17 a 40 N

b 60 N

c $4\frac{1}{3}$ m

18 a $\frac{g}{10}$

b $\frac{11}{5}g$

c $\frac{36}{5}g$

19 a i $\frac{7}{6}Mg$ ii $\frac{Mg}{3}$

b i 1.5 ii $\frac{5Mg}{2}$

20 a $\frac{2g}{5}$

b $\sqrt{\frac{4g}{5}}$

c $\frac{1}{2}$ m

Assessment chapters 18-19: Mechanics

1 C $X = \frac{3\sqrt{3}}{2}\,\text{N}, Y = \frac{3}{2}\,\text{N}$

2 A 1225 m

3 a

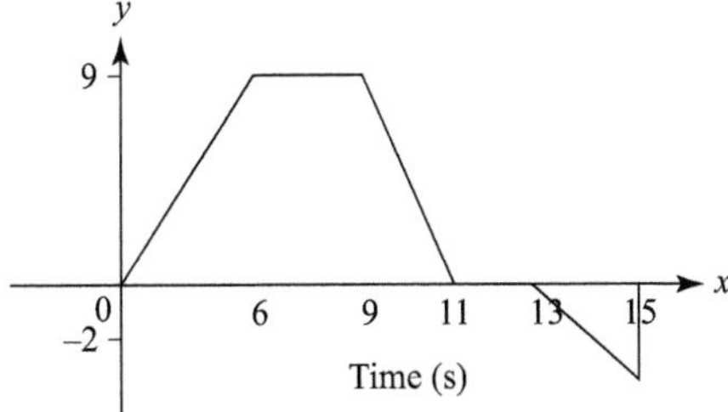

b 61 m

4 a $\frac{9}{8}g$

b $\frac{3}{2}g$

c $\frac{9}{8}g$

5 a i $42g$ or 411.6 N ii $0.8g$

b 0.4 m

6 a i $0.4g$ or $3.92\,\text{m s}^{-2}$ ii $\frac{12}{25}g$ or 4.71 N

b i String has no mass and does not stretch.

ii There is no frictional force.

7 13.3

8 a $\begin{pmatrix} -1 \\ 2 \end{pmatrix}$

b 5.59 m

9 a $20.6\,\text{m s}^{-1}$

b 22.0 s

10 a −1.73

b $\frac{t^2}{2} + \frac{1}{2}\sin 2t - \frac{\pi^2}{2}$

11 a $6\mathbf{i} + 2\mathbf{j}$

b $\left(\frac{t^3}{3} + 3\right)\mathbf{i} + (t^2 - 5)\mathbf{j}$

12 a i $6.18\,\text{m s}^{-1}$ ii $5.10\,\text{m s}^{-2}$

b 54 m

13 a $\theta = 52.6, P = 17.1$

b $\theta = 72.1, P = 14.3$

14 a i $F = 30\,\text{N}$ ii $F = 48.1\,\text{N}$

b $1.7\,\text{m s}^{-2}$

15 a $R_D = 2.4\,\text{N}, R_C = 0.6\,\text{N}$

b Beam is light so has no mass.

16 a At A: $R = 173\,\text{N}$

At B: $R = 238\,\text{N}$

b Weight acts at the centre of the rod.

17 $\sqrt{27}$

18 a $\begin{pmatrix} -1 \\ 0.375 \end{pmatrix}\text{m s}^{-2}$

b $4.27\,\text{m s}^{-1}$

c 291°

19 a $3.46\,\text{m s}^{-1}$

b $\begin{pmatrix} -4 \\ 12\sqrt{3} \end{pmatrix}$

20 a 1.06 s

b 1.38 m

c 3.18 m

d Ball is a particle/no spin. No air resistance/wind.

21 0.618

22 a 7.99 N

b $1.31\,\text{m s}^{-2}$

23 a i $R_B = 7g$ or 68.6 N
$R_A = 14g$ or 137.2 N
ii 55.7 cm
b The value of x will decrease as centre of mass will be 2.5 cm from end of the shelf.
24 $T_2 = 1.50$ N, $T_1 = 6.19$ N
25 2.87 m
26 a i $\frac{25}{1.5\cos\theta}$ **ii** $\theta = 27.7°$
b 17.7 m s^{-1} at 19.7° above the horizontal
27 a i 37880 N **ii** 10823 N
b 2.85 s
c 22130 N
28 a i 8.1°, 81.9° **ii** 0.61 s, 4.24 s
b Would travel less distance due to air resistance.
29 a $T_B = 26g$, $T_A = 4g$ **b** 3.47 m
30 72.9 m
31 a $14.9° < \theta < 86.1°$
b $14.9° < \theta < 18.6°$ or $69.8° < \theta < 86.1°$
32 90 m

Chapter 20

Exercise 20.1A Fluency and skills

1 a $\frac{1}{2}$ **b** $\frac{1}{8}$ **c** $\frac{4}{11}$

2

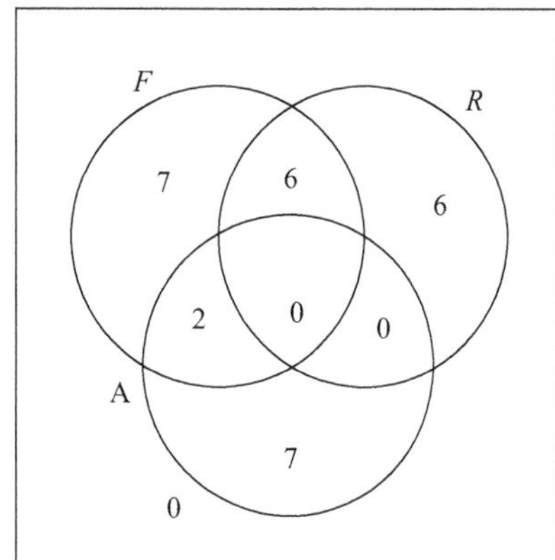

a $\frac{3}{7}$
b $\frac{19}{28}$
c $\frac{2}{9}$

3 a $\frac{1}{6}$
b $\frac{1}{2}$
c $\frac{1}{3}$

4 $\frac{48}{91}$

5 a $\frac{3}{20}$
b $\frac{5}{8}$

6 a $\frac{4}{7}$
b $\frac{1}{3}$
c $\frac{3}{7}$
d $\frac{2}{3}$

7 a $\frac{1}{2}$
b $\frac{1}{6}$
c $\frac{1}{3}$
d $\frac{1}{4}$

8 a $\frac{7}{20}$
b 0.45

Exercise 20.1B Reasoning and problem-solving

1 a

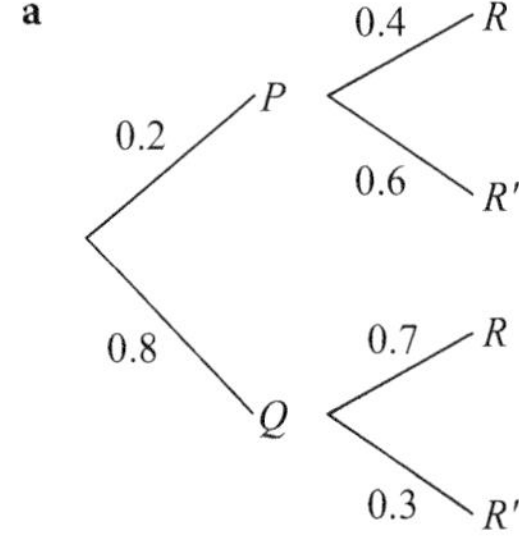

P(R) = 0.64

b $\frac{1}{8}$

2 a 0.3194
b 0.844 (3 dp)

3 a 0.215
b 0.377 (3 dp)

4 a 0.231 (3sf)
b 0.512 (3sf)

5 a 0.271 (3sf)
b 0.303 (3sf)
c 0.90 (2sf)

6 a $\frac{2}{3}$
b $\frac{1}{2}$

7 a $\frac{35}{72}$
b $\frac{2}{7}$

8 a $\frac{5}{39}$
b $\frac{5}{13}$

9 $\frac{101}{201}$

Exercise 20.2A Fluency and skills

1 a $\mu = 3.025$, $s = 1.102$ (3dp)

b Using $X \sim B(n, p)$, mean = 3, variance = 1.2
Using sample, mean = 3.025 ≈ 3, s = 1.102 ≈ 1.2. The values given by the sample suggest this model is a good fit for the data.

2 $X \sim B\left(300, \frac{1}{3}\right)$. Proportion of scores greater than 2 is $\frac{1}{3}$.
A binomial model is appropriate because there is a fixed number of independent trials (300), each trial has two possible outcomes ('greater than 2' or 'less than or equal to 2'), and the same dice is used each time so the probabilities should remain constant.

3 Using B(3, 0.2), $P(X = x) = 0.512, 0.384, 0.096, 0.008$ for $x = 0, 1, 2, 3$
Good fit to data ⇒ good model

Exercise 20.2B Reasoning and problem-solving

1 a $X \sim B(30, 0.5)$. A binomial model is appropriate because there is a fixed number of independent trials (30), each trial has two possible outcomes ('head' or 'not head') and the same coin is used each time so the probability should remain constant. Heads and not heads are assumed to be equally likely which gives $p = 0.5$.

b The coin may not be fair, in which case $p = 0.5$ would be inaccurate.

2 a i 0.5

ii 0.75

Purchasing patterns might be different in the town during 2015 due to e.g. variation in levels of disposable income of residents, number of shops in the town, convenience of purchasing mineral water (e.g. number of small shops or vending machines in the town).

b From the 2015 data $Q_1 = 160$, $Q_2 = 175.5$ (1dp), $Q_3 = 191.7$ (1dp). The median, lower quartile and upper quartile are all about 5 ml higher in 2015 than 2001 but the interquartile range is approximately the same at 32 to the nearest integer in both years. This suggests residents bought more mineral water on average in 2015 but the variation between purchased quantities did not change.

3 a Sian assumes that the later pages in her notes will take the same amount of time to review as the earlier pages, but these are likely to cover more challenging material so may take longer. Sian assumes she will work at the same rate over 8 hours that she did over 90 minutes, but it's likely that over a longer period she will need more breaks or get tired and work more slowly. Sian assumes a very small standard deviation but over a period of three weeks a number of unexpected events might happen which could change her working pattern, such as illness or family responsibilities.

b Sian's mean will reduce to account for a slower working pace. Sian's standard deviation will increase if she considers the high likelihood of unexpected events.

Exercise 20.3A Fluency and skills

1 a 0.8413

b 0.0918

c 0.8186

2 a 0.7734

b 0.2902

c 0.8186

3 a 0.65

b −2

c 0.65

4 a $\frac{5}{4}$

b 2

c 0

d $\frac{25}{17}$

e $-1\frac{7}{23}$

f −10

5 10.46 (2dp)

6 16.95 (2dp)

7 0.7499 (2dp)

Exercise 20.3B Reasoning and problem-solving

1 a $\mu_R = 13.43$ (2dp)

b $\sigma_S = 7.90$ (2dp)

c $\mu_T = 12.92$
$\sigma_T = 3.96$ (2dp)

2 a 6. Let T be a random variable for the time taken, in seconds, for a randomly chosen runner. $T \sim N(730, 80^2)$

b 10 minutes 27 seconds

3 a
$$\begin{aligned} X \sim N(\mu, \sigma^2) \Rightarrow P(\mu - 2\sigma < X < \mu + 2\sigma) &= P(-2 < Z < 2) \\ &= 2P(0 < Z < 2) \\ &= 2(0.9772 - 0.5) \\ &= 0.95 \end{aligned}$$

b For a normal population, mean ≈ mode ≈ median and ~95% of data lies within 2 standard deviations of the mean.
For this data 275 ≈ 270 ≈ 265 and 93% ≈ 95%
Reaction data gives very good fit to these statistics.
Therefore, evidence suggests that a normal distribution would provide a good model.

4 a $\mu = 8.25$, $\sigma = 2.91$

b 69.8%

c This distribution could be modelled by a Normal distribution as close to 68% of values lie within one standard deviation of the mean, the distribution is roughly symmetrical and the mean 8.25 is close to the modal class $6 \le m < 8$.

5 $\mu = 9.31$ (2dp)

6 a i 0.1587 ii 0.1587

iii
$$\begin{aligned} P(W > \mu + \sigma) &= P\left(Z > \frac{(\mu + \sigma) - \mu}{\sigma}\right) \\ &= P(Z > 1) \\ &\approx P(1 < Z < \mu + 5\sigma) \\ &= P(1 < Z < 5) \\ &= 0.1587 \end{aligned}$$

iv
$$\begin{aligned} P(W < \mu - n\sigma) &= P\left(Z < \frac{(\mu - n\sigma) - \mu}{\sigma}\right) \\ &= P(Z < -n) \end{aligned}$$

b 1.1

7 a 0.0194 b 0.943 (3dp)

8 T – waiting time in minutes. $T \sim N(\mu, \sigma^2)$

a
$$\begin{aligned} P(T > 5) &= P\left(Z > \frac{5 - \mu}{\sigma}\right) \\ &= 0.14 \end{aligned}$$

$\frac{5-\mu}{\sigma}=1.0801$

$P(T<1)=P\left(Z<\frac{1-\mu}{\sigma}\right)$

$=0.08$

$\frac{1-\mu}{\sigma}=-1.4065$

$\mu=3.26,\ \sigma=1.61$

$P(T>6)=4.439\%>2.5\%$

Requirement not met.

b Maximum mean waiting time is 2.84 minutes (2dp)

Exercise 20.4A Fluency and skills

1 a i $P(22.5<Y<23.5)$

ii $P(X=23)=0.0611$

$P(22.5<Y<23.5)=0.0604$

$0.0611\approx 0.0604$

b i $P(Y<50.5)$

ii $P(X\le 50)=0.999...$

Lower bound $=27-5\times\sqrt{14.85}=7.73$

$P(Y<50.5)\approx P(7.73<Y<50.5)=0.999...$

$P(X\le 50)\approx P(7.73<Y<50.5)$

c i $P(Y>11.5)$

ii $P(X\ge 12)=1-P(X\le 11)$

$=0.999986...$

Upper bound $=27+5\times\sqrt{14.85}=46.27$

$P(11.5<Y<46.27)=0.999970...$

$0.999986...\approx 0.999970...$

d i $P(31.5<Y<51.5)$

ii $P(32\le X\le 51)=P(X\le 51)-P(X\le 31)$

$=1-0.8783...$

$=0.12165...$

$P(31.5<Y<51.5)=0.12145...$

$0.12165...\approx 0.12145...$

e i $P(X>17)=P(X\ge 18)$

$P(Y>17.5)$

ii $P(X>17)=1-P(X\le 17)$

$=1-0.00608...$

$=0.99391...$

Upper bound $=27+5\times\sqrt{14.85}=46.27$

$P(17.5<Y<46.27)=0.99315...$

$0.99391...\approx 0.99315...$

f i $P(X<40)=P(X\le 39)$

$P(Y<39.5)$

ii $P(X<40)=P(X\le 39)$

$=0.99941...$

Lower bound $=27-5\times\sqrt{14.85}=7.73$

$P(Y<39.5)\approx P(7.73<Y<39.5)=0.99940...$

$0.99941...\approx 0.99940...$

2 a 11.8 and 7.6 (accept 12 and 7.5)

b $\mu=9.7$ (accept 9.75)

c $\sigma=2.1$ (accept 2.25)

3 0.608

4 a A: $X\sim B\left(300,\ \frac{1}{3}\right)$. A binomial model is appropriate because there is a fixed number of independent trials (300), each trial has two possible outcomes ('hot meal' or 'not hot meal') and the probability can be assumed to be constant between passengers.

B: $Y\sim N\left(100, 66\frac{2}{3}\right)$, n is large and p is close to $\frac{1}{2}$

b $P(X>115)=0.0299$

$P(Y\ge 115.5)=0.0288$

To 2 dp both answers are 0.03, but to 3dp the binomial model gives 0.030 and the Normal model gives 0.029

5 0.428

Exercise 20.4B Reasoning and problem-solving

1 Median value (20) far from mean.

Large standard deviation suggesting negative lifetimes.

2 n is not large and p is not close to 0.5 so Ava's model is not suitable.

3 a If coin is fair, $P(7)=0.1172$

No reason to suggest coin is biased.

b X – number of heads

If fair coin, $X\sim B(10000,\ 0.5)$

$Y\sim N(5000,\ 2500)$

$P(Y\ge 5100)=0.023$

Coin is probably biased.

4 a From 2001 school milk was classified as 'eating out'.

Welfare milk was discontinued in 2009.

b i Most people consume some form of milk. However, a significant number of people tend to use either skimmed or whole milk, possibly giving more zeros than would be expected under a Normal distribution.

ii 0.2375

iii 0.31 (2dp)

5 Y – vegetable content in g

Assume $Y\sim N(98,100)$ so $P(Y>98)=0.5$

X – number of pies out of 10 with vegetable content over 98 g

$X\sim B(10,0.5)$

Mean of X is $10\times 0.5=5$, variance $10\times 0.5\times 0.5=2.5$

Data has mean 3.1 and variance $2.12=4.41$

Discrepancy suggests normal model is not a good one.

6 X – number of passengers who turn up

$X\sim B(N,0.94)$ where N is the number of tickets sold

$Y\sim N(0.94N,\ 0.0564N)$

P(accommodate all passengers) $=P(X\le 300)=P(Y<300.5)$

$=P\left(Z<\frac{300.5-0.94N}{\sqrt{0.0564N}}\right)>0.99$

$\frac{300.5-0.94\times N}{\sqrt{0.0564\times N}}>2.326$

Test $N=309$ and $N=310$

$\frac{300.5-0.94\times 309}{\sqrt{0.0564\times 309}}=2.405$

$\frac{300.5-0.94\times 310}{\sqrt{0.0564\times 310}}=2.176$

Therefore $N=309$

Review exercise 20

1 a $\frac{1}{2}$

b $\frac{23}{45}$

2 a 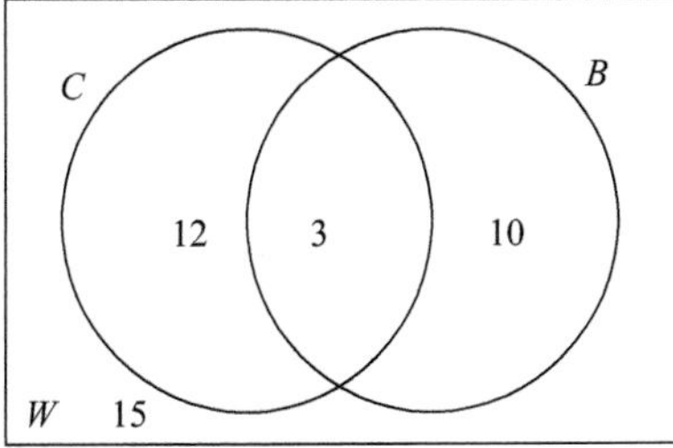

b i $\frac{3}{40}$

ii $\frac{13}{25}$

3 $\frac{5}{8}$

4 a

x	0	1	2	3	4
$P(X = x)$	0.3164	0.4219	0.2109	0.0469	0.0039

b

x	0	1	2	3	4
expected f	25.312	33.752	16.872	3.752	0.312

Good comparison with data suggests spinner is fair.

5 a i 0.1241 ii 0.83427

b $a = 16.675$

6 0.74988

7 Median = 279 s

Mode, median and mean approximately equal and about $\frac{2}{3}$ of all observations within one standard deviation of the mean so Normal is a good model.

8 0.2973 (4dp)

Assessment 20

1 D

2 C

3 a 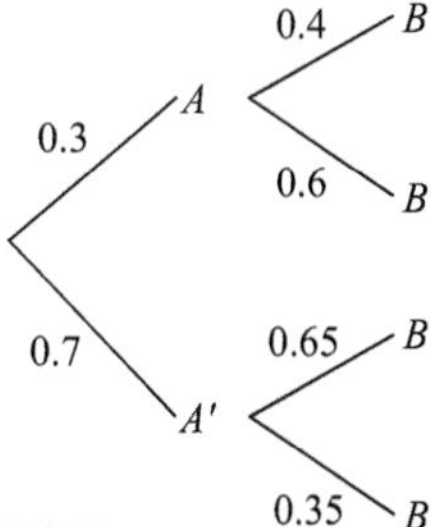

b 0.365

c 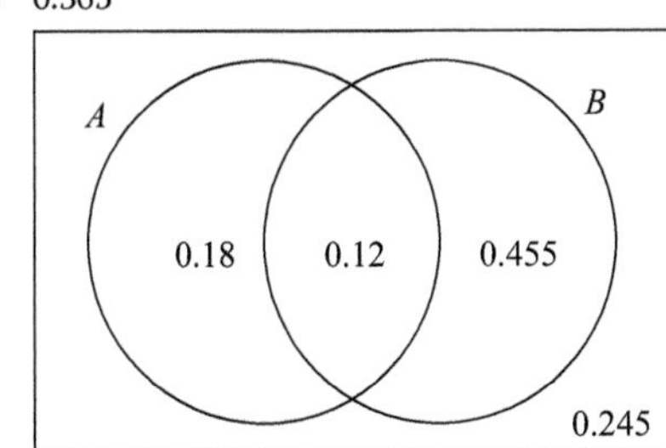

4 a i $\frac{9}{14}$

ii $\frac{2}{9}$

iii $\frac{5}{14}$

iv $\frac{2}{5}$

b The probability that the amount of sugar and preserves consumed exceeds 130g given that the amount of cheese consumed exceeds 115g is only $\frac{2}{9}$, so this is not a sensible assumption.

5 a 6 minutes

b 1

c It's not reasonable as delays are likely inevitable. The assumption that the buses never run late is almost definitely a false one.

6 a 0.163

b 0.223

c 0

7 a 0.775

b i 0.841

ii 0.159

iii 0.225

8 a 0.236 b 0.432

c No as there are weather conditions and psychological factors to consider

OR

Yes as a competition archer is likely to be reasonably consistent between each arrow being fired

d 2.16

9 a $\mu = 0.99, \sigma^2 = 0.07$

b $\mu - \sigma = 0.99 - 0.26 = 0.73$

$\mu + \sigma = 0.99 + 0.26 = 1.25$

$\frac{0.8 - 0.73}{0.8 - 0.7} = 0.7$

$0.7 \times 5 = 3.5$

masses in the $0.7 \le m < 0.8$ category lie within one standard deviation of the mean.

$\frac{1.25 - 1.2}{1.3 - 1.2} = 0.5$

$0.5 \times 6 = 3$

masses in the $1.2 \le m < 1.3$ category lie within one standard deviation of the mean.

$3.5 + 13 + 15 + 3 = 34.5$ masses lie within one standard deviation of the mean.

$\frac{34.5}{50} = 69\%$

This distribution could be modelled by a Normal distribution as close to 68% of values lie within one standard deviation of the mean, the distribution is roughly symmetrical and the mean 0.99 is close to the modal class $1.0 \le m < 1.2$

c i 0.5153 ii 0.8677 iii 0.9226

d (0.21, 1.77)

10 a i $\frac{5}{3}$ ii $\frac{25}{18}$ iii 3.25, 0.0855 iv 0.930

b $S \sim N\left(\frac{n}{6}, \frac{5n}{36}\right)$

c i 0.8344 ii 0.8246 iii 0.8207

Chapter 21

Exercise 21.1A Fluency and skills

1 Reject H_0: the p-value is less than the significance level, the actual result is more surprising than we need.

2 a The jewellery auction house is investigating to see if there is any correlation present. This is a two-tailed test.

b Accept H_0: the p-value is more than the significance level, so the actual result is not surprising enough.

3 Reject H_0: the PMCC for the sample is more negative than the critical value.

4 Accept H_0: the p-value corresponds to 35.5%. Since 35.5% > 5% (the result is not as extreme as required) the result is not significant.

5 Reject H_0: the critical value for this test is 0.2605. Since the PMCC for the sample is larger in size than the critical value, the result is significant.

Exercise 21.1B Reasoning and problem-solving

1 Accept H_0, no correlation between levels of taxation and the amount individuals give to charity.

2 Reject H_0, positive correlation between the Maths and English scores of primary school students.

3 **a** 5% chance of rejecting the null hypothesis if it is actually true.

b One flavour.

c Since the result exceeds the critical value for positive correlation, the scientists may conclude that there is positive correlation between that one flavour and being carcinogenic.

d As you expect one of the 20 flavours to show positive correlation it is not surprising when one of them actually does. It would not be reasonable to reach the conclusion that any of the flavours are carcinogenic without conducting further tests.

4 The critical value is smaller than the test statistic, so there is sufficient evidence at the 5% level to suggest that there is positive correlation between the average purchased quantities of infant milks and baby food per person per week in England.

5 **a** The critical value is larger than the test statistic, so there is insufficient evidence at the 5% level to suggest that there is any correlation between the amount of milk products (excluding cheese) and the amount of cheese consumed per person per week.

b The East consumes relatively average amounts of milk but a high amount of cheese, whereas Yorkshire and the Humber drinks a lot of milk and eats relatively little cheese.

6 The p-value for the test statistic (38.2%) is much larger than the significance level of the test so the result is not significant. There isn't sufficient evidence to reject the null hypothesis. You conclude that there is no correlation between the geographical location of a school and the likelihood that a student there will get at least one A at A level.

7 **a** There is evidence to suggest that A and B are positively correlated.

b There is evidence to suggest that B and C are positively correlated.

c There is no evidence to suggest that C and A are positively correlated.

d All three can be true simultaneously unless there is perfect correlation between the variables. It does not appear that there is perfect correlation between these variables. It doesn't follow that A and B being correlated and B and C being correlated makes C and A correlated.

Exercise 21.2A Fluency and skills

1 $H_0: \mu = 19$
$H_1: \mu \neq 19$

2 $H_0: \mu = 43$
$H_1: \mu \neq 43$

3 **a** $H_0: \mu = 32$
$H_1: \mu \neq 32$

b −2

c ±1.645

d Reject H_0: the test statistic is larger in size than the critical value.

4 **a** $H_0: \mu = 25$
$H_1: \mu \neq 25$

b 1.697

c 1.645

d Reject H_0: the test statistic is greater than the critical value.

5 **a** 1.714 **b** 0.0432

c Reject H_0: since the p-value is smaller than the significance level, the result is significant.

6 **a** −1.304 **b** 0.0961

c Reject H_0: the p-value is smaller than the significance level.

Exercise 21.2B Reasoning and problem-solving

1 **a** $H_0: \mu = 1845$, $H_1: \mu \neq 1845$, where μ is the average amount of milk and cream consumed per person in the East.

b The test statistic is $\dfrac{1824 - 1845}{\sqrt{\dfrac{4288.3}{48}}} = -2.222$. The critical values are ±1.960.

c Since the critical value is smaller than the test statistic, there is sufficient evidence at the 5% level to reject the null hypothesis. You conclude that the East consumes a different amount of milk and cream per person per week than England as a whole.

d The East does tend to consume much more than England as a whole. This is the conclusion of the test, but has arisen from a result which is, in fact, less than that of England's. The result should be treated with some scepticism given the previously-known differences.

2 **a** $H_0: \mu = 125$, $H_1: \mu > 125$

b $z = 3.175$ (4 sf)
The p-value is 0.1%

c You reject the null hypothesis as the p-value is less than the significance level.

d You conclude that there is sufficient evidence to suggest that the mean mass of a banana that summer is greater than 125 g.

e It is not reasonable as we've only found evidence that it's greater than 125 g. 131 g would likely be a reasonable guess but it's very unlikely to be exactly that amount.

3 **a** $H_0: \mu = 13.6$, $H_1: \mu < 13.6$ where μ is the mean amount of margarine purchased per person per week in London.

b The test statistic is −0.472, which has a p-value of 31.86%.

c For a fixed value of the sample mean, a larger sample size gives a test statistic with a decreased p-value, so at some point the result will be significant.

d The p-value is larger than the significance level, so there is insufficient evidence to reject the null hypothesis. You conclude that the mean amount of margarine purchased per person per week in London is not lower than 13.6 g.

4 **a** Statistician one claims $H_0: \mu = 16$ and $H_1: \mu \neq 16$
Statistician two claims $H_0: \mu = 17$ and $H_1: \mu \neq 17$

b The result is not significant for either statistician, and both conclude that they were correct. Increasing the sample size would give a better chance of yielding a result that favours one view over the other.

Review exercise 21

1 $H_0: \rho = 0$
$H_1: \rho \neq 0$

2 a $H_0: \rho = 0$
$H_1: \rho > 0$

b Reject H_0: the PMCC is greater than the critical value.

3 a Accept H_0: the p-value is greater than the significance level.

b There is no negative correlation between the amount of the chemical and the number of bacteria present in a petri dish.

4 a $H_0: \rho = 0$
$H_1: \rho > 0$

b The PMCC is less than the critical value so there is insufficient evidence to reject the null hypothesis. You conclude that there is no positive correlation between levels of annual household income and the amount spent per year on books.

5 Reject H_0: the p-value is lower than the significance level. Conclusion: the long jump and high jump scores achieved by heptathletes are correlated.

6 $H_0: \mu = -5$
$H_1: \mu > -5$

7 a $H_0: \mu = 30$
$H_1: \mu \neq 30$

b Accept H_0: the p-value is larger than the significance level.

c It is reasonable to believe that the mean length of a stick of rock in the population the sample is drawn from is 30 cm.

Assessment 21

1 a C
b B

2 $H_0: \mu = 40$; $H_1: \mu < 40$
Reject H_0; evidence supports mean is less than 40

3 No reason to reject H_0; evidence supports mean equals 22

4 a 2020 hours
b No reason to reject H_0; evidence supports mean equals 1930

5 Reject H_0; evidence suggests that the mean is less than 12

6 a $H_0: \rho = 0$
$H_1: \rho \neq 0$

b $-0.6319 < 0.532 < 0.6319$, so accept H_0. There is no evidence to support correlation between cheese and pickle purchases.

7 a The test statistic is further from zero than the critical value, reject the null hypothesis.

b The test statistic is further from zero than the critical value, reject the null hypothesis.

c The mean is not −13.6 but one is larger and one is smaller. The scientists should do a further test with a larger sample.

8 Reject H_0: evidence supports mean volume greater than 568 ml.

9 No reason to reject H_0

10 a $H_0: \rho = 0$
$H_1: \rho > 0$

b $0.532 > 0.631$, so reject H_0. There is evidence to support correlation between white and brown bread purchases.

11 No reason to reject H_0

12 Reject H_0 and the manufacturer's claim.

13 a $\bar{x} > \mu_0 + 2.054\frac{\sigma}{\sqrt{n}}$

b Reject H_0; evidence supports mean breaking point greater than 80 kN.

Assessment chapters 20–21: Statistics

1 D

2 a A b D

3 a (left to right, top to bottom): 12, 187, 126, 313, 5, 140

b i $\frac{32}{350}$ or $\frac{16}{175}$

ii $\frac{140}{350}$ or $\frac{2}{5}$

iii $\frac{187}{350}$

c $\text{P(mixed handed)} \times \text{P(man)} = \frac{5}{350} \times \frac{3}{5}$
$= \frac{3}{350}$
$= \text{P(mixed handed and man)}$

So yes

4 a Parameter is the underlying mean for the whole population of the town. The statistic is the mean of the sample that we are calculating.

b Could be biased due to location being similar/ people being out at work/non-response etc

c 4

5 a i 26.5 g ii 10.2 g (or 10.3)
iii 25.79 g iv 14.32 g

b Median, as not effected by outliers.

6 a 0.4

b

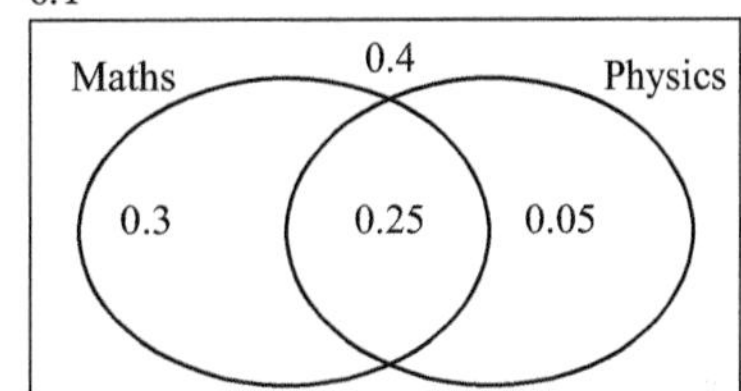

c $\frac{5}{6}$

d No; P(Maths | Physics) ≠ P(Maths)

7 a i $\frac{12}{20}$ or $\frac{3}{5}$ ii $\frac{8}{12}$ or $\frac{2}{3}$

b

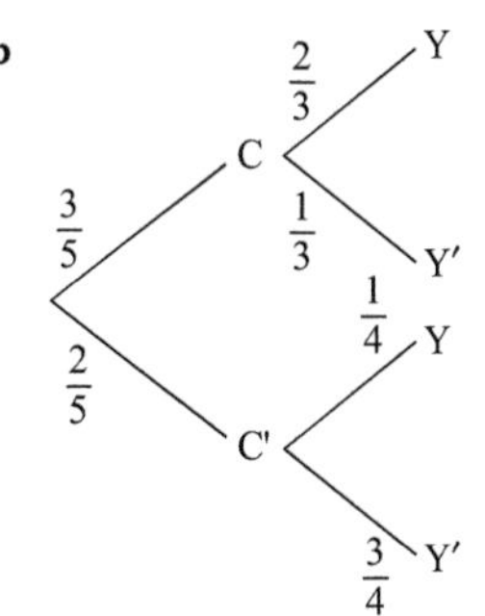

8 a i $\frac{1}{35}$ ii $\frac{6}{35}$ iii $\frac{4}{19}$

b i $\frac{19}{105}$ ii $\frac{7}{15}$

9 a i 0.045 **ii** 0.046

b

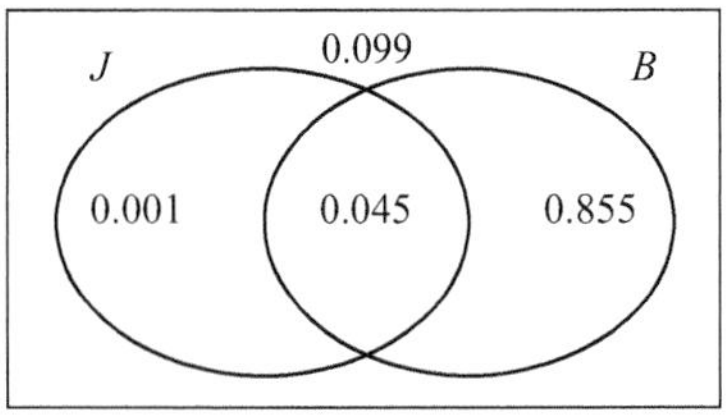

J	0.099	B
0.001	0.045	0.855

10 a i 0.9772 **ii** 0.1587 **iii** 0.2297

b 2.893

11 H_0: $\rho = 0$, H_1: $\rho > 0$

Reject H_0

12 a H_0: $\rho = 0$, H_1: $\rho < 0$

Reject H_0

b $r < -0.4226$

13 H_0 will be rejected when $r < -0.4438$ or $r > 0.4438$

14 a $\frac{11}{30}$ (or 0.36...) **b** $\frac{1}{12}$ (or 0.083...)

15 a 14 meerkats

b i 27.1 cm **ii** 3.4 cm

c Any one of: Continuous data; Approximately symmetrical; All data lies within 3 sd of the mean; Approximately 67% lies within 1 sd of the mean

d i 0.8743 **ii** 0.2692

e $n \geq 15$

16 a 11.6 **b** 10.4, 12.8

17 a 0.049 **b** 0.2485 **c** 0.0392

18 a Yes, data is continuous and the distribution is almost symmetric.

b i 0.7665

ii According to the box and whisker diagram, 75% of households bought less than 2400, so this is quite close to our result of 76.65%.

c i 0.3594 **ii** 0.7073 **iii** 0.2646

19 a Could be modelled by binomial, $X \sim B(8, 0.5)$

b No; not a fixed number of trials.

c No; probability is not constant/trials are not independent.

20 H_0: $\rho = 0$, H_1: $\rho < 0$

Accept H_0

21 a H_0: $\mu = 1.4$, H_1: $\mu < 1.4$

Accept H_0

b $z < -1.2816$

22 a $\sigma = 5.41$; $\mu = 132$ **b** 3.30

23 a 2.041 litres

b 0.2337

c i 0.2642 **ii** 0.6749

24 a i 0.8708 **ii** 0.459

b Normal distribution is appropriate since n is large and np and $nq > 5$

25 a $\sigma = 6.02$ g; $\mu = 9.35$ g **b** 0.1736 **c** 0.9319

26 H_0: $\mu = 30$, H_1: $\mu < 30$

Reject H_0 / there is enough evidence to support the customer's claim.

27 a H_0: $\mu = 13$, H_1: $\mu > 13$, Critical value is 1.96

b 13

Index

A
absolute value 10–11, 14
acceleration
    constant 198–201, 209, 237
    dynamics 236–8
    forces 212–13, 215–16
    variable 202–5
    see also Newton's second law
algebra 3–32
    algebraic fractions 20–22, 37–8
    functions 8–15
    parametric equations 16–19
    partial fractions 24–7
    proof 4–7
algebraic division 150
algebraic fractions 20–23
    binomial series 37–8
    factor theorem 21–23
    long division method 21
    useful formulae 20
alternative hypotheses 284, 286, 288, 290
ANS calculator key 40, 43, 174
approximation
    binomial distribution 274–7
    binomial series 36–7
    Newton-Raphson method 174, 176
    symbol 37
area between two curves 137–8
area of trapezium 178
area under curve 134, 178
arithmetic sequences 44–9
    arithmetic series 44–8
    common differences 44
    definition 44
    *n*th terms 44–8
    real-life situations 47–8
arithmetic series 44–8
    real-life situations 47–8
    sigma notation 45–6
    sum of first *n* terms 44–8
atmospheric modelling 163
axes in 3D 224

B
binomial distribution 274–7
binomial models 266–7
binomial series 34–9
    $(1 + x)^n$ expansion 34
    approximations 36–7
    partial fractions 37–8
    range of validity 35–6

C
Cartesian equations 16–19
celestial mechanics 133
centre of mass 241
chain rule 110–13, 114, 116, 120, 140, 152
change of sign method 164, 166, 170–1, 176
change of variable 140
circles 16, 64, 66
cobweb diagrams 168–9
coefficient of friction 229, 232, 237
coefficients, equating 225
column vectors 224
common differences 44
common ratios 50
composite functions 9, 110, 112
compound angles 74–80
concave curves 92–3, 100
conditional probability 260–5
constant acceleration 198–201, 209, 237
constant of integration 152, 203
continuity correction 275
continuous functions 164, 166, 268
continuous random variables 259–82
contradiction, proof by 4–6
convergent sequences 41, 168–72, 174, 176
convex curves 92–3, 96–7, 100
correlation 283, 284–7
cosecant (cosec) 68–9
cosine (cos)
    $a\cos\theta +/- b\sin\theta$ 80–3
    compound angle formula 74–5
    double angle formula 75–6
    integrating $\cos^2 x$ 142–3
    inverse functions 69–70, 114
    reciprocal functions 68–9
    small values of *x* in radians 64
    standard integrals 135
    trigonometric functions 63, 68–73, 98–101
    trigonometric identities 63–90
cotangent (cot) 68–9, 78
counter examples 4, 74
critical values 284–6, 288–91
cubic functions 92
curves
    area between two curves 137–8
    area under curve 134, 178
    function shapes 92–7, 100

D
decreasing sequences 40–1, 44
definite integrals 134, 141, 178
degree symbol 64
denominators 22–5, 150, 261
dependent variables 116
derivatives 92–104
Dido's problem 91
differential equations 133, 152–62, 163, 238
differentiation 91–132
    chain rule 110–13, 114, 116, 120
    exponential functions 102–5
    function shapes 92–7
    $f(x)=a^x$ 102–3
    implicit differentiation 116–19
    inverse functions 114–15
    logarithmic functions 102–5
    parametric functions 120–3
    product rule 106–9, 116
    quotient rule 106–9
    trigonometric functions 98–101
    variable acceleration 202, 204
direct proof 4
displacement 198–200, 202–4, 206
division, algebraic 21
domains 8, 10, 69, 70
double angle formula 75–6, 98
dynamics 234–9

E
e (irrational number) 102–5, 135, 152
empty sets 260
encryption 3
equating coefficients 37, 225
equations
    Cartesian 16–19
    differential 133, 152–62, 163, 238
    motion 198–217, 237
    parametric 16–19
    simultaneous 24, 52, 148, 150, 272
equilibrium 229, 231, 235, 237, 241, 243–4
exhaustion, proof by 4
explicit differentiation 116
explicit functions 152
exponential functions 102–5, 135, 152

F
factor theorem 23–25
Fibonacci sequence 33
first derivatives 92–6, 98–100, 102–4
forces 212–17, 223–52
    dynamics 234–9
    moments 223, 240–5
    Newton's second law 212–13, 215
    statics 228–33
    2D motion 212–17
    vectors in 3D 224–7
fractions, algebraic 20–27
friction 229, 232, 237
functions 3, 8–15
    composite 9, 110, 112
    cubic 92
    curves rising/falling 92–3
    definition 8
    differentiation 92–105, 114–15, 120–3
    explicit 152
    exponential 102–5, 135, 152
    implicit 116–19, 152
    inverse 10, 13, 68–73, 114–15
    logarithmic 102–5
    many-to-one 8, 10, 70
    one-to-one 8, 10, 13, 70
    parametric 120–2
    rational 148–51
    reciprocal 68–73
    shapes 92–7, 100
    trigonometric 63, 68–73, 98–101

G
general solutions 152–5
geometric sequences 50–5
    common ratio 50
    definition 50
    geometric series 50–4
    *n*th terms 50–2
    real-life situations 53–4
    simultaneous equations 52

geometric series 50–4
infinite series 53
real-life situations 53–4
sigma notation 51
simultaneous equations 52
sum of first n terms 50–1
sum to infinity 53
golden ratio 33
gradients 92, 171
gravity 206–11
equation of path 208
horizontal components 206–10
maximum height 208
range of particle 207–8
time of flight 207–8
variable value of 206
vertical components 206–10

H
horizontal forces 228–9, 235, 237
horizontal motion 206–10
hypothesis testing 283–96

I
identities 25–6
see also trigonometric identities
implicit functions 116–19, 152
improper rational functions 150
increasing sequences 40–1, 44
indefinite integrals 135
independent events 261
independent variables 116
infinite series 34, 53
inflection, point of 93–6
initial conditions 152–5, 238
integers 34–6, 40, 42
integral calculator key 178
integration 133–62
differential equations 133, 152–62
numerical 178–83
by parts 144–7
rational functions 148–51
standard integrals 134–9
by substitution 140–3
variable acceleration 202–4
intersections of sets 260–1
intervals 166, 178–9, 181
inverse calculator functions 270–2
inverse functions 10, 13, 68–73, 114–15
irrational numbers 5, 102–5, 135
iterative formulae 40, 168–77

L
Lami's Theorem 231
Leibniz notation 92, 106
limiting equilibrium 229
limiting friction 229, 232
limits 41, 98, 102, 134
lines of symmetry 241, 268
logarithmic functions 102–5, 135
long division, algebraic 21

M
magnitudes 199, 204, 212, 224
many-to-one functions 8, 10, 70
mass
centre of 241
see also Newton's second law
maximum points 92–3
mean
binomial model testing 266–7
hypothesis testing 288, 290–1
Normal distribution 268–72, 276–7
median 268
minimum points 92–3
mode 268
modelling real life 43, 47–8, 53–4, 266–7
modulus (absolute value) 10, 14
modulus of vector 224
moments 223, 240–5
motion 197–222
constant acceleration 198–201, 237
equations 198–217, 237
forces 212–17
gravity 206–11
2D 197–222
variable acceleration 202–5
mutually exclusive events 264

N
natural logs 102–5, 135
natural numbers 6
nCr calculator key 34
negative correlation 284
newton-metres (Nm) 240
Newton-Raphson method 174–7
Newton's second law 204, 212–13, 215, 234–8
Nm *see* newton-metres
Normal distribution 259, 268–77
approximating the binomial 274–7
calculator functions 268–72, 275, 277
hypothesis testing 288–91
*n*th terms of sequences 40, 42, 44–8, 50–1
null hypotheses 284–6, 288–91
numerical integration 178–83
numerical methods 163–88
finding roots 164–77
iterative root finding 168–73
Newton-Raphson method 174–7
numerical integration 178–83
simple root finding 164–7

O
one-tailed tests 284, 288
one-to-many mappings 10
one-to-one functions 8, 10, 13, 70
open intervals 166
order of sequence 41

P
parametric equations 16–19
parametric functions 120–3
partial differential equations 163
partial fractions 22–5, 37–8, 148–50
particular solutions 152–5
parts, integration by 144–7
Pascal's triangle 34
Pearson's product-moment correlation coefficient (PMCC) 284–5
periodic sequences 40–1
perpendicular forces 231–2, 234, 236
PMCC (Pearson's product-moment correlation coefficient) 284–5
points of inflection 93–6
polynomials 20
population correlation coefficient 284–6
position vectors 198–200, 203–4, 225, 227
positive correlation 284, 286
prime numbers 3, 5
principal values 70
probability 259–82
conditional 260–5
continuous random variables 259–82
modelling real life 266–7
Normal distribution 268–77
probability density functions 268
product-moment correlation coefficient 284–5
product rule 106–9, 116, 144
proof 4–7
by contradiction 4–6
direct 4
by exhaustion 4
public-key encryption 3
p-values 285–6, 288, 290
Pythagoras' theorem 69, 228

Q
quotient rule 38, 106–9

R
radians 64–7
Newton-Raphson method 176
small values of x 64–5
symbols for 64
trigonometric functions 98
random variables 259–82
range
functions 8, 10, 69–70, 207
motion under gravity 207
validity of expansion 35–6
rates of change 92, 103–4, 154–5
ratio formula 225
rational functions 148–51
real-life modelling 43, 47–8, 53–4, 266–7
reciprocal functions 68–73
recurrence relations 40
see also iterative formulae
reflection 11, 70
remainders 25
resolving forces 212–13, 215, 228, 231–2, 234–8, 243
resultant forces 228–9, 234, 241, 243
right-hand set of axes 224
root finding 164–77
iterative 168–73
Newton-Raphson method 174–7
simple 164–7

S
sample spaces 260–1, 264
secant (sec) 68–9, 72
second derivatives 92–100
separating variables 152–5, 238
sequences 33–62
arithmetic 44–9
binomial series 34–9
convergent 41, 168–72, 174, 176
decreasing 40–1, 44
definition 40
Fibonacci 33
geometric 50–5
increasing 40–1, 44
*n*th terms 44–8, 50–2
periodic 40–1
set notation 260
sigma notation 45–6, 51, 134

significance levels 285–6, 288, 290
simplifying assumptions 266–7
simultaneous equations
  geometric sequences 52
  integration 148, 150
  Normal distribution 272
  partial fractions 24
sine rule 231
sine (sin)
  $a\cos\theta+/-b\sin\theta$ 80–3
  compound angle formula 74–5
  double angle formula 75–6, 98
  integrating $\sin^2 x$ 142–3
  inverse functions 69–70, 72, 114
  reciprocal functions 68–9
  sine rule 231
  small values of x in radians 64–5
  standard integrals 135
  trigonometric functions 63, 68–70, 98–101
  trigonometric identities 63–90
special triangles 65, 75
speed 199–200, 204, 210
  see also velocity
speed-time graphs 181
staircase diagrams 168
standard deviation 268, 272, 288
standard integrals 134–9
  area between two curves 137–8
  area under curve 134, 178
  definite integrals 134
  indefinite integrals 135
  list of 135
  rational functions 148
standardised Normal distribution 288
standard Normal density function 268
statics 228–33
stationary points 92–3, 96, 108–9
substitution, integration by 140–3
surds 65, 75
symmetry 241, 268, 271–2

T

TABLE calculator facility 164
tangent (tan)
  compound angle formula 74
  double angle formula 75–6, 98
  inverse functions 69–70, 114
  reciprocal functions 68–9
  small values of x in radians 64
  standard integrals 135
  trigonometric functions 68–70, 72, 98–101
  trigonometric identities 63–90
tends to, symbol 41
tension 216
terms of sequences 40
test statistics 288–91
three dimensional vectors 224–7
transformations 9, 80
trapdoor functions 3
trapezium rule 178–81
tree diagrams 260–1, 263–4
triangles, special 65, 75
trigonometric functions 63, 68–73, 98–101
trigonometric identities 16, 63–90
  $a\cos\theta+/-b\sin\theta$ 80–3
  compound angles 74–9
  quotient rule 106
  radians 64–7
  trigonometric functions 63, 68–73
trigonometry in statics 228, 231
turning forces *see* moments
turning points 96, 120
  see also stationary points
two-dimensional motion 197–222
two-tailed tests 284, 288
two-way tables 260, 263

U

unions of sets 260
unit vectors 224

V

variable acceleration 202–5
variables
  change of 140
  continuous random 259–82
  integration by substitution 140
  separating 152–5, 238
variance
  binomial model testing 266–7
  hypothesis testing 288, 290
  Normal distribution 268–72, 276–7
vectors
  in 3D 224–7
  statics 231
  2D motion 198–205, 207, 209–10, 213
velocity
  constant acceleration 198–200
  motion under gravity 206–8, 210
  variable acceleration 202–4
  see also speed...
Venn diagrams 260, 263
vertical forces 228, 232, 235, 237–8, 243
vertical motion 206–10

Z

zero correlation 284
z-tests 288–90
z-values 268–72